Lecture Notes in Artificial Intelligence 8655

Subseries of Lecture Notes in Computer Science

Petr Sojka Aleš Horák
Ivan Kopeček Karel Pala (Eds.)

Text, Speech, and Dialogue

17th International Conference, TSD 2014
Brno, Czech Republic, September 8-12, 2014
Proceedings

Volume Editors

Petr Sojka
Masaryk University
Faculty of Informatics
Department of Computer Graphics and Design
Brno, Czech Republic
sojka@fi.muni.cz

Aleš Horák
Ivan Kopeček
Karel Pala
Masaryk University
Faculty of Informatics
Department of Information Technologies
Brno, Czech Republic
E-mail: {hales; kopecek; pala}@fi.muni.cz

ISSN 0302-9743
e-ISSN 1611-3349
ISBN 978-3-319-10815-5
e-ISBN 978-3-319-10816-2
DOI 10.1007/978-3-319-10816-2

Springer Cham Heidelberg New York Dordrecht London

Library of Congress Control Number: 2014946617

LNCS Sublibrary: SL 7 – Artificial Intelligence

Typesetting: Camera-ready by author, data conversion by Scientific Publishing Services, Chennai, India

Printed on acid-free paper

Springer is part of Springer Science+Business Media (www.springer.com)

Preface

The annual Text, Speech and Dialog Conference (TSD), which originated in 1998, is in the middle of its second decade. So far more than 1,000 authors from 45 countries have contributed to the proceedings. TSD constitutes a recognized platform for the presentation and discussion of state-of-the-art technology and recent achievements in the field of natural language processing. It has become an interdisciplinary forum, interweaving the themes of speech technology and language processing. The conference attracts researchers not only from Central and Eastern Europe but also from other parts of the world. Indeed, one of its goals has always been to bring together NLP researchers with different interests from different parts of the world and to promote their mutual cooperation.

One of the ambitions of the conference is, as its title says, not only to deal with dialog systems as such, but also to contribute to improving dialog between researchers in the two areas of NLP, i.e., between text and speech people. In our view, the TSD Conference was successful in this respect in 2014 again.

This volume contains the proceedings of the 17th TSD Conference, held in Brno, Czech Republic, in September 2014. In the review process, 70 papers were accepted out of 143 submitted, an acceptance rate of 49%.

We would like to thank all the authors for the efforts they put into their submissions and the members of the Program Committee and reviewers who did a wonderful job in helping us to select the most appropriate papers. We are also grateful to the invited speakers for their contributions. Their talks provided insight into important current issues, applications, and techniques related to the conference topics.

Special thanks are due to the members of the Local Organizing Committee for their tireless effort in organizing the conference.

The TEXpertise of Petr Sojka resulted in the production of the volume that you are holding in your hands.

We hope that the readers will benefit from the results of this event and disseminate the ideas of the TSD Conference all over the world. Enjoy the proceedings!

July 2014

Aleš Horák
Ivan Kopeček
Karel Pala
Petr Sojka

Preface

Organization

TSD 2014 was organized by the Faculty of Informatics, Masaryk University, in cooperation with the Faculty of Applied Sciences, University of West Bohemia in Plzeň. The conference webpage is located at http://www.tsdconference.org

Program Committee

Nöth, Elmar, Germany, *General Chair*
Agirre, Eneko, Spain
Baudoin, Geneviève, France
Cook, Paul, Australia
Černocký, Jan, Czech Republic
Dobrišek, Simon, Slovenia
Evgrafova, Karina, Russia
Fiser, Darja, Slovenia
Garabík, Radovan, Slovakia
Gelbukh, Alexander, Mexico
Guthrie, Louise, UK
Hajič, Jan, Czech Republic
Hajičová, Eva, Czech Republic
Haralambous, Yannis, France
Hermansky, Hynek, USA
Hitzenberger, Ludwig, Germany
Hlaváčová, Jaroslava, Czech Republic
Horák, Aleš, Czech Republic
Hovy, Eduard, USA
Khokhlova, Maria, Russia
Kocharov, Daniil, Russia
Kopeček, Ivan, Czech Republic
Kordoni, Valia, Germany
Krauwer, Steven, The Netherlands
Kunzmann, Siegfried, Germany
Loukachevitch, Natalija, Russia
Matoušek, Václav, Czech Republic
McCarthy, Diana, UK
Mihelić, France, Slovenia
Ney, Hermann, Germany
Oliva, Karel, Czech Republic
Pala, Karel, Czech Republic
Pavesić, Nikola, Slovenia
Pianesi, Fabio, Italy
Piasecki, Maciej, Poland
Przepiorkowski, Adam, Poland
Psutka, Josef, Czech Republic
Pustejovsky, James, USA
Rigau, German, Spain
Rothkrantz, Leon, The Netherlands
Rumshinsky, Anna, USA
Rusko, Milan, Slovakia
Sazhok, Mykola, Ukraine
Skrelin, Pavel, Russia
Smrž, Pavel, Czech Republic
Sojka, Petr, Czech Republic
Steidl, Stefan, Germany
Stemmer, Georg, Germany
Tadić, Marko, Croatia
Varadi, Tamas, Hungary
Vetulani, Zygmunt, Poland
Wiggers, Pascal, The Netherlands
Wilks, Yorick, UK
Wołinski, Marcin, Poland
Zakharov, Victor, Russia

Additional Referees

Agerr, Rodrigo
Fedorov, Yevgen
Gonzalez-Agirre, Aitor
Grzl, František
Hana, Jirka
Hajdinjak, Melita
Hlaváčková, Dana
Holub, Martin
Jakubíček, Miloš
Otegi, Arantxa
Veselý, Karel
Veselovská, Kateřina
Wang, Xinglong
Waver, Aleksander

Organizing Committee

Aleš Horák *(Co-chair)*, Ivan Kopeček, Karel Pala *(Co-chair)*, Adam Rambousek *(Web System)*, Pavel Rychlý, Petr Sojka *(Proceedings)*

Sponsors and Support

The TSD conference is regularly supported by the International Speech Communication Association (ISCA). We would like to express our thanks to the Lexical Computing Ltd. and IBM Česká republika, spol. s r. o. for their kind sponsoring contribution to TSD 2014.

Table of Contents

Invited Papers

Text

Speech

Dialogue

Invited Papers

An Information Extraction Customizer

Ralph Grishman and Yifan He

Department of Computer Science, New York University
New York, NY, USA
{grishman,yhe}@cs.nyu.edu
http://nlp.cs.nyu.edu

Abstract. When an information extraction system is applied to a new task or domain, we must specify the classes of entities and relations to be extracted. This is best done by a subject matter expert, who may have little training in NLP. To meet this need, we have developed a toolset which is able to analyze a corpus and aid the user in building the specifications of the entity and relation types.

Keywords: information extraction, distributional analysis.

1 Introduction

Information extraction (IE) systems, by rendering selected types of relations and events in natural language text into a structured form, make it possible to access, analyze, and reason about this information. A major obstacle to the wider use of IE is the need to customize the IE system to the particular relations and events of interest. There has been considerable research over the past 20 years to reduce the effort required to customize a system, using semi-supervised methods, unsupervised methods, and most recently distant supervision [7]. While there has been substantial progress, there are still significant limits on the performance of systems produced through largely-automatic customization.

In consequence, a number of research groups have developed interactive tools to aid in this customization. This paper describes one such system, integrating tools for corpus analysis and linguistic model building. In particular, our long term objective is to see whether such a tool can be effectively used after limited training by someone who may be a subject matter expert but is not an expert in computational linguistics. To this end, we set the following design goals:

- interaction with the user should be in terms of text examples; use of NLP terminology and formalisms should be minimized;
- the system should have a flexible control structure: a user who knows what to do next should be free to do it; a user who is temporarily at a loss can be guided by the system;
- guidance and assistance should be provided through corpus analysis (in particular, distributional analysis) tools.

P. Sojka et al. (Eds.): TSD 2014, LNAI 8655, pp. 3–10, 2014.

2 Extraction System

The system which is being customized is the NYU JET [Java Extraction Toolkit] [8], which was initially developed both for instructional purposes and for participation in the ACE [Automatic Content Extraction] evaluations [1]. It includes general linguistic analysis tools (for part-of-speech tagging, chunking, name tagging, coreference resolution, etc.) and tools for extracting mentions of entities, relations, and events.

The entity extractor identifies noun phrases headed by names or common nouns and classifies them by semantic type and subtype. The base system uses the 7 main types from ACE (person, organization, geo-political entity, location, facility, vehicle, and weapon). Nouns are classified by look-up in a semantic dictionary; names are identified and classified using a Maximum Entropy Markov Model (MEMM). Both dictionary and MEMM are trained from the ACE corpora.

The relation extractor identifies semantic relations between pairs of entities. There are approximately 20 types and subtypes of relations defined by ACE; the details changed from year to year during the series of evaluations. ACE required that a relation mention link two entity mentions in the same sentence, so a relation extractor can be structured as a classifier over pairs of entity mentions in the same sentence. JET includes several such classifiers, including one reliant only on the lexicalized dependency path between the entity mentions and one based on a maximum entropy classifier involving several dozen features.

The linguistic analysis tools of JET are used for the corpus analysis functions of the customizer. The one major analysis component imported into JET is a fast dependency analyzer developed by Tratz and Hovy at ISI [15] based on Goldberg's Easy-First parsing procedure [6]. The parser produces trees with fine-grained dependency labels similar to those of the Stanford parser. As we shall see, it is used as the basis for both distributional analysis and feature generation for relations.

3 Customization Tool: Entities

The seven main entity types cover many of the entities in news stories. Collections on specific topics, however, are likely to involve additional types. Reports on border security will involve drugs and other contraband; reports on energy will involve types of fuel; business news will involve financial instruments. Given a new corpus, we will want to identify the new classes and assemble the entity sets for these classes.

Our basic approach is to leave the user in control but to provide as much support as possible. Typically, the user can provide a few examples of a phenomenon (entity or relation type) but has difficulty producing a comprehensive list; users will find it much easier to judge examples provided by the system than to come up with additional examples on their own. So we ask the user to provide some *seed* examples, let the system display additional examples based on some similarity metric, and ask the user to pass judgement on these. Since the user is in control, it is also important that the state of the system be relatively *transparent*, so that the user is able to inspect (and possibly correct) the current models.

To find good candidate seeds around which to build our new entity classes, we start with a simple word frequency tool. Words in the corpus are POS tagged and lemmatized; we provide a ranked list with the frequency of each lemma, or the relative frequency compared to a background corpus (e.g., a general news corpus). We can limit the list to terms not yet assigned a semantic type.

The list will also include common multi-word terms in the corpus. To identify multi-word terms, we first tag the corpus with a noun-phrase chunker. If the noun chunk has N adjectives and nouns preceding the head noun, we obtain $N + 1$ multi-word term candidates consisting of the head noun and its preceding i $(0 \leq i \leq N)$ nouns and adjectives. We count the absolute frequency of these candidates and rank them with a ratio score, which is the relative frequency compared to a background corpus. We use the ratio score S_t to measure the representativeness of term t with regard to the given domain.

The top-ranked terms will typically include good candidates to serve as seeds for building new entity sets. If a user has some acquaintance with the domain, they will pick out a few terms as a seed set.

Given a seed set, we compute the distributional similarity of all terms in the corpus with the centroid of the seeds, using the dependency analysis as the basis for computing term contexts. We represent each term with a vector that encodes its syntactic context, which is the label of the dependency relation attached to the term in conjunction with the term's governor or dependent in that relation.

Consider the entity set of *drugs* in a corpus that describes drug enforcement actions. Drugs will often appear in the dependency relations *dobj(sell, drug)* and *dobj(sieze, drug)* (where *dobj* is the direct object relation), thus members in the *drugs* set will share the features *dobj_sell* and *dobj_sieze*. We use PMI [pointwise mutual information] to weight the feature vectors and use a cosine metric to measure the similarity between two term vectors.

The terms are displayed as a ranked list, and the user can accept or reject individual members of the entity set. At any point the user can recompute the similarities and rerank the list (where the ranking is based the centroids of the accepted and rejected terms, following [11]). When the user is satisfied, the entity set can be saved within Jet, and the set of accepted terms will become a new semantic type for tagging further text.

If the user is exploring a new domain, the choice of suitable seeds may not be clear. In such cases the system can make a suggestion. We select seeds with an agglomerative clustering procedure, which consists of the following steps:

- **Intialization.** We assign each term to a cluster which contains only the term itself.
- **Clustering.** We perform agglomerative clustering for several iterations. In each iteration, we merge the two most similar clusters. Similarity is measured using the cosine similarity of the cluster centroids. We stop when we have a cluster that has more than six terms.
- **Cluster scoring.** For each cluster we obtain, we calculate a cluster score S_c which is log sum of its members. i.e. $S_c = \sum_{t \in c} \log S_t$
- **Seed selection.** We select the two terms that are closest to the centroid of the cluster which has the maximum cluster score.

If the suggested seeds seem reasonable, the user can then proceed to expand the entity set as described above.

4 Evaluating the Entity Tool

To test the effectiveness of our procedure for acquiring new entity sets, we measured its performance in creating new *drug* and *law enforcement agent* (*agents*) entity types from a collection of approximately 5,000 web news posts from the U.S. Drug Enforcement Administration[1]. We first extracted the terms in this corpus and manually produced two lists of terms: drug names and law enforcement agent mentions. We extracted 3,703 terms from this corpus and identified 119 drug names and 97 law enforcement agent mentions in our list (the "gold standard" sets).

We then ran a simulated version of our customizer in the following manner: 1) we provided the entity set expansion program with two seeds. For the *drug* set, we provided "methamphetamine" and "oxycodone" (frequent terms in the corpus), while for the *agents* set, we provided "special agents" and "law enforcement officers"; 2) the program produced a ranked list of terms; 3) in each iteration, we examined the top N ($N = 20$ in our setting) new terms that had not been examined in previous iterations; 4) if a term is in the gold standard set, we added it to the expander as a positive seed, otherwise, we added it as a negative seed; 5) we continued the expansion with the updated seed set, and stopped after the kth ($k = 10$ in our setting) iteration.

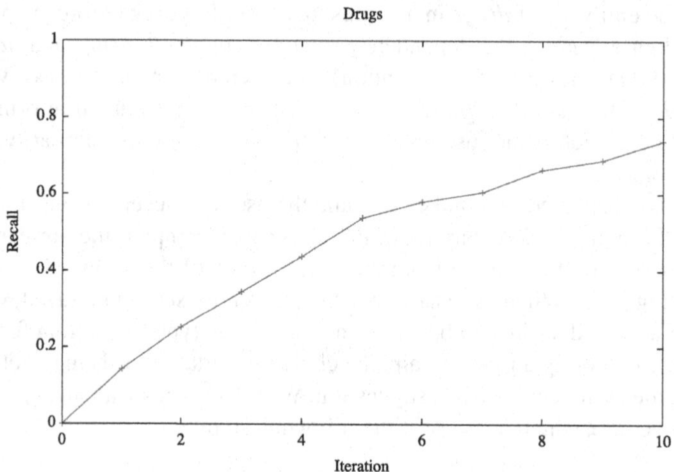

Fig. 1. Entity set expansion: recall at each iteration for the *drugs* set

We measured recall (fraction of terms found) at each iteration of our active learning process and show the result in Figures 1 and 2. Given two seeds, our active learning procedure can help the user to define a new entity set rapidly: for the *drugs* set, the user will be able to find more than 30% of all drugs after 3 iterations (i.e. reviewing 60 of the 3,703 terms); for the *agents* set, the initial 3 iterations will cover more than 40% of the agents. After 200 terms are reviewed in 10 iterations, the user will be able to build entity sets that cover more than 70% of the *drugs* or 80% of the *agents*.

[1] `http://www.justice.gov/dea/index.shtml`

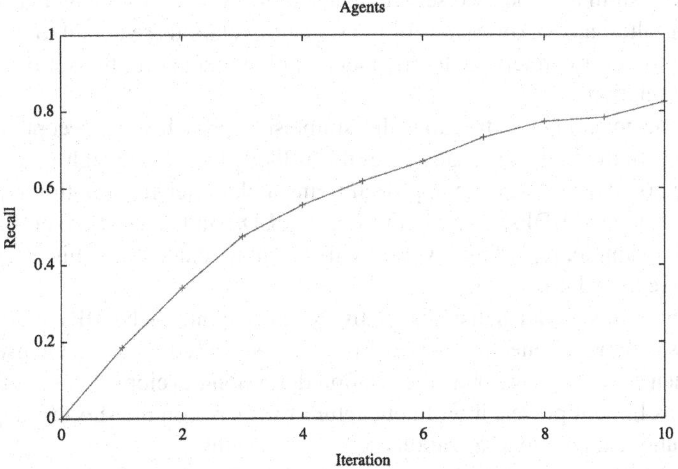

Fig. 2. Entity set expansion: recall at each iteration for the *agents* set

5 Customization Tool: Relations

The semantic relation (if any) between two entity mentions is generally (but not always) determined by the structure connecting the entities in the sentence, and in particular by the lexicalized dependency path between them. The lexicalized dependency path (LDP) includes both the labels of the dependency arcs and the lemmatized form of the lexical items along the path. For example, for the sentence "Cats chase mice." the path from "cats" to "mice" would be $nsubj^{-1}$ *chase dobj*, where the $^{-1}$ indicates that the *nsubj* arc is being traversed from dependent to governor. The LDP connecting two entity mentions, together with the semantic types of the two entities, has been shown to be a good predictor of the semantic relation (if any) between the entities [3].

We collect *all* the dependency paths connecting entity mentions in the corpus, and present them to the user ranked by frequency. As in the case of terms, we can present either the absolute frequencies or the relative frequency compared to a background corpus. This gives the user an overview of the most salient structures in the new corpus. In keeping with our general approach of hiding the linguistic notation from the user, we convert each dependency path back to a phrase which could have generated that path before displaying it to the user.

Just as the user has built up entity sets from seed terms, the user can now build up models of relations from seed LDPs. We are experimenting with two models.

The first model is simply a set of LDPs. The learner is a bootstrap learner: starting with a seed LDP, it gathers all the pairs of arguments (endpoints) which appear with this LDP in the corpus. It then collects all other LDPs which connect any of these pairs in the corpus, and presents these LDPs to the user for assessment. If the set of argument pairs connected by any of the seeds is S and the set of argument pairs of a candidate LDP x is $\mathcal{X}$, the candidate LDPs are ranked by $\mid \mathcal{S} \cap \mathcal{X} \mid / \mid \mathcal{X} \mid$, so that LDPs most distributionally similar to the seed set are ranked highest. The LDPs which are accepted by the user as alternative expressions of the semantic relation are added to the seed set. At any point the user can terminate the bootstrapping and accept the set of LDPs as a model of the relation.

In using the model for extraction, the simplest approach is to accept only exact matches to one of the LDP's. In many cases this is likely to produce high-precision, low-recall extractors. A more flexible approach trains a classifier from both the positively and negatively tagged LDPs, using both the exact LDP and a set of generalizations of the LDP (for example, replacing a word by its semantic class) [3]. This enables some trade-off of recall and precision.

The benefit of this model is that it is relatively transparent. Each LDP can be rendered to the user as a phrase connecting two entities of the specified types, so the user can see the list of such phrases growing as the relation definition develops. On the other hand, this model has limited power: it will not capture words not on the shortest dependency path, and cannot capture generalizations across such paths.

We have therefore also incorporated a second model, a maximum entropy model based on a rich set of features involving the context of the two entity mentions. Examples are presented to the user one-by-one using an extension of an active learning strategy termed *cotesting*; details are described in [13] and [5].

6 Related Work

Several groups have developed integrated systems for IE development:

The extreme extraction system from BBN [4] is similar in several regards: it is based on an extraction system initially developed for ACE, allows for the customization of entities and relations, and uses bootstrapping and active learning. However, in contrast to our system, it is aimed at skilled computational linguists.

The Language Computer Corporation has described several tools developed to rapidly extend an information extraction system to a new task [9,14]. Here too the emphasis is on tools for use by experienced IE system developers. Events and relations are recognized using finite-state rules, with meta-rules to efficiently capture syntactic variants and a provision for supervised learning of rules from annotated corpora.

A few groups have focused on use by NLP novices:

The WizIE system from IBM Research [10] is based on a finite-state rule language. Users prepare some sample annotated texts and are then guided in preparing an extraction plan (sequences of rule applications) and in writing the individual rules. IE development is seen as a rule programming task. This offers less in the way of linguistic support (corpus analysis, syntactic analysis) but can provide greater flexibility for extraction tasks where linguistic models are a poor fit.

The PropMiner system from T. U. Berlin [2] takes an approach more similar to our own. In particular, it is based on a dependency analysis of the text corpus and emphasizes exploratory development of the IE system, supported by search operations over the dependency structures. However, the responsibility for generalizing initial patterns lies primarily with the user, whereas we support the generalization process through distributional analysis.

7 Future Work

One limitation of the current system is the absence of a tool for capturing new event types (beyond those present in ACE). Some event types can be represented as binary relations, and further accomodation can be achieved by allowing time and location modifiers. However, to handle genuine 3-argument events ("X gave Y to Z") as well as binary events where one argument is omitted ("the American attack") a more general mechanism is required, such as the merging of alternative argument patterns described in [12].

A second limitation is the use of lists (rather than a statistical model) for names belonging to a new entity type. This can be addressed by annotating a single corpus using both the baseline statistical name tagger and the list of new names, training a new statistical model from this corpus, and then (optionally) refining this combined model through active learning.

The greatest challenge, however, will lie in creating a system which is useable by non-NLP experts. This will undoubtably require a number of iterations as we address the needs, concerns, and misunderstandings of users.

Acknowledgments. Supported by the Intelligence Advanced Research Projects Activity (IARPA) via Air Force Research Laboratory (AFRL) contract number FA8650-10-C-7058. The U.S. Government is authorized to reproduce and distribute reprints for Governmental purposes notwithstanding any copyright annotation thereon. The views and conclusions contained herein are those of the authors and should not be interpreted as necessarily representing the official policies or endorsements, either expressed or implied, of IARPA, AFRL, or the U.S. Government.

References

1. Automatic Content Extraction (ACE), `https://www.ldc.upenn.edu/collaborations/past-projects/ace`
2. Akbik, A., Konomi, O., Melnikov, M.: Propminer: A workflow for interactive information extraction and exploration using dependency trees. In: Proceedings of the 51st Annual Meeting of the Association for Computational Linguistics: Systems Demonstrations, pp. 157–162 (2013)
3. Bunescu, R., Mooney, R.: A shortest path dependency kernel for relation extraction. In: Proceedings of the Human Language Technology Conference and Conference on Empirical Methods in Natural Language Processing, pp. 724–731 (2005)
4. Freedman, M., Ramshaw, L., Boschee, E., Gabbard, R., Kratkiewicz, G., Ward, N., Weischedel, R.: Extreme Extraction – Machine Reading in a Week. In: Proceedings of the 2011 Conference on Empirical Methods in Natural Language Processing, pp. 1437–1446 (2011)
5. Fu, L., Grishman, R.: An Efficient Active Learning Framework for New Relation Types. In: Proceedings of International Joint Conference on Natural Language Processing (IJCNLP), Nagoya, Japan, pp. 692–698 (2013)
6. Goldberg, Y., Elhadad, M.: An efficient algorithm for easy-first non-directional dependency parsing. In: Human Language Technologies: The 2010 Annual Conference of the North American Chapter of the ACL, pp. 742–750 (2010)

7. Grishman, R.: Information Extraction: Capabilities and Challenges. The 2012 International Winter School in Language and Speech Technologies, Rovira i Virgili University, Spain (2012), http://cs.nyu.edu/grishman/survey.html
8. Java Extraction Toolkit, http://cs.nyu.edu/grishman/jet/jet.html
9. Lehmann, J., Monahan, S., Nezda, L., Jung, A., Shi, Y.: Approaches to Knowledge Base Population at TAC 2010. In: Proceedings of the 2010 Text Analysis Conference (2010)
10. Li, Y., Chiticariu, L., Yang, H., Reiss, F., Carreno-fuentes, A.: WizIE: A best practices guided development environment for information extraction. In: Proceedings of the 50th Annual Meeting of the Association for Computational Linguistics, pp. 109–114 (2012)
11. Min, B., Grishman, R.: Fine-grained entity refinement with user feedback. In: Proceedings of RANLP 2011 Workshop on Information Extraction and Knowledge Acquisition, pp. 2–6 (2011)
12. Shinyama, Y., Sekine, S.: Preemptive information extraction using unrestricted relation discovery. In: Proceedings of the Human Language Technology Conference of the NAACL, pp. 304–311 (2006)
13. Sun, A., Grishman, R.: Active learning for relation type extension with local and global data views. In: Proc. 21st ACM International Conf. on Information and Knowledge Management (CIKM 2012), pp. 1105–1112 (2012)
14. Surdeanu, M., Harabagiu, S.: Infrastructure for Open-Domain Information Extraction. In: Proceedings of the Second International Conference on Human Language Technology Research, HLT 2002, pp. 325–330 (2002)
15. Tratz, S., Hovy, E.: A fast, effective, non-projective, semantically-enriched parser. In: Proceedings of the 2011 Conference on Empirical Methods in Natural Language Processing, Edinburgh, Scotland, UK, pp. 1257–1268 (2011)

Entailment Graphs for Text Analytics in the Excitement Project

Bernardo Magnini[1], Ido Dagan[2], Günter Neumann[3], and Sebastian Pado[4]

[1] Fondazione Bruno Kessler, Trento, Italy
magnini@fbk.eu
[2] Bar Ilan University, Israel
dagan@cs.biu.ac.il
[3] DFKI, Germany
neumann@dfki.de
[4] Stuttgart University
pado@ims.uni-stuttgart.de

Abstract. In the last years, a relevant research line in Natural Language Processing has focused on detecting semantic relations among portions of text, including entailment, similarity, temporal relations, and, with a less degree, causality. The attention on such semantic relations has raised the demand to move towards more informative meaning representations, which express properties of concepts and relations among them. This demand triggered research on "statement entailment graphs", where nodes are natural language statements (propositions), comprising of predicates with their arguments and modifiers, while edges represent entailment relations between nodes.

We report initial research that defines the properties of entailment graphs and their potential applications. Particularly, we show how entailment graphs are profitably used in the context of the European project EXCITEMENT, where they are applied for the analysis of customer interactions across multiple channels, including speech, email, chat and social media, and multiple languages (English, German, Italian).

Keywords: Semantic inferences, textual entailment, text analytics.

1 Introduction

Textual entailment [4] suggests a long-term research direction where language understanding can take advantage from the capacity to resolve text-to-text semantic inferences. Among the others, the community has focused on two related aspects: the Recognizing Textual Entailment (RTE) shared task [3], aimed at capturing entailment between two portions of text, and knowledge acquisition, which aims at large-scale acquisition of entailment rules. On the application side, text-to-text semantic inferences may help in several scenarios where content analytics needs to be exploited. In this area taxonomy-based representations are currently widely used to model compactly large amounts of textual data. However, while current methods allow organizing knowledge at the lexical level (keywords/concepts/topics), there is an increasing demand to move towards more informative representations, which express properties of concepts and relations

P. Sojka et al. (Eds.): TSD 2014, LNAI 8655, pp. 11–18, 2014.

among them. This demand triggered our research on statement entailment graphs. In these graphs, nodes are natural language statements (propositions), comprising of predicates with their arguments and modifiers, while edges represent entailment relations between nodes. In this paper we report initial research that defines the properties of entailment graphs and their potential applications. Particularly, we show how entailment graphs can be profitably used for both knowledge acquisition and text exploration. Beyond providing a rich and informative representation, statement entailment graphs allow integrating multiple semantic inferences. So far, textual inference research focused on single, mutually independent, entailment judgments. However, in many scenarios there are dependencies among Text/Hypothesis pairs, which need to be captured consistently. This calls for global optimization algorithms for inter-dependent entailment judgments, taking advantage of the overall entailment graph structure (e.g. ensuring entailment graph transitivity).

From the applied perspective, we are experimenting with entailment graphs in the context of the EXCITEMENT project[1] industrial scenarios. We focus on the text analytics domain, and particularly on the analysis of customer interactions across multiple channels, including speech, email, chat and social media, and multiple languages (English, German, Italian). For example, we would like to recognize that the complaint *"they charge too much for sandwiches"* entails *"food is too expensive"*, and allow an analyst to compactly navigate through an entailment graph that consolidates the information structure of a large number of customer statements. Our eventual applied goal is to develop a new generation of inference-based text exploration applications, which will enable businesses to better analyze their diverse and often unpredicted client content.

The paper is structured as follows. Section 2 introduces the general aspects of entailment graphs as a tool for text analytics. Section 3 presents the architecture of the platform for textual inferences, which is at the core of the construction of entailment graphs. Finally, Section 4 provides an overview of the use and of the advantages of entailment graphs applied for customer interaction analysis.

2 Entailment Graphs

Recently, entailment graphs have been proposed ([1], [2]) as an efficient and informative organization of entailment rules automatically acquired from corpora. In this context, a node in a entailment graph is supposed to represent a simple statement composed by a predicate with its (possibly typed) arguments, while direct edges among nodes indicate an entailment relation. As an example, from [1], a node like "X-reduce-nausea" entails a node like "X-help-with-nausea", assuming the same instantiation for the X variable. In a entailment graph nodes are natural language statements (propositions), comprising of predicates with their arguments and modifiers, while edges represent entailment relations between nodes. Additionally, given that entailment is a directed and transitive relation, a well formed entailment graph should preserve transitivity among connected nodes.

In the EXCITEMENT project we consider entailment graphs based on more complex statements, where, in addition to a single predicate, we include grammatical modifiers

[1] `http://www.excitement-project.eu`

both of the predicate and of its arguments. For example, the sentence *"Lights in night trains are annoyingly bright"*, expresses a predication about the lights in a particular kind of trains. Within a statement we individuate the following basic elements: a top level predicate, usually corresponding to the root node in a dependency tree, expressed by the word "bright" in the example, one or more arguments of the predicate (i.e. the word "lights"), and a number of grammatical modifiers, either of the predicate (i.e. "annoyingly") or of the arguments of the predicate (i.e. "in night trains").

Grammatical modifiers, i.e. tokens which can be removed from the statement without affecting its comprehension, are represented as dependencies of the modifier from other tokens. As an example, given the following output of the Stanford Dependency Parser [6] for our statement:

```
nsubj(bright-7, Lights-1)
nn(trains-4, night-3)
prep_in(Lights-1, trains-4)
cop(bright-7, are-5)
advmod(bright-7, annoyingly-6)
root(ROOT-0, bright-7)
```

we can derive the following head-modifier dependencies: *trains* depends on *lights*, *night* on *trains* and *annoyingly* on *bright*, and the preposition-object dependencies *in-2* on *trains-4* and *trains-4* on *in-2*. Assuming that each head-modifier dependency indicates an entailment relation between the statement with the modifier (e.g. "Lights in night trains are annoyingly bright") and the statement without the modifier (e.g. "Lights are annoyingly bright"), we can recursively apply the procedure to build an entailment graph for the initial statement. The resulting graph is a Directed Acyclic Graph (DAG) generated over the partial order relation determined by the set inclusion over the modifiers in the statement. Figure 2 shows an example of entailment graph. In fact, the resulting DAG for a single statement is a lattice, with the maximal and the minimal statements being respectively the top and the bottom nodes [9], and where the entailment relations are ordered chains in the lattice.

While the construction of the entailment graph for a single statement requires syntactic knowledge (i.e. dependency relations), a more complex situation occurs when it is necessary to merge two entailment graphs, where a broader range of knowledge is necessary to compare two statements. This is addressed exploiting the potential of a text-to-text inference engine, described in the next Section.

3 The Excitement Open Platform (EOP)

A major result of the project is the release of the EXCITEMENT Open Platform (EOP). The goal of the platform is to provide functionality for the automatic identification of entailment relations among texts. The EOP is based on a modular architecture with a particular focus on *language-independent* algorithms. It allows developers and users to combine linguistic pipelines, entailment algorithms and linguistic resources within and across languages with as little effort as possible. The result is an ideal software environment for experimenting and testing innovative approaches for textual inferences.

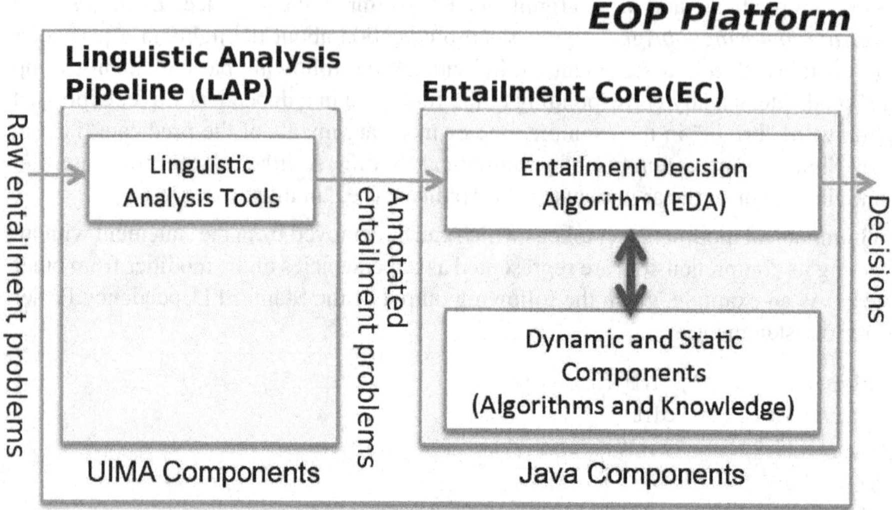

Fig. 1. EOP Architecture

The platform is distributed as an open source software[2] and its use is open both to users interested in using inference in applications and to developers willing to extend the current functionalities.

The EOP platform takes as input two text portions, the first called the Text (abbreviated with T), the second called the Hypothesis (abbreviated with H). The output is an entailment judgment, either "Entailment" if T entails H, or "NonEntailment" if the relation does not hold. A confidence score for the decision is also returned in both cases. The EOP architecture ([8], [7]) is based on the concept of modularization with pluggable and replaceable components to enable extension and customization. The overall structure is shown in Figure 1 and consists of two main parts. The Linguistic Analysis Pipeline (LAP) is a series of linguistic annotation components. The Entailment Core (EC) performs the actual entailment recognition. This separation ensures that (a) the components in the EC only rely on linguistic analysis in well-defined ways and (b) the LAP and EC can be run independently of each other. Configuration files are the principal means of configuring the EOP.

The Linguistic Analysis Pipeline is a collection of annotation components for Natural Language Processing (NLP) based on the Apache UIMA framework.[3] Annotations range from tokenization to part of speech tagging, chunking, Named Entity Recognition and parsing. The Entailment Core performs the actual entailment recognition based on the preprocessed text made by the Linguistic Analysis Pipeline. It consists of one or more Entailment Decision Algorithms (EDAs) and zero or more subordinate components. An EDA takes an entailment decision (i.e., "entailment" or "no entailment") while components provide static and dynamic information for the EDA. Scoring

[2] `http://hltfbk.github.io/Excitement-Open-Platform/`

[3] `http://uima.apache.org/`

Components accept a Text/Hypothesis pair as an input, and return a vector of scores. Their output can be used directly to build minimal classifier-based EDAs forming complete RTE systems.

The current version of the EOP platform ships with three EDAs corresponding to three different approaches to RTE: an EDA based on transformations between T and H, an EDA based on edit distance algorithms, and a classification based EDA using features extracted from T and H. Knowledge resources are crucial to recognize cases where T and H use different textual expressions (words, phrases) while preserving entailment. The EOP platform includes a wide range of knowledge resources, including lexical and syntactic resources, where some of them are grabbed from manual resources, like dictionaries, while others are learned automatically. Many EOP resources are inherited from pre-existing RTE systems migrated into the EOP platform, but now use the same interfaces, which makes them accessible in a uniform fashion.

Finally, the EOP infrastructure follows state-of-the-art software engineering standards to support both users and developers. In addition to communication channels, (e.g. mailing list, issue tracking, web site), the platform comprises a version control system, a rich documentation, an archive for storing results, and a package for continuous integration.

4 Analysing Customer Interactions

This Section provides details on the EXCITEMENT application scenario as well as how we are manually annotating datasets both for training and for evaluation.

Data are based on real customer interactions and business scenarios with high potential impact for the industrial partners of the project. The datasets cover three languages (English, German and Italian) and three communication channels (speech, email, and social media). All data comply with European and national privacy regulations and will be publicly distributed for research purposes under a Creative Commons license Attribution-NonCommercial-ShareAlike. Two different types of datasets - corresponding to the two main use cases addressed in the project were created performing different kinds of annotation. Section 4.1 describes the novel graph-based annotation, aimed at producing entailment graphs to be used for evaluation within Use Case 1 (text exploration), while Section 4.2 presents more traditional RTE-style entailment datasets, created to test entailment systems within Use Case 2 (information access).

The collected data (some hundreds of interactions for English and Italian) were anonymized where necessary, depending on the partners or customers restrictions. The following is an example (reported as it is, including orthographic and grammatical mistakes) of an anonymized interaction in English, where reasons for dissatisfaction in train service are reported:

> I would not recommend Quasigo as the the tickets are inflexible, I had to change at Moonport instead of Belville europe on the return journey, the food is terribly expensive and not good, there was a fight for luggage space, and on the return journey I did not get the table seats I had booked.

4.1 Use Case 1: Text Exploration

To create this dataset, we performed a novel graph-based annotation, which aims to build an entailment graph for a certain number of customer interactions pertaining to a given topic. A customer interaction can be a telephone call, a feedback received by email, a post on a social channel, while the topic gives a general reason for the interaction and can be a business event (e.g. express dissatisfaction, report billing problems, etc.) or any business case that the final user would like to explore.

The entailment graph creation starts from the customer interactions collected for a given topic and includes the following main steps:

- For each interaction, relevant text fragments are extracted (one or more fragment); each fragment contains a specific reason for complaining and corresponds to a single statement (see Section 2);
- For each fragment, a number of subfragments are extracted (see Section 2) and the corresponding entailment graph is manually created;
- All the fragment graphs are merged into the final entailment graph on the base of entailment judgments among statements provided by annotators.

Figure 2 shows an example of entailment graph. Statements within the same node are considered as equivalent (i.e. paraphrases); numbers on nodes indicate the frequency with which a certain statement occurs, while numbers on edges indicate the confidence of the entailment judgment. It is worth to notice that the statement-based representation provided by entailment graphs is much more informative (albeit still very compact) than labels currently provided by document categorization technology, resulting in a more effective tool for text analytics.

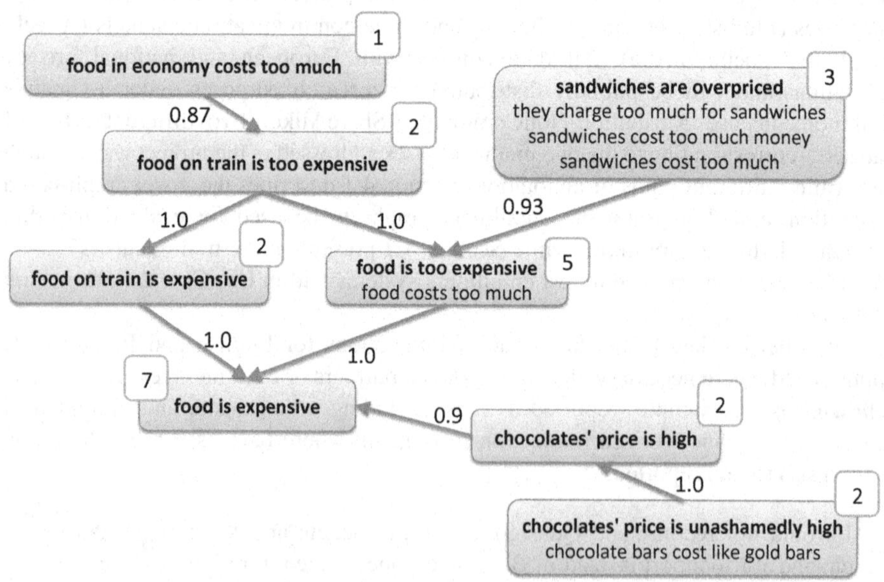

Fig. 2. Example of entailment graph for the text exploration use case

4.2 Use Case 2: Information Access

Use case 2 is based on a system used by agents in a support center for handling customers inquiries in email format. Customers issues are stored in a database along with their solutions after their handling. When a new email is received, the system maps the problems mentioned in the email with problems stored in the database and thus to automatically suggest a suitable solution to the agent. The mapping is done by ranking the relevance of each stored issue according to presence of keywords characteristic to the issue. Following this ranking process, the system displays the top ranked problems to the agent. In this scenario, the entailment engine (see Section 3 on the EOP) provides a more precise and effective statement-based ranking compared to the keyword-based ranking. Accordingly, a dataset is created following the RTE-style entailment annotation, where stand-alone sentence pairs composed of a text and a hypothesis are created and annotated with the corresponding entailment judgment. This annotation is used to evaluate the entailment engines in isolation, as well as the performance of the Use Case 2 about entailment-driven search at the statement level.

5 Conclusion

We have presented the main research lines that are being carried on in the context of the EXCITEMENT project. We are designing an innovative framework for a compact and effective representation of entailment relations among statements based on entailment graphs. We have shown how text-to-text entailment judgments are the core relations for building entailment graphs and we have presented the EOP, an open source platform that provides such text-to-text judgments. We are experimenting the use of entailment graphs and of the EOP for customer interaction analytics, where the relevant content of streams of interactions is detected and then organized in order to facilitate deeper analysis.

Several activities are still ongoing in the EXCITEMENT project. First, an extensive evaluation of the approach, based on manually annotated datasets, all of them publicly available, which we are using for both training and test purposes. A second research line involves a deep investigation of the strategies for automatically building entailment graphs with the EOP platform, including strategies for producing coherent graphs minimizing possible violations of transitivity of the entailment relation.

Acknowledgments. This work was partially supported by the EC-funded project EXCITEMENT (FP7ICT-287923).

References

1. Berant, J., Dagan, I., Goldberger, J.: Global Learning of Focused Entailment Graphs. In: Proceedings of the 48th Annual Meeting of the Association for Computational Linguistics, Uppsala, Sweden, pp. 1220–1229 (2010)
2. Berant, J., Dagan, I., Goldberger, J.: Global Learning of Typed Entailment Rules. In: Proceedings of the 49th Annual Meeting of the Association for Computational Linguistics: Human Language Technologies, Portland, Oregon, USA, pp. 610–619 (2011)

3. Dagan, I., Glickman, O., Magnini, B.: The Pascal recognising textual entailment challenge. In: Quiñonero-Candela, J., Dagan, I., Magnini, B., d'Alché-Buc, F. (eds.) MLCW 2005. LNCS (LNAI), vol. 3944, pp. 177–190. Springer, Heidelberg (2006)
4. Dagan, I., Dolan, B., Magnini, B., Roth, D.: Recognizing textual entailment: Rational, evaluation and approaches. Journal of Natural Language Engineering 15(04), i–xvii (2009)
5. Eads, P., McKay, B.: An Algorithm for generating subset of fixed size with a strong Minimal Change Property. Information Processing Letters 19, 131–133 (1984)
6. Klein, D., Manning, C.D.: Accurate Unlexicalized Parsing. In: Proceedings of the 41st Meeting of the Association for Computational Linguistics, pp. 423–430 (2003)
7. Magnini, B., Zanoli, R., Dagan, I., Eichler, K., Neumann, G., Noh, T.-G., Padó, S., Stern, A., Levy, O.: The Excitement Open Platform for Textual Inferences. To appear in Proceedings of the 52nd Meeting of the Association for Computational Linguistics, Demo papers (2014)
8. Padó, S., Noh, T.-G., Stern, A., Wang, R., Zanoli, R.: Design and Realization of a Modular Architecture for Textual Entailment. Journal of Natural Language Engineering (2014)
9. Wille, R.: Restructuring lattice theory: An approach based on hierarchies of concepts. In: Ferré, S., Rudolph, S. (eds.) ICFCA 2009. LNCS, vol. 5548, pp. 314–339. Springer, Heidelberg (2009)

Multi-lingual Text Leveling

Salim Roukos, Jerome Quin, and Todd Ward

IBM T. J. Watson Research Center, Yorktown Heights, NY 10598
{roukos,jlquinn,tward}@us.ibm.com

Abstract. Determining the language proficiency level required to understand a given text is a key requirement in vetting documents for use in second language learning. In this work, we describe our approach for developing an automatic text analytic to estimate the text difficulty level using the Interagency Language Roundtable (ILR) proficiency scale. The approach we take is to use machine translation to translate a non-English document into English and then use an English language trained ILR level detector. We achieve good results in predicting ILR levels with both human and machine translation of Farsi documents. We also report results on text leveling prediction on human translations into English of documents from 54 languages.

Keywords: Text Leveling, ILR Proficiency, Second Language Acquisition.

1 Introduction

As computerized tools for second language teaching become more widely available, the selection of content that is appropriate to a learner's proficiency in a language may benefit from the use of automatic text leveling tools. These tools can enable more automated content selection for personalized adaptive self-service language teaching systems. They can also support educators select more efficiently content that is contemporary and with the appropriate text difficulty level for their classes.

The Interagency Language Roundtable proficiency scale [1] of 11 levels ranges from level 0 to level 5 using half steps (0, 0+, 1, 1+, 2, 2+, 3, 3+, 4, 4+, 5). Level 0 indicates *no proficiency*, level 5 indicates *functional native proficiency* and level 3 indicates *general professional proficiency*.

In some contexts, there has been a significant investment in determining the ILR level of texts covering a variety of topics for multiple levels. However, updating the collection of documents to cover recent news, events, and topics can be a daunting task. The ILR text leveling guidelines are quite complicated, and the authors are not aware of any inter-annotator agreement studies for ILR level assignment. We report in this paper on some initial human ILR level annotation and the inter-annotator agreement reached in a preliminary annotation exercise. While we do not have access to expertly trained linguists, we attempted to assess the IAA that can be achieved with some training on ILR text difficulty level assessment. We report on these results in Section 2.

Recently, Shen et al [3] introduced their work to develop an automatic text leveling analytic for each of 4 languages. Their regression model is trained on about 2k documents from each language that have been annotated for ILR text difficulty level.

P. Sojka et al. (Eds.): TSD 2014, LNAI 8655, pp. 19–26, 2014.

The effort required to develop such a model for each new language is not scalable since it requires extensive training in ILR level labeling and applying it correctly for a new language. We propose an approach that relies on using machine translation from the source language to English and using an English-trained automatic text leveling analytic. This approach does not require text leveling annotation for the new language, though it requires a machine translation system. We report our initial results for Farsi documents.

Earlier work on text difficulty addressed the readability of a document based on the Flesch Reading Ease Formula, which uses two simple length features [2]: the average number of words per sentence and the average number of syllables per word. There has been various attempts at exploring weighing these features (linear regresion models) to improve the accuracy of predicting different readability levels. More recent work [4,5] used a richer feature set such as:

- average sentence length,
- average number of syllables per word,
- Flesch-Kincaid score,
- six out-of-vocabulary (OOV) rate scores,
- syntactic parse features, and
- 12 language model perplexity scores.

They also explored both classification and regression with SVMs to estimate the grade level (grades 2, 3, 4, and 5) of documents from the Weekly Reader newspaper. The richer models outperfomed the simpler Flesch-Kincaid score.

A similar feature set was used in a recent readability experiment conducted under the DARPA Machine Reading Program where *readability* was re-defined as the ease of reading of various styles of text as opposed to text level difficulty as addressed in earlier work. The range of documents cover various genres such newswire, transcriptions of conversational content, machine translation into English from other languages [6]. The methods used similar features ranging from parser-based features to n-gram language models.

Shen et al [3] used a corpus of 200 documents for each of seven levels (1, 1+, 2, 2+, 3, 3+, 4) for a given language. In their data, each of the texts was labeled by two independent linguists expertly trained in ILR level scoring. The ratings from these two linguists were then adjudicated by a third linguist. They did not provide inter-annotator agreement measures but took the adjudicated decision as the reference truth for both training and testing their system.

Using the fine grained ILR level training data, Shen et al developed regression models of text difficulty and proposed the use of mean square error (mse) metric where the plus-levels were mapped to the mid-point (e.g. 2+ is 2.5). They used a 80/20 split for training and test and built a separate regression model for each of the 4 languages. The best results were an mse of 0.2 for Arabic, 0.3 for Dari, 0.15 for English, and 0.36 for Pashto. These would correspond to a root mean square error (rmse) of 0.45, 0.55, 0.39, and 0.60 for each of the languages, respectively. Note a change of one level is an interval of 0.5 in their study.

They used two types of features: Length features and Word-usage features. The length features were three z-normalized length features:

1. average sentence length (in words) per document,
2. number of words per document, and
3. average word length (in characters) per document.

The Word-usage features were weighted word frequencies using TF-LOG weighted word frequencies on bag-of-words for each document. They compared length-based features which are not lexical to *Words-usage* features which are lexical items. The lexical features reduces the mse by 14% (Dari) to about 80% (Pashto). We are concerned about the word usage features. We surmise that the data used is more homogeneous than what is required for general second language acquisition (SLA) and may be influencing the significant performance improvement due to the *Words-usage* features since their leading examples of useful lexical features for English (which yielded a reduction of mse by 58%) appears to be topical. For example for level 3, the top ten lexical features, shown in Table 1, appear to be US politics centric.

Table 1. Top 10 Words-usage features and their weights for level 3

Word	Weight
obama	1.739
to	1.681
republicans	1.478
?	1.398
than	1.381
more	1.365
cells	1.355
american	1.338
americans	1.335
art	1.315

While it is hard to make solid claims about topicality without having access to the data, we are concerned about the robustness of the above results as we expect a sensitivity to topic change over time and geography for SLA content. For example, what would happen when the news is about French politics? Surely the names and parties will be different from the top indicators shown above.

In this work, we had access to data with single ILR level annotatio for 4 levels with coarser granularity spanning two consecutive levels (0+/1, 1+/2, 2+/3, 3+/4). The data had 5 broad topical areas and covered 54 languages. The texts were available in both the source language and its English human translation. We used these data to develop an ILR level detector based on English translations. While our work builds on Shen et al's results, we are different in 3 aspects: 1) we report initial measurements of ITA for human ILR text difficulty annotation, 2) our data set has coarser ILR annotation where a document was assigned a two level value (e.g 2+/3), and 3) our data have very broad variety of topics since it comes from 54 different languages.

The larger quantization interval of 1 versus an interval of 0.5 in the Shen et al study, implies that our mse error would be larger by 0.06 by definition, other factors being equal. Another aspect of our data is a skewed distribution of the levels with a severe under-representation of the 0+/1 level at 2% of the documents with the other categories at 23%, 58%, and 17%, respectively. We present, in Section 2, our work on ILR level annotation, in Section 3, the data set, in Section 4, our text leveling results, and in Section 5, our conclusions.

2 Text Leveling Annotation

We attempted to train a small pool of 5 annotators to perform the ILR text leveling of English documents. We had access to ILR level annotation to a set of documents and a multimedia course on ILR annotation that requires five to ten hours to go through. We performed our own training of the annotators by explaining the principles as best as we could. We conducted 5 rounds of annotations followed by feedback sessions comparing the annotation of the five annotators on the same set of about 40 English documents per round. These were human translations of various languages and covered content for both reading and listening comprehension.

Table 2. Text leveling: human annotator performance

	kt	mn	mr	rx
AVG TIME	0:02:31	0:07:24	0:03:14	0:03:30
AVG ABS ERROR	0.64	0.57	0.55	0.67
NUM ERRORS	34	28	28	36
LARGE ERR	3	3	4	7

We report our results on the fifth round of ILR Text leveling annotation (we dropped one of the annotators due to consistently poorer scores). We compared each annotator to the reference truth as provided in our data set and to the other annotators. We used 60 documents covering source languages: Dari, Persian, Somali, and Spanish. They covered 3 levels nominally 2, 3 and 4 (strictly speaking these should be represented as 1.75, 2.75, and 3.75 as our data was annotated by an intervals such as 1+/2 meaning a midpoint of 1.75). We show in Table 2, the average time an annotator took to perform the task per document, the average absolute error between the human and the reference, the number of documents that had a different label, and the number of documents where the error was more than one level (interval of 1) for each of our four annotators. We show, in Table 3, the mse and rmse comparing each annotator to the reference. The mse in this work by definition is 0.06 higher than the results of Shen et al due to the coarser granularity of our reference truth (1 unit interval instead of 0.5). On average all 4 annotators have a rmse of 0.72.

We also computed the Pearson correlation between the annotators as shown in Table 4. We computed the average correlation of one annotator to the other three, and found mr and rx to have the highest average correlation of 0.74 and 0.73 respectively. Computing the mse and rmse between mr and rx, we get 0.24 and 0.49, respectively

Table 3. Text leveling: human annotator performance

	kt	mn	mr	rx
mse	0.48	0.53	0.51	0.56
rmse	0.69	0.73	0.72	0.75

Table 4. Interannotator correlation

	mn	mr	rx
cor	mn	mr	rx
kt	0.59	0.64	0.68
mn		0.79	0.72
mr			0.79

which indicates interestingly a better agreement between the two annotators than with the reference where the average rmse for the 2 annotators is 0.73.

We were concerned that our annotators did not achieve a lower rmse than 0.72 relative to the reference and felt that the task is quite difficult for them. We decided not pursue our own annotation of text difficulty due to the larger investment required. The results with the human annotators' performance can be used as an indicator to assess how well our automatic text leveling analytic performs.

3 Text Leveling Data Set

Through our cooperation with the Center of Advanced Study of Language (CASL) at the University of Maryland, we were able to obtain a document collection with a single ILR text leveling annotation. The documents covered 5 broad topical areas and were evenly split between written (4.5k texts) and human transcribed genres (5k texts). The data were also provided with human English translations in addition to the spource language from 54 non-English languages. Table 5 shows the division by topic.

Table 5. Text leveling data set

Culture/ Society	Defense/ Security	Ecology/ Geography	Economics/ Politics	Science/ Technology
3,635	1,046	823	2,904	1,007

We received the data in two batches. The first one with about 2k documents and the second about 9k. Most of our results are based on the initial set of 2k documents. For the smaller condition we created a test set of 125 documents. We refer to the full set, as the larger condition, with a corresponding test set of 881 documents.

4 Experimental Results

We experimented with the following features:

- number of words in document length,
- average sentence length,
- average word length in characters,
- ratio of count of unique words (types) to total words,
- pronoun histogram,
- POS bigrams, and
- log term frequency.

We measure the performance by the classification accuracy, the mean square error (mse), and its nding root mean square (rms) error. We used a maximum entropy regression model. When we use the first three features, which are similar to the basic length features of earlier work, the level assignment accuracy is 66%, the mse is 0.37 with an rms of 0.60. Adding the remaining features listed above improves the accuracy to 77% and reduces the rmse to 0.51. Table 6 shows the confusion matrix for the full feature set.

Table 6. Confusion matrix between the 4 levels using the full feature set classifier

Level R/S	0.75	1.75	2.75	3.75
0.75	-	-	2	-
1.75	-	**15**	5	1
2.75	-	4	**70**	1
3.75	-	1	15	**11**

To evaluate the effect of machine translation on text leveling performance, we identified the largest subset of text material by source language in the smaller set which turned out to be Farsi. We had a Farsi test set of of 60 documents. We used a phrase-based Farsi-English translation system produce the machine translation version of the documents. We used the basic three feature set with the addition of ten binned vocabulary rank histogram. Table 7 compares human to machine translation in terms of accuracy, mse, and rms error. We can see that MT is relatively close to human translation though the rms on Farsi at 0.64 is higher than on the original set of 125 documents at 0.51.

Table 7. Performance with human and machine translation

	Accuracy	mse	rmse
Human translation	65%	0.41	0.64
Machine translation	57%	0.47	0.69

4.1 Experiments with the Larger Data Set

For the full set of 9k documents, we show the distribution by level in Table 8 which indicates the paucity of data for the first level and the dominance of the third level.

Table 8. Count of documents for each of the 4 levels

Level	0.75	1.75	2.75	3.75
Count	148	2,214	5,531	1,569

We compared the small and large training and test conditions. As can be seen in Table 9, the small trained model's rms error increases to 0.69 on the large test set. The large training set reduces the rms error from 0.69 to 0.54 on the large test set.

Table 9. RMSE using both the large and small training and test sets

Train/Test	small	large
small	0.63	0.69
large	0.58	0.54

5 Conclusion

We have built a text leveling system using an English training set of about 9k documents. The rms error of 0.54 achieved is comparable to the earlier work of Shen et al which had an average rms error across the 4 languages of 0.50 in spite of the larger quantization error in our data. Our approach depends on using machine translation instead of annotating for each new source language. Our results outperfom what our human annotators were able to achieve over 5 rounds of training annotations.

Acknowledgments. We gratefully acknowledge the help of the Center for the Advanced Study of Language at the University of Maryland. In particular, we want to thank Amy Weinberg, Catherine Daughty, and Jared Linck from CASL and Carol Van Ess-Dykema from DOD for their support in getting us access to the ILR annotated data set and the ILR annotation multimedia course.

References

1. Interagency Language Roundtable: ILR Skill Scale, `http://www.govtilr.org/Skills/ILRscale4.htm` (accessed June 15, 2014)
2. Flesch, R.: A new readability yardstick. Journal of Applied Psychology 32(3), 221–233 (1948)

3. Shen, W., Williams, J., Marius, T., Salesky, E.: A Language-Independent Approach to Automatic Text Difficulty Assessment for Second-Language Learners. In: Proceedings of the Workshop on Predicting and Improving Text Readability for Target Reader Populations, Sofia, Bulgaria, pp. 30–38 (2013)
4. Schwarm, S.E., Ostendorf, M.: Reading Level Assessment Using Support Vector Machines and Statistical Language Models. In: Proceedings of the 43rd Annual Meeting of the Association for Computational Linguistics (2005)
5. Petersen, S.E., Ostendorf, M.: A machine learning approach to reading level assessment. Computer Speech and Language 23, 89–106 (2009)
6. Kate, R.J., Xiaoqiang, L., Patwardhan, S., Franz, M., Florian, R., Mooney, R.J., Roukos, S., Welty, C.: Learning to predict readability using diverse linguistic features. In: Proceedings of the 23rd International Conference on Computational Linguistics, COLING 2010, pp. 546–554 (2010)

Text

"**Text**: a book or other written or printed work, regarded in terms of its content rather than its physical form: *a text which explores pain and grief.*"

NODE (The New Oxford Dictionary of English), Oxford, OUP, 1998, page 1998, meaning 1.

SuMACC Project's Corpus

A Topic-Based Query Extension Approach to Retrieve Multimedia Documents

Mohamed Morchid[1], Richard Dufour[1], Usman Niaz[2], Francis Bouvier[3], Clément de Groc[3], Claude de Loupy[3], Georges Linarès[1], Bernard Merialdo[2], and Bertrand Peralta[4,*]

[1] LIA - University of Avignon, Avignon, France
[2] Syllabs, Paris, France
[3] EURECOM, Sophia Antipolis, France
[4] WIKIO, Paris, France
{mohamed.morchid,richard.dufour, georges.linares}@univ-avignon.fr, {bouvier,groc,loupy}@syllabs.com, {usman.niaz,bernard.merialdo}@eurecom.fr, bertrand.peralta@ebuzzing.com

Abstract. The SuMACC project aims at automatically tracking new multimodal entities on Internet. The goal of the project is to propose robust multimedia methods that define relevant patterns allowing to automatically retrieve these entities. This paper describes the SuMACC corpus collected on video-sharing platforms using word-queries. Since concepts are limited to a single or few words, querying video-sharing platforms with the concept only can easily introduce irrelevant collected videos. In this paper, we propose to use an extended query obtained by mapping the initial concept into a topic space from a Latent Dirichlet Allocation (LDA) algorithm. This topic-based query extension approach allows to better retrieve videos related to the targeted concept. As a result, a corpus of 7,517 videos, extracted using the simple (*i.e.* concept only) and the extended queries, from 47 concepts, was obtained. Results show the effectiveness of the proposed thematic querying approach compared to the simple concept query in terms of relevance (+21%) and ambiguity (−4%). The annotation process as well as the corpus statistics are detailed in this paper.

Keywords: Multimedia corpus, Annotation, Latent Dirichlet Allocation, Topic modeling, Extended queries.

1 Introduction: The SuMACC Project

The search of a concept in multimedia database or on the Internet encounters major issues due to the diversity of concept representations, that may depend on one or several different modalities, such as pictures, video, speech, text, sounds... Typically, a

* This work was funded by the SuMACC project supported by the French National Research Agency (ANR) under contract ANR-10-CORD-007.

P. Sojka et al. (Eds.): TSD 2014, LNAI 8655, pp. 29–36, 2014.

concept such as *olympic games* may be mapped into videos of opening ceremony, in text documents focusing on a specific race, in radio shows... The SuMACC aproject[1] aims to develop models supporting these variabilities related to the multimedia contents, with a particular focus on the Web.

Methods for concept discovery and tracking in text documents have been largely studied in the last decades. These methods are now relatively mature and effective. Moreover, the video processing community produced great efforts to design methods for tracking concrete objects or object categories, especially in the TrecVid evaluation campaigns [8].

Nonetheless, multimodal approaches remain poorly developed, most of previous works proposing solutions for only one modality (video, audio or text). From a technological point of view, most of the identification methods are based on statistical models. To correctly estimate model parameters, a large amount of data is however mandatory. Collecting and annotating such large corpus is generally too costly, thus avoiding the emergence of multimedia approaches.

The SuMACC project addresses these two major issues related to the multimodal representation of concepts and to the training strategies that could enable a low-cost estimate of concept signatures. This paper describes the collected and annotated corpus used to propose multimodal searches and training strategies.

The next section presents the method followed to collect data, especially the query strategy based on a topic-representation of concepts. Section 3 describes the obtained corpus and its annotation protocol as well as a discuss about size, nature and quality of the collected database. Section 4 concludes and presents some future works.

2 Collecting Evaluation Data: Methodology

2.1 Motivation and Principle

To evaluate concept retrieval methods, plausible scenarios have to be simulated. These simulations require a set of realistic queries (*i.e.* that could be asked by users), a large video database in which targeted concepts will be searched, and tags that indicate if videos effectively contain the targeted concepts.

One of the major difficulty in collecting such a corpus is due to the fact that the data set is basically composed of videos obtained by requesting the search engines (SE) of video sharing platforms. The implicit video tagging performed by the SE can not be used as a ground truth: firstly, the SE makes errors; secondly, the collected database should be designed to enable simulation of a realistic information retrieval tasks. Consequently, the collected set has to contain not only the videos having the targeted concepts, but also ambiguous ones. Our proposal is to use a query extension method that allows the hit of videos related to the targeted concept and ambiguous ones.

The query generation process starts from the initial concept characterization. Each concept is expressed as a keyword (or a key expression) that could be used as a query to obtain concept-related videos. We expand this primary query by using closed keywords. In the context of information retrieval, most of the previous works proposed to expand

[1] http://sumacc.univ-avignon.fr/

Table 1. Statistics on the French Wikipedia dump for the topic space training

Characteristic	Statistic
Number of articles	3, 197, 395
Number of words ($\|D\|$)	898, 645, 071
Number of unique words (N)	13, 182, 180
Number of words per article	281.05
Number of unique words per article	4.12

the initial user query by using a vector space model [4], a word similarity matrix [6], the analysis of social networks [2] or visual descriptors [5]. We propose a query extension by mapping the concept into a topic space obtained by a Latent Dirichlet Allocation (LDA) method, presented in details in next section. By using such a topic space for query extension, we aim to introduce in the dataset not only negative examples, but ambiguous examples corresponding to ambiguous semantic contents.

2.2 LDA Approach

Latent Dirichlet Allocation (LDA) [3] is a generative probabilistic model which considers a document as a *bag-of-words* produced by a set of latent topics. Word occurrences are linked by latent variables that determines the distribution of topics in a document. This decomposition model of documents offers good generalization abilities compared to other generative models that are commonly used in automatic language processing such as Latent Semantic Indexing (LSI) or Probabilistic Latent Semantic Indexing (PLSI) [1,7].

A topic z, associated with a LDA class, is represented by a vector V_z, whose coefficients represent the probability of words w_i knowing the topic z:

$$V_i^z = P(w_i|z)$$

This method requires a large dataset to build a global model. Our training corpus D is composed by documents from the French Wikipedia dump (see Table 1), containing about 900 million words.

2.3 Building Query Set

Videos are collected by querying the Syllabs multimedia fetcher, that allows to search videos from four video sharing platforms: DailyMotion, Youtube, Vimeo and Flickr. The queries are composed of the initial concept (*i.e.* one keyword) and a set of extended keywords obtained with a LDA-based technique; this method consists in identifying the *n-best* words of the closest topic of the concept. The set of n words is considered as the first expanded query. This first extension step is followed by a second one, where we use the first expanded query to select the *k-best* concepts. Finally, a set of m words are extracted by cumulating conditional word probabilities belonging to the m best LDA classes. This step yields to get the m most relevant words for each initial concept. This second expanded query is submitted to the Syllabs multimedia fetcher to collect videos matching the final query. Therefore, the document retrieval process performs 4 steps:

- building an off-line thematic representation;
- mapping the concept into the topic space to extract the best topic z. Then, m words of the topic z are chosen to compose a first expanded query;
- mapping this first expanded query into the topic space to find its k closest topics. Then, scoring each word of $\mathbf{V}$ to find the final *expanded query*;
- sending each expanded query (and the initial one) to a multimedia fetcher to retrieve multimedia documents.

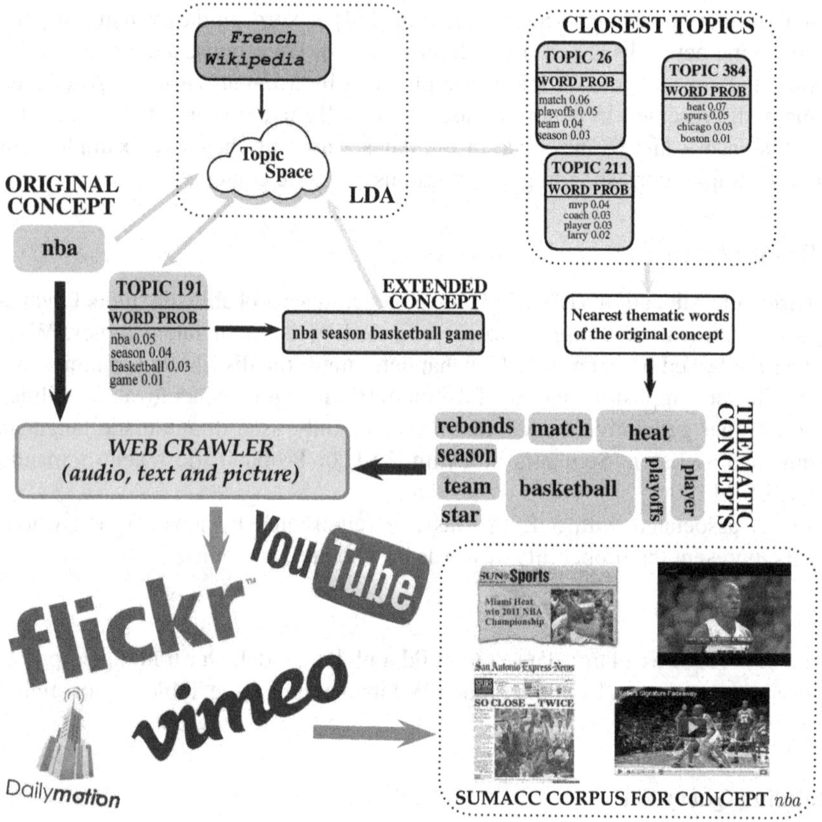

Fig. 1. Example of the initial query *nba* and its extension to retrieve documents

Figure 1 shows an example of an initial concept *NBA* processed in the document retrieval system (audio, text or picture documents).

3 The SuMACC Corpus

The manual corpus annotation consists in checking, in each video, the presence of the concept which was used to collect it (via the expanded queries).

Table 2. Statistics on the concepts of the SuMACC project

Concept	#documents	%	time	%
accident_de_la_route (*road_accident*)	535.0	7.117	1D-5:12:59	5.827
age_de_départ_à_la_retraite (*retirement_age*)	30.0	0.399	0:40:33	0.135
alpes	33.0	0.439	2:31:21	0.503
apple	114.0	1.517	7:0:45	1.399
applications_iphone	206.0	2.74	9:53:37	1.973
barack_obama	455.0	6.053	1D-6:49:42	6.148
barcelone-real_madrid	32.0	0.426	4:2:38	0.807
bnp_paribas	30.0	0.399	0:56:5	0.186
brad_pitt	455.0	6.053	1D-0:58:3	4.98
can_2012	30.0	0.399	1:50:56	0.369
cisjordanie	30.0	0.399	0:19:23	0.064
consoles_portables (*portable_game_consoles*)	30.0	0.399	0:56:24	0.187
cosmétique (*cosmetic*)	843.0	11.215	3D-2:38:17	14.886
dominique_strauss-kahn	341.0	4.536	14:29:41	2.891
françois_hollande	144.0	1.916	7:59:25	1.594
fukushima	33.0	0.439	1:14:43	0.248
galeries_d'art (*art_galleries*)	686.0	9.126	1D-10:52:20	6.955
gameplay	30.0	0.399	10:33:54	2.107
ground_zero	154.0	2.049	10:57:52	2.187
hôtel_de_ville (*city_hall*)	30.0	0.399	0:40:34	0.135
ipad_2	76.0	1.011	5:42:23	1.138
jacques_chirac	349.0	4.643	20:19:10	4.053
javier_pastore	30.0	0.399	0:58:44	0.195
jennifer_lopez	383.0	5.095	1D-4:49:39	5.749
kanye_west	38.0	0.506	0:33:22	0.111
liberté_d'expression (*freedom_of_expression*)	30.0	0.399	1:37:19	0.323
londres_2012	39.0	0.519	1:44:50	0.348
marché_financier (*financial_market*)	38.0	0.506	1:46:10	0.353
mouammar_kadhafi	112.0	1.49	6:6:38	1.219
nba	51.0	0.678	6:0:19	1.198
oscars	36.0	0.479	1:12:25	0.241
otan	30.0	0.399	2:58:42	0.594
paris_saint-germain	33.0	0.439	0:51:59	0.173
prix_nobel_de_la_paix (*nobel_peace_prize*)	42.0	0.559	1:19:41	0.265
présidentielle_2012 (*presidential_elections_2012*)	80.0	1.064	2:27:14	0.489
psn	30.0	0.399	2:47:37	0.557
racisme (*racism*)	155.0	2.062	11:44:10	2.341
real_madrid	30.0	0.399	0:24:7	0.08
renseignement_et_espionnage (*intelligence_and_espionage*)	34.0	0.452	3:50:59	0.768
semaine_de_la_mode (*fashion_week*)	568.0	7.556	2D-1:46:16	9.926
stade_de_france	30.0	0.399	4:38:35	0.926
steve_jobs	343.0	4.563	1D-0:54:10	4.967
tournages (*filming*)	30.0	0.399	0:41:0	0.136
vernissages_et_expositions (*openings_and_exhibitions*)	352.0	4.683	22:51:59	4.561
vitrolles	30.0	0.399	1:22:13	0.273
washington	180.0	2.395	20:49:14	4.153
zone_euro (*eurozone*)	127.0	1.69	6:25:40	1.282
Total	**7,517**	–	**20D-21:23:47**	–

Table 3. Comparison between the initial and the expanded queries after manual annotation

	Initial Queries			Expanded Queries		
Concept	Yes	No	Ambiguous	Yes	No	Ambiguous
accident_de_la_route (*road_accident*)	50	44	4	49	41	9
age_de_départ_à_la_retraite (*retirement_age*)	66	16	16	42	19	38
alpes	21	71	7	55	44	0
apple	6	86	6	17	68	13
applications_iphone	30	65	3	62	34	3
barack_obama	73	24	2	63	17	18
barcelone-real_madrid	11	77	11	30	25	45
bnp_paribas	16	83	0	68	31	0
brad_pitt	69	18	11	47	39	12
can_2012	33	50	16	12	87	0
cisjordanie	60	20	20	96	4	0
consoles_portables (*portable_game_consoles*)	20	60	20	**90**	9	0
cosmétique (*cosmetic*)	**87**	11	0	60	20	18
dominique_strauss-kahn	27	63	9	**88**	4	6
françois_hollande	66	9	23	74	17	8
fukushima	11	88	0	66	33	0
galeries_d'art (*art_galleries*)	40	31	28	54	28	16
gameplay	66	0	33	**96**	0	3
ground_zero	13	82	4	36	56	7
hôtel_de_ville (*city_hall*)	20	40	40	8	84	8
ipad_2	50	40	10	25	53	21
jacques_chirac	63	32	4	76	10	13
javier_pastore	20	70	10	15	78	5
jennifer_lopez	42	53	4	57	30	12
kanye_west	33	33	33	**82**	7	10
liberté_d'expression (*freedom_of_expression*)	33	44	22	64	29	5
londres_2012	30	46	23	64	20	16
marché_financier (*financial_market*)	25	66	8	70	20	8
mouammar_kadhafi	31	52	15	**81**	15	3
nba	33	0	66	65	18	16
oscars	25	62	12	44	55	0
otan	40	20	40	68	28	4
paris_saint-germain	16	50	33	**88**	7	3
prix_nobel_de_la_paix (*nobel_peace_prize*)	28	57	14	75	25	0
présidentielle_2012 (*presidential_elections_2012*)	14	57	28	54	33	12
psn	33	50	16	**91**	4	4
racisme (*racism*)	**89**	3	7	61	27	11
real_madrid	13	80	6	20	66	13
renseignement_et_espionnage (*intelligence_and_espionage*)	71	14	14	**80**	19	0
semaine_de_la_mode (*fashion_week*)	0	94	4	44	34	20
stade_de_france	25	62	12	61	28	9
steve_jobs	10	87	2	51	36	11
tournages (*filming*)	16	66	16	**83**	12	4
vernissages_et_expositions (*openings_and_exhibitions*)	64	10	25	76	10	12
vitrolles	28	42	28	21	65	13
washington	10	86	3	27	42	29
zone_euro (*eurozone*)	58	27	13	46	43	9
Total	**37%**	**47%**	**14%**	**58%**	**30%**	**10%**

The SuMACC concept list is composed of 47 concepts corresponding to different kinds of entities that may be searched on the Web, with respect to the project goals; therefore, some of them are very concrete (such as *Ipad2* or *Jennifer Lopez*), while others are much more abstract (such as *racism*). In Table 2, the number of videos retrieved by the initial queries and their expanded version is compared. We observe that the initial query allowed to retrieve 1,574 (21%) videos, the 5,943 (79%) others being extracted thanks to the thematic queries. Some concepts such as *can_2012* or *ipad_2* allow to retrieve more documents. This is mainly due to their popularity and to the fact that these precise concepts can better describe a video than a more general one such as *renseignement et espionnage (intelligence and espionage)* or *âge de départ à la retraite (retirement age)* which can be associated to a lot of heterogeneous videos.

The time duration of all documents related to a concept is detailed in Table 2. This table shows that the time duration is well distributed among the 47 concepts. The total number of hours of the corpus is about 89 days. This represents 2,162 hours of videos. Each concept contains 329 videos in average.

With this corpus of video documents, a set of video-related text documents that describes the extracted videos is added to the SuMACC corpus. Note that a description is not available for all the videos. Thus, 1,410 descriptions are collected for the 7,517 videos of the corpus. This is a real context fact: few videos have a textual document to describe their content. This set of text documents contains 9,692 unique words for a total of 56,474 running words.

As expected, Table 3 shows that the videos retrieved with the thematic queries are globally considered more relevant that the ones from the initial queries. Nonetheless, the videos retrieved with the thematic queries have a tendency to be much more variable than the one obtained with the initial queries. Indeed, if we take a look at the column *Ambiguous* of Table 3, we can notice that the total proportion of ambiguous videos is about 14% for the original queries, while only 10% is for the thematic queries.

Moreover, the proportion of videos considered as relevant (bold in column *Yes* in Table 3) is higher with extended queries than the initial ones (query containing the concept only). In details, 9 concepts among the 47 ones, have a proportion of relevant videos beyond 80% with the use of a topic space representation (only 2 for the not-extended queries).

The main issue of this video collecting task is to obtain a sufficient variability in the video content while remaining close to the query topic.

4 Conclusion

In this paper, the SuMACC project corpus was described as well as an unsupervised method to retrieve a large set of concept-related multimedia documents. This method expands simple requests in order to get a realistic sampling of videos returned by a search engine. Query extension is based on a 2-step mapping of keywords into a topic space estimated by a LDA approach. This method allows to add a necessary variability in the corpus while respecting a realistic ambiguity due to topic proximity. As a result, up to 23,000 videos from 47 concepts is obtained, 1,432 of them being annotated in a first annotation campaign. The corpus will be freely available under GPL license by the end of 2014.

References

1. Ando, R.K., Lee, L.: Iterative Residual Rescaling: An analysis and generalization of LSI. In: Proceedings of SIGIR, pp. 154–162 (2001)
2. Bertier, M., Guerraoui, R., Leroy, V., Kermarrec, A.M.: Toward personalized query expansion. In: ACM EuroSys Workshop on Social Network Systems (SNS), pp. 7–12 (2009)
3. Blei, D.M., Ng, A.Y., Jordan, M.I.: Latent Dirichlet Allocation. The Journal of Machine Learning Research 3, 993–1022 (2003)
4. Crouch, C.J., Crouch, D.B., Nareddy, K.R.: The automatic generation of extended queries. In: ACM SIGIR Conference on Research and Development in Information Retrieval, pp. 369–383 (1990)
5. Feng, B., Cao, J., Chen, Z., Zhang, Y., Lin, S.: Multi-modal query expansion for web video search. In: ACM SIGIR Conference on Research and Development in Information Retrieval, pp. 721–722 (2010)
6. Gauch, S., Wang, J., Rachakonda, S.M.: A corpus analysis approach for automatic query expansion and its extension to multiple databases. ACM Transactions on Information Systems 17(3), 250–269 (1999)
7. Hofmann, T.: Probabilistic latent semantic indexing. In: ACM SIGIR Conference on Research and Development in Information Retrieval, pp. 50–57 (1999)
8. Smeaton, A.F., Over, P., Kraaij, W.: Evaluation campaigns and trecvid. In: Proceedings of the 8th ACM International Workshop on Multimedia Information Retrieval, pp. 321–330 (2006)

Empiric Introduction to Light Stochastic Binarization

Daniel Devatman Hromada[1,2]

[1] Slovak University of Technology, Faculty of Electrical Engineering and Information Technology, Department of Robotics and Cybernetics, Ilkovičova 3, 812 19 Bratislava, Slovakia
[2] Université Paris 8, Laboratoire Cognition Humaine et Artificielle, 2, rue de la Liberté 93526, St Denis Cedex 02, France

Abstract. We introduce a novel method for transformation of texts into short binary vectors which can be subsequently compared by means of Hamming distance measurement. Similary to other semantic hashing approaches, the objective is to perform radical dimensionality reduction by putting texts with similar meaning into same or similar buckets while putting the texts with dissimilar meaning into different and distant buckets. First, the method transforms the texts into complete TF-IDF, than implements Reflective Random Indexing in order to fold both term and document spaces into low-dimensional space. Subsequently, every dimension of the resulting low-dimensional space is simply thresholded along its 50th percentile so that every individual bit of resulting hash shall cut the whole input dataset into two equally cardinal subsets. Without implementing any parameter-tuning training phase whatsoever, the method attains, especially in the high-precision/low-recall region of 20newsgroups text classification task, results which are comparable to those obtained by much more complex deep learning techniques.

Keywords: Reflective Random Indexing, unsupervised Locality Sensitive Hashing, Dimensionality Reduction, Hamming Distance, Nearest-Neighbor Search.

1 Introduction

In applied Computer Science one often needs to select from the database an object which most resembles the "query" object already at one's disposition. In order to do so, all members of the database are often transformed into ordered sequences of numeric values (i.e. vectors). Such vectors can be interpreted as points in the high-dimensional metric space allowing to calculate their distance to other points in the space. In such case, the resulting "most similar" entity is simply the entity whose vector has smaller distance to the vector representing the "query" entity than any other entity stored in the database, i.e is query's "nearest neighbor".

In Natural Language Processing (NLP), the nearest-neighbor search (NNS) is a widely-used approach applied for solving diverse problems. Seemingly trivial, NNS is nonetheless not an easy problem to tackle with, especially in the case of Big Data scenarios where database contains huge amount of highly-dimensional datapoints. In real-time scenarios where naive linear comparison of d-dimensional query vector with

P. Sojka et al. (Eds.): TSD 2014, LNAI 8655, pp. 37–45, 2014.

all *N* vectors stored in the database is simply not feasible due to its *O(Nd)* computational complexity. Thus, one is almost always obliged to take recourse in approximation or heuristic-based solutions.

One of the most common methods of reducing the complexity of the NN-search is by reducing the dimensionality of the database-representing vector space. Classical approach to do so is Latent Semantic Analysis [10] (LSA). Other family of more and more common approaches exploits so-called binary vectors as the ultimate means of entity's formalisation. Given the fact that contemporary computers are machines essentially -i.e. on the physical hardware level- always working with binary distinctions, the calculation of the distance between two binary vectors (i.e. Hamming distance – the number of times a bit in $vector_1$ has to be flipped in order to obtain the form of $vector_2$) can be indeed a very fast operation to realize, especially when implemented on the hardware level as a part of processor's instruction set.

Combination of dimensionality reduction and binarisation are basis for family of methods descending from the approach called Locally Sensitive Hashing (LSH) [11]. While concrete implementations often substantially differ – c.f. [13] for the state-of-the-art overview – the objective is always the same: to hash each object of the dataset into a concise binary vector in such a way that the objects which are similar shall end up in the same or similar bucket (i.e. shall be represented by same or similar binary vector) while the objects which are disparate shall end up in disparate buckets[1].

In order to attain stunningly good results, many of these methods have to be first trained. Such a tradeoff of high performance / complexity of training phase is the case, for example, in the "semantic hashing" (SH) approach of [1]. In SH one has to first learn the weights between different restricted Boltzmann machines in order to obtain a multi-layered "deep generative model" able to perform the hashing. But the SH has also certain non-negligeable disadvantages: the 1) training-related costs 2) need to work with restricted amount of features which shall enter the first layer of the network (e.g. 2,000 TF-IDF values in [1]) 3) possibility of over-fitting of the model etc.

In this article, we shall present approach, which could one take *vis-a-vis* the problem of "text hashing". Instead of founding our approach on a powerful supervised "deep learning" algorithm able to extract sophisticated combinations of regularities among restricted number of initial features, we shall exploit an algorithm so simple that it can easily integrate huge number of features in a very fast & frugal way. In fact, the algorithm presented here is completely unsupervised and does not need any training or feature-preselection at all in order perform the hashing process.

1.1 Reflective Random Indexing

Theoretically, our approach stems from the lemma of Johnson-Lindenstrauss stating that a small set of points in a high-dimensional space can be embedded into a space of much lower dimension in such a way that distances between the points are nearly preserved [9]. Practically, the JL-lemma was already implemented as so-called Random

[1] Note that the aim of hashing process as presented in this paper differs substantially from the aim of hashing algorithms like MD5 or SHA2 whose objective is to always hash objects into different buckets.

Projection or Random Indexing algorithms. Random Projection was already quite successfully proposed in relation to the hashing problem [12]. Its much simpler Random Indexing (RI) counterpart, however, was not.

Since a decade from its initial proposal in [4], RI has already proven its usefulness in regards to NLP problems as diverse as synonym-finding [4], text categorization [5], unsupervised bilingual lexicon extraction [6], implicit knowdledge discovery [2], automatic keyword attribution to scientific articles [3] or measurement of string distance metrics [8].

The basic RI algorithm is quite straightforward to both understand and implement: Given the set of N objects (e.g. documents) which can be described in terms of F features (e.g. occurence of the string in the document), to which one initially associates a randomly generated D-dimensional vector, one can obtain D-dimensional vectorial representation of any object X by summing up the vectors associated to all features F_1, F_2 observable within X. The original random feature vectors are generated in a way that out of D elements of vector, only S among them are set to either -1 or 1 value. Other values contain zero. Since the "seed" parameter S is much smaller than the total number of elements in the vector (D), i.e. S «D, initial feature vectors are very sparse, containing mostly zeroes, with occasional value of -1 or 1.

At the end of the process, one obtains vectorial characterisations of all documents which one can compare by means of cosine measure. Leaving aside some advantageous properties described elsewhere [8], RI as described here is nothing else than a randomly distorted variation on a "bag-of-words" theme. Consistently to other bag-of-word approaches, one can also weight the initial randomly generated feature vectors with feature's TFIDF [7] value.

But one is not obliged to stop the whole process after the calculation of initial document vectors. One can indeed "reflect" the whole process and proceed this time from object vectors toward feature vectors, forget the initial randomly generated feature vectors of 0th generation and obtain the feature vectors for feature F_X as a sum of vectors O_1,O_2 representing the objects within which one can observe the occurence of feature F_X. Subsequently, the object vectors can be once again calculated as a sum of feature vectors; feature vectors as a sum of object vectors etc. Such a multi-iterative approach whereby every iteration can be potentially followed by vector normalization is called Reflective Random Indexing (RRI).

While RRI keeps the advantageous properties of non-iterative RI like incrementality (i.e. it is very easy to enrich the model with new features or objects) and homogenity (both objects and features are points of the same space), it goes well-beyond simple bag-of-words properties of non-iterative RI. This is so because of certain symmetry in the algorithm where, in every iteration, not only features take part in the vectorial definition of objects but also objects help to construct the vectorial representations of features. Thus, RRI multiplies substantially the amount of mutually interacting forces within the generated metric space and allows for such usages as discovery of "implicit semantic inferences" within huge corpora [2].

Reflective Random Indexing

```
algorithm RRI ()
  #initial iteration is equivalent to plain Random Indexing
  foreach Feature
     Feature_Vectors[Feature] =  generate_Random_Vectors(Dimension, Seed)
     Feature_Vectors[Feature] *=  TFIDF_Weights[Feature] #optional
  foreach Object
     foreach Feature in Object2Feature[Object]
        Object_Vector[Object] += Feature_Vectors[Feature]
  normalize Feature_Vectors,Object_Vectors #optional
  #reflective iterations
  repeat
     foreach Feature
        foreach Object in Feature2Object[Feature]
           Feature_Vector[Feature] += Object_Vectors[Object]
     foreach Object
        foreach Feature in Object
           Object_Vector[Object] += Feature_Vectors[Feature]
     Iteration = Iteration + 1
     normalize Feature_Vectors,Object_Vectors  #optional
  until Iteration == MaxIterations
  return Feature_Vectors, Object_Vectors
```

2 Light Stochastic Binarization

Our hashing algorithm is a simple extension of Reflective Random Indexing. While the output of RRI are D-dimensional real-valued vectors, the output vectors of LSB are not real-valued but binary vectors of length D.

Transformation of RRI-generated real-valued vectors into binary vectors is a fairly straightforward process: after all object vectors are calculated by RRI, we simply determine the median value (i.e. 50th percentile[2]) for every dimension (i.e. column) D of the resulting Nd matrix. In such a way we obtain a threshold value for every dimension and we assign into d^{th} element of final binary representation of object *n* the 0 value if its real-valued coordinate along d^{th} dimension is smaller than the determined threshold and 1 if it is above the threshold. Rare tie situations are broken randomly. Result is a set of binary hashes cut in two equally cardinal subsets by every dimension-denoting bit. This binarization is the very last step of the indexing phase.

$$h_d(n) = \begin{cases} 0 & \text{if } n < \text{median}(D_d) \\ 1 & \text{if } n > \text{median}(D_d) \\ \text{rand} & \text{if } n == \text{median}(D_d) \end{cases}$$

Subsequently, during the query phase, can simply transform the query object is transformed into its binary vector by: 1)summing up the real-valued representations of the features observable within the query object 2) thresholding the resulting real-vectored by pre-determined medians. Resulting binary vector is subsequently considered to be the center of the Hamming-ball of radius R. Every binary vector

2 Determination of dimension's median value is the only nontrivial component of the process. Note that in case of particularly large and complex samples, law of large numbers shall push 50th percentile's value limitely close to 0.

contained within such Hamming-ball shall yield an index pointing to the bucket stored in the memory where we could look in order to find query's nearest-neighbor. In case $2^R<B$, i.e. in case when generated Hamming-ball can potentially contain more possibilities than is the total amount B of binary buckets generated during the indexing phase, we can calculate, in a linear fashion, query's Hamming distance H in regards to every bucket-denoting binary vector and subsequently select only those buckets for which H(query_hash,bucket_hash)<R. Radius R is the query-phase thresholding parameter by means of which one can trade precision with recall and vice versa.

3 Experiment

The aim of our preliminary experiment was to assess whether the LSB approach can be useful at all and, if yes, compare the information retrieval faculties of LSB with those of Semantic Hashing. Thus, our results shall be presented in terms of Precision-Recall curves as defined by [1]:

$$Recall = \frac{\text{Number of retrieved relevant documents}}{\text{Total number of all relevant documents}}$$

$$Precision = \frac{\text{Number of retrieved relevant documents}}{\text{Total number of retrieved documents}}$$

Analogicaly to the article with which we compare our data, the retrieved document is considered to be relevant to the query document when they have the same class label.

3.1 Corpus and Pre-processing

In this preliminary work we have confronted the LSB algorithm only with data contained in 20 newsgroups corpus [14]. The corpus contains 18,845 postings taken from the Usenet newsgroup collection divided into training set containg 11,314 postings, rest being the testing set. Both training and testing subsets are divided into 20 different newsgroups which correspond each to a distinct topic. Because our approach aims to introduce an unsupervised hashing scenario, we have left aside the training set and focused all evaluation solely on 7,531 postings contained in the testing set.

Words were extracted from postings by considering every non-word character as a word boundary – 93,591 words were thus extracted, among which 41,782 has occured in more than one posting. These words were considered as features by subsequent RRI. Data were not processed in any other way – no stop words were filtered away and all words which occured in more than one posting were taken into account.

3.2 Empiric Results

In a comlete analogy to simulations performed in [1], we have used every document from the test set as a query document which was compared to all other 7,530 documents. Precision and Recall values were calculated for every query and averaged over all 7,531 queries. Figure 1 compares the Precision – Recall curves of reflective and unreflective

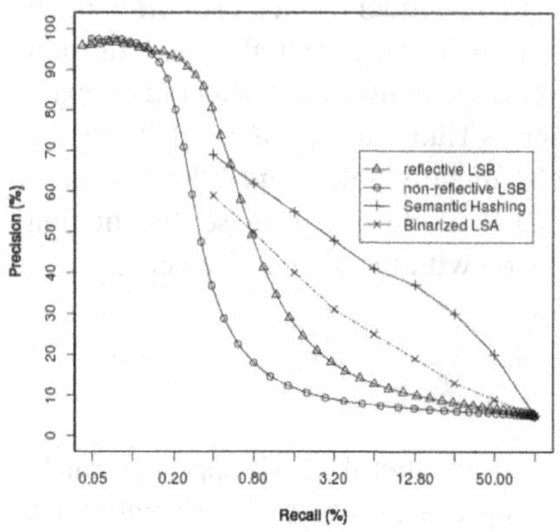

Fig. 1. Comparison of reflective LSB(I=2) and unreflective LSB(I=0) LSB with Semantic Hashing and binarized Latent Semantic Analysis

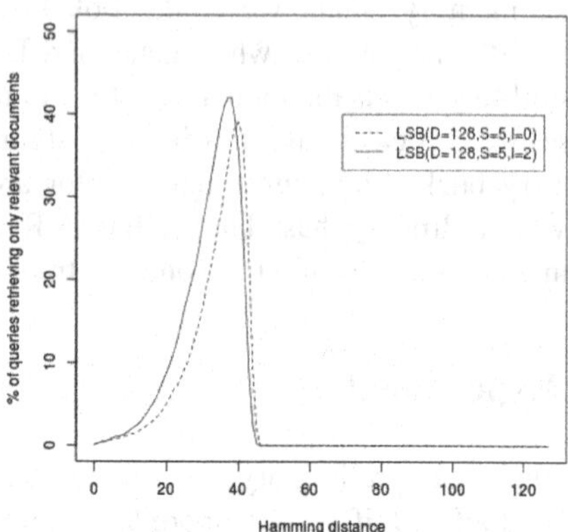

Fig. 2. More than 40% of queries are accompanied, within the Hamming ball of radius 38, only by neighbors belonging to the same newsgroup category

variants of LSB with both "Fine-tuned 128-bit Semantic Hashing" and 128-dimensional binarized Latent Semantic Analysis. Non-LSB results are reproduced from Figure 6 of study [1].

Figure 2 illustrates in closer detail the behaviour of both reflective and non-reflective variants of LSB in relation to the variation of the retrieval parameter (i.e. Hamming ball's radius R) which is plotted on the X axis. Y axis represents the number of queries which do not have – in their surrounding Hamming ball with radius R – any posting not having the same newsgroup label (i.e. they do not retrieve any false positive), but in the same time, they do retrieve at least one true positive (i.e. at least one object belonging to same newsgroup as query has the binary hash which is located within query's Hamming ball of radius R).

Notwithstanding the size of the theoretically possible hashing search space (2^{128}) which by large exceeds the number of 7,531 initial objects, LSB succeeded to create some collisions, hashing all articles in the 20newsgroup corpus into 7,526 binary buckets. While majority of such "collisions" are due to fairly trivial reposting of the same message, couple of somewhat more divergent messages from comp.graphics newsgroup was also hashed into the same 16-byte bucket. Displayed in the listing below is the difference of content between these two files, as produced by UNIX command diff.

```
< New since version of 2 May 1993:
<   * Added info on ImageViewer for NeXT.
---
> New since version of 18 April 1993:
>   * New version of XV supports 24-bit viewing for X Windows.
>   * New versions of DVPEG & Image Alchemy for DOS.
>   * New versions of Image Archiver & PMView for OS/2.
```

```
>    * New listing: MGIF for monochrome-display Ataris.
461,463c464,466
<      PMView 0.85: JPEG/GIF/BMP/Targa/PCX viewer.  GIF viewing very fast,
<      JPEG viewing roughly the same speed as the above two programs.  Has
<      image manipulation & slideshow functions.  Shareware, $20.
---
>      PMView 0.85: JPEG/GIF/BMP viewer.  GIF viewing very fast, JPEG viewing
>      fast if you have huge amounts of RAM, otherwise about the same speed
>      as the above programs.  Strong 24-bit display support.  Shareware, $20.
632,641d634
< NeXT:
<
< ImageViewer is a PD utility that displays images and can do some format
< conversions.  The current version reads JPEG but does not write it.
< ImageViewer is available from the standard NeXT archives at
< sonata.cc.purdue.edu and cs.orst.edu, somewhere in /pub/next (both are
< currently being re-organized, so it's hard to point to specific
< sub-directories).  Note that there is an older version floating around that
< does not support JPEG.
```

In spite of difference of their contents, files comp.graphics/39638 and comp.graphics/39078 of 20-newsgroup corpus, LSB assigned to them the same "10010011000010111001100010110010001111000011011101001001010011010111 0100000001001010011011000100100101010011010000101111110110011" hash

4 Discussion

Looking at the peak shown in Figure 2, one is tempted to state that when confronted with data from the testing set of 20newsgroups corpus, the reflective 128-dimensional LSB is able to retrieve, in 42% of cases (i.e. 3,166 out of 7,531), at least one relevant "neighbor" with maximal precision. It is indeed at the Hamming distance 38, where the method combining Reflective Random Indexing executed with parameters D=128, S=5, I=2 and followed by simple binary thresholding of every dimension, attains at overall recall rate 0.39%[3] to much higher precision (80.6%) than any method presented in the study of [1].

On the other hand, LSB performs much worse than compared methods in situations where one wants to attain higher recall. This is most probably due to almost complete lack of "training" – since with exception of 1) TFIDF weighting of initial randomly-generated feature vectors 2) the "reflection" procedure which aids us to characterize objects in terms of features and features in terms of objects 3) determination of binary thresholds (i.e. medians) – there is no kind of "learning" procedure involved.

But it might be the case that lack of any complex "deep learning" procedure shall prove itself to be a certain advantage. More concretely, the one who uses LSB is not obliged to drastically reduce the number of features by means of which all objects are

[3] We precise that when we mention 42% recall with 100% precision, we speak about NNS scenario where it is sufficient for a query to retrieve one relevant document. This scenario is documented on Fig. 2. On the other hand, when we mention attained recall rate 0.39%, we speak about much more difficult "classification" scenario where query, in order to attain maximal recall, must retrieve all >370 postings which belong to the same class. This scenario is documented on Figure 1.

characterized. Thus, in the case of the introductory experiment presented in this paper, we have represented every text as a linear combination of vectors associated to 41,782 features. We believe that it is indeed this ability to "exploit the minute details" (compare to 2,000 words with highest TFIDF score used by [1]) which allows the method hereby introuced to attain higher precisions in scenarios where just one relevant document is to be retrieved. It would be, however, unfair to state that can LSB "pefrorms better" than Semantic Hashing, because the goal purpose of SH was not to target the NN-search problem but to yield robust results in more exhaustive classification scenarios. Thus, comparison of LSB with other methods is needed.

It might also be the case that more advanced tuning of RRI's parameters could improve the performance. Another possible direction of research is to study the impact of strategies by means of which the initial random vectors are weighted. Due to the introductory nature of this paper, not much was unveiled about neither of two problems. Looking at the Figure 1, one can, however, assert that: 1) LSB seems to attain better results when its RI component involves more than one iteration, i.e. when it is "reflective".

In sum, we believe that the method hereby introduced is worth to be studied somewhat further. Not only because its dimensionality-reduction component – the RRI – is less costly and more opened to incremental addition of new data than, for example, LSA [10]. Not only because it is similar to LSH [11] in its ability to transform texts into hashes as big as concise as 16 ASCII characters and yet preserve the relations of similarity and difference held by original texts. But also because the algorithm is easy to comprehend, simple to implement and queries can be very fast to execute. That's why we label the method of binarization hereby presented as not only stochastic, but also light.

References

1. Salakhutdinov, R., Hinton, G.: Semantic hashing. International Journal of Approximate Reasoning 50(7), 969–978 (2009)
2. Cohen, T., Schvaneveldt, R., Widdows, D.: Reflective Random Indexing and indirect inference: A scalable method for discovery of implicit connections. Journal of Biomedical Informatics 43(2), 240–256 (2010)
3. El Ghali, A., Hromada, D., El Ghali, K.: Enrichir et raisonner sur des espaces sémantiques pour l'attribution de mots-clés. In: JEP-TALN-RECITAL 2012, p. 77 (2012)
4. Sahlgren, M., Karlgren, J.: Vector-based semantic analysis using random indexing for cross-lingual query expansion. In: Peters, C., Braschler, M., Gonzalo, J., Kluck, M. (eds.) CLEF 2001. LNCS, vol. 2406, pp. 169–176. Springer, Heidelberg (2002)
5. Sahlgren, M., Cöster, R.: Using bag-of-concepts to improve the performance of support vector machines in text categorization. In: Proceedings of the 20th International Conference on Computational Linguistics, p. 487 (2004)
6. Sahlgren, M.: An introduction to random indexing. In: Methods and Applications of Semantic Indexing Workshop at the 7th International Conference on Terminology and Knowledge Engineering, TKE, vol. 5 (2005)
7. Manning, C.D., Raghavan, P., Schütze, H.: Introduction to information retrieval, vol. 1. Cambridge University Press, Cambridge (2008)

8. Hromada, D.D.: Random Projection and Geometrization of String Distance Metrics. In: Proceedings of the Student Research Workshop Associated with RANLP, pp. 79–85 (2013)
9. Johnson, W.B., Lindenstrauss, J.: Extensions of Lipschitz mappings into a Hilbert space. Contemporary Mathematics 26(189–206), 1 (1984)
10. Landauer, T.K., Dumais, S.T.: A solution to Platos problem: The latent semantic analysis theory of acquisition, induction, and representation of knowledge. Psychological Review 104(2), 211–240 (1997)
11. Gionis, A., Indyk, P., Motwani, R.: Similarity search in high dimensions via hashing. In: VLDB, vol. 99, pp. 518–529 (1999)
12. Charikar, M.S.: Similarity estimation techniques from rounding algorithms. In: Proceedings of the Thiry-Fourth Annual ACM Symposium on Theory of Computing, pp. 380–388 (2002)
13. Andoni, A., Indyk, P.: Near-optimal hashing algorithms for approximate nearest neighbor in high dimensions. In: 47th Annual IEEE Symposium on Foundations of Computer Science, FOCS 2006, pp. 459–468 (2006)
14. 20 newsgroups, http://qwone.com/~jason/20Newsgroups/

Comparative Study Concerning the Role of Surface Morphological Features in the Induction of Part-of-Speech Categories

Daniel Devatman Hromada[1,2]

[1] Université Paris 8, Laboratoire Cognition Humaine et Artificielle, 2, rue de la Liberté 93526, St Denis Cedex 02, France

[2] Slovak University of Technology, Faculty of Electrical Engineering and Information Technology, Department of Robotics and Cybernetics, Ilkovičova 3, 812 19 Bratislava, Slovakia

Abstract. Being based on English language, existing systems of part-of-speech induction prioritize the contextual and distributional features "external" to the word and attribute somewhat secondary importance to features derived from word's "internal" morphologic and orthotactic regularities. Here we present some preliminary empirical results supporting the statement that simple "internal" features derived from frequencies of occurrences of character n-grams can substantially increase the V-measure of POS categories obtained by repeated bisection k-way clustering of tokens contained in Multext-East corpora. Obtained data indicate that information contained in suffix features can furnish c(l)ues strong enough to outperform some much more complex probabilist or HMM-based POS induction models , and that this can especially be the case for Western Slavic languages.

Keywords: part-of-speech induction, development of morphology, clustering, surface features, suffix.

1 Introduction

Part-of-speech (POS) induction is a constructivist process aiming to converge to the mechanism able to attribute the POS category (e.g. "verb", "noun", "adjective" etc.) membership information to any word of the language under study. Because "syntactic category information is part of the basic knowledge about language that children must learn before they can acquire more complicated structures" [15] POS induction (POS-i) is often considered to be the first step in a more complex process of grammar induction and language acquisition in general.

Given such an important place of POS-i in NLP studies, it is of no surprise that while first computational models of POS-i were proposed decades ago [3,6,15] the problem of unsupervised POS-label attribution still attracts attention of many computational linguists. Thus, dozens of POS-i systems exist, among which those based on class-based word n-grams [5], graph clustering [2] or diverse extensions to Hidden Markov Models [9,8,1] are compared in the [4] comparative study which suggests that "some of the oldest (and simplest) systems stand up surprisingly well against more recent approaches".

P. Sojka et al. (Eds.): TSD 2014, LNAI 8655, pp. 46–52, 2014.

Aims of this article are 1) to elucidate a superior peformance of Clark [5] and Berg-Kirkpatrick [1] models with the statement: "Their models perform better because they use better features" 2) to precise that for many languages, such features can be morphological ones. We precise that what shall be called "morphological feature" (MF) in the rest of this article is any feature "internal" to the word WITHIN which it occurs and as such can be opposed to contextual or distributional features "external" to the word under study (i.e. opposed to features which describe word's relation to other words and not its internal composition).

By focusing upon the role of such "orthotactic" MFs in diverse languages represented in the Multext-East corpus [7] we shall try to persuade the reader that while the "syntax-in-word-order paradigm" could (and did) yield useful models and tools for description of English language, the uncritical acceptation of such paradigm could turn to be somewhat contra-productive if one tends to develop POS-i models for highly flectional & morphology-rich languages.

2 Corpus

All analyses were effectuated with texts contained in the 4th version of Multext-East corpus [7] . Bulgarian (bg), Czech (cs), English (en), Estonian (et), Farsi (fa), Hungarian (hu), Polish (pl), Romanian (ro), Serbian (sr), Slovak (sk) and Slovene (sl) transcription of Orwell's 1984 were analysed. Quantitative descriptions of different corpora are present in the Table 1.

Corpus	Types	Tokens	Tags$_{POS}$
bg	17305	117238	13
cs	22341	100368	13
en	11160	134832	12
et	18911	111305	12
fa	13009	124823	12
hu	20642	132196	13
pl	24019	115185	14
ro	16220	135055	15
sk	23015	103452	13
sl	20597	112278	13
sr	21540	126611	13

3 Method

Every word from the corpus was described by a vector of features whose values were obtained by application of feature filters described below. Vectors were subsequently clustered into groups.

3.1 Feature Extraction

All tokens, punctuation marks included, were extracted as such from the corpus. Word characters were transcribed into lower case. In order to mark the word boundaries, ^

and $ characters were prefixed, respectively suffixed, to extracted tokens. Following features were then extracted from tokens:

Length [L] – yields only one feature whose value equals the character length of the token, i.e. 6 for word "^good$". Baseline.

Character n-grams of length X [N_x] – every feature encodes the number of occurrences of the character n-gram of length L within the token. Thus, if X=1, the word "^ good$" can be encoded by vector of features [1, 1, 2, 1, 1] whose second element denotes the number of "g" present in the word, third feature the number of "o" etc. If X=2, the vector could be [1, 1, 1, 1, 1], its first element representing the frequency of occurrence of "^ g" character bigram, second of "go" bigram, third of "oo" bigram etc.

Character fragments whose length <X [F_X] – this approach takes into account all n-gram fragments BELOW the specified length X. Thus if X=3, the word "^good$" could be represented by the vector [1, 1, 2, 1, 1, 1, 1, 1, 1, 1, 1, 1, 1] whose last four elements encode the presence of trigrams "^go", "goo", "ood" and "od$"; composition of first 10 elements is explained above.

All fragments [A] – same as above but X is equal to word's length. Word's vector thus encodes occurrences of all 1gram, 2gram, 3gram ... X-gram character sequences present within the word. Yields biggest number of features.

Prefixes of length X [P_X]– same as N_X but fragments of length X were extracted only from word's beginning

Suffixes of length L [S_X]– same as P_X but fragments of length L were extracted only from word's beginning Word's circumference n-grams of length X [B_X] – boundary n-gram feature is a conjunction of a prefix and suffix feature, e.g. the B_2 feature for the word good can be matched by regular expression /ĝ.+d$/ and its occurrence would be also observed in the words like "god" or "gold"

Word's circumference [C_X]– Conjunction of P_X and S_X,i.e. feature is defined by combination of prefix and suffix both of length X.

Word's root [R_X]– for the purpose of this article, we define the root feature "as all that rests in the token when its circumference n-grams of length X are removed"

Token's co-occurrence neighborhood of length L [O_L] – this is the only feature "external" to the token under study. Every co-occurrence of the definiens-token (column) maximum L words aside to the left or right from definiendum-token (row) augments the value by 1.

If the definiens does not co-occur aside the definiendum word or if a fragment (column) does not occur within the word, or a feature-representing pattern (column) does not match the word (row), then the value in the final vector is, of course, zero.

3.2 Clustering

Since our objective is to evaluate the (non)relevance of diverse sets of surface features for POS-i in different languages, and not to evaluate the subsequent grouping machinery, we have decided to use a simple (& fast) repeated bisection k-way clustering as is implemented in the clustered tool CLUTO [12]. Columns of the word x feature matrix were scaled according to inverse-document frequency paradigm, cosine function was used for the calculation of the similarity metrics.

4 Evaluation

For the purposes of this article we had decided to present our simulations principially in terms of V-measure. More theoretical [13] and empiric [4] reasons being explained elsewhere, our choice was partially motivated by the form of V-measure score equations:

$$h = 1 - \frac{H(T|C)}{H(T)}$$

$$c = 1 - \frac{H(C|T)}{H(C)}$$

$$V = \frac{(1+\beta)hc}{(\beta h) + c}$$

which strongly resembles the F-measure score often used in evaluation of classification problems. The homogenity (h) and completeness (c) were designed in order to be analogic to precision, respectively recall. Given its elegance, stability in regards to growing number of clusters but also certain "strictness" (note that even the best state-of-the-art present in [4] comparative study rarely surpass the V>0.6 limit), we consider the Vmeasure to be very valuable quantitative measure of performance of clustering POS-i algorithms.

Table 1. V-measures obtained after clustering different corpus according to different features. The most performant feature of every corpus is marked.

	L	**N_1**	**N_2**	**N_3**	**N_4**	**F_2**	**F_3**	**F_4**	**A**	**P_2**	**P_3**	**S_2**	**S_3**	**C_2**	**C_3**	**R_2**	**R_3**	**O_1**
bg	4.3	5.6	13.1	17.0	11.9	8.5	14.4	14.7	14.6	6.7	5.0	**18.9**	16.5	3.8	2.3	3.4	3.0	12.5
cs	5.4	9.2	**25.2**	20.7	11.6	23.1	24.8	23.9	24.3	7.4	7.1	**25.2**	18.7	4.7	3.1	3.7	3.4	7.9
en	3.8	6.5	14.1	15.3	9.4	10.4	14.9	16.1	14.7	3.9	3.6	**20.5**	19.7	2.4	1.7	2.9	2.2	14.4
et	4.2	4.0	12.2	14.2	11.9	5.8	6.92	9.38	7.24	4.2	6.0	14.2	**16.1**	3.6	2.8	3.4	3.3	6.77
fa	2.6	6.8	15.4	15.52	12.2	12.0	15.51	15.3	**15.55**	11.7	14.5	14.4	12.0	6.4	4.6	2.8	3.2	14.3
hu	2.3	4.3	6.1	10.7	9.4	5.2	6.26	6.58	5.65	5.4	5.7	**17.1**	14.2	3.0	1.8	2.4	2.0	7.1
pl	4.7	8.0	21.1	20.1	13.7	18.5	20.3	19.7	15.6	5.3	6.5	**25.1**	22.7	4.0	3.0	3.3	2.9	7.9
ro	4.6	7.1	11.1	13.6	9.5	8.23	11.3	11.8	10.9	6.5	5.9	**15.8**	14.8	3.1	1.9	2.5	2.4	15.6
sr	5.2	5.5	13.3	14.8	10.5	5.67	8.06	8.82	5.95	6.1	6.4	**19.1**	16.5	4.6	3.0	4.7	3.5	9.4
sk	5.9	11.2	26.9	21.0	14.0	23.8	24.9	24.2	22.5	8.2	5.8	**27.5**	21.3	4.8	3.5	3.6	3.5	8.7
sl	4.5	4.8	12.2	17.1	12.8	7.39	8.42	14.3	7.5	6.8	6.0	**21.6**	19.3	5.2	2.4	3.3	3.4	9.1

Table above shows V-measure*100 values obtained by clustering of words characterized by length (L), character n-gram fragments of fixed (N_2, N_3, N_4) length or n-gram fragments shorter than certain length (F_2, F_3, F_4) as well as of clusters created by considering all fragments (A).

The best results (i.e. highest V-measures) were observed in case of Western Slavic languages which have all attained >0.2 of V-measure performance when clustered according to features representing character bigram occurrences. Southern Slavic languages along with Romanian, Hungarian and Estonian performed the best when character trigrams were taken into account. English attained the 0.16 performance when all bigrammata, trigrammata and tetragrammata were taken into account while Farsi was clustered the best when all n-gram character fragments were taken into account.

Further results presented in the table below point in the same direction. Highest V-measure score was attained by Slovak, Czech and Polish when simple extractor of suffix features of length 2 was applied. In fact the same extractor yielded highest scores in case of all languages with exception of Estonian where somewhat longer suffixes tend to facilitate the POS-i, and in case of Farsi whereby prefixal features seem to be at least as important as suffixal features. Word circumference features C_2 and C_3 as well their "negation", the word root features R_2 and R_3 do not seem to bring any information relevant to the categorization process – in fact they seem to perform even worse than the baseline feature L.

Members of set of "external" distributional features (O_1), which represent the trivial frequency of occurrence of the feature-word to the left or right from the target word, performed worse in all cases, English included, than S_2.

5 Discussion

POS-i system comparative study of [4] indicates that POS-i models involving morphological features perform better than models which do not. However both in Clark's [5] probabilist model as well as in morphology-enriched HMM-derived [1] model, morphological features seem to play rather a role of a performance-increasing "cherry added to the top of the cake" than that of model's cornerstone.

Results presented in this paper suggest that focusing upon the phenomena occurring within the token, if the token's transcription allows it[1], seem to yield quite strong c(l)ues for subsequent clustering of tokens into their respective syntactic categories. It may be the case that especially the character bigrams occuring at word's offset position – suffixes – seem to play an important role in word→ POS category attribution. It is also worth noting that suffixes augment the performance of POS-i not only for Indo-European languages but also for Uralic languages like Estonian or Hungarian.

It is also worth reiterating that POS-i within Western Slavic languages tends to be much more sensitive to character N-gram and suffix-derived features than other languages compared in this study. Because the research presented hereby was based only on one particular litteral corpus (Orwell's 1984) and the results obtained may thus represent not the properties of languages as such, but rather a certain translation style, it would be somewhat hors propos to postulate that a kind of overall statistic property – labeled hereby as "word offset flectivity" – is more marked in Western Slavic languages than, for example, in Southern Slavic or Uralic languages. But given the fact that it was only Slovak, Czech and Polish whose V>0.25 when clustered according to outputs of S_2 feature-extracting prism, we believe that subsequent analyses involving more corpora and more languages may be worth the effort. Verily only more exhaustive comparative studies could assess the impact of morphology of word X upon the attribution of

[1] For example, an "internal" feature-oriented approach would hardly yield any interesting results if applied on Chinese logograms but could be of certain theoretic interest when applied upon pinyin transcription.

syntactic function to the very word X. And since syntax is often bound with semantics – for example by means of thematic relations – such studies, if ever they would verify and not falsify the results presented hereby, could possibly result in a partial revision of a canonical "signifiant is independent from signifié" paradigm [14].

To emit such a call was, however, not a motivation behind the redaction of this paper. Nor had we aimed to outperform existing distributional&probabilist models – for it may seem quite unprobable that one would outperform the "heavy Markovian artillery" with such a simple computational machinery as k-way clustering. Thus, it has been of certain surprise to us that the comparison of data presented on Figure 4 in [4] with our results indicated that for some Slavic corpora, our simplistic morphology-driven geometrically-clustered model has attained higher or more or less equal V-mesure scores than models presented in [11,9]. Our approach can also dispose of certain advantages when it comes to computational complexity – while some models like that of [2] have sometimes problems to converge to result in reasonable time, none of our 198 analyses whose results are presented above have lasted more than few seconds on an average desktop computer.

This being said, we believe that it may be the case that POS-i induction of systems of next generation could not only take into account but shall rather be based on word's "internal" morpho(phono)logical or even prosodic and metric features. While sufficient evidence exists for stating that in order to have a highly performant and robust POS-i model, one MUST take into account the distributional and contextual information "external" to the word under question, we believe that especially in case of highly flectional languages, the complexity of the whole POS-i clustering proccess could be significantly reduced if ever the process shall be "seeded" (i.e. initiated) with token's "internal" features. Since the performance-augmenting and complexity-reducing effects of such seeding are the principal topic of our ongoing work, we conclude that what we believe to be the ultimate advantage of such a model could be its "cognitive plausibility" [10].

At last but not least, by underlining the importance of suffixal features for POS-induction process, our results may well point in the same direction as hypothesis that "one of the first operating principles employed in the ontogenesis of grammar [is that] grammatical realizations in the form of suffixes or postpositions will be acquired earlier than realizations in the form of prefixes or prepositions" [16]. Thus, without an intention to do so[2] we ultimately find the results of our purely empiric study to be consistent with more general psycholinguistic theories of grammar induction and language development.

References

1. Berg-Kirkpatrick, T., Bouchard-Côté, A., de Nero, J., Klein, D.: Painless unsupervised learning with features. In: Human Language Technologies: The 2010 Annual Conference of the North American Chapter of the Association for Computational Linguistics, pp. 582–590 (2010)

[2] Both during conception and realization of our study, we have been utterly unaware neither of Slobin's "operating principle A", nor of amount of scientific evidence already associated with it.

2. Biemann, C.: Unsupervised part-of-speech tagging employing efficient graph clustering. In: Proceedings of the 21st International Conference on Computational Linguistics and 44th Annual Meeting of the Association for Computational Linguistics: Student Research Workshop, pp. 7–12 (2006)
3. Brown, P.F., Desouza, P.V., Mercer, R.L., Della Pietra, V.J., Lai, C.: Class-based n-gram models of natural language. Computational Linguistics 18(4), 467–479 (1992)
4. Christodoulopoulos, C., Goldwater, S., Steedman, M.: Two Decades of Unsupervised POS induction: How far have we come? In: Proceedings of the 2010 Conference on Empirical Methods in Natural Language Processing, pp. 575–584 (2010)
5. Clark, A.: Combining distributional and morphological information for part of speech induction. In: Proceedings of the Tenth Conference on European Chapter of the Association for Computational Linguistics, vol. 1, pp. 59–66 (2003)
6. Elman, J.L.: Representation and structure in connectionist models (1989)
7. Erjavec, T.: MULTEXT-East: Morphosyntactic resources for Central and Eastern European languages. Language Resources and Evaluation 46(1), 131–142 (2012)
8. Goldwater, S., Griffiths, T.: A fully Bayesian approach to unsupervised part-of-speech tagging. In: Annual Meeting of Association of Computational Linguistics, vol. 45, p. 744 (2007)
9. Graca, J., Ganchev, K., Taskar, B., Pereira, F.: Posterior vs. parameter sparsity in latent variable models. In: Advances in Neural Information Processing Systems, vol. 22, pp. 664–672 (2009)
10. Hromada, D.D.: Conditions for cognitive plausibility of computational models of category induction. In: Laurent, A., Strauss, O., Bouchon-Meunier, B., Yager, R.R. (eds.) IPMU 2014, Part II. CCIS, vol. 443, pp. 93–105. Springer, Heidelberg (2014)
11. Johnson, M.: Why doesnt EM find good HMM POS-taggers. In: Proceedings of the 2007 Joint Conference on Empirical Methods in Natural Language Processing and Computational Natural Language Learning (EMNLP-CoNLL), pp. 296–305 (2007)
12. Karypis, G.: CLUTO-a clustering toolkit (2002)
13. Rosenberg, A., Hirschberg, J.: V-measure: A conditional entropy-based external cluster evaluation measure. In: Proceedings of the 2007 Joint Conference on Empirical Methods in Natural Language Processing and Computational Natural Language Learning (EMNLP-CoNLL), vol. 410, p. 420 (2007)
14. de Saussure, F.: Cours de linguistique générale. Payot, Paris (1922)
15. Schütze, H.: Part-of-speech induction from scratch. In: Proceedings of the 31st Annual Meeting on Association for Computational Linguistics, pp. 251–258 (1993)
16. Slobin, D.: Cognitive prerequisities for acquisition of grammar. In: Studies of Child and Language Development, pp. 175–208 (1973)

Automatic Adaptation of Author's Stylometric Features to Document Types

Jan Rygl

Natural Language Processing Centre,
Faculty of Informatics, Masaryk University, Brno, Czech Republic
xrygl@fi.muni.cz

Abstract. Many Internet users face the problem of anonymous documents and texts with a counterfeit authorship. The number of questionable documents exceeds the capacity of human experts, therefore a universal automated authorship identification system supporting all types of documents is needed. In this paper, five predominant document types are analysed in the context of the authorship verification: *books, blogs, discussions, comments and tweets*. A method of an automatic selection of authors' stylometric features using a double-layer machine learning is proposed and evaluated. Experiments are conducted on ten disjunct train and test sets and a method of an efficient training of large number of machine learning models is introduced (163,700 models were trained).

Keywords: authorship verification, feature selection, machine learning, stylome, stylometric features.

1 Introduction

The Internet has become an integral part of our lives, it is used in the interpersonal communication (e-mails, forums, chat rooms, etc.) and the promotion of goods and events (product reviews, e-shops, corporate portals,...). Since the Internet is built as an anonymous source of information, many users have faced the problem of anonymous documents and texts with a counterfeit authorship. The sources of serious problems come in different forms, from anonymous threats and menaces [2] through fictitious product reviews and forged e-mail headers to anonymously or spuriously published illegal documents.

As the number of unsigned or unofficially signed documents increases rapidly, it is not possible to solve the problem of the authorship recognition with the capacity of human experts. Instead, automated methods for authorship detection are developed and applied.

Because long texts tend to be consistent and reveal sufficient information about the author's style, most methods have been developed for an analysis of books [11]. The emergence of the field of the authorship identification dates back to the antiquity and was motivated by the literature for a long period of time [10].

Nowadays, the authorship recognition methods are predominantly solved by **machine learning** (ML) methods [14] and stylometric analysis [13] that provide the best results when the input is formed with a sufficiently long and coherent text. However,

P. Sojka et al. (Eds.): TSD 2014, LNAI 8655, pp. 53–61, 2014.

the current situation of electronic means of communication calls for a universal automated authorship identification system that supports all types of documents. The results of experiments we conducted suggest that the performance of individual stylometric features depends on the length and type of documents. Therefore, we present an algorithm to supplement authorship recognition systems with automatic stylometric feature selection, depending on the type and the content of examined documents.

The fundamental problem of the authorship recognition field is the task of *authorship verification*. All other tasks, such as authorship attribution or clustering, can be converted to the verification problem. The verification problem can be defined in two different forms:

1. *Strict authorship verification:* Confirm or deny a document authorship by a single known author [18, p. 1].
2. *Authorship verification with corpora:* Given a set of documents written by a suspect along with a document dataset collected from the sample population, we want to determine whether or not an anonymous document is written by the suspect [7].

2 Authorship Verification Approaches

For common stylometric approaches to the authorship verification, we are given two texts A, B and two lists of stylometric features $s(A)$ and $s(B)$. Two predominant machine learning approaches to the authorship verification are presented in the following paragraphs:

1. **One classifier per each known author:** We are given $n \gg 1$ texts written by the author of the text A (documents $A_1, A_2, \ldots, A_n$). For each text, the stylomes $s(A_1), s(A_2), \ldots, s(A_n)$ are extracted, and a model M describing the style of the author of texts A_i is trained by ML methods. If the model is trained only from instances representing the same authorship (pairs A_i, A_j), one-class ML approach is used. However, one-class ML performs generally worse than two-class ML [8]. Training instances can be extended by data representing a different authorship (the second class) by adding texts of other authors, distinct from the author of the document A. In both cases, the author's stylomes are used as attributes (value of each stylometric feature is an attribute) for the ML process. If we are given only one text from the known author of A (or B), this method cannot be used.
2. **A classifier from the corpus:** We are given two texts (A and B) and a collection of texts with similar length to the documents A and B written by authors with known authorships: $C^1_{a_1}, C^2_{a_1}, \ldots, C^n_{a_1}, C^1_{a_2}, \ldots$. Each document can be compared with each other, therefore the problem is defined as a two-class ML (same authorship: $C^m_{a_i}, C^n_{a_i}$; different authorship: $C^m_{a_i}, C^n_{a_j}, a_i \neq a_j$). Training instances are generated from pairs of documents ($C^m_{a_i}, C^n_{a_j}$) in the following manner: stylomes $s(C_{a_i}), s(C_{a_j})$ are extracted and their normalized absolute difference $sim = 1 - |s(C_{a_i}) - s(C_{a_j})|$ represents a similarity vector corresponding to similarity between two texts ($sim \rightarrow 0 \equiv$ *same style*, $sim \rightarrow 1 \equiv$ *different style*).
The ML model M is created from the training instances and the similarity of texts A, B is defined as a probability of the same authorship predicted by M:

$sim \equiv M(1 - |s(A) - s(B)|)$. If we are given more texts by the author(s) of texts A and B, the training instances can be extended by pairs A_i–A_j, B_i–B_j (same authorships) and pairs A_i–$C^m_{a_j}$, B_i–$C^m_{a_j}$ (different authorships).

Since specialized webs as discussion forums, blog servers and libraries contain large text collections with known authorships, the second approach was selected for our experiments. The selected ML implementation supports probability estimation $\langle 0, 0.5)$ for different authorships, $\langle 0.5, 1\rangle$ for same authorships.

3 Author's Stylometric Features – Stylome

Each author has a unique vocabulary, popular phrases, stylistic preferences and a bias that characterizes the author. The author's features are important for the authorship recognition if they allow to distinguish the author from the majority of the population, or if the features can be consistently observed in author's texts.

For the purpose of authorship verification, a vector of decimal numbers (*feature vector*) is extracted from each author's document by a *quantification* of these features. Such vector, when averaged through all author's documents, summarizes all stylometric features of one author. The resulting vector is called a *stylome* [1] – it provides information that defines the style of the author.

The stylome can be analysed via statistical and machine learning methods to discover the author's key features (unique and consistent). The values of key features are used as attributes for ML to recognize whether two documents were written by the same author or not.

The selection of features for the stylome is certainly not fixed. In our experiments conducted on Czech texts, the stylome quantification contains the following 14 categories of features (each category contains tens to thousands of features):

1. **Punctuation analysis** – relative frequencies of punctuation marks are counted [4,12]. We have extended the punctuation analysis with the information about relative positions of punctuation marks in sentences (beginning, middle, or end).
2. **Sentence length distribution** *has been applied with some success and justification when used in conjunction with other attribution tests* [3, p. 17]. Again, we use extended version of the method in experiments – instead of comparing raw differences of frequencies, the data are preprocessed not to penalize differences in close sentence lengths.
3. **Syntactic analysis** – the availability of fast and accurate natural language parsers allows for serious research into syntactic stylometry [5]. SET [9] produces 3 types of sentence parsing trees: a dependency format, a constituent format and a hybrid format. The information from the syntactic trees such as the tree depth, the node count and relative frequency of particular non-terminal nodes are used [16].
4. ***N*-gram syntactic analysis** – instead of relative frequency of non-terminal nodes used in the previous method, the relative frequency of the most common syntactic *n*-grams is used (syntactic *n*-grams consist of interdependent non-terminal nodes).

5. **Analysis of morphological categories** – morphological tagging and disambiguation are required to perform analysis of morphological categories in the input text. The Czech tagger *Desamb* was used to annotate texts and extract relative frequencies of morphological categories that are used as features.
6. **Analysis of morphological tags** uses same tools as the analysis of morphological categories. For each word in a corpus, a normalized morphological-tag information is extracted (set of all important tags assigned to the word). The most frequent morphological tags are used as features and their relative frequencies are counted for each text.
7. **Frequency of word classes** uses fourteen most significant features from the analysis of morphological categories.
8. **Frequency of word-class bigrams** – the number of n-grams of all morphological categories is too big for $n \geq 2$, therefore bigrams of word classes are used as features (14^2 combinations).
9. **Frequency of stop words** – the main advantage of stop word analysis is the topic-independence. The differences of relative stop word frequencies in a large corpus and relative stop word frequencies in the examined text are used as attributes. Lemmatized tokens and stop words are used in our implementation.
10. **Word repetition analysis** – in particular, word lemmata are used here instead of words because Czech is an inflectional language. A separate numeric feature is defined for each word class. If a lemma is repeated in a sentence, value of a feature corresponding to the word class of the lemma is increased.
11. **Simpson's vocabulary richness** – if we manage to collect enough data, we can derive the characteristic of the author on the basis of his or her active vocabulary [6, p. 334]. Simpson in [17] presents a low metric score, which indicates high vocabulary richness: $D = \sum (V_i \cdot \frac{i}{N} \cdot \frac{i-1}{N-1})$ where V_i denotes the number of words with frequency i and N is the number of words in the text.
12. **Analysis of typographic errors** – sum of absolute differences of relative frequencies of selected typographic errors. Each frequency is computed as a ratio of error occurrences in the document to the number of characters in the document.
13. **Frequency of emoticons** – common emoticons were divided to three groups (28 positive, 42 negative, 58 neutral) and relative frequencies of each emoticon and group were used as features.
14. **Analysis of usage of capital letters** – 3 categories of words were distinguished: lowercase words, uppercase words, title words. Relative frequencies of categories and bigrams of two words with the same category were also used as features.

4 Experiments

4.1 Methodology

The evaluation of each category of stylometric features and combination of feature categories is required to discover the best subset of all stylometric features, as two high-quality features can describe similar aspects of the text and learning them in one set achieves worse results than two inferior features that inform about unique trait of the author.

For the above described fourteen feature categories, there are in all combinations $\sum_{i=1}^{14} \binom{14}{i} = 16{,}383$ subsets. Learning a standard machine learning model for thousands of features (categories contains tens to thousands features) is time-consuming (up to hours per model), therefore it is impossible to learn a full model for each subset. We decided to use the double-layer ML approach [15] that creates a model for each feature category (the first layer) and predictions of the first layer models are used as instances for a final model (the second layer):

1. For each category c, a model M_c is created from training instances (14 tasks with tens to thousands features). The models output probabilities of the same authorship of a document pair (a training instance). Train and test data are evaluated separately for each model and each training instance (a document pair) is converted to vector of 14 values (each value corresponds to one feature category).
2. For each subset of feature categories, train and test data are filtered to contain only values that correspond to features in the subset. Since all problems are composed of 14 or less attributes, 16 369 models are trained very fast (tens of seconds per model).

4.2 Data Sources

Texts from five data sources were downloaded and used for experiments (all documents were written by Czech authors), see Table 1. Documents were organized into train and test collections (document pairs were randomly generated). For each data source s, two pairs of collections were built:

1. **unbalanced collections** of productive authors with 5 documents by each of 20 authors – $train_s^{20\times5}$ (2 000 diff. inst., 200 same. inst.) and $test_s^{20\times5}$ (210 diff. inst., 210 same. inst.), where the training instances of the different authorship predominate over instances of the same authorship, and
2. **balanced collections** of 10 documents of each of 10 authors – $train_s^{10\times10}$ (495 diff. inst., 495 same. inst.) and $test_s^{10\times10}$ (473 diff. inst., 472 same. inst.), where the training instances are evenly distributed.

Table 1. Data sources used in experiments

type	doc. #	auth. #	avg. doc. length	source
books	801	79	3 970 words	various
blog articles	26 143	3 352	786 words	http://blog.idnes.cz/
posts from forums	3 664	101	358 words	http://www.filosofie.cz/forum/
comments under blogs	5 519	3 352	151 words	http://blog.idnes.cz/
tweets	1 621	20	33 words	http://www.klaboseni.cz

4.3 Results

Each feature category (FC) was evaluated on train data using cross-validation and on test data. The accuracy is an average accuracy of ML models using that FC. The score is a normalized ranking (a model with the best accuracy has rank 1, the second rank 2, etc.) of models using that FC:

$$score_{FC} = \sum_{i=1}^{|models|} \frac{1}{rank(model_i)} \qquad \text{if } FC \in model_i; \text{ else } 0$$

A higher score corresponds to a better feature category. In the following tables divided per document types, the FC score is shown.

Blogs. *Sentence length distribution*, *Frequency of stop words* and *Typographic errors* were the most effective methods and were used by the two best models. Models using *Frequency of emoticons*, *Vocabulary richness*, *Syntactic analysis* and *Word repetition analysis* have not achieved good results. The highest model accuracy for blogs on test data was 72.14 %. For detailed results, see Table 2.

Books. *Frequency of word classes* and *Word repetition analysis* were the most effective methods and both the best models have used them. Models using *Vocabulary richness* and *Freq. of word-class bigrams* have not achieved good results. The book authorship verification models expectedly provided the highest model accuracy among all document types at 75 % (see Table 3).

Discussions and User Comments. *Frequency of word classes*, *Punctuation analysis* and *Typographic errors* were the most effective methods for these short texts and were again used by the two best models. Models using *Word repetition analysis*, *Morphological categories*, *Usage of capital letters*, *N-gram syntactic analysis* and *Syntactic analysis* have not achieved good results. The best model accuracy on test data for this document type was 72.90 %. For detailed results (see Table 4).

Forums. *Morphological categories*, *Typographic errors* and *Frequency of emoticons* were the most effective methods, which both were used by the two best models. Models using *Usage of capital letters* and *Frequency of word classes* have not achieved good results. The best Forums model has reached 73.81 % accuracy (see Table 5).

Tweets. For the shortest texts in our experiments, authorship verification works best with *Morphological tags*, *Punctuation analysis*, *Frequency of stop words*, *Word repetition analysis*, *Vocabulary richness* and *Usage of capital letters* as the predominant feature categories used by the best two models. We can see that, as the overall accuracy decreases, there are more feature categories that can help to classify the results. Models using *Morphological categories* have not achieved good results. The overall best accuracy on test data for tweets was 66.43 % (see Table 6).

Table 2. Blog analysis

Feature category	Train$^{10\times10}$	Test$^{10\times10}$		Train$^{20\times5}$	Test$^{20\times5}$		Total	
	Acc.	Acc.	Score	Acc.	Acc.	Score	Score	Rank
Usage of capital letters	53.0 %	61.53 %	4.59	47.0 %	65.52 %	8.81	13.40	4
Sentence length distribution	54.0 %	61.59 %	8.30	53.0 %	65.21 %	7.78	16.08	2
Frequency of emoticons	50.0 %	61.42 %	5.02	48.0 %	65.14 %	4.23	9.25	11
Morphological tags	54.0 %	61.53 %	4.94	51.0 %	65.17 %	5.07	10.01	9
Morphological categories	63.0 %	61.62 %	5.22	56.0 %	65.44 %	3.59	8.81	13
Punctuation analysis	57.0 %	61.71 %	8.60	55.0 %	65.75 %	4.27	12.87	5
Frequency of stop words	61.0 %	63.11 %	10.27	55.0 %	66.96 %	10.19	20.45	1
Syntactic analysis	53.0 %	61.40 %	4.48	48.0 %	65.19 %	4.69	9.17	12
Typographic errors	48.0 %	61.50 %	7.77	48.0 %	65.32 %	8.15	15.91	3
N-gram syntactic analysis	50.0 %	61.36 %	5.24	45.0 %	65.25 %	5.69	10.93	8
Vocabulary richness	58.0 %	61.64 %	6.25	58.0 %	65.53 %	3.24	9.48	10
Word repetition analysis	59.0 %	61.37 %	4.08	58.0 %	65.20 %	1.75	5.83	14
Frequency of word classes	63.0 %	61.48 %	3.47	60.0 %	65.66 %	8.86	12.33	6
Freq. of word-class bigrams	60.0 %	61.94 %	6.45	56.0 %	65.58 %	5.54	11.99	7
Best Accuracy	70.00 %	66.23 %		76.00 %	72.14 %			

Table 3. Book analysis

Feature category	Train$^{10\times10}$	Test$^{10\times10}$		Train$^{20\times5}$	Test$^{20\times5}$		Total	
	Acc.	Acc.	Score	Acc.	Acc.	Score	Score	Rank
Usage of capital letters	72.0 %	62.30 %	2.27	65.0 %	71.35 %	8.50	10.77	5
Sentence length distribution	72.0 %	62.43 %	5.94	65.0 %	70.59 %	5.13	11.07	4
Frequency of emoticons	59.0 %	62.32 %	3.90	58.0 %	70.65 %	7.63	11.53	3
Morphological tags	70.0 %	62.29 %	5.40	65.0 %	70.59 %	4.88	10.27	6
Morphological categories	70.0 %	62.24 %	4.12	65.0 %	70.29 %	1.74	5.87	11
Punctuation analysis	71.0 %	62.20 %	2.30	64.0 %	70.90 %	7.07	9.37	8
Frequency of stop words	70.0 %	62.29 %	2.43	65.0 %	70.43 %	1.93	4.35	13
Syntactic analysis	69.0 %	62.12 %	1.43	63.0 %	70.57 %	6.72	8.15	10
Typographic errors	49.0 %	62.32 %	6.03	58.0 %	70.59 %	4.19	10.22	7
N-gram syntactic analysis	70.0 %	62.29 %	4.82	65.0 %	70.59 %	4.45	9.27	9
Vocabulary richness	60.0 %	62.22 %	2.29	57.0 %	70.25 %	1.66	3.95	14
Word repetition analysis	72.0 %	62.56 %	8.98	66.0 %	71.20 %	7.09	16.07	2
Frequency of word classes	71.0 %	62.44 %	6.70	66.0 %	71.56 %	9.77	16.46	1
Freq. of word-class bigrams	70.0 %	62.04 %	0.97	65.0 %	70.56 %	4.26	5.23	12
Best Accuracy	75.00 %	64.87 %		69.00 %	75.00 %			

Table 4. Discussion analysis

Feature category	Train$^{10\times10}$	Test$^{10\times10}$		Train$^{20\times5}$	Test$^{20\times5}$		Total	
	Acc.	Acc.	Score	Acc.	Acc.	Score	Score	Rank
Usage of capital letters	60.0 %	68.53 %	2.57	57.0 %	65.77 %	3.23	5.80	13
Sentence length distribution	55.0 %	68.47 %	3.97	52.0 %	65.81 %	8.67	12.65	6
Frequency of emoticons	60.0 %	68.99 %	9.80	59.0 %	65.90 %	4.29	14.09	4
Morphological tags	60.0 %	68.63 %	6.08	56.0 %	65.58 %	4.87	10.95	8
Morphological categories	64.0 %	69.02 %	4.64	63.0 %	65.55 %	3.90	8.54	9
Punctuation analysis	67.0 %	69.30 %	6.74	67.0 %	66.85 %	9.97	16.71	2
Frequency of stop words	59.0 %	68.62 %	4.41	57.0 %	65.66 %	2.35	6.76	12
Syntactic analysis	62.0 %	68.48 %	2.42	59.0 %	65.26 %	1.85	4.27	14
Typographic errors	57.0 %	68.72 %	7.29	62.0 %	66.80 %	9.91	17.20	1
N-gram syntactic analysis	55.0 %	68.50 %	3.34	54.0 %	65.44 %	4.37	7.71	11
Vocabulary richness	58.0 %	68.41 %	4.57	57.0 %	65.56 %	6.64	11.21	7
Word repetition analysis	58.0 %	68.59 %	3.94	60.0 %	65.70 %	3.93	7.87	10
Frequency of word classes	64.0 %	68.78 %	6.09	60.0 %	65.50 %	6.99	13.08	5
Freq. of word-class bigrams	62.0 %	68.87 %	8.96	63.0 %	65.62 %	5.75	14.71	3
Best Accuracy	70.00 %	72.90 %		74.00 %	71.43 %			

Table 5. Forum analysis

Feature category	$Train^{10\times10}$	$Test^{10\times10}$		$Train^{20\times5}$	$Test^{20\times5}$		Total	
	Acc.	Acc.	Score	Acc.	Acc.	Score	Score	Rank
Usage of capital letters	55.0 %	67.93 %	2.12	59.0 %	66.72 %	4.98	7.10	14
Sentence length distribution	55.0 %	68.17 %	7.82	62.0 %	66.51 %	4.07	11.89	7
Frequency of emoticons	49.0 %	68.12 %	5.85	50.0 %	66.77 %	7.88	13.72	4
Morphological tags	59.0 %	68.15 %	5.22	59.0 %	66.55 %	8.00	13.21	6
Morphological categories	59.0 %	68.39 %	8.33	65.0 %	66.95 %	8.67	17.01	2
Punctuation analysis	58.0 %	68.09 %	3.75	55.0 %	67.44 %	9.55	13.30	5
Frequency of stop words	57.0 %	68.58 %	8.35	61.0 %	66.38 %	3.01	11.36	10
Syntactic analysis	59.0 %	68.17 %	6.77	62.0 %	66.55 %	2.98	9.76	12
Typographic errors	52.0 %	68.95 %	10.02	61.0 %	67.99 %	10.13	20.16	1
N-gram syntactic analysis	52.0 %	67.99 %	4.78	51.0 %	66.57 %	6.64	11.43	9
Vocabulary richness	57.0 %	68.10 %	6.54	55.0 %	66.70 %	3.67	10.21	11
Word repetition analysis	57.0 %	68.56 %	8.25	66.0 %	66.52 %	6.55	14.81	3
Frequency of word classes	59.0 %	68.18 %	3.30	63.0 %	66.66 %	4.45	7.74	13
Freq. of word-class bigrams	60.0 %	68.51 %	9.24	66.0 %	66.39 %	2.54	11.78	8
Best Accuracy	72.00 %	72.48 %		81.00 %	73.81 %			

Table 6. Twitter analysis

Feature category	$Train^{10\times10}$	$Test^{10\times10}$		$Train^{20\times5}$	$Test^{20\times5}$		Total	
	Acc.	Acc.	Score	Acc.	Acc.	Score	Score	Rank
Usage of capital letters	56.0 %	62.93 %	10.28	62.0 %	61.78 %	9.27	19.55	1
Sentence length distribution	53.0 %	60.63 %	6.37	58.0 %	60.99 %	6.98	13.35	5
Frequency of emoticons	55.0 %	60.82 %	6.44	53.0 %	60.98 %	6.18	12.61	6
Morphological tags	49.0 %	60.62 %	6.02	48.0 %	60.74 %	5.20	11.22	10
Morphological categories	55.0 %	60.58 %	2.84	52.0 %	60.72 %	4.08	6.92	14
Punctuation analysis	57.0 %	61.15 %	5.47	60.0 %	61.29 %	6.84	12.31	7
Frequency of stop words	50.0 %	60.78 %	7.35	51.0 %	61.19 %	8.82	16.17	2
Syntactic analysis	55.0 %	60.38 %	1.69	53.0 %	61.04 %	6.51	8.20	13
Typographic errors	53.0 %	60.84 %	5.33	54.0 %	60.92 %	6.09	11.42	8
N-gram syntactic analysis	54.0 %	60.61 %	4.81	51.0 %	60.91 %	6.10	10.91	11
Vocabulary richness	55.0 %	60.91 %	7.17	56.0 %	61.43 %	7.53	14.70	4
Word repetition analysis	51.0 %	61.07 %	8.74	52.0 %	60.95 %	6.88	15.63	3
Frequency of word classes	53.0 %	60.57 %	2.44	51.0 %	61.03 %	6.90	9.34	12
Freq. of word-class bigrams	53.0 %	60.58 %	5.99	49.0 %	60.86 %	5.26	11.25	9
Best Accuracy	65.00 %	66.02 %		70.00 %	66.43 %			

5 Conclusions and Future Work

The experiments indicate that accuracies of stylometric features achieved in cross-validation tests on train data do not correlate with results obtained on test data. If all stylometric features are used, the authorship recognition performs worse than when using only the optimal feature set. Therefore, we recommend to use the double-layer machine learning technique to select the best performing stylometric features.

We have also provided a methodology for selection of the most effective stylometric features for five predominant document types when not enough training data or training time is available.

In the following research, we plan to conduct experiments on other document types and combinations of different document types, which are still a challenge in the field of authorship verification.

Acknowledgements. This work has been partly supported by the Ministry of the Interior of CR within the project VF20102014003.

References

1. Daelemans, W.: Explanation in computational stylometry. In: Gelbukh, A. (ed.) CICLing 2013, Part II. LNCS, vol. 7817, pp. 451–462. Springer, Heidelberg (2013)
2. Fitzgerald, J.R.: FBI's communicated threat assessment database: History, design, and implementation. FBI: Law Enforcement Bulletin 76, 6–9 (2007)
3. Grieve, J.W.: Quantitative authorship attribution: A history and an evaluation of technique. Master's thesis. Simon Fraser University (2005)
4. Hilton, O.: Scientific examination of questioned documents. Callaghan (1956)
5. Hollingsworth, C.: Using dependency-based annotations for authorship identification. In: Sojka, P., Horák, A., Kopeček, I., Pala, K. (eds.) TSD 2012. LNCS, vol. 7499, pp. 314–319. Springer, Heidelberg (2012)
6. Holmes, D.I.: The Analysis of Literary Style – A Review. Journal of the Royal Statistical Society 148(4), 328–341 (1985)
7. Iqbal, F., Khan, L.A., Fung, B.C.M., Debbabi, M.: e-mail authorship verification for forensic investigation. In: Proceedings of the 2010 ACM Symposium on Applied Computing, SAC 2010, pp. 1591–1598. ACM Press, New York (2010)
8. Koppel, M., Schler, J.: Authorship verification as a one-class classification problem. In: Proceedings of the Twenty-first International Conference on Machine Learning, ICML 2004, p. 62. ACM, New York (2004)
9. Kovář, V., Horák, A., Jakubíček, M.: Syntactic analysis using finite patterns: A new parsing system for czech. In: Vetulani, Z. (ed.) LTC 2009. LNCS (LNAI), vol. 6562, pp. 161–171. Springer, Heidelberg (2011)
10. Love, H.: Attributing Authorship: An Introduction. Cambridge University Press (2002)
11. Luyckx, K., Daelemans, W.: Authorship attribution and verification with many authors and limited data. In: Proceedings of the 22nd International Conference on Computational Linguistics COLING 2008, vol. 1, pp. 513–520. Association for Computational Linguistics, Stroudsburg (2008)
12. McMenamin, G.R., Choi, D.: Forensic Linguistics: Advances in Forensic Stylistics. Crc Press (2002)
13. Morton, A.Q., Michaelson, S.: The Q-Sum Plot. Technical report, Department of Computer Science, University of Edinburgh, CSR-3-90 (1990)
14. Pearl, L., Steyvers, M.: Detecting authorship deception: a supervised machine learning approach using author writeprints. LLC 27(2), 183–196 (2012)
15. Rygl, J., Horák, A.: Authorship Attribution: Comparison of Single-layer and Double-layer Machine Learning. In: Sojka, P., Horák, A., Kopeček, I., Pala, K. (eds.) TSD 2012. LNCS, vol. 7499, pp. 282–289. Springer, Heidelberg (2012)
16. Rygl, J., Zemková, K., Kovář, V.: Authorship Verification based on Syntax Features. In: Proceedings of Sixth Workshop on Recent Advances in Slavonic Natural Language Processing, RASLAN 2012, 1st edn., Tribun EU, Brno, Czech Republic, pp. 111–119 (2012)
17. Simpson, E.H.: Measurement of diversity. Nature 163, 688 (1949)
18. van Halteren, H.: Linguistic profiling for author recognition and verification. In: Proceedings of the 42nd Annual Meeting on Association for Computational Linguistics, ACL 2004. Association for Computational Linguistics, Stroudsburg (2004)

Detecting Commas in Slovak Legal Texts

Róbert Sabo[1] and Štefan Beňuš[1,2]

[1] Institute of Informatics of Slovak Academy of Sciences, Bratislava, Slovakia
[2] Constantine the Philosopher University in Nitra, Nitra, Slovakia
robert.sabo@savba.sk, sbenus@ukf.sk

Abstract. This paper reports on initial experiments with automatic comma recovery in legal texts. In deciding whether to insert a comma or not, we propose to use the value of the probability of a bigram of two words without a comma and a trigram of the words with the comma. The probability is determined by the language model trained on sentences with commas labeled as separate words. In the training database one sentence corresponds to one line. The thresholds of bigrams and trigrams probability were experimentally determined to achieve the best balance of precision and recall. The advantage of the proposed method is its high precision (95%) at a relatively satisfactory recall (49%). For judges as potential users of an ASR system with an automatic comma insertion function, precision is particularly important.

Keywords: automatic speech recognition, Slavic languages, judicial domain.

1 Introduction

With the advance of the automatic speech recognition technology, ASR systems began to be deployed in many varied areas. One such area in Slovakia is the judicial domain. Currently, judges can dictate legal texts which are transcribed by an ASR engine with the accuracy of about 95% [1]. In this state of the art the overall quality of transcribed texts is more and more important. Correct text formatting or punctuation plays an important role in the everyday use of ASR systems.

In studies focused on punctuation recovering, authors typically detect commas and sentence ends [2,3]. Within the sentence ends some authors further distinguish between periods, question marks and possibly also exclamation marks [4,5]. In this paper we concentrate only on comma recovering. Our aim is to offer a text which should corresponds to the desired text as much as possible. The presence of the correct punctuation in texts generated by the ASR system deployed for dictating legal documents should shorten the time necessary for final corrections by the user.

Most previous punctuation prediction techniques, developed mostly by the speech processing community, exploit both lexical and prosodic cues. There is comparatively little work exploiting lexical features exclusively [6,7,8,9]. In this paper, we tackle the task of predicting punctuation symbols from a standard text processing perspective without relying on additional prosodic features such as pitch and pause duration.

The proposed method of commas recovering is trained and tested on the Slovak language. Slovak is similar to other Slavic languages in having a rich inflectional and

P. Sojka et al. (Eds.): TSD 2014, LNAI 8655, pp. 62–67, 2014.

derivational morphology which causes additional problems with language modeling such as the need for much larger vocabulary or greater difficulties in part-of-speech tagging. Slovak also has a relatively free word order, which degrades the performance of N-gram language models. The rules for writing commas in Slovak are stricter than in English. Commas separate:

- subordinate clauses from main clauses;
- all co-ordinate constituents unless they are connected by copulative conjunctions *a, i, alebo* (eng. 'and', 'as well', 'or');
- all independent constituents that are inserted into a sentence (parentheses, complements, explanations, etc.).

The rest of this paper is organized as follows. Section 2 briefly introduces the text corpus. Section 3 presents the evaluation metrics that were employed. Section 4 describes the proposed method of comma detection. In Section 5, we report experimental results and Section 6 presents our conclusions and future work.

2 Text Database

To train the language model we used a text corpus of legal text consisting of 17M word tokens. In this corpus 1,035k words were followed by a comma and 437k by sentence ends.

To minimize the perplexity of the trained language model we introduced named entities of three categories. We replaced the numbers by tags <num> and proper names (words beginning with a capital letter) by tags <pn>. Some words in the original legal texts, such as the names of defendants, witnesses, and companies, were anonymised by court employees for security reasons. These anonymised words were replaced by special tag <anon>. Names of some persons e.g. minutes clerk, judges and people who were not parties to court proceedings were omitted from the anonymisation process. These words, as well as all other words beginning with a capital letter, were replaced by tag <pn>.

Commas were separated by spaces and replaced by tag <com>. Sentences were divided into separate lines as it is required for n-gram training by SRILM toolkit [10]. The database was divided into the training and testing sets. The testing database comprises about 1 percent of the training set.

3 Evaluation Metrics

Precision, recall and F-measure are well known performance measures widely used in punctuation tasks. In our task, precision expresses the percentage of correctly inserted commas of all commas inserted by the system. It is important for this number to be as high as possible because users (the judges) prefer a lower number of correctly inserted commas over a higher number of wrongly inserted ones.

The precision and recall are defined in (1) and (2) below.

$$P = \frac{C}{C + F} \tag{1}$$

$$R = \frac{C}{C + M} \tag{2}$$

In these equations C is the number of correctly inserted commas, F is the number of falsely inserted commas and M is the number of missing commas. To express the system performance by a single number, it is possible to use a harmonic mean of P and R called the F-measure (3).

$$F = \frac{2PR}{P + R} \tag{3}$$

4 Proposed Method

Our proposed method is based on comparing the probability of bigrams and trigrams obtained from the language model trained with the SRILM toolkit [10] and saved in the ARPA format. In the language model, for each line the logarithm (base 10) of conditional probability of each N-gram [11] is given; in this work it is labeled as p2 for bigrams and p3 for trigrams. The training database was formatted as a file containing one sentence per line.

In the first step we find the probability p3 of the trigrams (a pair of words separated by a comma). If the trigram (word1 <com> word2) is more probable than a defined threshold, the second step follows. In this step, the probability p2 of the bigram of these words without a comma (word1 word2) is evaluated. If the bigram (word1 word2) is less probable than a selected threshold, the script places a comma between word1 and word2. The thresholds have been examined for different combinations of p3 and p2 values shown in Figures 1 and 2 and discussed in the following section.

5 Experimental Results

The testing database size was about 1 percent of the training set and contained 14,7k words. The quality of the testing database is affected by the fact that the original text material contains a number of errors and ambiguities in comma inserting, which significantly influenced the resulting precision and recall values. The results were degraded by 1–2% due to these errors. The most frequent error was the absence of a comma. The ambiguities typical for the Slovak language involve commas in front of conjunctions *a, i, aj, ani, alebo, či* in graduative, adversative or disjunctive semantic relationships between the connected clauses. In these cases commas preceding the conjunctions are optional. The best results for precision, recall and the F-measure are shown in Table 1. The table shows two different settings of probability thresholds *p3* for trigrams and *p2* for bigrams. The first line shows the best possible precision and the second line the best possible recall.

Figures 1 and 2 illustrate precision and recall values for different thresholds for bigram (p2) and trigram (p3) probabilities. The precision values are consistently high when $p3 > -2.2$ and $p2 < -2.9$. In contrast, recall achieves the best results by $p3$ between -1.8 and -2.4 and by $p2$ between -2.9 and -3.3.

Table 1. Precisions, recalls and F-measures for detecting commas using different models

	Precision	Recall	F-measure
Best precision	97.33	44.12	60.72
Best recall	95.31	49.64	65.28

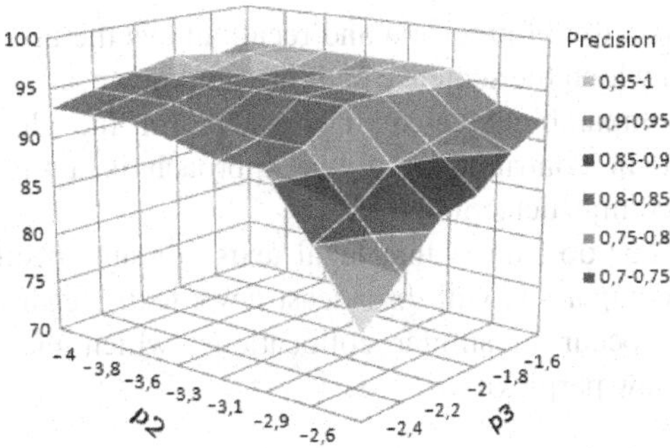

Fig. 1. Precision values (vertical axis) for different thresholds for bigram (p2) and trigram (p3) probabilities

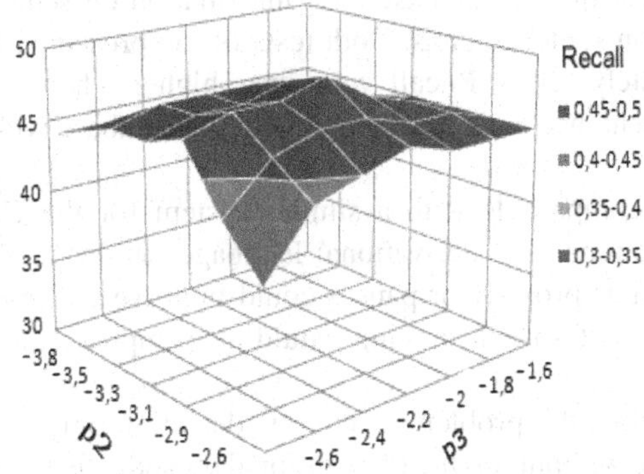

Fig. 2. Recall values (vertical axis) for different thresholds for bigram (p2) and trigram (p3) probabilities

The best precision is achieved when $p3 = -2$ and $p2 = -3.1$. The best recall is achieved by $p3 = -2.2$ and $p2 = -2.9$. We obtained the best F-measure with $p3 = -2.2$ and $p2 = -2.9$.

We also performed similar tests on the test database without the information about sentence ends. The results are shown in Table 2.

Table 2. Precisions, recalls and F-measures for detecting commas using different models [%] without information on sentence ends

	Precision	Recall	F-measure
Best precision	96.75	42.84	59.38
Best recall	95.14	47.64	63.48

The minimal degradation of precision and recall allows the proposed method to be applied in the tasks in which the sentence ends are not known.

The results are comparable to a similar system for the closely related Czech language [2]. The main contribution of our approach is a high precision and a satisfactory recall of comma detection.

The advantage of our domain is that legal texts contain specific vocabulary and stylistic features. In comparison with broadcast news texts, legal texts contain many repetitive words that occur in similar collocations, which facilitates training the language model with low perplexity.

6 Conclusions and Future Work

Results for detecting commas in Slovak legal texts were presented. In deciding whether to insert a comma or not we follow the value of the probability of a bigram of two words without a comma and a trigram of the words with a comma. We perform two tests. First, we test on the database with information on sentence ends and then without information on sentence ends. Both tests of the presented method show high precision (approximately 95%). Recall value was higher when testing the database with information on sentence ends (49%) than with the database without sentence ends (47%).

These results are comparable with a similar system for the Czech language [2] that is a highly inflectional and derivational language similar to Slovak. The use of acoustic features such as prosody or pauses could improve the results. It is plausible that implementing part of speech tagging would also improve the results of comma detection.

Our paper approached the problem of comma detection only from a standard text processing perspective without relying on additional prosodic features such as pitch and pause duration. We take it as a first, but very important, step towards solving the comma detection problem in Slovak. We expect that including prosodic cues may improve the results but at the same time, their influence might not be very significant in this domain, which is supported also in other studies [2,4].

Acknowledgement. This publication is the result of the project implementation: Technology research for the management of business processes in heterogeneous distributed systems in real time with the support of multimodal communication, ITMS 26240220064 supported by the Research & Development Operational Program funded by the ERDF. This publication was also supported by the grant of Slovak Academy of Sciences MVTS - GAMMA.

References

1. Rusko, M., Juhár, J., Trnka, M., Staš, J., Darjaa, S., et al.: Slovak automatic transcription and dictation system for the judicial domain. In: 5th Language & Technology Conference on Human Language Technologies as a Challenge for Computer Science and Linguistics, pp. 365–369. Fundacja Uniwersytetu Im, A. Miczkiewicza (2011)
2. Kolář, J., Švec, J., Psutka, J.: Automatic punctuation annotation in Czech broadcast news speech. In: SPECOM 2004, Saint-Petersburg, pp. 319–325 (2004)
3. Batista, F., Caseiro, D., Mamede, N., Trancoso, I.: Recovering capitalization and punctuation marks for automatic speech recognition: Case study for the Portuguese broadcast news. Speech Communication 50(10), 847–862 (2008)
4. Huang, J., Zweig, G.: Maximum entropy model for punctuation annotation from speech. In: Proceedings of International Conference on Spoken Language Processing, Denver, pp. 917–920, (2002)
5. Christensen, H., Gotoh, Y., Renals, S.: Punctuation annotation using statistical prosody models. In: Proc. ISCA Workshop on Prosody in Speech Recognition and Understanding, pp. 35–40 (2001)
6. Wei, L., Hwee, T.N.: Better punctuation prediction with dynamic conditional random fields. In: Proceedings of the 2010 Conference on Empirical Methods in Natural Language Processing, Cambridge, pp. 177–186 (2010)
7. Gravano, A., Jansche, M., Bacchiani, M.: Restoring punctuation and capitalization in transcribed speech. In: Proceedings of The International Conference on Acoustics, Speech, and Signal Processing, Dallas, pp. 4741–4744 (2009)
8. Stolcke, A., Shriberg, E., Bates, R., Ostendorf, M., Hakkani, D., Plauche, M., Tur, G., Lu, Y.: Automatic detection of sentence boundaries and disfluencies based on recognized words. In: Proc. of ICSLP 1998 (1998)
9. Jakubíček, M., Horák, A.: Punctuation Detection with Full Syntactic Parsing. Research in Computing Science, Special issue: Natural Language Processing and its Applications 46, 335–343 (2010)
10. Stolcke, A.: SRILM – An Extensible Language Modeling Toolkit. In: Proc. of ICSLP 2002, Denver, pp. 901–904 (2002)
11. `http://www.speech.sri.com/projects/srilm/manpages/ngram-format.5.html`

Detection and Classification of Events in Hungarian Natural Language Texts

Zoltán Subecz

College of Szolnok, Department of Economic and Analysis Methodology,
Tiszaligeti sétány 14, 5000 Szolnok, Hungary
subecz@szolf.hu

Abstract. The detection and analysis of events in natural language texts plays an important role in several NLP applications such as summarization and question answering. In this study we introduce a machine learning-based approach that can detect and classify verbal and infinitival events in Hungarian texts. First we identify the multiword noun + verb and noun + infinitive expressions. Then the events are detected and the identified events are classified. For each problem, we applied binary classifiers based on rich feature sets. The models were expanded with rule-based methods too. In this study we introduce new methods for this application area. According to our best knowledge ours is the first result for detection and classification of verbal and infinitival events in Hungarian natural language texts. Evaluating them on test databases, our algorithms achieved competitive results as compared to the current English results.

Keywords: Information extraction, Event detection, Event classification.

1 Introduction

The detection and analysis of events in natural language texts plays an important role in several NLP applications such as summarization and question answering. In this paper we deal with the detection and classification of events that occur in natural language texts.

Though other parts of speech (e.g. noun, participle) can also denote events, the most events belong to verbs in texts; therefore we deal with verbal and infinitival events in this study. e.g. *A tanár **bement** a terembe. (The teacher **went** into the room.)* However not all verbs and infinitives can be considered as event-indicator (e.g. auxiliaries), thus special attention is needed to filter out them. *e.g. Haza **akarok** menni. (I **want** to go home.)* Several events are expressed with two words, e.g. *döntést hoz (make a decision)*, these require distinct treatment. Some studies have dealt with multiword verbal expressions in detail [10,11], we utilized their results here.

The input of our system is a token-level labeled training corpus. The task was divided into three parts. First the single- and multiword verbal and infinitival expressions were picked out. Then from them the events were detected. Finally, the identified events were classified.

In our study we introduce new methods for this area. According to our best knowledge, this is the first result for detection and classification of verbal and infinitival events in Hungarian natural language texts. Evaluating them on test databases, our algorithms achieved competitive results as compared to the current English results.

P. Sojka et al. (Eds.): TSD 2014, LNAI 8655, pp. 68–75, 2014.

2 Related Work

Several papers are concerned with event detection. The most studies focused only on particular events (e.g. business events). In our present work we dealt with the detection and classification of **all** verbal and infinitival events. Most studies engaged in detection and classification of verbal events for only English texts.

Bethard [1] detected events with statistical features. He took into account multi-word expressions too. The model achieved an F-measure of 88.3 for detection and 70.7 for classification.

Llorens et al. [6] applied a CRF model with the application of semantic rules for event detection and classification. The model achieved an F-measure of 91.33 for detection and 73.51 for classification.

Marsaic [7] focused only on verbal event detection and classification. For detection the model achieved an F-measure of 86.49.

The previous studies were designed for the English language.

Bittar [2] detected events in French texts, and achieved an F-measure of 88.8.

Kata et al. [5] applied a clustering algorithm for Hungarian verbs.

Subecz et al. [9] detected and classified verbal events in Hungarian texts, but their methods and results were simplified, not well-worked-out, and was published only in Hungarian language.

Our approach detects and classifies the events with machine learning techniques, which were expanded with rule-based methods. In our system we applied the Hungarian WordNet [8] for the semantic characterization of the examined words, and we disambiguated the polysemic inspected words with the Lesk algorithm [4].

3 The Corpus and Applied Software Packages

In our application we used one part of the Szeged Corpus [3], which contains 5,000 sentences from the following domains: business and financial news, fictions, legal texts, newspaper articles, compositions of pupils. From each of the five domains we selected the first 1,000 sentences.

The sentences were annotated by two annotators with the help of a linguist expert for the detection and classification. The inter-annotator agreement for detection was 87% and for classification it was 81% (simple percentage).

The J48 decision tree algorithm of the Weka 3 data mining suite was employed for machine learning. For the linguistic processing of Hungarian texts the Magyarlanc 2.0 [12] toolkit was used.

4 The Detection of Verbal and Infinitival Events

In this module we detected the verbal and infinitival events. Binary classification was performed for this task, which we expanded with rule based methods. For this module a separate classifier was created, where the event candidates were the verbs and infinitives.

The 5,000 sentences contain 10,628 verbs and infinitives, which were used as event candidates. The annotators labeled 6,479 of them as event.

4.1 Feature Set

The following features were defined for each event candidate.

- **Surface features**: bigrams and trigrams: The character bigrams and trigrams of the beginning and end of the examined words. Besides them: word length, lemma length and the word position within the sentence.
- **Lexical features**: binary feature: Is the examined word a copula or an auxiliary verb? Two lists were created with copulas and auxiliary verbs. These features indicate the presence of the lemma in these lists. Since the eventive nature of a word could be determined by the presence of a copula or an auxiliary verb before or after the word, these four binary features were used.
- **Morphological features**: Since the Hungarian language has rich morphology, therefore several morphology-based features were defined. We defined the MSD codes (morphological coding system) of the event candidates, using the next morphological features: type, mood(Mood), case(Cas), tense(Tense), person of possessor (PerP), number(Num), definiteness (Def). The following features were also defined: the verbal prefix, the examined word, the POS code and the POS codes of the previous and the subsequent words.
- **Syntactic features**: We defined the syntactic labels of the children of the examined event candidate (e.g. Subject, Object…)
- **Semantic features**: The Hungarian WordNet was used here, which contains 3,611 verbal synsets out of the all 42,292 synsets. The semantic relations of the WordNet hypernym hierarchy were used. We applied the following method, *which is new compared to the previous studies*. A separate model was created that without human interaction picked out synsets that are typically in the hypernym chains of events, or have an important role in the decision of the eventive nature. One of the advantages of our method is the automatic collection of the suitable synsets. Otherwise, finding all the required synsets with a simple method would be a complicated task because the events do not belong to some specific synsets in the diverse hypernym relation system of the WordNet. The second advantage of our method is that it can be applied generally, without modification, also to similar problems where it is necessary to find common hypernym intersections, relations for the group of given words in the WordNet hierarchy. It was applied also for the event classification. First we created a model, to which we collected the hypernyms of each event candidate as features during the training phase. On the basis of the features of the decision tree, the model picked out those synsets that are typically in the hypernym chains of events, or have an important role in the decision of the eventive nature. It picked out 95 synsets out of the 3,611 verbal synsets into a list. Then for the main model, these 95 binary features were added to the feature set. At the evaluation phrase we checked whether the event candidate belongs to the hyponyms of any of the collected synsets. Since several meanings can belong to a word form in the WordNet, therefore we performed word sense disambiguation (WSD) between the particular senses with the Lesk algorithm. [4]: Definition and illustrative sentences belong to the synsets in the WordNet. In the case of polysemic event candidates, we counted how many words from the syntactic environment of

the event candidate can be found in the definition and illustrative sentences of the particular WordNet synset (neglecting stopwords). That sense was chosen which contained the highest number of common words.

- **Frequency features**: This feature group was applied as a *new method* as compared to previously published papers. As one of the features, we counted for each event candidate the rate of the cases when the particular word's lemma is an event in the training set. As the second feature a similar rate was counted for the verbal prefix + lemma pair of each event candidate.

The number of features in each group: • Surface: 7, Lexical: 6, Morphological: 10, Syntactic: 4, Semantic: 1–10, Frequency: 2

We completed our machine learning technique also with a rule based method. There were several expressions in the legal texts where the verb usually indicates event in other contexts, but not in the legal context. For example: *A törvény **kimondja**, hogy... (The law **states** that...)* We defined rules for such cases. An example for such a rule: If Subject = "law" And Candidate = "state" Then Candidate ≠ Event.

In the course of evaluation of event detection and classification, the precision, recall and F-measure metrics were used. We examined the significance of the particular feature groups too, then the model's performance on the five subcorpora separately.

Two baseline solutions were applied. At the first one, every verb and infinitive was treated as event. At the second one, only those verbs and infinitives were treated as event that is not copulas or auxiliary verbs.

5 Results – Event Detection

The following experiments on event detection were performed with 10 fold cross validation.

Our first baseline method achieved an F-measure of 79.45, the second one 84.37.

With only the WordNet feature used independently, the model achieved an F-measure of 91.84.

With the whole feature set, the model achieved the following scores: precision: 94.76, recall: 96.20 and F-measure: 95.48.

We examined the efficiency of the particular feature groups with an ablation analysis. In this case the particular feature groups were left out from the whole feature set, and we trained on the basis of the residual features. The results can be found in Table 1. According to the results the Semantic and Frequency features proved to be the most useful ones. The best result was achieved without the Surface features, therefore our further experiments were performed without them.

Then we tested our model on verbs and without the rule based method. We got an F-measure of 94.75 with focusing only on verbs. We got an F-measure of 95.20 without the rule based method. Henceforward the rule based method was used together with focusing on verbs and infinitives.

We examined the model's performance on each subcorpus by randomly splitting the particular subcorpus into training/evaluation sets in a 9/1 ratio. These results can be seen in Table 2. The model achieved the best performance on the Business news domain, and the lowest performance on the Legal corpus.

Table 1. Results of the ablation analysis - Event detection

Left out features	**Precision**	**Recall**	**F-measure**	**Difference**
Surface	94.52	96.50	***95.50***	**+0.02**
Lexical	94.67	96.16	*95.41*	−0.07
Morphological	94.74	96.17	*95.45*	−0.03
Syntactic	94.80	95.99	*95.39*	−0.09
Semantic	94.63	96.06	***95.34***	**−0.14**
Frequency	92.70	96.26	***94.45***	**−1.03**

Table 2. Performance on the subcorpora – Event detection

Corpus	**Precision**	**Recall**	**F-measure**
Compositions	96.08	98.00	*97.03*
Legal	89.74	86.42	***88.05***
Fictions	95.45	97.35	*96.39*
Business news	97.86	98.56	***98.21***
Newspaper articles	96.71	97.35	*97.03*

5.1 Additional Experiments for Event Detection

These experiments did not improve the results for event detection.

The feature set was extended with bag-of-words features. First the lemmas of the syntactic dependents of the particular event candidate were used as bag-of-words. The extended model achieved an F-measure of 95.33 with 10 fold cross validation.

Then similar to the previous case, the lemmas of the syntactic dependents of the particular event candidate together with the relationship type were used as bag-of-words. For example: *OBJ-book*. This extended model achieved an F-measure of 95.39 with 10 fold cross validation.

6 The Classification of Verbal and Infinitival Events

After the detection of verbal and infinitival events we classified them. The classification was performed considering multiple aspects. First, we investigated the main verb types: actions, occurrences, existence and states. Out of them the action and occurrence categories are mostly related to events, therefore these two categories were focused on. **Examples** Action: *A postás **hoz** egy csomagot. (The postman **brings** a package.)* Occurrence: *A levél **leesett** a fáról. (The leaf has **fallen** from the tree.)* Within the 5,000 sentences, among the 6479 events there were 4,158 actions and 1,752 occurrences.

The actions and occurrences together constitute the main part of the events. We wanted to test our model, independently from the former classification, on smaller, but frequent categories. Hence for the second experiment two smaller categories were chosen: movement and communication. **Examples** Movement: *A gyerek **elment** az iskolába. (The child **went** to the school.)* Communication: *Tegnap telefonon **beszélgettünk**. (We **talked** on the phone yesterday.)* In the corpus there were 586 movement and 1,120 communication events.

The same feature set and feature selection methods were used as for the event detection.

Our machine learning technique was extended in the case of movements with a rule based method. Several expressions can be found that denote movement in most contexts, but in some cases they do not. For example: *Az árak szűk sávban **mozogtak**. (The prices **moved** in a narrow range.)* We defined rules for such cases. An example for such a rule: If Subject = "price" And Candidate = "move" Then Candidate ≠ Movement. We created baseline models for classifications too.

7 Results – Event Classification

The following experiments on event detection were performed with 10 fold cross validation.

In the action-occurrence classification task, the baseline model treated all events as action. The model achieved an F-measure of 78.38. In the movement and communication classification task, for the baseline model we selected 11 frequent verbs that denote movement and 16 frequent verbs that denote communication events. The model treated only these events as belonging to the particular category. The model achieved an F-measure of 49.15 for movement and 45.07 for communication.

Henceforward the following abbreviations indicate the given categories:

A: Action, **O**: Occurrence, **M**: Movement, **C**: Communication

With only the WordNet feature used independently, the model achieved F-measures of **A**: 86.63; **O**: 66.00; **M**: 65.64; **C**: 81.24

With the whole training set, the model achieved F-measures of **A**: 87.06; **O**: 73.43; **M**: 68.51; **C**: 81.57

We examined the significance of the particular feature groups with an ablation analysis. In this case the particular feature groups were left out from the whole feature set, and we trained on the basis of the residual features. The results can be found in Table 3. According to the results, the Semantic and Frequency features proved to be the most useful ones. According to the average differences, the best results were achieved without the Morphological features, therefore our further experiments were performed without them.

Table 3. The results of ablation - F-measure - Event classification

Left out features	**Action**	**Occurrence**	**Movement**	**Communication**	**Difference**
Surface	87.02	73.58	68.40	81.13	−0.04/+0.15/−0.11/−0.44
Lexical	86.90	73.09	68.37	80.32	−0.16/−0.34/−0.14/−1.25
Morphological	87.10	73.66	68.72	82.34	+0.4/+0.23/+0.21/+0.77
Syntactic	85.58	73.54	68.54	80.74	−1.48/+0.11/+0.03/−0.83
Semantic	**86.21**	**72.52**	**66.02**	**80.22**	**−0.85/−0.91/−2.49/−1.35**
Frequency	**85.58**	**71.16**	**60.76**	**79.93**	**−1.48/−2.27/−7.75/−1.64**

We examined the model's performance on each subcorpus by randomly splitting the particular subcorpus into training/evaluation sets in a 9/1 ratio. These results can be seen

Table 4. Performance on the sub-corpora - F-measure - Event classification

Corpus	Action	Occurrence	Movement	Communication
Compositions	85.32	56.67	86.96	75.68
Legal	84.40	71.43	66.67	84.85
Fictions	85.71	60.32	70.27	72.34
Business news	**88.89**	**92.86**	**62.37**	**85.71**
Newspaper articles	83.09	47.76	58.22	70.18

in Table 4. According to the average results, the model achieved the best performance on the Business news domain, and the lowest performance on the Newspaper articles corpus.

7.1 Additional Experiments for Event Classification

In the next two paragraphs we marked bold the improved results compared to the outcome of the ablation analysis.

The feature set was extended with bag-of-words features. First, the lemmas of the syntactic dependents of the particular event candidate were used as bag-of-words. The extended model achieved F-measures of **A: 87.18; O: 74.01; M: 69.20;** C: 81.61 with 10 fold cross validation.

Then similar to the previous case, the lemmas of the syntactic dependents of the particular event candidate together with the relationship type were used as bag-of-words. For example: *SUBJ-teache*r. This extended model achieved F-measures of **A: 87.63; O: 74.04; M: 68.92;** C: 81.69 with 10 fold cross validation.

8 Discussion, Conclusions

In this paper, we introduced our machine learning approach based upon a rich feature set, which can detect verbal and infinitival events in Hungarian texts and classify the identified events. We solved the problem in 3 steps. First, we identified the multiword noun + verb or noun + infinitive expressions. Then we detected the events, and classified the identified events. We tested our methods on 5 domains of the Szeged Corpus.

We applied for each problem binary classifiers based on rich feature sets. We expanded the models with rule based methods too. In this study we introduced new methods for this application area. According to our best knowledge ours is the first result for detection and classification of verbal and infinitival events in Hungarian natural language texts. We tested the model's feature set with an ablation analysis, then the model's performance on 5 subcorpora. Evaluating them on test databases, our algorithms achieved competitive results as compared to the current English results. An F-measure of 95.5 was achieved for detection and F-measure of 87.63; 74.04; 69.20 and 82.34 for the four classifications.

References

1. Bethard, S.J.: Finding Event, Temporal and Causal Structure in Text: A Machine Learning Approach. Computer Science. Boulder, CO, University of Colorado (2002)
2. Bittar, A.: Annotation of Events and Temporal Expressions in French Texts. In: ACL-IJCNLP 2009 Proceedings of the Third Linguistic Annotation Workshop, pp. 48–51 (2009)
3. Csendes, D., Csirik, J.A., Gyimóthy, T.: The Szeged Corpus: A POS Tagged and Syntactically Annotated Hungarian Natural Language Corpus. In: Sojka, P., Kopeček, I., Pala, K. (eds.) TSD 2004. LNCS (LNAI), vol. 3206, pp. 41–47. Springer, Heidelberg (2004)
4. Jurafsky, D., Martin, J.H.: Speech and Language Processing: An Introduction to Natural Language Processing. In: Computational Linguistics, and Speech Recognition. Prentice-Hall, Upper Saddle River (2000)
5. Kata, G., Héja, E.: Clustering Hungarian verbs on the basis of complementation patterns. In: Proceedings of the 45th Annual Meeting of the ACL: Student Research Workshop, pp. 91–96 (2007)
6. Llorens, H., Saquete, E., Navarro-Colorado, B.: TimeML Events Recognition and Classification: Learning CRF Models with Semantic Roles. In: Proceedings of the 23rd International Conference on Computational Linguistics, COLING 2010, pp. 725–733 (2010)
7. Marsic, G.: Temporal processing of news: annotation of temporal expressions, verbal events and temporal relations. PhD thesis, University of Wolverhampton (2011)
8. Miháltz, M., Hatvani, C., Kuti, J., Szarvas, G., Csirik, J., Prószéky, G., Váradi, T.: Methods and Results of the Hungarian WordNet Project. In: Tanács, A., Csendes, D., Vincze, V., Fellbaum, C., Vossen, P. (eds.) Proceedings of the Fourth Global WordNet Conference, GWC 2008, pp. 311–320. University of Szeged, Szeged (2008)
9. Subecz, Z., Nagyné, C.É.: Igei események detektálása és osztályozása magyar nyelvű szövegekben. In: X. Magyar Számítógépes Nyelvészeti Konferencia, Szeged, pp. 237–247 (2014)
10. Vincze, V.: Félig kompozicionális főnév + ige szerkezetek a Szeged Korpuszban. In: VI. Magyar Számítógépes Nyelvészeti Konferencia, Szeged, pp. 390–393 (2009)
11. Vincze, V., Zsibrita, J., Nagy, T.I.: Dependency Parsing for Identifying Hungarian Light Verb Constructions. In: Proceedings of International Joint Conference on Natural Language Processing, pp. 207–215 (2013)
12. Zsibrita, J., Vincze, V., Farkas, R.: Magyarlanc 2.0: szintaktikai elemzés és felgyorsított szófaji egyértelműsítés. In: IX. Magyar Számítógépes Nyelvészeti Konferencia, Szeged, pp. 368–374 (2013)

Generating Underspecified Descriptions of Landmark Objects

Ivandré Paraboni, Alan K. Yamasaki, Adriano S.R. da Silva, and Caio V.M. Teixeira

School of Arts, Sciences and Humanities, University of São Paulo (EACH / USP)
Av. Arlindo Bettio, 1000 - São Paulo, Brazil
ivandre@usp.br

Abstract. We present an experiment to collect referring expressions produced by human speakers under conditions that favour landmark underspecification. The experiment shows that underspecified landmark descriptions are not only common but, under certain conditions, may be largely preferred over minimally and fully-specified descriptions alike.

Keywords: Natural Language Generation, Referring Expressions.

1 Introduction

In Natural Language Generation (NLG), Referring Expression Generation (REG) [3,9] is the computational task of providing adequate referring expressions (e.g., pronouns, definite descriptions, proper names etc.) for discourse entities.

Let us consider the issue of selecting the semantic contents that a referring expression should convey.[1] For instance, suppose a domain containing two identical desks side-by-side, and a single man behind the right desk. The following are possible examples of uniquely identifying reference to the target man.

a The man
b The tall man
c The man behind the desk
d The tall man behind the desk
e The tall man behind the desk, on the right side

In this paper we will focus on the problem of generating non-ambiguous *relational* referring expression between a target r and a landmark object o, as in (c-e) above. More specifically, we will discuss the amount of information used by human speakers to produce the landmark (desk) description $L(o)$ as part of a larger reference to the main target (man).

We will say that the reference to a landmark object is *underspecified* when $L(o)$ does not fully distinguish o from all other objects in the same context, that is, when target and landmark descriptions are meant to mutually disambiguate each other as in (c) and

[1] For surface realisation issues, see, e.g., [14,15,11].

P. Sojka et al. (Eds.): TSD 2014, LNAI 8655, pp. 76–83, 2014.

(d). Conversely, we will say that $L(o)$ is *fully-specified* when $L(o)$ denotes o and no other object in the context as in (e).

Underspecified landmark descriptions as in (c-d) above are likely to be felicitous in simpler, visual domains as those discussed in, e.g., [2]. Allowing target and landmark descriptions to disambiguate each other in these cases may arguably demand little cognitive effort from either speaker or hearer. In more complex situations, by contrast, underspecified landmark descriptions seem to be best avoided. For instance, 'the man behind the door' may demand considerable search effort if, e.g., there is a large number of potential landmarks (i.e., doors) to be inspected [13].

Leaving aside the issue of what kinds of domain may favour landmark underspecification, we will focus on simple situations of reference in which landmark underspecification is most likely frequent, and we will ask under which circumstances this may be preferred over full-specification. In addition to that, since the situations under consideration are simple, we will also investigate whether minimal descriptions are common – in line with studies such as [4] – or not [13].

The present investigation will be carried out as a controlled experiment to collect referring expressions under conditions that favour landmark underspecification. The resulting data set shows that underspecified landmark descriptions are not only common but, under certain conditions, may be largely preferred over minimally and fully-specified descriptions alike.

2 Related Work

Underspecification may be a natural way of referring to landmark objects. This seems to be the case at least in simple visual scenes as discussed in [2]. Accordingly, a number of REG algorithms focused on brevity or minimality may produced underspecified landmark descriptions, although this seems to be largely a side-effect of the main reference strategy, e.g., [4,2,10].

Finding empirical evidence of landmark reference underspecification is however difficult. Referring expression corpora are ubiquitous in NLP (e.g., for anaphora resolution as in [1]), but they lack the necessary semantic 'transparency' [7]. On the other hand, there are few publicly available, semantically annotated REG corpora conveying relational descriptions, but these generally do not seem to offer support to our current investigation.

One possible source of relational descriptions is the GIVE-2 corpus of instructions in virtual environments [6]. In a set of 992 definite descriptions of button object extracted from GIVE-2, 467 (47.1%) descriptions were found to use some kind of relation to a landmark object (e.g., doors, pieces of furniture etc.) However, landmark objects in GIVE-2 have few referable attributes, and most objects do not present variation in size, colour or shape. As a result, these objects are usually referred to by making use of the *type* attribute alone (e.g., 'the door'), with little variation in the way they may be (under or fully) specified.

Closer to our present interests, there is the case of the GRE3D and GRE3D7 corpora of referring expressions [5,17]. Both GRE3D and GRE3D7 are fully annotated and represent situations of reference in visual scenes in which the use of spatial relations was likely to occur (e.g., 'the cube next to the large sphere'). There are 224 relational descriptions in GRE3D, and further 600 in GRE3D7. In GRE3D, the relation between target and landmark is always unique, whereas in GRE3D7 it is not.

We performed a brief analysis of the number of underspecified landmark descriptions in GRE3D and GRED7. Results are summarized in Table 1. As we shall focus on landmark underspecification, the top row – representing target underspecification – is presented for completeness only.

Table 1. Underspecified relational descriptions in the GRE3D and GRE3D7 data

Underspecification	GRE3D		GRE3D7		Overall	
target only	21	9.4%	3	0.5%	24	2.9%
landmark only	15	6.7%	36	6.0%	51	6.2%
target and landmark	22	9.8%	18	3.0%	40	4.9%
total	58	25.9%	57	9.5%	115	14.0%

From these results we observe that landmark underspecification (as would be produced by an algorithm such as [2]) is not infrequent. Over 15% of the descriptions in GRE3D show landmark underspecification, and even in the more complex scenes from GRE3D7 landmark underspecification is at 9%. This contrasts, for instance, studies such as [12,13], in which this kind of underspecification is shown to hinder identification in larger or structurally complex domain structures. In absolute numbers, however, both GRE3D and GRE3D7 corpora still lack sufficient evidence to support a study on landmark underspecification. For that reason, we decided to design a controlled experiment to collect referring expressions of this kind.

3 Current Work

3.1 Experiment Design

We designed a simple within-subjects experiment to investigate the generation of underspecified and minimal descriptions of landmark objects. The experiment makes use of near-identical pairs of visual scenes (kept as simple as possible to encourage minimal and/or underspecified descriptions without the identification risks discussed in [13]). Two examples are provided in Figure 1. The difference between the two is that on the left scene the landmark $q6$ has a uniquely distinguishing colour (white), whereas on the right side it does not.

In all scenes, a reference to the target object pointed by an arrow (e.g., $e2$) will most likely require a reference to the nearest landmark object (i.e., the box $q6$). Even when there are several similar landmark distractors (i.e., boxes) available, an unambiguous underspecified landmark description as in 'the star next to the box' is always possible, although with different degrees of difficulty.

From a REG perspective, the two scenes offer different choices in case the speaker decides to fully specify $q6$. Following the hierarchy of cognitive effort of spatial relations in [8], in the first scene the speaker may arguably prevent landmark

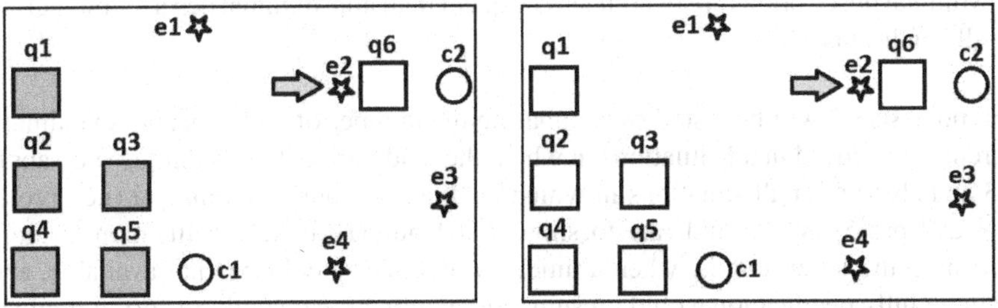

Fig. 1. Two situations of reference to a target (e2) via landmark object (q6)

underspecification simply by making use of an *absolute* reference form (e.g., colour).[2] This, according to [8], may require little cognitive effort. In the second case, by contrast, avoiding landmark underspecification may require a *projective* reference form (e.g., 'on the right'), or some less accessible attribute (e.g., ordinal reference as in 'the second box from left to right' etc.). This, also according to [8] may require more cognitive effort.

In our experiment setting, the left side of Figure 1 is an example of potential *absolute* (*abs*) reference, and the right side is an example of potential *projective* (*proj*) reference. The main goal of the experiment is to investigate in which of these two situations landmark underspecification (e.g., 'next to the box') is more common. In addition to that, given that we make use of simple visual scenes that are likely to encourage underspecification, we will also investigate whether minimal descriptions are common at all.

A landmark description is assumed to be underspecified only if conveying the single attribute *type* as in 'the box'. Any reference to colour, position or any other information that helps ruling out landmark distractors represent full-specification. Thus, typical underspecified landmark descriptions include 'the star *next to the box*' or 'the white star *next to the box*', since in our experiment setting the target colour is never uniquely distinguishing. By contrast, in 'the star *on the right side, next to the box*' the reference to the screen position, although unnecessary, helps landmark disambiguation, and it is therefore assumed to be fully-specified.[3] Minimal descriptions are always as in 'the star next to the box' or involving a similar spatial relation.

To gain further insight on these issues, both situations will be tested in three context sets of different sizes, conveying 1, 3 or 5 landmark distractors each. This gives rise to (2 situations * 3 context sizes =) 6 experiment conditions hereby called abs1, abs3, abs5, proj1, proj3 and proj5. For instance, the left side of Figure 1 represents the abs5 condition, and the right side represents the proj5 condition. Our two research hypotheses are as follows:

h1 Landmark underspecification is more frequent than full-specification when it is not possible to distinguish the landmark from other objects of the same type by means of an absolute reference form (e.g., colour).

[2] For a discussion on the semantics of the colour attribute see [18].

[3] Or, to be more precise, not underspecified.

h2 Minimal descriptions are overall less frequent than non-minimal descriptions across all conditions.

Hypothesis $h1$ will be tested by comparing the number of underspecified landmark references produced in all situations in which the landmark colour is unique (i.e., abs1, abs3 and abs5) with all situations in which landmark colour is not unique (i.e., proj1, proj3 and proj5). We would like to show that landmark underspecification is more frequent in the latter, that is, when a uniquely identifying colour is not available, and obtaining full-specification would presumably require more cognitive effort to produce and interpret. Conversely, we would like to show that landmark underspecification is less frequent when a uniquely distinguishing colour is available, presumably because full-specification is easier.

Hypothesis $h2$ will be tested by comparing the number of minimal and non-minimal descriptions in all situations and context sizes. We would like to show that minimal descriptions are less frequent that non-minimal descriptions even when the context set is quite simple as in our experiment setting.

3.2 Subjects

We recruited 73 Information Systems students who replied to an invitation made by email, and who agreed to volunteer. Participants were on average 20.9 years old and mostly male (62, or 84.9%). All participants were native speakers of Brazilian Portuguese.

3.3 Procedure

Participants were asked to run an executable (jar) file attached to the invitation e-mail and follow the instructions on screen. Upon execution, a brief instruction page was presented, informing the participant that she was about to volunteer anonymously for an experiment on Computer Science. A student id number was required as a means to collect gender and average age information.

Participants were informed that the task consisted of completing the sentence ‘The object pointed by the blue arrow is the...’ in a series of images, and that they should do so as naturally as possible, as if talking to a friend about the objects seen on screen, and free of ambiguity. In order to prevent biased answers, no examples were provided.

Each screen showed a different stimulus in random order, the command sentence and a text field accompanied by a ‘Next’ button. Simple cases of ambiguity as in ‘the star’ were automatically checked and, when necessary, an error message as in ‘I do not know which star you are talking about. Please be more specific” was displayed. Other kinds of ambiguity were not automatically verified.

By providing an answer in the text field and pressing the ‘Next’ button, the next stimulus was displayed. At the end of the experiment, an encrypted file containing the participant’s answers was produced, and the participant received instructions on how to e-mail her answer file back to the researchers in charge.

3.4 Materials

We used purpose-built software for presenting the experiment instructions and the stimuli (in random order), collecting the participant's answers and saving the data onto an encrypted file. The stimuli consisted of 11 images (6 representing our present research questions, and 5 fillers). The relatively large number of fillers was necessary to prevent answer patterns, as all 6 research questions could be in principle answered with the same (underspecified) description as in 'the star next to the box'.

The actual images used in the experiment are similar to the two scenes seen in Figure 1, but without object labels (presently added for ease of discussion). The target is always a star, and it is accompanied by three identical distractors, which forces subjects to add information for disambiguation. Describing the target will usually involve referring to the nearest landmark object, which is always a box.

3.5 Results

We collected 803 descriptions produced by 73 participants, who took on average 5.5 minutes each to complete the task. Since the task was performed without supervision and the participants did not receive examples of the expected description, the collected data set was manually verified for correctness. As a result, nine participants (12%) were identified as outliers. This was the case of participants who provided highly ambiguous descriptions, as in 'a white star', and who most likely misunderstood the task.

After removing the data produced by the outliers, our final data set contained 704 descriptions produced by 64 participants. For the purpose of the present study, a subset of 320 descriptions represents filler situations. These descriptions are not considered in the analysis to follow, which is solely based on the subset of 384 descriptions produced in the six situations of interest (abs and proj, with 1, 3 or 5 distractor landmarks each, cf. Section 3.1).

3.6 Analysis

We use χ^2 to compare description counts. The number of distractor landmarks (1, 3 or 5) had no significant effect on either $h1$ or $h2$. For that reason, in what follows we will consider mean frequencies for each condition group (abs 1/3/5 versus proj 1/3/5). Table 2 shows descriptive statistics for both hypotheses.

Most participants used both under- and fully-specified landmark descriptions. Only 8 participants (12.5%) always underspecified, and only 10 participants (15.6%) always fully-specified. According to Table 2 (left), landmark underspecification is less frequent

Table 2. Descriptive statistics for hypotheses h1 (left) and h2 (right)

	Underspec.		Fully-spec.			Minimal		Non-min.		
Cond.	mean	sdv	mean	sdv		mean	sdv	mean	sdv	
abs	11.7	1.5	50.3	3.8	31.0	7.3	0.6	55.0	2.6	31.2
proj	43.3	4.0	18.0	3.6	30.7	18.7	1.5	43.0	1.0	30.8
	27.5		34.2			13.0		49.0		

than full specification when a uniquely distinguishing landmark colour is available (abs), and more frequent otherwise (proj). The difference is highly significant ($\chi^2 = 98.5, df = 1, p < 0.0001$). This confirms hypothesis $h1$.

Minimally distinguishing descriptions were overall rare. A total of 39 participants (60.9%) never produced minimal descriptions, and only 8 participants (6.3%) always produced them. As seen in Table 2 (right), minimally descriptions are less frequent than non-minimally distinguishing ones in both condition groups (abs and proj). The difference is highly significant ($\chi^2 = 18.12, df = 1, p < 0.0001$). This confirms hypothesis $h2$.

3.7 Further Issues

In addition to testing hypotheses $h1$ and $h2$, we performed a post-hoc analysis of description lengths and attribute usage across experiment conditions. Regarding description length, each description contained on average 3.1 attributes besides *type*, and there was no significant variation across experiment conditions. Since participants were free to complete each sentence with or without a noun (e.g., 'the star'), the use of *type* cannot be taken as indicative of a particular reference strategy, and is not presently analysed.

As for other attributes, there were only two significant difference across experiment conditions: first, the use of screen position attributes (e.g., 'the star *on the top-right corner*') increases three-fold when no absolute reference form is available (proj) ($\chi^2 = 17.09, df = 1, p < 0.0001$), and the use of landmark attributes (e.g., 'the star next to the *white* box') has a 28% increase when an absolute reference form is available (abs) ($\chi^2 = 11.35, df = 1, p < 0.0008$). Both results were to be expected as the absence of an absolute landmark attribute calls for an alternative (e.g., referring to the relative screen position) and, conversely, the presence of a discriminatory colour enables this reference strategy.

4 Conclusions

This paper described a controlled experiment to collect referring expressions under conditions that favour landmark underspecification. Results show that, at least in the simple visual scenes under consideration, the use of underspecified landmark descriptions is highly frequent, and that underspecification is preferred when full-specification seems more difficult to obtain (e.g., when the landmark object does not have a distinguishing colour).

As future work, we intend to apply some of these insights to the design of a more informed REG algorithm to produce definite descriptions conveying the appropriate level of information (e.g., under versus fully-specification). Preliminary results regarding this issue are presented in [16].

Acknowledgments. The authors acknowledge support by USP and FAPESP.

References

1. Cuevas, R.R.M., Paraboni, I.: A machine learning approach to portuguese pronoun resolution. In: Geffner, H., Prada, R., Machado Alexandre, I., David, N. (eds.) IBERAMIA 2008. LNCS (LNAI), vol. 5290, pp. 262–271. Springer, Heidelberg (2008)
2. Dale, R., Haddock, N.J.: Content determination in the generation of referring expressions. Computational Intelligence 7, 252–265 (1991)
3. Dale, R., Reiter, E.: Computational interpretations of the gricean maxims in the generation of referring expressions. Cognitive Science 19 (1995)
4. Dale, R.: Cooking up referring expressions. In: Proceedings of the 27th Annual Meeting of the Association for Computational Linguistics, pp. 68–75 (2002)
5. Dale, R., Viethen, J.: Referring expression generation through attribute-based heuristics. In: Proceedings of ENLG 2009, pp. 58–65 (2009)
6. Gargett, A., Garoufi, K., Koller, A., Striegnitz, K.: The GIVE-2 corpus of giving instructions in virtual environments. In: Proceedings of LREC 2010 (2010)
7. Gatt, A., van der Sluis, I., van Deemter, K.: Evaluating algorithms for the generation of referring expressions using a balanced corpus. In: ENLG 2007 (2007)
8. Kelleher, J.D., Costello, F.J.: Applying computational models of spatial prepositions to visually situated dialog. Computational Linguistics 35(2), 271–306 (2009)
9. Krahmer, E., van Deemter, K.: Computational generation of referring expressions: A survey. Computational Linguistics 38(1), 173–218 (2012)
10. de Lucena, D.J., Pereira, D.B., Paraboni, I.: From semantic properties to surface text: The generation of domain object descriptions. Inteligencia Artificial. Revista Iberoamericana de Inteligencia Artificial 14(45), 48–58 (2010)
11. de Novais, E.M., Paraboni, I.: Portuguese text generation using factored language models. Journal of the Brazilian Computer Society, 1–12 (2012)
12. Paraboni, I.: Generating references in hierarchical domains: the case of Document Deixis. Ph.D. thesis, University of Brighton (2003)
13. Paraboni, I., van Deemter, K.: Reference and the facilitation of search in spatial domains. Language and Cognitive Processes Online (2013)
14. Pereira, D.B., Paraboni, I.: A language modelling tool for statistical NLP. In: Proceedings of TIL 2007, pp. 1679–1688 (2007)
15. Pereira, D.B., Paraboni, I.: Statistical surface realisation of Portuguese referring expressions. In: Nordström, B., Ranta, A. (eds.) GoTAL 2008. LNCS (LNAI), vol. 5221, pp. 383–392. Springer, Heidelberg (2008)
16. Teixeira, C.V.M., Paraboni, I., da Silva, A.S.R., Yamasaki, A.K.: Generating relational descriptions involving mutual disambiguation. In: Gelbukh, A. (ed.) CICLing 2014, Part I. LNCS, vol. 8403, pp. 492–502. Springer, Heidelberg (2014)
17. Viethen, J., Dale, R.: GRE3D7: A corpus of distinguishing descriptions for objects in visual scenes. In: Proceedings of UCNLG+Eval 2011, pp. 12–22 (2011)
18. Viethen, J., Goudbeek, M., Krahmer, E.: The impact of colour difference and colour codability on reference production. In: CogSci 2012, pp. 1084–1098 (2012)

A Topic Model Scoring Approach for Personalized QA Systems

Hamidreza Chinaei[1], Luc Lamontagne[1], François Laviolette[1], and Richard Khoury[2]

[1] Department of Computer Science and Software Engineering, Université Laval, Québec, Canada
[2] Department of Software Engineering, Lakehead University, Thunder Bay, Canada

Abstract. To support the personalization of Question Answering (QA) systems, we propose a new probabilistic scoring approach based on the topics of the question and candidate answers. First, a set of topics of interest to the user is learned based on a topic modeling approach such as Latent Dirichlet Allocation. Then, the similarity of questions asked by the user to the candidate answers, returned by the search engine, is estimated by calculating the probability of the candidate answer given the question. This similarity is used to re-rank the answers returned by the search engine. Our preliminary experiments show that the reranking highly increases the performance of the QA system estimated based on accuracy and MRR (mean reciprocal rank).

Keywords: Personalized QA, User Modeling, Topic Modeling.

1 Introduction

In QA systems, the system automatically finds answers to questions asked by the user. For instance, the user may ask *How old is the oldest complete genome?* The question is processed linguistically and search phrases or keywords are extracted, which are then used to retrieve documents, snippets (the passages returned by the search engine), or sentences [3]. Then a short answer is constructed from the search results and returned to the users. For instance, for the above question, the answer would be *700,000 years old*.

In personalized QA systems, the user may ask several questions either to complete a task or to learn several facts about a topic of interest. The user may need to find as much information as possible say about the *oldest complete genome*. Each time, the system returns an answer and may give extra information or suggestions for other questions of potential interest to the user.

Through its interaction with the user, the personalized QA system can learn incrementally user models based on the user browsed contents and from questions submitted to the system. We assume that a personalized QA system supports some user tasks. While accomplishing their tasks, the users should consult some documents (e.g., news articles). The learned models are then reused in the topic model function that we propose in this paper to estimate the probability of a candidate answer given a question. In this way, the system results can be assessed and prioritized using the learned user model.

Schlaefer [8] introduced the Ephyra question answering engine, a modular and extensible framework that allows to integrate multiple approaches to question answering

P. Sojka et al. (Eds.): TSD 2014, LNAI 8655, pp. 84–92, 2014.

in one system. We use an open version of the Ephyra engine, i.e., *openEphyra*[1], and build our personalized QA system on top of it. Our QA system includes for instance a logger that collects the documents read by user, a topic modeler that models the topics of interest to the user, etc.

In particular, we propose a reranking function to increase the evaluation of the passages, returned by the search engine, that may contain the answer. This is because the QA systems have a bigger chance to extract the correct answers if the top retrieved page./nippets contain the answers [6].

To this end, we propose a new probabilistic scoring approach based on the topics of the questions and the candidate answers. First, the set of interesting topics to the user is learned by applying a topic modeling approach, such as Latent Dirichlet Allocation (LDA) [1], on passages of interest to the user. Then, the similarity between questions asked by the user and candidate answers returned by the system is estimated by calculating the probability of the answer given the question. The probabilities are then estimated using the models learned by LDA. This topic model function is added to the set of QA functions (so called filters) that contribute to extracting and ranking the candidate answers to the question.

In the rest of this paper, we describe topic modeling. A brief introduction to the LDA method is presented in Section 2. We then explain in Section 4 our proposed method of scoring the candidate answers based on the topics of the question and those of the candidate answers. In Section 5, we briefly describe the architecture of our personalized QA system. We then explain in Section 6 our experiments and results. Finally, we conclude this work in Section 7.

2 Topic Modeling

Topic modeling techniques learn a set of possible topics from text documents of a corpus. The Latent Dirichlet Allocation (LDA) model is the pioneer topic modeling approach with interesting results. LDA is a latent Bayesian topic model which is used for discovering the hidden topics of documents [1]. In this model, a document can be represented as a mixture of the hidden topics, where each hidden topic is represented by a distribution over words occurred in the document.

Suppose we have the sentences shown in Table 1. Using the LDA method, we can automatically discovers the topics that are contained in the given collection (data). For example, given n asked topics, the LDA model learns the topics and the assignment of each topic to each sentence in the given data. The learned topics are represented using the words and their probabilities of occurring within each topic. The topics are presented in Table 2 with their top-10 words. The topic representation for topic A illustrates that this topic is about horse, dna, genome, etc. The topic representation for topic B illustrates that this topic B is about film, festivals, stars, etc. For any given snippet, then the algorithm produces in output a probability distribution over the set of all topics. This distribution is interpreted as the probability that the snippet belong to each of the learned topic, as presented in Table 3. Note that the examples shown in this section are from our collection as well as the results of the experiments of this work.

[1] `https://mu.lti.cs.cmu.edu/trac/Ephyra/wiki/OpenEphyra`

Table 1. The sample passage used to learn topic models with LDA

Snippet 1:	Scientists have sequenced the oldest genome to date —and shaken up the horse family tree in the process. Ancient DNA derived from a horse fossil that's between 560,000 and 780,000 years old suggests that all living equids—members of the family that includes horses, donkeys, and zebras—shared a common ancestor that lived at least 4 million years ago, approximately 2 million years earlier than most previous estimates. ...
Snippet 2:	Cameron Diaz is ready to make it a hard-knock life for Annie. The actress has just secured the role of Miss Hannigan in the long-simmering Annie remake.Prior to nabbing the part, Sandra Bullock had been the rumored frontrunner for the mean-spirited, often drunken matron who runs the orphanage in the story. ...
Snippet 3:	Riyaj Shamsudeen offers an in-depth look at Oracle Database 12c , which he calls a 'true cloud database', bringing a new level of efficiency and ease to database consolidation....
...	...

Table 2. Topics learned by LDA from the passage presented in Table 1

Topic A:	dna years horse drone genome ago year time previous million
Topic B:	film jolie festival angelina brad vice year baalbek season trip
Topic C:	food pizza place top made water sugar high set side
Topic D:	space earth vision satellite yankees closer stereo images bridgeman alex
...	...

Table 3. The LDA topic assignments to the passages presented in Table 1

Topics	A	B	C	D	...
Snippet 1:	**44%**	3%	25%	2%	...
Snippet 2:	4%	**65%**	11%	5%	...
Snippet 3:	4%	5%	2%	5%	...
...	...	...	...	...	...

Formally, given a document in the form of $d = (w_1, \dots, w_M)$ in a document corpus D, and given a parameter N that represents the total number of different topics to be found by the algorithm; the LDA model learns two parameters:

1. The parameter θ which is generated from the Dirichlet prior α.
2. The parameter β which is generated from Dirichlet prior η.

The first parameter, θ, is a vector of size N for the distribution of the hidden topics z. The second one, β, is a matrix of size $M \times N$ in which column j stores the probability of each word given the topic z_j.

Figure 1 shows the LDA model in the plate notation in which the boxes are plates that represent replicates. The shaded nodes are the observation nodes, i.e., the words w. The unshaded nodes z represent hidden topics. To build a generative model, LDA performs as follows:

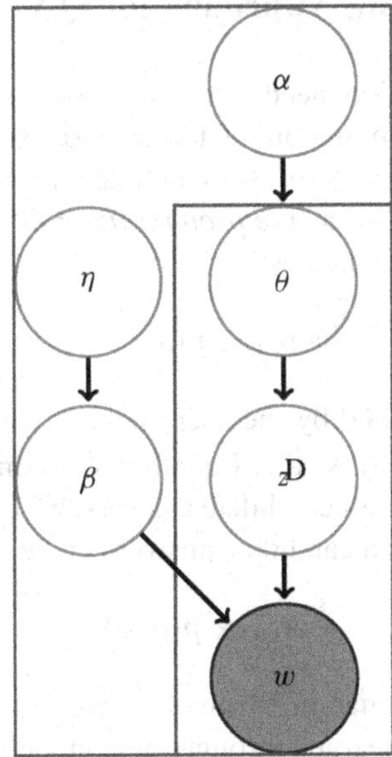

Fig. 1. Latent Dirichlet allocation. The boxes are plates that represent replicates. The shaded nodes represent observations and the unshaded nodes represent hidden topics.

1. For each document d, a parameter θ, is drawn for the distribution of hidden topics based on multinomial distribution with the Dirichlet parameters α.
2. For each document set D, parameter β is learned for the distribution of words given topics. Given each topic z, the vector β_z is drawn based on multinomial distribution with the Dirichlet parameters η.
3. Generate $w_{i,j}$, the jth word in the document i, as follows:
   (a) Draw a topic $z_{i,j}$ based on the multinomial distribution with the parameter θ_i.
   (b) Draw the word $w_{i,j}$ based on the multinomial distribution with the parameter $\phi_{z_{i,j}}$.

3 Related Work

Topic modeling has been used in few QA systems [3,2,4]. The closest to our research effort is the work of [3]. They proposed an LDA-based similarity measure for QA. Their proposed degree of similarity DES(q, a), for a question q and a candidate answer a, based on the similarity of the topic distribution of the question and that of the candidate answers. The degree of similarity is also based on the distributions of question's words, given each topic, and the candidate answers' words given each topic. The similarity is then calculated using an information-theoretic similarity metric, called transformed information radius (IR), which is based on Kullback-Liebler (KL) divergence. Note that in IR measurement, the divergence is transformed into a distance measure [5].

4 A Topic Model Scoring Approach for QA

To account for user information needs in a personalized QA, we propose a new probabilistic scoring approach based on the topics of questions and candidate answers. In a probabilistic approach, the score of a candidate answer, the snippet returned by a search engine, can be estimated as *the probability of the answer given the question*. Formally, the score can be calculated as:

$$\text{score}_a = p(a|q) \tag{1}$$

where q is the question submitted by the user and a is the candidate answer returned by the search engine. Using Bayes rule, Equation 1 becomes $\text{score}_a = \frac{p(a,q)}{p(q)}$. Since the denominator is the same for all candidate answers, we can remove the denominator. Thus, the topic model score for a candidate answer can be approximated as $p(a,q)$:

$$\text{score}_a \propto p(a,q) \tag{2}$$

To learn the expression in Equation 2 from data, we should consider the information need of the user (which can be learned through topic modeling). To do so, using the law of total probabilities, we reformulate the topic model score:

$$p(q,a) = \sum_z p(a,q,z)$$

where z is the topic learned from training data. Then, using conditional rule, we can insert the user topic models in the estimation function:

$$\begin{aligned} &= \sum_z p(a,q|z)\, p(z) \qquad (3)\\ &= \sum_z p(a|z)\, p(q|z)\, p(z) \end{aligned}$$

In the last equality, we assume that given the topic the answer is independent from the question.

We estimate three probability distributions in Equation 3 from topic models learned by LDA. Note that LDA learns a set of topics from the passages of interest to the user. Given a topic z, and question q, then $p(q = (w_1, \ldots, w_{|q|})|z)$ is calculated based on the bag of word assumption that is considered in LDA. Thus, we assume that $p(q = (w_1, \ldots, w_{|q|})|z) = p(w_1|z) \ldots p(w_{|q|})$. Similarly, we calculate $p(a = (w_1, \ldots, w_{|a|})|z)$, where a is a candidate answer. Note however that longer answers will result to smaller probabilities; thus, we normalize Equation 3 by taking it to the power of $1/|a|$ similar to [7].

Each $p(w_i|z)$ is estimated based on the learned model by LDA stored in β (where w_i is a word in the question or the answer). If some words do not occur in the training data, no probabilities are considered for them. In addition, $p(z)$, the learned distributions over topics are learned by LDA, stored in the vector θ. For the passages that do not occur in training data, the distribution over topics can be inferred using LDA inference method [1], as we have done in our experiments.

Our work is similar to that of [3] in using LDA in QA systems. However, [3] used their degree of similarity (that is an information-theoretic similarity metric) particularly as a feature of an SVM classifier. In contrast, we proposed probabilistic scoring approach using topic modeling that is used as a ranking function. This ranking function is added to the set of QA answer extractio./election functions. The topic model function contributes to ranking the candidate answers to the question. For each extracted answer, its score is calculated from Equation 3. In particular, we use the topic distributions, $p(z)$, as a prior probability over the user's topic of interest, whereas [3] use the topic distributions as a factor in their measure of degree of similarity.

5 QA Architecture

Figure 2 shows the architecture of our personalized QA system. Our personalized QA system supports user tasks. While accomplishing theirs tasks, the users should consult some documents (e.g., news articles). The document, read by the user, are logged to be used as input to learn the user model. The learned models are then reused in our proposed topic model function to estimate the probability of a candidate answer given a question. Our personalized QA system is built using open source resources. We developed a graphical user interface (GUI) allowing the users to submit their queries through. The GUI also displays the topics learned for the users based on their questions and browsed snippets/documents. We use RSSOwl for the news feed reader, which is a free news reader software[2]. The core of the QA system is built on the top of the OpenEphyra QA system, an open source QA system developed at CMU. In addition, we use Lucene as our search engine, which is an open source text search engine[3]. The user model is learned using Mallet, an open source machine learning software[4].

6 Experiments

For our experiments, our users collected 1,872 news articles from their topics of interests. These articles were used as our collection in the QA system. Our users then asked 100 factoid questions. The answers for the questions have to occur in the documents read by them (the 1,872 logged news articles). For testing, we manually extracted the answers from the collection (one to five answers based on the question). We then made patterns of each answer (as regular expressions), to be used in the automated evaluation module. This module verifies if each answer returned by the QA system includes the answer's pattern.

For each question, we collected the snippets returned by the Lucene search engine. We then regarded the question and all the returned snippets as a document to be used in LDA, so called *the question document*. In this case, the prior probability of topic

[2] http://www.rssowl.org/

[3] https://lucene.apache.org/core/

[4] http://mallet.cs.umass.edu/

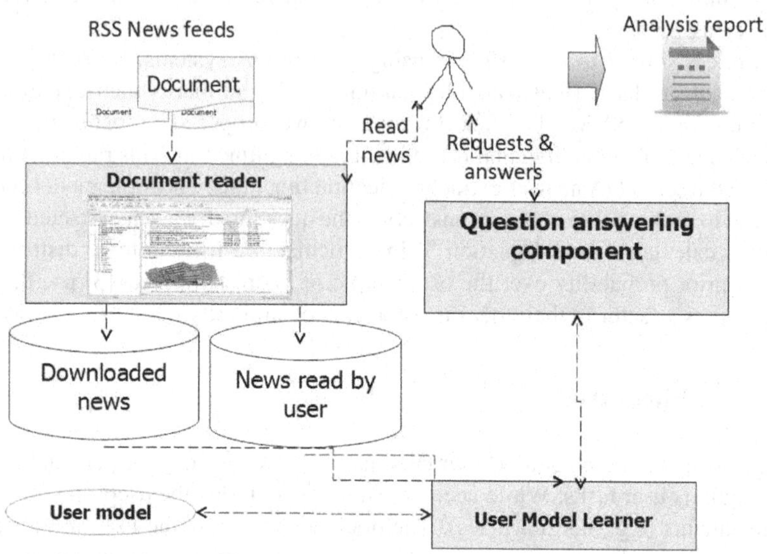

Fig. 2. The personalized QA System architecture

distributions, $p(z)$, for each question is the learned topic distribution of the question document. Specifically, we used LDA to learn 10, 20, 40, 60, 80, and 100 topics using parameters $\alpha = 50$ and $\beta = 1$. For each learned model, the probability of each word given each topic was learned by Mallet for the total of 34,669 distinct words.

We then implemented the topic model function for our QA system. In particular, the topic model function takes the top 10 results, returned by Lucene. For each result, if it includes the question's target, the topic model function scores the result based on the topic model scoring in Equation 3. Then, the topic model score is normalized over all results that got a topic model score (the top 10 results returned by Lucene that contained the question's target). The score for all other results remains minus infinity (the default value in openEphyra). Then, all results returned by Lucene are passed to the result of default functions (filters) in openEphyra, so that the answer extraction and selection be completed.

The topic model performance is evaluated based on the QA system performance. That is, the system performance is calculated in the presence and absence of topic model function. We calculated the following measures: MRR, top1-accuracy, top5-accuracy, and top10-accuracy. The MRR is the mean reciprocal rank. The top1-accuracy, top5-accuracy, and top10-accuracy are respectively the accuracy for the top1, top5, and top10 results [9].

The results are shown in Table 4. The results demonstrate that the topic model function significantly improves the QA system's performance with any number of topics. The table shows consistent results for MRR, top1-accuracy, top5-accuracy, and top10-accuracy. The best performance is achieved when 20 topics are used where MRR drastically increases from 0.20 to 0.30 and top-10-accuracy from 0.27 to 0.40, i.e., 48% improvement.

Table 4. The QA system's performance on the news data

	default functions	10 topics	20 topics	40 topics	60 topics	80 topics	100 topics
MRR	0.20	0.28	**0.30**	0.27	0.29	0.25	0.28
top1-acc	0.16	0.23	**0.24**	0.22	0.24	0.20	0.23
top5-acc	0.25	0.34	**0.37**	0.34	0.35	0.31	0.34
top10-acc	0.27	0.38	**0.40**	0.37	0.38	0.35	0.36
all-acc	0.27	0.38	**0.40**	0.37	0.38	0.35	0.36

Table 5. The QA system's performance on the news data, using two-fold cross validation

	default functions	20 topics
MRR	0.22	**0.29**
top1-acc	0.17	**0.24**
top5-acc	0.27	**0.38**
top10-acc	0.29	**0.38**
all-acc	0.29	**0.38**

We also performed a two-fold cross validation on the same data set using 20 topics (the rest of parameters remain the same). We used half of the data for training, i.e., learning the LDA model, particularly the vector β. The other half was used for testing. In particular, the topic distributions were inferred for the testing data, i.e., the vector θ used as the prior in Equation 3.

Specifically, we randomly took half of the questions and made their question document, similarly we made the question documents for the second half to make two folds. We then used one of the folds for training, and the second half for testing. The prior topic distributions, $p(z)$, were inferred from the question document of the tested questions (using the inference approach introduced in LDA [1]). We then calculated the MRR and top1-accuracy, top5-accuracy, top10-accuracy, and accuracy. We calculated the same measures on the first half while we used the second half for training. The results summarized in Table 5 are averaged over the two folds. These experiments show consistent results with the results shown in Table 4. That is, the topic model filter highly increases the QA system performance on the news dataset.

7 Conclusions

In this paper, we proposed a new probabilistic scoring approach based on the topics of the question and the candidate answers particularly useful for personalized QA systems. In particular, the score of a candidate answer is estimated by calculating the probability of the candidate answer given the question. Our preliminary experiments show that the reranking based on the user topics highly increases the performance of the QA system. In the future work, we will run our experiments on the TREC QA dataset. In addition, we are going to use hierarchical topic modeling approach in the place of flat LDA.

References

1. Blei, D.M., Ng, A.Y., Jordan, M.I.: Latent Dirichlet allocation. Journal of Machine Learning Research 3, 993–1022 (2003)
2. Cai, L., Zhou, G., Liu, K., Zhao, J.: Learning the latent topics for question retrieval in community qa. In: Proceedings of 5th International Joint Conference on Natural Language Processing, Chiang Mai, Thailand (2011)
3. Celikyilmaz, A., Hakkani-Tur, D., Tur, G.: LDA based similarity modeling for question answering. In: The Conference of the North American Chapter of the Association for Computational Linguistics, NAACL HLT 2010, Workshop on Semantic Search, SS 2010, pp. 1–9. Association for Computational Linguistics, Los Angeles (2010), http://dl.acm.org/citation.cfm?id=1867767.1867768
4. Ji, Z., Xu, F., Wang, B., He, B.: Question-answer topic model for question retrieval in community question answering. In: Proceedings of the 21st ACM International Conference on Information and knowledge Management, Maui, Hawaii, USA (2012)
5. Manning, C.D., Schütze, H.: Foundations of Statistical Natural Language Processing. MIT Press, Cambridge (1999)
6. Pala Er, N., Cicekli, I.: A factoid question answering system using answer pattern matching. In: Proceedings of the 6th International Joint Conference on Natural Language Processing, IJNLP 2013, Nagoya, Japan (October 2013), http://www.aclweb.org/anthology/I13-1106
7. Qu, M., Qiu, G., He, X., Zhang, C., Wu, H., Bu, J., Chen, C.: Probabilistic question recommendation for question answering communities. In: Proceedings of the 18th International Conference on World Wide Web (WWW 2009), Madrid, Spain (2009), http://doi.acm.org/10.1145/1526709.1526942
8. Schlaefer, N., Gieselmann, P., Schaaf, T., Waibel, A.: A pattern learning approach to question answering within the ephyra framework. In: Sojka, P., Kopeček, I., Pala, K. (eds.) TSD 2006. LNCS (LNAI), vol. 4188, pp. 687–694. Springer, Heidelberg (2006), http://dx.doi.org/10.1007/11846406_86
9. Voorhees, E.M.: Overview of the TREC 2004 Question Answering Track. In: Text REtrieval Conference (TREC). Special Publication 500-261. National Institute of Standards and Technology, NIST (2004)

Feature Exploration for Authorship Attribution of Lithuanian Parliamentary Speeches

Jurgita Kapočiūtė-Dzikienė[1], Andrius Utka[1], and Ligita Šarkutė[2]

[1] Vytautas Magnus University, K. Donelaičio 58, LT-44248, Kaunas, Lithuania
[2] Kaunas University of Technology, K. Donelaičio 73, LT-44029 Kaunas, Lithuania
j.kapociute-dzikiene@if.vdu.lt, a.utka@hmf.vdu.lt, ligita.sarkute@ktu.lt

Abstract. This paper reports the first authorship attribution results based on the automatic computational methods for the Lithuanian language. Using supervised machine learning techniques we experimentally investigated the influence of different feature types (lexical, character, and syntactic) focusing on a few authors within three datasets, containing transcripts of the parliamentary speeches and debates. Due to our aim to keep as many interfering factors as possible to a minimum, all datasets were composed by selecting candidates having the same political views (avoiding ideology-based classification) from the overlapping parliamentary terms (avoiding topic classification task).

Experiments revealed that content-based features are more useful compared with the function words or part-of-speech tags; moreover, lemma n-grams (sometimes used in concatenation with morphological information) outperform word or document-level character n-grams. Due to the fact that Lithuanian is highly inflective, morphologically and vocabulary rich; moreover, we were dealing with the normative language; therefore morphological tools were maximally helpful.

Keywords: Authorship attribution, supervised ML, Lithuanian.

1 Introduction

Authorship attribution is a process based on an "writeprint" notion. Due to this view, each individual possess his/her own unique idiosyncratic way to express thoughts, which can distinguish him among the others. Authorship attribution analysis is relevant to such applications as author verification (based on the decision if the text is written by the certain author), plagiarism detection (based on finding similarities between the texts written by the different authors), author characterization (based on the extraction of meta information about the authors: i.e. his/her gender, age, education, personality, emotional state, etc.), etc.

But typically authorship attribution task (for review see [18]) is formulated as a task of assigning a text of unknown authorship to one of the candidate authors, when the text samples of those candidates are available. Hence, from the machine learning perspective this problem can be assumed as a supervised multi-class single-label text classification task [17].

The early works on the authorship attribution done for the Lithuanian language date 1971, when the concept of idiolect (individual's distinctive and unique use of

P. Sojka et al. (Eds.): TSD 2014, LNAI 8655, pp. 93–100, 2014.

language) was disputed for the first time by Pikčialingis [16]. Since then lots of descriptive linguistic studies have been done in this field focusing on more specific applications such as forensic linguistics or the analysis of e-mail messages [21]. Despite these important linguistic works, we still need more results on the automatic methods. Consequently, this research is the first attempt at finding a good method to perform authorship attribution task on the Lithuanian texts. In this work we focus on the exploration of the various feature types (lexical, character and syntactic) used together with the supervised machine learning methods. The task is complex due to the couple of reasons. First, the datasets used in our task contain text transcripts of the Lithuanian parliamentary speeches and debates. Since the task is referred to the task of predicting the authorship according to the speaking style, popular orthographic and typographic stylometric features are not valid. Besides, we have to deal with the Lithuanian language, which is highly inflective; ambiguous (47% of all words and their forms are ambiguous); has rich morphology, vocabulary (0.5 million headwords), and word derivation system (e.g. 78 suffixes for diminutives and hypocoristic words).

2 Methodology

2.1 The Datasets

Our experiments were carried out on three datasets to make sure that findings generalize over the different domains. All datasets were composed of the text transcripts of the Lithuanian parliamentary speeches and debates [9], thus represent formal spoken, but normative Lithuanian language. The text transcripts are from the regular parliamentary sessions and cover the period of 7 parliamentary terms starting from March 10, 1990 and ending with December 23, 2013.

Due to the fact that we want to find the best set of features for the authorship attribution, the other interfering factors that determine variation in the text and could facilitate our task must be kept to a minimum, therefore:

- Selected author candidates were from the overlapping (but not necessary from absolutely the same) parliamentary terms (thus avoiding topic classification task).
- Selected author candidates had the same political views (thus avoiding ideology-based classification task).

There is no consensus about the minimum text length appropriate for the authorship attribution. Some of the researchers argue that 2,500 words is the optimal length independent of the random noise [12], some of them obtain promising results with the text fragments of 500 words [6], the others achieve reasonable results with the extremely short texts (with an average length of only 60 words) [10], [15].

The majority of the parliamentary speeches are rather short, thus we could hardly follow the recommendations of e.g. 2,500 words. Besides, we wanted to test the robustness of our methods dealing with the short texts, therefore 150 words was chosen as the minimum text length and all shorter text samples were filtered out.

After previously described pre-processing steps three datasets were created (see Table 1)[1]. Each of them refers to the different Lithuanian party groups (*Social Democrats*,

[1] We did not balance our datasets, because it can cause negative influence on the results [13].

Conservatives, and *Liberals*) and contains parliamentary speeches and debates of three different parliamentarians (e.g. *Social Democrats* contains speeches and debates of parliamentarians with id *SD*1, *SD*2, and *SD*3).

Table 1. Statistics about all datasets: categories; number of texts; number of (distinct) tokens, and lemmas; average text lengths; random and majority baselines

Dataset	Category	Numb. of texts	Numb. of tokens	Numb. of dist. tokens	Numb. of dist. lemmas	Avg. text length	Random baseline	Majority baseline
Social Democrats	SD1	3,932	925,864	58,673	19,263	235.47	0.3685	0.4863
	SD2	2,130	451,212	36,621	12,524	211.84		
	SD3	2,024	400,807	35,719	12,987	198.03		
	TOTAL	8,086	1,777,883	84,484	26,845	219,87		
Conser-vatives	C1	5,093	1,075,880	57,728	18,616	211.25	0.3738	0.4974
	C2	2,654	455,712	29,612	11,893	171.71		
	C3	2,491	484,446	32,220	12,609	194.48		
	TOTAL	10,238	2,016,038	77,885	25,986	196.92		
Liberals	L1	2,539	456,501	31,239	11,371	179.80	0.4602	0.6209
	L2	930	187,501	24,594	8,599	201.61		
	L3	619	125,523	20,995	7,863	202.78		
	TOTAL	4,088	769,525	50,566	16,581	188.24		

2.2 Experimental Setup

Before the experiments all three datasets were preprocessed:

- *Tokenized* (see Table 1),
- *Lemmatized* (see Table 1). Documents were processed using Lithuanian morphological analyzer and lemmatizer "Lemuoklis" [20,3], which replaces all numbers with the special tag and transforms generic words into the lowercase. This preprocessing technique reduced the number of tokens by ~32–33%.
- *Part-of-speech tagged*. Documents were processed using "Lemuoklis", which also performs coarse-grained (identifies main 18 part-of-speech categories, such as noun, verb, etc.) and fine-grained (identifies 12 morphological categories, such as case, gender, tense, etc.) part-of-speech tagging.

We explored the wide range of the individual and compound features that covered lexical, character and syntactic levels[2]:

- *fwd* – the content-free lexical feature which involves only function words[3]. This feature type by consensus is considered as the topic-neutral and was proved to be a relatively good identifier of the author writing style [1].

[2] All typographic and orthographic features were skipped, because we are dealing with the speaking (not writing) style of the author.

[3] Considering the Lithuanian language specific, the function words are: prepositions, pronouns, conjunctions, particles, interjections, and onomatopoeias.

- *lex* – the most popular content-based lexical feature which involves word unigrams or their interpolation (up to *n*=3 in our experiments).
- *lem* – content-based lexical feature, especially recommended for the highly inflective languages, which involves lemmas based on the word unigrams or their interpolation (up to *n*=3 in our experiments).
- *chr* – character feature which involves document-level character n-grams (*n*=[2, 7] in our experiments) – i.e. successions of *n* characters including spaces and punctuation marks. This feature type was proved to surpass other types for Dutch in authorship attribution [11]. Moreover, it gave the most accurate results for the Lithuanian topic classification task [5].
- *pos* – content-free syntactic feature which involves coarse-grained part-of-speech tags based on the word unigrams or their interpolation (up to *n*=3 in our experiments). This feature type is not among the most accurate, but it usually selected for the comparison purposes or used in concatenation with the lexical features.
- *lexpos, lempos, lexmorf, lemmorf* – aggregated features which involves unigrams of concatenated lexical and syntactic features or their interpolation (up to *n*=3 in our experiments) (e.g. *žodis_dktv* (word_noun) is the example of *lexpos* feature). Feature *morf* – determines single string of the concatenated fine-grained morphological category values (e.g. *morf*(*esanti* (existent))="*Noun_Nominative_Feminine_Singular_Present_Active_Non-reflexive_Non-pronominal_Positive*").

Taking into account inflective character of the Lithuanian language and the fact that the normative language texts are used in all our datasets, we formulated our main hypothesis which states that authorship attribution should significantly benefit from the morphological information, in particular, from lemmatization. Besides, the results should be even more improved using as much morphological information as possible, in particular, lexical features in concatenation with the syntactic.

2.3 Classification

In this paper we focus on the supervised machine learning techniques [7] applied to the text categorization [17]) and used for the authorship attribution [18].

The aim of our task is to find a method, which could distinguish authors from each other by creating a model the best approximating the "writeprints" of each individual author speaking style.

We explored the features described in the previous Section using two different machine learning approaches:

- *Support Vector Machine* (SVM) [2] – discriminative instance-based approach is the most popular technique for the text classification, because it can cope with the high dimensional feature spaces (e.g. 84,484 word features in *Social Democrats* dataset; 77,885 in *Conservatives* dataset; and 50,566 in *Liberals* dataset) and the sparseness of the feature vector (only ~220 non-zero feature values among 84,484 in *Social Democrats* dataset instances; ~197 in among 77,885 in *Conservatives*; and ~188 among 50,566 in *Liberals*).
- *Naive Bayes Multinomial* (NBM) [8] – generative profile-based approach was selected for the comparison purposes, in particular, because it is very fast, performs especially well when the number of features is large, and sometimes surpass SVM.

In our experiments we used chi-squared feature extraction method, SMO polynomial kernel (it gave the highest accuracy in our preliminary control experiments) with SVM and NBM implementations in WEKA [4] machine learning toolkit, version 3.6 [19]. All remaining parameters were set to their default values.

3 Results

We performed experiments based on the stratified 10-fold cross-validation with SVM and NBM methods using feature types, described in Section 2.2, but for clarity reasons (not to overload with the information) only the best results of each type are reported (see Table 2 and Table 3). All obtained results are reasonable, because they outperform random and majority baselines.

Table 2. Accuracies, macro-averaged and micro-averaged F-scores for all datasets with different feature types and SVM. The best obtained authorship attribution results are underlined.

Feature types	**Social Democrats**			**Conservatives**			**Liberals**		
	acc.	**microF**	**macroF**	**acc.**	**microF**	**macroF**	**acc.**	**microF**	**macroF**
fwd	0.8260	0.8250	0.8150	0.7931	0.7930	0.7783	0.8312	0.8300	0.7773
chr3	0.9129	0.9130	0.9090	0.9204	0.9200	0.9173	0.9261	0.9260	0.9027
lex1	0.9374	0.9370	0.9347	0.9302	0.9300	0.9277	0.9354	0.9350	0.9193
lex2	0.9375	0.9370	0.9347	0.9363	0.9360	0.9340	0.9291	0.9290	0.9100
lem1	0.9440	0.9440	0.9407	0.9400	0.9400	0.9380	0.9511	0.9510	0.9357
lem2	0.9489	0.9490	0.9470	0.9384	0.9380	0.9360	0.9550	0.9550	0.9417
pos3	0.8167	0.8160	0.8080	0.8272	0.8270	0.8160	0.8378	0.8380	0.7963
lexpos3	0.9387	0.9390	0.9353	0.9318	0.9320	0.9290	0.9359	0.9360	0.9187
lempos3	0.9535	0.9530	0.9513	0.9334	0.9330	0.9310	0.9496	0.9500	0.9367
lexmorf2	0.9400	0.9400	0.9373	0.9355	0.9360	0.9330	0.9293	0.9290	0.9103
lexmorf3	0.9389	0.9390	0.9360	0.9326	0.9330	0.9300	0.9247	0.9250	0.9040
lemmorf2	0.9436	0.9440	0.9410	0.9326	0.9330	0.9300	0.9418	0.9420	0.9270
lemmorf3	0.9427	0.9430	0.9397	0.9333	0.9330	0.9310	0.9406	0.9410	0.9260
Random baseline:	*0.3685*			*0.3738*			*0.4602*		
Majority baseline:	*0.4863*			*0.4974*			*0.6209*		

4 Discussion

Zooming into the results presented in Table 2 and Table 3, allows as to report the following statements.

The content information is very important to achieve high authorship attribution accuracy: i.e. content-free feature types (based on the function words or part-of-speech tags) are easily beaten.

Document-level character n-grams reaching the peak at n=3 are surpassed by the content-based lexical features. Thus, we can assume that for this authorship attribution task character n-grams are not robust enough to capture the patterns of the Lithuanian inflection system intrinsically as it can be done with the lemmas. Lemmas, reducing

Table 3. Authorship attribution results with NBM

Feature types	**Social Democrats**			**Conservatives**			**Liberals**		
	acc.	**microF**	**macroF**	**acc.**	**microF**	**macroF**	**acc.**	**microF**	**macroF**
fwd	0.7857	0.7870	0.7780	0.7125	0.7090	0.6893	0.7774	0.7820	0.7353
chr3	0.7833	0.7840	0.7770	0.7232	0.7210	0.7087	0.7432	0.7520	0.7240
lex1	0.8865	0.8850	0.8773	0.7744	0.7750	0.7730	0.7740	0.7830	0.7763
lex2	0.8910	0.8900	0.8823	0.7719	0.7720	0.7677	0.7630	0.7740	0.7660
lem1	0.8700	0.8700	0.8640	0.7740	0.7760	0.7753	0.7833	0.7910	0.7787
lem2	0.8922	0.8910	0.8850	0.7701	0.7700	0.7670	0.7725	0.7820	0.7713
pos3	0.6838	0.6840	0.6787	0.7131	0.7110	0.6950	0.7277	0.7390	0.6993
lexpos3	0.8909	0.8900	0.8823	0.7678	0.7670	0.7620	0.7561	0.7660	0.7547
lempos3	0.8957	0.8950	0.8887	0.7690	0.7690	0.7647	0.7605	0.7720	0.7613
lexmorf2	0.8918	0.8900	0.8823	0.7712	0.7710	0.7670	0.7625	0.7730	0.7617
lexmorf3	0.8925	0.8910	0.8833	0.7680	0.7670	0.7620	0.7571	0.7670	0.7563
lemmorf2	0.8957	0.8940	0.8867	0.7762	0.7760	0.7727	0.7605	0.7720	0.7593
lemmorf3	0.8978	0.8960	0.8890	0.7757	0.7750	0.7697	0.7544	0.7670	0.7550
Random baseline:	*0.3685*			*0.3738*			*0.4602*		
Majority baseline:	*0.4863*			*0.4974*			*0.6209*		

number of features by ~32–33%, also reduce the sparseness of the feature vector which, in turn, results in a more robust classification model creation.

The best results with NBM are obtained using concatenated lemmas & fine-grained part-of-speech tags based on the interpolation of unigrams to trigrams with *Social Democrats* dataset; concatenated lemmas & fine-grained part-of-speech tags based on the interpolation of unigrams & bigrams with *Conservatives* dataset; and lemmas based on the token unigrams with *Liberals* dataset. SVM method is much more accurate compared to NBM. Thus, the best results with SVM and the best overall results are obtained using concatenated lemmas & coarse-grained part-of-speech tags based on the interpolation of unigrams & bigrams with *Social Democrats* dataset; lemmas based on the token unigrams with *Conservatives* dataset; and lemmas based on the interpolation of token unigrams & bigrams with *Liberals* dataset. McNemar [14] test with one degree of freedom applied on the results with the different feature types revealed that differences between features based on lemmas or used in concatenation with lemmas in most of the cases are not statistically significant, but mostly are statistically significant when compared with the other feature types. Hence, it allows as to confirm our hypothesis that lemmatization and morphological information indeed improves the results. Moreover, the analysis of words unrecognized by the lemmatizer allows assuming that the results would probably be even better, if lemmatizer could cope with the shortened endings (often used in the spoken Lithuanian language) and deal with a bigger set of the international words and named entities.

5 Conclusions and Future Work

In this paper we report the first authorship attribution results on the normative Lithuanian language texts (transcripts of the Lithuanian parliamentary speeches and debates) obtained with the supervised machine learning techniques.

We formulated and experimentally confirmed our hypothesis, that such highly inflective and morphologically rich language as Lithuanian (especially when dealing with the normative texts) mostly benefits from the morphological information, in particular, lemmatization; besides lemmas supplemented with the part-of-speech information can even more boost the performance.

In the future research we are planning to expand the number of authors in the datasets; to experiment with the different domains (e.g. blog data, tweets, etc.) and language types (not only normative Lithuanian).

Acknowledgments. This research was funded by a grant (No. LIT-8-69) from the Research Council of Lithuania.

References

1. Argamon, S., Levitan, S.: Measuring the usefulness of function words for authorship attribution. In: 2005 Joint Conference of the Association for Computers and Humanities and the Association for Literary and Linguistic Computing, pp. 1–3 (2005)
2. Cortes, C., Vapnik, V.: Support-vector networks. Machine Learning 20(3), 273–297 (1995)
3. Daudaravičius, V., Rimkutė, E., Utka, A.: Morphological annotation of the Lithuanian corpus. In: Proceedings of the Workshop on Balto-Slavonic Natural Language Processing: Information Extraction and Enabling Technologies (ACL 2007), pp. 94–99 (2007)
4. Hall, M., Frank, E., Holmes, G., Pfahringer, B., Reutemann, P., Witten, I.H.: The WEKA Data Mining Software: An Update. SIGKDD Explorations 11(1), 10–18 (2009)
5. Kapočiūtė-Dzikienė, J., Vaassen, F., Daelemans, W., Krupavičius, A.: Improving Topic Classification for Highly Inflective Languages. In: 24th International Conference on Computational Linguistics (COLING 2012), pp. 1393–1410 (2012)
6. Koppel, M., Schler, J., Bonchek-Dokow, E.: Measuring Differentiability: Unmasking Pseudonymous Authors. Journal of Machine Learning Research 8, 1261–1276 (2007)
7. Kotsiantis, S.B.: Supervised Machine Learning: A Review of Classification Techniques. Informatica 31, 249–268 (2007)
8. Lewis, D.D., Gale, W.A.: A sequential algorithm for training text classifiers. In: 17th Annual International ACM-SIGIR Conference on Research and Development in Information Retrieval (SIGIR 1994), pp. 3–12 (1994)
9. Lithuanian Parliament official page, `http://www3.lrs.lt/pls/inter/w5_sale.kad_ses`
10. Luyckx, K.: Authorship Attribution of E-mail as a Multi-Class Task – Notebook for PAN at CLEF 2011. In: Petras, V., Forner P., Clough P. (eds.) Cross-Language Evaluation Forum (Notebook Papers/Labs/Workshop) (2011)
11. Luyckx, K., Daelemans, W.: The effect of author set size and data size in authorship attribution. Literary and Linguistic Computing 26(1), 35–55 (2011)
12. Maciej, E.: Does size matter? Authorship attribution, small samples, big problem. In: Literary and Linguistic Computing (2013)
13. Manning, C.D., Schütze, H.: Foundations of Statistical Natural Language Processing. MIT Press, Cambridge (1999)
14. McNemar, Q.M.: Note on the sampling error of the difference between correlated proportions or percentages. Psychometrika 12(2), 153–157 (1947)
15. Mikros, G.K., Perifanos, K.: Authorship identification in large email collections: Experiments using features that belong to different linguistic levels – Notebook for PAN at CLEF 2011. In: Petras, V., Forner P., Clough P. (eds.) Cross-Language Evaluation Forum (Notebook Papers/Labs/Workshop) (2011)

16. Pikčilingis, J.: Kas yra stilius (What is style?). Vaga, Vilnius (1971) (in Lithuanian)
17. Sebastiani, F.: Machine Learning in Automated Text Categorization. ACM Computing Surveys 34, 1–47 (2002)
18. Stamatatos, E.: A Survey of Modern Authorship Attribution Methods. Journal of the Association for Information Science and Technology 60(3), 538–556 (2009)
19. WEKA Machine Learning Toolkit, `http://www.cs.waikato.ac.nz/ml/weka/`
20. Zinkevičius, V.: Lemuoklis morfologinei analizei (Morphological analysis with Lemuoklis). In: Gudaitis, L. (ed.) Darbai ir Dienos, vol. 24, pp. 246–273 (2000) (in Lithuanian)
21. Žalkauskaitė, G.: Idiolekto požymiai elektroniniuose laiškuose (Idiolect signs in the e-mails). PhD dissertation, Vilnius University, Lithuania (2012)

Processing of Quantitative Expressions with Measurement Units in the Nominative, Genitive, and Accusative Cases for Belarusian and Russian

Yury Hetsevich and Alena Skopinava

The United Institute of Informatics Problems of the National Academy of Sciences of Belarus, Minsk, Belarus
{yury.hetsevich,skelena777}@gmail.com

Abstract. This paper outlines an approach to the stage-by-stage solution of the computer-linguistic problem of the processing of quantitative expressions with measurement units by means of the linguistic processor NooJ. The focus is put on the nominative, genitive, and accusative cases for Belarusian and Russian. The paper gives a general analysis of the problem providing examples not only for Belarusian and Russian, but also for English.

Keywords: NooJ, text-to-speech synthesis, text processing, Belarusian, Russian, quantitative expressions, measurement units.

1 Introduction

In order to make text interfaces more "natural", systems of human computer interaction should be able to voice electronic texts. High-quality text-to-speech synthesis cannot be achieved without excellent performance of text processing. There are plenty of objects in texts which require a specific way of treatment: tables, formulas, program codes, etc. This research concentrates on the processing of quantitative expressions with measurement units (QEMUs).

Let us take the following sentence in Belarusian as an example: *Маса Зямлі складае* 5.9736×10^{24} *кг* 'The mass of the Earth is equal to 5.9736×10^{24} kg'. Obviously, there is one quantitative expression with a measurement unit, namely 5.9736×10^{24} кг '5.9736×10^{24} kg'. The purpose is to develop algorithms and resources which will allow expanding this expression into an orthographic form: *пяць цэлых дзевяць тысяч семсот трыццаць шэсць дзесяцітысячных на дзесяць у дваццаць чацвёртай ступені кілаграмаў* 'five point nine thousand seven hundred thirty-six multiplied by ten raised to the twenty-fourth degree kilograms'. The problem is not easy to solve due to the enormous variety of ways in which QEMUs are expressed in writing. Plenty of these ways differ within various language systems. The main difficulty lies in the need to correctly define the categories of case, number, and gender, and to coordinate all the elements of the expression. Both in Belarusian and Russian, six cases, two numbers, and three genders are grammatically possible, which is different from English in

P. Sojka et al. (Eds.): TSD 2014, LNAI 8655, pp. 101–107, 2014.

which there are only two cases, and the ending of the word after the numeral depends only on the number. By now the authors of this paper have focused on the nominative, genitive and accusative cases.

Previously in 2009 a team of Croatian linguists developed algorithms which identify dimensional expressions of length, square and numerical ranges for Croatian and English [1]. Much has been achieved by European developers of the Numeric Property Searching service [2], and Quantalyze semantic annotation and search service [3]. Later in 2013, Belarusian researchers demonstrated finite-state automata which identify, analyse, and classify QEMUs for the Belarusian and Russian languages [4]. The paper's authors have decided to go in another research direction, in particular turning QEMUs (namely combinations of digits, symbols, and letters) into orthographically correct sequences of words which agree in gender, number, and case according to the grammar rules of the Belarusian and Russian languages.

2 Characterizing the Problem of the Processing of QEMUs

At first it is important to decide which tokens should be viewed as measurement units generally. Every scientific field has its own terminological apparatus, objects, subject area, methods of research, etc. For instance a survey is one of research methods in sociology [5]. Sometimes results of surveys are represented with the help of the well-known unit *percent* but often sociologists use such words as *a person, a male, a child, a woman, a student, an employee, an American, etc.*, for example: *5 млн. чал* '5 mln ppl' (should be turned into *пяць мільёнаў чалавек* 'five million people'). At the same time these "units" are not relevant for astronomy or mathematics. This research aims to develop a system which processes QEMUs used as standards for measuring certain physical quantities according to the International System of Units (SI) [6]. At the same time some frequently used and often abbreviated words, such as Belarusian/Russian *штука* or *шт* 'piece' or 'pc' are not going to be ignored either.

Another difficulty is the variety within one and the same language system. For example, the Belarusian language possesses one more system of spelling, which is called Taraškievica (or Belarusian classical orthography). Nowadays the modern and classical systems co-exist, so it is important to take both of them into consideration. Thus, the full list of variants for the Belarusian word секунда 'second' (the SI base measurement unit of time) will be the following: *с, сек, сэк, секунда, секунды, секундзе, секунду, секундай, секундаю, секундзе, секунд, секундаў, секундам, секундамі, секундах, сэкунда, сэкунды, сэкундзе, сэкунду, сэкундай, сэкундаю, сэкундзе, сэкунд, сэкундаў, сэкундам, сэкундамі, сэкундах* – 27 variants. By analogy, there is American English, British English, etc. Thus, we have American *meter-meters*, and British *metre-metres* [5].

It is important to note that the problem of the QEMUs processing can be complicated not only by language peculiarities, but also by the human factor (orthographic mistakes, misprints), and the machine properties (the encoding of power signs, software limitations, etc.).

To sum up, the problem of QEMUs processing is not as easy as it may seem in the beginning. At the same time it requires to be solved because QEMUs can be found in electronic texts of almost any thematic domain in various spheres of everyday life: starting from culinary recipes, and ending with scientific data from space satellites and probes.

3 Creating the Algorithm for the Processing of QEMUs

To solve the problem, by the moment we have developed a syntactic grammar (Fig. 1) consisting of over 350 graphs which can be applied to any electronic texts in Belarusian and Russian. The grammar is represented by the finite-state automaton which has been created by means of the visual graphic editor built in the linguistic processor NooJ [7]. For the present the grammar covers three cases out of six: nominative (the graphs with *Nom*), genitive (the graphs with *Gen*), and accusative (the graphs with *Acc*).

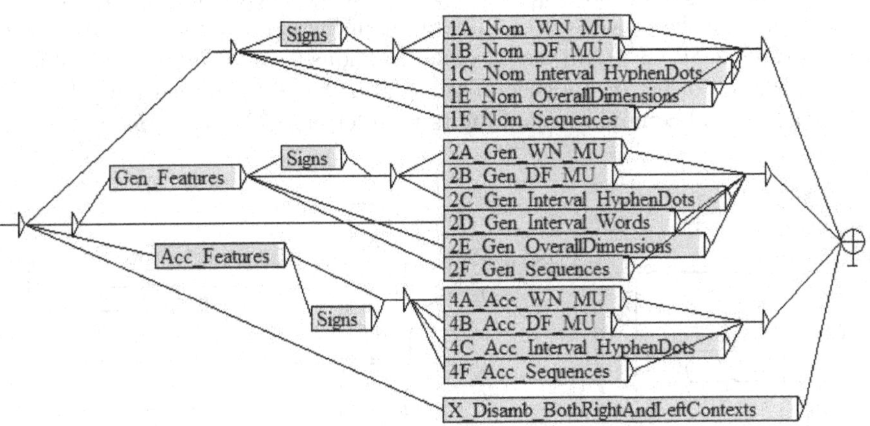

Fig. 1. The main graph of the algorithm for the processing of QEMUs in Belarusian/Russian

As one can see, the names of some graphs start with numbers followed by capital Latin letters. Since there are over 350 graphs we had to foresee a convenient way of the graphs ordering. There are six cases in Belarusian and Russian according to the modern grammar rules of these languages. Traditionally they are listed in the following order: 1st nominative, 2nd genitive, 3rd dative, 4th accusative, 5th instrumental, 6th prepositional. That is why we have put 1 before *Nom*, 2 before *Gen*, and 4 before *Acc*. Latin letters are used as codes for phenomena described by one or another graph. The phenomena are also specified at the end of the graphs' names. The *A*-graphs are for QEMUs in which numeral descriptors are expressed by whole numbers: *678 мм* '678 mm'. The *B*-graphs are used when numeral descriptors are decimal fractions: *18.0005 кілапаскаля* '18.0005 kilopascals'. The *C*-graphs work out for intervals connected either with a hyphen or dots: *125. . . 1000 метров* '125. . . 1000 meters'. The graph with *D* processes intervals formed with certain prepositions: *ад 5 да 6 мілірэнтгенаў* 'from 5 to 6 milliroentgens'. The algorithmic branches with *E* process expressions which describe overall dimensions: *240x707x1500 дм* '240x707x1500 dm'. The F-graphs are necessary for QEMUs in which there are homogeneous numeral descriptors: *около 2, 3, 5 ампер* 'nearly 2, 3, 5 amperes'. As one can notice in Fig. 1, there are also branches named *Signs*, *Gen_ Features*, and *Acc_ Features*. The *Signs*-branch works out when the algorithm finds one of the following sequences: -+± *плюс мінус* (for Belarusian); -+± *плюс минус* (for Russian). Under *Features* we mean word or symbol indicators of one or another case. For instance, after Belarusian *больш* за 'more than' or Russian *более чем* the expression is used in the accusative case, for example: *больш за 6 A* will transform into *больш за шэсць_ампераў*. At the same time Belarusian *больш* 'over' or Russian *более* will make the algorithm behave in a different way, and the phrase will take the genitive case: *больш 6 A* will be turned into *больш шасці ампераў*.

Within the subgraphs the algorithm fulfils plenty of checking operations. At first it focuses on the processing of numeral quantifiers, including not only whole numbers (*99837680*, ± *2*) but also decimal fractions (*678,0000009*, ≈*567.7800*), numbers raised to one or another power ($1{,}43128 \times 10^{15}$), dimensions (Fig. 2) (*10*×*20*×*60*), and combinations of digits with letters (*3892 тыс*, *90 млрд*).

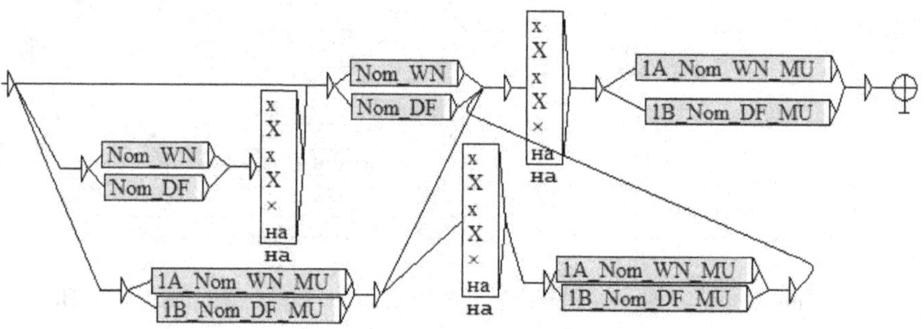

Fig. 2. The subgraph for the processing of QEMUs which denote overall dimensions for Belarusian and Russian

What concerns the graphs for measurement units, they process SI base units, SI derived units, some units which belong to other systems, and some tokens which can be used for measuring or even counting in general: *штука* 'piece', *диоптрия* 'dioptre', *цаля* 'inch', etc. They have been collected separately within subgraphs which contain *Extra* in their names (Fig. 3). So it is possible "to switch them off" if there ever arises the necessity to do this. We have also added

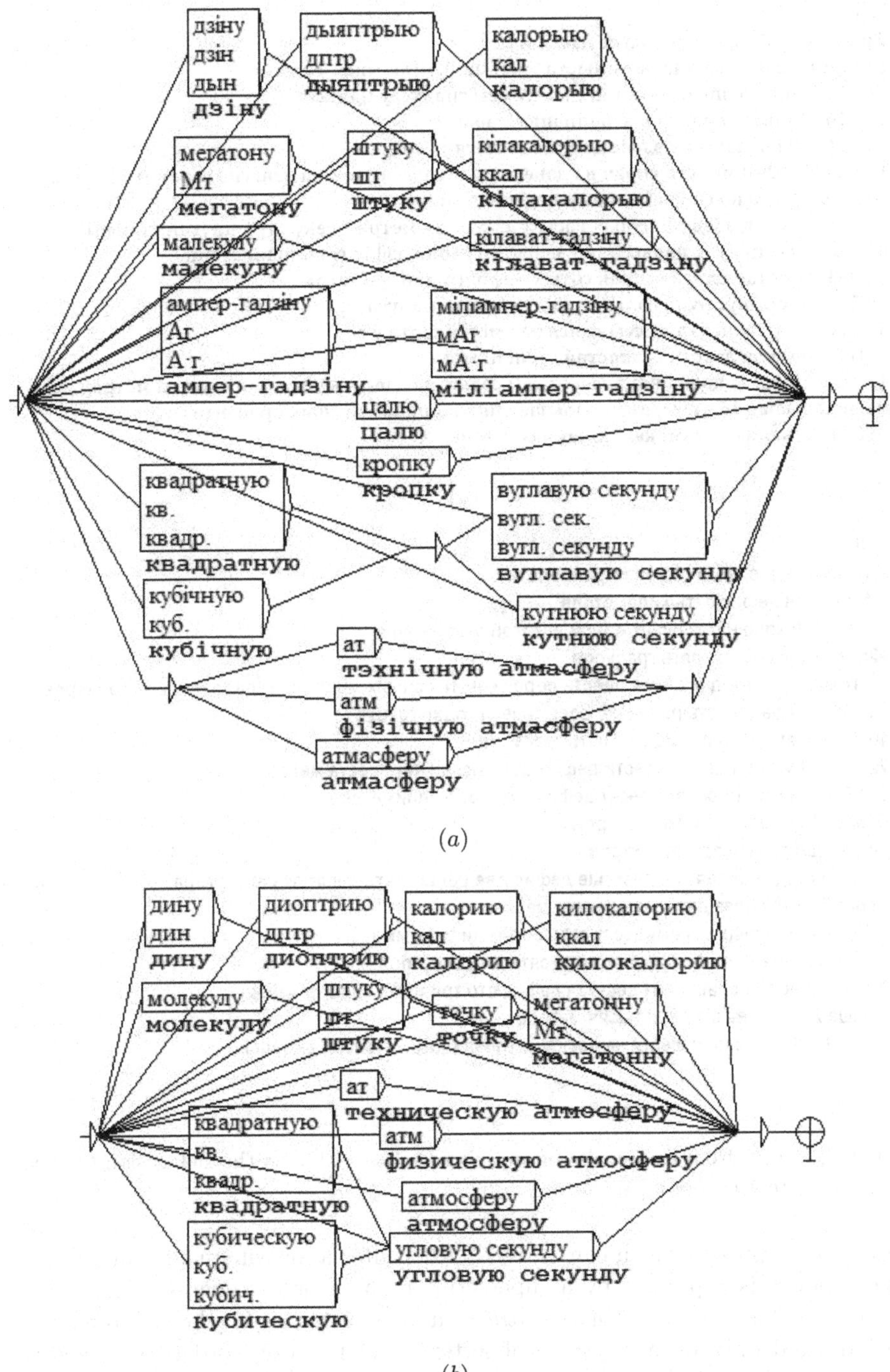

Fig. 3. The subgraph for the processing of QEMUs which denote overall dimensions for Belarusian and Russian

Seq.
ў радыусе 10 міль/ў радыусе дзесяці міль
1430 мАг/адна тысяча чатырыста трыццаць міліампер-гадзін
-5.. -7°C/мінус пяць шматкроп'е мінус сем градусаў Цэльсія
12 і 16 Гб/дванаццаць і шаснаццаць гігабайт
200-20000 Гц/дзвесце дэфіс дваццаць тысяч Герц
140 x 257 x 240 мм/сто сорак на дзвесце пяцьдзесят сем на дзвесце сорак міліметраў
100x40 м^2/сто на сорак метраў у другой ступені
~9,8 м/с^2/каля дзевяці цэлых васьмі дзясятых метра ў секунду ў другой ступені
0,1 Гц-300 кгц/нуль цэлых адна дзясятая Герца дэфіс трыста кілагерц
1,40–3 мкм/адна цэлая сорак сотых дэфіс тры мікраметры
±0,3°/плюс-мінус нуль цэлых тры дзясятыя градуса
0,7 мзв/год/нуль цэлых сем дзясятых мілізіверта ў год
7*10^6 м/сем на дзесяць у шостай ступені метраў
1,57*10^6 м/адна цэлая пяцьдзесят сем сотых на дзесяць у шостай ступені метраў
не перавышае 15 ат. %/не перавышае пятнаццаць атамных працэнтаў
~107 К/с/каля ста сямі кельвінаў у секунду

(*a*)

Seq.
~ 1 ГэВ/около одного гигаэлектронвольта
1-5 км/один дефис пять километров
~ 0,1 МДж/около нуля целых одной десятой мегаджоуля
+25°/плюс двадцать пять градусов
1,247 млн. кв. км/одна целая двести сорок семь тысячных миллиона квадратных километров
74-81°/семьдесят четыре дефис восемьдесят один градус
10-35°/десять дефис тридцать пять градусов
2200-2400 м/две тысячи двести дефис две тысячи четыреста метров
380-500 мм/триста восемьдесят дефис пятьсот миллиметров
более 40 га/более сорока гектаров
до 18 м/до восемнадцати метров
1,3-2,5 см/одна целая три десятые дефис две целые пять десятых сантиметра
Через 29 мес/через двадцать девять месяцев
0,85 кг/ноль целых восемьдесят пять сотых килограмма
20-210 мм/двадцать дефис двести десять миллиметров
0,1-130 кг/ноль целых одна десятая дефис сто тридцать килограммов
свыше 2000 м/свыше двух тысяч метров
около 6·10^{13} Дж/около шести на десять в тринадцатой степени джоулей

(*b*)

Fig. 4. Excerpts from the results of applying the algorithm to the Belarusian (*a*) and Russian (*b*) text corpora

indicators of powers which can be used either after a measurement unit (in this case a power is expressed by a superscript number or a sequence of letters: [1] *кубічны* ‘cubic’, *кв* ‘sq’, *квадратный* ‘square’), or before it (then a power can be indicated only by a sequence of letters, and the word order is reverse: *5 метраў кубічных*). The analysis of text corpora has shown that most often powers are used with measurement units of length or distance. So it has been decided not to overload the algorithm with unnecessary checks.

It is important to note that our finite-state automaton (Fig. 1) represents both the working model of the NooJ grammar [7], and the algorithm at once, so it is unnecessary to perform additional programming transformations in order to apply it in practice. Fig. 4 illustrates some results which have been received after the processing of Belarusian (*a*) and Russian (*b*) texts by our finite-state automata.

We have performed an evaluation test of the algorithm for Belarusian on the material of the text corpus with 100,000 word usages. They cover various scientific and non-scientific thematic domains: astronomy, biology, botany, chemistry, cosmology, culinary, defence technology, geography, history, law, mineralogy, news, physics, space travel science, etc. The corpus belongs to the specific NooJ-format *.noc*. Linguistic experts have counted the total number of QEMUs in the corpus: $N = 1765$. The quantity of all expressions processed by the algorithm is $L = 1430$; the number of those which have been correctly processed is $M = 1464$. The calculations have showed the following results: precision ≈ 83 %, recall ≈ 81 %, and F-score ≈ 82 %.

4 Conclusion

The main goal of the paper – to develop the algorithm for the processing of quantitative expressions with measurement units in the nominative, genitive, and accusative cases in Belarusian and Russian electronic texts – has been achieved. On the whole the algorithm processes whole numbers ($\pm$999 999 999 999), decimal fractions ($\pm 999999999999, 999999999$) combined with 120 measurement units which belong to the classification of the International Bureau of Weights and Measures, units which belong to other systems, and some tokens which can be conditionally called units. In order to construct and test the algorithm, the computer-linguistic development environment NooJ has been used. The evaluation testing of the developed finite-state automaton has proved its rather high quality. At present we continue working in this direction, and further we plan to cover the dative, instrumental, and prepositional cases.

References

1. Bekavac, B., Agić, Ž., Šojat, K., Tadić, M.: Units of Measurement Detection Module for NooJ. In: Mesfar, S., Silberztein, M. (eds.) NooJ 2009 International Conference and Workshop. Finite State Language Engineering, pp. 121–127. Centre de Publication Universitaire, Tunisia (2010)
2. Numeric Property Searching in Derwent World Patents, `http://www.stn-international.com/numeric_property_searching.html`
3. Quantalyze service, `https://www.quantalyze.com/en/`
4. Hetsevich, Y., Skopinava, A.: Identification of Expressions with Units of Measurement in Scientific, Technical & Legal Texts in Belarusian and Russian. In: Proceedings of the Workshop on Integrating IR Technologies for Professional Search (2013), `http://ceur-ws.org/Vol-968/irps_6.pdf`
5. American Heritage Science Dictionary, `http://dictionary.reference.com`
6. International System of Units (SI), `http://www.bipm.org/en/si/`
7. NooJ, `http://www.nooj4nlp.net/pages/nooj.html`

Document Classification with Deep Rectifier Neural Networks and Probabilistic Sampling

Tamás Grósz and István Nagy T.

Department of Informatics, University of Szeged, Hungary
{groszt,nistvan}@inf.u-szeged.hu

Abstract. Deep learning is regarded by some as one of the most important technological breakthroughs of this decade. In recent years it has been shown that using rectified neurons, one can match or surpass the performance achieved using hyperbolic tangent or sigmoid neurons, especially in deep networks. With rectified neurons we can readily create sparse representations, which seems especially suitable for naturally sparse data like the bag of words representation of documents. To test this, here we study the performance of deep rectifier networks in the document classification task. Like most machine learning algorithms, deep rectifier nets are sensitive to class imbalances, which is quite common in document classification. To remedy this situation we will examine the training scheme called probabilistic sampling, and show that it can improve the performance of deep rectifier networks. Our results demonstrate that deep rectifier networks generally outperform other typical learning algorithms in the task of document classification.

Keywords: deep rectifier neural networks, document classification, probabilistic sampling.

1 Introduction

Ever since the invention of deep neural nets (DNN), there has been a renewed interest in applying neural networks (ANNs) to various tasks. The application of a deep structure has been shown to provide significant improvements in speech [5], image [7], and other [11] recognition tasks. As the name suggests, deep neural networks differ from conventional ones in that they consist of several hidden layers, while conventional shallow ANN classifiers work with only one hidden layer. To properly train these multi-layered feedforward networks, the training algorithm requires modifications as the conventional backpropagation algorithm encounters difficulties ("vanishing gradient" and "explaining away" effects). In this case the "vanishing gradient" effect means that the error might vanish as it gets propagated back through the hidden layers [1]. In this way some hidden layers, in particular those that are close to the input layer, may fail to learn during training. At the same time, in fully connected deep networks, the "explaining away" effects make inference extremely difficult in practice [6].

As a solution, Hinton et al. presented an unsupervised pre-training algorithm [6] and evaluated it for an image recognition task. After the pre-training of the DNN,

P. Sojka et al. (Eds.): TSD 2014, LNAI 8655, pp. 108–115, 2014.

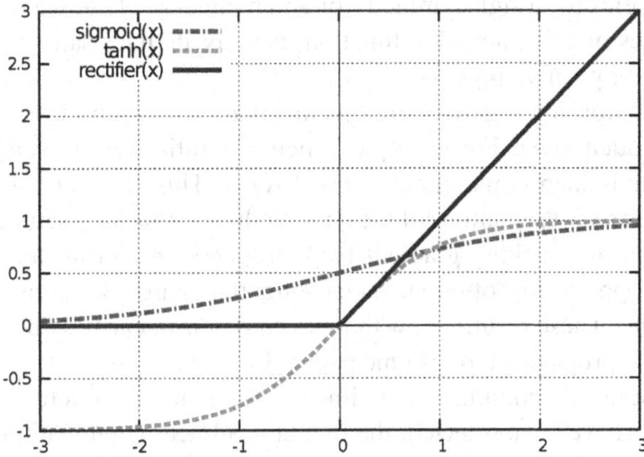

Fig. 1. The rectifier activation function and the commonly used activation functions in the neural networks, namely the logistic sigmoid and hyperbolic tangent (tanh)

the backpropagation algorithm can find a much better local optimum of the weights. Based on their new technique, a lot of effort has gone into trying to scale up deep networks in order to train them with much larger datasets. The main problem with Hinton's pre-training algorithm is the high computational cost. This is the case even when the implementation utilizes graphic processors (GPUs). Several solutions [4,10,2] have since been proposed to alleviate or circumvent the computational burden and complexity of pre-training, one of them being deep rectifier neural networks [2].

Deep Rectifier Neural Networks (DRNs) modify the neurons in the network and not the training algorithm. Owing to the properties of rectified linear units, the DRNs do not require any pre-training to achieve good results [2]. These rectified neurons differ from standard neurons only in their activation function, as they apply the rectifier function ($max(0,x)$) instead of the sigmoid or hyperbolic tangent activation. With rectified neurons we can readily create sparse representations with true zeros, which seem well suited for naturally sparse data [2]. This suggests that they can be used in document classification, say, where the bag of words representation of documents might be extremely sparse [2]. Here, we will see how well DRNs perform in the document classification task and compare their effectiveness with previously used successful methods. To address the problem of unevenly distributed data, we combine the training of DRNs and ANNs with a probabilistic sampling method, in order to improve their overall results.

2 Deep Rectifier Neural Networks

Rectified neural units were recently applied with success in standard neural networks, and they were also found to improve the performance of Deep Neural Networks on tasks like image recognition and speech recognition. These rectified neurons apply the rectifier function ($max(0,x)$) as the activation function instead of the sigmoid or hyperbolic tangent activation. As Figure 1 shows, the rectifier function is one-sided, hence it does not enforce a sign symmetry or antisymmetry. Here, we will examine the two key properties of this one-sided function, namely its hard saturation at 0 and its linear behaviour for positive input.

The hard saturation for negative input means that only a subset of neurons will be active in each hidden layer. For example, when we initialize the weights uniformly, around half of the hidden units output are real zeros. This allows rectified neurons to achieve truly sparse representations of the data. In theory, this hard saturation at 0 could harm optimization by blocking gradient back-propagation. Fortunately, experimental results do not support this opinion, suggesting that hard zeros can actually help supervised training. These results show that the hard non-linearities do no harm as long as the gradient can propagate along some path [2].

For a given input, the computation is linear on the subset of active neurons. Once the active neurons have been selected, the output is a linear combination of their input. This is why we can treat the model as an exponential number of linear models that share parameters. Based on this linearity, there is no vanishing gradient effect [2], and the gradient can propagate through the active neurons. Another advantage of this linear behaviour is the smaller computational cost: there is no need to compute the exponential function during the activation, and the sparsity can also be exploited. A disadvantage of the linearity property is the "exploding gradient" effect, when the gradients can grow without limit. To prevent this, we applied L1 normalization by scaling the weights such that the L1 norm of each layer's weights remained the same as it was after initialization. What makes this possible is that for a given input the subset of active neurons behaves linearly, so a scaling of the weights is equivalent to a scaling of the activations.

Overall, we see that Deep Rectifier Neural Networks use rectified neurons as hidden neurons. Owing to this, they can outperform pre-trained sigmoid deep neural networks without the need for any pre-training.

3 Probabilistic Sampling

Most machine learning algorithms – including deep rectifier nets – are sensitive to class imbalances in the training data. DRNs tend to behave inaccurately on classes represented by only a few examples, which is sometimes the case in document classification. To remedy this problem, we will examine the training scheme called probabilistic sampling [12].

When one of the classes is over-represented during training, it might cause that the network will favour that output and label everything as the most frequent class. To avoid this, it is necessary to balance the class distribution by presenting more examples taken from the rarer classes to the learner. If we have no way of generating additional samples from any class, then resampling is simulated by repeating some of the samples of the rarer classes.

Probabilistic sampling is a simple two-step sampling scheme: first we select a class, and then randomly pick a training sample from the samples belonging to this class. Selecting a class can be viewed as sampling from a multinomial distribution after we assign a probability to each class. That is,

$$P(c_k) = \lambda \frac{1}{K} + (1 - \lambda) Prior(c_k), \tag{1}$$

where $Prior(c_k)$ is the prior possibility of class c_k, K is the number of classes and $\lambda \in 0, 1$ is a parameter. If λ is 1, then we get a uniform distribution over the classes; and with $\lambda = 0$ we get the original class distribution.

4 Experimential Setup

In our experiments, the Reuters-21,578 dataset was used as our training and testing sample set. This corpus contains 21,578 documents collected from the Reuters newswire, but here just the 10 most frequent categories were taken from the 135. For each category, 30% of the documents were randomly selected as test documents and the rest were employed as the training sets. In the evaluation phase, one category was employed as the positive class, and the other nine categories were lumped together and treated as the negative class; and each category played the role of the positive class just once. The documents were represented in a tf-idf weighted vector space model, where the stopwords and numeric characters were ignored.

4.1 Baseline Methods

In order to compare the performance of our method with that for the other machine learning algorithms, we also evaluated some well-known machine learning methods on our test sets.

First, we applied C4.5, which is based on the well-known ID3 decision tree learning algorithm [9]. This machine learning method was a fast learner as it applied axis-parallel hyperplanes during the classification. We trained the J48 classifier of the WEKA package [3], which implements the decision tree algorithm C4.5. Decision trees were built that had at least two instances per leaf, and used pruning with subtree raising and a confidence factor of 0.25.

Support Vector Machines (SVM) [13] were also applied. SVM is a linear function having the form $f(x) = w^t x + b$, where w is the weight vector, x is the input vector and $w^t x$ denotes the inner product. SVM is based on the idea of selecting the hyperplane that separates the space (between the positive and negative classes) while maximizing the smallest margin. In our experiments we utilized LibSVM[1] and the Weka **SMO** implementation.

4.2 Neural Network Parameters

For validation purposes, a random 10% of the training vectors were selected before training. Our deep networks consisted of three hidden layers and each hidden layer had 1,000 rectified neurons, as DRNs with this structure yielded the best results on the development sets. The shallow neural net was a sigmoid net with one hidden layer, with the same number of hidden neurons (3,000) as that for the deep one.

---

[1] `http://www.csie.ntu.edu.tw/~cjlin/libsvm/`

Table 1. The F-score results got from applying different machine learning algorithms (DRN: Deep Rectifier Network, ANN: Shallow Neural Network, SMO, LibSVM: Support Vector Machine, J48: Decision Tree) on the Reuters Top 10 classes

Task	DRN	ANN	SMO	LibSVM	J48
ship	88.20	87.12	87.65	**88.61**	83.15
grain	**96.40**	95.11	94.77	93.1	95
money-fx	93.52	**94.06**	88.56	78	86.13
corn	83.22	76.80	86.9	78.12	**91.78**
trade	**95.74**	93.38	94.41	91.04	85.52
crude	**94.62**	91.21	91.23	90.63	86.36
earn	**98.74**	98.31	98.46	98.52	96.43
wheat	87.12	81.97	**92.49**	86.42	91.86
acq	**97.54**	97.13	96.76	96.86	91.83
interest	94.46	**96.00**	89.96	77.25	82.71
micro-avg	**96.22**	95.42	92.18	87.64	87.86

The output layer for both the shallow and the deep rectifier nets was a softmax layer with 2 neurons – one for the positive class and one for the negative class. The softmax activation function we employed was

$$softmax(y_i) = \frac{e^{y_i}}{\sum_{j=1}^{K} e^{y_j}}, \tag{2}$$

where y_i is the ith element of the unnormalised output vector y. After applying the *softmax* function on the output, we simply select the output neuron with the maximal output value, and this gives us the classification of the input vector. For the error function, we applied the cross entropy function.

Regularization is vital for good performance with neural networks, as their flexibility makes them prone to overfitting. Two regularization methods were used in our study, namely early stopping and weight decay. Early stopping regularization means that the training is halted when there is no improvement in two subsequent iterations on the validation set. The weight decay causes the weights to converge to smaller absolute values than they otherwise would.

The DRNs were trained using semi-batch backpropagation, the batch size being 10. The initial learn rate was set to 0.04 and held fixed while the error on the development set kept decreasing. Afterwards, if the error rate did not decrease in the given iteration, then the learn rate was subsequently halved. The λ parameter of the probabilistic sampling was set to 1, which means that we sampled from a uniform class distribution.

5 Results

Table 1 lists the overall performance we got from training the different machine learning methods on the Reuters dataset. Here, F-scores were used to measure the effectiveness of the various classifiers and we applied the micro-average method [8] to calculate an

Table 2. Neural networks results got with and without probabilistic sampling (P.S.), on the three most unbalanced tasks

	ship			corn			wheat		
Method	F-score	Prec.	Recall	F-score	Prec.	Recall	F-score	Prec.	Recall
DRN	88.20	94.67	82.56	83.22	80.52	86.11	87.12	93.42	81.61
DRN+ P.S.	**90.48**	92.68	88.37	**87.50**	87.50	87.50	**89.89**	87.91	91.95
ANN	87.12	92.21	82.56	76.80	90.57	66.67	81.97	78.13	86.21
ANN+ P.S.	90.36	93.75	87.21	85.29	90.63	80.56	85.56	80.00	91.95

overall F-score. Micro-averaging pools per-document decisions across classes, and then computes an effectiveness measure on the pooled contingency table.

As can be seen, the DRN method outperformed the other methods in general, but it performed poorly (F-score below 90) on three classes. From among the baseline algorithms, the best one was the SMO, with a micro-average score of 92.18. Compared to the other two baseline methods, which yielded approximately the same micro-average score, the SMO achieved a better overall score of 4.5. To make a sense of the relative effectiveness of the neural nets, we decided to compare their perfomance with that for the SMO – the best one of the baseline methods. The micro-average score of the DRN is 96.22, which is 4.04 higher than that for the SMO. The ANN achieved an average F-score of 95.42, which is 3.24 higher than that for the micro-average score of the SMO. This means that the average effectiveness of DRNs is competitive with classifiers like SVMs and decision trees. However, on small classes ('ship', 'corn' and 'wheat'), which were represented with fewer than 200 positive examples in the training set, DRNs and ANNs performed much worse. Interestingly, on these rare classes the baseline algorithms performed quite differently. On the 'ship' class LibSVM yielded the best result, but on the 'corn' class J48 was the best and for the 'wheat' class the SMO achieved the best result.

Next, we investigated the three tasks on which the neural networks approach was outperformed by the other methods. These tasks were the most under-represented classes, so to improve the results we applied probabilistic sampling. In Table 2, we see the improvements got for the deep and the shallow networks after applying it. For the DRNs, the improvement was 3.11 on average, while for the ANNs it was 5.1; but the DRNs yielded better results for all three classes.

With probabilistic sampling, DRNs outperformed LibSVM on all three tasks, and the SMO was better only on the 'wheat' class. The J48 results were still better on the 'corn' and the 'wheat' classes, but the DRNs performed much better on the other eight classes.

6 Discussion

Deep Rectifier Neural Networks outperformed our baseline algorithms, which probably tells us that they are suitable for document classification tasks. However, they face difficulties when some of the classes are underrepresented.

The results of our experiment show that probabilistic sampling greatly improves the F-scores for the DRNs and the ANNs on the underrepresented classes. To understand precisely how probabilistic sampling helps the training of neural networks on these classes, we investigated the effects it produced. The most important one is that after probabilistic sampling balanced the distribution of positive and negative examples, the recall values increased here. The reason behind this is quite simple: the neural networks get more positive examples during training. As the neural nets get more positive samples, the proportion of negative samples decrease. This sometimes caused a drop in the precision score. However this reduction was much smaller than the increase in the recall score, as the negative samples were still well represented.

Comparing the results of the DRNs with those got using ANNs, we can say that the DRNs are not only better but their training and evaluation phases are faster too. To support this opinion, we should mention that the shallow sigmoid network had approximately 1.5 times more parameters. The ANN had 2,000×3,000 connections between input units and hidden units and 3,000×2 weights for the output layer, while the DRN had only 2,000×1,000 input-hidden, 1,000×2 hidden-output, and 2×1,000×1,000 hidden-hidden connections. Thanks to the greater number of parameters, ANNs were able to learn a better model for the 'money-fx' and 'interest' classes. On the other eight classes, the DRNs yielded better results, and this suggests that deep structures are better than shallow ones, for the tasks described earlier.

7 Conclusions

In this paper, we applied deep sparse rectifier neural nets to the Reuters document classification task. Overall, our results tell us that these DRNs can easily outperform SVMs and decision trees if the class distribution is reasonably balanced. With extremely unbalanced data, we showed that probabilistic sampling generally improves the performance of neural networks.

In the future, we would like to investigate a semi-supervised training method for DRNs, so they could be applied on such tasks where we have only a small number of labelled examples and a large amount of unlabelled data.

Acknowledgment. Tamás Grósz were funded in part by the European Union and the European Social Fund through the project FuturICT.hu (TÁMOP-4.2.2.C-11/1/KONV-2012-0013).

References

1. Glorot, X., Bengio, Y.: Understanding the difficulty of training deep feedforward neural networks. In: Proc. AISTATS, pp. 249–256 (2010)
2. Glorot, X., Bordes, A., Bengio, Y.: Deep sparse rectifier networks. In: Proc. AISTATS, pp. 315–323 (2011)
3. Hall, M., Frank, E., Holmes, G., Pfahringer, B., Reutemann, P., Witten, I.H.: The WEKA data mining software: an update. SIGKDD Explorations 11(1), 10–18 (2009)
4. Hinton, G.E., Srivastava, N., Krizhevsky, A., Sutskever, I., Salakhutdinov, R.R.: Improving neural networks by preventing co-adaptation of feature detectors. CoRR. 1207.0580 (2012)

5. Hinton, G.E., Deng, L., Yu, D., Dahl, G.E., Rahman Mohamed, A., Jaitly, N., Senior, A., Vanhoucke, V., Nguyen, P., Sainath, T.N., Kingsbury, B.: Deep neural networks for acoustic modeling in speech recognition: The shared views of four research groups. IEEE Signal Process. Mag. 29(6), 82–97 (2012)
6. Hinton, G.E., Osindero, S., Teh, Y.W.: A fast learning algorithm for deep belief nets. Neural Computation 18(7), 1527–1554 (2006)
7. Krizhevsky, A., Sutskever, I., Hinton, G.E.: Imagenet classification with deep convolutional neural networks. In: Proc. NIPS, pp. 1106–1114 (2012)
8. Manning, C.D., Raghavan, P., Schütze, H.: Introduction to Information Retrieval. Cambridge University Press (2008)
9. Quinlan, J.R.: C4.5: Programs for Machine Learning. Morgan Kaufmann Publishers Inc., San Francisco (1993)
10. Seide, F., Li, G., Chen, X., Yu, D.: Feature engineering in context-dependent deep neural networks for conversational speech transcription. In: Proc. ASRU, pp. 24–29 (2011)
11. Srivastava, N., Salakhutdinov, R.R., Hinton, G.E.: Modeling documents with a deep Boltzmann machine. In: Proc. UAI, pp. 616–625 (2013)
12. Tóth, L., Kocsor, A.: Training HMM/ANN hybrid speech recognizers by probabilistic sampling. In: Duch, W., Kacprzyk, J., Oja, E., Zadrożny, S. (eds.) ICANN 2005. LNCS, vol. 3696, pp. 597–603. Springer, Heidelberg (2005)
13. Vapnik, V.N.: Statistical learning theory. Wiley (September 1998)

Language Independent Evaluation of Translation Style and Consistency: Comparing Human and Machine Translations of Camus' Novel "The Stranger"

Mahmoud El-Haj[1], Paul Rayson[1], and David Hall[2]

[1] School of Computing and Communications, Lancaster University, UK
{m.el-haj,p.rayson}@lancaster.ac.uk
[2] UK Data Archive, The University of Essex, UK
djhall@essex.ac.uk

Abstract. We present quantitative and qualitative results of automatic and manual comparisons of translations of the originally French novel "The Stranger" (French: L'Étranger). We provide a novel approach to evaluating translation performance across languages without the need for reference translations or comparable corpora. Our approach examines the consistency of the translation of various document levels including chapters, parts and sentences. In our experiments we analyse four expert translations of the French novel. We also used Google's machine translation output as baselines. We analyse the translations by using readability metrics, rank correlation comparisons and Word Error Rate (WER).

1 Introduction

Translation in general is a complex task and it is more challenging when translating novels [7], especially ones written by Nobel Prize winners such as the French author, journalist and philosopher, Albert Camus. In a talk by [7], translator of a volume of Camus' Combat editorials, he called it "nonsense" to believe that "good translation requires some sort of mystical sympathy between author and translator" and instead he compared translating to playing the piano, saying that "The sinews and reflexes that translation requires are capable of development through exercise.". [7] believed that "it is different when translating novels, where translators need to have the eyes, ears, and wits to savour its beauties, and that they are obliged to preserve as many of them as they can.". [4] defined the task of the translator as consisting of "finding that intended effect upon the language into which he is translating which produces in it the echo of the original". He used this feature to differentiate translation from poet's work as the latter is never directed at the language but solely at specific contextual aspects.

These different opinions indicate that translation is a complex task and that there are many factors and features that could play a big role in defining translation quality. In our work we focus on the use of statistical metrics to judge translation consistency, comparing our results to qualitative analysis by expert readers for both the Arabic and English translations of the French novel, "The Stranger" (L'Étranger) by Albert Camus[1]. In our study we used four translations, male and female professional Arabic

[1] http://en.wikipedia.org/wiki/Albert_Camus

P. Sojka et al. (Eds.): TSD 2014, LNAI 8655, pp. 116–124, 2014.

translations and male and female professional English translations. We also used Google's machine translation to automatically translate the French novel into Arabic and English, using these as baseline translations to allow us to judge the efficiency of our approach across both human and machine translation.

2 Related Work and Our Hypothesis

[13] demonstrate that readability features can be used in statistical machine translation to produce simplified text translations that could be useful, for example, for language learners and others who want to have a feel for the major content in highly domain-specific documents written in a foreign language.

The study by [6] measures the relationships between linguistic variation and reader perceptions by analysing 74 linguistic features in a set of 80 introductory academic textbooks. The goal was to study whether it is possible to assess student perception of effectiveness, comprehensibly and organisation in the textbooks. The statistical test showed three ("Elaboration and Involvement", "Colloquial Style", and "Academic Clarity") of the 74 linguistic features of variation were significant predictors of perception.

Previous research on translation quality has not taken into account the variability or consistency of the translation text. In our work we propose a novel approach to evaluate translations without the need for a reference translation (gold standard). The approach uses language-independent readability metrics in combination with statistical rank correlation comparisons. The approach overcomes problems found with BLEU [10] and other n-gram based scores. Those problems include the unreliability of the metrics when it comes to evaluating on an individual sentence level, caused by data sparsity, and the dependency of n-gram metrics on word order [9].

Our approach focuses on checking whether the translation process (human or machine) has correctly preserved the variation in style and complexity in the original language at various document levels including chapters and parts. We hypothesise that the readability scores for each block of text in the original and translated versions should be similarly ranked if the translation quality is good. A poor translation would not preserve the original variation in style and readability. Our approach is therefore product-driven making explicit use of the structure of the original novel in our case study.

3 Data Collection and Pre-processing

In our work we analyse the translation of four human expert translators. Two translations into Arabic and two translations into English with a male and female translator for each language.

The original French novel is divided into two parts with 6 and 5 chapters respectively. We followed the same order with the Arabic and English translations when running the experiments. The analysis and experiments were carried out at three levels: document, part, and chapter. Our approach does not require alignment of sentence level. The original novel and the translations differ in size. Table 1 shows the number of words and sentences for each language.

Table 1. Data Collections Statistics

Language	sentences	words	unique words
French	2,204	30,867	4,928
Arabic Male	951	24,129	6,808
Arabic Female	1,945	24,608	7,363
English Male	2,110	33,583	4,420
English Female	2,131	31,293	3,651

4 Readability

In order to detect consistency in the translation style we use readability as a proxy for style and then consider how it varies both within and between translations. The readability metrics used in our experiments are Laesbarheds-Index (LIX) [5] and Automated Readability Index (ARI) [12]. LIX and ARI have been used to measure readability of Arabic and French languages and found to correlate with measuring the readability for the English version [2,3,14,1]. We calculated LIX and ARI readability metrics for each part and chapter in addition to the full text of the original novel and the four translations. Table 2 shows the LIX and ARI readability scores for Part 1 and 2 of the novel in addition to the full text for the three languages. The lower the score the easier to read.

Table 2. LIX and ARI Readability Scores

	LIX			ARI		
Language	Part 1	Part 2	Full Text	Part 1	Part 2	Full Text
French	33.91	38.85	36.33	6.15	7.79	6.94
Arabic Male	24.48	27.39	25.91	6.21	7.36	6.76
Arabic Female	18.54	19.24	18.88	3.85	3.57	3.69
Arabic Google	26.11	28.11	26.97	6.98	7.51	7.20
English Male	27.39	31.07	29.22	4.70	5.90	5.29
English Female	25.43	29.59	27.51	3.78	5.12	4.44
English Google	33.06	37.44	35.31	7.16	9.05	8.12

Table 3 shows the LIX readability score for each chapter in Parts 1 and 2 for the three languages. The LIX readability scores show consistency across chapters for each language. Taking the French text readability scores, we can see the writer's style is consistent across chapters. Using this finding to judge the translation quality, chapters with readability scores close to the original text are considered to be high quality translations. As an example, for Part 1, Chapter 1, the English male translator produces a higher quality than the other human translators.

5 Rank Correlation and Kendall Tau Comparisons

To identify and test the strength of the readability relationship lists shown in Section 4, we used Spearman's Rank Correlation Coefficient and Kendall Tau statistical methods.

Table 3. Part 1 & 2 Chapters LIX Readability (*1_1 stands for Part 1 Chapter 1 and so on.*)

	Part 1						Part 2				
Language	1_1	1_2	1_3	1_4	1_5	1_6	2_1	2_2	2_3	2_4	2_5
French	31.5	32.3	29.3	30.3	31.6	31.1	34.8	32.5	37.8	36.6	38.2
Arabic Male	25.0	28.4	21.9	24.7	22.9	25.0	30.2	26.9	29.8	28.7	23.7
Arabic Female	18.5	22.0	16.7	17.4	20.8	17.9	20.1	19.1	19.9	19.5	18.1
Arabic Google	25.0	28.1	26.3	27.2	30.0	24.7	27.9	27.6	30.0	33.8	27.7
English Male	27.4	29.3	25.0	26.0	28.5	28.7	31.1	29.8	32.4	33.1	28.1
English Female	25.5	27.9	23.9	24.8	25.4	25.9	28.0	27.8	31.7	32.2	27.9
English Google	33.0	39.2	31.2	36.4	38.7	36.2	36.7	35.7	44.4	41.0	37.2

Table 4 shows the Spearman's correlation scores for the LIX readability metric. We only report the Spearman's correlation results as the results we observed using Kendall Tau showed the same trends when compared to Spearman's.

Table 4. Spearman's for LIX scores

	Arabic Male	Arabic Female	Arabic Google	English Male	English Female	English Google
French	0.49	0.29	0.58	0.66	0.47	0.53
Arabic Male	–	0.61	0.49	0.89	0.81	0.54
Arabic Female	–	–	0.74	0.71	0.58	0.70
Arabic Google	–	–	–	0.70	0.70	0.89
English Male	–	–	–	–	0.92	0.69
English Female	–	–	–	–	–	0.72

The correlation scores shown in Table 4 support our consistency hypothesis in Section 4 where we considered translations with readability scores consistent with the original text to be of higher quality. The scores also show the closeness in translation and style consistency between the translations. Taking the English translation as an example we can see the closeness between the male and female translation style as indicated by consistency. Similarly we found male translations to be more consistent with the original French text when compared to the female translations. The Spearman's scores in the table are consistent with the readability scores shown in tables 2 and 3. As we expected, Google's Arabic and English translations were found to be very close and consistent across chapters.

6 Evaluation

To judge the translation performance, two types of evaluation were carried out: automatic and manual. To evaluate the translation performance for both the human translators and Google machine translations we used the Word Error Rate (WER) metric [11]. WER is derived from Levenshtein distance [8], working at the word level instead of the phoneme level. In addition and to ensure quality, we used domain experts to judge the comprehensibility and readability of the four Arabic and English translation. For the experiments in this paper we used one Arabic and one English native speaker and reader. The Arabic speaker read the Arabic translations in addition to the English translations. The speaker extracted some examples for terms that can be considered dialectical and are not used in Modern Standard Arabic (MSA) (Sample in Table 7). The English speaking participant read the English translations in addition to parts of the original French script and made his judgement based on comparing the translation and by taking some examples where the translators used idioms and metaphors (Sample in Table 8). The speakers were asked to judge the texts' readability and comprehensibility by writing short paragraphs describing how they thought the translations compared. Both speakers have PhD degrees and are experts in Arabic or English literature and are expert readers/writers and critics.

Tables 5 and 6 show the WER scores and the number of substitutions, insertions and deletions needed to transform the hypothesis translation (Example: Google English) into the reference one (Example: Human English). The tables also show the number of words in the reference and the hypothesis translations in addition to the number of correct matches. Table 5 suggests the Arabic male and female translations to be closer to each other than to Google translation. Comparing the male and female scores we can see that each translation contained around 25% correct matches from the other. The percentage is higher, *c*34% between the English male and female translations (Table 6).

Table 5. Arabic Translations WER and Levenshtein Stats

Arabic Full Text	WER	Reference	Hypothesis	Correct	Substitutions	Insertions	Deletions
Female vs. Male	0.85	24,867	23,969	6,131	15,424	2,414	3,312
Male vs. Female	0.88	23,969	24,867	6,131	15,433	3,306	2,408
Male vs. Google	1.03	23,969	27,028	3,839	18,820	4,369	1,310
Female vs. Google	0.96	24,867	27,028	4,852	18,438	3,738	1,577

Table 6. English TranslationsWER and Levenshtein Stats

English Full Text	WER	Reference	Hypothesis	Correct	Substitutions	Insertions	Deletions
Female vs. Male	0.86	31,460	35,012	10,801	17,732	6,479	2,927
Male vs. Female	0.77	35,012	31,460	10,801	17,726	2,931	6,483
Male vs. Google	0.81	35,012	31,379	9,599	18,912	2,868	6,501
Female vs. Google	0.73	31,460	31,379	12,544	14,856	3,979	4,060

Google Arabic baseline translations were not close to the human translation references, with less than 20% considered correct matches. Google English did considerably better with *c*40% and *c*30% correct matches to the female and male reference translations. This suggests translating French to English using Google to be more accurate than translating into Arabic. The Arab expert reader found the male translation to be easier to comprehend, but the use of long sentences made the readability harder. In contrast the female translation was easier to read with the use of short sentences, but difficult to comprehend with the frequent use of dialect mainly used in Levant countries (i.e. Jordan, Lebanon etc.). Table 7 shows the use of dialect by the Arab female translator, referred to by [DL]. The translator also used transliteration [TL] and foreign words [FN] when translating from French in contrast to the Arab male translator who avoided the use of dialect and transliteration. But the male translator did use some common foreign words, e.g. "Billiard". The male translator used simplification (a tendency to produce simpler and easier-to-follow text) when translating, and this is also noticeable with the amount of skipped words (see example in Table 7).

The right most column in the table showing the Arabic Google translation shows a sample of wrong translations [WT] that have been found by the expert reader. The table also gives an insight into the writing style and use of vocabulary between the female and male translators, both translators using a rich vocabulary as seen in the unique words column in Table 1. The English expert reader found the male translator to be more accurate in places. He also found the male translator to be more literal and much more straightforward. Take for example the first example in Table 8 where the word 'audience' is more literal than the female's 'spectators'. This makes the male translation easier to understand, but the language is more dated due to a gap of over 60 years between translations. This raises questions such as: whether the male translator was more familiar with contemporaneous French idiom, or whether the female translator is interpreting the work for a modern audience? The male translator may indeed have been more in tune with the general philosophic and artistic feelings of the time, when he translated the French novel in 1946, only 4 years after it was first published.

Table 7. Keywords Human Comparisons (*M: Male, F: Female*)

Arabic (M)	Arabic (F)	English (M)	English (F)	French	English Google	Arabic Google
يغلق فمه	[DL] يسد بوزه	shut his trap	Shut your trap	fermer sa gueule	keep his mouth shut	يبقي فمه مغلقا
[FN] البلياردو	[FN] البليار	billiards	billiards	billard	billiards	[FN] البلياردو
حقل الملاهي	[TL] الشان دو مانوفر	Parade Ground	Parade Ground	Champ de Manoeuvres	Field Labourers	[WT] عمال الحقل
skipped	ملك الفرار	Handcuff King	King of the Escape Artists	le Roi de l'évasion	King of Escape	ملك الهروب

7 Results

The results in Table 5 show that the Arabic human translators (male and female) are closer to each other than to Google. But the high WER scores between the male and female translations (and vice versa) suggest a big difference between the translations vocabulary, which is consistent with what has been reported by the human expert reader (see Section 6). The expert reader found the two translations to be using a rich vocabulary (see Table 1). The automatic and manual evaluation results are also consistent with the readability and the rank correlation scores shown in the Tables 2, 3 and 4 in Sections 4 and 5, which show the Arabic female translation to be easier to read with low and consistent readability scores.

The rank correlation comparison scores in Table 4 show that the readability scores are consistent across parts and chapters in addition to the full text. The results strengthen our assumption that translations with readability scores close to those of the original French novel are of better quality considering readability and style consistency. The results show that the Arabic male translation readability scores are closer to the original French than the female translation. This was supported by the rank correlation scores that suggest the Arabic male translation to be closer and more consistent with the French original.

Tables 7 and 8 show the differences between the two Arabic translations in the vocabulary usage and the big difference when translating metaphors and idioms. Which explains the low WER scores (Table 5), which found only 20% similarity between the two translations.

Table 8. Idioms and Metaphors Human Comparisons (English/Arabic) and their meaning *Pt: Part, Ch: Chapter, M: Male, F: Female*

Pt	Ch	English/Arabic (F)	English/Arabic (M)	Original Meaning
1	2	let a wave of spectators out	disgorged their audiences	those attending the cinema came out
		موجة من الشاهدين	بحشود من المتفرجين	
1	2	we thrashed them	we licked them	defeated the opposition in a comprehensive way
		لقد انتصرنا عليهم	لقد هزمناهم	

Table 6 shows that the English female translation is closer to the male and Google translations with slightly more towards the latter. But when looking at the rank correlation scores (Table 4) we can see that the scores show high correlation between the English male and female translations, indicating that the male and female translations are more consistent across chapters than is Google.

Table 7 shows through examples the similarity and closeness between the two translation, which most of the time are also consistent with Google translate. This is consistent with the statistics shown in Table 1, which found these translations to be using nearly the same number of words and sentences. Table 8 shows the differences between both translations when it comes to translating idioms and metaphors in sentences. This suggests the two translations are close on a word level rather than on a sentence level. The readability scores in Tables 2 and 3 are consistent with the expert reader judgements. The reader found the female translation to be more readable when compared to the male translation finding the language of the male translation to be a bit dated with over 60 year gap. The results suggest the male translation to be more consistent with the French novel. The results also show the Arabic and English male translations to be consistent when compared to each other.

8 Conclusion and Future Work

The paper shows language-independent readability metrics and rank correlation comparisons can be used to evaluate the translation quality and style consistency without the need for human translation references. To evaluate the results we used human expert readers and the Word Error Rate (WER) metric. The results show that using readability scores in combination with rank correlation comparison is consistent with human judgements about translation quality. As shown in our experiments, we were able to directly compare the Arabic and English human or machine translations to the French original text. The results also suggest that the Arabic and English male translation were closer, based on manual and automatic evaluation. We have shown that consistent readability scores across parts and chapters between original and translated text are indicators of good quality translations. Since it is usually hard to find enough human gold-standard references for any particular text, especially in under-resourced languages such as Arabic, we believe our method moves towards being an alternative economical solution for machine-translation evaluation. The current evaluation tools usually require several gold-standard references to provide a performance score and they measure the closeness between the hypothesis translation and the references without referring to the original text that is in a different language. Other evaluation metrics rely on the quality of the translation references, while our technique relies completely on the closeness of the translation to the consistency of style of the original text.

References

1. Altamimi, A.K., Jaradat, M., Aljarrah, N., Ghanem, S.: Aari: Automatic Arabic readability index iajit first online publication (2013)
2. Anderson, J.: Analysing the readability of English and non-English texts in the classroom with lix. In: The Australian Reading Association Conference. ERIC Institute of Education Sciences (1981)
3. Anderson, J.: Lix and rix: Variations on a little-known readability index. Journal of Reading 26(6), 490–496 (1983)
4. Benjamin, W.: Illuminations. Houghton Mifflin Harcourt (1968), `http://books.google.co.uk/books?id=mV06rdTclagC`
5. Björnsson, C.H.: Läsbarhet. In: Stockholm: Liber (1968)
6. Egbert, J.: Student perceptions of stylistic variation in introductory university textbooks. Linguistics and Education 25, 64–77 (2014)
7. Goldhammer, A.: Translating subtexts: What the translator must know. In: Talk Delivered to Brandeis University Translation Seminar. CUNY Conference on Translation. Association for Computational Linguistics, Waltham (1994), `http://www.people.fas.harvard.edu/ agoldham/articles/WhatMust.htm`
8. Levenshtein, V.I.: Binary codes capable of correcting deletions, insertions, and reversals. Tech. Rep. 8 (1966)

9. Padó, S., Galley, M., Jurafsky, D., Manning, C.: Robust machine translation evaluation with entailment features. In: Proceedings of the Joint Conference of the 47th Annual Meeting of the ACL and the 4th International Joint Conference on Natural Language Processing of the AFNLP, ACL 2009, vol. 1, pp. 297–305. Association for Computational Linguistics, Stroudsburg (2009)
10. Papineni, K., Roukos, S., Ward, T., Zhu, W.J.: Bleu: A method for automatic evaluation of machine translation. In: Proceedings of the 40th Annual Meeting on Association for Computational Linguistics, ACL 2002, pp. 311–318. Association for Computational Linguistics, Stroudsburg (2002)
11. Popović, M., Ney, H.: Word error rates: Decomposition over pos classes and applications for error analysis. In: Proceeding of the Second Workshop on Statistical Machine Translation Held within ACL 2007, pp. 48–55 (2007)
12. Smith, E., Senter, R.: Automated Readability Index. AMRL-TR-66-220, Aerospace Medical Research Laboratories (1967)
13. Stymne, S., Tiedemann, J., Hardmeier, C., Nivre, J.: Statistical machine translation with readability constraints. In: Proceedings of the 19th Nordic Conference of Computational Linguistics (NODALIDA 2013). NEALT Proceedings Series 16, pp. 375–386. LiU Electronic Press (2013)
14. Uitdenbogerd, A.L.: Readability of french as a foreign language and its uses. In: Proceedings of the 10th Australasian Document Computing Symposium (2005)

Bengali Named Entity Recognition Using Margin Infused Relaxed Algorithm

Somnath Banerjee, Sudip Kumar Naskar, and Sivaji Bandyopadhyay

Department of Computer Science and Engineering,
Jadavpur University, India
s.banerjee1980@gmail.com,
{sudip.naskar, sbandyopadhyay}@cse.jdvu.ac.in

Abstract. The present work describes the automatic recognition of named entities based on language independent and dependent features. Margin Infused Relaxed Algorithm is applied for the first time in order to learn named entities for Bengali language. We used openly available annotated corpora with twelve different tagset defined in IJCNLP-08 NERSSEAL shared task and obtained 91.23%, 87.29% and 89.69% precision, recall and F-measure respectively. The proposed work outperforms the existing models with satisfactory margin.

1 Introduction

Named entities (NEs) have a special status in Natural Language Processing (NLP) because of their distinctive nature which other elements of human languages do not have, e.g. NEs refer to specific things or concepts in the world and are not listed in the grammars or the lexicons. Automatic identification and classification of NEs benefits in text processing due to their significant presence in the text documents. Named Entity Recognition (NER) is a task that seeks to locate and classify NEs in a text into predefined categories such as the names of persons, organizations, locations, expressions of times, quantities, etc. The NER task can be viewed as a two stage process: a) Identification of entity boundaries, b) Classification into the correct category. For example, if *"Sachin Tendulkar"* is a named entity in the corpus, it is essential to identify the beginning and the end of this entity in the sentence. Following this step, the entity must be classified into the predefined category, which is PERSON (Named Entity Person) in this case.

The NER task has important significance in many NLP applications such as Machine Translation, Question-Answering, Automatic Summarization, Information Extraction etc. The task of building an NER for Indian languages (ILs) presents various challenges related to their linguistic characteristics. Some of them are: no capitalization, unavailability of large gazetteer, relatively free word order, spelling variation, rich inflection, ambiguity, etc. In this work, we identified suitable language independent and dependent features for the Bengali NER task and used *Margin Infused Relaxed Algorithm* (MIRA) to develop NER system for Bengali.

P. Sojka et al. (Eds.): TSD 2014, LNAI 8655, pp. 125–132, 2014.

2 Related Work

Bandyopadhyay[1] stated that the computational research aiming at automatically identifying NEs in texts forms a vast and heterogeneous pool of strategies, techniques and representations from hand-crafted rules towards machine learning approaches. Most of the previous NER systems are based on one of the following approaches:

- Linguistic approaches
- Machine Learning(ML) based approaches
- Hybrid approaches

The linguistic approaches based NER systems ([2, 3, 4]) typically use hand-crafted grammatical rules written by linguistics. On the other hand, ML based NER systems use learning algorithms that require large annotated datasets for training and testing [5]. ML methods such as Hidden Markov Model (HMM) [6], Conditional Random Field (CRF) [7], Support Vector Machine (SVM) [8], Maximum Entropy (ME) [9] are the most widely used approaches. Besides the above two approaches, Hybrid approaches based NER systems [10] combines the strongest point from both the Rule based and statistical methods.

Mainly, ML and hybrid approaches were used successfully in NER for Bengali language. The survey by [11] on NER for ILs detailed the various approaches used for Bengali NER by researchers.

ML-based: [12],[13],[14],[15],[16],[17],[18].

Hybrid-based: [10],[19],[20].

Though a few use of MIRA was noted for English [21], but it has not been used in NER for any Indian languages till date. This is one of the reasons to use MIRA in this work.

3 Margin Infused Relaxed Algorithm

Crammer and Singer [22] reported *Margin Infused Relaxed Algorithm* is a machine learning algorithm for multiclass classification problems. It is designed to learn a set of parameters (vector or matrix) by processing all the given training examples one-by-one and updating the parameters according to each training example, so that the current training example is classified correctly with a margin against incorrect classifications at least as large as their loss. The change of the parameters is kept as small as possible. MIRA is also called passive-aggressive algorithm (PA-I), is an extension of the perceptron algorithm for online machine learning that ensures that each update of the model parameters yields at least a margin of one. The flow of the MIRA is depicted in Figure 1.

Suppose sequence $(\bar{x}^1, y^1), ..., (\bar{x}^t, y^t), ...$ is the instance-label pairs. Each instance $\bar{x}^t$ is in $\mathbb{R}^n$ and each label belongs to a finite set Y of size k. It can be assumed without loss of generality that $\mathrm{Y} = \{1, 2, ..., k\}$. A multiclass classifier is a function $H(\bar{x})$ that maps instances from $\mathbb{R}^n$ into one of the possible labels in Y. The classifier is in the form $H(\bar{x}) = \text{argmax}_{r=1}^{k}\{\bar{M}_r.\bar{x}\}$, where **M** is a $k \times n$ matrix over the reals and $\bar{M}_r \in \mathbb{R}^n$ denotes the r$^t h$ row of **M**. The inner product of $\bar{M}_r$ with the instance $\bar{x}$ is called the *similarity-score* for class r. Thus, the considered classifiers set the label of an instance to be the index of the row of **M** which achieves the highest *similarity-score*.

Initialize: Set $\mathbf{M} = 0$ $(\mathbf{M} \in \mathbb{R}^{k \times n})$.
Loop: For $t = 1, 2, \ldots, T$

- Get a new instance $\vec{x}^t \in \mathbb{R}^n$.
- Predict $\hat{y}^t = \arg\max_{r=1}^{k} \{\bar{M}_r \cdot \vec{x}^t\}$.
- Get a new label y^t.
- Set $E = \{r \neq y^t : \bar{M}_r \cdot \vec{x}^t \geq \bar{M}_{y^t} \cdot \vec{x}^t\}$.
- If $E \neq \emptyset$ update **M** by choosing any $\tau_1^t, \ldots, \tau_k^t$ that satisfy:
    1. $\tau_r^t \leq 0$ for $r \neq y^t$ and $\tau_{y^t}^t \leq 1$.
    2. $\sum_{r=1}^{k} \tau_r^t = 0$.
    3. $\tau_r^t = 0$ for $r \notin E \cup \{y^t\}$.
    4. $\tau_{y^t}^t = 1$.
- For $r = 1, 2, \ldots, k$ update: $\bar{M}_r \leftarrow \bar{M}_r + \tau_r^t \vec{x}^t$.

Output : $H(\bar{x}) = \arg\max_r \{\bar{M}_r \cdot \bar{x}\}$.

Fig. 1. Algorithm (Crammer and Singer [22])

On round t the learning algorithm gets an instance $\bar{x}^t$. Given $\bar{x}^t$, the learning algorithm outputs a prediction, $\hat{y} = \arg\max_{r=1}^{k} \{\bar{M}_r . \bar{x}^t\}$. It then receives the correct label y^t and updates its classification rule by modifying the matrix **M**. It can be said that the algorithm made a (multiclass) prediction error if $\hat{y}^t \neq \bar{y}^t$. The goal is to make as few prediction errors as possible.

We used *miralium*[1] which is the open source java implementation of MIRA.

4 Features

The success of any machine learning algorithm depends on finding an appropriate combination of features. This section outlines language dependent and language independent features.

4.1 Language Independent Features

Language independent features can be applied to any language including ILs, e.g., Bengali, Hindi, Tamil, Punjabi, etc. The following language independent features are applied to this work.

Window of words: Preceding or following words of the target word might be used to determine its category. The previous m words and next n words along with target word are considered to build the window. But, it has been observed that majority of research works used $m = n$. Following a few trials we found that a suitable window size is five with $m = 2$ and $n = 2$.

Word Suffix: Target word suffix information is very helpful to identify NEs especially for highly inflectional languages like ILs. Though the stemmer or morphological analyzer recognizes the suffix properly, but in the absence of those fixed length suffix

[1] `https://code.google.com/p/miralium/`

can be used as a feature. We used four fixed length suffixes of length 5, 4, 3 and 2 respectively.

Word prefix: Target word prefix also can be used like prefix feature. We used four fixed length prefixes of length 5, 4, 3 and 2 respectively.

First word: First word of a sentence can be used as a feature because in most of the languages the first word is the subject of the sentence.

Word length: It has been observed that short words are rarely NEs. So, length of the word may be used as a feature.

Part of Speech (POS): The POS of the target word and surrounding words may be useful feature for NER. Since NEs are noun phrases, the noun tag is very relevant.

Presence of Digit: Presence of digit in the target word is a very useful feature. This feature is very helpful to identify time expression, measurement and numerical quantities. Most of the cases, digit combines with symbols make NEs, e.g., 12/10/2014, 55.44%, 22/-, etc.

4.2 Language Dependent Features

Language dependent features are increasing the accuracy of the NER systems. So, in most of the experiments they are used along with language independent features.

Clue words: Clue words play a useful role to determine NEs. They occur before or after the NEs. For example, *'Mr.'* is most likely present before starting a person name. Similarly, *'Limited'* is most likely present after an organization. List of clue words can be prepared for NER. In this work, we prepared two clue word lists, namely person clue list and organization clue list under human supervision from the archive (100 documents) of an online available widely used Bengali newspaper. Person and organization clue word lists contain 39 and 53 words. This feature is used as binary feature. If the target word present in the lists, then the value is set to 1, otherwise 0.

Gazetteers list: Lists of names of various types are helpful for NER. We manually prepared four lists, namely names of months, names of sessions, Days of a week, names of units.

5 Experiments

This section describes our study of NER using language independent and dependent features applying MIRA and comparative study with the existing Bengali NER models. We considered the same baseline system (i.e., name finder tool) reported in [23] which is an open source, maximum entropy based and part of OpenNLP[2] package. At first, experiments were performed with language independent features only. Then we used language independent and dependent features together. It has been observed that the use of language dependent features increase the overall F-measure (3.12%).

[2] http://opennlp.sourceforge.net/

5.1 Corpus and Tagset

We used IJCNLP-08 NER on South and South East Asian Languages (NERSSEAL) *shared task data*[3]. The shared task data is tagged with the Tagset[4] which consists of 12 NE tags. Corpus statistics, tagset and tagset statistics are shown in Table-1 and Table-2 respectively.

Table 1. Corpus Statistics

	Training	Testing
Sentences	6030	1835
Words	112845	38708
NEs	5000	1723

Table 2. NERSSEAL NE Tagset and Statistics

Tag	Name	Example	Training	Testing
NEP	Person	Bob Dylan	1299	728
NED	Designation	President, Chairman	185	11
NEO	Organization	State Government	264	20
NEA	Abbreviation	NLP, I.B.M.	111	9
NEB	Brand	Pepsi, Windows	22	0
NETP	Title-Person	Mahatma, Dr., Mr.	68	57
NETO	Title-Object	American Beauty	204	46
NEL	Location	New Delhi, Paris	634	202
NETI	Time	10th July, 5 pm	285	46
NEN	Number	3.14, Fifty five	407	144
NEM	Measure	three days , 5 kg	352	146
NETE	Terms	Horticulture	1165	314

5.2 Results

The performance of the system is evaluated in terms of the standard precision, recall, and F-Measure as follows:

Precision: $P = \frac{c}{r}$

Recall: $R = \frac{c}{t}$

F-Measure: $F_{\beta=1} = \frac{2 \times P \times R}{P+R}$

where c is the number of correctly retrieved (identified) NEs, r is the total number of NEs retrieved by the system (correct plus incorrect) and t is the total number of NEs in the test data. Using language independent features, we obtained 89.26%, 82.99% and 86.01% precision, recall and F-measure respectively. Then using language independent and dependent features, we obtained 91.20%, 87.17% and 89.13% precision, recall and F-measure respectively. NE tags specific results is shown in Table-3.

[3] http://ltrc.iiit.ac.in/ner-ssea-08/index.cgi?topic=5

[4] http://ltrc.iiit.ac.in/ner-ssea-08/index.cgi?topic=3

Table 3. Evaluation for specific NE tags. (NP: Not present in reference data).

	Lang. Independent			Lang. Dependent		
Tag	P	R	$F_{\beta=1}$	P	R	$F_{\beta=1}$
NEP	92.86	89.29	91.04	94.20	91.48	92.82
NED	66.67	36.36	47.06	70.00	63.64	66.67
NEO	73.68	70.00	71.79	78.95	75.00	76.92
NEA	37.50	33.33	35.29	42.86	33.33	37.50
NEB	NP	NP	NP	NP	NP	NP
NETP	82.35	73.68	77.78	83.93	82.46	83.19
NETO	80.95	73.91	77.27	84.09	80.43	82.22
NEL	89.67	81.68	85.49	91.94	84.65	88.14
NETI	84.21	69.57	76.19	88.37	82.61	85.39
NEN	88.89	77.78	82.96	92.86	81.25	86.67
NEM	88.28	77.40	82.48	90.85	88.36	89.58
NETE	87.00	83.12	85.02	88.60	86.62	87.60

Table 4. Comparative Evaluation Results

Model	**F-Measure**
Baseline ([23])	12.30%
Karthik et al., 2008 ([20])	40.63%
Ekbal et al., 2008a ([17])	59.39%
Saha et al., 2008 ([10])	65.95%
Ekbal and Bandyopadhyay, 2010 ([24])	84.15%
MIRA	**89.13%**

5.3 Comparisons with Existing Systems

The existing Bengali NER systems reported in Table-4 used the same corpus and evaluation metrics as described in this work; i.e., NERSSEAL shared task data and evaluation metrics. The obtained results confirms that the proposed system outperforms the existing models based on CRF, ME, HMM and the best performing existing SVM-based system by 4.98%. The reasons behind the superior performance of the proposed system are the better optimization technique of MIRA and its ability to handle the overlapping features efficiently than the existing systems. Basically MIRA does not explicitly optimize any function, so there is no involvement of probabilistic interpretation. Due to unavailability of experimental data reported in [13, 14, 15] and [18], we were unable to compare this work with those systems.

6 Conclusions

This paper presents a system based on MIRA for an IL namely Bengali using both language-independent and language-dependent features. The results show that the proposed system outperforms the existing systems based on CRF, ME, HMM and

SVM. But this system was unable to identify NEA (i.e., Abbreviation) properly due to our assumption that short words are rarely NEs which is not true in this case. Post-processing with heuristic patterns may be applied for that. As MIRA has been applied to neither Bengali nor other ILs, so besides improving accuracy one of the notable contributions of this work is to incorporate MIRA in the NER task of one of the ILs. MIRA may be used for other ILs to enhance the performance of state-of-the-art NER systems.

The performance of this work may be enhanced further by applying post-processing with a set of heuristics and Ensemble approaches. Also one of the extension of MIRA, e.g., AdaGrad may be used to improve further performance of NER systems.

Acknowledgements. We acknowledge the support of the Department of Electronics and Information Technology (DeitY), Ministry of Communications and Information Technology (MCIT), Government of India funded project *"CLIA System Phase II"*.

References

1. Bandyopadhyay, S.: Multilingual Named Entity Recognition. In: Proceedings of the IJCNLP 2008 Workshop on NER for South and South East Asian Languages, Hyderabad, India (2008)
2. Ralph, G.: The New York University System MUC-6 or Where's the syntax? In: Proceedings of Message Understanding Conference (1995)
3. McDonald, D.: Internal and external evidence in the identification and semantic categorization of proper names. In: Boguraev, B., Pustejovsky, J. (eds.) Corpus Processing for Lexical Acquisition, pp. 21–39 (1996)
4. Takahiro, W., Gaizauskas, R., Wilks, Y.: Evaluation of an algorithm for the recognition and classification of proper names. In: Proceedings of COLING (1996)
5. Hewavitharana, S., Vogel, S.: Extracting parallel phrases from comparable data. In: Proceedings of the Workshop on Building and Using Comparable Corpora, ACL, Portland, Oregon, pp. 61–68 (2011)
6. Bikel, D.M., Scott, M., Richard, S., Ralph, S.: Nymble: A High Performance Learning Name-finder. In: Proceedings of Applied Natural Language Processing, Hyderabad, India, pp. 194–201 (1997)
7. Wei, L., Andrew, M.: Rapid Development of Hindi Named Entity Recognition using Conditional Random Fields and Feature Induction. ACM Transactions on Computational Logic (2004)
8. Hiroyasu, Y., Kudo, T., Matsumoto, Y.: Japanese Named Entity Extraction using Support Vector Machine. Transactions of IPSJ 43(1), 44–53 (2002)
9. Andrew, B.: A Maximum Entropy Approach to Named Entity Recognition. Ph.D. Thesis, New York University (1999)
10. Saha, S.K., Chatterji, S., Dantapat, S., Sarkar, S., Mitra, P.: A Hybrid Approach for Named Entity Recognition in Indian Languages. In: NERSSEAL-IJCNLP 2008, Hyderabad, India, pp. 17–24 (2008)
11. Sharma, P., Sharma, U., Kalita, J.: Named Entity Recognition: A Survey for the Indian Languages. In: Parsing in Indian Languages, pp. 35–39 (2011)
12. Ekbal, A., Haque, R., Das, A., Bandyopadhyay, S.: Language Independent Named Entity Recognition in Indian Languages. In: Proceedings of the NERSSEAL-IJCNLP 2008, Hyderabad, India, pp. 33–40 (2008)

13. Ekbal, A., Saha, S.: Weighted Vote Based Classifier Ensemble Selection Using Genetic Algorithm for Named Entity Recognition. In: Hopfe, C.J., Rezgui, Y., Métais, E., Preece, A., Li, H. (eds.) NLDB 2010. LNCS, vol. 6177, pp. 256–267. Springer, Heidelberg (2010)
14. Ekbal, A., Saha, S.: Classifier Ensemble using Multiobjective Optimization for Named Entity Recognition. In: European Conference on Artificial Intelligence (ECAI 2010), Lisbon, Portugal, pp. 783–788 (2010)
15. Ekbal, A., Saha, S.: Maximum Entropy Classifier Ensembling using Genetic Algorithm for NER in Bengali. In: International Conference on Language Resources and Evaluation (LREC 2010), Malta (2010)
16. Ekbal, A., Bandyopadhyay, S.: Maximum Entropy Approach for Named Entity Recognition in Bengali. In: Proceedings of International Symposium on Natural Language Processing (SNLP 2007), Thailand, pp. 1–6 (2007)
17. Ekbal, A., Bandyopadhyay, S.: Bengali Named Entity Recognition using Support Vector Machine. In: NERSSEAL-IJCNLP 2008, Hyderabad, India, pp. 51–58 (2008)
18. Ekbal, A., Bandyopadhyay, S.: Voted NER System using Appropriate Unlabeled Data. In: Named Entities Workshop: Shared Task on Transliteration (NEWS 2009), ACL-IJCNLP, Singapore, pp. 202–210 (2009)
19. Chaudhuri, B., Bhattacharya, S.: An Experiment on Automatic Detection of Named Entities in Bangla. In: NERSSEAL-IJCNLP 2008, Hyderabad, India, pp. 75–82 (2008)
20. Gali, K., Surana, H., Vaidya, A., Shishtla, P., Sharma, D.M.: Aggregating Machine Learning and Rule Based Heuristics for Named Entity Recognition. In: NERSSEAL-IJCNLP 2008, Hyderabad, India, pp. 25–32 (2008)
21. Ganchev, K., Pereira, F., Mandel, M., Carroll, S., WhiteCrammer, P., Singer, Y.: Semi-automated named entity annotation. In: Proceedings of the Linguistic Annotation Workshop, pp. 53–56. ACL (2007)
22. Crammer, K., Singer, Y.: Ultraconservative Online Algorithms for Multiclass Problems. Journal of Machine Learning Research, 951–991 (2003)
23. Singh, A.K.: Named Entity Recognition for South and South East Asian Languages: Taking Stock. In: NERSSEAL-IJCNLP 2008, Hyderabad, India (2008)
24. Ekbal, A., Bandyopadhyay, S.: Named entity recognition using support vector machine: A language independent approach. International Journal of Electrical, Computer, and Systems Engineering 4(2), 155–170 (2010)

Score Normalization Methods Applied to Topic Identification*

Lucie Skorkovská and Zbyněk Zajíc

University of West Bohemia, Faculty of Applied Sciences
New Technologies for the Information Society
Univerzitní 22, 306 14 Plzeň, Czech Republic
{lskorkov,zzajic}@ntis.zcu.cz

Abstract. Multi-label classification plays the key role in modern categorization systems. Its goal is to find a set of labels belonging to each data item. In the multi-label document classification unlike in the multi-class classification, where only the best topic is chosen, the classifier must decide if a document does or does not belong to each topic from the predefined topic set. We are using the generative classifier to tackle this task, but the problem with this approach is that the threshold for the positive classification must be set. This threshold can vary for each document depending on the content of the document (words used, length of the document, ...). In this paper we use the Unconstrained Cohort Normalization, primary proposed for speaker identification/verification task, for robustly finding the threshold defining the boundary between the correc and the incorrect topics of a document. In our former experiments we have proposed a method for finding this threshold inspired by another normalization technique called World Model score normalization. Comparison of these normalization methods has shown that better results can be achieved from the Unconstrained Cohort Normalization.

Keywords: topic identification, multi-label text classification, Naive Bayes classification, score normalization.

1 Introduction

Multi-label classification is increasingly required in modern categorization systems, especially in the fields of newspaper article topic identification, social network comments classification, web content topical organization or email routing, where each "document" (either newspaper article or email) can belong to many topics (or keywords or tags) selected from a large set of possible labels. Usually, the multi-label classification is handled through a set of binary classifiers, one for each label, deciding if a document does or does not belong to a specified topic. The issue with this approach is that for each topic the classifier must be trained and the threshold for the positive classification must be set. This may not be a problem for a classification task with a small set of topics (ten topics for example), where for each one of them a sufficient amount of training data is available, but in a real application the set of topics is usually quite large (450 topics in our case) and for some of them very little training data can be obtained.

* The work was supported by the Ministry of Education, Youth and Sports of the Czech Republic project No. LM2010013 and University of West Bohemia, project No. SGS-2013-032.

P. Sojka et al. (Eds.): TSD 2014, LNAI 8655, pp. 133–140, 2014.

Possible alternative is to use a single generative classifier like Naive Bayes (NB) classifier [1][8], which outputs a distribution of probabilities (or likelihood scores) of the document belonging to the topics from the topic set. In this approach only a single threshold defining the boundary between the "correct" and the "incorrect" topics of a document has to be set. The problem addressed in this paper is how to process the distribution of topics and select this threshold. Since it may vary depending on the content of each document, it can not be fixed for the whole document collection, but a dynamically set threshold is needed.

In our former experiments we have proposed a General Topic Model Normalization (GTMN) method [14] for finding the threshold inspired by the World Model score normalization technique used in speaker identification/verification task. Since this method has shown promising results, in this paper we try to propose advanced technique for the threshold selection based on another technique used in speaker identification area - Unconstrained Cohort Normalization (UCN).

The score normalization methods are used to improve the newspaper topic identification results in a real-life application for language modeling data filtering [17], where the topics are chosen from a quite extensive hierarchy - it contains about 450 topics.

2 Multi-label Text Classification

The multi-label classification methods can be divided into two main categories - *data transformation methods* and *algorithm adaptation methods*. The methods of the first group transform the problem into the single-label classification problem and the methods in the second group extend the existing algorithms to handle the multi-label data directly. In [16] a detailed overview of the existing *data transformation methods* is presented: The easiest way is to transform the multi-label data set into single-label by either selecting only one label from the multiple labels for each data instance or by discarding every multi-label data instance from the set. Another option is to considers each set of labels as one label together [8].

The most common option is to train a binary one-vs.-rest classifier for each class. The labels for which the binary classifier yields a positive result are then assigned to the tested data item. The disadvantage of this method is that you have to transform the data set into $|L|$ data sets, where L is the set of possible labels, containing only the positive and negative examples. The second disadvantage is that you have to find the threshold for each binary classifier. This method was used for example in [4][18].

Another possibility is to decompose each training data with n labels into n data items each with only one label. One generative classifier with the distribution of likelihoods for all labels is learned from the transformed data set. The distribution is then processed to find the correct labels of the data item. This approach is used in the work [3][8] and also in our experiments.

2.1 Threshold Definition for Generative Classifiers

A related work on the problem how to select the set of correct topics from the output distribution of the generative classifiers is presented in this section. A straightforward approach is to select the labels for which the likelihood is greater than a specific threshold or select a predefined number of topics. In the work [1] only the one best label is assigned to each news article. In our later work, we selected 3 topics for each article [15]. In the work [8] this problem is bypassed by creating a mixture topic model from all possible topic subsets and then choosing the subset for which the corresponding mixture model has achieved the maximum likelihood.

To our knowledge, the only work concerning the finding of a threshold for choosing the correct topics in the distribution output of a classifier is described in [3]. A dynamic threshold is set as the mean plus one standard deviation (MpSD) of the topic likelihoods. The assumption is that topics that have a likelihood greater than this threshold are the best choices for the article. In our former experiments [14] we have proposed a General Topic Model Normalization (GTMN) method for finding the threshold inspired by the World Model score normalization technique and compared it to the related methods. The results obtained from the comparison can be seen in Section 4.4 in Table 1.

3 Score Normalization Applied to Multi-label Topic Identification

A topic identification problem is quite similar to the open-set text-independent speaker identification (OSTI-SI) problem. Similarly as in the speaker identification, the multi-label document classification can be described as a twofold problem: First, the speaker model best matching the utterance has to be found and secondly it has to be decided, if the utterance has really been produced by this best-matching model or by some other speaker outside the set. The difficulty in this task is that the speakers are not obliged to provide the same utterance that was the system trained on. The document classification problem can be described in the same way: First, we need to find the topic models which have the best likelihood score for the tested document and second, we have to choose only the correct topic models which really generated the document. The only difference in topic identification is that we try to find more than one correct topic model. The normalization methods from OSTI-SI can be used in the same way, but have to be applied to all topic models likelihoods.

3.1 Naive Bayes Classification

For the first phase of the topic identification the multinomial Naive Bayes classifier is used, which is formally equal to the language modeling based approach in the information retrieval [7]. Each topic is defined by its unigram language model and a probability of a document A being generated by a topic model T is a conditional model $P(T|A)$. Using the Bayes' theorem and leaving out the prior probability of an article $P(A)$, the following equation can be written:

$$P(T|A) \propto \frac{P(T)p(A|T)}{P(A)} \propto p(A|T), \tag{1}$$

where $P(T)$ is the prior probability of a topic T, which can be estimated as a relative frequency of a topic in training data, or considered uniform and be left out as in our case [14]. The distribution of topic likelihoods $p(A|T)$ is then used to find the most likely topics of an article. Under the "naive" conditional independence assumption $p(A|T)$ can be computed in the following way:

$$p(A|T) = \prod_{t \in A} p(t|T), \qquad \hat{p}(t|T) = \frac{tf_{t,T}}{N_T}, \tag{2}$$

where $p(t|T)$ is a conditional probability of a term t given the topic T. This probability is estimated by the maximum likelihood estimate as the relative frequency of the term t in the training data of the topic T. The uniform prior smoothing was used in the estimation of $p(t|T)$.

3.2 Score Normalization

Now that we have the distribution of topic likelihoods $p(A|T)$ we have to find the threshold for selection the correct topics of an article. A score normalization methods have been used to tackle the problem of the compensation for the distortions in the utterances in the second phase of the open-set text-independent speaker identification problem [12]. In the topic identification task, the likelihood score of a topic obtained from the classifier is dependent on the characteristics of the document (words used, length of the document, ...).

Similarly as in the OSTI-SI [12] we can define the decision formula:

$$P(T_C|A) > P(T_I|A) \rightarrow A \in T_C \quad \text{else} \quad A \in T_I, \tag{3}$$

where $P(T_C|A)$ is the score given by the correct topic model and $P(T_I|A)$ is the score given by the incorrect topic model. By the application of the Bayes' theorem, formula (3) can be rewritten as:

$$\frac{p(A|T_C)}{p(A|T_I)} > \frac{P(T_I)}{P(T_C)} \rightarrow A \in T_C \quad \text{else} \quad A \in T_I, \tag{4}$$

where $l(A) = \frac{p(A|T_C)}{p(A|T_I)}$ is the normalized likelihood score and $\theta = \frac{P(T_I)}{P(T_C)}$ is a threshold that has to be determined. Setting a threshold θ a priori is a difficult task, since we do not know the prior probabilities $P(T_I)$ and $P(T_C)$. Similarly as in the OSTI-SI the topic set is open - an article belonging to a topic not contained in our set can easily occur.

A frequently used form to represent the normalization process is the following [12]:

$$L(A) = \log p(A|T_C) - \log p(A|T_I). \tag{5}$$

The score $\log p(A|T_C)$ is affected by the document characteristics as well as the score $\log p(A|T_I)$. Thus, the distance between them should stand constant for various documents and finding the threshold experimentally for the whole collection of documents can be achieved.

Since the normalization score $\log p(A|T_I)$ of an incorrect topic is not known, there are several possibilities how to approximate it:

General Topic Model Normalization (GTMN). The unknown model T_I can be approximated by the General topic model G [14] which was created as a language model from all documents in the training collection. This technique was inspired by the World Model normalization [11]. The normalization score of a topic model T_I is defined as:

$$\log p(A|T_I) = \log p(A|G) \tag{6}$$

Unconstrained Cohort Normalization (UCN). In this method [2], for every topic model a set (cohort) of N similar models $C = \{T_1, ..., T_N\}$ is chosen. These models in the set C are the most competitive models with the reference topic model, i.e. models which yield the next N highest likelihood scores. The normalization score is given by:

$$\log p(A|T_I) = \log p(A|T_{UCN}) = \frac{1}{N}\sum_{n=1}^{N} \log p(A|T_n). \tag{7}$$

Even when we have the topic likelihood score normalized, we still have to set the threshold θ in (4) for verifying the correctness of each topic in the list. Selecting a threshold defining the boundary between the correct and the incorrect topics in a list of normalized likelihood is more robust, because the normalization removes the influence of the various document characteristics. In our former experiments with GTMN [14] we have selected only the topics which are better scoring than the general topic model and we have defined the threshold as 80% of the normalized score of the best scoring topic. The topics which achieved better normalized score are the "correct" topics to be assigned. The threshold selected in this way has experimentally proven to be robust, the change in the range of percents does not influence the result of the topic identification. For the UCN normalization, we have chosen the same threshold - 80% of the best scoring topic, and we have performed experiments with N - size of the set C to be chosen.

4 Performed Experiments

In this section the experiments with the UCN score normalization method are presented. All experiments were performed with the topic identification module which is a part of the System for acquisition and storing data [17] designed to gather the training data for the estimation of the parameters of statistical language models for natural language processing. For the topic identification experiments the most important parts of the system are the text preprocessing modules. On each article a *tokenization*, *text normalization*, *vocabulary-based substitution* and *decapitalization* algorithms are applied. Automatic *text lemmatization* [6] is also applied in our work, since it has been shown to improve the results when dealing with sparse data [5] [10] in highly inflected languages.

4.1 Topic Identification Module

Since it has been shown that not only the size of the training data is important, but also the right scope of the language models training texts is needed [9], the topic identification algorithm is used for large scale language modeling data filtering [15].

The topic identification module uses a multinomial Naive Bayes classifier, since based on the nature of our application (every day more than 600 new articles are downloaded containing more than 130 new topic training articles) we needed the topic identification algorithm which will be fast and can use the easily updatable statistics stored in the database tables as the trained classifier data. The motivation for choosing the NB classifier is more addressed in [14][15].

The topics are chosen from a hierarchical system, which is built in a form of a topic tree and is based on our expert findings in topic distribution in the articles on the Czech favorite news servers. Totally it contains about 450 topics and topic categories. The advantage of the hierarchical organization of the topics is currently used only for the selection of documents to be used as the training data for the estimation of the parameters of statistical language models but not for the topic identification. For the classification all topic are used only as the set of topics on an equal level (all 450 topics). This is caused by the nature of the training data since we use as training data the real articles from the different news servers and we do not want to change it in any way. The authors of these articles to our knowledge do not use any topic hierarchy, or at least not strictly and easily readable from the data. Sometimes the articles have assigned also the more general topic for some detailed topic, but mostly it does not (for example the article with the topic `soccer` mostly does not have also the topic `sports`.

4.2 Evaluation Metrics

The evaluation of the results of the multi-label classification requires different metrics than those used in the single-label classification. The commonly used metric is somewhat similar to the evaluation used in the field of information retrieval (IR), where each article undergoing classification is considered to be a query in IR and precision and recall is computed for the answer topic set. Similar measures was used in [4]. For the article set D and the classifier H precision ($P(H, D)$) and recall ($R(H, D)$) is computed:

$$P(H,D) = \frac{1}{|D|}\sum_{i=1}^{|D|}\frac{T_C}{T_A}, \qquad R(H,D) = \frac{1}{|D|}\sum_{i=1}^{|D|}\frac{T_C}{T_R}, \tag{8}$$

where T_A is the number of topics assigned to the article, T_C is the number of correctly assigned topics and T_R is the number of relevant reference topics. The $F_1(H, D)$-measure, which is used for straightforward comparison of methods, is then computed from the $P(H, D)$ and $R(H, D)$ measures:

$$F_1(H,D) = 2\frac{P(H,D)\cdot R(H,D)}{P(H,D)+R(H,D)}. \tag{9}$$

These metrics used for the evaluation of multi-label classification express also the partial match of the classification result, so they have to be understood slightly different than those used in single-label classification. For each data item being classified (news article in this case) we obtain a precision and a recall values expressed as a percentage of the full match between the correct topics set and the assigned topics set. The metrics computed for the whole set of articles ($P(H, D)$, $R(H, D)$ and $F_1(H, D)$-measure) therefore express how the classification of an article is "good" on average (e.g. the result $F_1(H, D) = 0.66$ means that on average the classification of an article is 66% good - 2 correct and 1 incorrect topics was assigned to an article with 3 relevant topics).

4.3 Test Data

For the experiments a smaller collection containing 31 419 articles from the news server *ČeskéNoviny.cz* separated from the whole corpus was used [13]. The collection contains articles published in the year 2011(January to October) and is divided into 27 000 training and 4 419 testing articles. The articles have not been rearranged in any way, therefore all the test articles was published after the training articles and may contain events not described in the training set. This reflects the real situation in our system, where we need to identify the topics of each newly downloaded article.

Table 1. Comparison of different threshold finding methods

metric / method(H)	1 topic	3 topics	MpSD	GTMN	UCN
$P(H, D)$	0.8123	0.5859	0.0554	0.5916	**0.6650**
$R(H, D)$	0.3191	0.6155	0.9611	0.6992	**0.6311**
$F_1(H, D)$	0.4582	0.6003	0.1048	0.6409	**0.6476**

4.4 Results

The results of the UCN method applied to the topic identification score for robustly finding the threshold are compared to the results of the GTMN method proposed in [14]. The results are also compared to the previously used selection of 3 topics for each article and also to selection only one topic [1] and setting the threshold as the mean plus one standard deviation (MpSD) of the topic likelihoods [3]. For UCN the size of the cohort was selected experimentally $N = 80$. The experiments (see Table 1) with score normalization techniques from speaker identification domain has shown significantly better results than other techniques used for threshold selection in multi-label document classification. The newly proposed UCN method yields even better results than previously tested GTMN method.

5 Conclusions

This article has proved that score normalization techniques are very useful in topic identification task. The score normalization methods are not time consuming, therefore they can be used even in real-life application like ours. Although we still have to set the threshold for verifying the correctness of the topics, selecting a threshold defining the boundary between the correct and the incorrect topics is more robust, because the normalization removes the influence of the various document characteristics. The proposed UCN technique achieved 1% relative improvement compared to the GTMN method and 7.9% relative improvement compared to the selection of fixed number of topics. The same evaluation was repeated on a different collection of documents separated from our database and the results has shown the same trend.

References

1. Asy'arie, A.D., Pribadi, A.W.: Automatic news articles classification in indonesian language by using naive bayes classifier method. In: Proc. of the 11th Int. Conf. iiWAS 2009, pp. 658–662. ACM, New York (2009)
2. Auckenthaler, R., Carey, M., Lloyd-Thomas, H.: Score normalization for text-independent speaker verification systems. Digital Signal Processing 10(13), 42–54 (2000)
3. Bracewell, D.B., Yan, J., Ren, F., Kuroiwa, S.: Category classification and topic discovery of japanese and english news articles. Electron. Notes Theor. Comput. Sci. 225, 51–65 (2009)
4. Godbole, S., Sarawagi, S.: Discriminative methods for multi-labeled classification. In: Dai, H., Srikant, R., Zhang, C. (eds.) PAKDD 2004. LNCS (LNAI), vol. 3056, pp. 22–30. Springer, Heidelberg (2004)
5. Ircing, P., Müller, L.: Benefit of Proper Language Processing for Czech Speech Retrieval in the CL-SR Task at CLEF 2006. In: Peters, C., Clough, P., Gey, F.C., Karlgren, J., Magnini, B., Oard, D.W., de Rijke, M., Stempfhuber, M. (eds.) CLEF 2006. LNCS, vol. 4730, pp. 759–765. Springer, Heidelberg (2007)
6. Kanis, J., Müller, L.: Automatic lemmatizer construction with focus on oov words lemmatization. In: Matoušek, V., Mautner, P., Pavelka, T. (eds.) TSD 2005. LNCS (LNAI), vol. 3658, pp. 132–139. Springer, Heidelberg (2005)
7. Manning, C.D., Raghavan, P., Schütze, H.: Introduction to Information Retrieval. Cambridge University Press, New York (2008)
8. McCallum, A.K.: Multi-label text classification with a mixture model trained by em. In: AAAI 1999 Workshop on Text Learning (1999)
9. Psutka, J., Ircing, P., Psutka, J.V., Radová, V., Byrne, W., Hajič, J., Mírovský, J., Gustman, S.: Large vocabulary ASR for spontaneous Czech in the MALACH project. In: Proceedings of Eurospeech 2003, Geneva, pp. 1821–1824 (2003)
10. Psutka, J., Švec, J., Psutka, J.V., Vaněk, J., Pražák, A., Šmídl, L., Ircing, P.: System for fast lexical and phonetic spoken term detection in a czech cultural heritage archive. EURASIP J. Audio, Speech and Music Processing 2011 (2011)
11. Reynolds, D.A., Quatieri, T.F., Dunn, R.B.: Speaker verification using adapted gaussian mixture models. In: Digital Signal Processing. p. 2000 (2000)
12. Sivakumaran, P., Fortuna, J., Ariyaeeinia, M.A.: Score normalisation applied to open-set, text-independent speaker identification. In: Eurospeech, Geneva, pp. 2669–2672 (2003)
13. Skorkovská, L.: Application of lemmatization and summarization methods in topic identification module for large scale language modeling data filtering. In: Sojka, P., Horák, A., Kopeček, I., Pala, K. (eds.) TSD 2012. LNCS, vol. 7499, pp. 191–198. Springer, Heidelberg (2012)
14. Skorkovská, L.: Dynamic threshold selection method for multi-label newspaper topic identification. In: Habernal, I. (ed.) TSD 2013. LNCS, vol. 8082, pp. 209–216. Springer, Heidelberg (2013)
15. Skorkovská, L., Ircing, P., Pražák, A., Lehečka, J.: Automatic topic identification for large scale language modeling data filtering. In: Habernal, I., Matoušek, V. (eds.) TSD 2011. LNCS, vol. 6836, pp. 64–71. Springer, Heidelberg (2011)
16. Tsoumakas, G., Katakis, I.: Multi-label classification: An overview. Int. J. Data Warehousing and Mining 2007, 1–13 (2007)
17. Švec, J., Hoidekr, J., Soutner, D., Vavruška, J.: Web text data mining for building large scale language modelling corpus. In: Habernal, I., Matoušek, V. (eds.) TSD 2011. LNCS, vol. 6836, pp. 356–363. Springer, Heidelberg (2011)
18. Zhang, M.L., Zhou, Z.H.: A k-nearest neighbor based algorithm for multi-label classification. In: 2005 IEEE International Conference on Granular Computing, vol. 2, pp. 718–721 (2005)

Disambiguation of Japanese Onomatopoeias Using Nouns and Verbs

Hironori Fukushima[1], Kenji Araki[1], and Yuzu Uchida[2]

[1] Hokkaido University, Sapporo, Japan
{hukupoyo,araki}@media.eng.hokudai.ac.jp
[2] Hokkai Gakuen University, Sapporo, Japan
yuzu@hgu.jp

Abstract. Japanese onomatopoeias are very difficult for machines to recognize and translate into other languages due to their uniqueness. In particular, onomatopoeias that convey several meanings are very confusing for machine translation systems to distinguish and translate correctly. In this paper, we discuss what features are helpful in order to automatically disambiguate the meaning of onomatopoeias that have two different meanings. We used nouns, adjectives, and verbs extracted from sentences as features, then carried out a machine learning classification analysis and compared the accuracy of how well these features differentiate two meanings of ambiguous onomatopoeias. As a result, we discovered that employing a combination of machine learning with nouns and verbs as a feature achieved accuracy of above 80 points. In addition, we were able to improve the accuracy by excluding pronouns and proper nouns and also by limiting verbs to those that are modified by onomatopoeias. In future, we plan to concentrate on dependency between verbs that are modified by onomatopoeia and nouns, as we believe that this approach will help machine translation to translate Japanese onomatopoeias correctly.

1 Introduction

In 2020, the Tokyo Olympics [1] will be held and many non-Japanese speakers will visit Japan. Not only these visitors, but many other non-Japanese will face a necessity to understand Japanese during their stay due to the lack of translated information in other languages. However, Japanese is one of the most difficult languages to understand because of unique expressions such as onomatopoeia [2]. There are even some onomatopoeias with multiple meanings. In addition, although these words must not be ignored in order to understand the meaning of Japanese more clearly and precisely, such onomatopoeias are very difficult to translate into other languages and recognize using machines because the meaning differs according to the context [3]. Therefore, it is very important to distinguish the meanings of onomatopoeias. In our study, we conducted a consideration of which features need to be extracted from a sentence in order to distinguish the meanings of onomatopoeias with two meanings. Furutake [4] et al. proposed a method for an automatic acquisition system of alternative expressions for onomatopoeia. Their method paraphrases an onomatopoeia as an adjective typically modified by a verb which is typically modified by the onomatopoeia. The meaning of

P. Sojka et al. (Eds.): TSD 2014, LNAI 8655, pp. 141–149, 2014.

the onomatopoeia greatly depends on the verb which it modifies, as demonstrated by the fact that the accuracy of this method achieved 80.6 points. Moreover, Uchida et al. [5] showed that the most often used parts of speech which the onomatopoeia modifies are commonly verbs, nouns, and adjectives, in descending order. Therefore, these parts of speech are very useful for distinguishing the meanings of onomatopoeias.

With the recent proliferation of social media, there are many short sentences and colloquial expressions on the Web. As a result, it can be assumed that there will be cases of Japanese sentences that have no verbs, or where it is difficult to find the part of speech that the onomatopoeia modifies. An example Tweet in Japanese, "*Yokatta-, yuki sarari(*'_'*.)*"[1] , meaning "I'm glad the snow is flowing" has no verbs according to the result of morphological analysis by MeCab [7], although the meaning "touch, feel" can be understood due to the existence of "*sarari*" with the noun "*yuki*", meaning "snow". Consequently, words which onomatopoeia do not modify can be useful as features for extraction, and their effectiveness should be considered. In our study, we extract verbs, nouns, and adjectives from sentences which include an onomatopoeia with two meanings. We extract these words as features and conduct machine learning using SVM-Light [8] to measure the accuracy of semantic disambiguation. Moreover, we subsequently changed the range of extracted nouns and used verbs modified by onomatopoeias as features to improve accuracy.

2 Experiment Preparation

2.1 Defining Target Onomatopoeia

First, we extracted onomatopoeias and mimetic words which have multiple meanings from 'Basic Word Usage Dictionary for Foreigners' [9]. These can be classified into two types depending on whether or not the parts of speech are the same when the onomatopoeia is used in each meaning. When the onomatopoeia has meanings that differ according to the part of speech, we need only distinguish the part of speech of the onomatopoeia in order to disambiguate the meanings. The meanings of such onomatopoeias are distinguished easily. Therefore, in our study, we define onomatopoeias with meanings that are difficult to distinguish as being "onomatopoeias that are used as the same part of speech in two different meanings". An example of this is "*garari*". When we use "*garari*" as an adverb, it has the following two meanings: (1) Opening the door vigorously. (e.g., "*Genkan no to wo "garari" to akeru*", He open the front door "*garari*".), (2) Changing an attitude suddenly. (e.g., " *Hanashi no tochuu de kareno taido ga "garari" to kawatta.*", His attitude changed "*garari to*" in the middle of talking.)

We found 21 onomatopoeias with two meanings which match the above definition in 'Basic Word Usage Dictionary for Foreigners'. This research is based on the concept that onomatopoeias that are used widely and regularly should be studied. Therefore, we targeted the two onomatopoeias that rank highest in terms of number of hits in a Google search [10]. As a result, "*sarari to*" and "*gatan to*" were selected. Both of these consist of the particle "*to*" and an onomatopoeia. However, it was found that onomatopoeias

[1] Extracted from Twitter [6].

often appear in blogs accompanied by the particle "*to*" [11]. In order to reduce errors in morphological analysis we retained the particle "*to*" in this research, using these onomatopoeias in the forms of "*gatan to*"and "*sarari to*".

2.2 Definition of Meanings of Onomatopoeias

It is necessary to provide original definitions of the meanings of "*sarari to*" and "*gatan to*", because there are some sentences that are impossible to classify using only the meanings described in 'Basic Word Usage Dictionary for Foreigners'.

The meanings of "*sarari to*" and "*gatan to*" that we defined are shown in Table 1 and 2. we define "Meaning 1 " as "m1" and "Meaning 2 " as "m2".

Table 1. The meanings of "*sarari to*" that we defined

Meaning 1 (ml)	**A touch, a feel, a flavor.** Wind blowing "*sarari to.*"(English) "*sarari to" shita kaze ga huku.*(Japanese)
Meaning 2 (m2)	**With good grace, good decisiveness.** He said difficult things "*sarari to*".(English) *Kare ha muzukashii koto wo "sarari to" itta.*(Japanese)

Table 2. The meanings of "*gatan to*" that we defined

Meaning 1 (ml)	**Appearance situations or sounds of heavy things faling or crushing heavy things, crush heavy things.** Things fall "*gatan to*" from a shelf.(English) *tana kara monoga "gatan to" ochiru.*(Japanese)
Meaning 2 (m2)	**A sudden drop in price, performance etc.** Sales fell "*gatan to*".(English) *uriage ga "gatan to" ochiru.*(Japanese)

2.3 Collection of Sentences That Include Onomatopoeia

We collected sentences from Ameba blog [12] and Twitter [6] due to the assumption that such blogs contain a high number of onomatopoeia that are difficult to distinguish, and have many colloquial and everyday expressions. Any noisy sentences were removed manually. Example sentences which were removed are shown in Table 3.

Example 1 contains sentences consisting only of onomatopoeias. Example 2 is a sentence which can not be generally understood. Example 3 contains an onomatopoeia used as a proper noun in the sentence. In Example 4, features of the string which construct the onomatopoeia are repeated directly before or after the onomatopoeia.

Table 3. Examples of manually removed sentences

Ex1	*"Gatanto."*, *"Sarari sarari to."*
Ex2	*Nama de mireru nante sasu "ga tanto" n san.* - I'm proud to see Tanton in real life (English)
Ex3	*"Sarari to" shita umeshu* - *Sarari to* plum liquor. (Product name)
Ex4	*Otonari no akishitsu kara "gatan gatan" to to ga aku oto.* - The sound of the door opening in the vacant room next to mine.(English)

2.4 Constructing Correct Data

We constructed a corrected version of the data to input in SVM-Light. We asked ten Japanese participants to judge which meaning is correct for each onomatopoeia in each sentence. We determined a meaning to be correct if eight or more people gave the same answer. Sentences with "*sarari to*" were checked by ten participants, consisting of five women and five men. Sentences with "*gatan to*" were checked by another ten participants, consisting of six women and four men. We collected approximately 200 sentences for each onomatopoeia. Of the "*sarari to*" sentences, we collected 78 sentences with meanings of m1 (defined in 2.2), and 124 sentences with meanings of m2. Of the "*gatan to*" sentences, we collected 100 sentences each of m1 and m2. The number of sentences given correct meanings for "*sarari to*" is 70 sentences as m1, and 119 as m2. For "*gatan to*" 98 sentences were given correct meanings as m1, and 94 as m2. There were 13 sentences which more than three people judged to have different meanings among 202 sentences featuring "*sarari to*". The rate of occurrence is 6.4 percent. On the other hand, there were eight such sentences among 200 sentences containing "*gatan to*", an occurrence of 4.0 percent. As a result, it was demonstrated that there are cases of different interpretations of the meanings of Japanese onomatopoeia even among native Japanese speakers.

3 Meanings Distinction Experiment

3.1 Onomatopoeia Semantic Disambiguation Method

Firstly, we conducted morphological analysis on the sentences using MeCab [7] and extracted verbs, nouns, and adjectives. Secondly, we constructed the data in sets of words (verbs, nouns and adjectives) and the number of occurrences. Finally, we classified the meanings of the onomatopoeias using SVM-Light. We compared the accuracy using 10-fold cross-validation, as the amount of data is small. There are seven patterns of features in SVM-Light : only verbs; only nouns; only adjectives; nouns and adjectives; verbs and adjectives; nouns and verbs; and nouns, verbs and adjectives. Our definition of accuracy is as the equation (1). The meaning given is the correct meaning.

$$accuracy = \frac{TP + TN}{TP + FP + TN + FN} \tag{1}$$

TP = The meaning given "m1" and its output is "m1"
FP = The meaning given "m2" and its output is "m1"
TN = The meaning given "m2" and its output is "m2"
FN = The meaning given "m1" and its output is "m2"
("m1" and "m2" are defined in **2.2**)

3.2 Result of Experiment

We performed 10-fold cross-validation twelve times and calculated the average after omitting the highest and lowest values for each onomatopoeia. The calculated categorie are accuracy(defined in **3.1** (1)). These results are shown in Table 4.
n : noun, v : verb, a : adjective.

Table 4. Accuracy for "*sarari to*" and "*gatan to*"

POS	"*sarari to*" [%]	"*gatan to*" [%]
n	75.66	75.70
a	67.06	52.60
v	79.81	72.59
n and a	76.07	74.10
n and v	83.50	80.50
a and v	78.09	73.32
n,a and v	83.66	80.77

3.3 Consideration of Results

For both "*sarari to*" and "*gatan to*", the feature pattern of "nouns, adjectives and verbs" gave the highest accuracy. The feature pattern of "nouns and verbs" gave the second highest accuracy. The difference in accuracy between "nouns, adjectives and verbs" and "nouns and verbs" is very small. Therefore, we demonstrated that adjectives have poor potential for improving accuracy. On the other hand, we found some cases where adjectives did make a contribution to improving accuracy. For example, "nouns and adjectives" has higher accuracy than "nouns" in the results for "*sarari to*". Thus, adjectives cannot be said to be completely noisy. This fact is related to the frequency of co-occurrence with the onomatopoeia. Furthermore, when we compare the features "verbs", "nouns" and "adjectives", "verbs" has the highest accuracy for "*sarari to*" , whereas "nouns" has the highest accuracy for "*gatan to*". Therefore, it was demonstrated that in the semantic disambiguation of onomatopoeia, the most significant parts of speech are dependent on the type of onomatopoeia. The conclusion of our consideration is that "nouns" and "verbs" are highly significant for distinguishing meanings of onomatopoeia, and combining them as features makes a major contribution to semantic disambiguation.

4 Improvement of Features

4.1 Method of Improvement

Firstly, we conducted a consideration of verbs. From related research, we know that the verbs that the onomatopoeia modifies are very important in order to distinguish meanings. Therefore, we extracted only verbs that the onomatopoeia modifies. We used CaboCha [13] to determine dependency relations. Furthermore, we extracted only verbs that belong to independent words using MeCab with the exception of suffixes. A suffix includes the "*ori*" of " ...*orimasu*" or "*i*" of " ...*imasu*" (English: both forms of "to be"). Secondly, we conducted a consideration of nouns. There are proper nouns, numerals, and pronouns in sentences. For example, names such as "*Ogawa*" and "*Darvish*", and the pronoun "*watashi*" ("I"). These belong to the category of nouns, but they do not affect semantic disambiguation. Therefore, we removed them. Moreover, nouns which belong to the category of non-independent words were also removed. For example, "*suru*" of "...*suru koto ni natta*" (English: "(I) ended up doing "). Therefore, as a method of improvement, we extracted nouns belonging to the categories of "general", "sahen-connection", "nai-adjectivestem" and "quoted string". In this way, we improved the features selection method, and performed 10-fold cross-validation again.

4.2 Results and Consideration after Improving the Features

As the purpose of this study is improvement of accuracy(defined in 3.1 (1)), we compared the results of experiments after improving the features selection method with the previous results. These results and a comparison shown in Table 5, Table 6, Table 7 and Table 8. n : features are nouns. v : features are verbs. n': features are only nouns, with the exception of proper nouns and pronouns, etc. v': features are only verbs modified by the onomatopoeia.

Table 5. Comparison of "nouns"

POS	"*sarari to*"[%]	"*gatan to*" [%]
n	75.66	75.70
n'	72.70	80.73
Change	-2.96	5.03

Table 6. Comparison of "verbs"

POS	"*sarari to*"[%]	"*gatan to*"[%]
v	79.81	72.59
v'	85.23	81.67
Change	5.42	9.08

Table 7. Results after improving features

POS	"*sarari to*"[%]	"*gatan to*"[%]
n'	72.70	80.73
v'	85.23	81.67
n' and v'	83.87	82.52

Table 8. Combination of nouns and verbs

POS	"*sararito*"[%]	"*gatanto*"[%]
n and v	83.50	80.50
n' and v'	83.87	82.52
Change	0.37	2.02

Firstly, we will discuss nouns. For "*gatan to*", accuracy increased by 5.03 points. On the other hand, for "*sarari to*" accuracy decreased by 2.96 points. When we change the types of extracted nouns, there are 19 sentences that have no nouns for "*sarari to*", and nine sentences without nouns for "*gatan to*". These sentences were removed from the data. Therefore, the target sentences for "*sarari to*" were reduced to 170 sentences, and "*gatan to*" to 183 sentences. The number of distinct nouns that appeared was decreased to 327 from 456 for "*sarari to*", and to 341 from 500 for "*gatan to*". It can be considered that the frequency of occurrence of proper nouns and pronouns was an influential factor on the increased rate of accuracy. For example, the pronoun "*watashi*" ("I") appeared in six sentences of 183 sentences for "*sarari to*", whereas it appeared in only three sentences of 170 sentences for "*gatan to*". This difference is about 1.5 points. This affected the increased rate of accuracy. Moreover, failures of morphological analysis affected the decreasing rate of accuracy. For example, "*Darvish ga watashi no me no mae wo "sarari to" tootte kandou* " (I was impressed when Darvish (a baseball player) "*sarari to*" passed right in front of me). The extracted words from this sentence are "*Darvish*", "*me*", "*kayou*" and "*kandou*". The point of interest here is "*Darvish*". The word "*Darvish*" belongs to the category of "proper nouns" , but it was analyzed by MeCab in this sentence as belonging to the category of "general nouns". As a result , the word can not be removed as a proper noun. This failure causes a reduction in accuracy.

Secondly, we will discuss verbs. In sentences containing "*sarari to*", accuracy increased by 5.42 points, and for "*gatan to*", accuracy increased by 9.08 points ; a major increase for both features. This shows that the verbs modified by onomatopoeias are very important for distinguishing the meanings. However, the number of features is insufficient, because a Japanese onomatopoeia basically only modifies one verb. Therefore, there are few features to distinguish the meaning. There are a few sentences in which the meanings are different even though the features are the same. Such sentences are very difficult to disambiguate clearly. The combination of nouns and verbs increased accuracy by 2.02 points for sentences containing "*gatan to*", and 0.37 points for "*sarari to*". When "v' and n' " were combined and used as a feature, the highest accuracy was achieved for "*gatan to* ", although this was not the case for "*sarari to*". The reason why the accuracy for "*gatan to*" is high is that the accuracy of the feature "n' " is high. The accuracy of "*gatan to*" is 8 points higher than '*sarari to*" with the feature "n' ". This shows that "n' " has a positive influence on improvement of accuracy. On the other hand, for "*sarari to*", the feature "n' and v' " is 1.36 points lower than "v' " , because the accuracy of "n' " was low. Below, we analyze examples of failed semantic disambiguation. These examples are classified into three patterns. These patterns and an example show in Table 9, Table 10, and Table 11. We conducted analysis for each pattern.

Example of pattern (1), there are three forms of the verb "*suru*" ("to do") in this sentence. The other nouns extracted from this sentence nouns do not exist in any other sentences in our data. Therefore, the only feature for disambiguation is "*suru*". Furthermore, there are 75 sentences which include "*suru*", of which 45 sentences were judged as m1(defined in 2.2), and 30 sentences as m2. The proportion of sentences judged as m1 is larger than those judged as m2. Therefore, there is a trend to wards output as m1 although correct meaning is m2.

Example of pattern (2), the nouns and verbs in this sentence do not occur in other sentences. Therefore, there are insufficient features for disambiguation. The output meaning of this sentence is m2 though the correct meaning is m1.

Table 9. Pattern (1) : In case that semantic disambiguation depends on non-important verbs

Example sentence	*Watashi mo gochisou shitari purezento shitari suru no ga suki dakedo kono "sarari to" to iu no ga otokomae.* (I like treating people to meals and giving presents too, but the way he does it so smoothly ("*sarari to*") is really manly.)
(Words : frequency)	(gochisou : 1), (purezento : 1), (otokomae:1), (suru : 3)

Table 10. Pattern (2) : In case that there is a lack of information to distinguish meanings

Example sentence	*Hontou, futshuu no ro-ru ke-ki nan desu ga, suponji ha fuwafuwa de, kuri-mu ha tottemo nameraka de, shita no ue de "sarari to" tokemasu.* (It is a secret that I freaked out a bit the cream is so smooth it melts "*sarari to*" on the tongue.)
(Words : frequency)	(shita : 1), (suponji : 1), (hontou : 1), (kuri-mu : 1), (tokeru : 1)

Example of pattern (3), the output meaning of this sentence is m2(defined in **2.2**) though the correct meaning is m1. Both "*yuka*" and "*ochiru*" exist in other sentences in our data: two sentences include "*yuka*" and 57 sentences include "*ochiru*". Therefore, if we consider the proportions, "*yuka*" has very poor potential to distinguish meanings even though there are tendencies for "*yuka*" to be an important word for disambiguation and to exist disproportionately in either meaning.

Table 11. Pattern (3) : In case that the difference in frequency causes a failure

Example sentence	*Tenjo ya yuka ga "gatan to" ochite karuku tenpatta no ha naisho* (I like treating people to meals and giving presents too, but because the ceiling and floor came crashing ("*gatan to*") down)
(Words : frequency)	(yuka : 1),(tenjo : 1),(naisho : 1),(ochiru : 1)

5 Conclusions and Future Works

Our research evaluated a selection method for features used to semantically disambiguate onomatopoeias with two meanings. As a result, using nouns and verbs as features achieved a stable accuracy of over 80 points. Moreover, using verbs that are modified by onomatopoeias as a feature improved the accuracy without depending on the type of onomatopoeia. In addition, according to types of onomatopoeia, the combination of verbs modified by onomatopoeias and nouns further improved the accuracy. Therefore, these features highly effective for semantic disambiguation. However, there remains a problem in the handling of nouns. The frequency of word occurrence has a major influence on semantic disambiguation. Consequently, nouns whose frequency of occurrence is low are not emphasized in disambiguation, even when the noun is important in order to distinguish the meaning. Therefore, we need to apply weighting to words according to frequency of co-occurrence of "nouns and onomatopoeias" or "verbs modified by onomatopoeias and nouns". In future work, we will incorporate this co-occurrence as a feature and aim for further improvements in accuracy.

References

1. Tokyo Olympics, http://tokyo2020.jp/
2. Shimizu, Y., Doizaki, R., Sakamoto, M.: A System to Estimate an Impression Conveyed by Onomatopoeia. Transactions of the Japanese Society for Artificial Intelligence 29(1), 41–52 (2014) (in Japanese)
3. Asaga, C., Yusuf, M., Watanabe, C.: ONOMATOPEDIA: Onomatopoeia Online Example Dictionary System Extracted from Data on the Web. In: Proceeding of the 10th Asia Pacific Web Conference (APweb), pp. 8b-1 (2008)
4. Furutake, Y., Sato, S., Komatani, K.: Onomatope wo iikaeru hyougen no jidoushushu (Automatic Collection of Expression in Other Words of Onomatopoeias). In: Proceeding of the 17th.Annual Meeting of the Association for Natural Language Processing, pp. 904–907 (2011) (in Japanese)
5. Uchida, Y., Araki, K., Yoneyama, J.: Semantic Ambiguity of Onomatopoeia Extracted from Blog Entries. In: Proceeding of the 27th Fuzzy System Symposium, pp. 853–856 (2011) (in Japanese)
6. Twitter, https://twitter.com/
7. Kudo, T., et al.: MeCab, Yet Another Part-of-Speech and Morphological Analyzer, http://mecab.sourceforge.net/
8. Joachims, T.: SVM-Light, http://svmlight.joachims.org/
9. Agency for Cultural Affairs, gaikokujin no tame no kihongoyourei jitenn (Basic Word Usage Dictionary for Foreigners), National Printing Bureau (1990)
10. Google, https://www.google.co.jp/
11. Uchida, Y., Araki, K., Yoneyama, J.: Affect Analysis of Onomatopoeia Sentences Extracted from Blog Entries. In: Proceeding of the 10th Forum on Information Technology, pp. 274–279 (2011) (in Japanese)
12. Ameba, http://www.ameba.jp/
13. CaboCha, Yet Another Japanese Dependency Structure Aalyser, http://code.google.com/p/cabocha/

Continuous Distributed Representations of Words as Input of LSTM Network Language Model

Daniel Soutner and Luděk Müller

University of West Bohemia, Faculty of Applied Sciences
New Technologies for the Information Society
Univerzitní 22, 306 14 Plzeň, Czech Republic
{dsoutner,muller}@ntis.zcu.cz

Abstract. The continuous skip-gram model is an efficient algorithm for learning quality distributed vector representations that are able to capture a large number of syntactic and semantic word relationships. Artificial neural networks have become the state-of-the-art in the task of language modelling whereas Long-Short Term Memory (LSTM) networks seem to be efficient training algorithm.

In this paper, we carry out experiments with a combination of these powerful models: the continuous distributed representations of words are trained with skip-gram method on a big corpora and are used as the input of LSTM language model instead of traditional 1-of-N coding. The possibilities of this approach are shown in experiments on perplexity with Wikipedia and Penn Treebank corpus.

Keywords: language modelling, neural networks, LSTM, skip-gram, word2vec.

1 Introduction

Language modelling is a crucial task in many areas of natural language processing. Speech recognition, optical character recognition and many other areas heavily depend on the performance of the language model that is being used. Each improvement in the language modelling may also improve the particular job where the language model is used.

In a lot of applications we are not able to obtain enough amount of the in-domain data for a language model; for example in case of specific style of text or spontaneous speech. Sometimes, adding more out-domain data do not cover our performance demands.

We have focused in our work whether the state-of-the-art models could be improved by adding some information from an out-domain corpus. More specifically, we are exploring the possibilities of two currently favourite LM techniques: *word2vec* architecture for training word distributed representations and LSTM neural networks.

Recurrent neural networks (RNN) have attracted much attention in last years among other types of language models (LM) caused by their better performance [1] and their ability to learn on a smaller corpus than conventional n-gram models. Skip-gram and continuous bag of words (CBOW) [2] – sometimes deonted as *word2vec* – are recently developed technique for building a neural network that maps words to real-number vectors, with the desideratum that words with similar meanings will be mapped to the similar vectors.

P. Sojka et al. (Eds.): TSD 2014, LNAI 8655, pp. 150–157, 2014.

There are described LSTM neural network language model and *word2vec* architecture for training continuous distributed representations of words in the Section 2; the introduction of our proposed language model which combines both of them follows. Section 3 deals with experiments and results and Section 4 summarizes the results and draws some conclusions.

2 Model Structure

In this Section the two main models we were working with are described: LSTM language model and the *word2vec* model for training continuous distributed representations of words from text. Afterwards we proposed our fused model, where we are using word vectors trained from *word2vec* as the input to LSTM neural net instead of standard 1-of-N coding inputs.

2.1 LSTM Language Model

The long-short term memory neural networks were successfully introduced to the field of language modelling by [3] and the basic scheme is shown in Figure 1. LM is based (similarly to conventional recurrent neural network) on these principals:

- The input vector is a word encoded in 1-of-N coding.
- By the training the output vector is also in 1-of-N coding.
- There is a softmax function used in output layer to produce normalized probabilities.
- The cross entropy is used as training criterion.

The vanishing gradient seemed to be problematic during the training of recurrent neural networks as is shown in [4]. This led authors to re-design the network unit, in

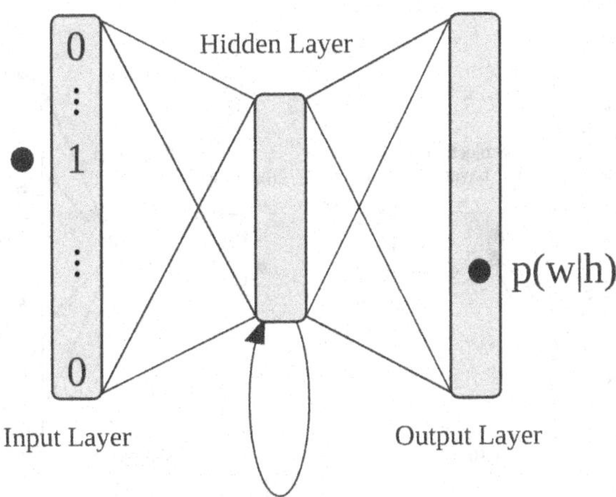

Fig. 1. LSTM language model scheme

LSTM called as a *cell*. Every LSTM cell contains *gates* that determine when the input is significant enough to remember, when it should continue to remember or forget the value, and when it should output the value.

Regular back-propagation could be effective at training LSTM cell due to this specific topology of LSTM, especially because of a constant error flow. This leads to easier model training and in LM application these networks capture even a longer context then standard RNN as it is showed in [3] and perform better on dynamic evaluation [5].

2.2 Distributed Representations of Words

Representation of text is very important for performance in many real-world applications. The most commonly used techniques in these days are *n*-grams, bag-of-words, 1-of-N coding (which all belongs to local representations). Continuous representations of the words, which represent words as a vector with real numbers are for example LDA, LSA or distributed representations. Distributed representations of words can be obtained from various NN-based language models.

Mikolov et al. in [2] introduced new architecture for training distributed word representations which belongs to class of methods called "neural language models". Using a scheme that is much simpler than previous work in this domain, where neural networks with many hidden units and several non-linear layers were normally constructed (e.g., [6]), *word2vec* constructs a simple log-linear classification network [7]. There were proposed two architectures: continuous bag-of-words (cbow) predicts the context given the word, continuous skip-gram predicts the current word given the context. The basic scheme of model is shown in Figure 2. The input words are encoded in 1-of-N coding, the model is trained with hierarchical softmax, a context in interval 5-10 word is usually considered.

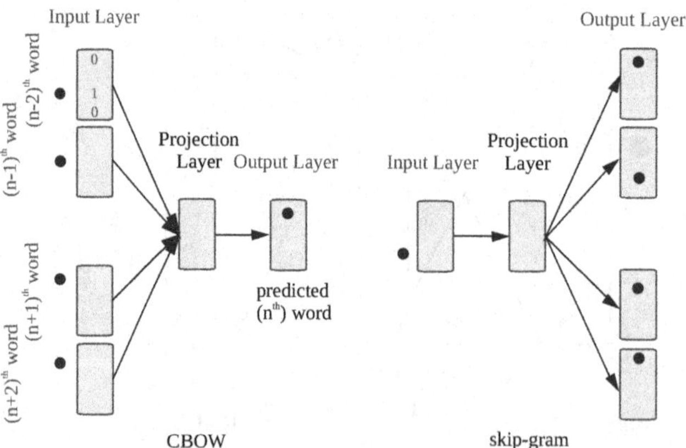

Fig. 2. The CBOW architecture predicts the current word based on the context, and the Skip-gram predicts surrounding words given the current word.

This model provides more efficient word vectors than neural networks, the resulting distributed representations of words contain surprisingly a lot of syntactic and semantic information. There are multiple degrees of similarity among words:

- KING is similar to QUEEN as MAN is similar to WOMAN
- KING is similar to KINGS as MAN is similar to MEN

Simple vector operations (addition and subtraction) with the word vectors provide very intuitive results, $vector(KING) - vector(MAN) + vector(WOMAN) \approx vector(QUEEN)$ [8].

Word vectors from neural networks were previously criticized for their poor performance on rare words – performance of skip-gram on rare words has good results for the nearest neighbours rare words which are more correlated with human ratings [9]. For the reasons and properties described above we choose this model for our experiments, further we work only with skip-gram.

2.3 Our Architecture

The idea was to replace the word vectors (sometimes called as projections or feature vectors) which computes the recurrent neural network itself with better ones computed on more data, even out-of-domain. As it was described in Section 2.2, the skip-gram provides us – in some point of view – better word vectors. Moreover, the vectors form skip-gram are computationally cheaper. The LSTM neural net architecture provides very stable learning (we have already done some experiments with modifications of the input vector in [10]). The LSTM network should not have any difficulties with learning from this skip-gram word vectors.

The word vectors obtained by training skip-gram model are trained on bigger corpora and afterwards they are used as an input of LM. It means that on the input of the LSTM language model we do not put words in 1-of-N coding, but words represented by skip-gram word vectors. This architecture should allow to language model to discover more regularities in a language.

3 Experiments

We performed the experiments to evaluate the contribution of our model to the current state-of-the-art were performed on Penn Treebank and Wikipedia corpus. Continuous distributed representations of words were obtained with *word2vec* toolkit[1], training of LSTM model has been provided with our own toolkit.

3.1 Text Data

To maintain comparability with the other experiments, we chose well-known Penn Treebank (PTB) [11] portion of the Wall Street Journal corpus for testing our models. The following standard preprocessing was applied to the corpora: the vocabulary was

[1] `https://code.google.com/p/word2vec/`

short-listed to 10k most frequent words, all numbers were unified into N tag and a punctuation was removed. The corpus was divided into 3 parts (training, development and test) with 42k, 3.3k and 3.7k tokens.

Continuous distributed representations of words need to be computed on big corpora, so we decided for free and easy available Wikipedia corpus (Wiki) [12]. In our experiments we used database dump from the year 2009, which has been preprocessed and the vocabulary was limited to 225k words (words that occurs more than 50 times), it contains 930M tokens. This data could be assumed as out-domain data if referred to PTB.

As the in-domain data we have used Wall Street Journal (WSJ) corpus [13] with 38M tokens.

3.2 Evaluated Models

All LSTM models used in our experiments were with 100 cells in the hidden layer. The input vocabulary was short-listed to 10k words in case of PTB, to 230k in case of PTB+Wiki; the output vocabulary is short-listed to 10k words from PTB.

Models denoted as **KN5** is a standard 5-gram language model with Knesser-Ney discounting [14], the **LSTM-100** means conventional LSTM neural network models with 100 cells in the hidden layer and **LSTM-100&skip1200** means our model fusion where word vectors of the width 1200 are on input of LSTM network.

For network training we used the PTB-development dataset as validation data and as test data is always used the PTB-test set.

3.3 Results

First, we evaluated how the amount of in-domain data impacts the resulting perplexity. This was done by adding increasing part of WSJ portion (by step of 10%) to the skip-gram model. The word vectors by *skip-gram* were computed from combined PTB and portion of WSJ corpus, the LSTM model was trained on PTB-train. As it could be seen in Figure 3, the model performs better with bigger amount of data for continuous distributed representations of words. But for word vectors computed only on PTB (for *portion of WSJ = 0* in the graph), model performs worse (perplexity 211) then KN5 baseline (perplexity 141.4, ($ID = 4$) in Table 1).

Figure 4 shows the influence of the word vector width on the results, the best results we achieved with the width 1200. The skip-gram word vectors were computed from combined PTB and Wiki corpus, the LSTM model was trained on PTB-train.

Table 1 shows the results for various models, all tested on PTB-test. It could be seen, that Wiki corpus is, as we expected, out-of-domain for PTB-test ($ID = 2$). The performance of LSTM-100 and KN5 are comparable, when trained on PTB, resulting perplexities are 145.0 and 141.4 ($ID = 3$ and $ID = 4$). As the baseline model we have chosen linear combination of KN5 models from PTB-train and Wiki corpus (tuned with EM-algorithm on PTB-dev data) with perplexity 139.5 ($ID = 5$). Our proposed model, even with word vectors trained on out-of-domain Wiki corpus, notable outperforms the baseline models with perplexity 130.4 ($ID = 6$).

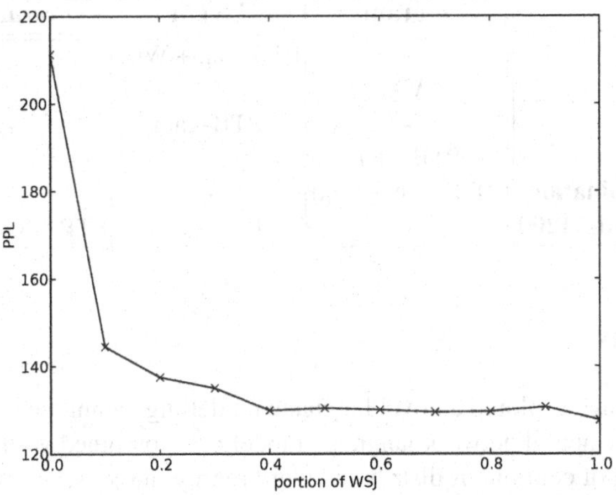

Fig. 3. Perplexity results with an increasing portion of WSJ added to PTB for training word2vec, measured on PTB-test

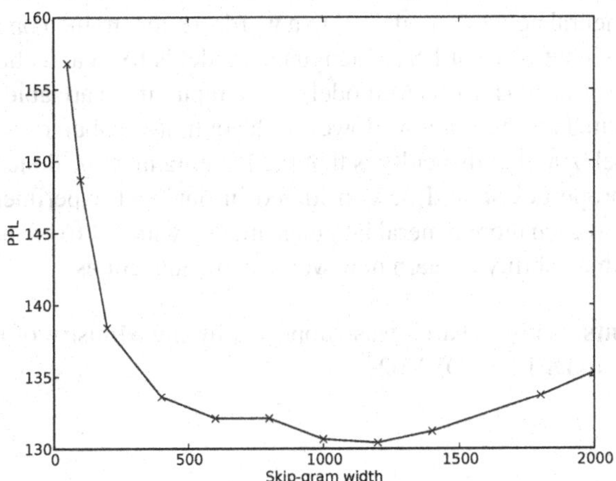

Fig. 4. Perplexity results with a various width of word vectors computed on PTB+Wiki corpus, measured on PTB-test

Table 1. Perplexity results with various models, measured on PTB-test

ID	Model Type	Train data			PPL
		n-gram	LSTM	*word2vec*	
1	**LSTM-100**	-	PTB-train+ Wiki	-	308.0
2	**KN5**	Wiki	-	-	188.0
3	**LSTM-100**	-	PTB-train	-	145.0
4	**KN5**	PTB-train	-	-	141.4
5	**KN5 (lin.combination)**	PTB-train + Wiki	-	-	139.5
6	**LSTM-100&skip-1200**	-	PTB-train	PTB-train + Wiki	**130.4**

4 Conclusions

In this paper, we have explored a novel language modelling technique where the long-short term memory neural network language model was combined with the skip-gram method of training of continuous distributed word representations.

We empirically evaluated the proposed design against the conventional LSTM model with perplexity on Penn Treebank corpus. The experiments revealed that the combination of LSTM and skip-gram word vectors trained on huge out-of-domain data from Wikipedia notable outperforms baseline models.

The advantages of the LSTM&skip solution are that we are able to improve performance of a neural network model using a words vectors trained on an out-domain text. Secondly, the input layer of LSTM language model is fixed and allows eventually adding a new unseen word (for LSTM model) to the input, if we are able to compute its continuous distributed representation. However, the output vocabulary is fixed as in the conventional model. Another difficulty is that the learning must be done in two phases.

These model properties should be considered in our next experiments, which we would like to lead to even more general language model, with not fixed input and output dictionary and with an ability to learn new words and connections.

Acknowledgements. This research was supported by the Ministry of Culture Czech Republic, project No.DF12P01OVV022.

References

1. Mikolov, T., Kombrink, S., Deoras, A., Burget, L., Černocký, J.: RNNLM - Recurrent Neural Network Language Modeling Toolkit (2011)
2. Mikolov, T., Chen, K., Corrado, G., Dean, J.: Efficient Estimation of Word Representations in Vector Space. In: Proceedings of Workshop at ICLR (2013)
3. Sundermeyer, M., Schlüter, R., Ney, H.: LSTM Neural Networks for Language Modeling. In: INTERSPEECH (2012)
4. Hochreiter, S., Schmidhuber, J.: Long Short-term Memory. Neural Computation 9(8), 1735–1780 (1997)
5. Graves, A.: Generating sequences with recurrent neural networks. arXiv:1308.0850 (cs.NE) (2013)

6. Bengio, Y., Ducharme, R., Vincent, P., Janvin, C.: A neural probabilistic language model. J. Mach. Learn. Res. 3, 1137–1155 (2003)
7. Mnih, A., Hinton, G.: Three new graphical models for statistical language modelling. In: Proceedings of the 24th International Conference on Machine Learning, ICML 2007, pp. 641–648. ACM, New York (2007)
8. Mikolov, T., Yih, W., Zweig, G.: Linguistic Regularities in Continuous Space Word Representations. In: Proceedings of NAACL HLT (2013)
9. Mikolov, T., Sutskever, I., Chen, K., Corrado, G., Dean, J.: Distributed Representations of Words and Phrases and their Compositionality. In: Proceedings of NIPS (2013)
10. Soutner, D., Müller, L.: Application of LSTM Neural Networks in Language Modelling. In: Habernal, I. (ed.) TSD 2013. LNCS, vol. 8082, pp. 105–112. Springer, Heidelberg (2013)
11. Charniak, E., et al.: BLLIP 1987-89 WSJ Corpus Release 1, Linguistic Data Consortium, Philadelphia (2000)
12. Wikimedia Foundation: Wikipedia, The Free Encyclopedia (2009), http://en.wikipedia.org
13. Garofalo, J., et al.: CSR-I (WSJ0) Complete, Linguistic Data Consortium, Philadelphia (2007)
14. Kneser, R., Ney, H.: Improved backing-off for M-gram language modeling. Acoustics, Speech, and Signal Processing 1, 181 (1995)

NERC-fr: Supervised Named Entity Recognition for French

Andoni Azpeitia[1], Montse Cuadros[1], Seán Gaines[1], and German Rigau[2]

[1] HSLT, IP Department - Vicomtech-IK4
Donostia-San Sebastián, Spain
{aazpeitia,mcuadros,sgaines}@vicomtech.org
http://www.vicomtech.org/en
[2] IXA NLP Group - UPV/EHU, Donostia-San Sebastián, Spain
german.rigau@ehu.es
http://ixa.si.ehu.es/Ixa

Abstract. Currently there are only few available language resources for French. Additionally there is a lack of available language models for for tasks such as Named Entity Recognition and Classification (NERC) which makes difficult building natural language processing systems for this language. This paper presents a new publicly available supervised Apache OpenNLP NERC model that has been trained and tested under a maximum entropy approach. This new model achieves state of the art results for French when compared with another systems. Finally we have also extended Apache OpenNLP libraries to support part-of-speech feature extraction component which has been used for our experiments.

1 Introduction

The Named Entity Recognition and Classification (NERC) task consists of detecting lexical units that refer to specific entities, in a sequence of words, and determining which kind of entity the unit refers to (e.g. person, location, organisation). NERC consists of two steps that can be approached either in sequence or in parallel. The first step is the detection of named entities in a text, while the second is the correct classification of the detected named entities, using a set of predefined categories.

Currently there are fewer publicly available NLP tools for French than, for instance, English or Spanish. Our first goal is to overcome this problem building a state of art NLP component for this language, and make it available to the research community[1]. One of the first problems faced when building a NERC is the lack of publicly available labelled datasets. For our task, we have used the ESTER corpus, a NE annotated corpus with 1.2 million words[2]. Section 3 provides a detailed description of the corpus.

Among the possible methods to determine and classify named-entities, we opted for a machine learning (ML) approach, as it provides language-independent core

[1] Both the NERC model for OpenNLP and the extensions made to the OpenNLP library will be made available here:
http://opener-project.github.io/documentation/ner.html

[2] http://catalog.elra.info/product_info.php?products_id=999

P. Sojka et al. (Eds.): TSD 2014, LNAI 8655, pp. 158–165, 2014.

algorithms for the development of state of the art language processing modules. The system we present extends a supervised approach to NERC implemented in the Apache OpenNLP library[3].

Previous research on support vector machines has shown that part of speech (PoS) information improves the overall performance of NERC systems [6]. Although OpenNLP integrates various feature extractors, it does not yet support PoS information extraction. We thus extended the OpenNLP library to include PoS feature extraction and integrated this information into our NERC system for French.

This paper is organised as follows: Section 2 presents related work; section 3 describes the dataset used to train system; section 4 contains the description of the system; section 5 presents evaluation results; section 6 describes and evaluates the system's performance; finally, section 7 draws conclusions from this work and presents suggestions for future research.

2 Related Work

A remarkable amount of work has been carried on in recent years on named entity recognition and classification. The main approaches to NERC can be categorised into Knowledge-based, Supervised, Semi-supervised and Unsupervised. Briefly, **Knowledge-based systems** were developed in the early stages of NERC research. Methods in this approach are essentially based on finite state machines and rule sets (see, e.g. [2,8]). Research has continued along these lines, with e.g., [5] reaching an F1 measure of approximately 70% for the recognition task. These tools are however costly to develop, as they require the development of knowledge-based resources which are difficult to port to other languages.

Supervised learning is currently the most widely used approach for the NERC task [9]. Different techniques have been used such as Hidden Markov Models [3], Decision Trees [14] and Maximum Entropy Models [4]. The main problem with Supervised Learning techniques is that a large amount of tagged data is needed to implement an effective system, and the accuracy of the models in a given domain is dependent on the training corpus. For this reason, some more recent references have experimented with the inclusion of knowledge-based techniques [11,12].

With **Semi-supervised systems**, a first classifier is learned, which is then improved using unlabelled data. The most effective systems are based on linguistic features. In recent approaches for French, the use of large lexical resources like Wikipedia are commonly used [13,10].

In 2002 and 2003[4] very well known evaluation champaings called CONLL were one of the major research efforts to join community and work towards an improvement of the NERC performance.

[3] `http://opennlp.apache.org`

[4] `http://www.cnts.ua.ac.be/conll2002/ner/`, `http://www.cnts.ua.ac.be/conll2003/ner/`

3 Datasets

To create the Apache opennlp French NERC model, we used the ESTER corpus. This corpus is based on more than 1700 hours of Broadcast News data (from 6 French radio channels), out of which 100 hours were manually transcribed. The corpus contains 1.2 millions of words for a vocabulary of 37,000 words and 74,082 named entities (15,152 unique NEs), labelled with a tagset of about 30 categories folded into 8 main types: persons (pers), locations (loc), organisations (org), geo-socio-political groups (gsp), amounts (amount), time (time), products (prod) and facilities (fac).

We divided the data into 3 sets: a training set (77.8% of the total corpus), a development set (9.7%) and a test set (12.5%). There is a 6 month time difference in the occurrence between data in the training/development set and in the test set: the training set contains Broadcast News ranging from 2002 to December 2003 and the test corpus was recorded in October 2004. The test set also contained data collected from 2 news radio channels which were not among the sources of the training data. One of the main characteristics of the ESTER corpus is the size of the NE tagset and the high ambiguity rate amongst NE categories (e.g. administrative regions and geographical locations): 40% of the matched corpus entities in the training corpus, and 32% of the unmatched ones, are ambiguous [7].

In order to use the ESTER corpus to train and test the models, we first converted it to the OpenNLP format placing each sentence on a separate line, sentence initial words were capitalised, and all sentences were tokenised. The two tagsets being different, we reduced the 30 named entity categories defined in ESTER corpus into the 6 categories needed for the task, along the lines described as follows:

- The geo-social-political tags were divided into three subcategories: *gsp.pers*, *gsp.loc* and *gsp.org*. These subcategories were then placed under *person*, *location* and *organisation*, respectively.
- The *amount* category, and its *amount.cur* subcategory, were categorised under *money*.
- *product* named entities were not used in our system.
- *person*, *location*, *organization*, *time* and *date* types were maintained as is.

4 System Description

The system developed for the experiment described in this paper implements a supervised approach. We used a maximum entropy framework and a classifier that detects each NE candidate given certain features. The system's properties are described in the next sub-sections.

4.1 OpenNLP Library Extension

To build the NERC model, we used the Apache OpenNLP library, which is a machine learning based toolkit for natural language processing. It supports the most common NLP tasks, and in particular provides a maximum entropy-based framework for NERC.

As previously mentioned, [6] demonstrated the usefulness of PoS information for NERC, when training models based on support vector machines. For that reason, we decided to extend the OpenNLP library for PoS feature extraction.

To extract PoS information, we trained a PoS tagger using annotated data in the French Treebank [1]. The tagger is MaxEnt-based and was developed using OpenNLP's functionality.

4.2 Estimating Parameters

Although OpenNLP provides default parameters for feature selection, we estimated optimal parameters in order to achieve better results. Parameters were tested along the lines given as follows:

1. **Sentence Boundaries:** the impact of beginning and end of sentence features was measured.
2. **Neighbouring tokens:** contextual features were extracted to measure the impact of neighbouring tokens. We defined a unit w as a token, a token pattern and token class, estimating optimal values for the i and j parameters within different w_{-i}, w, w_{+j} windows.
3. **Bigram window:** we extracted bigrams of neighbouring words, also estimating optimal window parameters for bigram sequences.
4. **Prefixes and suffixes window:** for any word w, we extracted i prefixes and suffixes on the left, and j prefixes and suffixes on the right. The maximum prefix and suffix char length was set to 4.
5. ***charngram* length parameter:** *charngram* concerns the features covering the minimum and maximum length of entities character-level *ngrams*.
6. **PoS window:** the system extracts features from the neighbouring PoS tags of the w-th word. We will provide results with and without this step in what follows.
7. **Cutoff:** cutoff specifies the minimal number of times a feature must be seen to be selected.
8. **Number of iterations:** this part estimates the number of iterations for the Generalised Iterative Scaling (GIS) procedure.

4.3 Named Entity Categories

To create our NERC model, we opted to recognise and classify six different named entity categories: **person**, **location**, **organisation**, **date**, **time** and **money**.

5 Experiments

To evaluate the performance of the NERC model at different stages, we used the development corpus included in the ESTER corpus and the three standard measures of $precision, recall$ and F_1.

As described in section 4, we made use of the following features to improve our NERC model: default parameters, sentence boundaries, neighbouring tokens, bigram window, prefixes and suffixes window, *charngram* length parameter, PoS window, cutoff and number of iterations. Table 1 shows the results with and without the PoS feature.[5]

Table 1. Best performing features, with and without PoS

Steps	Without PoS			With PoS		
	Precision	*Recall*	F_1-*score*	*Precision*	*Recall*	F_1-*score*
1. Default parameters	90,47	82,87	86,5	90,47	82,87	86,5
2. Sentence boundaries	90,47	82,87	86,5	90,47	82,87	86,5
3. Neighbouring tokens	90,8	83,35	86,91	90,8	83,35	86,91
4. Bigram window	90,91	83,68	87,14	90,91	83,68	87,14
5. Prefixes and suffixes window	91,2	85,32	88,16	91,2	85,32	88,16
6. charngram	91,54	85,09	88,2	91,54	85,09	88,2
7. Pos window	-	-	-	91,35	84,6	87,85
8. Cutoff	91,54	85,09	88,2	91,35	84,6	87,85
9. Iterations	91,45	85,23	88,23	91,5	85,39	88,34

The best results were obtained with the parameters below:

- **Sentence Boundaries**: the best sentence boundaries features were sentence start indicators.
- **Neighbouring tokens**: the best window scores for tokens, token classes and token patterns were 2-1, 1-2 and 0-0, respectively.
- **Bigram window**: the best results were achieved with a bigram window of 1-0.
- **Prefixes and suffixes window**: for prefixes and suffixes, the best window sizes were 1-0 and 1-1, respectively.
- ***charngram* length parameter**: *charngram* best length was 6.
- **PoS window**: 0-3 was the best PoS window size.
- **Cutoff**: both with and without PoS features, the optimal cutoff was 5.
- **Number of iterations**: without PoS features 380 iterations were needed to achieve the best scores; including PoS features, 260 iterations were needed.

As shown in table 1, the best performances in terms of F_1-*measure* were 88.34% and 88.23%, with and without PoS features, respectively. The most performing features for our NERC model are Bigram window and the Prefixes and Suffixes window. Both show an improvement of the performance in terms of F-score.

Table 2 shows the results obtained on the ESTER corpus development and test datasets. The 6 months difference and sources for the data that made up both sets, described in section 3, might be a reason for the differences in scores that were observed.

Table 3 presents the evaluation results for each entity category in the test set. Although the overall detection was good, the low recall value for the *money* entity category has a negative impact on the final results.

As an overall conclusion, despite PoS improves our overall performance it is not stadistically significant in comparison with the results without PoS.

[5] *Default parameters* denotes the default OpenNLP parameters.

Table 2. Performance results on the ESTER corpus test and dev sets

Dataset	Without PoS			With PoS		
	Precision	*Recall*	F_1*-score*	*Precision*	*Recall*	F_1*-score*
Development	91.45	85.23	88.23	91.5	85.39	88.34
Test	86.15	75.69	80.59	86.2	75.85	80.69

Table 3. Evaluation results per entity category

Categories	Without PoS			With PoS		
	Precision	*Recall*	F_1*-score*	*Precision*	*Recall*	F_1*-score*
Location	88.58	86.04	87.29	88.56	85.84	87.18
Person	88.71	80.5	84.41	88.59	80.84	84.54
Time	89.54	81.07	85.09	89.61	81.66	85.45
Date	84.01	74.48	78.96	84.03	74.59	79.03
Organization	76.53	54.99	64.00	76.79	55.42	64.38
Money	70.31	24.59	36.44	74.19	25.14	37.55
Total	86.16	75.69	80.59	86.20	75.85	80.69

6 Discussion

For languages such as Spanish or English, many resources are available for NERC. For French, few systems are readily available due to the absence of publicly available entity annotated datasets which could be used to create NERC supervised models. The two NLP tools described below perform NER for French and were used to compare systems performance:

Unitex CasEN[6] uses lexical resources and local description of patterns, transducers which act on text insertions, deletions and substitutions. The CasEN tool requires a certain knowledge of the language, as the NERC component is based on regular expressions which would need to be adapted for other languages.

LingPipe[7] is a toolkit for text processing which has been used for French in HLT 2005 [7]. In this paper, an automatic speech recognition system (ASR) for NERC tasks is presented, which uses a text-based NERC model trained with the LingPipe tool. To train and evaluate their NERC model, the ESTER corpus was used, which allows for a direct comparison with our system. LingPipe comes with a license that makes the code freely available for research purposes, with additional constraints for its integration in a commercial product. The NERC component we have developed will be distributed under the Apache License v.2 and will thus be useful for both research and industrial applications.

Table 4 shows the performance of our model against CasEN and Lingpipe. As can be seen, our model showed better performance overall, although our results are close to those obtained with LingPipe.

[6] `http://tln.li.univ-tours.fr/Tln_CasEN.html`

[7] `http://alias-i.com/lingpipe/`

Table 4. NERC tools comparison

Systems	*Precision*	*Recall*	F_1-*score*
CasEN	68.86	40.63	50.99
LingPipe	-	-	79.00
NERC-fr	86.2	75.85	80.69

7 Conclusion and Future Work

In this paper, we presented a module for Named Entity Recognition and Classification in French. This tool has been developed using OpenNLP, extended to extract PoS information features. The component was used to train several NERC models, using the ESTER corpus. The optimal model built under our approach showed slightly better performance than comparable tools, without requiring the development of language-dependent resources beyond annotated corpora. Despite the simplicity of the approach, the system showed better performance than the CasEN rule-based system and slightly better results than a supervised system developed with LingPipe.

When compared to NERC in other languages, the results for French are lower on average, in terms of precision and recall, which means that there is still a lot of research to be carried in this field and in this language.

In future work, we plan to extend the system with additional linguistic features, for example integrating syntactic information and evaluate their impact on the performance of the NERC module we described in this paper. Additionally we plan to train well-known tools such as LBJ of Illinois[8] or Stanford Core NLP[9] with ESTER corpus to evaluate the performance against OpenNLP.

Acknowledgements. This work is part of the OpeNER project, funded by the European Commission 7th Framework Programme (FP7), under grant agreement no 296451. The authors would like to thank Thierry Etchegoyhen for proof reading the paper and providing suggestions for improvement.

References

1. Abeillé, A., Clément, L., Toussenel, F.: Building a treebank for french. In: Treebanks, pp. 165–187. Springer (2003)
2. Appelt, D.E., Hobbs, J.R., Bear, J., Israel, D., Kameyama, M., Martin, D., Myers, K., Tyson, M.: Sri international fastus system: Muc-6 test results and analysis. In: Proceedings of the 6th Conference on Message Understanding, pp. 237–248. Association for Computational Linguistics (1995)
3. Bikel, D.M., Miller, S., Schwartz, R., Weischedel, R.: Nymble: A high performance learning name-finder. In: Proceedings of the 5th Conference on Applied Natural Language Processing, ANLP, Washington DC (1997)

[8] http://cogcomp.cs.illinois.edu/page/software_view/11

[9] http://www-nlp.stanford.edu/software/index.shtml

4. Borthwick, A.: A maximum entropy approach to named entity recognition. Ph.D. thesis, New York University (1999)
5. Budi, I., Bressan, S.: Association rules mining for name entity recognition (2003)
6. Ekbal, A., Bandyopadhyay, S.: Named entity recognition using support vector machine: A language independent approach. International Journal of Computer Systems Science & Engineering 4(2) (2008)
7. Favre, B., Béchet, F., Nocéra, P.: Robust named entity extraction from large spoken archives. In: Proceedings of HLT-EMNLP, pp. 491–498. Association for Computational Linguistics (2005)
8. Mikheev, A., Moens, M., Grover, C.: Named entity recognition without gazetteers. In: Proceedings of the 9th EACL, pp. 1–8 (1999)
9. Nadeau, D., Sekine, S.: A survey of named entity recognition and classification. Lingvisticae Investigationes 30(1), 3–26 (2007)
10. Nothman, J., Ringland, N., Radford, W., Murphy, T., Curran, J.R.: Learning multilingual named entity recognition from wikipedia. Artificial Intelligence 194, 151–175 (2013)
11. Petasis, G., Vichot, F., Wolinski, F., Paliouras, G., Karkaletsis, V., Spyropoulos, C.D.: Using machine learning to maintain rule-based named-entity recognition and classification systems. In: Proceedings of the 39th Annual Meeting on Association for Computational Linguistics, pp. 426–433. Association for Computational Linguistics (2001)
12. Poibeau, T.: The multilingual named entity recognition framework. In: Proceedings of the tenth conference on European Chapter of the Association for Computational Linguistics, vol. 2, pp. 155–158. Association for Computational Linguistics (2003)
13. Richman, A.E., Schone, P.: Mining wiki resources for multilingual named entity recognition. In: ACL, pp. 1–9 (2008)
14. Sekine, S.: Nyu: Description of the japanese NE system used for met-2. In: Proc. Message Understanding Conference (1998)

Semantic Classes and Relevant Domains on WSD

Rubén Izquierdo[1], Sonia Vázquez[2], and Andrés Montoyo[2,*]

[1] Computational Lexicology and Terminology Lab
Vrije University of Amsterdam, The Netherlands
ruben.izquierdobevia@vu.nl

[2] Department of Software and Computing Systems, University of Alicante, Spain
{svazquez,montoyo}@dlsi.ua.es

Abstract. Language ambiguities are a problem in various fields. For example, in Machine Translation the major cause of errors is ambiguity. Moreover, ambiguous words can be confusing for Information Extraction algorithms. Our purpose in this work is to provide a new approach to solve semantic ambiguities by dealing with the problem of the fine granularity of sense inventories. Our goal is to replace word senses with Semantic Classes that share properties, features and meanings. Also another semantic resources, Relevant Domains, is used to extract extract semantic information and enrich the process. The results obtained are evaluated in the Evaluation Exercises for the Semantic Analysis of Text (SensEval) framework.

1 Introduction

WSD is considered to be one of the most difficult problems in Artificial Intelligence and is also considered as an intermediate task within the Natural Language Processing (NLP) field [16]. Rather than an isolated problem, WSD often forms part of other NLP task. One of the most successful approaches in the last years has been Supervised Machine Learning, and in concrete *supervised learning from examples*. In this approach, Machine Learning models are induced from semantic annotated examples [8] usually using WordNet [3] as sense repository. Despite this wide use of WordNet, it has been criticized because the sense distinctions it provides are often too fine-grained for higher applications such as Machine Translation or Information Extraction, and represent too narrow distinctions that are not informative for NLP [9].

In this work we present a new supervised approach based on Semantic Classes in order to solve the problem of semantic ambiguity. Our goal is to provide a method with which obtain a hierarchy of semantic labels that represents a set of equivalent senses and use them into a supervised system. One of the main purposes of this work is to tackle the problems caused by semantic resources that are too fine-grained using another type of semantic indicators instead.

* This research work has been partially funded by the University of Alicante, Generalitat Valenciana, Spanish Government and the European Commission through the projects, "Tratamiento inteligente de la informacion para la ayuda a la toma de decisiones" (GRE12-44), ATTOS (TIN2012-38536-C03-03), LEGOLANG (TIN2012-31224), SAM (FP7-611312), FIRST (FP7-287607) and ACOMP/2013/067.

P. Sojka et al. (Eds.): TSD 2014, LNAI 8655, pp. 166–172, 2014.

Generally speaking, a semantic class is a concept that groups sub-concepts and word senses that share properties, features and meanings, such as VEHICLE with sub-concepts such as *car* or *helicopter* or BUILDING with *school* or *church* as sub-concepts. We will explore the performance of a WSD system when different types of semantic classes are used to build the semantic features. Firstly, we propose a basic system using traditional features, mainly contextual features. This system will be used as a reference to compare its results with other experiments. In order to improve this first approach, different semantic classes will be employed to generate the semantic features and the performance of this extended system will be analyzed.

2 State of the Art

Many researchers have focused their efforts on developing efficient WSD systems. However, WSD has been a challenging problem since [15] first introduced the question of ambiguity in the late 1950s, and it continues to be so. As last SensEval and SemEval[1] competitions have shown, there has been a very small improvement in the performance of the WSD systems.

Various approaches have been followed to tackle the problem of the excessively fine granularity of word senses. Some research has been focused on deriving different word–sense groupings to overcome the fine-grained description of word senses in WordNet [1,11,12]. However, these approaches only attempt to group senses of the same word, thus producing a coarse-grained repository of word senses, but only taking advantage of the reduction of the polysemy.

Like our approach, some research has been focused on using a predefined set of semantic classes for learning class–based classifiers for WSD [14,2]. Most of these approaches use the lexicographic files of WordNet (also known as SuperSenses (SS) as coarse-grained sense distinctions, but less attention has been paid to learning class-based classifiers from other available sense groupings.

The use of semantic classes rather than word senses in the development of a WSD system can also be found in unsupervised approaches, for instance in [7,4] which study the relations between meanings and domains.

3 Semantic Resources

In this section we explain the semantic resources used in our work. These can be divided into two groups: semantic classes and domains. A **semantic class** is a concept that subsumes other concepts that share common characteristics. The link between concepts and subconcepts is modelled by the relation “is-a”, for example a *car* “is-a” *vehicle*, (*vehicle* is a semantic class). A **domain** is a specific area of interest for humans. Each domain has a set of related words known as a terminology. The link between these words and their domain is modelled by the relation “belongs-to". For example, the word *soccer* “belongs-to" the SPORTS domain.

[1] http://www.senseval.org

There are different **Semantic Classes** repositories, each one groups the word senses in a different way, resulting in sets with a different granularity. We have used two types of Semantic Classes derived from WordNet: *Basic Level Concepts* (BLC) and *SuperSenses*. BLC are concepts that result from the compromise between representing as many concepts and features as possible. We have used the BLC sets created by [5]. SuperSenses are the name given to the WordNet Lexicographer Files within the framework of WSD. WordNet synsets are organized into 45 SuperSenses, based on syntactic categories (nouns, verbs, adjectives and adverbs) and logical groupings, such as person, phenomenon, feeling, location, etc.

On the other hand, we have used *WordNet Domains* (WND) and *Relevant Domains* (RD) as **domains** resources. WND [6] is a hierarchy of 165 domain labels which have been used to tag all WN synsets. This set of labels is organized into a taxonomy by following the Dewey Decimal Classification System[2].

WND is used as a basis to develop the resource called Relevant Domains (RD). The aim of this resource is to collect information from WordNet glosses in order to obtain new relations among words and domains. This resource has been previously evaluated in a WSD knowledge-based system [13]. The idea is to use the Mutual Information formula to calculate association ratios between words (or senses) and domains.

4 Semantic Class-Based WSD System

In order to use the resources and features previously explained, we have followed a supervised Machine Learning approach to develop a set of class-based WSD taggers. Our systems use an implementation of a Support Vector Machine[3] algorithm to train the classifiers (one per class) in semantic annotated corpora in order to acquire positive and negative examples of each class and in the definition of a set of features to represent these examples. The system decides and selects from among the possible semantic classes defined for a word. In the sense approach, one classifier is generated for each word sense, and the classifiers choose between the possible senses of the word. The examples used to train a single classifier for a concrete word are all the examples of this word sense. In the class-based approach, one classifier is generated for each semantic class. Thus, when we wish to label a word, our system obtains the set of possible semantic classes for this word, and then launches each of the semantic classifiers related to these semantic categories. The most likely category is then selected for the word.

We believe that this semantic class-based approach has several advantages. First, semantic classes reduce the average polysemy degree of words (some word senses are grouped together within the same semantic class). Moreover, the problem of the acquisition bottleneck in supervised machine learning algorithms is attenuated, because the number of examples for each classifier is increased.

---

[2] It was selected because it provides a good coverage, is freely available and widely used to classify written data. `http://www.oclc.org/dewey`

[3] SVMLight.

SemCor [10] was used for training while the corpora from the English all-words tasks of SensEval-2 (SE2)[4], SensEval-3 (SE3)[5] and SemEval–1(SEM1)[6] were used for testing. We also considered the SemEval-2007 coarse-grained task corpus for testing, but this dataset was discarded because this corpus is also annotated with clusters of word senses which are not comparable with Semantic Classes.

We have defined a set of features in order to represent the examples according to previous works in WSD and the nature of class-based WSD. Features widely used in literature have been selected. In particular, our systems use the following **basic features. Word-forms and lemmas** in a window of 10 words around the target word **PoS**: the concatenation of the preceding/following three/five PoS **bigrams and trigrams** formed by lemmas and word-forms and obtained in a window of 5 words. We use all tokens regardless of their PoS to build bi/trigrams. The target word is replaced with *X* in these features in order to increase their generalization for the semantic classifiers. We have also defined a set of **Semantic Features** with which explode different semantic resources in order to enrich the set of basic features: (1) the most frequent semantic class over SemCor for the target word, and the semantic classes of monosemic words around the target word.

Several types of semantic classes have been considered during the creation of these features, in particular, WordNet Domains, BLC–20[7] and Relevant Domains. We have selected these sets of semantic classes because they have been created by following different approaches and each one therefore has specific characteristics and a certain level of abstraction.

4.1 Experiments

We have defined several experiments focused on analyzing the influence of the semantic features in our semantic class-based system. In all the experiments, the classifiers have been built using BLC–20 semantic classes. In other words, one classifier is trained for each of the semantic classes within the BCL–20 set (in this set there are 649 semantic classes for nouns and 616 for verbs). Our classifiers therefore assign the proper semantic class to each ambiguos noun or verb in a text, as defined in the semantic-based WSD approach. Since the traditional evaluation in international WSD competitions (mainly SensEval and SemEval) is performed at a word sense level, we need to adapt the output of our system, and transform each semantic class assigned to a word into a word sense. We follow the same approach used in previous related work: our classifiers propose a Semantic Class, and the first sense for the word according to WordNet that belongs to that class is selected.

The evaluation was performed on SE2, SE3 and SEM1 All-Words tasks. We computed the standard measures for the evaluation of WSD systems (precision, recall

[4] `http://www.sle.sharp.co.uk/senseval2`

[5] `http://www.senseval.org/senseval3`

[6] `http://nlp.cs.swarthmore.edu/semeval`

[7] The value 20 refers to the threshold of minimum number of synsets that a possible BLC must subsume to be considered as a proper BLC. These BLC sets were built using all kind of relations.

and $F_{\beta=1}$ of each experiment on the three corpora. We have also included the baseline based on the most frequent sense ($F_{\beta=1}$ measure), and the result of the best system of the oficial participants in the task (also the $F_{\beta=1}$).

Table 1 contains the F1 value of our classifiers, when only one semantic class was used to build the semantic features (the type of this semantic class is represented in the header of the columns where RelDomW is Basic features and semantic features built with Relevant Domains based on words and RelDomS is Basic features and semantic features built with Relevant Domain based on word senses)

Table 1. F1 value over test corpora

Corpus	Basic	WND	BLC20	RelDomW	RelDomS	Baseline	Best
SV2	0.667	0.665	0.658	0.627	0.626	0.570	0.685
SV3	0.641	0.644	0.642	0.637	0.638	0.620	0.652
SEM1	0.511	0.511	0.543	0.511	0.521	0.514	0.591

It will first be observed that in three cases with different test corpus, the most frequent baselines are exceeded. As stated previously, these kinds of baselines are usually very hard to attain in WSD tasks. Moreover, despite the fact that our classifiers do not attain the result of the best participating systems in each task (SE2, SE3 and SEM1), in some cases our classifiers show a similar performance to the best system. In general, the results when semantic features are used are higher than if no semantic information is included, except in the case of SE2, in which the best result is obtained by the *Basic* experiment. This could be explained if we look at the domain of the evaluation set from SE2, which was built from three different sources, a mystery novel, a medical report and a paper on children education.

Furthermore, upon comparing *RelDomW* and *RelDomS* it will be noted that the performance is similar. This indicates that the statistic information provided by words, without considering senses, is sufficient to model the domains of a certain context.

Table 2 shows the performance of the system when pairs of semantic classes are combined to generate the semantic features. The aim of these experiments was to combine different sources of semantic information, in order to analyze whether they can provide complementary information.

Table 2. F1 value over test corpora

Corpus	Basic	BLC20+WND	BLC20+RelDomS	WND+SS	Baseline	Best
SV2	0.667	0.658	0.659	0.666	0.570	0.685
SV3	0.641	0.642	0.642	0.641	0.620	0.652
SEM1	0.511	0.541	0.543	0.519	0.514	0.591

On the one hand, we can see that the baselines are again exceeded. On the other hand, the results of these experiments when combining two semantic classes are worse than when only one class is used to build the semantic features. In general, it would appear that the information provided by semantic features can help up to a certain point, and that no combination of semantic classes improves these results. The information provided by different semantic classes therefore appears not to be complementary, and there is a kind of overlapping between the semantic features according to different semantic classes. This is normal if we consider that the main source from which all the semantic classes were derived was the same: WordNet.

5 Conclusions and Discussion

In this work we have presented a supervised approach to WSD using semantic classes rather than word senses. The goal of our proposal is to avoid the problem of the fine granularity of sense inventories. We propose to solve semantic ambiguities and to maintain semantic coherence by grouping together those concepts that share certain properties. This proposal has been evaluated in the SensEval framework and compared with the results of the best systems. According to the results obtained, we have demonstrated that semantic information extracted from resources based on WordNet leads to an improvement in the performance of our WSD system. However, there is a critical point at which the results are not improved and remain steady. One of the reasons for this behaviour is that the information added is redundant, despite the fact that it comes from different sources.

After the evaluation analysis and with regard to the results obtained, we can observe that the performance with the use of semantic attributes works better or worse depending on the evaluation corpora utilized. For instance, the use of WND as semantic attributes provides the best results in SensEval-3. However, BLC classes work better in SemEval. Again, we can observe the influence of domain specific corpora during the WSD process. Related to this question, although the resources should be adapted to a specific domain to improve the results, it is important to highlight that in several cases, the results obtained using semantic attributes built with an automatic process (BLC and Relevant Domains) work similarly or even better than manual attributes (WND). This fact supposes an advantage, since these kinds of attributes could be generated automatically and adapted to a new domain. On the other hand, the manual development of a resource for this purpose would be time consuming. Furthermore, our results are very close to the best systems in each competition, and in all cases exceed the baseline based on the most frequent sense.

References

1. Agirre, E., Calle, O.L.D.L.: Clustering Wordnet word senses. In: Proceedings of RANLP 2003, Borovets, Bulgaria, pp. 121–130 (2003)
2. Ciaramita, M., Altun, Y.: Broad-coverage sense disambiguation and information extraction with a supersense sequence tagger. In: Proceedings of the Conference on Empirical Methods in Natural Language Processing (EMNLP 2006), pp. 594–602. ACL, Sydney (2006)
3. Fellbaum, C. (ed.): WordNet. An Electronic Lexical Database. The MIT Press (1998)
4. Gliozzo, A.M., Strapparava, C., Dagan, I.: Unsupervised and supervised exploitation of semantic domains in lexical disambiguation. Computer Speech & Language 18(3), 275–299 (2004)

5. Izquierdo, R., Suárez, A., Rigau, G.: An empirical study on class-based word sense disambiguation. In: Proceedings of the 12th Conference of the European Chapter of the Association for Computational Linguistics, EACL 2009, pp. 389–397. Association for Computational Linguistics, Stroudsburg (2009)
6. Magnini, B., Cavaglià, G.: Integrating subject field codes into Wordnet. In: Proceedings of LREC, Athens, Greece (2000)
7. Magnini, B., Pezzulo, G., Gliozzo, A.: The role of domain information in word sense disambiguation. Natural Language Engineering 8, 359–373 (2002)
8. Màrquez, L., Escudero, G., Martínez, D., Rigau, G.: Supervised corpus-based methods for WSD. In: Agirre, E., Edmonds, P. (eds.) Word Sense Disambiguation: Algorithms and applications. Text, Speech and Language Technology, vol. 33. Springer (2006)
9. Mccarthy, D.: Relating Wordnet senses for word sense disambiguation. In: Proceedings of the ACL Workshop on Making Sense of Sense, pp. 17–24 (2006)
10. Miller, G., Leacock, C., Tengi, R., Bunker, R.: A Semantic Concordance. In: Proceedings of the ARPA Workshop on Human Language Technology, pp. 303–308 (1993)
11. Navigli, R.: Meaningful clustering of senses helps boost word sense disambiguation performance. In: ACL-44: Proceedings of the 21st International Conference on Computational Linguistics and the 44th Annual Meeting of the Association for Computational Linguistics, pp. 105–112. Association for Computational Linguistics, Morristown (2006)
12. Snow, R., et al.: Learning to merge word senses. In: Proceedings of Joint Conference on Empirical Methods in Natural Language Processing and Computational Natural Language Learning (EMNLP-CoNLL), pp. 1005–1014 (2007)
13. Vázquez, S., Montoyo, A., Rigau, G.: Método de desambiguación lingüistica basada en el recurso semántico dominios relevantes. In: Proceedings of the XIX Congreso de la Sociedad Española para el Procesamiento del Lenguaje Natural (SEPLN 2003), Alcalá de Henares, Spain, pp. 141–149 (2003)
14. Villarejo, L., Màrquez, L., Rigau, G.: Exploring the construction of semantic class classifiers for wsd. In: Proceedings of the 21th Annual Meeting of Sociedad Española Para el Procesamiento del Lenguaje Natural, SEPLN 2005, Granada, Spain, pp. 195–202 (September 2005) ISSN 1136-5948
15. Weaver, W.: Translation. In: Locke, W.N., Booth, A.D. (eds.) Machine Translation of Languages, pp. 15–23. MIT Press (1957)
16. Wilks, Y., Stevenson, M.: The grammar of sense: Is word sense tagging much more than part-of-speech tagging? Tech. rep., CS-96-05, University of Sheffield, UK (1996)

An MLU Estimation Method for Hungarian Transcripts

György Orosz[1,2] and Kinga Mátyus[3]

[1] Pázmány Péter Catholic University, Faculty of Information Technology and Bionics
50/a Práter street, 1083 Budapest, Hungary
[2] MTA-PPKE Hungarian Language Technology Research Group
50/a Práter street, 1083 Budapest, Hungary
oroszgy@itk.ppke.hu
[3] MTA Research Institute for Linguistics
33. Benczúr street, 1068 Budapest
matyus.kinga@nytud.mta.hu

Abstract. Mean length of utterance (MLU) is an important indicator for measuring complexity in child language. A generally employed method for calculating MLU is to use the CLAN toolkit, which includes modules that enable the measurement of utterance length in morphemes. However, these methods are based on rules which are only available for just a few languages not involving Hungarian. Therefore, in order to automatically analyze and measure Hungarian transcripts adequate methods need to be developed. In this paper we describe a new toolkit which is able to estimate MLU counts (in morphemes) while providing morphosyntactic tagging as well. Its components are based on existing resources; however, many of them were adapted to the language of the transcripts. The toolchain performs the annotation task with a high precision and its MLU estimates are correlated with that of human experts.

1 Introduction

Mean length of utterance (MLU) is a metric that has been widely used for measuring linguistic productivity of children for almost a hundred years. Utterance lengths are usually calculated in morphemes for morphologically complex languages, while in the case of analytical languages counting only words is also a feasible solution. Concerning the calculation of MLU in morphemes (MLUm), several tools are available which employ natural language processing methods and could be used to boost this labor-intensive task. Nonetheless, none of them is able to deal with Hungarian transcripts. However, existing resources with slight modifications can be used to estimate MLUm for transcripts of Hungarian child language.

2 Background

Ever since the complexity of child language has been measured, several methods have been developed. While manual counting prevailed for decades, automatic counting tools have been sought for in the past years.

P. Sojka et al. (Eds.): TSD 2014, LNAI 8655, pp. 173–180, 2014.

2.1 Measuring Morphosyntactic Complexity

Several studies (e.g. [3]) showed that MLUm indicates language development for normal children, especially at very early stages. In contrast, MLUw was shown to be highly correlating [7,16] with MLUm in the case of analytical languages such as English or Irish. Therefore, some studies concur that MLUw is a reliable measure as opposed to MLUm, where researchers often need to make ad hoc decisions on what (not) to count [4].

However, Crystal also points out [4] that MLUm is a good way to measure morphologically complex languages (see e.g. [2]). Hungarian is an agglutinative language, thus MLUm can be considered to be a more reliable indicator of language development than MLUw (similarly to Turkish [20]). Moreover, previous studies which measured language development in Hungarian manually [18,23] mostly employed MLUm as a metric.thus we used their work as bases.

In the case of corpora which follow the CHAT guidelines [9], MLU values (including MLUm) can be calculated with the CLAN [8] toolkit. This system is widely used, since it contains components that perform the necessary preprocessing steps. One of its modules is MOR, which is a morphological analyzer specially designed for spoken language corpora. A subsequent component is POST, which does the morphological disambiguation for such texts. With these components, the number of morphemes can be calculated in a corpus of transcribed spoken data in a number of languages. However, they lack rules for Hungarian, thus none of them can be used for processing such transcripts.

2.2 Tagging Spoken Language

One of the pioneers in tagging spoken language was Eeg-Oloffson [6], who used manually annotated transcripts to train a statistical tagger. Others employ and adapt statistics that derive from written language corpora [11,13,15]. Furthermore, building domain specific rules could also lead to satisfactory taggers (e.g. [12]), while combination of such systems with stochastic tools [1] yields effective algorithms as well.

Based on the studies above, a proper morphological annotation system aiming to process transcripts must be able to handle the following types of difficulties: *i*) existence of new morphosyntactic tags which are missing from the tagset of the training data, *ii*) occurrence of tokens with non-standard orthography in the texts, *iii*) the number of words unknown to a statistical tagger are increased compared to written language corpora, *iv*) in the case of stochastic taggers, if probability estimates are derived from a written language training corpus, the model learnt can become non-representative (e.g. the distribution of PoS tags may significantly differ in written and spoken language).

2.3 Tagging (Spoken) Hungarian

Beside Pos tagging, finding the roots of words is an indispensable task for estimating MLUm. There are numerous studies investigating the tagging performance of machine learning methods on Hungarian. Only two taggers are freely available and perform lemmatization as well. These tools are usually trained on the Szeged Corpus [5] (SZC), since it is the only linguistic resource for Hungarian that is manually annotated.

PurePos. [14] is an open-source full morphological disambiguator system which incorporates the Humor [17] morphological analyzer (MA). The tool is based on statistical trigram-tagging algorithms, but it is extended to employ rule-based components effectively.

magyarlanc. [24] is a freely available language processing chain for parsing Hungarian. Its morphological tagging component is based on the Stanford tagger [21] and incorporates a MA which relies on morphdb.hu [22].

We are not aware of any research investigating the tagging of spoken Hungarian. Moreover, there is no study aiming to calculate MLUm for Hungarian transcripts automatically. Therefore this article investigates methods for analyzing transcripts of Hungairan child speech.

3 The HUKILC Corpus

The contemporary Hungarian Kindergarten Language Corpus (HUKILC) [10] was selected as a base of our research, since there were no morphosyntactically annotated transcripts available. This corpus has been compiled predominantly for child language variation studies. It contains 62 interviews with 4.5–5.5 year-old kindergarten children from Budapest, recorded in spring 2012. The interviews are 20–30 minutes long, and consist of different types of story-telling tasks. Its transcription was carried out using the Child Language Data Exchange System (CHILDES) [9], following its guidelines. The corpus consists of about 39,000 utterances with 140,000 words.

As for morphological annotation of the corpus, general tagging principles were established first. We chose the morphosyntactic labels and lemmata of the Humor analyzer to represent morphological analyses. Next, an annotation manual was developed for human annotators to guide their work in the morphological disambiguation of the corpus. Finally, 6 interviews with about 1,000 utterances were labelled manually. This gold standard corpus was split into two sets of equal sizes: a development and a test set containing 509 and 449 lines respectively.

4 Tagging the Transcripts

The morphological tagging algorithm employed here is a hybrid one, composed of a morphological analyzer, a stochastic tagger tool and domain-specific disambiguation rules as well. Since the tagset of Humor was chosen to be used for the annotation, a plausible solution was to employ this analyzer. Further on, PurePos was used to disambiguate between the morphological tags, and Szeged Corpus was employed as a training corpus for the tagger.

In order to apply a MA prepared to analyze written texts, its analyses had to be adjusted for the transcripts. Thus, rules adapting its analyses – based on regular expressions and domain-specific word lists – were developed. Their formulation could be done with high confidence, since most of the transcripts contained controlled conversation covering only a few topics.

As a first step, morphological analyses of about 40 words typical of spoken language were created manually. These tokens were mostly interjections not used in written language (such as *hűha* 'wow'), while some adverbs were regarded as utterance words[1] in the corpus (e.g. *komolyan* 'seriously'). Furthermore, those tokens that are written in one word in transcripts but are spelled as two words in formal texts were also added to the lexicon. An example is *légyszíves* 'please' which is written formally as *légy szíves*. Finally, diminutive analyses were also provided where it was necessary. E.g. *kutyus* 'doggy' was also analysed as N.DIM with the lemma *kutya* 'dog' beside the old label N and the *kutyus* 'doggy' root. This process was carried out by investigating the lemmata produced by Humor: if the deletion of the derivational affix resulted in a root that was enumerated in a domain-specific list, a new diminutive analysis was created as well.

Concerning the disambiguation process, PurePos was extended with rules in order to customize its knowledge to the target domain. First, the tagger was forced to assign diminutive analyses when a related label had previously been selected by the disambiguator. Then further enhancements were carried out by investigating the common mistakes of the chain on the development dataset.

A frequent error of the chain was the mistagging of *akkor* 'when' and *azért* 'in order to'. These words are pronouns and can be categorized as either adverbial, noun phrase level or demonstrative ones, and can also behave as pronomial adjectives. Generally, when *akkor* is followed by *amikor* 'when' (as in *Akkor érkezett meg, amikor mentem* 'He arrived, when I left') and when *azért* is followed by *mert* 'because' (as in the sentence *Azért eszik, mert éhes* 'He eats, because he is hungry') these pronouns are demonstrative ones. Furthermore, such co-occurances are more frequent in the transcripts than in the Szeged Corpus, since they are usually used for reasoning or telling a story. As these long-term dependencies could not be learnt by the trigram tagger used, rules were employed to tag these tokens correctly.

The next issue was the case of the word *utána* 'afterwards, then; after him/her/it'. It can either be used as an adverb of time (as in the sentence *Utána elindultunk* 'Then we left') and as a postpositional phrase meaning 'after him/her/it, following him/her/it' (as in *Elindultunk utána* 'We went after him'). The former usage is much more common in spoken language: when this word is directly followed by conjunctions such as *meg* 'and' or *pedig* 'however', it is always an adverb. Therefore *utána* was tagged as an adverb in the transcripts when it is followed by one of these trigger words.

The last rule introduced deals with *meg*, which may function as a verbal prefix or as a conjunction. Moreover, it is commonly used as an expletive in spoken language. Therefore, the conjunctive label was assigned to the word when there was not any verb in its two token window.

5 MLU Estimation

As a first step, general principles of counting morphemes were established. This was based on the work of Brown [3], Retherford [19], Wéber [23] and Réger [18], with

[1] Annotation schemes for Hungarian distinguish utterance words and interjections. An utterance word forms a sentence or an utterance alone by interrupting or managing the communication. In contrast, interjections are either onomatopoeic or used to indicate emotions.

some necessary modifications. The basic principles were: *1*) only meaningful words were analyzed, thus fillers (filled pauses such as *ööö* 'er'), punctuation marks and repetitions are not counted in the utterances; *2*) phatic expressions (e.g. *igen* 'yes, mhm') serving to maintain communication and not conveying meaning were omitted; *3*) inflectional suffixes and lemmata were each counted as one unit; *4*) derivational morphemes (including diminutives) were not counted as separate ones, *5*) reciprocal and indefinite pronouns (e.g. *minden#ki* 'everybody') and compound words (such as *kosár#labda* 'basketball') were counted as one morpheme.

In a language with such a rich derivational system as Hungarian, it is often very complicated to determine the lemmata. This is even more difficult in our case, since no common methodology exists to determine the boundary of productivity in child language. Following the work of Brown [3], proper names (such as *Nagy Béla*, *Sári néni* 'Miss Sári') and lexicalized expressions (e.g. *Jó napot* 'Good morning'), which are frequently used in speech, were also considered as one unit. Their identification was based on capitalization rules and a domain specific list.

As for the automatization of rules, they were implemented relying on the morphological annotation of the corpus. First, each item on the list of fillers was eliminated. Afterwards, tagged words known to the MA were split into morphemes by the Humor analyzer. If more than one analysis was created for a word, the least complex one was chosen, since in the majority of the cases the analyses only differed in the number of derivative tags and compound markers (which we previously decided not to count). As the labels of the annotation scheme were composed of morphemic properties, the estimation for unknown words could be based on the morphosyntactic labels. This calculation was carried out by counting only the inflection markers in the guessed tags.

6 Evaluation

First of all, the morphosyntactic tagging performance of the system was investigated. For this, we calculated its accuracy following the work of Orosz et al. [14]. Therefore, full analyses – containing both the lemmata and the tag – were compared to the gold standard data, not counting punctuation marks and hesitation fillers.

Table 1. Accuracy of the different tagging chains

Method	Tagging accuracy	
	Token	Sentence
Baseline	91.97%	68.37%
DIM	94.92%	79.96%
CONJ	95.53%	81.74%
The full chain	96.15%	83.96%

For measuring the individual advances of the enhancements presented, three different settings were evaluated on the test set. The first of them was a baseline that used the raw analyses of Humor disambiguated by PurePos. The second system (DIM) employed the extended vocabulary and handled the diminutive analyses as described in Section 4.

The next one – marked with CONJ – used further rules aiming to tag *azért* and *amikor* correctly. Finally, the last system presented contains all the enhancements detailed above. Measurements in Table 1 show that in contrast with the baseline tool which tagged erroneously 3 out of 10 sentences, the accuracy of the full chain is comparable with that of the tagging methods for written corpora [24]. Furthermore, each of the enhancements improved the overall performance significantly.

As for the MLU estimation task, two metrics were used for the evaluation. Mean relative error[2] (MRE) was calculated to show the average relative deviation of the estimated morpheme counts from the one of human annotators.

In addition, Pearson's correlation coefficient was employed to measure the correlation between the output of the processing chain and the counts of human annotators. Since both metrics required a gold dataset, MLUm was manually calculated for 300 utterances in the test set.

Table 2. Evaluation of the MLU estimation algorithm

Tagged utterances	MRE	Correlation
Gold standard	0.0279	0.9933
The full chain	0.0449	0.9901

Table 2 presents the evaluation of the morpheme count estimation algorithm. Both the gold standard data and the output of the tagging tool were used as an input for the estimator, thus enabling detailed comparison of the components. The latter values confirm that the overall performance of the methodology described in this study is satisfactory. Therefore, the estimation algorithm introduced can be used to measure the morphosyntactic complexity of Hungarian spoken language in practice.

7 Conclusion

In this study, methods for measuring the morphosyntactic complexity in a corpus of Hungarian spoken child language were investigated. First, an annotation scheme for the HUKILC corpus was created, then a morphological tagging chain was developed. Although the components of our method use resources created for analysing written corpora, its most typical errors could be located and fixed. Further on, principles for counting MLUm for the corpus were laid down and got implemented. The tool developed is suitable for the estimation task, and its morphological disambiguation performance reaches the accuracy of taggers created for written language. In addition, the pipeline architecture of the system allows its modification, thus it can be used as a basic resource in the research of child language.

The contribution of our study is threefold. First, guidelines for Hungarian transcript annotation were created. Further on, MLUm calculation principles for Hungarian were collected, adapted for the HUKILC corpus and got automatized. Finally, a tool is

[2] $MRE = \sum_{i=1}^{n} \frac{|a_i - p_i|/a_i}{n}$, where a_i marks the manual morpheme count of the ith utterance and p_i stands for the ith prediction.

presented that not just performs morphological tagging with a high accuracy, but is an adequate one to be used in practice for estimating morpheme counts. Therefore, the labor-intensive manual calculation could be replaced by the execution of a tool, radically shortening the time required for measuring MLUm.

References

1. Bick, E., Mello, H., Panunzi, A., Raso, T.: The annotation of the C-ORAL-BRASIL oral through the implementation of the Palavras Parser. In: Proceedings of the Eight International Conference on Language Resources and Evaluation (LREC 2012), pp. 3382–3386. ELRA, Istanbul (2012)
2. Bowerman, M.: Early syntactic development: A cross-linguistic study with special reference to Finnish. Cambridge University Press (1973)
3. Brown, R.: A first language: The early stages. Harvard University Press (1973)
4. Crystal, D.: Review of R. Brown 'A first language'. Journal of Child Language 11, 289–307 (1974)
5. Csendes, D., Csirik, J.A., Gyimóthy, T.: The Szeged Corpus: A POS Tagged and Syntactically Annotated Hungarian Natural Language Corpus. In: Sojka, P., Kopeček, I., Pala, K. (eds.) TSD 2004. LNCS (LNAI), vol. 3206, pp. 41–47. Springer, Heidelberg (2004)
6. Eeg-Olofsson, M.: Probabilistic Tagging of a Corpus of Spoken English. University of Goteborg: Department of Computational Linguistics (1991)
7. Hickey, T.: Mean length of utterance and the acquisition of Irish. Journal of Child Language 18(3), 553–569 (1991)
8. MacWhinney, B.: The childes project: Tools for analyzing talk. Child Language Teaching and Therapy 8(2), 217–218 (1992)
9. MacWhinney, B.: CHAT manual (1996)
10. Mátyus, K., Orosz, G.: MONYEK: morfológiailag egyértelműsített óvodai nyelvi korpusz. Beszédkutatás (in press, 2014)
11. Mendes, A., Amaro, R., do Nascimento, M.F.B.: Morphological tagging of a spoken Portuguese corpus using available resources. In: Branco, A., Mendes, A., Ribeiro, R. (eds.) Language Technology for Portuguese: Shallow Processing Tools and Resources, pp. 47–62. Colibri, Lisboa (2004)
12. Moreno, A., Guirao, J.M.: Tagging a spontaneous speech corpus of Spanish. In: Proceedings of Recent Advances in Natural Language Processing (RANPL 2003), Borovets, Bulgaria, pp. 292–296 (2003)
13. Nivre, J., Grönqvist, L., Gustafsson, M., Lager, T., Sofkova, S.: Tagging spoken language using written language statistics. In: Proceedings of the 16th conference on Computational Linguistics, vol. 2, pp. 1078–1081. Association for Computational Linguistics (1996)
14. Orosz, G., Novák, A.: PurePos 2.0: a hybrid tool for morphological disambiguation. In: Proceedings of the International Conference on Recent Advances in Natural Language Processing (RANLP 2013), pp. 539–545. INCOMA Ltd., Shoumen (2013)
15. Panunzi, A., Picchi, E., Moneglia, M.: Using PiTagger for Lemmatization and PoS Tagging of a Spontaneous Speech Corpus: C-ORAL-ROM Italian. In: 4th Language Resource and Evaluation Conference (LREC), pp. 563–566 (2004)
16. Parker, M.D., Brorson, K.: A comparative study between mean length of utterance in morphemes (MLUm) and mean length of utterance in words (MLUw). First Language 25(3), 365–376 (2005)
17. Prószéky, G.: Industrial applications of unification morphology. In: Proceedings of the Fourth Conference on Applied Natural Language Processing, p. 213. Association for Computational Linguistics, Morristown (1994)

18. Réger, Z.: Mothers' speech in different social groups in Hungary. Children's Language 7, 197–222 (1990)
19. Retherford, K.S.: Guide to analysis of language transcripts. Thinking Publications University (1993)
20. Saygın, A.P.: A Computational Analysis of Interaction Patterns in the Acquisition of Turkish. Research on Language and Computation 8(4), 239–253 (2010)
21. Toutanova, K., Klein, D., Manning, C., Singer, Y.: Feature-rich part-of-speech tagging with a cyclic dependency network. In: Hearst, M., Ostendorf, M. (eds.) Proceedings of the 2003 Conference of the North American Chapter of the Association for Computational Linguistics on Human Language Technology, pp. 173–180. Association for Computational Linguistics (2003)
22. Trón, V., Halácsy, P., Rebrus, P., Rung, A., Vajda, P., Simon, E.: Morphdb.hu: Hungarian lexical database and morphological grammar. In: Proceedings of the Fifth Conference on International Language Resources and Evaluation, pp. 1670–1673 (2006)
23. Wéber, K.: Rejtelmes kétféleség. – A kétféle igeragozás elkülönülés a magyar nyelvben. Ph.D. thesis, University of Pécs, Pécs, Hungary (2011)
24. Zsibrita, J., Vincze, V., Farkas, R.: Magyarlanc: A Toolkit for Morphological and Dependency Parsing of Hungarian. In: Proceedings of Recent Advances in Natural Language Provessing 2013, pp. 763–771 (2013)

Using Verb-Noun Patterns to Detect Process Inputs

Munshi Asadullah, Damien Nouvel, and Patrick Paroubek

Laboratoire d'Informatique pour la Mécanique et les Sciences de l'Ingénieur (LIMSI)
Rue John von Neumann, Campus Universitaire d'Orsay, Bât 508,
91405 Orsay cedex, France
{munshi.asadullah,damien.nouvel,pap}@limsi.fr

Abstract. We present the preliminary results of an ongoing work aimed at using morpho-syntactic patterns to extract information from process descriptions in a semi-supervised manner. The experiments have been designed for generic information extraction tasks and evaluated on detecting ingredients from cooking recipes in French using a large gold standard corpus. The proposed method uses bi-lexical dependency oriented syntactic analysis of the text and extracts relevant morpho-syntactic patterns. Those patterns are then used as features for different machine learning methods to acquire the final ingredient list. Furthermore, this approach may easily be adapted to similar tasks since it relies on mining generic morpho-syntactic patterns from the documents automatically. The method itself is language independent, considering language specific parsers being used. The performance of our method on the DEFT 2013 data set is nevertheless satisfactory since it significantly outperforms the best system from the original challenge (0.75 vs 0.66 MAP).

1 Process Inputs in Specification Document

We are looking for a generic method to extract clearly defined target information from specification documents written in natural language. Specification in this context is the generic class of documents written to explain how a set of input elements are to be used or intended to be used in the processes described in the document. For the purpose of clarity we shall exemplify all the formal discussion within the cooking recipe domain.

1.1 Specification

The dictionary definition of "Specification" can be; "a detailed description of work to be done or materials to be used in a project; an instruction that says exactly how to do or make something"[1]. A large number of domains can crawl under this umbrella definition (e.g. cooking recipes, software specification, description of experiments, instruction manuals etc.). We are not suggesting that specification is a category of sentences, a taxonomical class (specification sentences), coined by Higgins [1]. We are in fact in total agreement with Heycock [2] in this aspect that a "specification" class for sentences is rather unnecessary.

[1] Source: Merriam-Webster Online Dictionary.

P. Sojka et al. (Eds.): TSD 2014, LNAI 8655, pp. 181–188, 2014.

However, it is reasonable to assume that the type of text to be found in a specification document uses a finite number of sentence patterns to express the action descriptions. We thus hypothesize that a finite variation of morpho-syntactic pattern must exist to express process-object interaction. We then verified the hypothesis on the cooking recipe corpus described in Section 2.1. First, we need to establish the theoretical basis of events, processes and their interaction with objects.

1.2 Objects, Events and Processes

We are not trying to provide the philosophical or ontological basis for these concepts, rather present the extent of these concepts that have been used in our research. A detail study on objects and events and process can be found in [3], and we shall present most of the definition according to this work. However, the granularity of details for these concepts required for our research is much coarse and it shall be reflected in the following discussion.

Objects are spatial elements i.e. something that occupies physical space, that can change over time (e.g. from egg to boiled egg), but do not have any temporal part (e.g. an egg after one hour is still an egg). According to [3] objects can have spatial parts, but those parts are not the same object, rather matters or different objects (e.g. half of an egg is not an egg). However, This relation has little impact in our work (e.g. one may need only egg white for a recipe but the ingredient may still be an egg). Therefore, we restrict our focus to normalized concrete objects and their interaction in a process or an event.

Events and processes are difficult concepts to put one's fingers on precisely. Both events and processes has been described in [3] as an action over time, where processes describe an action without defined temporal boundary and events, with defined temporal boundaries i.e. start and end. From the cooking recipe a definitive example can be "boiling an egg for 15 minutes". By definition, this is an event i.e. start boiling an egg and finished after 15 minutes. Any intervals in between, by definition are not events, thus are processes.

Another property of the events and processes is that they can have spatial components i.e. involvement of objects in the action. In the cooking recipes all the processes and events are thus expected be informative. For our research these stringent definitions have little impacts, thus events and processes have been used as interchangeable through out the whole article. However, these definitions can be useful in the adaptation of the method for some other tasks. For the ingredient extraction task we are only interested in the inputs of a process or an event.

1.3 Why Focusing on Process Inputs?

Following Faure and Nédellec [4], we have assumed that verbs play a fundamental role in process detection and extract objects that are connected to them as direct objects, where subject dependency is optional, in particular for cooking recipes where verbs are often in imperative mood. From a syntactic and semantic point of view, we are looking for common structures involving a verb and their arguments, among which we are less interested in the syntactic subject and more in the semantic patients and

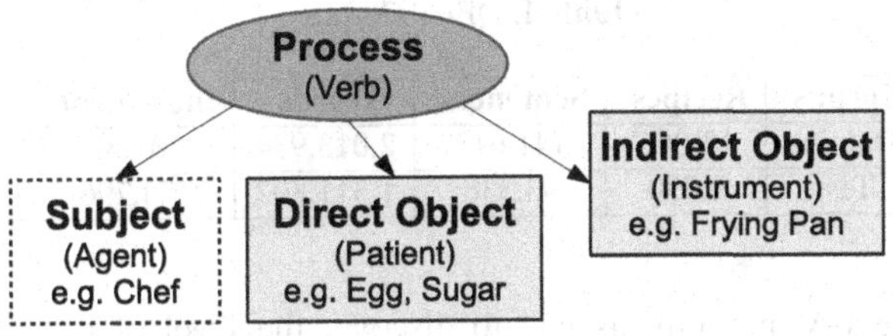

Fig. 1. Generic Verb-Argument Semantic

recipients arguments of a verb. For instance, in the following sentences, the presence of interesting verbs and their arguments are strong clues for recognizing processes and process inputs:

> **Put** the *flour*, *yeast*, *sugar* and *butter* in a food processor with a pinch of *salt*. [...] **Add** the *butter* and *egg white*, pulse to a paste, then **fold in** the *chocolate* and set aside. [...] **Sprinkle** with the remaining *almonds* and a little *caster sugar* and bake for 40–50 mins until puffed up and golden."[2]

Detection of the process input is a task specific requirement, thus, the adaptation of the method would require necessary changes at this level. We look for specific relations to identify ingredients, but one can look for the actors or instruments of a process for a different set of tasks (e.g. actors might be more interesting for named entity recognition).

2 Extracting Ingredient

From the original DEFT Challenge [5], we chose to evaluate on the task of extracting the ingredients from a recipe using a list of ingredients. This particular task was chosen because of the large amount of gold standard data. We shall give an overview of the task, performance of the participants of the original challenge and our approach in the following sub-sections.

2.1 The DEFT Challenge

The DEFT challenge is an annual French text mining evaluation workshop. Inspired by the Computer Cooking Contest[3], the 9th edition of this challenge was focused on the analysis of recipes written in French. There were 4 tasks from 2 main category,

1. Document Classification (Task 1–3)
2. Information Retrieval (Task 4)

[2] From http://www.bbcgoodfood.com/recipes/3466/double-chocolate-easter-danish

[3] http://computercookingcontest.net

Table 1. DEFT Corpus

Corpus	Recipes	Sentences	Words	Ingredients
Training	13,866	141,613	2,013,934	101,563
Test	9,230	93,338	1,311,802	74,796

For the first category, participants had to discover the level of difficulty (4 levels) for a recipe, the type of dish (starter, main dish or desert) a recipe most likely to be and identifying the best title for a recipe from a list of possible titles. For the information retrieval task, participants had to identify the ingredients for each recipe from a normalized list of possible ingredients. The details of the DEFT Challenge corpus, has been presented in Table 1. It is important to note that the normalized list of possible ingredients, contains at least all the ingredients to be found in the recipes of the corpus. There is also the possibility that the actual ingredient name is not in the recipe text directly (e.g. "Fry egg" implies the presence of oil) and ingredients that are not present in the training corpus at all [6].

2.2 Evaluation and Results of the Challenge

There were 6 teams participated in the challenge, of which 2 industrial participants and the rest from the academic arena. All the results were processed using an evaluation system designed specifically for the challenge. We also used this evaluation platform to evaluate our system's performance. The evaluation metric used for the task was Mean Average Precision (MAP). Let us consider that there are N recipes and for any recipe R_i there are n_i ingredients (i.e. $\{I_i^1 \dots I_i^j \dots I_i^{n_i}\}$) and P be the precision, then the MAP metric is,

$$MAP = \frac{1}{N}\sum_{i=1}^{N}\frac{1}{n_i}\sum_{j=1}^{n_i} P\left(I_i^j\right)$$

The final MAP score of the top 5 teams are listed in Table 2. "*Celi France*" performed the best with a MAP of **0.6622**. They presented a system that uses a hybrid approach to solve the given problem [6]. They used a rule-based system to identify the potential ingredient and then filter them using a classifier on the basis of the type of the recipe (from task 2). Their system relied strongly on the lexical features (mostly lemma). We shall try to present the contrast between this method and our approach in the following section.

Table 2. DEFT Challenge Result (MAP Score)

Team	LIM&Bio	GREYC	LIA	Celi Fr.	Wikimeta
Run #1	0.4115	0.4881	*0.6287*	**0.6662**	0.5675
Run #2	0.4170	0.5074	0.6218	—	*0.6428*
Run #3	0.4649	0.5556	0.6191	—	—
Rank	5	4	3	1	2

2.3 Dependency Sub-tree Patterns

After careful revision of the approaches used in the original challenge [5], we were convinced that lexical features (regardless of the type of the approach i.e. rules, regular expressions or machine learning) are not effective for this particular problem. Thus, we decided to use document level features. However, we attempted several methods using lexical features before using pattern features. Searching the ingredient tokens from the global list in a recipe has been used as baseline. Several improvements has been attempted for the lexical approaches (e.g. using token, lemma, token+lemma etc.). We then decided to use morpho-syntactic patterns at document level. Our patterns are dependency sub-trees obtained by generating sub-trees of all possible depth using either or both lemma and POS features. The use of sub-trees for text representation can be found in [7]. the significant features of our sub-tree patterns are,

1. All possible depth are explored.
2. Each element of a pattern can represent lemma or POS or both.
3. Each element of a given pattern is concrete i.e. no variable element exists.
4. For a verb we explore the direct object and the prepositional phrases.

These patterns were extracted (as in [8]) automatically and since we focused on the dependency structure of a sentence, less significant tokens (e.g. prepositions) were discarded. It in turn gave use more information rich patterns. Example of some patterns can be found in Table 3, Among these patterns we found the longer and the verb centred patterns to be information rich. Different document level heuristics of these patterns have been used in our system as features for the ingredient extraction. The following sections will present an overview of our experiments.

Table 3. Top Patterns for Some Ingredients

Sel(Salt)	Œuf(Egg)	Courgette
N/poivre	N/sucre	N/courgette
N/huile/N/olive	N/chocolat	**N/huile/N/olive**
N/tomate	N/oeuf	**N/dés/N/légume**
N/pomme/N/terre	N/gâteau	**N/grains/N/moitié**
N/viande	V/battre	**N/rondelle/N/courgette**
N/oignon	**N/jaune/N/oeuf**	**V/couper/N/courgette**
N/poulet	N/crème	N/poivre
N/ail	A/vanillé	N/ail
N/courgette	N/vanille	**N/blanquette/N/veau**

3 Data Preprocessing

The recipes were parsed using statistical dependency parser for French[4]. Partly developed by ANR-SEQOIA Project[5], BONSAI is a collection of resources for French

[4] http://alpage.inria.fr/statgram/frdep/fr_stat_dep_parsing.html
[5] https://sites.google.com/site/anrsequoia/home

parsing, namely 3 statistical parsers. This resources have been presented in [9], that also reported 86.8% accuracy for dependency output using Berkeley parser [10] for French. We used the Berkeley parser for the preprocessing that establishes the dependencies between two tokens in FTB-DEP [11] formalism.

FTB Surface Dependency Annotation Guide [12] lists the basic annotation guideline for FTB-DEP formalism. FTB-DEP is based on the Dependency Grammar (DG) [13] formalism and like any DG based formalism, FTB-DEP adapted the relation types according to the target language and domain. There are 12 relations to annotate the relations of a token with the verbal governors (e.g. suj, obj etc.) and 8 to annotate the relations with non verbal governors (e.g. mod, coord etc.). There are 8 more specific relations reserved for manual annotation (e.g. mod_loc etc.). There is a virtual "*ROOT*" element for each sentence in the FTB-DEP, which is the hierarchical nucleus of a sentence and a natural extension for many formalisms of the DG family. Among all the relations only coordination required some normalization.

The coordination dependency is represented as a chain rather than separate relations, i.e. *N* coordinated tokens are represented using a combination of two dependencies, "*COORD*", connects the first conjunct with the first coordinator and "*DEP–COORD*" connect the next coordinator. Consecutive conjuncts are connected in a chain with the "*COORD*" relation. We had to resolve the "*COORD*" to a flatter representation for some of the experiments. All the parsers output in an adapted **CoNLL**[6] data format. Once we have the parsed output[7] we can extract the patterns and start experimenting.

4 Experiments and Results

We experimented with two machine learning methods, logistic regression and perceptron. For the pattern based experiments, we have to map each pattern at document level for each ingredient. Thus, in the training set if we have n documents $D = \{d_1, d_2 \cdots d_n\}$ and m ingredients $I = \{i_1, i_2 \cdots i_m\}$ then we can have $D^x \subset D$, where the ingredient $i^x (1 < x < m)$ is present. All the features for the machine learning methods had calculated from this subset. The results from out experiments are listed in Table 4,

Table 4. Experimental Results

System Features	MAP	P(5)	P(10)	P(100)	R(5)	R(10)	R(100)
identify-tokens	0.3564	0.4960	0.3556	0.0375	0.3607	0.4930	0.5114
identify-lemmas	0.4355	0.5402	0.4193	0.0456	0.4000	0.5893	0.6262
identify-tokens+lemmas	0.4430	0.5375	0.4249	0.0469	0.3990	0.5986	0.6428
learn-lemmas	0.7196	0.7409	0.5306	0.0695	0.5420	0.7446	0.9493
learn-lemmas+mine	0.7362	0.7565	0.5432	0.0695	0.5538	0.7615	0.9487
learn-lemmas+mine+coord	0.7364	0.7555	0.5431	0.0695	0.5532	0.7610	0.9490
learn-lempos	0.7182	0.7414	0.5305	0.0695	0.5423	0.7446	0.9495
learn-percept	**0.7500**	0.7588	0.5547	0.0706	0.5545	0.7779	0.9648

[6] `http://nextens.uvt.nl/depparse-wiki/DataFormat`

[7] script available at `https://github.com/eldams/ConLL-SimpleReader`

All the systems with the prefix "identify" uses string matching and represent the baseline for the task. All the system with the prefix "learn" is a machine learning system. Except for "learn-percept" and "learn-percept-rank" all the machine learning system uses logistic regression. In Table 4, the suffix of a system name state the features used (e.g. tokens, lemma etc.). the suffix element "mine" means, the patterns had been used as features. Although it is not explicitly mentioned, the perceptron based system uses the patterns as features. The suffix element "coord" refers to the fact that the data has been normalized, following the discussion in Section 3 and then used in the extraction process.

The primary reason for the lexical features to fail is found to be the similar reasons discussed in Section 2.3. For example if "*fry*" is mentioned in a recipe that implicitly considering "*oil*" as an ingredient. If "*oil*" never appears in the recipe, there shall be no lexical map between "*oil*" and "*fry*". But by document level mapping, any pattern may be linked to any ingredient e.g. the verb "*fry*" to "*oil*". As it can be seen clearly, even the logistic regression produces higher scores when patterns are used in conjunction with lexical features. Although the performance of the "perceptron" algorithm is very good, the first significant improvement was seen when document level mapping was used ("learn-lemmas" shows about 30% improvement) where as using the patterns improves the MAP between 2% and 4%. All the results using document level features shows better performance than the top system from the original challenge.

5 Conclusion

We have presented our preliminary experimental results in search of a semi-supervised information retrieval System. The results are never the less inspiring. We can be confident of the performance of the system once tested with data from different domains. We are also looking forward to test the system for some other information retrieval task, that might require us to push the boundary of our current systems. Introducing document level semantic information is also a possible future direction of the research.

References

1. Higgins, F.R.: The pseudo-cleft construction in English. PhD thesis. Massachusetts Institute of Technology (MIT), Cambridge (1973)
2. Heycock, C.: Specification, equation, and agreement in copular sentences. The Canadian Journal of Linguistics / La Revue Canadienne De Linguistique 57, 209–240 (2012)
3. Galton, A., Mizoguchi, R.: The water falls but the waterfall does not fall: New perspectives on objects, processes and events. Applied Ontology 4, 71–107 (2009)
4. Faure, D., Nédellec, C.: Knowledge acquisition of predicate argument structures from technical texts using machine learning: The system asium. In: Fensel, D., Studer, R. (eds.) EKAW 1999. LNCS (LNAI), vol. 1621, pp. 329–334. Springer, Heidelberg (1999)
5. Grouin, C., Zweigenbaum, P., Paroubek, P.: DEFT2013 se met à table: présentation du défi et résultats. In: Actes de DEFT 2013: 9e DÉfi Fouille de Textes, Les Sables d'Olonne, France, pp. 1–14 (2013)

6. Dini, L., Bittar, A., Ruhlmann, M.: Approches hybrides pour l'analyse de recettes de cuisine DEFT, TALN-RECITAL 2013. In: Actes de DEFT 2013: 9e DÉfi Fouille de Textes, Les Sables d'Olonne, France, pp. 53–65 (2013)
7. Pak, A., Paroubek, P.: Text representation using dependency tree subgraphs for sentiment analysis. In: Xu, J., Yu, G., Zhou, S., Unland, R. (eds.) DASFAA Workshops 2011. LNCS, vol. 6637, pp. 323–332. Springer, Heidelberg (2011)
8. Nouvel, D., Antoine, J.Y., Friburger, N.: Pattern mining for named entity recognition. LNCS/LNAI Series 8387 (post-proceedings LTC 2011) (2014)
9. Candito, M., Nivre, J., Denis, P., Henestroza Anguiano, E.: Benchmarking of statistical dependency parsers for french. In: Proceedings of the 23rd International Conference on Computational Linguistics: Posters, COLING 2010, Beijing, China, pp. 108–116 (2010)
10. Petrov, S., Barrett, L., Thibaux, R., Klein, D.: Learning accurate, compact, and interpretable tree annotation. In: Proceedings of the 21st International Conference on Computational Linguistics and the 44th Annual Meeting of the Association for Computational Linguistics, ACL-44, pp. 433–440. Association for Computational Linguistics, Stroudsburg (2006)
11. Candito, M., Crabé, B., Pascal, D.: Statistical french dependency parsing: Treebank conversion and first results. In: Proceedings of the Seventh International Conference on Language Resources and Evaluation (LREC 2010), pp. 1840–1847. European Language Resources Association (ELRA), Valletta (2010)
12. Candito, M., Crabbé, B., Falco, M.: Dépendances syntaxiques de surface pour le français – Schéma d'annotation pour un corpus en dépendances obtenu par conversion du FrenchTreebank (2011)
13. Tesnière, L.: Éléments de Syntaxe Structurale. Éditions Klinksieck, Paris (1959)

Divergences in the Usage of Discourse Markers in English and Mandarin Chinese

David Steele and Lucia Specia

Department of Computer Science, The University of Sheffield, UK
{dbsteele,l.specia}@sheffield.ac.uk

Abstract. Statistical machine translation (SMT) has, in recent years, improved the accuracy of automated translations. However, SMT systems often fail to deliver human quality translations especially with complex sentences and distant language pairs. Current SMT systems often focus on translating single sentences with clauses being treated in isolation, leading to a loss of contextual information. Discourse markers (DMs) are vital contextual links between discourse segments and this paper examines the divergences in their usage across English and Mandarin Chinese. We highlight important structural differences in composite sentences extracted from a number of parallel corpora, and show examples of how these cases are dealt with by popular SMT systems. Numerous significant divergences, such as contextual omissions, were observed, which can lead to incoherent automatic translations. Our objective is to use these findings to guide a framework proposal to address divergences in DM usage in order to improve SMT output quality.

1 Introduction

In general "discourse" is used to signify an arbitrary length of coherent language-based communication consisting of either phrases, sentences or utterances [1]. With respect to natural language processing (NLP), and more specifically, Statistical Machine Translation (SMT) – our application of interest, discourse is mainly concerned with both written text and spoken dialogue consisting of some connected sequential units.

On a fundamental level discourse is linked in a meaningful way (lexical cohesion) by discourse markers (DMs, also known as discourse connectives), which separate the discourse into discourse segments or language structures, such as words, phrases, clauses or composite sentences [2], each of which contain a local coherence and context. However, DMs cover a range of connectives, conjunctions, conjunctives and other cue words and can be difficult to define precisely [3].

Despite the important role DMs have in terms of lexical cohesion, current SMT systems do not explicitly address DM constructions as such, and therefore translations can often lack the cohesive cues that DMs provide. Indeed, DMs are often translated into the target language in ways that differ from how they are used in the source language [4,5]. While recent developments in SMT potentially allow the modelling of discourse information across sentences [6], no efforts have been dedicated to address DMs in particular. Additionally it has been shown that single DMs can signal numerous

P. Sojka et al. (Eds.): TSD 2014, LNAI 8655, pp. 189–200, 2014.

discourse relations depending on where they occur and current SMT systems are unable to adequately distinguish between each of the relations during the translation process [7].

This paper examines the usage of a set of frequent DMs in Chinese[1] and English, highlighting some natural and common divergences observed in parallel corpora, and some of the problems that arise when the contextual information that surrounds them is not utilised by SMT systems. The focus is on Chinese into English translations. The results were produced from inspecting four corpora of various genres, domains and sizes, comparing given DMs in Chinese sentences against DMs in the English parallel human translation. Only DMs within compound sentences, rather than across discourse segments, were used for the analysis. The study shows that the parallelism in the usage of DMs in the two languages varies significantly across corpora. It also shows substantial divergences in the usage of DMs in a large proportion of the cases. This evidences the problem of using such parallel corpora as a source of information to build SMT systems without special treatment of DMs.

Popular online SMT systems were also used to translate the Chinese, with the resulting automated translation being compared to the given human translation, hence illustrating their limitations. The results show that these SMT systems are often unable to deal with the complex changes in word order and, because of DMs, struggle with contextual omissions, even across closely linked sentential clauses. As the sentences become more complex the problems are further compounded and more errors occur in the automated translations, ultimately suggesting that too much information is lost when the context carried by DMs is not utilised by SMT systems.

2 Discourse Markers in Chinese

Chinese and English stem from two very different language families (Sino-Tibetan and Indo-European respectively) which can be a chief cause of translation difficulty [9]. For example, Chinese is logographic and does not use inflection, relying on generating meaning through word order, which can often be quite flexible. Moreover, the positioning and order of connective markers is very fluid and syntactically they can take many positions including: "the initial position, the predicate-initial position, and the final position" [10]. English, on the other hand, has an alphabet and uses a degree of inflection with a relatively fixed word order where DMs can, for the main, only be placed in the initial position [10].

Defining DMs is not necessarily a trivial task. Chinese uses a rich array of DMs to link parts of speech in both simple and complex sentences [2]. Chinese conjunctions appear in two main types: those linking words or short phrases (simple conjunctions) such as: 和 (hé – and), 跟 (gēn – and/with), 或 (huò – or) as in 刀和叉 (dāo hé chā – knife and fork), and those that link clauses (composite conjunctions). Conjunctives are also used often appearing in the main (usually second) clause of a sentence and link back to the previous clause [11]. Additionally, there are instances where clauses may be linked in a sentence without the use of any DM (zero connective structures). In

[1] For the purposes of this paper the term Chinese is used to mean Mandarin or Mandarin Chinese, considered to be the main standardised language of China [8].

these cases the meaning or context is strongly inferred across the clauses, leading to the creation of sentences that have natural omissions, which can cause problems for current SMT systems.

3 Settings: Corpora and SMT Systems

We used four well known corpora for gathering the data necessary for observing DM frequency and pertinent translations:

• Basic Travel Expression Corpus (BTEC): This corpus is primarily made up of short simple phrases and utterances that occur in travel conversations. For this study, 44016 sentences in each language were processed with over 250000 Chinese characters and over 300000 English words [12].

• Foreign Broadcast Information Service (FBIS) corpus: This corpus uses a variety of news stories and radio podcasts in Chinese. For this study 302996 parallel sentences were used containing 215 million Chinese characters and over 237 million English words.

• Ted Talks corpus (TED): This corpus is made up of approved translations of the live Ted Talks presentations[2]. This corpus contains over 300.000 Chinese characters and over 2 million English words [13] spread across 156805 parallel sentences.

• Multi-UN corpus (UN): This is a parallel corpus (for 6 languages) using data extracted from the United Nations Website. It includes over 220 million words in English and over 629 million Chinese characters in 8.8 million parallel sentences [14].

The SMT systems used to produce the automatic translations are Google Translate[3] and Bing Translator[4]. Whilst these are specific commercial translation tools and they may not represent the best quality translation systems for Chinese-English, they are good representatives of statistical translation approaches, known to use state of the art techniques and achieve reasonable translation quality. In addition they are freely available, making it possible to reproduce and expand the analysis presented here.

4 Analysis of Chinese Discourse Markers

In this section we examine the main types of Chinese DMs, including conjunctions for composite sentences, sequential paired conjunctions and zero connectives [11,15,16,17,18,19]. Our first step was a simple quantitative analysis to identify the most commonly used DMs in our corpora, so that we could select a few cases of interest to analyse in more detail. Table 1 shows the proportion of sentences containing the ten most frequent disyllabic DMs in the four different corpora. It also shows one or more frequent English translations for each DM, but we note that variants of these translations are possible.

While the percentages of sentences containing specific DMs in Table 1 may seem small at first, overall DMs are present in a significant proportion of sentences. The

[2] `http://www.ted.com`

[3] `http://translate.google.com`

[4] `http://www.bing.com/translator`

Table 1. Ten most frequently occurring DMs in the four corpora

TED	UN
因为 (4.72%) : because	因此 (1.70%) : so/therefore
如果 (4.32%) : if	以便 (1.42%) : so that
所以 (4.05%) : so/therefore	因为 (1.24%) : because
但是 (3.58%) : but	由于 (1.22%) : due to/as a result of
或者 (1.68%) : or	如果 (1.05%) : if
还有 (1.59%) : furthermore	而且 (1.04%) : moreover
那么 (1.59%) : then/in that case	为了 (0.88%) : in order to
而且 (1.47%) : moreover	但是 (0.81%) : but
并且 (1.34%) : and also	并且 (0.73%) : and also
因此 (1.24%) : so/therefore	虽然 (0.62%) : although
FBIS	**BTEC**
因为 (1.39%) : because	如果 (1.18%) : if
如果 (1.30%) : if	但是 (1.10%) : but
因此 (1.19%) : so/therefore	那么 (0.44%) : then/in that case
为了 (1.13%) : in order to	还是 (0.39%) : or
由于 (1.10%) : due to/as a result of	所以 (0.29%) : so/therefore
但是 (1.01%) : but	因为 (0.25%) : because
而且 (0.85%) : moreover	或者 (0.23%) : or
虽然 (0.80%) : although	并且 (0.17%) : and also
然而 (0.79%) : however/but	只有 (0.17%) : only
甚至 (0.72%) : even	而且 (0.13%) : moreover

frequency analysis highlights certain trends, for instance 如果 (rúguǒ – if) and 因为 (yīnwèi – because) have a relatively high frequency in all four corpora. 因为 (yīnwèi) is classed as one of the high frequency (causal) connectives [20] and is considered to have a strong correlation in usage with 'because'. In what follows we pinpoint some of the divergences in the use of these markers through examples of constructions, and connect these divergences to the behaviour of SMT systems when faced with such constructions.

Ex (1) shows the 因为 (yīnwèi) DM being used in a relatively short causal sentence, and it is clear that the SMT system has problems with the DM, dropping it completely from its position before the comma.

Ex (1)[5] 他因为病了，没来上课。
he because ill, not come class.
Because he was sick, he didn't come to class. [18]
He is ill, absent. (Bing)

[5] Each example in this paper has the following format: Line 1 is the correct Chinese in characters; line 2 is a literal word-for-word translation; line 3 is the given translation and line 4 is (usually) the best translation returned by the SMT system. In some cases more than one automated translation is given for comparison purposes.

In Ex (1) the two parts of the sentence appear to have a very weak link in the translation as the DM is simply not used at all in the automated translation. The information after the comma (in the Chinese sentence) is correct and as Chinese does not use inflection, a sentence segment similar to ‘did not come to class’ should appear in the translation rather than simply having 'absent'. In Ex (2) the problem seems to be the reverse. The 因为 (yīnwèi – because) being present in the Chinese sentence causes problems for the SMT system as it tries to force ‘because’ into the translation (rather than omitting it) and by doing so significantly alters the meaning.

Ex (2) 你因为这个在吃什么药吗?
you because this (be) eat what medicine [MA]
Have you been taking anything for this? (BTEC)
What are you eating because of this medicine? (Google)

The automated translation gives the impression that the person has changed their diet due to having medicine, rather than their being required to take medicine for an ailment.

4.1 Sequential Constructions: Paired Conjunctions/Conjunctives

Paired DMs are frequently used in Chinese [21] and feature in many translations of complex sentences. Some paired constructions are formed using two conjunctions, but other formations are also possible such as: ‘conjunction …conjunctive’. Typical conjunctives include: 才 (cái – only/only if/ not unless), 就 (jiù – then/that), 却 (què – but/yet/while) and are often treated as connecting referential adverbs [11]. Conjunctions tend to appear in both clauses, while conjunctives frequently appear in just the second clause. They represent even more challenging problems for both human and machine translation.

Table 2 shows (for each corpus) the proportion of sentences that contain at least one occurrence of the given paired marker patterns. The main outcome of this frequency analysis is that for each corpus the …一…就…(…yī…jiù…) pattern appears with the highest frequency. However, manual inspection of a random sample of sentences showed that the …一…就…(…yī…jiù…) was only being used as a sequential paired marker construction in around one quarter of the cases.

Chinese does not have a specific word which maps one-to-one exactly with ‘then’ and so 就 (jiù) and 那么 (nàme – so) are often utilised to perform a similar function [18]. It is difficult to categorise 就 (jiù) on its own as it serves numerous functions. Many other characters such as 来 (lái) and 的 (de) can also be difficult to categorise for a similar reason, but perhaps none more so than the character ‘一’ (yī – one/single/ whole/same…) which covers six pages in the Oxford Chinese dictionary. By themselves 一 (yī) and 就 (jiù) can be ambiguous, but as a sequential construction they work together as a pair in a specific pattern with a relatively fixed meaning. Ex (3) shows a short five-character sentence that uses the …一…就…(…yī…jiù…) pattern as a sequential paired construction to mean: ‘…no sooner…than…’; ‘the moment…’; ‘as soon as…’; ‘once…’

Table 2. Ten most frequently occurring paired DMs in the four corpora

TED	UN
…一…就…(3.67%) : once/as soon as, (then)	…一…就…(0.92%) : once/as soon as, (then)
…如果…就…(1.33%) : if,(then)	…越…越…(0.30%) : more, more
…如果…那…(0.95%) : if, (then)	…由于…因…(0.24%) : due to, because
…也…也…(0.49%) : also, and	…如果…就…(0.22%) : if,(then)
…越…越…(0.49%) : more, more	…不仅…而且…(0.21%) : not only, but also
…从…开始…(0.48%) : starting from…	…从…起…(0.17%) : starting from…
…是…还是…(0.48%) : [be], or	…从…开始…(0.14%) : starting from…
…如果…那么…(0.34%) : if, (then)	…是…还是…(0.14%) : [be], or
…不是…而是…(0.29%) : not, but(is)	…虽然…但是…(0.12%) : although, but
…从…起…(0.27%) : starting from…	…也…也…(0.11%) : also, and
FBIS	**BTEC**
…一…就…(2.20%) : once/as soon as, (then)	…一…就…(0.28%) : once/as soon as, (then)
…越…越…(0.63%) : more, more	…如果…就…(0.22%) : if, (then)
…也…也…(0.40%) : also, and	…从…开始…(0.15%) : starting from…
…从…起…(0.38%) : starting from…	…如果…那…(0.10%) : if, (then)
…如果…就…(0.36%) : if,(then)	…从…起…(0.09%) : starting from…
…从…开始…(0.35%) : starting from…	…是…还是…(0.06%) : [be], or
…不仅…而且…(0.30%) : not only, but also	…只要…就…(0.06%) : as long as, (then)
…是…还是…(0.27%) : [be], or	…又…又…(0.05%) : both, and
…既…又…(0.25%) : both, also	…越…越…(0.03%) : more, more
…既…也…(0.24%) : both, also	…的话…就…(0.03%) : …if, (then)

Ex (3) 他一学就会。

he as soon as study then can.

He learned it (the trick) in a jiffy. [22]

He learn. (Google)

In Ex (3) it is clear that very little concrete information can be extracted from the five characters alone, and there is a lot of inference such as the speed in which the person learned to do something (in this case – a trick). To identify both the 'trick' and 'speed' would require additional contextual information. The overarching pattern for the …一…就…(…yī…jiù…) construct is fairly simple: …一 VP^a 就 VP^b

The 一 (…yī…) should come immediately before the prepositional phrase and/or verb or verb phrase [18], although it can have some subject information that precedes it. In the case of Ex (3) a pronoun is used for the subject.

It is possible that by itself the sentence in Ex (3), while grammatically correct, has too much inference for an SMT system to manage and sentences that contain more information may produce better translations. The actions in the structure do not have to be related and the subjects in each clause do not have to be the same, but it is often the case that the second action is as a direct result of the first.

Ex (4) 一有空位我们就给你打电话。

As soon as have space we then give you make phone.

We'll call you as soon as there is an opening. (BTEC)
A space that we have to give you a call. (Google)

In Ex (4) the SMT system tries to remain closer to the actual order of the given sentence, but once again misses the ‘as soon as’. If the word order is to be kept close to the original then a sentence similar to ‘as soon as we have a vacancy (then) we will give you a call’ could be used.

4.2 Linking Clauses Without Discourse Markers (Zero Connectives)

The zero connective [11] is often used to link closely set clauses where the meaning of the second clause is contextually implied by the meaning of the first clause. This can be done through repetition, answering or qualifying conditions as in Ex (5) or for rhythmic balance [17].

Ex (5) 东西太贵, 我不买。
things too expensive, I not buy
If things are too expensive, I won't buy them. [17]
Too expensive, I do not buy it. (Google)

In this case, the SMT system, appears to translate the Chinese word for word, but loses some meaning. The gist of the condition is evident, but the translation is not adequate. Manual insertion of two standard DMs into the sentence is actually required for the SMT system to produce a better output.

Ex (6) 如果东西太贵，我就不买 (了)。
If things too expensive, I then not buy(le).
If something is too expensive, I do not buy it. (Google)

5 Analysis of Chinese and English Discourse Markers in Parallel Corpora

In this Section we perform a quantitative analysis on the usage of DMs in both Chinese and English (human translation). SMT systems learn translation models primarily from parallel corpora with examples of translations aligned at the sentence level. The goal of this analysis is to study whether Chinese markers and their corresponding English markers appear in sentences that are aligned in parallel corpora. For a given DM, a high percentage of aligned sentences containing the marker in both Chinese and English could be an indication that learning the translation of such a marker from the corpus is potentially feasible. On the other hand, a low percentage of aligned sentences containing both Chinese and English markers could be an indication that the markers might be dropped or translated using different linguistic constructs, making the learning of SMT models a more difficult task.

Given that we start the analysis with Chinese DMs, a question that arises is how to find their corresponding English DMs. Each of the given DMs (Tables 1 and 2) are

Table 3. Frequencies of six Chinese DMs and their corresponding translations in parallel corpora

Chinese Marker	Occurrence rate in Chinese (%)				Occurrence rate in human translation (%)				Appear in both the Chinese and English translation (%)			
	BTE	FBIS	UN	TED	BTE	FBIS	UN	TED	BTE	FBIS	UN	TED
因为 (because)	0.25	1.39	1.24	4.72	0.20	1.01	0.48	3.92	80	73	39	83
如果 (if)	1.18	1.30	1.05	4.32	1.15	1.09	0.76	3.84	89	84	72	89
因此 (consequently)	0.02	1.19	1.70	1.24	0.02	0.83	1.09	1.07	100	70	64	86
但是 (but)	1.10	1.01	0.81	3.58	1.07	0.89	0.54	3.19	97	88	67	89
而且 (moreover)	0.13	0.85	1.04	1.47	0.13	0.59	0.69	1.15	100	69	66	78
虽然 (although)	0.02	0.80	0.16	0.36	0.02	0.65	0.15	0.15	100	81	94	42

relatively common, but can naturally have variance in the associated translations. For example, a strong link has already been suggested between 因为 (yīnwèi) and ‘because’, but there are numerous comparable ways of uttering or writing ‘because’ such as: ‘in light of’, ‘for this reason’, ‘as a result of’ [23,24]. For this paper, interchangeable values are classed as variance rather than ambiguity. Ambiguity is taken to mean a word that has numerous different functions as per the individual characters ‘一’ yī and ‘就’ jiù discussed in Section 4.1.

Table 3 shows the occurrence percentages of six frequently used Chinese DMs in the four corpora. The first column shows the Chinese DM with its commonly associated English equivalent. Column two shows the occurrence rate of the Chinese marker in sentences across the corpora. Column three shows the occurrence rate where a directly equivalent English DM (with variance included) is used in the parallel translations (e.g. 因为 = ‘because’ or a variant of ‘because’); that is, for each set of sentences with a given Chinese DM, a subset is formed from the parallel translations of the sentences. The percentages in column three show the size of the resulting subsets compared to the size of the whole corpus. The final column shows the percentages of sentences that contain, within a set, both the Chinese DM along with the equivalent usage of an English DM in the translation. The percentages in the fourth column can be used as general measure of the strength of correlation.[6]

We note that the source language of our corpora is not always Chinese. For TED it is English, while for UN it could be any of the six languages. BTEC and FBIS however consist of segments originally in Chinese, and their translations into English. Therefore the implications of the numbers in Table 3 will be different for different corpora.

Overall, the numbers show that in short everyday sentences (BTEC) the main DMs are used as expected (e.g. 因为 maps closely to ‘because’ or a strong variant of ‘because’). As the sentences become more complex and are used at a higher level (FBIS and TED), then the way DMs are used becomes more fluid. The markers appear to be

[6] It must be noted that whilst the percentages show trends, there is still a small degree of error where less common variant phrases may have possibly been used in the parallel translation (e.g. because = this is down to). Detailed discussion of further variance is beyond the scope of this paper and can be considered in future work. The given percentages are considered to offer a close enough approximation for the related discussion.

increasingly omitted or absorbed into the general meaning of a clause rather than translated directly. As expected with the UN corpus, where complex language is used and discourse is divided into subsections, addenda and annexes, there is even less need for certain markers and there are inevitably fewer occurrences of items such as 'if' and 'but'.

Ex (7) 这将是一次规模盛大, 而且受到广泛国际关注的聚会.
This will be one scale grand, moreover receive wide international attention [DE] meeting.
This will be a grand gathering with wide international concern. (FBIS)
This will be a grand scale, but widespread international concern gatherings. (Google)

In Ex (7), the 而且 (érqiě – moreover) is serving as a link that brings together the qualities of the meeting; that is, it will be on a 'grand scale' and will receive 'wide international attention'. Clearly the human translation is very succinct and does away with the need for the 'moreover' or 'furthermore' type link.

For an SMT system to reach a similar translation it would need to be aware of when to drop the marker, and how to reorder the sentence accordingly. Additionally the [DE] adds complication as grammatically it implies that the described qualities (scale and attention) belong to the meeting, which is not necessarily an easy connection to automatically recognise.

6 Related Work

In the last few years much work has gone into improving machine translation from Chinese into English, including major efforts as part of the DARPA GALE program [25]. A number of useful parallel corpora and wordlists have been developed. Additionally, due to the contextual information connected to DMs, there has been a shift to working with them to improve MT. Initially projects such as the Penn Discourse (Chinese) Treebank (PDTB) [26] started identifying DMs according to type for parts of speech (POS) tagging. The Chinese Discourse Treebank (CDTB) [21] was designed to add a discourse annotation layer to the PDTB.

Efforts have also been made to improve identification of Chinese DMs through applying machine learning [2] and indeed categorising them in terms of relationship (e.g. causal and conditional). Additional work has gone into identifying the meaning of DMs [15] to ascertain their type (e.g. concession or contrast) and improve classification techniques. More recently, word reordering of grammatical structures around DMs, a known translation difficulty, is also being explored [27] and tools such as the Stanford Parser have been built.

Further work has gone into cross-lingual identification of DMs and disambiguation [28], which builds on information from bi-lingual dictionaries, the PDTB and parallel corpora. There is now a fresh trend with a focus on lexical and grammatical cohesion as well as the disambiguation of connectives [29,30,31] and recognising the variety of discourse relations attached to DMs [7].

7 Conclusions and Future Work

Chinese and English both belong to very different language families leading to numerous structural differences between the two languages including differing word order and the use of DMs. DMs in particular provide a level of lexical cohesion between phrases and clauses, but are not always utilised during the MT process. This means that sentential positioning is often incorrect and words are frequently omitted leading to unclear translations with a loss of context and information.

In many cases Chinese discourse has significant subject inference carried across clauses and sentences leading to contextual omission of many items (often pronouns) within a sentence. Ex (9) shows a modified version of Ex (2) where the pronoun and second marker have been manually inserted into the Chinese sentence. With the extra information Bing returns a better translation, highlighting the importance of preserving DM/contextual information.

Ex (8) 他因为病了，所以他没来上课。(modified version of Ex (2))
he because ill, so he not come class. (extra *he* and *so* in the 2nd clause)
Because he was sick, he didn't come to class.
He is ill, so he did not come to class. (Bing)

In the case of paired DMs, especially with the 一 (yī) and 就 (jiu) structure, the SMT systems struggled with inference and disambiguation, often failing to spot the 'as soon as' relation. The main focus of this paper has been on Chinese to English translation. A positive next step would be to analyse DM usage and translation patterns for English to Chinese translation, which would enable comparisons of DMs in both directions. A detailed analysis of the comparisons would look at the relative sentential positioning of DMs and examine where direct equivalents do and do not exist. Additionally, where DMs are used, it will be important to examine the changes in word order that are required to accommodate the respective DMs in the target language. Part of this will include an in depth analysis of contextual omissions (e.g. dropped pronouns) within a sentence, but also will examine the distance that context can be carried in discourse through larger discourse segments that go beyond sentence level.

It is also expected that analysis of translations in both directions will produce more data detailing the variance that can be applied to individual DMs and hence work can go into developing improved recognition of such variance. The results of this further analysis of the corpora will provide insights to help develop a framework to model discourse markers in SMT between Chinese and English.

References

1. Zuffery, S., Degand, L.: Annotating the Meaning of Discourse Connectives in Multilingual Corpora. Corpus Linguistics and Linguistic Theory, 1–24 (2013)
2. Tsou, B., Gao, W., Lai, T., Chan, S.: Applying Machine Learning to Identify Chinese Discourse Markers. In: International Conference on Information, Intelligence and Systems, Chania Crete, Greece (1999)

3. Hussein, M.: Two Accounts of Discourse Markers in English. University of Damascus, Syria (2002)
4. Hardmeier, C.: Discourse in Statistical Machine Translation: A Survey and a Case Study. In: Discours – Revue de linguistique, psycholinguistique et informatique, Caen, Presses Universitaires de Caen (2012)
5. Meyer, T., Webber, B.: Implicitation of Discourse Connectives in (Machine) Translation. In: Workshop on Discourse in Machine Translation (DiscoMT), Sofia, Bulgaria, pp. 19–26 (2013)
6. Hardmeier, C., Stymne, S., Tiedemann, J., Nivre, J.: Docent: A Document-Level Decoder for Phrase-Based Statistical Machine Translation. In: 51st Annual Meeting of the ACL, Sofia, Bulgaria, pp. 193–198 (2013)
7. Hajlaoni, N., Popsecu-Belis, A.: Translating English Discourse Connectives into Arabic: a Corpus-based analysis and an Evaluation Metric. In: CAASL4 Workshop at AMTA (Fourth Workshop on Computational Approaches to Arabic Script-based Languages), San Diego, CA, pp. 1–8 (2013)
8. Swan, M., Smith, B.: Learner English, 2nd edn. Cambridge University Press, Cambridge (2004)
9. Chang, P., Jurafsky, D., Manning, C.: Disambiguating "DE" for Chinese-English Machine Translation. In: 4th Workshop on SMT, Athens, Greece, pp. 215–223 (2009a)
10. Li, Y.: Sensitive Positions and Chinese Complex Sentences: A Comparative Perspective. Journal of Chinese Language and Computing 18(2), 47–59 (2008)
11. Po-Ching, Y., Rimmington, D.: A Comprehensive Grammar. Routledge, London (2004)
12. Takezawa, T., Sumita, E., Sugaya, F., Yamamoto, H., Yamamoto, S.: Toward a broad-coverage bilingual corpus for speech translation of travel conversations in the real world. In: LREC, Las Palmas, Spain, pp. 147–152 (2002)
13. Cettolo, M., Girardi, C., Federico, M.: WIT3: Web Inventory of Transcribed and Translated Talks. In: EAMT, Trento, Italy, pp. 261–268 (2012)
14. Eisele, A., Chen, Y.: MultiUN: A Multilingual Corpus from United Nation Documents. In: 7th Conference on International Language Resources and Evaluation, Pages, La Valletta, Malta, pp. 2868–2872 (2010)
15. Hutchinson, B.: Acquiring the Meaning of Discourse Markers. In: 42nd Meeting of ACL, Main Volume, Barcelona, Spain, pp. 684–691 (2004)
16. Po-Ching, Y., Rimmington, D.: Chinese: Intermediate Chinese, A Grammar and Workbook. Routledge, London (1998)
17. Po-Ching, Y., Rimmington, D.: Chinese: An Essential Grammar, 2nd edn. Routledge, London (2010)
18. Ross, C., Sheng Ma, J.: Modern Mandarin Chinese Grammar. Routledge, London (2006)
19. The Conjunction (2010), `http://www.chineseteachers.com/blog/resource_content.jsp?id=142`
20. Wang, C., Huang, L.: Grammaticalisation of Connectives in Mandarin Chinese: A Corpus-Based Study. Language and Linguistics 7(4), 991–1016 (2006)
21. Xue, N.: Annotating Discourse Connectives in the Chinese Treebank. In: ACL Workshop on Frontiers in Corpus Annotation 2: Pie in the Sky (2005)
22. Oxford Chinese Dictionary: English-Chinese Chinese-English. Oxford University Press, UK (2009)
23. Macmillan Publishers Limited 2009–2014, `http://www.macmillandictionary.com/thesaurus-category/british/`
24. Thesauraus.com. Roget's 21st Century Thesaurus, 3rd edn., `http://thesaurus.com/`
25. Olive, J., Christianson, C., McCary, J.: Handbook of Natural Language Processing and Machine Translation. Springer, New York (2011)

26. Xia, F.: The Part-Of-Speech Tagging Guidelines for the Penn Chinese Treebank (3.0). Technical Reports, IRCS Report 00-07. Pennsylvania (2000)
27. Chang, P., Tseng, H., Jurafsky, D., Manning, C.: Discriminative Reordering with Chinese Grammatical Relations Features. In: 3rd Workshop on Syntax and Structure in Statistical Translation at NACCL HTL, Boulder, Colorado (2009b)
28. Zhou, L., Gao, W., Li, B., Wei, Z., Wong, K.: Cross-lingual Identification of Ambiguous Discourse Connectives for Resource-Poor Language. In: 24th International Conference on Computational Linguistics (COLING), Mumba, India (2012)
29. Tu, M., Zhou, Y., Zong, C.: Enhancing Grammatical Cohesion: Generating Transitional Expressions for SMT. In: 52nd Annual Meeting of the ACL, Baltimore, USA, June 23-25 (2014)
30. Guilou, L.: Analysing Lexical Consistency in Translation. In: Workshop on Discourse in Machine Translation (DiscoMT), Sofia, Bulgaria, pp. 10–18 (2013)
31. Wong, B., Kit, C.: Extending machine translation Evaluation Metrics with Lexical Cohesion to Document Level. In: 2012 Joint Conference on Empirical Methods in Natural Language Processing and Computational Natural Language Learning, Jeju Island, Korea, pp. 1060–1068 (2012)

Sentence Similarity by Combining Explicit Semantic Analysis and Overlapping N-Grams

Hai Hieu Vu[1], Jeanne Villaneau[1], Farida Saïd[2], and Pierre-François Marteau[1]

[1] IRISA, Université de Bretagne Sud (UBS), France
{hai-hieu.vu,jeanne.villaneau,
pierre-francois.marteau}@univ-ubs.fr
[2] LMBA, Université de Bretagne Sud, France
farida.said@univ-ubs.fr

Abstract. We propose a similarity measure between sentences which combines a knowledge-based measure, that is a lighter version of ESA (Explicit Semantic Analysis), and a distributional measure, ROUGE. We used this hybrid measure with two French domain-orientated corpora collected from the Web and we compared its similarity scores to those of human judges. In both domains, ESA and ROUGE perform better when they are mixed than they do individually. Besides, using the whole Wikipedia base in ESA did not prove necessary since the best results were obtained with a low number of well selected concepts.

1 Introduction

Measuring the similarity between sentences is an important task in a variety of applications in natural language processing and related areas, such as information retrieval [3], paraphrase detection [4], text categorization [13] and text summarization [9].

If much work has been presented in the literature for measuring the similarity between long texts and documents, few address the characterization of the similarity between short terms or sentences. P. Achananuparp et al. [1] investigated the efficiency of sentence similarity measures, which are split into three categories by the authors: Word overlap measures, *Tf-Idf* measures and Linguistic Measures as semantic relations and word semantic similarity scores. According to their results, linguistic measures are superior when a low-complexity data-set is considered but, in the presence of "hard" test pairs, for example when one of both sentences can be inferred from the other, most sentence similarity measures do not produce a satisfactory result. More recently, different approaches were tested in the Semantic Textual Similarity task [2] of SEMEVAL 2013, including mixed approaches [5] or UNL (Universal Networking Language) [6].

Measuring similarity requires to define which type of similarity is involved in the corresponding application [17]. In our area of interest, multi-document summarization systems rely widely on similarity measures to both extract the most informative sentences from the original texts and to tackle the issue of semantic redundancy. Since we are interested in extracting the important events of a given domain, the type of similarity between sentences we are seeking for, should focus on the common information and events described in the sentence pairs.

P. Sojka et al. (Eds.): TSD 2014, LNAI 8655, pp. 201–208, 2014.

In Section 2, we present a new method for measuring sentence similarity which combines a word co-occurence method and a knowledge-based method. Section 3 provides the results of an annotation task involving 7 human participants who scored the similarities in two domain specific datasets composed with 60 selected sentence pairs each. These results are used in Section 4 to evaluate our similarity method. Finally Section 5 summarizes the presented work, draws some conclusions and proposes future perspectives.

2 Proposed System

We propose a similarity measure between sentences which makes use of the knowledge-based model ESA [10,8] together with cosine similarity and the lexical metric ROUGE [16]. The similarity score between two sentences is set as a linear combination of ROUGE and ESA scores:

$$Score = (1 - p) \times Score_{\text{ROUGE}} + p \times Score_{\text{ESA}} \quad (1)$$

where $0 \leq p \leq 1$ is a tuning parameter.

2.1 Lexical Similarity

ROUGE (Recall-Orientated Understudy for Gisting Evaluation) is a family of metrics, based on the counts of overlapping units, which was introduced in 2003 [16] to measure the quality of automatically produced summaries through comparison with ideal summaries [15]. We used them in this paper as measures of similarity between any pair of sentences.

We tested four ROUGE measures during our experimentations. ROUGE1 counts overlapping unigrams, ROUGE2 counts overlapping bigrams, ROUGE-S enables at most 1 unigram to be skipped inside bigram components and ROUGE-SU takes into account both overlapping unigrams and skip-bigrams.

We obtained the best results with ROUGE-SU which will be refered to as ROUGE in the following.

2.2 Semantic Relatedness

The concept of Explicit Semantic Analysis (ESA) was introduced in 2007 by Gabrilovitch and Markovitch [10,8], to compute the semantic relatedness of words or texts. Since then, it has shown to be highly effective in various applications, among which information retrieval [18], cross-lingual information retrieval [20,21] and text categorization [12,7].

The ESA representation of a word is a vector whose entries reflect its degree of affinity with all documents of a collection, called index collection. A text is represented as the centroid of the vectors representing its words and cosine similarity is used to assess the semantic relatedness between texts. Gabrilovich and Markovitch used Wikipedia articles (concepts) as index documents and a standard *Tf-Idf* weighting scheme for the entries of the vectors.

In [11], Gottron et al. reformulated ESA as a variation of the generalized vector space model (GVSM) [23,22] and showed that ESA essentially captures term correlation information from the index collection. Larger index collections obviously provide more reliable correlation information. However, stability is reached at some point, so there is no need for excessively large index collections. Furthermore, for applications in a specific domain, there is benefit to take an index collection from the same topic domain while a general topic corpus introduces noise.

In order to build domain-orientated index collections, and prior to ESA, we have retrieved from Wikipedia the K-best concepts that are related to key words of the domain. We use a link mining method, *Pf-Ibf* (Path frequency - Inversed backward link frequency) [19], to score the relatedness of a key word's concept to other concepts in the Wikipedia graph. While *Tf-Idf* analyzes relationships to neighbor articles (only single hop neighbors are considered), *Pf-Ibf* analyzes the relations among nodes in n-hop range. In our experimentations, we considered a 4-hop range neighborhood.

3 Evaluation

We tested our system on two French Web corpora extracted from Wikipedia. The first corpus is about "epidemics" and the second one is about "space conquest".

In the Wikipedia graph, the best related concepts to the key words "space conquest" are mostly named entities which refer to people, countries or names of space vehicles. For the key word "epidemics", we find mostly common nouns, such as names of scientists or diseases.

3.1 Pairs of Sentences

In each corpus, a preliminary human scoring of all pairs of sentences showed a skewed distribution towards 0 (unrelated sentences) and 1 (same topic). In order to construct a set which would reflect a uniform distribution over the range of similarities (0 to 4), we selected a set of sixty sentences as follows: ten sentences, called reference sentences, were selected: they are informative sentences, which contain various important informations of the tested domains. Each of them was associated with six sentences chosen so as to respect the uniform distribution of similarity scores.

A reference sentence (in bold) along with its six associated sentences are presented in Table 1.

3.2 Manual Annotation

We adopted the same annotation procedure as in Li et al. [14]. The participants were asked to rate the similarity between sentences on a scale of 0.0 to 4.0, according to the following definitions:

4: the sentences are completely equivalent;
3: the sentences are mostly equivalent, but some unimportant details differ;
2: the sentences are not equivalent, but they share some parts of information;
1: the sentences are not equivalent, but they are on the same topic;
0: the sentences are unrelated.

Table 1. A reference sentence (in bold) with its associated sentences and their mean similarity scores

(1) Mars est l'astre le plus étudié du système solaire, puisque 40 missions lui ont été consacrées, qui ont confirmé la suprématie américaine - des épopées Mariner et Viking aux petits robots Spirit et Opportunity (2003 et 2004).
(Mars is the most studied celestial body in the solar system, since 40 missions were dedicated to it, which confirmed the American ascendancy from Marinate and Viking epics to the small robots Spirit and Opportunity (on 2003 and 2004).)
(2) Le 28 novembre 1964, la sonde Mariner 4 est lancée vers Mars, 20 jours après l' échec de Mariner 3.
(On November 28th, 1964, the probe Mariner 4 is launched towards Mars, 20 days after the failure of Mariner 3.)
(3) Les robots Spirit et Opportunity , lancés respectivement le 10 juin 2003 et le 8 juillet 2003 par la NASA , représentent certainement la mission la plus avancée jamais réussie sur Mars.
(Robots Spirit and Opportunity, launched respectively on June 10th, 2003 and July 8th, 2003 by the NASA, represent certainly the most advanced successful mission on Mars.)
(4) Le bilan de l'exploration de Mars est d'ailleurs plutôt mitigé : deux tiers des missions ont échoué et seulement cinq des quinze tentatives d'atterrissage ont réussi (Viking 1 et 2, Mars Pathfinder et les deux MER).
(The assessment of the exploration of Mars is rather mitigated: two thirds of the missions failed and only five out of fifteen landing attempts succeeded (Viking 1 and 2, Mars Pathfinder and both SEA))
(5) Le 6 août 2012, le rover Curiosity a atterri sur Mars avec 80 kg de matériel à son bord.
(On August 6th, 2012, the rover Curiosity landed on Mars with 80 kg of material aboard.)
(6) Arrivé sur Mars en janvier 2004 comme son jumeau Spirit, et prévu comme lui pour fonctionner au moins trois mois, Opportunity (alias MER-B) roule encore et plusieurs de ses instruments répondent présents.
(Arrived on Mars in January 2004 with its twin Spirit, and planned to work at least three months, Opportunity (alias MER-B) still runs and several of its instruments are still operational.)
(7) Mars est mille fois plus lointaine que la Lune et son champ d'attraction plus de deux fois plus intense : la technologie n'existe pas pour envoyer un équipage vers Mars et le ramener sur Terre.
(Mars is a thousand times more distant than the Moon and its gravitation field is more than twice as intense: the technology to send a crew on Mars and get it back on Earth, does not exist .)

pairs	(1)-(2)	(1)-(3)	(1)-(4)	(1)-(5)	(1)-(6)	(1)-(7)
mean score	1.49	2.06	1.86	1.19	1.57	1.1

The participants worked independently and with no time constraint on a web application which was developed to ease the annotation task. For each reference sentence chosen at random, its associated sentences were randomly and successively presented to the annotator. A history of the similarity scores was available and the annotators were free to modify them at any time.

Seven human volunteers, aged 18 to 60, were involved in the annotation task and three of them were experts.

Table 2. Pearson correlation scores between one annotator and the rest of group

Annotators	1	2	3	4	5	6	7
Correlation (space conquest)	0.872	0.869	0.844	0.941	0.886	0.815	0.855
Standard Deviation (space conquest)	0.586	0.640	0.714	0.364	0.624	0.671	0.568
Correlation (epidemics)	0.862	0.904	0.903	0.931	0.846	0.846	0.806
Standard Deviation (epidemics)	0.544	0.514	0.622	0.367	0.651	0.580	0.617

To investigate the inter-annotator agreement, we compared the scores of each annotator to the averaged scores of the rest of the group. The resulting Pearson correlation scores and standard deviations are given in Table 2. They show that the human raters largely agreed on the definitions used in the scale, even if they found the annotation task quite hard.

4 Results

RESA, that mixes ESA and ROUGE scores, was evaluated using the Pearson correlation coefficient between the system scores and the human scores, as customary in text similarity.

Tables 3 and 4 give Pearson correlations for different values of the tuning parameter p (cf. the formula 1 in Section 2) and for different sizes of the index collection in ESA. The number of reported concepts corresponds to the actual dimension of the representation space, once the ESA inner inverted index has been trimmed. For instance, in Table 3, we selected the 2,000 best related concepts to "space conquest" and we ended up with 1,492 concepts.

The main outcome of this experimentation is that, whatever is the size of the index collection, ESA and ROUGE perform better when they are combined than they do individually. However, there is a good chemistry to find between them. ROUGE always benefits from ESA which in turn, does not benefit from "too much" of ROUGE.

Table 3. Pearson correlations between the group of annotators and the system for the "space conquest" dataset. *(p=0: ROUGE and p=1: ESA).*

$p=$	0	0.10	0.15	0.175	0.20	0.25	0.3	0.35	0.4	0.6	0.8	1
1492 concepts	.800	.814	.819	.821	.8231	.8257	**.8265**	.8251	.821	.777	.682	.554
1857 concepts	.800	.813	.818	.820	.8221	.8247	**.8256**	.8246	.821	.779	.687	.558
3349 concepts	.800	.813	.819	.821	.8225	.8251	**.8259**	.8247	.821	.778	.684	.555

Table 4. Pearson correlations between the group of annotators and the system for the "epidemics" dataset. *($p = 0$: ROUGE and $p = 1$: ESA).*

$p=$	0	0.10	0.15	0.175	0.20	0.25	0.3	0.35	0.4	0.6	0.8	1
1,016 concepts	.751	.770	.7732	**.7734**	.7726	.768	.760	.749	.735	.666	.591	.525
1,721 concepts	.751	.769	**.7722**	.7720	.7708	.765	.757	.745	.730	.661	.588	.525
3,002 concepts	.751	.770	.7734	**.7735**	.7727	.768	.760	.749	.736	.668	.597	.533
5,094 concepts	.751	.770	.7729	**.7729**	.7720	.767	.759	.748	.734	.666	.595	.531

With the "space conquest" dataset, the highest correlation score of RESA is 0.8265 and it is achieved for a mixing parameter $p \simeq 0.3$ and about 1,500 concepts. The corresponding scores of the sole ROUGE and ESA are 0.800 and 0.554 respectively.

With the "epidemics" dataset, the best performance of RESA is 0.7735 and it is achieved for a mixing parameter $p \simeq 0.175$ and about 3,000 concepts. The corresponding scores of the sole ROUGE and ESA are 0.751 and 0.533 respectively.

The best value for the mixing parameter p does not seem to depend on the dimension of ESA representation space but rather seems to depend on the datasets. Indeed, "space conquest" sentences share generally more common words than "epidemics" sentences, which leads in the "epidemics" dataset, to a higher contribution of ROUGE in the RESA similarity scores.

5 Conclusion

We proposed a similarity measure, (RESA), between sentences which takes advantage of the knowledge-based model ESA and the lexical measure ROUGE. The similarity scores provided by RESA are linear combinations of ESA and ROUGE scores with p and $1-p, 0 \leq p \leq 1$ respective weights. When tested on two French datasets collected from the Wikipedia, RESA proved its efficiency and outperforms ESA for p values not exceeding a threshold which seems to depend on the datasets.

Further work is needed to understand what features of the dataset may influence the mixing parameter p (domain, language, lengths of the sentences, their complexity...), and to assess the generalization performance of the RESA measure. Currently, other experiments are conducted on SEMEVAL 2013 data (English language) and in other domains (French language).

The approach presented in this paper is a part of an ongoing work on summarization of domain-oriented French documents. It remains to evaluate the performance of RESA relatively to this specific task.

References

1. Achananuparp, P., Hu, X., Shen, X.: The evaluation of sentence similarity measures. In: Song, I.-Y., Eder, J., Nguyen, T.M. (eds.) DaWaK 2008. LNCS, vol. 5182, pp. 305–316. Springer, Heidelberg (2008)
2. Agirre, E., Cer, D., Diab, M., Gonzalez-Agirre, A., Guo, W.: *sem 2013 shared task: Semantic textual similarity. In: Second Joint Conference on Lexical and Computational Semantics (*SEM). Proceedings of the Main Conference and the Shared Task: Semantic Textual Similarity, vol. 1, pp. 32–43. Association for Computational Linguistics, Atlanta (2013), http://www.aclweb.org/anthology/S13-1004
3. Balasubramanian, N., Allan, J., Croft, W.B.: A comparison of sentence retrieval techniques. In: Kraaij, W., de Vries, A.P., Clarke, C.L.A., Fuhr, N., Kando, N. (eds.) SIGIR, pp. 813–814. ACM (2007)
4. Barzilay, R., Elhadad, N.: Sentence alignment for monolingual comparable corpora. In: Proceedings of the 2003 Conference on Empirical Methods in Natural Language Processing, EMNLP 2003, pp. 25–32. Association for Computational Linguistics, Stroudsburg (2003), http://dx.doi.org/10.3115/1119355.1119359

5. Buscaldi, D., Le Roux, J., Garcia Flores, J.J., Popescu, A.: Lipn-core: Semantic text similarity using n-grams, wordnet, syntactic analysis, esa and information retrieval based features. In: Second Joint Conference on Lexical and Computational Semantics (*SEM). Proceedings of the Main Conference and the Shared Task: Semantic Textual Similarity, vol. 1, pp. 162–168. Association for Computational Linguistics, Atlanta (2013), http://www.aclweb.org/anthology/S13-1023
6. Dan, A., Bhattacharyya, P.: Cfilt-core: Semantic textual similarity using universal networking language. In: Second Joint Conference on Lexical and Computational Semantics (*SEM). Proceedings of the Main Conference and the Shared Task: Semantic Textual Similarity, vol. 1, pp. 216–220. Association for Computational Linguistics, Atlanta (2013), http://www.aclweb.org/anthology/S13-1031
7. Dasari, D.B., Rao, V.G.: A text categorization on semantic analysis. International Journal of Advanced Computational Engineering and Networking 1(9) (2013)
8. Egozi, O., Markovitch, S., Gabrilovich, E.: Concept-based information retrieval using explicit semantic analysis. ACM Trans. Inf. Syst. 29(2), 8:1–8:34 (2011), http://doi.acm.org/10.1145/1961209.1961211
9. Erkan, G., Radev, D.R.: Lexrank: Graph-based lexical centrality as salience in text summarization. J. Artif. Intell. Res. (JAIR) 22, 457–479 (2004)
10. Gabrilovich, E., Markovitch, S.: Computing semantic relatedness using wikipedia-based explicit semantic analysis. In: Proceedings of the 20th International Joint Conference on Artifical Intelligence, IJCAI 2007, pp. 1606–1611. Morgan Kaufmann Publishers Inc., San Francisco (2007), http://dl.acm.org/citation.cfm?id=1625275.1625535
11. Gottron, T., Anderka, M., Stein, B.: Insights into explicit semantic analysis. In: CIKM 2011: Proceedings of 20th ACM Conference on Information and Knowledge Management (2011), http://dl.dropbox.com/u/20411070/Publications/2011-CIKM-Gottron-AS.pdf
12. Gupta, R., Ratinov, L.: Text categorization with knowledge transfer from heterogeneous data sources. In: Proceedings of the 23rd National Conference on Artificial Intelligence, AAAI 2008, vol. 2, pp. 842–847. AAAI Press (2008), http://dl.acm.org/citation.cfm?id=1620163.1620203
13. Ko, Y., Park, J., Seo, J.: Automatic text categorization using the importance of sentences. In: Proceedings of the 19th International Conference on Computational Linguistics (COLING 2002), pp. 65–79 (2002)
14. Li, Y., McLean, D., Bandar, Z.A., O'Shea, J.D., Crockett, K.: Sentence similarity based on semantic nets and corpus statistics. IEEE Trans. on Knowl. and Data Eng. 18(8), 1138–1150 (2006), http://dx.doi.org/10.1109/TKDE.2006.130
15. Lin, C.: Rouge: a package for automatic evaluation of summaries, pp. 25–26 (2004)
16. Lin, C.Y., Hovy., E.: Automatic evaluation of summaries using n-gram co-occurrence statistics. In: Proceedings of 2003 Language Technology Conference (HLT-NAACL 2003), Edmonton, Canada (May-June 2003)
17. Lin, D.: An information-theoretic definition of similarity. In: Proceedings of the 15th International Conference on Machine Learning, pp. 296–304. Morgan Kaufmann (1998)
18. Müller, C., Gurevych, I.: A study on the semantic relatedness of query and document terms in information retrieval. In: Proceedings of the 2009 Conference on Empirical Methods in Natural Language Processing, EMNLP 2009, vol. 3, pp. 1338–1347. Association for Computational Linguistics, Stroudsburg (2009), http://dl.acm.org/citation.cfm?id=1699648.1699680
19. Nakayama, K., Hara, T., Nishio, S.: Wikipedia mining for an association web thesaurus construction. In: Proceedings of IEEE International Conference on Web Information Systems Engineering, pp. 322–334 (2007)

20. Potthast, M., Barrón-Cedeño, A., Stein, B., Rosso, P.: Cross-language plagiarism detection. Lang. Resour. Eval. 45(1), 45–62 (2011), http://dx.doi.org/10.1007/s10579-009-9114-z
21. Sorg, P., Cimiano, P.: Cross-lingual information retrieval with explicit semantic analysis. In: Working Notes for the CLEF 2008 Workshop (2008), http://www.aifb.kit.edu/images/7/7c/2008_1837_Sorg_Cross-lingual_I_1.pdf
22. Tsatsaronis, G., Panagiotopoulou, V.: A generalized vector space model for text retrieval based on semantic relatedness. In: Proceedings of the 12th Conference of the European Chapter of the Association for Computational Linguistics: Student Research Workshop, EACL 2009, pp. 70–78. Association for Computational Linguistics, Stroudsburg (2009), http://dl.acm.org/citation.cfm?id=1609179.1609188
23. Wong, S.K.M., Ziarko, W., Wong, P.C.N.: Generalized vector spaces model in information retrieval. In: Proceedings of the 8th Annual International ACM SIGIR Conference on Research and Development in Information Retrieval, SIGIR 1985, pp. 18–25. ACM, New York (1985), http://doi.acm.org/10.1145/253495.253506

Incorporating Language Patterns and Domain Knowledge into Feature-Opinion Extraction

Erqiang Zhou[1,2], Xi Luo[1], and Zhiguang Qin[1]

[1] School of Computer Science and Engineering, University of Electronic Science and Technology of China, No.2006, Xiyuan Ave, Chengdu 611731, Sichuan, P.R. China
{zhoueq,qinzg}@uestc.edu.cn, xiluouestc@gmail.com

[2] Guangdong Key Laboratory of Popular High Performance Computers, Shenzhen Key Laboratory of Service Computing and Applications, No. 3688, Nanhai Ave, Shenzhen 518060, Guangdong, P.R. China

Abstract. We present a hybrid method for aspect-based sentiment analysis of Chinese restaurant reviews. Two main components are employed so as to extract feature-opinion pairs in the proposed method: domain independent language patterns found in Chinese and a lexical base built for restaurant reviews. The language patterns focus on the general knowledge which is implicit contained in Chinese, thus can be used directly by other domains without any modification. The lexical base, on the other hand, targets for particular characteristics of a given domain and acts as a plug-in part in our prototype system, thus does not affect the portability when applying the proposed approach in practice. Empirical evaluation shows that our method performs well and it can gain a progressive result when each component takes into effective.

Keywords: Opinion Mining, Sentiment Analysis, Restaurant Review.

1 Introduction

As a kind of user generated content, online reviews become more important whenever for the consumers and the sellers of e-commerce or for a service provider and its user. Based on the reviews of online products or services, consumers or users can evaluate the aspects with which they are concerned in detail in addition to the overall scores; sellers or service providers can know which aspects need to be improved for attracting more clients.

Realizing the importance of extracting feature-opinions for practical requirements, much efforts [2,5,10,11,16] have been devotes into this field since the pioneer work done by Hu and Liu [4]. However, automatic acquiring feature opinion pairs from texts is still a challenging problem.

In fact, in addition to a few researches [3,12,14,15], most of existing work in this research field focused on English reviews and has not attempted to handle texts written in other languages. Furthermore, to the best of our knowledge, the corpus of Chinese restaurant reviews, which is the target of our work, has not been fully investigated in the field of sentiment analysis.

P. Sojka et al. (Eds.): TSD 2014, LNAI 8655, pp. 209–216, 2014.

Although the domain of restaurant reviews share main characteristics with other domains for feature-opinion extraction, it has particular distinguishing points. For example, some Chinese dish names not only tell the ingredients of certain dishes, but also imply how they are cooked or which regional style they may have, such as '蒜苗回锅肉' (garlic sprouts with twice-cooked pork), '水煮牛肉' (water cooked beef), '重庆辣子鸡' (Chongqing spicy deep-fried chicken), etc. This fact leads to the difficulty of identifying such kind of entity names. Another fact which distinguishes the corpus of restaurant from product or movies reviews is that the quality of dishes and services provided by restaurants is time sensitive. In other words, the features of a product or a movie does not change once it is put on the market, while foods and services are hand work thus a recent opinion is more important than the viewpoint of a long time ago.

Based on all these facts, a fully investigation of Chinese restaurant reviews has both practical and scientific values. In this paper, we study the problem of feature-opinion extraction from online reviews of Chinese restaurants. We address the problem by employing following two steps. Firstly, we build a knowledge base for the domain of restaurant reviews. The base consists of two parts: 1) the feature words and their potential aspects, which can be classified into following five categories: taste, environment, service, price and overall evaluations; 2) the possible opinion words for remarking restaurant features. Secondly, we develop a rule sets for capturing the general patterns that are used when people expressing their opinions.

Our experiments are conducted on a real life dataset, which consists of 5,251 restaurants and each restaurant has 26 reviews in average. The overall extracted review texts is 125M bytes, we randomly selected 5 restaurants and manually labelled the feature-opinion pairs of all their reviews for testing. It should be noted that our test standard is whether the automatically extracted feature-opinion pairs match the manually labelled ones. Experiments show that the proposed method can reach a reasonable and acceptable result: the average Precision, Recall and F-score is 0.55, 0.83 and 0.66 respectively.

2 Related Work

There are two main tasks involved in the research of extracting feature-opinion: feature identification and opinion extraction. Hu and Liu [4] utilized frequent item to identify product features and considered adjectives which modifies feature words as the opinion words. Zhuang *et al.* [16] integrated WordNet, statistical analysis and movie knowledge into a multi-knowledge base so as to extract feature opinion pairs and summarize movie reviews. Somprasertsri and Lalitrojwong [11] extracted product feature and opinion by considering syntactic and semantic information, that is by applying dependency relations and ontological knowledge with probabilistic based model.

It is worth notice more and more ontologies or knowledge bases are used for the tasks of opinion mining recently. For example, O'Leary [8] utilized knowledge gained from tags on blogs for blog mining and showed that there is a need for domain specific terms so as to capture a richer understanding of mood of a blog. Peñalver-Martínez *et al.* [9] proposed a method for improving feature-based opinion mining by employing ontologies in the selection of features. Freitas and Vieira [3] identified the polarity of Portuguese reviews according to features described in domain (movie and hotel) ontologies. Ittoo and Bouma [5] showed that open-domain corpora, like Wikipedia, can be exploited as knowledge bases for extracting causal relations from domain-specific texts so as to overcome data sparsity issues. Yin *et al.* [14] presented a linguistic model, which is based on a automatically constructed ontology, for identifying the basic appraisal expression in Chinese product reviews.

In addition to methods of using ontologies, several kinds of rules are designed in the field of opinion mining. Zhai *et al.* [15] presented an unsupervised approach to identify people's opinions on topics and their aspects (the so called evaluative opinions) by proposing several rules. Jiao *et al.* [6] employed Chinese dependency grammar to set several rules for extracting candidate feature-opinion word pairs.

Statistical topic models are also often employed for tackling opinion mining tasks. Brody and Elhadad [1] applied the LDA model to the unsupervised features extract: they treated one sentence as a document and adjusted the parameters of LDA so as to make it suit the sentence-document model. After finding feature words in some topic, Brody and Elhadad treated the nearest adjective words as the opinion words. Li *et al.* [7] introduced a dependency-sentiment-LDA, which relaxes the sentiment independent assumption and is an extension of a joint sentiment and topic model, Sentiment-LDA, which is also proposed by the authors. Xu *et al.* [13] propose a generative topic model, the Joint Aspect/Sentiment (JAS) model, to jointly extract aspects and aspect-dependent sentiment lexicons from online customer reviews.

However, according to our experiments, the LDA-based topic models fail to find the feature words from our Chinese restaurant reviews. One possible reason may be the so called data sparsity issue: the dish names and the recipe of foods are candidates words of restaurant features, but they usually varies in restaurant reviews. We handle this problem by semi-automatically building an lexical base for the domain of Chinese restaurant reviews.

3 Opinion-Feature Extraction

Figure 1 gives an overview of the proposed method. Three main steps involves in the identification of feature-opinion pairs: 1) pre-processing; 2) lexical base building and; 3) feature-opinion pair extraction. As to the first step, we use ICTCLAS[1] to POS-tag all reviews, and then each review is split into sentences for further processing. The second step and the third step will be explained in Section 3.1 and Section 3.2 respectively.

3.1 Lexical Base Building

The lexical base consists three parts: aspect words, dish names, words that can match with aspect words and dish names. Accordingly, the building process can be decomposed into three tasks: 1) to collect dish names; 2) to identify aspect words and; 3) to extract the matching words by designing rules.

As to the first task, a crawling spider is developed to obtain candidate dish names from the website of *www.dianping.com* since it allows users to write their own recommending dishes for each restaurant. However, some irrelevant phrases or opinions (such

[1] `http://www.ictclas.org/ictclas_download.aspx`

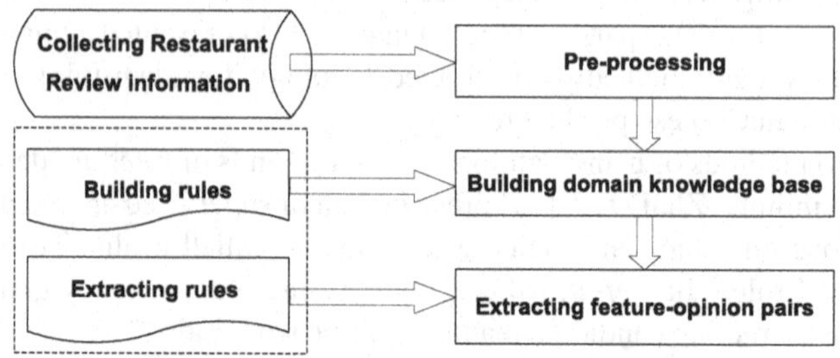

Fig. 1. Overview of the proposed method

as: 都好吃/delicious, 不喜欢/dislike, etc.) were also given by users, thus a cleanup step is needed for removing such kind of noises. In our research, we treat the word that are rarely used in dish name as seed words for identifying the noise phrases. Finally we collect 6433 dish names altogether.

As to the second task, we define a noun as a *frequent-noun* if its *sentence appear ratio*, which is defined as a division between the number of sentences which has the *frequent-noun* and the number of all review sentences, is great than 1%. The generated *frequent-nouns* are treated as candidate aspect words in this paper.

To perform the third task, following two rules are used:

Rule 1: {(F, N)—(A, V)}. This rule means that if a noun N or a dish name F appears near a word which was labeled as an adjective A or a verb V in one *short sentence*[2], the noun term N or dish name F and adjective A or verb V are considered as a candidate pair of matching words.

Rule 2: {F—N}. If a food term F appears with a noun term N in a short sentence, F and N are considered as a candidate pair of matching words. The noun N has certain chance of being an aspect of the food.

We define the number when term N or dish name F and adjective A or verb V co-occurs as the *Matched Degree* (MD). The bigger an MD is, the more chance of being a pair the term N (or F) and the term A (or V) have. A global threshold value of MD is used in our research.

The rules are quite intuitive, and they are mainly used for finding the matching words. For example, in POS-tagged sentence '环境/n (environment) 和/c (and) 味道/n (taste) 很/d (very) 不错/a (nice)', '环境' (environment) and '味道' (taste) are both noun terms, '不错' (nice) is an adjective word which co-occurs with '环境' (environment) and '味道' (taste). So the MD of '环境(environment)—不错(nice)' and '味道(taste)—不错(nice)' will all increase one number each.

3.2 Extracting Feature-Opinion Pairs

The extraction of feature-opinion pairs is performed by exploring following rules:

[2] Short sentences are obtained by splitting a natural sentence by punctuations.

Rule 3: if a noun N appears with and an adjective A (or a verb V), and there is no other[3] nouns, adjectives or verbs between N and A, then we make use of the MD of matching pair to decide whether N and A can be treated as a feature-opinion pair.

Rule 4: if a dish name F and an adjective A (or a verb V) appear in a *short sentence*, decide whether F and A can be treated as a feature-opinion pair according to the global threshold value of MD.

Rule 5: if a frequent-noun follows a dish name and an adjective comes along with the frequent-word, the frequent-noun might be an attribute of the dish. Then we check the MD of pair (dish name, frequent-noun), if the MD satisfies the global threshold, we extract the pair (dish name, frequent-noun) as a feature-opinion pair; otherwise the following two pairs (dish name, adjective) and (frequent-noun, adjective) should be checked.

For example, in POS-tagged sentence '回锅肉/ny (Double cooked pork) 和/c (and) 环境/n (environment) 很/d (very) 好/a (good)', '回锅肉(Double cooked pork)' is a dish name since it has a tag 'ny', and '环境(environment)' is a frequent-noun. So we first check whether the MD of pair (回锅肉(Double cooked pork), 环境(environment)) satisfy the global threshold value, if not, the two pairs (回锅肉(Double cooked pork, 好(good)) and (环境(environment), 好(good)) should be check. In our lexical base, these two pairs both satisfy the global threshold, therefore, they are treated as feature-opinion pairs.

4 Empirical Evaluation

The dataset that we used for evaluating the proposed was crawled from one of the most popular reviewing website (www.dianping.com) in China. The dataset consists of two parts: The first part includes all restaurant reviews of Chengdu city (about 125MB). The overall number of restaurants is 22579 while only 5251 ones have been reviewed more than 10 times. The reviews of these 5251 restaurants is used for building our lexical base.

As to the second part, we firstly filter restaurants which have more than 100 reviews, then the most recent 100 reviews of 5 random selected restaurants are chosen as the second part of our dataset. The five restaurants are 陈麻婆豆腐(CMP),大嘴霸王排骨(DZ),好伦哥(HLG), 红杏酒家(HX), 陋室茶居烧烤五花肉(LS). Two students are employed to annotate all sentences and find the feature-opinion pairs. The Kappa scores for inter-rater agreements range from 0.75 to 0.79, which indicate good agreement. The details of the labeled feature-opinions pairs are given in Table 1.

4.1 Experiment Settings

Our method RuleMD is compared with three baseline methods: SimpleNoun, SimpleVerb, RuleNoMD which are described as follows.

SimpleNoun: This method only uses the *frequent-nouns* as feature words. If an adjective appears with a *frequent-noun* in a *short sentence*, the adjective is regarded as an opinion word. If there are more than one adjective words, only the closest one is regarded as the opinion word.

[3] Words like adverbs, conjunction and auxiliary words can appear between N and A.

Table 1. Restaurant details

Shop	Labeled-Pairs	Kappa
CMP	284	0.776
DZ	330	0.752
HLG	308	0.754
HX	386	0.791
LS	265	0.784

Table 2. F-score of restaurants

Index	F-score					
Shop	CMP	DZ	HLG	HX	LS	Arg
SimpleNoun	0.49	0.55	0.53	0.58	0.46	**0.52**
SimpleVerb	0.51	0.60	0.60	0.65	0.56	**0.58**
RuleNoMD	0.53	0.64	0.65	0.68	0.61	**0.62**
RuleMD	0.56	0.67	0.68	0.71	0.65	**0.66**

Table 3. Comparison Results

Index	Precision						Recall					
Shop	CMP	DZ	HLG	HX	LS	Arg	CMP	DZ	HLG	HX	LS	Arg
SimpleNoun	0.40	0.46	0.43	0.48	0.36	**0.43**	0.62	0.69	0.67	0.73	0.66	**0.67**
SimpleVerb	0.40	0.48	0.46	0.51	0.40	**0.45**	0.69	0.81	0.84	0.91	0.88	**0.83**
RuleNoMD	0.43	0.53	0.52	0.53	0.46	**0.49**	0.72	0.82	0.86	0.93	0.90	**0.85**
RuleMD	0.48	0.57	0.57·	0.59	0.52	**0.55**	0.67	0.80	0.85	0.89	0.88	**0.83**

SimpleVerb: This method extends SimpleAdj by considering the verbs which have a 'v' POS-tag. A verb is considered as an opinion words if it is a *frequent-verb*, in other words, its *sentence appear ratio* is great than 1%.

RuleNoMD: This method extends SimpleVerb by incorporating the modified rules described in Section 3.2. The modification is made by disregarding the global threshold value of MD for further selecting the extracted feature-opinion pairs. That is to say, all word pairs that conform to the rule patterns will be treated as feature-opinion pairs. The purpose of testing this method is to check whether the rule pattern can improve performance than SimpleVerb, and furthermore, to check whether the lexical base has effects in feature-opinion extraction.

RuleMD: This is our proposed method and it extends SimpleVerb by incorporating the rules described in Section 3.2.

4.2 Evaluation Results

The comparison results are shown in Table 2 and Table 3, where 'Arg' represents the average result of the five restaurants. Below we discuss some detailed observations:

1: On F-Score, SimpleVerb is better than SimpleNoun. The reason is that certain verbs, such as '喜欢' (like), '失望' (disappoint), directly express the users' opinions. From Table 3 we can see that SimpleVerb's recall is much higher (16%) than SimpleNoun, and SimpleVerb's precision is also improved by 2%. This fact shows that the adding of verbs plays an important role in opinion extractions. Accordingly, the F-score is improved by 6%.

2: On F-Score, RuleNoMD is better than SimpleVerb. In addition, the precision improves by 4% and recall has 2% advances. The reason is that certain feature opinion pairs are not adjacent in a sentence, and the rules given in Section 3.2 is effective for capturing such kind of pairs.

3: On F-Score, RuleMD is better than RuleNoMD because of the using of our lexical base. It is worth notice that while the precision improves by 6%, recall decrease by 2%. This result shows that while we improve the precision by filtering some extracted feature-opinion pairs, a small part of correct pairs are also affected, consequently, ther recall of RuleMD is lower than RuleNoMD.

The proposed method has one importance parameter: the global threshold value of MD. Figure 2 shows the influences of this parameter on the overall performance. In Figure 2, the proposed method achieves the best average performance when the threshold is around 5.

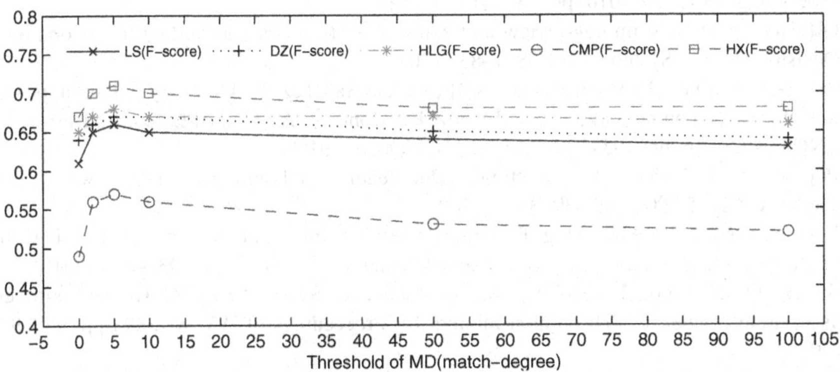

Fig. 2. Influence of the global threshold value of MD

5 Conclusion

This paper presented a hybrid approach that incorporates domain independent language patterns of Chinese and a lexical base for addressing the problem of feature-opinion extraction. Experiments showed that our method is effective and can gain a reasonable result. Currently, the introduced method has only been tested on the domain of restaurant reviews. In future work, we will further adapt our method (such as developing an adaptive architecture, refining the designed rules) so that it can be ported to other domain with minimal efforts.

Acknowledgements. This work was jointly supported by the National Natural Science Foundation of China (No. 61133016, No. 61300094) and the Fundamental Research Funds for the Central Universities (No. ZYGX2013J070).

References

1. Brody, S., Elhadad, N.: An unsupervised aspect-sentiment model for online reviews. In: Proceeding of HLT 2010, pp. 804–812 (2010)
2. Bross, J., Ehrig, H.: Automatic construction of domain and aspect specific sentiment lexicons for customer review mining. In: Proceedings of CIKM 2013, pp. 1077–1086 (2013)

3. Freitas, L.A., Vieira, R.: Ontology based feature level opinion mining for portuguese reviews. In: Proceedings of WWW 2013 Companion, pp. 367–370 (2013)
4. Hu, M., Liu, B.: Mining and summarizing customer reviews. In: Proceedings of SIGKDD 2004, pp. 168–177 (2004)
5. Ittoo, A., Bouma, G.: Minimally-supervised learning of domain-specific causal relations using an open-domain corpus as knowledge base. Data & Knowledge Engineering 85(0), 57–79 (2013)
6. Jiao, F., Dong, G., Li, Q., Zhu, J.: Opinminer: Extracting feature-opinion pairs with dependency grammar from chinese product reviews. In: Proceedings of WISA 2012, pp. 217–222 (2012)
7. Li, F., Huang, M., Zhu, X.: Sentiment analysis with global topics and local dependency. In: Proceedings of AAAI 2010, pp. 1371–1376 (2010)
8. O'Leary, D.E.: Blog mining-review and extensions:"from each according to his opinion". Decision Support Systems 51(4), 821–830 (2011)
9. Peñalver-Martínez, I., Valencia-García, R., García-Sánchez, F.: Ontology-guided approach to feature-based opinion mining. In: Muñoz, R., Montoyo, A., Métais, E. (eds.) NLDB 2011. LNCS, vol. 6716, pp. 193–200. Springer, Heidelberg (2011)
10. Popescu, A.M., Etzioni, O.: Extracting product features and opinions from reviews. In: Proceedings of HLT 2005, pp. 339–346 (2005)
11. Somprasertsri, G., Lalitrojwong, P.: Mining feature-opinion in online customer reviews for opinion summarization. Journal of Universal Computer Science 16(6), 938–955 (2010)
12. Su, Q., Xu, X., Guo, H., Guo, Z., Wu, X., Zhang, X., Swen, B., Su, Z.: Hidden sentiment association in chinese web opinion mining. In: Proceedings of WWW 2008, pp. 959–968 (2008)
13. Xu, X., Cheng, X., Tan, S., Liu, Y., Shen, H.: Aspect-level opinion mining of online customer reviews. Communications, China 10(3), 25–41 (2013)
14. Yin, P., Wang, H., Guo, K.: Feature pinion pair identification of product reviews in chinese: a domain ontology modeling method. New Review of Hypermedia and Multimedia 19(1), 3–24 (2013)
15. Zhai, Z., Liu, B., Zhang, L., Xu, H., Jia, P.: Identifying evaluative sentences in online discussions. In: Proceedings of AAAI 2011, pp. 933–938 (2011)
16. Zhuang, L., Jing, F., Zhu, X.Y.: Movie review mining and summarization. In: Proceedings of CIKM 2006, pp. 43–50 (2006)

BFQA: A Bengali Factoid Question Answering System

Somnath Banerjee, Sudip Kumar Naskar, and Sivaji Bandyopadhyay

Department of Computer Science and Engineering,
Jadavpur University, India
s.banerjee1980@gmail.com,
{sudip.naskar,sbandyopadhyay}@cse.jdvu.ac.in

Abstract. Question Answering (QA) research for factoid questions has recently achieved great success. Presently, QA systems developed for European, Middle Eastern and Asian languages are capable of providing answers with reasonable accuracy. However, Bengali being among the most spoken languages in the world, no factoid question answering system is available for Bengali till date. This paper describes the first attempt on building a factoid question answering system for Bengali language. The challenges in developing a question answering system for Bengali have been discussed. Extraction and ranking of relevant sentences have also been proposed. Also extraction strategy of the ranked answers from the relevant sentences are suggested for Bengali question answering system.

Keywords: BFQA, Question Answering (QA), Bengali Factoid QA.

1 Introduction

A QA system is an automatic system capable of answering natural language questions in a human-like manner: with a concise, precise answer. Generally questions can be classified into five broad categories ([1,2]): factoid questions, list questions, definition questions, complex questions and speculative questions.

As there exists no Bengali QA system till date and the development of Bengali QA system is at its nascent stage, our aim for the work reported here was to address the factoid questions for the following factors:- (i) a considerable percentage of the questions actually submitted to a search engine belongs to factoid questions. (ii) The percentages of factoid questions are increased each year in TREC due to frequent occurrences in daily usage. (iii) Sophisticated state-of-the-art approaches to open-domain QA use named entity recognition as a core process for detecting candidate answers.

2 Related Work

Designing a QA system for European languages particularly for English is not new in natural language processing. A number of QA systems have been developed since the 1960s. Two such early QA systems were BASEBALL[3] and LUNAR[4].The most notable QA system available to date is IBM Watson [2] which was developed under

P. Sojka et al. (Eds.): TSD 2014, LNAI 8655, pp. 217–224, 2014.

IBM's DeepQA project. Research in QA received significant boost when a shared task on factoid QA was included in the 8thText REtrieval Conference (TREC).

A number of QA systems were developed for European languages particularly for English ([5,6,7]), Middle Eastern languages ([8,9,10]) and Asian languages, e.g., Japanese ([11,12]) Chinese ([13,14]), etc. The aforesaid systems are capable of providing answers with reasonable accuracy. However, for Bengali, which is a widely spoken language in India and among the most spoken languages in the world, very little work [12,13,14] have been reported so far in QA research like other Indian languages.

3 Challenges

To the best of our knowledge, there exists no QA system till date for Bengali. Developing a QA system for low resource language is very much challenging. Several issues were confronted for developing the system which includes-

- *Presence of many interrogatives*: Unlike English there are many interrogatives present in the Bengali. A study [15] identified a total of twenty six interrogatives and classified them into three categories – Unit Interrogative (UI), Dual Interrogative (DI) and Compound/Composite Interrogative (CI).
- *Interrogative position*: A Bengali interrogative can appear in all potential positions, i.e., three positions (first, in between, last) of a question text [12]. This makes it difficult to propose rule-based question analysis.
- *Resource scarcity*: The language processing tools for Bengali are either under development phase or not developed yet. Even a fully-fledged parser has not been developed yet and no NER system is publicly available for Bengali. Besides, gold-standard corpora for QA research are not developed yet.

4 BFQA Architecture

Our proposed factoid QA system for Bengali language, named BFQA, has a pipeline architecture having three components, namely question analysis, sentence extraction and answer extraction. The question analysis module accepts natural language question in Bengali as input posed by the user. The question analysis step processes in five stages, namely question type (QType) identification, expected answer type (EAT) identification, named entity identification, question topical target (QTT) identification and keyword identification. The valid keywords are 'AND'ed together to form the query. Sentences are extracted from the passages based on the query and are ranked based on the answer score value. Finally, extracted answers are validated using the EAT module. The architecture of the proposed model is depicted in Figure 1

5 Question Analysis

Question analysis plays a crucial role for an automatic QA system. Acquiring the information embedded in a question is a primary task that allows the QA system

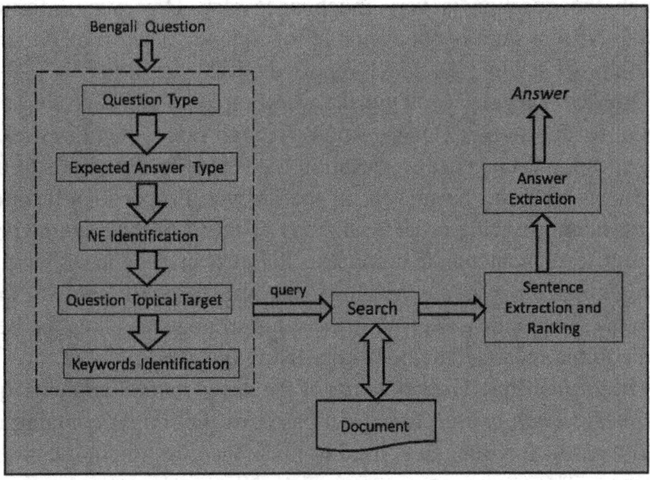

Fig. 1. BFQA Architecture

to decide the appropriate strategy in order to provide the correct answer to the question. [18] stated that when the question analysis module fails, it is hard or almost impossible for a QA system to provide correct answer. In this work, question analysis task is divided into five parts: question type identification, inferring the expected answer type, obtaining the question topical target, named entity (NE) extraction and identification of keywords.

Question Type: Question Type (QType) identification or question classification is an important component of every Question Answering System (QAS). Depending on the classification strategy, the task of a question classifier is to assign one or more predefined class labels to a given question written in natural language. The set of predefined categories which are considered as question classes is usually referred to as question taxonomy or answer type taxonomy. The single layer taxonomy with nine coarse-grained classes proposed by [12] is the only standard taxonomy available in Bengali QA research. We followed the approach of [14] which is the best reported work (91.65% accuracy) yet for Bengali QType identification task.

Expected Answer Type: Prager ([16]) defined the Expected Answer Type (EAT) as the class of object or rhetorical type of sentence required by the question. Another way EAT may be defined is as the semantic category associated with the desired answer, chosen from a predefined set of labels. A set of hand-crafted rules written as regular expressions (RE) were used by many sophisticated QA systems ([17],[18],[19]) for finding the EAT. Every RE is associated with an EAT that is to be assigned to a question if it matches its pattern. Though hand-crafted rules are very specific to the particular domain, but those are very useful at the initial stage of the development. In this work, hand-written rules were used to determine EAT of a question.

Named Entity Identification: NEs play a crucial role in question answering. NEs present in the question are believed to be present in the sentence that contains the expected answer. Again, the answer of a factoid question is a NE. Therefore, NE identification in the question is very much necessary. However, unfortunately any openly available NER system is not available for Bengali. Therefore, we implemented the NER system proposed by [20] which reported 65.95% accuracy.

Question Topical Target: Knowing the question type alone is not sufficient for finding answers to questions [21]. Question Topical Target (QTT) (sometimes also referred to as question focus [22], or question topic [16]) corresponds to a noun or a noun phrase that is likely to be present in the answer. Proper identification of QTT benefits the question answering process, since QTT terms or their synonyms are likely to appear in a retrieved sentence that contains the answer. Due to varying position of Bengali interrogatives, it is very challenging to separate the QTT when the interrogative appears within the noun phrase. To follow the same strategy for all cases, Bengali QTT is defined as a comma separated named entity list in this work.

Keywords Identification: The keywords of the posed natural language question are used to form query which is used to extract passages that might contain the expected answer. Shallow parser for Bengali is used to parse the question in the absence of any full-fledged parser. The nouns, proper nouns i.e., NEs, verbs and adjectives are taken as legitimate keywords and other words are considered as stop words. As the NEs are already included in QTT, query is formed by merging QTT with nouns, verbs and adverbs present in the question. The valid keywords are 'AND'ed together to form query.

6 Sentence Extraction and Ranking

The developed corpus for this work has the constraint that all the questions are related to a particular document only. So, we imposed some constraints on the searching technique: the answer to the posed question is searched from a single document and each of the paragraphs of the document is treated as individual documents while searching. The task of this module is to extract the relevant sentences that contain the expected answer from different paragraphs of a single document. The relevant sentences are searched using the query. The relevant sentences are searched taking into account the imposed constraints.

The extracted sentences that contain the expected answer are initially passed to the same NE system that tagged the NEs appearing in the question text. The extracted sentences are then ranked on the basis of answer score. EAT and similarity score are used to calculate the answer score. The answer score is calculated based on – (i) syntactic similarity, (ii) name proportions, and (iii) paragraph relevancy.

Syntactic Similarity: if Q_t is the natural language (NL) question and is composed of n words, then the question text Q_t can be expressed as:

$Q_t = Q_1 Q_2 Q_3 Q_4 ... Q_{n-2} Q_{n-1} Q_n$

Let $V_Q = Q_1, Q_2, Q_4, Q_7, Q_8, ... Q_n$, and $V_{Stop} = Q_3, Q_5, Q_6, Q_9, ...$;

Then $Q_t = V_Q \cup V_{Stop}$ and $V_Q \cap V_{Stop} = \emptyset$; where V_Q and V_{Stop} are the two word vectors, namely content and stop words respectively.

V_S is the sentence vector which contains all the sentences in the document of p sentences. i.e. $V_S = V_{S_1}, V_{S_2}, V_{S_3}, V_{S_4}, ... V_{S_{P-1}}, V_{S_P}$; where V_{S_k} represents the kth sentence in the document and contains c words.

$V_{S_k} = w_1, w_2, w_3, w_4, ..., w_{c-1}, w_c$

In similarity measure, we only consider the part-of-speech (POS): verb (VB), noun (NN), adjective (ADJ) and proper noun (NE). Let {VB, NN, ADJ, NE ∈ POS}. Four weights λ_{vb}, λ_{nn}, λ_{adj} and λ_{ne} have been defined corresponding to the verb, noun, adjective and named entity, respectively. We have set $\lambda_{vb} = 0.2$, $\lambda_{np} = 0.3$, $\lambda_{adj} = 0.1$ and $\lambda_{ne} = 0.4$ (so that the four weights add up to 1).

i.e., $\sum_{Pos \in \{vb,np,adj,ne\}} \lambda_{Pos} = 1$

So, the similarity of a NL question Q_t and a sentence S_l is calculated by the following formula: $Similarity_{(Q_t,S_l)} = \sum_{K=1}^{n} frequency_Q.w_K$;

where, $w_K = Q(\lambda_{Pos})$ and *frequency*$_Q$ is the number of occurrence of question word Q in the sentence S_l.

Name proportions(nprop): Jaccard similarity coefficient is used to measure name proportion. Jaccard similarity coefficient is a similarity measure that compares the similarity between two feature sets. In name proportion measure, it is defined as the size of the intersection of the named entities in the question and a sentence normalized by the size of the union of the named entities in the question and the sentence.

Paragraph relevancy: Relevancy of a paragraph to a question is measured by counting the presence of query words in that paragraph. The synonyms of the query words are also considered during relevancy count. To distinguish between original query word and synonymous words a *relevancy weight* is assigned to each appearance.

$$relevancy\ weight(r_w) = \begin{cases} 1.0 & \text{if original query term appears in the paragraph} \\ 0.9 & \text{if synonym appears in the paragraph} \\ 0.0 & \text{neither query word nor any synonym} \end{cases}$$

Each of the words in the paragraph is considered for paragraph relevancy calculation. Paragraph relevancy of a word is calculated as follows.

$$R_w = frequency \times relevancy\ weight$$
$$= f_w \times r_w$$

Therefore, if a paragraph contains k distinct words, then the *paragraph relevancy* for that paragraph can be measured using the following formula.

$$R_p = f_{w_1} \times r_{w_1} + f_{w_2} \times r_{w_2} + f_{w_3} \times r_{w_3} + \ldots + f_{n_1} \times r_{w_n}$$
$$= \sum_{i=1}^{k} f_{w_i} \times r_{w_i}$$

= sum of the paragraph relevancy for each distinct word in the paragraph

Finally, the score for the three metrics are summed up to arrive at the answer score.

answer score = syntactic similarity + name proportion + paragraph relevancy

7 Answer Extraction

Answer Extraction is the final module of the QA pipeline architecture. After the extracted sentences are ranked based on answer score, the answer to the natural language question is determined by the NE which is suggested by EAT in the question analysis module. Here, three cases are possible -

i) Only a single word in the sentence is of suggested NE type by EAT.
ii) Multiple words having the same NE tag suggested by EAT.
iii) No word in the sentence having NE tag suggested by EAT.

In the first case, the answer extraction is trivial and the NE word suggested by EAT is the answer to the question. However, the second case is a bit ambiguous and we need to apply some extra effort. We use a novelty factor to solve this ambiguity, i.e., choose the NE as candidate answer which is not present in the query. In the third case, the QA system simply fails to answer the question.

8 Experiments

As mentioned earlier, the work presented here is focused on factoid questions. Also we had to build our own corpus for experimentation. Corpus preparation and experimentation are described in the following subsections.

Corpus: As per our knowledge, there is no Bengali corpus available for QA research. So we had no other choice but to build our own corpus for experimentation. Fourteen documents from the geography and agriculture domains were acquired from the Wikipedia. Twenty Bengali language experts were involved in this small corpus development work. A total of 184 factoid questions were prepared and annotated according to three question answering based levels, namely Question Class (kappa - 0.91), Expected Answer Type (kappa - 0.85) and Question Topical Target (kappa - 0.89).

Results: Mean Reciprocal Rank (MRR) metric is used to evaluate the QA system. MRR is formulated as follows:

$$\text{MRR} = \frac{1}{|Q|}\sum_{i=1}^{|Q|}\frac{1}{rank_i}$$

where, $rank_i$ represents the best rank of the correct answer of the ith question and $|Q|$ is the number of test questions. Corpus statistics and evaluation results are shown in Table 1 and Table 2, respectively.

Table 1. QType Statistics

Type	Geography	Agriculture	Overall
Person	22	0	22
Organization	4	6	10
Location	6	11	17
Temporal	47	13	60
Numerical	60	15	75

Table 2. Corpus Statistics and System Evaluation

Domain	#Documents	#Questions	MRR
Geography	10	139	0.34
Agriculture	4	45	0.31
Overall	14	184	0.32

9 Conclusion

This paper presents the first attempt to build a factoid QA system for Bengali. We propose an architecture to address the scenarios common to low-resource languages particularly for Indian languages. Also, we discussed the major challenges of developing a QA system for Bengali. We proposed a sentence ranking strategy for the BFQA system. However, it was observed from the experiments that the accuracy of the system is not at par with those for the European languages. The probable reasons for the somewhat poor performance of the system can be attributed to the low accuracies of the shallow parser and the NER system as the accuracy of factoid QA system is largely dependent on the performance of the NER component and the parser. In absence of any gold standard test set, we also prepared our own test set to evaluate the system.

Acknowledgements. We acknowledge the support of the Department of Electronics and Information Technology (DeitY), Ministry of Communications and Information Technology (MCIT), Government of India funded project *"CLIA System Phase II"*.

References

1. Zheng, Z.: AnswerBus question answering system. In: The Second International Conference on HLT Research, pp. 399–404. Morgan Kaufmann Publishers Inc. (2002)
2. Ittycheriah, A., Franz, M., Zhu, W.J., Ratnaparkhi, A., Mammone, R.J.: IBM's Statistical Question Answering System. In: TREC (2002)
3. Green Jr, B.F., Wolf, A.K., Chomsky, C., Laughery, K.: Baseball: An automatic question-answerer. In: Western Joint IRE-AIEE-ACM Computer Conference, pp. 219–224. ACM (1961)
4. Woods, W.A., Kaplan, R.M., Nash-Webber, B.: The lunar sciences natural language information system: Final report. BBN Report 2378, Bolt Beranek and Newman Inc. (1972)
5. Mohammed, F.A., Nasser, K., Harb, H.M.: A knowledge based Arabic question answering system (AQAS). ACM SIGART Bulletin 4(4), 21–30 (1993)
6. Kanaan, G., Hammouri, A., Al-Shalabi, R., Swalha, M.: A new question answering system for the Arabic language. American Journal of Applied Sciences 6(4), 797 (2009)
7. Hammo, B., Abu-Salem, H., Lytinen, S.: QARAB: A question answering system to support the Arabic language. In: ACL 2002 Workshop on Computational Approaches to Semitic Languages, pp. 1–11 (2002)
8. Sakai, T., Saito, Y., Ichimura, Y., Koyama, M., Kokubu, T., Manabe, T.: ASKMi: A Japanese Question Answering System based on Semantic Role Analysis. In: RIAO, pp. 215–231 (2004)
9. Isozaki, H., Sudoh, K., Tsukada, H.: NTT's japanese-english cross-language question answering system. In: Proceedings of the NTCIR Workshop 5 Meeting, pp. 186–193 (2005)
10. Yongkui, Z.H.A.N.G., Zheqian, Z.H.A.O., Lijun, B.A.I., Xinqing, C.H.E.N.: Internet-based Chinese Question-Answering System. Computer Engineering 15, 34 (2003)
11. Sun, A., Jiang, M., He, Y., Chen, L., Yuan, B.: Chinese question answering based on syntax analysis and answer classification. Acta Electronica Sinica 36(5) (2008)
12. Banerjee, S., Bandyopadhyay, S.: Bengali Question Classification: Towards Developing QA System. In: SANLP-COLING 2012, IIT,Mumbai,India (2012)
13. Banerjee, S., Bandyopadhyay, S.: An Empirical Study of Combining Multiple Models in Bengali Question Classification. In: IJCNLP, Nagoya, Japan (2013)

14. Banerjee, S., Bandyopadhyay, S.: Ensemble Approach for Fine-Grained Question Classification in Bengali. In: PACLIC-27. National Chengchi University, Taiwan (2013)
15. Moldovan, D., Pasca, M., Harabagiu, S., Surdeanu, M.: Performance issues and error analysis in an open-domain question answering system. ACM Trans. Inf. Syst. 21, 133–154 (2003)
16. Prager, J.: Open-Domain Question-Answering. In: Foundations and Trends in Information Retrieval. Now Publishers (2007)
17. Moll Aliod, D.: AnswerFinder in TREC 2003. In: TREC, pp. 392–398 (2003)
18. Chen, J., Diekema, A.R., Taffet, M.D., McCracken, N., Ozgencil, N.E., Yilmazel, O., Liddy, E.D.: Question answering: CNLP at the TREC-10 question answering track. In: TREC (2001)
19. Hovy, E.H., Gerber, L., Hermjakob, U., Junk, M.: Chin-Yew Lin.: Question Answering in Webclopedia. In: TREC (2000)
20. Saha, S.K., Chatterji, S., Dantapat, S., Sarkar, S., Mitra, P.: A Hybrid Approach for Named Entity Recognition in Indian Languages. In: IJCNLP 2008 workshop on NER for South and South East Asian Languages, Hyderabad, India, pp. 17–24 (2008)
21. Moldovan, D., Harabagiu, S., Pasca, M., Mihalcea, R., Girju, R., Goodrum, R., Rus, V.: The structure and performance of an open-domain question answering system. In: ACL, pp. 563–570 (2000)
22. Monz, C.: From Document Retrieval to Question Answering. Ph.D. thesis, University of Amsterdam (2003)

Dictionary-Based Problem Phrase Extraction from User Reviews

Valery Solovyev[1] and Vladimir Ivanov[1,2,3]

[1] Kazan Federal University, Kremlevskaya St., 18, Kazan, Russia
http://www.kpfu.ru

[2] National University of Science and Technology "MISIS", Leninskiy Pr., 4, Moscow, Russia
http://www.misis.ru

[3] Institute of Informatics, Tatarstan Academy of Sciences,
Levoboulachnaya St., 36a, Kazan, Russia

Abstract. This paper describes a system for problem phrase extraction from texts that contain users' reviews of products. In contrast to recent works, this system is based on dictionaries and heuristics, not a machine learning algorithms. We explored two approaches to dictionary construction: manual and automatic. We evaluated the system on a dataset constructed using Amazon Mechanical Turk. Performance values are compared to a machine learning baseline.

Keywords: natural language processing, information extraction.

1 Introduction

User reviews are an important element of feedback to the company from its customers. Most works in this area are devoted to analysis of sentiments expressed in reviews [8] and [6]. Problem detection and extraction of problem phrases from texts are less studied. Gupta in [4] and [3] studies extraction of problem phrases with AT&T products and services from English Twitter messages. In [1] authors study Japanese texts with arbitrary problems from the Web. Both efforts try to find (i) a problem phrase and (ii) an artifact the problem is related to. Both these works make use of machine learning models trained on annotated corpora. In contrast, our system is based on dictionaries and templates constructed by experts. Our system analyzes reviews about Hewlett-Packard products from the company's website (http://reviews.shopping.hp.com). The system pays more attention to recall than to precision, because it is much more important to find every single (and potentially serious) problem.

We will detect problem phrases at the sentence-level only. Consider the following sentences extracted from user reviews that represent three possible cases. Example 1 contains a problem phrase. Example 2 mentions a problem that is implicit. Example 3 does not contain problems.

Example 1. *My printer indicates that the cartridge has a problem.*

Example 2. *Edges around the laptop are sharp in some areas.*

Example 3. *Battery life is really good.*

In theory it is very hard to formally define why Example 2 contains a problem; however, it is almost obvious for a human expert. The system was evaluated on a dataset of

P. Sojka et al. (Eds.): TSD 2014, LNAI 8655, pp. 225–232, 2014.

1,669 sentences constructed using Amazon Mechanical Turk (http://www.mturk.com/). Performance results are slightly better than a simple machine learning baseline. The main research questions we study here are the following. How does quality of problem extraction change when switching domain and extraction methods (here we compare results with [4])? How does performance of our system depend on dictionaries' sizes and construction methods? How does performance depend on difficulties with understanding an input sentence?

The main feature of problem extraction is the lack of a clear definition for a «problem». An evaluation of well-known methods for text classification is the first step we made to solve this unstudied task. The novelty of the work is in adaptation of existing approaches.

In this paper, we study a lightweight dictionary-based approach and compare it to a weak machine learning baseline. The main rationale behind this decision is the following: we show that even a simple method may lead to good results, and we suppose that better results can be reached only by much more sofisticated approaches. In a closely related work [4], Gupta had also evaluated a rule-based approach, but he applied it to another domain (telecommunication) and Twitter messages. In our work, we adopt similar ideas and apply them to reviews of consumer goods (computers), so our results are comparable to the results shown in [4]. The second reason for evaluation of the rule-based method in this paper is a clarity of reasons for a classification decision to the end user. Most machine learning approaches deal with weights and/or parameters that are hard to interpret (if something goes wrong inside the system, one can only optimize the parameters for a given data set). Thus the reason for a system's solution usualy unclear to the end user. In our future work, we will focus on machine learning approaches (including generation of feature sets from the dictionaries).

The rest of the paper is organized as follows. In Section 2 we describe dictionaries and simple heuristics for problem extraction. Section 3 represents evaluation settings and system performance. Section 4 contains the conclusion and future work.

2 Dictionaries and Heuristics for Problem Extraction

A common approach to problem phrase extraction is to use problem indicators (i.e., words or multiword expressions) that indicate a problem in a sentence. In [4] and [3], a list of about 40 indicators were collected from a few hundred tweets. In [1] a dictionary was generated automatically. Starting with one (seed) word, "trouble", authors discover nouns that represent synonyms and hyponyms of the seed word using templates X "similar" Y, X "called" Y, X "like" Y, etc. After this step, they construct templates consisting of discovered nouns and a postposition word that means "because". Our system uses five different dictionaries: ProblemWord, Action, Contradiction, NegativeWord, and Negation. Three of them (Contradiction, NegativeWord, and Negation) play auxiliary roles. The Contradiction dictionary contains words like "but", "after", "when", "despite", etc. These words indicate a contrast that may be related to a problem. The Contradiction dictionary contains 43 words. The NegativeWord dictionary contains negative-sentiment words and has been obtained from a well-known opinion lexicon [5]. The dictionary contains about 4,800 words. Finally, the Negation dictionary contains common expressions for negations (e.g., "not", "n't", "cannot"). The Action dictionary contains all verbs denoting some action, because a problem phrase may be represented as a negation of an action. All verbs found in the reviews have been included in this dictionary, except auxiliary verbs ("to have", "to do", "to be", etc.). This dictionary contains about 7,600 words.

The ProblemWord dictionary includes problem indicators. Initially the dictionary had very few problem indicators, such as "problem", "error", "failure", "malfunction", "defect", "damage", "deficiency", "weakness", "mistake", "fault". We explore two approaches to extend this dictionary: manual vs. automatic. In the manual approach, we collected synonyms for problem indicators and also added to the dictionary a few multiword expressions like "shut down", "have to replace", "need to exchange", "could be", "should have", etc. We found that WordNet is only a partially useful resource in collecting synonyms, because some problem indicators are neither synonyms nor fall under the same hierarchy of WordNet. The manually created dictionary (PWM) contains about 300 terms.

An automatic population of the ProblemWord dictionary makes use of Google Books NGram Viewer dataset (http://books.google.com/ngrams/) in the following manner. Starting with an initial set of seed words (pw_i), we collect all 4-grams of two kinds: "a pw_i or X" and "X or a pw_i". Then we extract words that fill the position of X, keeping nouns and verbs only. Finally, we remove terms that occur in the Action dictionary and include remaining terms into the ProblemWord dictionary. Thus, starting with ten initial indicators, we end up with 282 terms added to the dictionary (we call this dictionary PWA).

Both dictionaries, PWM and PWA, have only 47 terms in common, but as we show further, these (core) problem indicators cover most problem phrases in a test set. Core terms for problem extraction are provided in Table 1. Seed terms for PWA construction are presented in bold.

Table 1. Core terms for problem extraction

defect	**deficiency**	injury	issue	worry	remove	breakdown
error	disappointment	**failure**	**fault**	**problem**	lack	accident
loss	**malfunction**	flaw	absence	break	confusion	complaint
insufficiency	crack	failed	problems	refusal	puzzle	shortcoming
obstacle	**damage**	emergency	fail	gap	lesion	misfortune
inability	miss	overload	rejection	risk	trouble	disability
setback			warning	collapse	inadequacy	struggle

To extract problem phrases, we use three heuristics: PW – problem indicator presence; AC – negation of some action (i.e., a negation followed by an action word, from the Action dictionary); CN – a contradiction word (from the Contradiction dictionary) followed by a negative word (from the NegativeWord dictionary). The following texts show how these heuristics work:

PW: *"Worked fine first time around. when i turn printer off, next time i turn it on, it tells me there is a **problem** with the new print head..."*

AC: *"While plugging in USB 3 from dock station, the battery power does **not switch** over to use USB 3 dock power supply ..."*

CN: *"I recently received a shipment of this presentation paper that was not the same presentation paper that I always receive from hp (same product number, **but** suddenly an entirely different, **poorer** quality paper)..."*

Further we briefly describe an algorithm for problem phrase extraction. To decide whether a sentence (s) contains a problem or not, the algorithm exploits the following dictionaries: Action, ProblemWord, Contradiction, NegativeWord, and Negation.

Step 1. If the sentence (s) mentions a problem word (pw) from the ProblemWord dictionary, then

check whether the sentence (s) contains a negation word (neg) from the Negation dictionary. If the negation is related to the mention of the problem word, go to Step 2 (there is no explicit problem in the sentence s).

if there is no negation word related to the pw, mark the sentence s as a problem sentence and extract from its parse tree[1] a node of type S, which contains a pw and does not contain any other node of type S.

Else go to Step 2.

Step 2. If the sentence (s) contains mention of an action word (a) from the Action dictionary along with a related negation word (neg) from the Negation dictionary (i.e., the sentence contains a couple Negation+Action), then

mark the sentence s as a problem sentence and extract from its parse tree a node of type S that contains both words and does not contain any other node of type S.

Else go to Step 3.

Step 3. If the sentence (s) contains mention of a contradiction word (c) from the Contradiction dictionary followed by a negative word (nw) from the NegativeWord dictionary (i.e., the sentence contains a couple Contradiction + NegativeWord), then the sentence (s) contains a problem.

3 Evaluation of Problem Extraction

To carry out evaluation of the system, we have created a corpus consisting of 1,669 sentences. Class labels for each sentence were acquired by using Amazon Mechanical Turk's service. MTurk workers were told to assign one of three labels to each sentence: "a problem is indicated in text", "a problem is implicit", and "no problem in text". The Fleiss' kappa [2] measured after four separate runs was 0.44 (we treat each run as a pseudoexpert). After merging two labels – "a problem is indicated in text" and "a problem is implicit" – into a single label, the Fleiss' kappa was 0.58, which may be treated as a fair inter expert agreement. However, there are still 161 sentences with two positive and two negative labels, which we excluded from the evaluation set. The final evaluation corpus contains 1,508 sentences. The distribution of labels is presented in Table 2.

[1] In the parsing step we use the Stanford Parser.

Table 2. Distribution of positive and negative labels in the evaluation corpus after four MTurk runs

ID	Count of positive marks	Class label	Sentence count
S0	0	-	643
S1	1	-	209
S2	2	?	161
S3	3	+	258
S4	4	+	398

3.1 A Machine Learning Baseline

We implemented a simple machine learning baseline. The baseline approach to problem phrase extraction is based on a Naïve Bayes classifier from WEKA machine learning toolkit (http://www.cs.waikato.ac.nz/ml/weka/). WEKA supports textual data via its transformation to the attribute-relation file format (arrf) using a text directory loader converter. Once data is transformed and loaded into WEKA, the StringToWordVector filter converts unstructured documents into feature vectors that can be used to train a number of WEKA's classifiers. We enabled stemming while converting to vectors what resulted in 1,780 features. We then ranked these features according to chi-square test and selected the best 500 features (this number was identified experimentally). In calculating performance metrics, we use the most common measures: precision (P), recall (R), F1 measure, and common approaches to averaging, represented as follows.

$$P_i = \frac{TP_i}{TP_i + FP_i}; \quad R_i = \frac{TP_i}{TP_i + FN_i}$$

$$P_{\text{micro}} = \frac{\sum_{i=1}^{|C|} TP_i}{\sum_{i=1}^{|C|} TP_i + FP_i}; \quad R_{\text{micro}} = \frac{\sum_{i=1}^{|C|} TP_i}{\sum_{i=1}^{|C|} TP_i + FN_i}$$

$$P_{\text{macro}} = \frac{1}{|C|}\sum_{i=1}^{|C|} \frac{TP_i}{TP_i + FP_i}; \quad R_{\text{macro}} = \frac{1}{|C|}\sum_{i=1}^{|C|} \frac{TP_i}{TP_i + FN_i}$$

The distribution of instances in our dataset was as follows: 852 instances belong to the "no problem" class and 656 instances to the "problem" class. The simplest classifier ZeroR, which determines the most common class and gives the lower-bound performance estimates, produces the following metrics: accuracy 56.5%, weighted precision 31.9%, weighted recall 56.5%, and weighted F1 measure 40.8%. Naïve Bayes classifier considered by us as a baseline approach to problem phrase extraction using tenfold cross validation produces the following performance metrics: accuracy 69.2%, weighted precision 69.8%, weighted recall 69.2%, and weighted F1 measure 69.3%. Baseline performance metrics without weighting (for a problem sentences class): macro precision 69%, macro recall 69.4%, and macro F1 measure 69%.

3.2 Evaluation of Dictionary-Based Extraction

We evaluated different settings of the dictionary-based system with respect to the ProblemWord dictionary contents:

PWM – ProblemWord dictionary created manually

PWA – ProblemWord dictionary generated automatically

$PWA \cap PWM$ – ProblemWord dictionary is an intersection of manual and automatic versions

$PWA \cup PWM$ – ProblemWord dictionary is a union of manual and automatic versions

The performance metrics for these settings are provided in Table 3. All metrics are calculated on a set of 1,508 sentences (S0+S1+S3+S4). The metrics for the PWM setting for each subset are presented in Table 4.

Table 3. Evaluation of problem extraction with different dictionaries of problem indicators

Dictionary	TP	FP	P	R	F1
$PWA \cap PWM$	440	205	.68	.67	.68
PWM	521	225	.70	.79	.74
PWA	486	366	.57	.74	.65
$PWA \cup PWM$	552	382	.59	.84	.69
Naïve Bayes	453	202	.69	.69	.69
ZeroR	374	790	.32	.57	.41

Table 4. Distribution of system (based on the PWM dictionary) decisions over the evaluation corpus

ID	Sentences	Positive labels	Negative labels
S0	643	174	469
S1	209	51	158
S2	161	92	68
S3	258	186	72
S4	398	335	63

Table 5. Analysis of problem extraction heuristics

Heuristic name	TP fraction	FP fraction
PW	258 (49.7%)	69 (30.7%)
AC	209 (39.9%)	120 (53.3%)
CN	54 (10.4%)	36 (16%)

Finally, we evaluated influence of each heuristic on "true positives" and on "false positives" respectively. These results are presented in Table 5.

3.3 Discussion

First of all, we can see that the system divided the subset (S2) into almost equal classes, but the problem class is a little bigger, which reflects a system's trade-off between precision and recall. This point is also illustrated by a comparison of positive labels in subsets S3 and S4 and negative labels from subsets S0 and S1 (Table 4). The latter case system shows a better performance. It is interesting, that more than a half of all problems in the test set were discovered using few (core) problem indicators (Table 3). Given this and the fact that core indicators are included in an automatically constructed dictionary, one could adapt a dictionary-based system to another domains using Google Books NGram Viewer dataset. An error analysis of false negatives revealed two additional sources of problem phrases:

1. The sentence may not contain a problem indicator but may contain indirect features, like "sent it back", "to return the tablet", "contacted hp product support", etc.
2. Some multiword expressions may indicate problem, even if they do not include problem indicators (e.g., "too small", "too many", "little low").

4 Conclusion and Future Work

It is hard to define the notion of a problem. We consider problem extraction as an information extraction task. Performance metrics comparison confirms this idea. The value of the F1 measure (about 75%) is less than the F1 measure for Named Entity Recognition (more than 90%), but it is slightly better than the best F1 values for event extraction, which is about 60% [7]. Thus, problem extraction holds an intermediate position. In order to improve performance, we will develop additional dictionaries, mentioned in the previous section. This will allow to find all sentences that may potentially contain problems (even if there is no problem indicator). We suppose that improving a recall higher than 85% will be difficult (using dictionaries and our current lightweight approach). Probably, the extraction of a problem sentence not found by the dictionary approach requires deeper semantic analysis (e.g., extending a bag-of-words model and using advanced machine learning).

The performance metrics of our system are very close to those that have been shown in [4] and [3] and close to the machine learning baseline. This may indicate some limit for state-of-the-art information extraction methods. Our future work will focus on developing interactive algorithms for lexical feature extraction (including WordNet utilization) and adaptation of the system to new domains. We also plan to extend the corpus in order to carry out accurate evaluations. In particular, we will annotate sentences from full texts, not only single sentences separated from texts.

Acknowledgments. We are grateful to Sergey Serebryakov for their support of this research, useful discussions and help with our approaches. We are grateful to reviewers for their insightful and precise comments.

References

1. De Saeger, S., Torisawa, K., Kazama, J.: Looking for Trouble. In: Proceedings of the 22nd International Conference on Computational Linguistics, vol. 1, pp. 185–192. Association for Computational Linguistics (2008)
2. Fleiss, J.L.: Statistical Methods for Rates and Proportions. Wiley Series in Probability and Mathematical Statistics. Applied Probability and Statistics. Wiley, New York (1981)
3. Gupta, N.: Extracting Descriptions of Problems with Products and Services from Twitter Data. In: Proceedings of the Third Workshop on Social Web Search and Mining, Beijing (2011)
4. Gupta, N.: Extracting Phrases Describing Problems with Products and Services from Twitter Messages. Technical report, Conference on Intelligent Text Processing and Computational Linguistics (2013)
5. Hu, M., Liu, B.: Mining and Summarizing Customer Reviews. In: Proceedings of the Tenth ACM SIGKDD International Conference on Knowledge Discovery and Data Mining, KDD 2004, pp. 168–177. ACM, New York (2004a)
6. Hu, M., Liu, B.: Mining Opinion Features in Customer Reviews. In: Proceedings of the 19th National Conference on Artificial Intelligence (2004b)
7. Indurkhya, N., Damerau, F.J.: Handbook of Natural Language Processing, vol. 2. CRC Press (2010)
8. Liu, B., Hu, M., Cheng, J.: Proceedings of the 14th International Conference on World Wide Web, WWW 2005, pp. 342–351. ACM, New York (2005)

RelANE: Discovering Relations between Arabic Named Entities

Ines Boujelben, Salma Jamoussi, and Abdelmajid Ben Hamadou

Miracl, University of Sfax, Tunisia
{Boujelben_ines@yahoo.fr,jamoussi}@gmail.com,
adelmajid.benhamadou@isimsf.rnu.tn

Abstract. In this paper, we describe the first tool that detects the semantic relation between Arabic named entities, henceforth RelANE. We use various supervised learning techniques to predict the word or the sequence of terms that can highlight one or more semantic relationship between two Arabic named entities.

For each word in the sentence, we use its morphological, contextual and semantic features of entity types. We do not integrate a relation classes predefined in order to cover more relations that can be presented in sentences. Given that free Arabic corpora for this task are not available, we built our own corpus annotated with the required information.

Plenty of experiments are conducted, and the preliminary results proved the effectiveness of our process that allows to extract semantic relation between Arabic NEs. We obtained promising results in terms of F-score when applied to our corpus.

Keywords: relation, named entity, supervised method, Arabic language.

1 Introduction

The extraction of Relations involving Named Entities (RNE) task is seen as a step towards a more structured model of the text meaning. Therefore, it presents a fundamental task for the many Natural Language Processing (NLP) and information extraction tasks, such as question answering and automatic summarization. Hence, the NLP community shows a great interest concerning this issue. This interest in RNE is shown for English and European languages. However, a few works are done for the Arabic language. This is due to the complexity of this language morphology and the lack of available resources, notably annotated corpus. Given that the Arabic language is a rich morphological language, the RNE applying to Arabic language task has to be challenging.

In literature, RNE is the task of finding pre-defined semantic relations between two entities or entity mentions from text (e.g. [1,5]). This task requires two main subtasks: relation detection and relation classification. In our work, we aimed at finding all binary relations without any restriction to relation classes. Our main goal is to detect a set of words that predicts relation between NEs. In a subsequent step, we will assign a specific class to each extracted trigger word.

P. Sojka et al. (Eds.): TSD 2014, LNAI 8655, pp. 233–239, 2014.

The rest of the paper is organized as follows. We firstly introduce the entity relation extraction task. Then, we survey previous work on relation extraction. The ensuing section is devoted to describe our data in which we depict our annotation guidelines. In the last section, we present the different experiments from which we discuss the reported results.

2 Relation Extraction

In our case, a relation can be expressed directly through one word or a sequence of words which is very common for some family (e.g. ابن العم/cousin, ابن عمه/ his cousin) or functional relations (e.g أمين عام/Secretary-General), , respectively. These words can be depicted in the same context (before, between NEs or after NEs), or each of them can be located in different contexts (e.g. طلب أحمد يد سلمى من أبيها للزواج /Ahmed ask Salma's father for her hand in marriage). We assume that support words can offer helpful information to recognize the semantic relations holding between Arabic NEs.

Unlike some recent researches which focused on semantic relation classes, we assume that detecting an infinite number of relations (independently of semantic relations classification) poses a more challenging problem.

3 Related Work

Several methods have been proposed to extract semantic relation between NEs. Some of them are based on linguistic method, which relies on rules that are usually implemented in the form of regular expressions or finite-state transducers. [2,4] have been elaborated local grammars under the linguistic platform NooJ[1] to discover relation between Arabic NEs. In order to automate this task, some researchers have been oriented towards machine learning (ML) methods. We distinct three main methods: (i) the unsupervised methods that make use of massive quantities of unlabeled text. They focalize almost on clustering techniques and similarities between features or context words [8,12]. To overcome the problems encountered by unsupervised approach, some researchers are oriented towards semi-supervised learning methods which rely on a small set of initial seeds. These latter can be depicted as a sample of linguistic patterns or some target relation instances for the purpose to acquire more basics until discovering all target relations such as [14,11]. A last method under the ML methods is the supervised technique which considers the relation extraction as a classification task. This method requires fully labeled corpus. An early attempt to extract relation between Arabic NEs is carried out by [1], who used MaxEnt classifier. Based only on morphological and POS information, his system achieves satisfactory results when applied to ACE 2005 Multilingual training data[2]. In [5], the author combined two supervised techniques which are simply decision trees (DT) and PART decision lists algorithms to extract

[1] NooJ local grammars are typically used in order to describe sequences of words that present meaningful units or entities. They represent a set of rules by means of transducers.

[2] Available on http://www.nist.gov/speech/tests/ace/

three semantic relations (role, social and location) between NEs. The author focused on the part-of-speech tag of the context before and between the two entities only, without considering the context after NEs and he reported an F-score of 81.2% when applied to I-CAB[3] data. Through the above study of different works, we decide to rely on supervised techniques regarding their promising results. Therefore, we examined a set of supervised techniques used in prior work to reach a conclusion. We aimed at discovering trigger words that can explicitly predict a relation between NEs. Consequently, an infinite number of relations will be detected, without being dependent on predefined relations classes.

4 Data

As far as we know, Arabic misses lexical resources, especially free resources available for a RNE purposes. It is mandatory to point out the existence of an Arabic corpus which is annotated with the relation between NEs, namely ACE multilingual training data. Unfortunately, it is not freely accessible. This is the reason why, we have to contract our own corpus to carry out this task. After wards, in sub step, we tend to share our corpus with other researches who are interested to make a comparative study on the RNE task in Arabic language.

Our corpus consists of 870 heterogeneous articles. They were gathered from various sources of Arabic electronic newspapers such as "البيان /AlbyAn", "الجزيرة /Ajazeera", "الشروق /Al$rwq", "الحياة /AlHyAp" and from Wikipedia[4]. The present work focuses on the possible relations between couple of NEs from Person (PERS), Location (LOC) Organization (ORG) and Date (DATE). The choice of these types of NEs is motivated by their high frequency in the majority of electronic texts. Our corpus is composed of a set of 1,245 sentences containing at least a pair of NEs. These sentences were automatically annotated with:

- Morpho-syntactic analyzer which provides the Part of speech (POS) tagging of each word. This information is retrieved using linguistic resources elaborated by (Mesfar, 2008) like Verbs, nouns and adjectives dictionaries as well as some lexical and syntactic grammars elaborated through the linguistic platform NooJ.
- Clauses splitter which is a cascade of finite-state transducers elaborated by [9], proceeds to split long sentences into a set of clauses.
- Arabic NE recognition [10] that is presented as a set of syntactic grammars to recognize different types of NEs.

Otherwise, segmentation, NE tagging and morphological annotation errors were manually rectified in order to obtain an efficient relation recognizer. The NEs distribution along the different types and the characteristics of our corpus is presented in Table 1.

At the relation annotation stage, we just identify the word that can predict a semantic relation between NEs presented within a clause. Therefore, any word on the sentence should be annotated as one of the following tags:

---

[3] Italian Content Annotation Bank: an Italian corpus annotated with temporal expressions and four named entity types (person, organization, location and geo-political entity).

[4] Available on http://www.wikipedia.org/

Table 1. Gold corpus characteristics

Types	Sentences	Words	PERS	LOC	ORG	REL	P-REL
Number	1245	10234	966	764	257	1709	923

- Rel= A word expresses a relation between NEs pair.
- PRel= A word expresses only a part of a relation between two NEs in a given clause.
- N= A word doesn't enclose a relation.

For the sentences that contain more than two NEs, we treated each related pair of entities separately. Unlike proposed ACE annotation guidelines, the negatively defined relations will be taken into account in our annotation (e.g. أحمد ليس في كندا/Ahmed is not in London) and as the output, these relations will be deduced as negatively defined relationship. Then, three Arabic linguistic experts were asked to predict which word or a sequence of words can define semantic relation between NEs within a clause. We provided them with detailed description of our relation extraction task as well as our main goal. The inter-annotator agreements are computed from which we get a promising Cohen kappa of 79%. The focal disagreements came from some examples in which a relation cannot be predicted directly from words. Moreover, some relations are expressed through more than one word, which poses little disagreements between our linguistic experts.

5 Features

Each word in a given clause is assigned to a set of learning features, in order to built the learn data base. We investigated:

- The POS information of a word.
- The POS of the three words before and after this word.
- Grammatical structure of clause: To simplify the relation entity extraction task, we focused on a clause rather than a sentence. We added the grammatical structure of of a clause which is provided by the Stanford tagger [6].
- The semantic features concerned the NE tags which can be PERS, ORG LOC and DATE.
- Numeric features included the position of a word in the clause, the position according to the first NE as well as the second NE, the number of words in the sentence, number of words before, between and after the second NEs and the number of characters of each word.

Once these features assignments are done, we build 10,234 instances which is presented as a set of pairs (attribute or feature and its corresponding value) and a class label. We enclosed three classes: ***Rel, PRel*** and ***N***.

6 Experiments and Results

To avoid the over fitting, all the reported experiments are done with 10-fold cross-validation on our entire data-set by means of standard evaluation metrics. In order to analyze how the learning procedure can be influenced by the instances number, we have computed a learning curve, by dividing our corpus into different learning sets. For each set, we apply the DT algorithm.

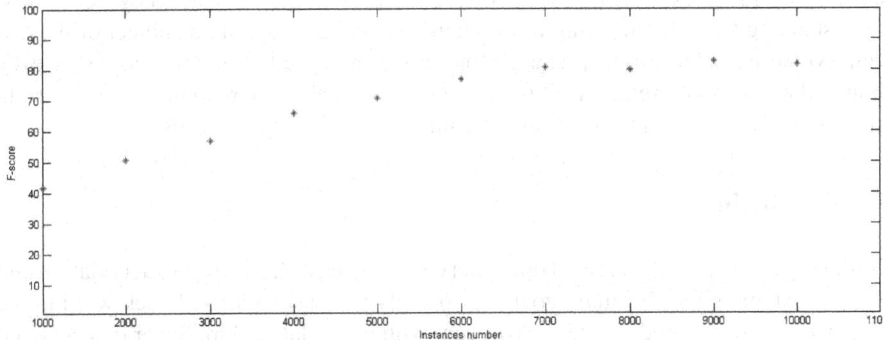

Fig. 1. F-score behavior for each instances number (using Adaboost)

The F-score curve shows that the curve grows regularly between 0 and 7,000 instances while it seems to plateau between 6,000 and 10,000 instances. We can thus conclude that the addition of more than 10,000 instances will only slightly increase the performance of the relation extraction task. Four our learning model, we investigated six ML techniques. They have been examined individually in order to choose the best technique to be applied on our Arabic non standard corpus. Firstly, we adopted the MaxEnt technique since it has been successful used not only in the RNE [1] but in many other NLP tasks, peculiarly NE recognition [3]. Similarly, we applied the DT and PART algorithms which are used also in [5] and the support vector machine (SVM) which is used in [7]. Other algorithms in literature are used to make a comparative study. The DT (C4.5), SVM, Adaboost (with DT), PART and Naïve Bayes are available in WEKA[5]. While, the MaxEnt is available on NLP Stanford tools[6]. Table 2 shows the system's performance in terms of precision, recall and F-measure when applied to our gold corpus.

Table 2. Results of different supervised algorithms

MaxEnt	PART	Decision	Tree	Adaboost	Naïve Bayes	SVM (SMO)
Precision	57.4	82.1	82.36	84.43	53.9	86.5
Recall	64.4	75.3	72.26	80.16	71.2	84
F-score	60.7	78.2	76.56	82.13	58.1	85.23

According to the empirical results illustrated in this table, the used algorithms obtained very competitive scores. The highest performance of our system is accomplished

[5] http://www.cs.waikato.ac.nz/ml/weka/

[6] http://nlp.stanford.edu/software/classifier.shtml

by PART and Adaboost classifiers. Experimental study exhibits that SVM significantly outperforms other algorithms in term of precision. To conclude, SVM and Adaboost performed well on the entity relation detection task.

When examining the output of our process, we can deduce that some relations are difficult to be extracted from a word or a sequence of words. They need to be understanding from the meaning of previous sentences, or the main subject of the text from which this clause is extracted. Moreover, some words are not recognized even though they express a relation. This can be caused by the non recognition of the right category of NEs or the ambiguity in determining the POS tag of words.

7 Conclusion

Through this paper, we described our supervised process relANE to extract relationship between Arabic NEs. Our main goal was to study various features of each word in the sentence in orderto predict which term can explicit a relationship. Several supervised algorithms were applied.But mainly, the SVM and Adaboost techniques proved to be efficient for the NE relation task.

For a future work, we obviously intend to classify supported words into adequate level of semantic relationship classification. Similarly, we seek to evaluate our process on a standard corpus such as ACEdatawhich is not available yet. We are also considering the possibility of investigating other features to boost the overall performance of our system.

Acknowledgments. We thank our developers Sana Trigui and Marwa Ben Hammouda for their efforts and their feedback to design our extractor interfaces.

References

1. Alotayq, A.: Extracting Relations between Arabic Named Entities. In: Habernal, I., Matousek, V. (eds.) TSD 2013. LNCS (LNAI), vol. 8082, pp. 265–271. Springer, Heidelberg (2013)
2. Hamadou, A.B., Piton, O., Fehri, H.: Multilingual Extraction of functional relations between Arabic Named Entities using NooJ platform. hal-00547940- version 1 (2010)
3. Benajiba, Y., Rosso, P., BenedíRuiz, J.M.: ANERsys: An Arabic Named Entity Recognition system based on Maximum Entropy. In: Gelbukh, A. (ed.) CICLing 2007. LNCS, vol. 4394, pp. 143–153. Springer, Heidelberg (2007)
4. Boujelben, I., Jamoussi, S., Ben Hamadou, A.: Rules based approach for semantic rela-tions extraction between Arabic named entities. In: NooJ 2012, in INALCO-Paris (2012)
5. Celli, F.: for Semantic Relations between Named Entities in I-CAB (2009), technical report available at `http://clic.cimec.unitn.it/fabio`
6. Green, S., Manning, C.: Better Arabic Parsing: Baseline, Evaluations and Analysis. In: 23rd International Conference on Computational Linguistics COLING 2010, Beijing, China (2010)
7. Hong, G.: Relation Extraction Using Support Vector Machine. In: Dale, R., Wong, K.-F., Su, J., Kwong, O.Y. (eds.) IJCNLP 2005. LNCS (LNAI), vol. 3651, pp. 366–377. Springer, Heidelberg (2005)

8. Hasegawa, T., Sekine, S., Grishman, R.: Discovering relations among named entities from large corpora. In: Proceedings of Association for Computational Linguistics, Morris-town, NJ, USA (2004)
9. Keskes, I., Benamara, F., Belguith, L.: Clause-based Discourse Segmentation of Arabic Texts. In: Language Resources and Evaluation LREC, Istanbul, pp. 2826–2832 (2012)
10. Mesfar, S.: Analyse morpho-syntaxique automatique et reconnaissance des entités nommées en Arabe standard, University of Franche-Comté (2008)
11. Zhang, Z.: Weekly supervised relation classification for information extraction. In: CIKM 2004, Washington D.C., USA (2004)
12. Zhang, M., Su, J., Wang, D., Zhou, G., Tan, C.-L.: Discovering Relations between Named Entities from a Large Raw Corpus Using Tree Similarity-Based Clustering. In: Dale, R., Wong, K.-F., Su, J., Kwong, O.Y. (eds.) IJCNLP 2005. LNCS (LNAI), vol. 3651, pp. 378–389. Springer, Heidelberg (2005)
13. Zhao, S., Grishman, R.: Extracting Relations with Integrated Information Using Kernel Methods. In: 43rd ACL Meeting, Ann Arbor (2005)
14. Zhou, G., Qian, L., Zhu, Q.: Label propagation via bootstrapped support vectors for semantic relation extraction between named entities. Computer Speech & Language 23(4), 464–478 (2009)

Building an Arabic Linguistic Resource from a Treebank: The Case of Property Grammar

Raja Bensalem Bahloul[1], Marwa Elkarwi[1], Kais Haddar[1], and Philippe Blache[2]

[1] Multimedia Information Systems and Advanced Computing Laboratory, Higher Institute of Computer Science and Multimedia, Sfax, Tunisia
raja_ben_salem@yahoo.com, marwaelkarwi89@gmail.com, kais.haddar@fss.rnu.tn

[2] Laboratoire Parole et Langage, CNRS, Université de Provence, France
pb@lpl.univ-aix.fr

Abstract. This paper presents a survey of Arabic treebanks to facilitate their reuse for the building of new linguistic resources. In our case, we created from a treebank an automatically induced Property Grammar (GP). So, we discussed characteristics of these treebanks to choose the appropriate one. To build our resource, we adopted an automatic technique, acquiring first a context-free grammar (CFG) from the chosen treebank, and second, inducing a GP by generating relations between grammatical units described in the CFG.

Keywords: treebanks, Arabic language, reuse, property grammar.

1 Introduction

Treebanks, as rich corpora with annotations, are used to build other linguistic resources, such as extensional and intentional lexicons, context-free grammars (CFG), property grammars (GP), bilingual dictionaries, etc. This promotes their reuse as well as makes explicit their implicit information. Also, treebanks have many advantages: they are not only developed and validated by linguists, but also submitted to consensus, which promotes their reliability. Having such resource makes it possible to generate automatically and in a very controlled basis, new and wide coverage resources on other formalisms, inheriting the original treebank qualities, while gaining on construction time. For Arabic, treebanks are scarce while the most important are: the Penn Arabic Treebank (ATB) [6], the Prague Arabic Dependency Treebank (PADT) [5], the Columbia Arabic Treebank (CATiB) [4] and the Quranic Arabic Dependency Treebank (QADT) [3]. But, what resource can we choose to build, particularly a GP in our case? This depends on several factors to consider such as the size of the corpus, its richness with different types of annotations, the annotation granularity level reached, the representation format of annotation suitable and easy to manipulate, the syntactic representation structure, the used grammar, etc. And even if we find the appropriate treebank, understanding of its categories describing linguistic units is not an easy task. Another difficulty may be encountered when we build a GP namely, the induction of properties connecting categories, which can be easily deduced, or require heuristics.

P. Sojka et al. (Eds.): TSD 2014, LNAI 8655, pp. 240–246, 2014.

In this paper, the building process of our GP consists of two tasks: The first one induces a CFG from the treebank. The second one specifies relations between categories of each syntactic unit from the CFG rules. In view of the defined types of properties in this paper, the technique of GP induction we adopted is purely automatic and independent of any language and of the source treebank formalism. This promotes its reuse. In addition, this technique produces properties by providing changes of different granularity levels of grammatical categories. To the best of our knowledge, the obtained GP induced from a treebank, represents the first test product for Arabic. This product can be used by several other resources to extract their implicit information.

This paper is organized as follows: Section 2 is devoted to comparing the different Arabic treebanks under various criteria. Section 3 gives the reasons for the selection of appropriate treebank. Section 4 explains our approach. Experimental results will be presented and discussed in section 5. Section 6 gives a conclusion and perspectives.

2 Comparison among Arabic Treebanks

Many criteria distinguish the ATB, PADT, CATiB, and QADT treebanks, namely:

- **Source corpus:** The ATB source corpus is composed of newswires [6]. It was divided into more than 12 sets of texts (called divisions); comprising about 750K tokens [2]. It proved its effectiveness in a large number of works in various fields, and its divisions were converted by other treebanks to their syntactic representations, like PADT and CATiB [8,4]. As against, QADT treats the Holy Quran, from which it annotates 11K words and represents them on syntactic dependency graphs [3].
- **Used grammar:** PADT and ATB annotations follow the MSA theories suitable to their source texts [5,6]. CATiB facilitates annotation, based on traditional Arabic genre [4]. QADT follows, also, the same grammar suitable to its source text [3].
- **Syntactic representation:** The phrase structure, used by ATB, is a tree representation in which the words of a sentence appear as leaves and the categories as non-terminal nodes. Dependency structure, used by the 3 other treebanks, has also a tree representation except that the words of the sentence are the nodes of the tree.
- **POS tags:** ATB uses more than 400 tags and specifying different morphological features of Arabic words and includes empty pronouns [6]. PADT has a more complex morphology than ATB, including more detail in the features [5]. CATiB uses only 6 tags [4]. QADT uses 44 tags based on the 3 the main lexical categories (noun, verb and particle) of traditional Arabic grammar [3].
- **Syntactic and semantic relations:** ATB uses about 20 dashtags to represent syntactic and semantic features [6]. But, CATiB marks only syntactic functions [4]. PADT uses about 20 detailed tags deeper than CATiB, and QADT uses 43 syntactic and semantic relations based on dependency links of traditional Arabic grammar [3].

3 Choice of Treebank

The choice of source treebank depends on the goal we want to achieve. Indeed, we aim to induce a GP from an Arabic treebank. This grammar must describe, for each syntactic category, all grammatical categories contributing in its construction and the relations existing between them. Syntactic categories are intermediate representations of the constituents of a sentence from the source treebank. This representation corresponds to a syntactic phrase structure but not to a dependency structure. Only ATB, described in Section 2, represents its annotations according to this structure. Several other issues led us to use ATB, namely: its rich POS tags and syntactic and semantic relations, its grammar adapted to MSA, the syntactic relevance of its source documents (converted to several other treebank representations), the variable category granularity offered, and the availability of a parentheses simple format to manipulate.

4 Proposed Approach

Our goal is to automatically induce a GP from ATB. This mechanism consists of two automatic tasks: The first one is to induce a CFG from ATB. The second task is to infer for each syntactic unit, the various relations (called properties) between its constituents from the rules described in the CFG. A constituent can be a syntactic or a lexical unit. The application of these two tasks allows us to obtain a GP.

Fig. 1. Property grammar induction approach

As shown in Figure 1, the two outputs generated by the process tasks are not similar. Indeed, the first output represents each syntactic unit by a set of rules combining its various constituents (categories). But, the second task represents each syntactic unit in terms of its constituents (const), and of its properties. These properties have these different types described in the following figure:

Table 1. Functions of properties in the GP

Properties	Symbols	Functions
Uniqueness (unic)	Uniq	Set of constituents that cannot be repeated in a syntactic unit
Obligation (oblig)	Oblig	Set of possible heads of a syntactic unit
Linearity (lin)	$\prec$	Linear precedence relations between constituents of a syntactic unit
Requirement (req)	$\Rightarrow$	Mandatory co-occurrence between constituents
Exclusion (excl)	$\otimes$	Cooccurrence restriction between constituents

The contribution of the obtained GP focuses on the representation form of linguistic information in its formalism. Unlike the CFG, information is represented independently of its type or its position [1]. Thus, the GP also describes incomplete, partial and non-canonical information, which enhances its flexibility and robustness.

5 Experiments and Results

In a first step, we used an annotated corpus extracted from ATB to induce a CFG. From this, we induced a GP by deducing the set of properties describing the categories.

5.1 The Used Annotated Corpus

We adopted the Part 2 of ATB (ATB2 v 3.1) consisting of 501 news, and including POS tags, morpho-syntactic structures at many levels and gloss [7]. It is available in various formats: The "sgm" format refers to source documents. The "pos" format gives information about each token as fields before and after clitic-separation. The "xml" format contains the "tree token" annotation after clitic-separation. The "penntree" format generates a Penn Treebanking style. Each terminal is on the form of ("POS tag" "word"). And the "integrated" format brings together information about the source tokens, about the tree tokens, and the mapping between them and the tree structure.

After the presentation of these formats, we should choose the format to use as input in the CFG induction step. Since the CFG describes only the category level, our choice depends on 3 criteria: the simplicity of representation, the presence of a tree structure and the annotation at the syntactic level. The "penntree" format was the only one selected because it meets all these criteria. More specifically, we used the vowelized version of this format to avoid ambiguities related to the absence of vowels in Arabic.

5.2 Obtained Grammar

With this approach, we obtained successively two different grammars: The CFG and the GP. Their size depends on the granularity level of categories it describes. A category specifies many features like mood, gender, number, etc. The higher this level, the more these grammars are complex, but the more it respects the language and vice versa.

Table 2. Frequency of the rule "PREP NP" describing the PP subcategories at the highest granularity level in the CFG

Phrases	PP	PP-CLR	PP-PRP	PP-TMP	PP-LOC	PP-PRD	PP-MNR	PP-DIR	Others
Σ# Rules	50	44	15	15	13	13	12	9	–
#Occ of "PREP NP"	12,834	3,025	445	754	1511	762	246	154	–
Σ# Occ of rules	13,814	3,781	684	805	1537	805	286	165	222

The obtained CFG contains sets of rules describing each non-terminal unit XP. The rule form is an ordered list of grammatical categories describing a syntactic category. Table 1 shows information about the obtained CFG at the highest granularity level, describing the category PP (Prepositional Phrase) and its subcategories (e.g. PP-MNR and PP-TMP), which include more details than PP [7]. In this level, there are 263 rules of different types. According to what we observed in Table 1, we note that the highest granularity does not make a big difference for some subcategories of PP. "PREP NP" remains the most frequent rule whatever the PP subcategory it describes. Other rules are not frequent, sharing the rest of occurrences. For example, the subcategory PP-LOC is described, in addition, by a set of rules, needless to represent, each one not exceeding 10 occurrences and bringing together only 19 occurrences. Futhermore, the sign "#" assigned to some parameters denotes their cardinality. But, if we observe the CFG, illustrated in part in Table 2, we can note that regardless of the granularity level, the occurrences of "PREP NP" represents 90% of all rules in the treebank, often making unnecessary the increase of the granularity level. By reducing this granularity to 0, we obtain a CFG for PP more compact comprising only 59 rules incorporating mostly the rule "PREP NP". Generally, for all syntactic categories, we noticed that the granularity of categories affects also the size of all the grammar. Indeed, the number of rules in the CFG is divided by 6 at the lowest level compared to the highest one (2998/14452).

Table 3. Excerpt from the CFG at the lowest granularity level describing the category PP

Rules	#Occ	Rules	#Occ	Rules	#Occ
PREP NP	19886	PREP ADVP	32	PP PREP NP	10
PREP SBAR	1346	NP PREP NP	28	PREP NP PUNC	10
PREP S	237	PREP PUNC NP	22	PREP UCP	8
PP CONJ PP	126	PP PP	20	PUNC PREP SBAR PUNC	7
-NONE-	87	PUNC PREP NP	20	PREP NP PP	6
PRT PREP NP	63	ADVP PREP NP	19	14 rules	≤ 5
PP PUNC CONJ PP	48	PREP PUNC NP PUNC	18	25 Other rules	1
PUNC PREP NP PUNC	42	Σ# Occurrences			22099

The GP generated at a given granularity level describes, for each syntactic category, all of its constituents and the properties which connect these constituents. Fig. 3 illustrates an excerpt of the obtained GP at the highest granularity level for the category PP. Thanks to the GP formalism, implicit information in the treebank are made explicit. It induces different types of properties connecting the various constituents. For example, in Fig. 3, we have "PREP $\prec$ S-NOM" as linearity property describing the subcategory PP-DIR and indicating that if the category PREP (Preposition) appears with the category S-NOM (Nominative clause) in the same construction, it will always directly or indirectly proceed S-NOM. Such information is not explicit in the treebank. But, with the highest granularity level of categories, a lot of implicit information may be repeated for several subcategories, increasing the GP size and making its run more difficult. In Fig. 3, this is so for the properties connecting the categories PREP and NP, which are repeated at least 6 times in the grammar for the indicated subcategories. The GP at the lowest granularity level is very different. It becomes much more compact, the categories are simpler and the properties are not repeated. This is because these categories were generalized and factored. But, this lack of precision may lead to a loss of information. The linearity property "PRON_3MS $\prec$ DET+ADJ+ CASE_DEF_NOM" describing the subcategory NP-ADV-1 is an example, among others, proving this idea. After its generalization, normally, its precision should be reduced, and should be converted to the following rule "PRON ≺ ADJ" to describe the basic category NP. But this did not happen. This is explained by the fact that the validity of this property has not been guaranteed for all NP subcategories. The absence of several properties due to generalization can increase the error rate.

PP-DIR	Const	{PP, PREP, NP, ADVP, S-NOM, SBAR, PRT}
	Uniq	{PREP, NP, ADVP, S-NOM, SBAR, PRT}
	Lin	PP ≺ {PREP, NP}; PREP ≺ {NP, ADVP, S-NOM, SBAR}; PRT ≺ {PREP, NP}
	Req	{NP, ADVP, S-NOM, SBAR}⇒ PREP; PRT ⇒ {PREP, NP}
	Excl	PP ⊗ {ADVP, S-NOM, SBAR, PRT}; NP ⊗ {ADVP, S-NOM, SBAR} ADVP ⊗ {S-NOM, SBAR, PRT}; S-NOM ⊗ {SBAR, PRT};SBAR ⊗ {PRT}

PP-DTV	Const	{PREP, NP}
PP-TPC	Uniq	{PREP, NP}
PP-LOC	Oblig	{PREP, NP}
PP-MNR	Lin	PREP ≺ NP
PP-PRD	Req	NP ⇒ PREP; PREP ⇒ NP

Fig. 2. Excerpts from the GP at the highest granularity level describing the category PP

According to the results, we note that the granularity level has a major impact on the complexity of the GP inducing problem. Indeed, with a high granularity level, we have representative and detailed properties, because of the high number of categories in the CFG. But, this produces particularly an over-generation for exclusion properties, which increases the complexity of the problem. On the opposite, with a low granularity level, we obtain a more reduced number of categories in the CFG, which makes the properties safe but very general, losing thus information. We should control the granularity level to make a compromise between the quantity and the quality of these properties.

6 Conclusion and Perspectives

We proposed in this paper an approach of building a GP from ATB. In this new resource, we made explicit different types of implicit information by inducing properties connecting various grammatical categories of ATB. The result is a resource on a wide coverage provided at different granularity levels, and inheriting ATB qualities among which its reliability, its submission to consensus and its rich annotation. This resource was built in an internship at the laboratory LPL (Aix-en-provence). The technique that we adopted to build this resource is generic. Indeed, it is independent of any language as well as of the source formalism, since properties are directly generated from the CFG. In addition, by applying only the types of properties defined so far, this technique is purely automatic, which promotes its reusability.

As perspectives, we will show a control of the granularity level of categories in future works. Besides, in order to offer a very precise representation of syntactic information, we can enrich or modify the relations presented in the GP. In future works, this grammar can also be used to enrich the Arabic treebank with a property-based representation to improve its quality. To optimize this enrichment, several control mechanisms can be integrated into determining syntactic categories and evaluability of their linguistic properties. The ATB size can also be enriched by converting the source documents annotated by other Arabic Treebanks (like PADT and CATiB) to the ATB representation as they have already done. So, having a corpus format unifying the different representations is useful, offering thus interoperability of annotated data and extending the applicability of divergent NLP tools in different research contexts.

References

1. Blache, P.: Les Grammaires de Propriétés: Des contraintes pour le traitement automatique des langues naturelles. Hermès Sciences Publications (2001)
2. Diab, M.T., Habash, N., Rambow, O., Roth, R.: LDC Arabic Treebanks and Associated Corpora: Data Divisions Manual. Columbia University. Technical Report, Center for Computational Learning Systems (2013)
3. Dukes, K., Buckwalter, T.: A Dependency Treebank of the Quran using traditional Arabic grammar. Institute of Electrical and Electronics Engineers (2010)
4. Habash, N., Faraj, R., Roth, R.: Syntactic Annotation in the Columbia Arabic Treebank. In: Conference on Arabic Language Resources and Tools, Cairo, Egypt (2009)
5. Hajič, J., Smrž, O., Zemánek, P., Snaidauf, J., Beska, E.: Prague Arabic Dependency Treebank: Development in Data and Tools. In: Proceedings of the NEMLAR International Conference on Arabic Language Resources and Tools (2004)
6. Maamouri, M., Bies, A., Buckwalter, T.: The Penn Arabic Treebank: Building a Large-scale Annotated Arabic Corpus. In: Proceedings of the Network for Euro-Mediterranean Language Resources Conference on Arabic Language Resources, Cairo, Egypt (2004)
7. Maamouri, M., Bies, A., Krouna, S., Gaddeche, F., Bouziri, B.: Penn Arabic Treebank guidelines v4.8. Technical report, LDC, University of Pennsylvania (2009)
8. Smrž, O., Bielický, V., Kouřilová, I., Kráčmar, J., Hajič, J., Zemánek, P.: Prague Arabic Dependency Treebank: A Word on the Million Words. In: Proceedings of the Workshop on Arabic and Local Languages (LREC 2008), Marrakech, Morocco, pp. 16–23 (2008)

Aranea: Yet Another Family of (Comparable) Web Corpora

Vladimír Benko[1,2]

[1] Slovak Academy of Sciences, Ľ. Štúr Institute of Linguistics
Panská 26, SK-81101 Bratislava, Slovakia
[2] Comenius University in Bratislava, UNESCO Chair in Translation Studies
Šoltésovej 4, SK-81334 Bratislava, Slovakia
vladob@juls.savba.sk
http://www.juls.savba.sk/~vladob

Abstract. Our paper deals with an on-going Project in the framework of which, by means of open-source and free tools, a family of web corpora is being created that would (to a large extend) deserve the designation of being "comparable". A summary of results after the first stage of the Project is given, and experiences with the tools are commented.

Keywords: web-based corpora, web data filtration and deduplication, universal PoS tagset, compatible sketch grammar.

1 Introduction

In spring 2013, a new web crawler, *SpiderLing*, has been released. This tool was the last missing stone in the mosaic of open-source and free tools for effective creation and annotation of web corpora. Thanks (mostly) to Computational Linguistics Departments at the Masaryk University in Brno and the University of Stuttgart, this mosaic now contains the following elements:

- *SpiderLing* [17] (a specialized crawler for downloading textual data from the web)
- *chared* (Python module for web page encoding detection, taking into consideration the expected language of the web page)
- *trigrams* (Python module for web page language detection)
- *jusText* [13] (utility for boilerplate removal)
- *Onion* [13] (deduplication utility based on n-grams)
- *Tree Tagger* [16] (tokenization and PoS tagging tool with parameter files for many languages
- *NoSketch Engine* [14] (corpus manager)

In our Project, by means of all the tools mentioned, we decided to create a family of web corpora that would (to a large extend) deserve the designation of being "comparable", i.e. the data would be downloaded at (approximately) the same time, they would contain similar (web-specific) composition of text types, genres and registers, would be of the same size, and would be available at one place via the unified

P. Sojka et al. (Eds.): TSD 2014, LNAI 8655, pp. 247–256, 2014.

access mechanism. The project can be described as Slovak-centric, as it should (in the first phase) cover the languages used and/or taught in Slovakia and the neighbouring countries, i.e. Slovak, Czech, German, Hungarian, Polish, Ukrainian and English, French, Spanish, Italian and Russian.

2 Corpus Design Decisions

Why we need other corpora? Besides our interest in testing the new corpus-building tools, the motive for starting our Project was the lack of suitable corpora that could be used by students of foreign languages and translation studies at our University. The existing corpora presented in [2] and [15] do not cover all languages needed. As for corpora described in [8] and hosted at the Sketch Engine web site,[1] they (1) are not available for downloading, (2) are typically too large for classroom use, and (3) have too different sketch grammars, which makes them difficult to use in a mixed-language classroom.

We expect that a family of corpora for several languages of equal size and built by standardized methodology can not only be used for teaching purposes, but also in linguistic research (contrastive studies) and in lexicography (both mono- and bilingual).

The names: For our corpora, we have decided to use "language-neutral" Latin names denoting the language of the texts and their size. The whole corpus family is called *Aranea*,[2] and the respective members bear the appropriate language name, e.g. *Araneum Anglicum*, *Araneum Germanicum*, *Araneum Russicum* for English, German and Russian, respectively, etc.

The sizes: Each corpus will exists in several editions, differing by their sizes. The basic medium-sized version, *Maius* ("greater"), will contain approximately 1.2 billion of tokens. This size is expected to contain at least 1 billion words, and can be reached relatively quickly for all participating languages. For the "large" ones with plenty of web data available it usually takes just one or two days to download the source data. The 10% random sample of *Maius*, called *Minus* ("smaller"), is to be used for teaching purposes. A 1% sample, *Minimum* ("minimal"), is not intended to be used directly by the end users, and is utilized in debugging of the processing pipeline and tuning the sketch grammars. And lastly, the largest *Maximum* ("maximal") edition will contain as much data as can be downloaded from the web for the particular language, and its size is mostly determined by the configuration of the server.

3 Crawling and Preprocessing

All source data acquisition is being performed by means of *SpiderLing*, a web crawler optimized for collecting textual data from the web. The system contains an integrated character encoding (*chared.py*) and language recognition (*trigrams.py*) module, as well as a tool for boilerplate removal (*jusText*).

The input seed URLs have initially been harvested by various methods. At present, the procedure has been standardized to consist of these steps:[3] (1) Take first two

[1] `http://www.sketchengine.co.uk`

[2] *Araneum* (pl. *aranea*, n.) is the Latin expression both for spider and (spider)web.

[3] The procedure has been partially inspired by [6].

paragraphs of the documents as follows: (a) Universal Declaration of the Human Rights, (b) Bible (John 1:1), (c) Wikipedia article for a concrete noun ("bicycle"), and (4) Wikipedia article for an abstract noun ("love"); (2) Tokenize and deduplicate the resulting wordlist, sort randomly; (3) Use the wordlist in several steps as seed for BootCAT [1], collect list of URLs (do not download the web pages themselves); (4) Deduplicate and filter the resulting URL list.

Using this method, we were quickly able to get several thousands of URLs that were subsequently used as seed for *SpiderLing*.

Several input parameters of the crawling process can (or must) be set by the user, most notably the language name, a file containing sample text in the respective language (to produce a model for language recognition), language similarity threshold (a value between 0 and 1 (default 0.5), number of parallel processes, and the crawling time.

In our processing, we usually crawled in 24-hour slots (the process could be later restarted) with all other values set to defaults. The only exception was crawling for Slovak and Czech, where we crawled in 7-day slots, as the process was much slower for these languages. The language similarity threshold had also to be changed in case of Slovak and Czech. As these languages are fairly similar, the trigram method did not seem to be able to distinguish between them sufficiently. We have therefore increased the threshold value to 0.65 (saving many "good" documents, and causing many "wrong" ones to pass the filter) and removed the unwanted texts by subsequent filtration based on character frequencies.[4]

Table 1 shows the share of eight most frequent top-level domains (TLDs) in documents for the respective languages (in percents).

Table 1. TLD distribution (in %)

de		en		es		fr		pl		ru		sk	
.de	71.34	.com	53.35	.com	45.52	.com	38.18	.pl	81.31	.ru	71.78	.sk	86.64
.com	10.55	.org	19.06	.es	20.11	.fr	33.41	.com	6.16	.com	10.95	.com	4.76
.at	5.26	.uk	6.67	.org	8.92	.org	9.86	.eu	4.69	.ua	6.42	.eu	3.99
.ch	3.78	.edu	5.10	.net	5.67	.net	5.45	.net	2.07	.org	2.97	.net	1.53
.net	3.34	.net	3.57	.ar	5.32	.ca	4.99	.info	1.80	.net	2.92	.cz	1.42
.org	2.76	.au	2.31	.mx	3.19	.be	3.09	.org	1.61	.info	2.56	.org	0.88
.info	1.57	.ca	1.85	.cl	2.92	.ch	2.20	.biz	0.51	.by	1.12	.info	0.54
.eu	1.18	.gov	1.53	.info	0.94	.info	1.59	.sk	0.36	.su	1.10	.rs	0.06
other	0.22	*other*	6.54	*other*	7.40	*other*	1.21	*other*	1.49	*other*	0.18	*other*	0.17

Quite consistent with our expectation, the national TLDs prevail in all languages spoken predominantly in a single country, and the "other" item is really significant only for languages spoken in many countries (English and Spanish).

Filtration: Besides the standard cleanup provided by the *SpiderLing* itself, we made use of some filters originally developed for our older Slovak web corpus, most notably to normalize representation of white space and special graphic characters, and to remove

[4] The idea is (conceptually) based on counting frequencies of graphemes present in Slovak ("ä", "ľ", "ô"), and Czech ("ě", "ř", "ů") only, respectively.

documents with misinterpreted encoding and/or having non-standard distribution of punctuation and uppercase characters (two few punctuation and/or too many uppercase chars usually mean that a page does not contain a "discursive" text). We also performed segmentation of the text on sentence boundaries by means of a rather rudimentary procedure (this segmentation was later used in deduplication).

Table 2 shows some statistics on the downloaded and preprocessed web data.

Table 2. Data downloaded, filtered and normalized

	Domains	Docs	Tokens	Docs per domain	Tokens per doc
de	80,722	2,332,921	1,200,000,087	28.9	514.4
en	23,968	1,163,007	1,200,048,075	48.5	1031.8
es	22,343	1,439,567	1,049,739,252	64.4	729.2
fr	48,398	1,780,315	1,233,336,202	36.8	692.8
pl	58,338	1,783,411	1,110,120,825	30.6	622.5
ru	37,200	1,034,734	1,216,800,424	27.8	1176.0
sk	33,037	1,724,512	1,200,003,757	52.2	695.9

Deduplication: The whole procedure (described in more detail in our recent paper [3]) will finally consists of three stages. The first stage detects the near-duplicate documents by means of the Onion utility (similarity threshold 0.95), and the duplicate documents are deleted. The second stage deduplicates the remaining text at the paragraph level using the same procedure and settings. The tokens of the duplicate paragraphs, however, are not deleted but rather they are marked to make them "invisible" during corpus searches, while they can be displayed as context at the boundary of non-duplicate and duplicate text. In the last stage, we make use of our own tool based on the fingerprint method (with ignoring punctuation, special graphics characters and digits) to deduplicate the text at the sentence level. The tokens of duplicate sentences are marked similarly to the previous stage. This last step can "clean up" the duplicities among the short segments that fail to be detected as duplicates by Onion [13].

At present, only the fingerprint sentence deduplication has been used, and the whole procedure was postponed to later stages (to produce the upgraded version of our corpora). The results of the process can be seen in Table 3.

4 Linguistic Annotation

For all languages covered by parametric/dictionary files of *Tree Tagger* [16], this tagger has been used to annotate the respective corpora. For Polish, the *TaKIPI* [12], and for Czech, the *Morče* [7] taggers were used, respectively. The question of tools for PoS tagging of Hungarian and Ukrainian has not been resolved yet.

To simplify the creation of compatible sketch grammars, all native tagsets are mapped into the *Araneum Universal Tagset* (AUT) [4] (partially inspired by the Google Universal PoS Tagset [11]) creating a secondary layer of morphosyntactic annotation. The AUT PoS tags are shown in Table 4.

Table 3. Segmentation and deduplication

	Sentences	Sentences per doc	Tokens per sentence	% of tokens removed
de	71,964,893	30.8	16.7	29.61
en	56,922,473	48.9	21.1	24.93
es	43,301,352	30.1	24.2	26.69
fr	54,650,594	30.7	22.6	26.57
pl	67,992,427	38.1	16.3	35.98
ru	69,180,355	66.9	17.6	21.10
sk	68,380,608	39.7	17.5	47.16

Table 4. Araneum Universal Tagset

aTag	PoS	aTag	PoS
Dt	determiner/article	Ij	interjection
Nn	noun	Pt	particle
Aj	adjective	Ab	abbreviation/acronym
Pn	pronoun	Sy	symbol
Nm	numeral	Nb	number
Vb	verb	Xx	other (content word)
Av	adverb	Xy	other other (function word)
Pp	preposition/postposition	Yy	unknown/alien/foreign
Cj	conjunction	Zz	punctuation

Besides the traditional 11 word classes, AUT contains 7 more items to accommodate information provided by the individual native tagsets (that is being merged into single a tag by the Google Universal PoS Tagset). Table 5 shows the share (in percents) of the respective word classes in seven Aranea corpora:

Table 5. PoS distribution (in %)

	de	en	es	fr	pl	ru	sk
Dt	9.17	9.31	10.17	10.58	0.00	0.00	0.00
Nn	24.48	26.19	24.05	23.06	28.28	27.48	27.36
Aj	8.30	6.95	6.16	6.30	12.39	8.45	8.76
Pn	8.48	5.56	3.64	7.68	1.52	9.27	6.99
Nm	2.19	2.02	2.96	2.32	0.60	2.69	1.04
Vb	12.02	15.15	14.09	12.84	15.46	11.32	11.87
Av	5.21	4.83	2.97	4.86	2.10	3.64	2.32
Pp	9.06	10.44	12.72	15.43	10.20	9.22	9.23
Cj	5.12	3.42	7.72	4.18	5.87	6.18	5.85
Ij	0.02	0.05	0.00	0.06	0.00	0.07	0.04
Pt	1.71	0.34	3.11	0.00	5.32	2.68	2.62
Zz	13.64	12.50	11.82	12.11	5.02	18.99	14.55
other	0.61	3.24	0.58	0.57	3.22	0.00	9.38

We can see that the numbers for the respective languages are surprisingly similar, with the exception of the "other" value Slovak, caused by some peculiarities of the SNK tagset.[5]

The subsequent filtration fixes some known tagger issues for the respective languages, namely the misassigned tags for several punctuation and special graphic characters (that are often tagged as nouns, verbs, or adjectives). For some languages, an additional tag with masked subcategories for gender and number is created, that can be later used by some rules within the respective sketch grammars.

5 Corpus Access

The standard environment for users to access the corpus data is the open-source *NoSketch Engine* developed at the Faculty of Informatics of the Masaryk University in Brno [14]. It is a mature, stable and user-friendly corpus manager offering all traditional concordancing- and wordlist-related search and display functions with queries based on wordform, lemma or PoS tag with optional use of regular expressions and the powerful Corpus Query Language (CQL). For users having an account at the Sketch Engine site, the installed versions of the *Aranea* corpora with compatible sketch grammars offer full capabilities of that system [10]. The source versions of the corpus data can be made available for download (for research and educational purposes). Note, however, that the copyright status of the data is not clear and users from countries where this might cause legal problems will have to solve this issue themselves.

6 The Sketch Grammar

For all corpora, compatible sketch grammars have been written. Their main idea is having an equal number of gramrels (and word sketch tables displayed) for all word classes across all languages. The principles and main design decisions behind creation of compatible sketch grammars are discussed in our work [4]. The Appendix contains an example of compatible word sketches generated from two *Aranea* corpora.

7 Current State of the Project

At the time of writing this Paper (May 2014), the basic medium-sized *Maius* (as well as the smaller *Minus*) *Aranea* editions for seven languages (Russian, French, German, Spanish, Polish, English, and Slovak) have been created, and compatible sketch grammars have been written for all of them. For Slovak, the *Araneum Slovacum Maximum* (cca. 3 billion tokens) has also been compiled. Data for the Czech *Araneum Bohemicum* have been downloaded and filtered, and is being tagged at present. The downloading of data for the remaining languages of the "inner circle" (Hungarian, Ukrainian and Italian) will follow soon and the first stage of the Project is expected to complete by the end of 2014.

[5] The SNK [5] tagset assigns a word class of its own for reflexive formants ("sa", "si") and for participles, with both having fairly high frequencies in the corpus.

8 Conclusion and Further Work

The *Aranea* project has showed that by using the available open-source and free tools, billion-token web corpora can be created with minimal additional programming. After our processing pipeline has been tuned, a corpus for a new language (provided that a PoS tagger is available), including creation of a new sketch grammar, can typically be produced in some two weeks.

Our further activities are expected to follow several tracks. Firstly, based on the feedback from the users of our corpora, we would like to improve the data (filtration, tokenization, better deduplication, and tagging) of the existing corpora, and, where possible, to provide for alternative layer(s) of annotation, e.g. by using different taggers. Secondly, we want to include more languages into our *Aranea* corpus family, at least those taught as foreign languages at the Slovak universities (provided that suitable taggers exist for them). And lastly, we plan to compare the Aranea corpora among themselves and with other available web-based corpora for matching languages by means of methodology described in [9], and try to establish the degree of their mutual "comparability".

Acknowledgements. The presented results were partially obtained under the VEGA Grant Agency Project No. 2/0015/14 (2014–2016).

References

1. Baroni, B., Bernardini, S.: BootCaT: Bootstrapping corpora and terms from the web. In: Proc. 4th Int. Conf. on Language Resources and Evaluation, Lisbon (2004)
2. Baroni, M., Bernardini, S., Ferraresi, A., Zanchetta, E.: The WaCky Wide Web: A Collection of Very Large Linguistically Processed Web-Crawled Corpora. Language Resources and Evaluation 43(3), 209–226 (2009)
3. Benko, V.: Data Deduplication in Slovak Corpora. In: Slovko 2013: Natural Language Processing, Corpus Linguistics, E-learning, pp. 27–39. RAM-Verlag, Lüdenscheid (2013)
4. Benko, V.: Compatible Sketch Grammars for Comparable Corpora. In: Proc. XVI EURALEX Int. Congress, Bolzano (in print, 2014)
5. Garabík, R., Šimková, M.: Slovak Morphosyntactic Tagset. Journal of Language Modelling (1), 41–63 (2012)
6. Grefenstette, G.: Generating resources for the lexicography of under-resourced languages. Invited lecture at eLex 2013 Int. Conference, Tallinn (2013)
7. Hajič, J.: Disambiguation of Rich Inflection (Computational Morphology of Czech). Karolinum, Praha (2004)
8. Jakubíček, M., Kilgarriff, A., Kovář, V., Rychlý, P., Suchomel, V.: The TenTen Corpus Family. In: Proc. Int. Conf. on Corpus Linguistics, Lancaster (2013)
9. Kilgarriff, A.: Comparing Corpora. International Journal of Corpus Linguistics 6(1), 97–133 (2001)
10. Kilgarriff, A., Rychlý, P., Smrž, P., Tugwell, D.: The Sketch Engine. In: Proc. XI EURALEX Int. Congress, Lorient, pp. 105–116 (2004)
11. Petrov, S., Das, D., McDonald, R.: A Universal Part-of-Speech Tagset. In: Proc. 8th Int. Conf. on Language Resources and Evaluation, Istanbul (2012)

12. Piasecki, M.: Polish Tagger TaKIPI: Rule Based Construction and Optimisation. Task Quarterly 11, 151–167 (2007)
13. Pomikálek, J.: Removing Boilerplate and Duplicate Content from Web Corpora. Ph.D. thesis, Masaryk University, Brno (2011)
14. Rychlý, P.: Manatee/Bonito – A Modular Corpus Manager. In: 1st Workshop on Recent Advances in Slavonic Natural Language Processing, pp. 65–70. Masaryk University, Brno (2007)
15. Schäfer, R., Bildhauer, F.: Web Corpus Construction. Synthesis Lectures on Human Language Technologies. Morgan & Claypool Publishers (2013)
16. Schmid, H.: Probabilistic Part-of-Speech Tagging Using Decision Trees. In: Proceedings of International Conference on New Methods in Language Processing, Manchester (1994)
17. Suchomel, V., Pomikálek, J.: Efficient Web Crawling for Large Text Corpora. In: 7th Web as Corpus Workshop (WAC-7), Lyon, France (2012)

Appendix

Word sketches (collocation profiles) for the verb "drink" ("boire") generated by Sketch Engine from *Araneum Anglicum Maius* and *Araneum Francogallicum Maius*. The gramrel (table) names denote collocational relationships. The "X" symbol stand for the keyword, i.e. the lemma the word sketch is made for. The left-hand and right-hand collocates are indicated by the respective PoS abbreviations. The "Y" stands for collocates of any PoS, and the "Z" indicates a collocate of PoS not covered by the "explicit" rules (i.e. a "catch all" rule).

Sketch Engine

drink *(verb)* Alternative PoS: non-verb (34500)
Araneum Anglicum Maius (En Web 1.2.01) 1,20 G freq = 42609 (35.5 per million)

X/Y, X/Y	5,596	-0.1
smoke	149	4.74
thirst	8	4.3
eat	709	3.63
party	8	3.15
gamble	10	2.92
breathe	31	2.05
dance	25	1.97
proclaim	23	1.9
sleep	42	1.31
marry	41	1.23
taste	16	0.83
shop	8	0.67
laugh	13	0.3
relax	13	0.22

X/Y Cj X/Y	10,312	-0.1
smoke	295	5.69
eat	2,951	5.68
carouse	15	5.39
party	43	5.37
drug	29	4.87
bathe	12	3.98
gamble	22	3.91
bath	31	3.72
dance	65	3.32
feast	11	3.29
snack	8	3.11
dine	21	2.86
socialize	12	2.6
drive	338	2.4

YX	16,949	-0.0
binge	31	4.65
tritium	12	3.95
habitually	10	3.45
arsenic	13	3.38
quit	45	2.65
litre	11	2.65
contamination	18	2.58
Benefits	12	2.42
happily	18	2.4
contaminate	10	2.11
abstain	8	2.08
seldom	11	1.96
rarely	31	1.85
sport	71	1.66

XY	27,833	-0.1
alcohol	805	5.95
Kool-Aid	51	5.75
beer	597	5.66
responsibly	83	5.64
tea	459	5.4
bottled	69	5.39
soda	123	5.37
coffee	591	5.28
alcoholic	121	5.27
champagne	76	5.11
wine	626	5.02
Kool	26	4.7
Jiaogulan	22	4.56
excessively	38	4.54

Nn X	16,383	-0.0
binge	33	4.77
Ye	34	4.6
tritium	17	4.48
drinker	21	4.0
arsenic	16	3.7
fluoride	15	3.67
gall	9	3.25
litre	16	3.2
alcoholic	25	3.16
wine	158	3.06
beer	90	2.97
cup	112	2.95
tea	80	2.92
beverage	30	2.9

X Nn	33,413	-0.1
tea	991	6.49
beer	1,032	6.43
alcohol	1,040	6.3
soda	237	6.23
Kool-Aid	83	6.21
coffee	1,095	6.16
smoothie	148	6.07
wine	1,129	5.86
juice	418	5.75
beverage	238	5.71
cup	755	5.65
milk	621	5.58
champagne	103	5.44
water	4,558	5.27

Aj X	4,639	-0.0
thirsty	37	5.56
underage	12	3.88
unfit	10	3.64
bottled	8	3.09
safe	200	2.71
unsafe	10	2.52
okay	21	2.13
clean	50	1.79
pregnant	24	1.67
drunk	9	1.54
ready	86	1.52
pleasant	14	1.51
OK	10	1.5
sick	19	0.88

X Aj	9,473	-0.1
bottled	93	6.42
caffeinated	31	6.26
koolaid	18	5.89
alcoholic	141	5.77
iced	32	5.55
sugary	24	4.97
contaminated	51	4.97
alkaline	30	4.92
fizzy	12	4.88
thirsty	26	4.81
copious	23	4.72
spirituous	8	4.68
non-alcoholic	9	4.35
herbal	49	4.33

Vb X/X Vb	44,353	-0.0
thirst	28	4.14
carbonate	25	4.02
hydrate	21	3.52
intoxicate	27	3.4
water	37	3.39
contaminate	33	3.39
milk	23	3.37
poison	30	3.35
fluoridate	14	3.26
purify	28	2.99
flush	28	2.97
boil	52	2.97
decaffeinate	10	2.82
vinegar	10	2.81

Av X/X Av	18,106	-0.1
responsibly	92	6.07
excessively	47	5.14
moderately	47	4.91
heavily	205	4.49
greedily	12	4.22
habitually	15	3.98
unworthily	9	3.96
sensibly	13	3.89
regularly	109	3.16
too	913	2.92
eagerly	14	2.85
freely	45	2.82
abundantly	9	2.79
anymore	46	2.77

ZX	32,200	-0.1
whoever	17	2.33
who	1,516	1.69
I	3,747	1.56
he	1,594	1.47
she	550	1.35
you	3,115	1.24
they	1,368	1.04
&	157	0.92
we	1,356	0.91
those	280	0.78
him	286	0.7
TO	8	0.68
What	83	0.55
to	8,542	0.54

XZ	25,135	-0.1
eight	80	2.01
himself	113	1.61
every	273	1.52
8	120	1.02
some	423	0.66
herself	17	0.63
six	61	0.61
myself	47	0.6
it	2,103	0.58
themselves	60	0.42
half	62	0.33
neither	11	0.28
whatever	20	0.24
2	186	0.22

Sketch Engine

boire **Araneum Francogallicum Maius (Fr Web 1.2.02) 1,23 G freq = 48230** (39.1 per million)

X/Y , X/Y	7,445	-0.1
fumer	311	5.21
droguer	25	4.91
manger	864	4.48
papoter	9	3.36
pisser	10	3.05
danser	54	2.75
dormir	78	2.3
rigoler	16	2.25
respirer	20	1.67
chanter	49	1.56
pendre	21	1.46
discuter	54	1.14
cuisiner	10	0.94
laver	17	0.92

X/Y Cj X/Y	9,211	-0.2
manger	2,362	5.93
fumer	268	4.99
droguer	25	4.82
uriner	18	4.33
grignoter	28	4.33
festoyer	9	4.13
enivrer	16	3.89
déboire	16	3.85
trinquer	11	3.81
pisser	9	2.84
danser	51	2.66
baigner	20	2.1
rigoler	13	1.93
dormir	56	1.82

YX	26,598	-0.1
last	83	6.36
ossature	95	5.56
reputation	21	4.51
Serial	20	4.46
yogourt	23	4.31
yaourt	49	4.28
chaufferie	22	4.2
Last	21	4.14
serial	18	4.07
honte	100	3.88
nover	16	3.84
navette	48	3.71
pout	11	3.55
bardage	12	3.42

XY	31,428	-0.1
gorgée	335	7.67
bière	689	6.66
verre	1,653	6.42
thé	670	6.11
not	116	5.87
calice	85	5.86
tasse	265	5.79
café	995	5.78
alcool	552	5.56
champagne	172	5.49
tisane	71	5.45
coca	79	5.42
potion	77	5.09
pisse	39	4.99

NnX	17,569	-0.1
last	83	6.82
ossature	78	5.49
truck	30	5.19
universal	37	5.02
reputation	21	5.02
yogourt	23	4.7
chaufferie	22	4.57
yaourt	52	4.53
Last	21	4.53
pout	11	4.05
navette	56	4.04
honte	102	3.96
bardage	13	3.94
Rosny	9	3.73

XNn	31,427	-0.1
gorgée	647	8.62
bière	936	7.1
verre	2,210	6.83
least	108	6.66
thé	840	6.44
tasse	385	6.33
calice	109	6.22
alcool	845	6.18
café	1,310	6.17
champagne	263	6.1
coca	112	5.92
tisane	93	5.84
whisky	86	5.57
soda	71	5.5

AjX	3,627	-0.1
Serial	20	6.6
serial	18	5.6
Urban	15	4.17
rosé	15	3.95
préférable	20	2.07
agréable	61	1.97
potable	9	1.21
mixte	13	1.2
facile	70	1.11
prêt	73	0.77
mini	9	0.54
idéal	30	0.33
chaud	28	0.15
incapable	9	0.14

XAj	6,084	-0.1
cul-sec	16	6.3
also	19	5.85
gazeuse|gazeux	24	4.73
empoisonné	10	4.66
alcalin	9	4.15
rosé	16	3.94
tiède	25	3.74
chaud	285	3.49
sec	78	3.31
you	9	2.69
digestif	11	2.46
frais	154	2.46
potable	21	2.41
amer	11	2.26

Vb X/X Vb	40,042	-0.1
not	152	6.08
gorger	49	4.34
it	44	4.06
epervier	18	3.86
alcooliser	33	3.82
sucrer	67	3.55
hydrater	30	3.41
nover	16	3.39
assoiffer	21	3.25
Vera	15	3.2
is	39	3.15
granuler	14	3.15
be	28	3.13
désaltérer	13	3.02

Av X/X Av	19,221	-0.1
modérément	36	5.36
avidement	19	4.67
tranquillement	99	4.48
jeun	22	4.47
abondamment	35	4.28
goulûment	10	3.65
suffisamment	132	3.39
lentement	50	3.19
régulièrement	194	3.11
occasionnellement	9	2.91
excessivement	9	2.89
trop	765	2.76
beaucoup	872	2.67
raisonnablement	10	2.66

ZX	35,379	-0.1
Donne-moi	27	4.53
donne-moi	13	3.49
Pouvez-vous	16	3.3
Assurez-vous	16	3.3
Peut-on	16	2.75
quiconque	23	2.46
on	2,456	2.24
tu	505	2.19
RER	9	2.05
quel	349	2.04
je	3,335	1.75
la	232	1.27
quoi	121	1.26
il	3,500	1.23

XZ	27,144	-0.1
+	94	1.5
quelque	425	1.06
mon	579	1.05
ton	138	0.94
&	81	0.91
~	26	0.84
un	7,147	0.83
L	26	0.8
I	21	0.78
=	41	0.67
son	1,424	0.39
huit	17	0.23
toi	32	0.14
chaque	135	0.09

Towards a Unified Exploitation of Electronic Dialectal Corpora: Problems and Perspectives*

Nikitas N. Karanikolas[1], Eleni Galiotou[1], and Angela Ralli[2]

[1] Department of Informatics, Technological Educational Institute of Athens, GR-12210 Aigaleo, Athens, Greece
{nnk,egali}@teiath.gr

[2] Department of Philology, University of Patras, GR-26504 Rio, Patras, Greece
ralli@upatras.gr

Abstract. In this paper, we deal with the problem of storing and retrieving dialectal data in a unified framework. In particular, we discuss issues concerning the design and implementation of a multimedia database which will contain written and oral data from three Greek dialects in Asia Minor. At first, we describe the overall architecture of a system aiming at providing the user with the possibility to store audio recordings, text transcripts, and other annotations. Then we discuss the possibilities and limitations of a retrieval module aiming at combining different linguistic levels for a unified exploitation of oral and written corpora.

Keywords: Computational Dialectology, Electronic Corpora, Modern Greek Dialects, Multimedia databases.

1 Introduction

The use of computational techniques and the possibility to store oral and written dialectal data on electronic media has greatly contributed to the advancement of research in dialectal change and language contact. For example, the on-line tool for Dutch dialect syntax research DynaSAND (the Dynamic Syntactic Atlas of Dutch Dialects) provides a database, a search engine, a cartographic component, and a bibliography concerning the syntactic variation in dialects located in the Netherlands, Belgium and France [2]. Another interesting approach is that of LAMSAS (Linguistic Atlas of Middle and South Atlantic States) [18] where dialectal material from the Atlantic coast of the United States is comprised. As for Greek dialectal data, no results of a computational processing were reported until very recently, with the exception of an electronic dictionary of Cypriot Greek [21]. Greek dialects of Asia Minor constitute a quite interesting case in the scientific field of dialectology and language contact; although they are of the same Indo-European origin (Greek), they have gradually

* This research has been co-financed by the European Union (European Social Fund - ESF) and Greek national funds through the Operational Program "Education and Lifelong Learning" of the National Strategic Reference Framework (NSRF) - Research Funding Program: Thalis. Investing in knowledge society through the European SocialFund.

P. Sojka et al. (Eds.): TSD 2014, LNAI 8655, pp. 257–266, 2014.

diverged from one another partly under the influence of an Altaic language (Turkish) to such an extent that they are considered as different dialects. Moreover, Greek and Turkish belong to different typological groups (fusional vs. agglutinative). Therefore, a systematic study of Asia Minor Greek dialects would give useful insights as for the nature of language change within the domain of dialectal variation. The aforementioned task would be greatly facilitated by the availability of dialectal data on electronic media and the development of computational tools for their processing. Recently, an attempt to combine Informatics and Theoretical Linguistics in order to describe and analyze dialectal phenomena of three Greek dialects in Asia Minor has been undertaken in the course of the "Thalis" program: "Pontus, Cappadocia, Aivali: in search of Asia Minor Greek" (AMiGre). The aim of his project is twofold: (a) to provide a systematic and comprehensive study of Pontic, Cappadocian and Aivaliot, three Greek dialects of Asia Minor of common origin and of parallel evolution that are faced with the threat of extinction; (b) to digitize, archive and process a wide range of oral and written data thus contributing to the sustainability and awareness of this longwinded cultural heritage.

The computational component of the project comprises activities such as [8]: (a) design and development of a multimedia tri-dialectal dictionary which contains lemmata from the three dialects (Pontic, Cappadocian, Aivaliot) in a comparative way [11]; (b) design and development of a multi-media software and database for the archiving and processing of oral and written dialectical data.

In this paper, we deal the problem of storing and retrieving dialectal corpora on electronic media. In particular, we present the design of a multimedia software and database for archiving and processing of oral and written data from the three Asia Minor Greek dialects. In Section 2, we present the design principles and the modules of the multimedia software. Next we discuss the technical aspects and the structure of the database. In Section 4, we present the search and retrieve module. Finally, in Section 5, we draw conclusions and point to future work.

2 A Multimedia Software for Oral and Written Dialectal Resources

2.1 Design Principles

The Nature of the Data. The oral corpus of the AMiGre project in its current state consists of approx. 180 hours (i.e. 60 hours/dialect) of recorded raw data and was compiled in the Laboratory of Modern Greek Dialects of the University of Patras [19]. The raw data were annotated, abstracted and analyzed according to the 3A (annotation, abstraction, analysis) model [22]. Some 45 hours (15hours/dialect) of raw data were further processed resulting in a multimodal sub-corpus which combines raw data with transcription, translation, annotation and metadata. This multimodal sub-corpus is processed using ELAN, a software package for the creation of audio and video resources ([6]; [20]) while spoken data are further analyzed using Praat, a scientific software package for the analysis of Speech in Phonetics ([3,4]). The spoken data are annotated according to the speaker's turn-takings, utterances and intonation phrases. Phonological words, syllables and phonemes are also annotated

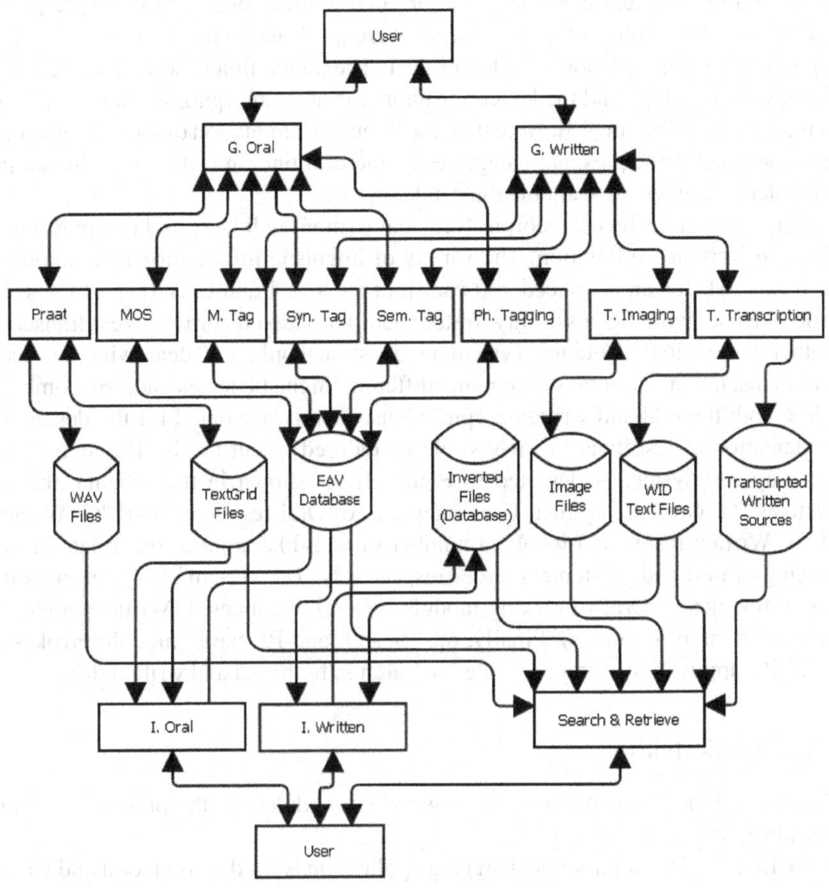

Fig. 1. The Architecture of the system

using the IPA (International Phonetic Alphabet) symbols [10]. Explicit representations of vowels, diphthongs, consonants and consonant clusters appear on different layers of representation (tiers). So, the output files of the processing with ELAN and Praat act as input files for our system. The written corpus consists of 1,000,000 words of digitized dialectal texts from primary written sources of the 19th and early 20th centuries [12]. The inclusion of a text in the corpus was based upon criteria such as representativeness (according to a dialect, a local sub-dialect, chronological period) and quality (closeness to the actual spoken language, consistent linguistic terminology or transcription system). A sub-corpus consisting of 200,000 words was transcribed using a custom-made transcription system based on SAMPA (Speech Assessment Methods Phonetic Alphabet) [23]. The use of the particular transcription system aims at: (a) facilitating further electronic elaboration by allowing transcription without the use of special diacritics on keyboard configurations; (b) representing all the special sounds included in the phonetic inventory of the dialects under investigation; (c) unify the disparate and inconsistent notation of the original written sources [15]. Finally, some 50,000 words of the transcribed sub-corpus were annotated with the use of a special tool developed for the purposes of this project. The annotation describes the levels of phonology, morphology and the lexicon. Annotation at the morphological level follows the principles described in [13]. Note that, the linguistic annotation of the written corpus follows the same principles and categories as the oral one, in order to enable unified searches across the whole available dialectal corpus.

Design. The whole dialectal corpus (oral and written) will be stored in a multimedia database for further exploitation. The variety of linguistic information and annotation types would ask for an advanced software tool such as Labb-CAT ([7], [14]) which provides the user with the possibility to store audio or video recordings, text transcripts and other annotations. Yet, the system in question could not deal with our basic requirements, i.e. (a) Annotations at many different linguistic levels and, (b) Combined search at both the oral and written corpus. Consequently, we opted for the design and implementation of a software which would be tailored to our needs. The architecture of the proposed system is depicted in Figure 1. As shown in the schema, the two subsystems "G. Oral" (Graphical User Interface for Oral resources) and "G. Written" (GUI for Written resources) invoke a number of web-like applications related to the processing of oral and written resources respectively. The system also comprises two indexing modules (I. Oral= Indexing module for oral resources, I. Written= Indexing module for written resources). Finally, the "Search and Retrieve" module invokes all the web-like applications for a combined research in both oral and written data.

2.2 The Applications

The aforementioned system comprises 8 web-like modules for the processing of oral and written resources:

Phon Tagger: The phonological tagging application is used on both oral and written resources. Annotation on the written resources is performed at the word level. In order to achieve a unified treatment, information on morphological word boundaries will be added to oral resources.

Morph Tagger: The morphological tagging application is also used on both oral and written resources. In all the resources annotation is performed at the word level. For each morphological word, information on part of speech, grammatical properties, and morphological phenomena such as derivation and compounding, is provided.

Synt Tagger: The syntactic tagger is also used on both oral and written resources. In the current state of the system, annotation is performed at the word level; each word is associated to at most one syntactic structure. The application provides also the possibility to perform annotations on a phrase or sentence level.

Sem Tagger: The semantic tagging application is also used on both oral and written resources. Annotation is performed on a sequence of words with values such as “loan”, “idiomatic phrase” etc.

Text Imaging: This application aims at the preview of pages of written resources.

Text Transcription: The application of transcription of a written resource provides two panels. The left panel contains the image of the page under processing and the right one contains the transcription of the page in the form of processable text.

MOS (Oral Metadata). This application provides the possibility to store and update the metadata of oral resources comprising information as age, sex, cultural background of the speaker. Note that, such information is not available for written resources.

PRAAT: It invokes the Praat software for phonetic analysis [4].

3 The Structure of the Database

3.1 Transcribed Written Documents Source

The Transcribed Written Documents source is a collection of text files encoded with UTF-8 (Unicode) encoding. Each file contains the transcription of a written document. The written documents are manuscripts, typescripts or printed material that is related with records of the Asia Minor Greek Dialects. The transcriptions are based on lowercase letters of the Greek alphabet and some other symbols (uppercase Greek letters, Latin letters and letter-couples). This is an endeavor to homogenize the various symbols used by different researchers of the Asia Minor Greek Dialects, in order to indicate the same phonological phenomenon [15].

3.2 Image Files Source

The Image Files source is a collection of digitized versions of the original documents (manuscripts, typescripts or printed materials that are related with records of the Asia Minor Greek Dialects). Each image file is usually a digitized version of a single page of one written document. The alternative solution to keep a single file (e.g. a multi-page tiff file) for each written document is considered inexpedient. This is because the system runs on the web and loading the pages can cause great delays when the user moves from one document to another. The overhead of the selected solution is to associate a list of Image File Paths (one path for each page) with any single written document (the primary record used for the written document). It is obvious that the values of the Image File Paths list should be accessible by the Text Imaging module.

3.3 WID Text Files Source

Each WID Text File is a collection of words that emanate from the tokenization of a Transcribed Written Document. Each word (token) is characterized by:

- WordID (an identifier of the word/token which is unique for the whole WID Text Files source),
- Word (the word as is in the transcribed text),
- PositionIndex (sequence number of the word/token in the transcribed written document),
- Location (starting byte and ending byte of the word/token in the transcribed written document),

- PartID (an identifier of the page containing the word/token, it is unique for the whole set of parts/pages constituting the written documents).

From the PartID we can deduce the source (the primary record used for the written document) and the page of the document (the sequence number of page into the document). A set of words/tokens (identified by a set of WordIDs coming from the same document) constitute a single WID Text File. The set of WID Text Files constitute the WID Text Files source.

3.4 TextGrid Files Source

This is a collection of files produced by the Praat software package for phonetic analysis [4]. Praat uses different tiers defined by the user to annotate an audio file. Annotations can vary and can comprise turn takings, transcription, intonation phrases, intonation words, syllables, phonemes, etc. It also supports point tiers that can be used for pitch (e.g. pitch accent, phrase accent, boundary tone), etc. In order to capture all necessary levels of annotation, we have also introduced a "morphological words" tier. In our system, each TextGrid file is associated with a relevant audio file (discussed in Section 3.5). We can make the abstraction that tokens/words in a WID Text File are the equivalents to the "morphological words" in a TextGrid file. The association of a TextGrid file with the relevant oral document (the primary record) imposes to keep the TexGrid File Path inside the primary record of the oral document.

3.5 WAV Files Source

This is a collection of audio files in the WAV (Wave Audio File) format. The purpose of an audio file has been discussed in Section 3.4. The association of an audio file with the relevant oral document (the primary record) is achieved by including a WAV File Path field into the primary record of the oral document.

3.6 EAV Database

Actually the EAV (Entity Attribute Value) database consists of five sub-schemas, namely:

- EAV Phon (for storing phonological phenomena of single words),
- EAV Syn (for storing syntactic phenomena regarding sequences of words),
- EAV Sem (for storing semantic phenomena regarding sequences of words),
- EAV Morpho (for storing morphological attributes characterizing single words),
- EAV Meta (for storing metadata for oral documents).

With the term "single words" we refer to tokens from the transcribed written documents or morphological words from the TextGrid files. With the term "sequences of words" we refer to sequences of tokens from the transcribed written documents or sequences of morphological words from the TextGrid files. The EAV Morpho sub-schema is the most complicated one because we have to store a sparse set of morphological attribute - value pairs for each morphological word. In addition to the sparse nature of morphological attributes, we have to deal with subschema evolution (the attributes are not fixed from the beginning) and also with hierarchical values in the attribute - value pair. A similar sparse nature and need for subschema evolution we have to cope with, is the EAV Meta subschema. These are the main reasons for adopting the EAV (Entity - Attribute - Value) technology ([1,9,16,17]) to design the above five sub schemata. An example of the EAV Morpho subschema is depicted in Tables 1, 2 and 3. The "Repetition" field in the "EAV Morpho" table indicates the number of morphological phenomena associated to a single word while, the "TypeId" field in the "EAV Morpho Properties" table indicates whether the value of a property supports single or multiple values and whether the values come from a predefined list or they are text values.

Table 1. EAV Morpho

WordId	PropertyId	Repetition	PropertyValueId
385	4	1	177
385	26	1	268
387	4	1	176
387	5	1	182
390	4	1	175
391	4	1	176
391	5	1	182

3.7 Inverted Files Database

Inverted Files [5] do not have the simple form used in Text Information Retrieval (couples of a word/lemma and a list of occurrences of the word/lemma in documents). In our case (an Electronic Multimedia Dialectal Corpus) there is a need for a form such as :

- a word/token/phenomenon,
- a list of quadruples, where each quadruple specifies the location of the occurrence of a word/token/phenomenon.

Each quadruple (location) contains:

- The identifier of the relevant database or source (type of information), a value from the set EAV Phon, EAV Syn, EAV Sem, EAV Morpho, EAV Meta, Text-Grid File, Transcribed Written Document,
- The identifier (the primary key or other unique name) that specifies a concrete instance among the collection of instances in the relevant database (file collection),
- The attribute (only in the case of inversion of EAV Morpho and EAV Meta data),
- The word/token sequence (a couple of the form (StartWordId, EndWordId)).

Table 2. EAV Morpho Properties

PropertyId	**Name**	**TypeId**
4	'PART OF SPEECH'	4
5	'GENDER'	4
6	'INFL. CLASS'	4
...	...	...
24	'ORIGIN OF BASE FORM'	4
25	'ORIGINAL LOAN WORD'	1
26	'PART OF SPEECH OF LOAN WORD'	4

Table 3. EAV Morpho Lookup

PropertyValueId	**Description**	PropertyId
175	'Adjective'	4
176	'Noun'	4
177	'Verb'	4
...	...	...
182	'Neuter'	5
183	'3-gender'	5
184	'Masculin'	5
185	'Feminin'	5
...	...	...
225	'Turkish'	15
226	'Italian'	15
227	'Roman Dialects'	15
228	'Greek'	15
...	...	...

4 Search and Retrieve and Other Advanced Modules

The Search & Retrieve module invokes the relevant application (among the ones explained in Section 2.2) using OLE (Object Linking and Embedding) Automation or other equivalent technology. The selection of the relevant application, among the available ones, is automatically decided, depending on the types of information that match the user defined criteria. It can also be defined by the user. The Search & Retrieve module provides a query builder to the user who can add and combine criteria. For each criterion, the user defines the requested value (word/token/phenomenon) and the location (a completely or partially defined quadruple) where the value should occur. Table 4 depicts a simple example of a query defined by the Search and Retrieve query builder:

The G. Oral (GUI for Oral sources) and G. Written (GUI for Written sources) are typical administrative applications that permit Add, Update, Delete and Browse facilities. The I. Oral (Indexing Module for Oral sources) and the I. Written (Indexing Module for Written sources) are mainly responsible for performing inversions.

Table 4. Query on the transcriptions of written documents where both phenomena of synaeresis and palatalization appear into a word/token window of size 10

Word/token/phenomenon	Location (Quadruple)			
synaeresis	EAV Phon	-	-	(X, X+10)
palatalization	EAV Phon	-	-	(X, X+10)
Output	Transcribed Written Document			

5 Conclusions and Future Work

In this paper, we have presented the design of a software application aiming at storing and exploiting oral and written dialectal corpora in a unified way. Our system takes into account alternative sources of information using the EAV technology and performing a combined inversion which is adapted to the complexity of linguistic dialectal data. The software is currently under development and the first implementation results are expected to contribute to our understanding and awareness of dialectal corpora processing at both linguistic and computational levels.

References

1. Anhoj, J.: Generic Design of Web-Based Clinical Databases. Journal Medical Internet Research 4 (2003)
2. Barbiers, S., et al.: Dynamic Syntactic Atlas of the Dutch dialects (DynaSAND). Meertens Institute, Amsterdam (2006), http://www.meertens.knaw.nl/sand/
3. Boersma, P.: The use of Praat in corpus research. In: Jacques Durand, J., Gut, U., Kristofferson, G. (eds.) Handbook of Corpus Phonology, OUP, Oxford (2012)
4. Boersma, P., Weenink, D.: Praat: Doing phonetics by computer (2013), http://www.praat.org
5. Buttcher, S., Clarke, C., Cormack, G.: Information Retrieval: Implementing and Evaluating Search Engines. MIT Press, Cambridge (2010)
6. ELAN: Max Planck Institute for Psycholinguistics, The Language Archive, Nijmegen, The Netherlands, http://tla.mpi.nl/tools/tla-tools/elan/
7. Fromont, R., Hay, J.: ONZE Miner: the development of a browser-based research tool. Corpora 3(2), 173–193 (2008)
8. Galiotou, E., Karanikolas, N., Manolessou, I., Pantelidis, N., Papazachariou, D., Ralli, A., Xydopoulos, G.: Asia Minor Greek: Towards a Computational Processing. In: Procedia: Social and Behavioral Science. Elsevier (in press, 2014)
9. Johnson, S.B., Chatziantoniou, D.: Extended SQL for manipulating clinical warehouse data. In: AMIA 1999, pp. 819–823 (1999)
10. IPA chart, http://www.langsci.ucl.ac.uk/ipa/ipachart.html
11. Karanikolas, N.N., Galiotou, E., Xydopoulos, G.J., Ralli, A., Athanasakos, K., Koronakis, G.: Structuring a Multimedia tridialectal dictionary. In: Habernal, I. (ed.) TSD 2013. LNCS, vol. 8082, pp. 509–518. Springer, Heidelberg (2013)
12. Koliopoulou, M., Markopoulos, T., Pantelidis, N.: Pontus, Cappadocia, Aivali: Challenges of a digital corpus of written material. In: The 11th International Conference of Greek Linguistics, Rhodes (September 2013) (in Greek)

13. Koutsoukos, N., Ralli, A.: From derivation to inflection: a process of grammaticalization. In: Morphology Meeting 2012. Leiden, the Netherlands (2012)
14. LaBB-CAT (formerly ONZE Miner), `http://onzeminer.sourceforge.net/`
15. Manolessou, I., Beis, S., Bassea-Bezantakou: The phonetic transcription of Modern Greek dialects. Lexicographicon Deltion 26, 161–222 (2012) (in Greek)
16. Nadkarni, P.: Clinical Patient Record Systems Architecture: An Overview. Journal of Postgraduate Medicine 46(3), 199–204 (2000)
17. Nadkarni, P.: An introduction to entity-attribute-value design for generic clinical study data management systems. Presentation in: National GCRC Meeting, Baltimore, MD (2002)
18. Nerbonne, J., Kleiweg, P.: Lexical distance in LAMSAS. Computers and the Humanities 37(3), 339–357 (2003)
19. Ralli, A., Papazachariou, D., Karasimos, A.: Laboratory of Modern Greek Dialects and the project GreeD. In: Ralli, A., et al. (eds.) Proc. 4th Int. Conf. of Modern Greek Dialects and Linguistic Theory (2010)
20. Sloetjes, H., Wittenburg, P.: Annotation by category - ELAN and ISO DCR. In: Proceedings of the 6th International Conference on Language Resources and Evaluation, LREC 2008 (2008)
21. Themistocleous, C., Katsogiannou, M., Armosti, S., Christodoulou, K.: Cypriot Greek Lexicography: An Online Lexical Database. In: Proceedings of Euralex, pp. 889–891 (2012)
22. Wallis, S., Nelson, G.: Knowledge discovery in grammatically analyzed corpora. Data Mining & Knowledge Discovery 5, 305–335 (2001)
23. Wells, J.C.: 'SAMPA computer readable phonetic alphabet'. In: Gibbon, D., Moore, R., Winski, R. (eds.) Handbook of Standards and Resources for Spoken Language Systems 1997, Part IV, section B, Mouton de Gruyter, Berlin (1997)

Named Entity Recognition for Highly Inflectional Languages: Effects of Various Lemmatization and Stemming Approaches

Michal Konkol and Miloslav Konopík

Department of Computer Science and Engineering,
Faculty of Applied Sciences,
University of West Bohemia,
Univerzitní 8, 306 14 Plzeň, Czech Republic
{konkol,konopik}@kiv.zcu.cz

Abstract. In this paper, we study the effects of various lemmatization and stemming approaches on the named entity recognition (NER) task for Czech, a highly inflectional language. Lemmatizers are seen as a necessary component for Czech NER systems and they were used in all published papers about Czech NER so far. Thus, it has an utmost importance to explore their benefits, limits and differences between simple and complex methods. Our experiments are evaluated on the standard Czech Named Entity Corpus 1.1 as well as the newly created 2.0 version.

Keywords: Named Entity Recognition, Lemmatization, Stemming.

1 Introduction

Named entity recognition (NER) is a standard natural language processing (NLP) task. A NER system detects phrases with interesting meaning and classifies them into predefined groups – typically persons, organizations, locations, etc. NER is used as a preprocessing for many other NLP tasks, including question answering [1], machine translation [2], or summarization [3]. Together with named entity disambiguation, it can be used to link entities from text to a knowledge base.

NER is mostly studied on English. In this paper, we study the NER task on Czech – a highly inflectional Slavic language. All the published results on Czech NER use lemmatization to reduce the overwhelming amount of different word forms.

In this paper, we explore various lemmatization and stemming approaches. These approaches range from very simple methods to very complex hybrid (combining rule-based and machine learning techniques) systems. We also experiment with a morphological analyzer, which does not disambiguate the lemmas as full lemmatizer.

Our main goal is to measure the difference between the simplest and the most complex systems, i.e. to answer the question: Is it worth using a complicated, computationally expensive lemmatization (resp. stemming) system?

2 Related Work

The first NER system for Czech was published together with the CNEC 1.0 corpus [4]. It was based on Decision Trees and achieved 68% F-measure. This system was

P. Sojka et al. (Eds.): TSD 2014, LNAI 8655, pp. 267–274, 2014.

outperformed by SVM-based NER system [5] with 71% F-measure. The following two systems were based on Maximum Entropy [6] (72.94% F-measure) and Conditional Random Fields [7] (58% F-measure). Both systems were evaluated by different methods and were not directly comparable to the previous systems and to each other. Another system based on Conditional Random Fields was published in [8] together with comparison of all previous systems. It achieved 74.08% F-measure on the CoNLL version of the CNEC 1.0 corpus and outperformed all of the previous systems. A system based on Maximum Entropy Markov Models was published at the same time [9]. It was evaluated on the original version of CNEC 1.0 corpus and achieved 82.82% F-measure. The last two systems form the current state-of-the-art for Czech NER. The former for the CoNLL versions of the CNEC corpora and the latter for the original (hierarchical) versions of the CNEC corpora.

All the published systems seem to use the PDT 2.0 corpus [10] for training of their lemmatizers. The first two systems [4,5] use directly the annotations that can be found in the CNEC corpus. In [6,8,7], they use a standard HMM tagger trained on PDT 2.0. The last system [9] uses a lemmatizer based on average perceptron sequence labeling [11].

3 Lemmatizers and Stemmers

This section briefly describes the lemmatization and stemming approaches we use in this paper. It should provide a basic idea about the quality and complexity of each method.

3.1 OpenOffice Based lemmatizer

This lemmatizer (proposed by authors of this paper) is inspired by the approach from [12]. It uses the dictionaries and rules created for error correction in the OpenOffice. These resources are meant to be used in a generative process – the words forms are created from the dictionary using the rules.

In our approach, we try to do an inverse operation. This operation is ambiguous and in many cases there are more possible lemmas. We simply choose the first one. The necessary OpenOffice resources are freely available for many languages, thus this approach can be used for many languages.

3.2 HPS Stemmer

The High Precision Stemmer[1] (HPS) is an unsupervised stemmer. It works in two phases. In the first phase, lexically similar words are clustered using MMI clustering [13]. The word similarity is based on the longest common prefix. The output of this phase are clusters which share a common prefix and have the minimal MMI loss. The method assumes that the common prefix is a stem and the rest is a suffix.

The second phase consists of training of a Maximum Entropy classifier. The clusters created in the first phase are used as the training data for the classifier. The classifier uses general features of the word to decide where to split the word into a stem and a suffix.

---

[1] `http://liks.fav.zcu.cz/HPS/`

3.3 HMM Tagger

The HMM (Hidden Markov Model) tagger represents a standard (pure) statistical approach to lemmatization [14]. Transition probabilities in HMM are estimated using 3-gram Kneser-Ney smoothing [15]. This approach can be easily reproduced using common machine learning libraries. It is trained on the PDT 2.0 data [16].

3.4 PDT 2.0 Lemmatizer

The PDT 2.0 lemmatizer [16] uses the most complex approach. The system as a whole is a hybrid system[2]. It is based on two main components – a morphological analyser and a tagger. The morphological analyser is rule-based. It is based upon a dictionary with 350,000 entries and derivation rules. The tagger is statistical (feature-based). The system also contains a statistical guesser for out-of-dictionary words.

3.5 Majka

Majka [17] is rule-based morphological analyser. It provides all possible word forms for a given word. It is not a tagger as it does not disambiguate the proposed lemmas and tags. The authors of Majka are currently working on the disambiguation tool, but it has not been released as a usable library yet.

For our use, we always select the most frequent lemma-tag pair. This is definitely not an optimal solution, but it will be interesting to compare it to the other lemmatization and stemming approaches. Keep in mind, that in this way, we do not use the full potential of Majka and the results would be probably better using some state-of-the-art tagging approach.

3.6 MorphoDiTa

MorphoDiTa[3] [18] is a state-of-the-art tool for morphological analysis, which is based on the averaged perceptron algorithm [11]. The algorithm is derived from standard HMMs, but the transition and output scores are given by a large set of binary features and their weights.

4 NER System

Our NER system is based on Conditional Random Fields (CRF) [19], which are considered as the best method for NER by many authors. We use Brainy [20] implementation of CRF.

All features are used in a window $-2, \ldots, +2$. We use the following feature set for our experiments:

[2] According to `http://ufal.mff.cuni.cz/pdt2.0/browse/doc/tools/machine-annotation/`

[3] `https://ufal.mff.cuni.cz/morphodita`

Word – Each word that appears at least twice is used as a feature.

Lemma – The lemmatization approaches are described in section 3. A lemma has to appear at least twice to be used as a feature.

Affixes – We use both prefixes and suffixes of the actual word. Their length ranges from 2 to 4. The affixes are based on lemmas and have to appear at least 5 times.

Bag of lemmas – Identical to bag of words, but uses lemmas instead of words. The lemma has to appear at least twice to be used as a feature.

Bi-grams – Bi-grams of lemmas have to appear at least twice to be used. Higher level n-grams did not improve the results, probably due to the size of the corpora.

Orthographic features – Standard orthographic features. Including *firstLetterUpper; allUpper; mixedCaps; contains ., ', –, &; upperWithDot; various number formats; acronym*

Orthographic patterns – Orthographic pattern [21] rewrites the word to a different representation, where every lower case letter is rewritten to `a`, upper case letter to `A`, number to `1` and symbol to `-`.

Orthographic word pattern – A compressed orthographic pattern is created for each word in the window. The combination of these patterns forms the orthographic word patter feature. Each combination has to appear at least five times.

Gazetteers – We use multiple gazetteers. They are acquired from publicly available sources such as list of cities from the Czech Ministry of Regional Devel- opment

5 Corpora

In this paper we use the Czech Named Entity Corpus (CNEC) versions 1.1 and 2.0 [4]. We use the older 1.1 version for smooth transition to the 2.0 version, i.e. we can directly compare the results on these corpora and use only the 2.0 version in the future work. We use the CoNLL versions [8] of both corpora.

The 1.1 version contains 5,868 sentences (149,538 tokens), the 2.0 version 8,993 sentences (199,216 tokens). The CNEC 1.1 has a higher density of named entities than common texts. This problem was addressed by adding 3,000 sentences with only a few entities in the new version. The CNEC 2.0 was also extended by some sentences that contains addresses and emails. Both CoNLL versions use seven types of named entities – time (T), geography (G), person (P), address (A), media (M), institution (I) and other (O).

6 Experiments

Our experiments are relatively straightforward. We train a NER model on the training data using each lemmatization (or stemming) approach. Then, we evaluate these models on the validation and test data. As we do not use the validation data for choosing any parameters of the system, they have the same information value as the test data. For all experiments, we use the feature set described in section 4. It consists of frequently used features with default parameters and should work very well in all our experiments.

Table 1. Results for the CNEC 1.1

		Strict			Lenient		
		Precision	Recall	F-measure	Precision	Recall	F-measure
Validation set	Baseline	69.67	66.18	67.87	76.65	72.41	74.47
	OO lemmatizer	76.16	72.74	74.41	82.93	78.79	80.81
	HPS stemmer	76.02	72.53	74.23	82.91	78.56	80.68
	HMM tagger	77.57	74.67	76.09	83.90	80.29	82.05
	PDT 2.0 lemmatizer	78.20	75.52	76.84	84.62	81.29	82.92
	Majka	77.03	73.65	75.30	83.51	79.43	81.42
	MorphoDiTa	78.57	75.63	**77.07**	84.58	80.99	82.79
Test set	Baseline	69.69	66.47	68.05	76.68	72.79	74.69
	OO lemmatizer	72.87	68.36	70.55	80.03	74.80	77.33
	HPS stemmer	73.13	69.10	71.05	80.62	75.70	78.09
	HMM tagger	75.40	71.09	73.18	81.75	76.79	79.19
	PDT 2.0 lemmatizer	76.16	72.40	**74.23**	82.06	77.73	79.84
	Majka	74.58	70.19	72.32	81.55	76.40	78.89
	MorphoDiTa	75.76	71.67	73.66	82.36	77.51	79.86

Table 2. Results for the CNEC 2.0

		Strict			Lenient		
		Precision	Recall	F-measure	Precision	Recall	F-measure
Validation set	Baseline	74.70	70.47	72.52	81.27	76.26	78.69
	OO lemmatizer	76.91	72.26	74.51	83.36	77.89	80.53
	HPS stemmer	75.66	72.06	73.82	82.33	77.83	80.02
	HMM tagger	77.42	74.34	75.85	83.93	80.12	81.98
	PDT 2.0 lemmatizer	78.06	75.24	**76.62**	84.91	81.39	83.11
	Majka	77.56	73.40	75.42	84.27	79.28	81.70
	MorphoDiTa	78.24	74.99	76.58	84.41	80.37	82.34
Test set	Baseline	73.40	69.14	71.20	80.39	75.35	77.79
	OO lemmatizer	73.99	69.34	71.59	81.33	75.88	78.51
	HPS stemmer	73.87	69.58	71.66	81.38	76.25	78.73
	HMM tagger	75.27	70.91	73.03	82.63	77.42	79.94
	PDT 2.0 lemmatizer	76.41	72.43	**74.37**	82.58	78.01	80.23
	Majka	74.99	70.86	72.87	82.27	77.36	79.74
	MorphoDiTa	76.39	72.33	74.31	82.87	78.17	80.45

We use two separate metrics – the strict CoNLL evaluation and the lenient evaluation used in GATE[4]. Both are based on the precision, recall and F-measure. The strict metric considers the marked entity as correct only if it agrees with gold data in both span and type. The lenient metric is a supplement to the strict metric and covers the cases, where the system guesses the correct type, but the span is partially wrong (e.g. two words of three are marked).

The results of our experiments are shown in Tables 1-2. An important finding is that even the simplest methods improve the results, even though the word-based baseline is much stronger on the CNEC 2.0.

The methods based on the standard tagging approaches significantly outperform the methods based on approximative techniques (OO lemmatizer, HPS). This also holds for our approach of using Majka, which in fact, do not use disambiguation but only a morphological analysis. The HMM tagger, MorphoDiTa and PDT 2.0 lemmatizer outperforms our Majka-based approach, but we believe that some combination of our HMM approach and Majka would perform better than both individual methods.

The best results were achieved using the PDT 2.0 lemmatizer with a slight edge over MorphoDiTa. The difference is probably caused by the OOV guesser as entities are more often OOV words than common words. Both significantly outperformed our basic HMM approach.

Generally, the more complex the method is, the better the result is achieved in our tests. This trend is much more obvious than we expected at the beginning.

7 Conclusion and Future Work

We have tested six different lemmatization and stemming approaches in the NER task. They range from very simple to complex state-of-the-art systems. We use a standard setting for Czech NER – the CNEC corpus and CoNLL evaluation.

The results show that supervised lemmatization approaches significantly outperform both the simple rule-based system and the unsupervised stemmer. All lemmatization and stemming approaches were better than the word-based baseline. The best results were achieved using the PDT 2.0 lemmatizer – 74.23% F-measure on the CNEC 1.1 corpus and 74.37% F-measure on the CNEC 2.0 (both in CoNLL format). These results outperform the current state-of-the-art on the CNEC 1.1 corpus and define it for the newly created CNEC 2.0 (as there are no previous results on this corpus).

Acknowledgements. This work was supported by grant no. SGS-2013-029 Advanced computing and information systems, by the European Regional Development Fund (ERDF). Access to the MetaCentrum computing facilities provided under the program "Projects of Large Infrastructure for Research, Development, and Innovations" LM2010005, funded by the Ministry of Education, Youth, and Sports of the Czech Republic, is highly appreciated.

[4] `http://gate.ac.uk/sale/tao/splitch10.html#x14-26900010.2`

References

1. Mollá, D., Van Zaanen, M., Smith, D.: Named entity recognition for question answering. In: Cavedon, L., Zukerman, I. (eds.) Proceedings of the 2006 Australasian Language Technology Workshop, pp. 51–58 (2006)
2. Babych, B., Hartley, A.: Improving machine translation quality with automatic named entity recognition. In: Proceedings of the 7th International EAMT Workshop on MT and other Language Technology Tools, Improving MT through other Language Technology Tools: Resources and Tools for Building MT. EAMT 2003, pp. 1–8. Association for Computational Linguistics, Stroudsburg (2003)
3. Kabadjov, M., Steinberger, J., Steinberger, R.: Multilingual statistical news summarization. In: Poibeau, T., Saggion, H., Piskorski, J., Yangarber, R. (eds.) Multilingual Information Extraction and Summarization. Theory and Applications of Natural Language Processing, vol. 2013, pp. 229–252. Springer, Heidelberg (2013)
4. Ševčíková, M., Žabokrtský, Z., Krůza, O.: Named entities in Czech: annotating data and developing NE tagger. In: Matoušek, V., Mautner, P. (eds.) TSD 2007. LNCS (LNAI), vol. 4629, pp. 188–195. Springer, Heidelberg (2007)
5. Kravalová, J., Žabokrtský, Z.: Czech named entity corpus and SVM-based recognizer. In: Proceedings of the 2009 Named Entities Workshop: Shared Task on Transliteration, NEWS 2009, pp. 194–201. Association for Computational Linguistics, Stroudsburg (2009)
6. Konkol, M., Konopík, M.: Maximum entropy named entity recognition for Czech language. In: Habernal, I., Matoušek, V. (eds.) TSD 2011. LNCS, vol. 6836, pp. 203–210. Springer, Heidelberg (2011)
7. Král, P.: Features for Named Entity Recognition in Czech Language. In: KEOD, pp. 437–441 (2011)
8. Konkol, M., Konopík, M.: Crf-based czech named entity recognizer and consolidation of czech ner research. In: Habernal, I. (ed.) TSD 2013. LNCS, vol. 8082, pp. 153–160. Springer, Heidelberg (2013)
9. Straková, J., Straka, M., Hajič, J.: A new state-of-the-art czech named entity recognizer. In: Habernal, I. (ed.) TSD 2013. LNCS, vol. 8082, pp. 68–75. Springer, Heidelberg (2013)
10. Hajič, J.: Disambiguation of Rich Inflection (Computational Morphology of Czech). Karolinum, Charles University Press, Prague (2004)
11. Collins, M.: Discriminative training methods for hidden markov models: Theory and experiments with perceptron algorithms. In: Proceedings of the ACL 2002 Conference on Empirical Methods in Natural Language Processing, EMNLP 2002, vol. 10, pp. 1–8. Association for Computational Linguistics, Stroudsburg (2002)
12. Kanis, J., Skorkovská, L.: Comparison of different lemmatization approaches through the means of information retrieval performance. In: Sojka, P., Horák, A., Kopeček, I., Pala, K. (eds.) TSD 2010. LNCS, vol. 6231, pp. 93–100. Springer, Heidelberg (2010)
13. Brown, P.F., de Souza, P.V., Mercer, R.L., Pietra, V.J.D., Lai, J.C.: Class-based n-gram models of natural language. Comput. Linguist. 18, 467–479 (1992)
14. Kupiec, J.: Robust part-of-speech tagging using a hidden markov model. Computer Speech & Language 6, 225–242 (1992)
15. Chen, S.F., Goodman, J.T.: An empirical study of smoothing techniques for language modeling. Technical report, Computer Science Group, Harvard University (1998)
16. Hajič, J., Panevová, J., Hajičová, E., Sgall, P., Pajas, P., Štěpánek, J., Havelka, J., Mikulová, M., Žabokrtský, Z., Ševčíková Razímová, M.: Prague dependency treebank 2.0, PDT 2.0 (2006)
17. Šmerk, P.: Fast morphological analysis of czech. In: Proceedings of the Raslan Workshop 2009. Masarykova Univerzita, Brno (2009)

18. Spoustová, D.J., Hajič, J., Raab, J., Spousta, M.: Semi-Supervised Training for the Averaged Perceptron POS Tagger. In: Proceedings of the 12th Conference of the European Chapter of the ACL (EACL 2009), pp. 763–771. Association for Computational Linguistics, Athens (2009)
19. Lafferty, J.D., McCallum, A., Pereira, F.C.N.: Conditional Random Fields: Probabilistic Models for Segmenting and Labeling Sequence Data. In: Proceedings of the Eighteenth International Conference on Machine Learning, ICML 2001, pp. 282–289. Morgan Kaufmann Publishers Inc., San Francisco (2001)
20. Konkol, M.: Brainy: A machine learning library. In: Rutkowski, L., Korytkowski, M., Scherer, R., Tadeusiewicz, R., Zadeh, L.A., Zurada, J.M. (eds.) ICAISC 2014, Part II. LNCS, vol. 8468, pp. 490–499. Springer, Heidelberg (2014)
21. Ciaramita, M., Altun, Y.: Named-Entity Recognition in Novel Domains with External Lexical Knowledge. In: Proceedings of Human Language Technologies: The 2009 Annual Conference of the North American Chapter of the Association for Computational Linguistics, Companion Volume: Short Papers (2005)

An Experiment with Theme–Rheme Identification

Karel Pala and Ondřej Svoboda

Natural Language Processing Centre,
Faculty of Informatics, Faculty of Arts,
Masaryk University,
Botanická 68a, 602 00 Brno, Czech Republic
{pala,xsvobo15}@fi.muni.cz

Abstract. In this paper we start from the theory of Functional Sentence Perspective developed primarily by Firbas [1], Svoboda [12] and also later by Sgall et al. [9].

We make an attempt to formulate and implement a procedure for Czech allowing to automatically recognize which sentence constituents carry information that is contextually dependent and thus known to an addressee (*theme*), constituents containing new information (*rheme*), and also constituents bearing non-thematic and non-rhematic information (*transition*).

The experimental implementation of the procedure uses tools developed in NLP Centre, FI MU, particularly the morphological analyzer Majka [17], disambiguator DESAMB [16] and parser SET [5].

As a starting data resource we use a small corpus of 120 Czech sentences, which at the moment does not include a free continuous text. This is motivated by the fact that we do not use syntactically pre-tagged text but perform syntactic analysis directly using the parser SET. Thus, we offer only a very basic evaluation, which captures the main FSP phenomena and shows that the task is feasible.

The toolset developed for the experiment consists of two parts: first, a chunker, which determines word-order positions from the parse tree of a sentence, second, an FSP tagger which is the implementation of the procedure. It labels the chunks with the tags of what is further called *functional elements* (e.g. theme proper, transition, rheme proper). An experimental version is available at `http://nlp.fi.muni.cz/~xsvobo15/fsp/fsp.html`.

Keywords: rule-based parsing, chunking, functional sentence perspective.

1 Introduction

The theory of Functional Sentence Perspective (FSP in the sequel) was proposed by V. Mathesius [6] and further elaborated by his pupil J. Firbas [1]. The term itself was created by J. Firbas as a more convenient English equivalent of Mathesius' Czech term *aktuální členění větné*.

The FSP theory naturally attracted other Czech researchers as well, particularly P. Sgall and E. Hajičová [9] who creatively introduced slightly different terminology: instead of FSP they started to use Topic-Focus Articulation (further TFA). The Czech

P. Sojka et al. (Eds.): TSD 2014, LNAI 8655, pp. 275–284, 2014.

equivalents of these terms are used rather interchangeably. In this paper we will prefer to use FSP as the original term as well as other terms like *theme* (topic in TFA), *rheme* (focus in TFA). In FSP also terms *transition* and *diatheme* [12] are used which do not seem to have straightforward counterparts in TFA. Then there are terms of the *context dependency* and *communicative dynamism*, introduced by J. Firbas [2]. They express the intuition that some information in a sentence is linked to the previous (verbal and also nonverbal) context and some is perceived as new.

In Firbas (and Svoboda) this is grasped in the following way: the sentence constituents bearing known or contextually dependent information are labelled as *themes*, then there are transitional elements – *transitions* and constituents carrying communicatively new (dynamic) information are called *rhemes*. Within thematic elements *themes proper* (ThPr) and *diathemes* (DTh) are further distinguished, which carry new information within the theme or refer to the new information from the previous text. *Transitions proper* (TrPr) and *rhemes proper* (RhPr) are also recognized among transitional and rhematic elements.

Some results by Karlík and Svoboda [4] offer a solution which inspired us to try a more formal formulation of the procedure able to automatically identify FSP elements in a sentence. They offer rules describing word order positions which can be occupied by individual sentence constituents and depending on their nature allowing to decide whether they can be labelled as thematic, transitional or rhematic. The first attempt to formulate the rules of Karlík and Svoboda as a formal procedure can be found in [8].

1.1 Early Experiments

There were attempts to propose an automatic procedure for TFA by Hajičová et al. [3] and Steinberger et al. [10] in the past. Steinberger's attempt was designed for German, Hajičová's proposal dealt with simple English sentences.

For both papers it is characteristic that they have an experimental nature and do not contain evaluation as we are used to it now. So it is not possible to asses at least approximately how successful the mentioned experiments were. This, however, is understandable if we take into consideration the time of their origin (almost 20 years ago).

1.2 Recent Development

Prague group members have published many papers related to the various aspects of the TFA theory recently, here we would like to mention especially the work related to the manual annotation of FSP (TFA) in Prague Dependency Treebank 3.0 (PDT), see [13].

PDT 3.0 contains annotation of the sentence constituents on three levels: morphological, analytical (syntactic) and tectogrammatical. We will touch here the tectogrammatical level, on which TFA elements (topic, focus and contrast) and communicative dynamism are manually annotated. Procedures for automatic topic/focus bipartition of sentences have been proposed and tested [15,14].

Initially, we considered to compare the PDT annotation obtained manually with our results. After a closer look at the PDT annotation we, however, came to the conclusion that this would be a completely separate task:

- first, apart from the terminological differences there are also differences in the notation that would require more detailed analysis,
- second, we, in fact, looked at some example sentences in the PDT Annotators Manual for T-level and found relevant terminological differences preventing us from trying to use PDT data for comparison in this paper, [7],
- third, TFA annotation in PDT is closely linked to the tectogrammatical level, which we do not work with,
- fourth, TFA annotation in PDT works with the terms *context (non)boundness* which we do not use in the same sense, and semantical relation of aboutness, which is difficult to grasp formally.

The mentioned points show that the more detailed comparison would be very stimulating but, as we hinted, it is a time consuming task for future. It would also require to create some reasonable test data, on which broader agreement could be hopefully reached. We also observe that within FSP theory it is not necessary to work with the tectogrammatical level.

2 Motivation

The task described above has been considered difficult but also challenging. Its successful solution will make it possible to obtain better insight into the information structure of utterances, which should allow for more accurate information extraction as well as meaningful understanding of the thematic progression in natural language texts [11]. Our ambition in this paper is to show that the automatic identification of themes and rhemes is feasible on the basic level at least. We concentrate on the basic aspects of the problem but are well aware of the wider context (e.g. anaphors or particles functioning as rhematizers). So far we work with some methodological constraints, see below.

Our approach is motivated by the fact that we try to answer simple questions first to gain firm ground for solving more complex parts of the problem in the next step. After having managed simple sentences we can come to complex clauses though the basic types of the Czech clauses are handled already.

3 Resources

Though the idea of using the PDT data as a resource for our experiments came to us as quite tempting we had to abandon it as we hinted above. One methodological decision was adopted: due to the experimental setup we decided not to work with free text yet, thus we have prepared a small corpus containing a collection of 120 sample Czech sentences representing various syntactic structures which allow us to test the FSP tagger and improve it step by step to be able to process free text, ultimately. The sentences in our experimental corpus are partly sample sentences displaying relevant syntactic structures and partly sentences taken randomly from online newspapers (such as iDNES or Lidové noviny).

4 Word Order Positions

The free word order in Czech makes it possible to combine sentence constituents quite freely. However, the internal word order within noun, adjective and adverbial phrases is practically fixed in Czech.

It can be observed that a finite verb takes the medium position in Czech sentences in approx. 60%. The morphosyntactic cases in Czech permit to have a direct object in accusative case or indirect object in dative case at the beginning of the sentence and subject in nominative case at the end frequently. The same can be said about adverbial constituents expressed either by adverbs or prepositional groups in various cases, most frequently in locative.

Following [4] we distinguish up to five word-order positions in Czech sentences: pre-initial (usually occupied by conjunctions), initial, post-initial (where enclitics follow Wackernagel's rule), medial and final. In this point we differ from TFA as it is annotated in PDT. The order of enclitic elements in Czech is strictly given: auxiliary forms of verb *být – to be*) are followed by reflexives (pronouns or particles), then by personal, adverbial and demonstrative pronouns.

Unlike the medial position, the initial and the final positions must always be present (even in the form of a merged initial-final position) and can contain only one sentence constituent. The initial, medial and final positions may be occupied by a noun phrase, an adverbial phrase or a verb. A conjunction or a particle may occur in the pre-initial position.

5 Levels of Analysis

To finally obtain labelled sentence constituents in their word order positions, several discrete steps, using automated tools, must be successfully performed. We will use the following sentence as an example:

Přijdu do školy, až napíšu ten text. which translates as *I will come to school when I finish writing the text.*

- First, tokenization (by `unitok.py`) of the input text takes place, yielding a basic vertical text.
- The vertical is extended with complex POS tags and lemmas by the morphological analyser Majka [17] and morphosyntactic disambiguator Desamb [16].
- Using an experimental grammar, partly written by one of the authors, the morphologically annotated input is unambiguously parsed by the SET parser to produce a dependency tree, see Figure 1.
- The chunker processes the obtained trees and segments the sentence(s) into word-order positions along with indication of the constituent type, e.g. a conjunction, a noun phrase or a relative clause.
- Finally, the FSP tagger labels the word order positions with functional elements, taking the information from the chunker and the position in a clause into account.

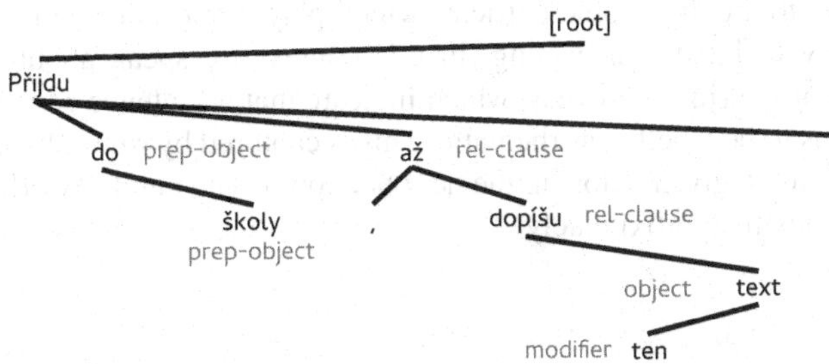

Fig. 1. Dependency tree created by SET with our grammar (note that the labels of edges are technical and unused currently)

The grammar is required to have the following properties:

- ability to recognize a coordination of top-level clauses within compound sentences,
- to link main types of subordinate clauses to their head words through subordinating expressions.

If these conditions are met, a single unit (e.g. a NP/PP, an adverb, a clause) can be extracted by the chunker from the parse tree to form, as a whole, a word-order position. Components of a verb phrase (VP) are, in contrast, separated into positions of their own.

To delimit individual sentences, the chunker first finds the root node. In accordance with the experimental grammar design, the root is either a finite verb (VF), or a coordinate conjunction with VFs as immediate children. The subtrees of VFs represent individual sentences.

Among a head VF's children, other parts of a verb phrase (e.g. auxiliars and modals) are found. Other subtrees of a VP (e.g. NPs represented by their head, a noun) then represent the rest of chunks, or sentence constituents. The chunks occupy word-order positions on their own. Even particles, not usually considered sentence constituents, are included.

The tagger's output format is the vertical text with tokens belonging to the word-order positions enclosed in XML notation. The XML opening tag (named after the position) contains morphosyntactic information added by the tagger and the FSP label explaining the choice, e.g.

```
<initial NP="k1gFnSc1" diatheme="NP in initial">
```

For testing, a web interface has been developed. Apart from plain text, vertical text may be passed to the tools in the pipeline.

The resulting XML is presented graphically after being XSL-transformed to HTML. The positions are shown as boxes, with the name above the content, and the labels underneath. Additional information is displayed in tooltips. Debugging output is also available in separate windows.

It has to be noted that in the course of the syntactic analysis we face usual problems with the ambiguity of prepositional phrase attachment. So far we have decided to work with the longest possible constituents but we are aware that this is just one of the heuristic solutions which has to be further tested.

We also have to mention some particles which play a relevant role in FSP tagging but are not easy to handle in parsing. In this context we speak about *rhematizers*, e.g. *jen* (only), *právě* (just), *i* (even), which indicate that a sentence constituent which follows them has to be labelled as rhematic. This is captured by rules of the FSP tagger (particularly in rule 4 given below in the next Section 6. The number of rhematizers in Czech is rather small, approximately 20.

```
<s>
   <initial PNE="p1nS" TME="eAaPmI" theme-proper="PNE"
            transition-proper="TME" verb="přijít">
Přijdu  přijít  k5eAaPmIp1nS  0  -1,
   </initial>
   <medial NP="k7c2" diatheme="NP">
do      do      k7c2          1  0,prep-object
školy   škola   k1gFnSc2      2  1,prep-object
   </medial>
   <final clause="až" rheme-proper="final position">
,       ,       kIx,          3  4,
až      až      k8xS          4  0,rel-clause
dopíšu  dopsat  k5eAaPmIp1nS  5  4,rel-clause
ten     ten     k3xDgInSc4    6  7,modifier
text    text    k1gInSc4      7  5,object
   </final>
.       .       kIx.          8  0,
</s>
```

Fig. 2. Vertical representation of word-order positions carrying FSP labels

6 FSP Tagger

The procedure first determines the nature of the chunks, matching them with word-order positions. The implemented procedure consists of the following main points:

1. If an input sentence is compound it is split into separate clauses by means of coordinate conjunctions or punctuation.
2. Some coordinate conjunctions, such as *a* (and) or *ale* (but), which stand in the front of a clause, create a pre-initial position (they can occupy no other position).
3. If a (group of) enclitics is found the chunks then form the post-initial position. This position is also optional.
4. The other chunks represent the initial, medial (optional) and final positions with one exception: if the post-initial position is the last in the clause the initial and final positions merge into one initial-final position.

Following the raising communicative dynamism scale, basic rules are applied to label the positions as thematic, transitional or rhematic:

1. All elements in the post-initial position and also finite verbs (for expressing the gender and number of a subject) in other positions are labelled as themes proper (ThPr).

2. Noun phrases (NP/PP), adverbs, infinitives, relative clauses and some particles are labelled as diathemes (DTh). Depending on the word-order position, some of them will be relabelled as transitional or rhematic in the next steps.
3. Finite verbs (bearing temporal and modal grammatical categories) and some particles are labelled as transitions proper (TrPr).
4. A position containing a rhematizer before a NP/PP is labelled as rheme proper (RhPr). The final position is labelled as transitional if it contains a finite verb. If a rhematizer was not found the position is labelled as RhPr.

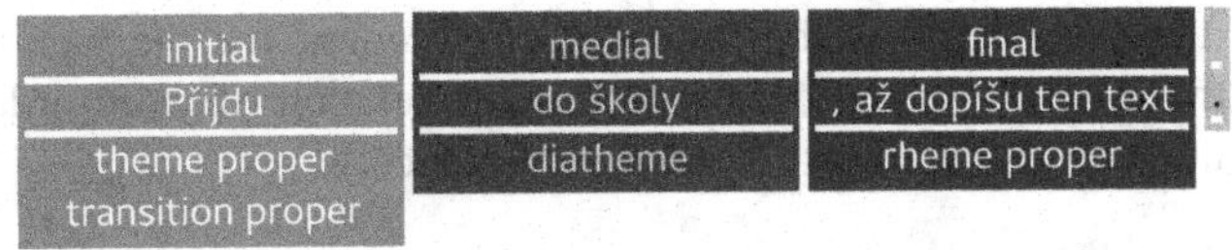

Fig. 3. Graphical output of the chunker and FSP tagger

6.1 Example

The procedure applied to the example sentence finds a single clause in the input. The full stop is left aside. No element is found to mark a pre-initial or post-initial position. The first, last and middle chunks fall into the initial, final and medial positions, respectively.

The labelling of themes proper is performed, marking the the verb *přijdu* (I will come), whose ending *-u* expresses the subject (1st person and singular number), as ThPr. A NP is labelled as a diatheme. The verb is labelled additionally as TrPr. Finally, the clause standing in the final position receives the RhPr label.

One can argue that the label of *do školy* (to school) should be a *rheme* (Rh) rather than a diatheme because it expresses a rather important argument of the verb. On the other hand, if the school had already been mentioned the labelling would have been correct. Working with context and exploiting verbal valencies is, however, a subject to further experimenting.

7 Results and Evaluation

Presently, we have performed a basic experiment, in which sentence constituents have been labelled automatically with the FSP tags. As we have said above we have decided to work with some methodological limitations, particularly:

- In the experiment we work with simple sentences which contain the basic types of the dependent clauses (relative, content and adverbial ones).
- We do not work yet with a continuous text but only with a collection of the sample sentences, each considered out of context, to create a baseline we can build upon.

We have developed two tools, a chunker which processes the output from the parser SET and provides identification of the word-order positions in sentences taken from the the sample corpus (120 sentences so far). The output from the chunker is then handled by the FSP tagger assigning FSP labels to the sentence constituents occuring in the corpus sentences, see Figure 2 and Figure 3).

The results in Table 1 are very basic by their nature, they indicate that the success in FSP labelling is 88%. The number of serious errors (incorrect assignment of the Rh label) can be considered acceptable. They are basically caused by the quality of the used grammar.

Sentences in the testing corpus were evaluated by authors and care was taken to treat them in the same way as the procedure to account for the lack of context-sensitivity of the original formal description of the relation of word-order positions and the distribution of FSP elements in a Czech sentence.

Table 1. The results of experimental FSP labelling

	Sentences total	120	100.0%
A	Correctly analyzed (incl. marginal errors)	106	88.3%
N	Fatal errors	14	11.7%
A−	Marginal errors	13	10.8%
	All errors	27	22.5%

The line A comprises sentences, in which the FSP labels have been assigned correctly. As to errors we can clearly distinguish two sorts of errors:

- N, the result is completely negative, i.e. the FSP tagger does not assign the labels Th/Tr/Rh at all or assigns them incorrectly,
- *A*−, partial errors, here the FSP tagger assigns the label Rh correctly but single errors may occurr with other labels (Th, Tr, Dth).

In our view, the distinction between errors of the type N and A- has to be made, we are convinced that sentences with partial errors can be still considered acceptable. Thus we can conclude that the situation with evaluation is not black and white and will require further analysis.

It has to be remarked that there are several language phenomenona that lower the success rate of chunking and tagging:

- semi-sentential infinitive constructions and, similarly, nominal valencies are not well recognized currently,
- PP attachment causing e.g. adverbial NPs to be connected to other constituents than they belong to,
- coordinated relative clauses.

8 Conclusions

We have been dealing with the task consisting of the identification of word-order positions and automatic theme-rheme tagging in Czech. Starting from the work of Karlík and Svoboda [4] we attempted to formulate formal rules capturing behaviour of the constituents in Czech sentences with regard to the word-order positions they occupy. On this ground the rules form an algorithm for labelling thematic, transitional and rhematic elements in Czech sentences. The first experimental version of the procedure has been implemented, consisting of the two modules:

- the chunker processing simple Czech sentences with canonical (standard) word order,
- the FSP tagger tagging sentence constituents as thematic, transitional and rhematic.

We are well aware of the experimental and modest character of the presented results but, in our view, they show that it makes sense to go in the indicated direction. In the further research we will pay attention to the phenomena that so far prevent the FSP tagger from handling the continuous text with reasonable success.

Acknowledgements. This work has been partially supported by the Ministry of Education of Czech Republic under the project Lindat-Clarin.

References

1. Firbas, J.: On the problem of non-thematic subjects in contemporary English (English summary of "k otázce nezákladových podmětů v současné angličtině", ib. pp. 22–42 and 165–173). Časopis pro moderní filologii 39, 171–173 (1957)
2. Firbas, J.: Functional sentence perspective in written and spoken communication. Cambridge University Press (1992) (reprinted 1995)
3. Hajičová, E., Sgall, P., Skoumalová, H.: An automatic procedure for topic-focus identification. Journal of Computational Linguistics 21(1), 81–94 (1995)
4. Karlík, P., Svoboda, A.: Skladba češtiny pro cizince (Czech Syntax for Foreigners). Univerzita J.E. Purkyně, Faculty of Arts, Brno (1982)
5. Kovář, V., Horák, A., Jakubíček, M.: Syntactic analysis using finite patterns: A new parsing system for Czech. In: Human Language Technology: Challenges for Computer Science and Linguistics, pp. 161–171 (2011)
6. Mathesius, V.: O tak zvaném aktuálním členění větném (on the so-called functional sentence perspective). Slovo a Slovesnost 5, 171–174 (1939)
7. Mikulová, M., Bémová, A., Hajič, J., Hajičová, E., Havelka, J., Kolářová-řezníčková, V., Kučová, L., Lopatková, M., Pajas, P., Panevová, J., Razímová, M., Sgall, P., Štěpánek, J., Urešová, Z., Veselá, K., Žabokrtský, Z.: Annotation on the tectogrammatical layer in the Prague Dependency Treebank. Tech. rep., ÚFAL MFF UK, Prague, Czech Republic (2005), `http://ufal.mff.cuni.cz/pdt2.0/doc/manuals/en/t-layer/html/index.html`
8. Pala, K., Svoboda, O.: Semi-automatic theme-rheme identification. In: Proceedings of the Raslan Workshop, pp. 39–48. Karlova Studánka (2013)
9. Sgall, P.: Towards a definition of focus and topic. Prague Bulletin of Mathematical Linguistics 31, 32, 3–25, 24–32 (1979, 1980)

10. Steinberger, R., Bennett, P.: Automatic recognition of theme, focus and contrastive stress. In: Proceedings of the Conference Focus and NLP (1994)
11. Svoboda, A.: České slovosledné pozice z pohledu aktuálního členění. Slovo a slovesnost 45, 22–34, 88–103 (1984), http://kramerius.lib.cas.cz/search/i.jsp?pid=uuid:c9de3a32-530d-11e1-1418-001143e3f55c
12. Svoboda, A.: Kapitoly z funkční syntaxe. In: Spisy pedagogické fakulty v Ostravě. vol. 66 (1989)
13. Veselá, K., Havelka, J.: Anotování aktuálního členění věty v pražském závislostním korpusu, ÚFAL/CKL TR-2003-20 (2003), http://ufal.mff.cuni.cz/pdt2.0/publications/VeselaHavelkaTR2003.pdf
14. Zikánová, Š., Týnovský, M.: Identification of topic and focus in czech: Comparative evaluation on prague dependency treebank. In: Studies in Formal Slavic Phonology, Morphology, Syntax, Semantics and Information Structure (Formal Description of Slavic Languages 7, pp. 343–353. Peter Lang, Frankfurt am Main (2009)
15. Zikánová, Š., Týnovský, M., Havelka, J.: Identification of topic and focus in czech: Evaluation of manual parallel annotations. The Prague Bulletin of Mathematical Linguistics (87), 61–70 (2007)
16. Šmerk, P.: Unsupervised learning of rules for morphological disambiguation. In: Sojka, P., Kopeček, I., Pala, K. (eds.) TSD 2004. LNCS (LNAI), vol. 3206, pp. 211–216. Springer, Heidelberg (2004)
17. Šmerk, P.: Majka – fast morphological analyzer. In: Proceedings of the Raslan Workshop, pp. 13–16. Masarykova Univerzita, Brno (2009)

Self Training Wrapper Induction with Linked Data*

Anna Lisa Gentile, Ziqi Zhang, and Fabio Ciravegna

Department of Computer Science, University of Sheffield, UK
{a.l.gentile,z.zhang,f.ciravegna}@dcs.shef.ac.uk

Abstract. This work explores the usage of Linked Data for Web scale Information Extraction, with focus on the task of Wrapper Induction. We show how to effectively use Linked Data to automatically generate training material and build a self-trained Wrapper Induction method. Experiments on a publicly available dataset demonstrate that for covered domains, our method can achieve F measure of 0.85, which is a competitive result compared against a supervised solution.

1 Introduction

Information Extraction (IE) is the process of transforming unstructured or semi-structured textual data into structured representation that can be understood by machines. IE is a crucial technique to deal with the continuously growing data published on the Web. Many websites use scripts to generate pages, which get populated with values from an underlying database. These automatically generated pages have high structural similarity. The technique to extract information from this kind of pages is commonly referred as Wrapper Induction (WI) and consists of identifying a set of rules which enables the systematic extraction of specific data records from the pages. In general, WI addresses the extraction of data from *detail* Web pages [3], which are pages corresponding to a single data record (or entity) of a certain type or *concept* (e.g. in a collection of detail pages about films, each page describes a single film).

An extensive range of work has been carried out to study WI, with a mainstream of research focusing on supervised WI [9,11,5], i.e. using a collection of manually annotated Web pages as training data to generate extraction patterns. To reduce the number of required annotations, porting techniques have been proposed to adapt wrappers learnt on one specific website to other websites of the same domain [14,7]. Completely unsupervised methods [4,1] overcome the need of manual annotation but the semantic of produced results is to be interpreted by the user. Hybrid methods [6] propose to automatically generate annotations (exploiting available resources, e.g semantic data on the Web) to be used in the learning phase.

Our solution adopts the methodology proposed in [6], which focus on the usage of *Linked Data*[1] (*LD*) to automatically generate training data to learn extraction patterns. We extend the method by focusing on *limiting noise* in the generation of annotations, and the selection of *reliable* patterns. Experiments report an an average F measure of 0.85.

* Part of this research has been sponsored by the EPSRC funded project LODIE: Linked Open Data for Information Extraction, EP/J019488/1

[1] A collection of interrelated datasets on the Web http://tinyurl.com/lbncdjl

P. Sojka et al. (Eds.): TSD 2014, LNAI 8655, pp. 285–292, 2014.

2 State of the Art

Using Wrapper Induction to extract information from structured Web pages has been studied extensively. Early studies focused on the DOM-tree representation of Web pages and learn a template that wrap data records in HTML tags, such as [9,12,13]. The challenges related to WI concern three main aspects: the need of training data in the case of supervised methods, the lack of semantic for extracted values in the case of unsupervised methods and the robustness problem in both cases. Supervised methods require manual annotation on example pages to learn wrappers for similar pages [9,11,5]. The number of required annotations can be drastically reduced by annotating pages from a specific website and then adapting the learnt rules to previously unseen websites of the same domain [14,7]. Completely unsupervised methods (e.g. RoadRunner [4] and EXALG [1]) do not require any training data, nor an initial extraction template (indicating which concepts and attributes to extract), and they only assume the homogeneity of the considered pages. If homogeneity is not assumed, cluster techniques can be used to obtain homogeneous pages [2]. The drawback of unsupervised methods is that the semantic of produced results is left as a post-process to the user. Hybrid methods [6] intend to find a tradeoff with these two limitations by proposing a supervised strategy, where the training data is automatically generated exploiting *LD*. The methodology proposed by [6], which we will adopt in this work, consists of three steps: (i) dictionary generation, (ii) annotation generation and (iii) pattern extraction. It suggests to build pertinent dictionaries from *LD* and then use them to automatically generate annotations of pages. These automatically generated annotations, potentially incomplete and imprecise, are used to discover the common structural patterns in the Web pages that encapsulate the target information. A limitation of this approach is that extremely noisy annotations can lead to learn incorrect patterns, and the strategy do not propose any check for the reliability of produced patterns.

3 Methodology

Our Wrapper Induction method has two inputs: a schema, defining objects to extract and set of homogenous Web pages. The schema specifies a set of *concepts* of interest $C = \{c_1, \ldots, c_i\}$ with their attributes $\{a_{i,1}, \ldots, a_{i,k}\}$. A set of homogenous Web pages W_{c_i} contains pages from the same website, describing entities of type c_i. They share a similar structure since they are generated using the same script.

We adopt a three steps methodology which consists of (i) dictionary generation, (ii) dictionary based annotation and (iii) pattern extraction, originally proposed by [6] and we introduce novel methods to implement each step. The intuition behind the methodology is that large scale dictionaries will help produce a good number of annotations. Dictionary based annotation is applied in a brute force way that can over-annotate, creating false positives. Good seeds in the dictionary will contribute to the creation of true positive annotations while bad seeds will create false annotations. If there are no good seeds in the dictionary only false positive annotations are created. The number of false positives ca be reduced, by reducing the number of bad seeds in the dictionaries. In the absence of true positives, we cannot generate a useful extraction pattern, but we propose a strategy to detect unreliable patterns and we rather do not propose any for the attribute than proposing an incorrect one. In remaining of this Section we will describe our solution in detail.

Dictionary generation. The goal of this step is to create sufficiently large gazetteers that are good representation of attributes of each concept. *LD* contains billions of facts for specific domains, and can be used as a large entity knowledge base in many tasks [8,10]. [6] propose to query available SPARQL endpoints[2], with a pertinent query for each concept c_i – attribute $a_{i,k}$ pair, thus obtaining a dictionary $d_{i,k}$ for each attribute $a_{i,k}$ of each concept c_i. We propose two simple strategies to limit noise in each dictionary. First, we classify each entry in the result set of the query according to very broad data types, i.e. *NUMBER*, *DATE*, *SHORT TEXT* and *LONG TEXT* using simple regular expressions. We assume the majority type to be the correct one for the dictionary and we discard all entries other than the majority type. After that, for each concept c_i we check for intersections in the dictionaries $d_{i,k}$ of all considered attributes $a_{i,k}$. Values present in more than one dictionary are likely to be either noise (as the wrong usage of properties is common on *LD* [6]) or ambiguous examples, therefore likely to generate misleading annotations. The dictionary generation process is completely independent from the data in W_{c_i}. No a priori knowledge about the data is introduced to this process and thus the dictionary $d_{i,k}$ is unbiased and universal for any extraction tasks concerning the pair c_i–$a_{i,k}$.

Web page annotation. We generate annotations for each pair c_i - $a_{i,k}$, using the dictionary $d_{i,k}$. Each W_{c_i} can contain a number of Web pages varying from a few hundreds to several thousand (e.g. W_{c_i} could be the set of pages on the website `http://www.imdb.com/` describing films). Instead of generating annotations for the whole W_{c_i} we propose a simple strategy to reduce the number of required annotations. Although the annotation process is automatic and no human intervention is needed, limiting the number of pages to annotate is crucial to speed up the process at Web scale.

Algorithm 1. *annotate*(W_{c_i}, $d_{i,k}$)

```
1: M ← xpathDensity(W_ci)
2: M_filt ← filterXpath(M)
3: M_match ← matchXpath(M_filt, d_i,k)
```

Algorithm 2. *xpahDensity*(W_{c_i})

```
1: trainingSize ← 0; M ← ∅; W_train ← ∅
2: while new nodes are added to M do
3:     W_train ← selectRandom(W_ci, n)
4:     trainingSize ← trainingSize + n
5:     M ← M + indexNodes(W_train)
6: end while
```

Algorithm 1 illustrates the main steps of the annotation procedure. The *xpahDensity* function takes as input all the Web pages in W_{c_i}, determines a sufficient subset $W_{train} \subseteq W_{c_i}$ and produces an index M of all text nodes in W_{train} and their values. In detail, *xpahDensity* (algorithm 2) starts by randomly selecting a small fixed number n of pages from W_{c_i} (e.g. $n = 15$). Each page is parsed into a DOM tree and, for

[2] A service to query a knowledge base via SPARQL `http://tinyurl.com/n9h3kce`

each leaf node containing text, *xpahDensity* extracts (i) its text value and (ii) its *xpath*[3], which are added to the index M (function *indexNodes*). M will contain all possible *xpaths* (identifying textual nodes) found in the pages and all different content values for them. The process of adding pages to M is repeated iteratively, on the the next n random pages from W_{c_i}, until no new *xpath* are added to M by the current set of pages W_{train}. The stopping criteria does not consider the values of the nodes, but only checks if new structural elements are introduced by the set of pages. The intuition behind this procedure is that once we cover all structural patterns in the pages, we do not need to generate more annotations. The stop criteria also provides a useful information: the number of pages used before covering all possible path in the website gives an idea of the complexity of structure of website itself. On simple websites, the structure will be covered after using a small number of pages (as little as 45 pages, i.e. 3 iteration with $n = 15$) while for more complex websites the number of pages used can go up to 1000. The *filterXpath* function (algorithm 1) implements simple heuristics to filter out non useful *xpath* in M. The goal of this function is to make sure that boilerplate in the pages does not contribute to the generation of spurious annotations. *FilterXpath* removes all *xpath* that have the same value on all the training pages: these are likely to be formatting and menu items that do not contribute to the extraction process. The *matchXpath* function (algorithm 1) is the one generating the annotation examples. This function simulates a human identifying which nodes in the page contain the information we are interested in. *MatchXpath* generates a positive example every time that the text contained by a particular node matches a value in the dictionary $d_{i,k}$. If there is an exact match between the text content of a node and any item in $d_{i,k}$, the annotation is saved for the page, as a pair $\langle$*xpath, text value*$\rangle$.

Pattern Extraction. The intuition beyond this step is that useful *xpaths* will be likely to match a bigger variety of dictionary entries, on a sufficiently large sample of web pages. Therefore the same *xpaths* producing different annotations over the set of Web pages are more likely to be useful than the ones with limited variety of annotations. For a particular attribute $a_{i,n}$ and a website collection W_{c_i}, starting from all generated annotations (M_{match}) and exploiting the *xpath* density (M_{filt}) we attribute a score to each *xpath* in M_{match}. Based on the hypothesis of structural consistency in W_{c_i}, we expect the majority of true positives to share the same or similar *xpath*. The collection of pages considered to calculate the score is the final set W_{train} in algorithm 2), from which M_{match} and M_{filt} are produced. For each *xpath*, using M_{match} and M_{filt}, the score takes into account three factors: (i) the number of different results produced and (ii) the number of matches with the dictionary, with respect to (iii) the number of pages in the collection. This is to favour *xpaths* which both produce values which are consistent with the dictionaries and apply to the majority of web pages, i.e. are likely to produce a different value for each page. The scoring function *score* is detailed in algorithm 3.

Using M_{filt} we count the number of different values $allVal_i$ corresponding to $xpath_i$, regardless if they match or do not match the dictionary. We then calculate what we call a *margin* for $xpath_i$, which is the ratio between $allVal_i$ and the total number number of pages in the collection W_{train}. The *margin* indicates if an *xpath* is (i) highly applicable,

[3] `http://www.w3.org/TR/xpath/`

Algorithm 3. *score*(M_{match}, M_{filt}, *trainingSize*)

for all $xpath_i \in M_{match}$ **do**
    $allVal_i \leftarrow |values(M_{filt}, xpath_i)|$
    $margin_i \leftarrow allVal_i / trainingSize$
    $matchVal_i \leftarrow |values(M_{match}, xpath_i)|$
    $score_i \leftarrow matchVal_i / allVal_i + margin_i$
    $X \leftarrow addPair(xpath_i, score_i)$
end for

produces a result for most pages, and (ii) informative , produces different results for different pages. Then we count the number $matchVal_i$ of different values corresponding to $xpath_i$ matching the dictionary, using M_{match}. The final $score_i$ is given by the linear combination of the margin $margin_i$ and the ratio between $matchVal_i$ and $allVal_i$.

We introduce a reliability strategy for extracted *xpaths*. We repeat Algorithm 1 multiple times, while applying the *score* function at each iteration (in this work we set the number of iterations to 20). In each iteration, we verify if the best scoring *xpath* and check if it converges; since pages for generating annotations are selected randomly by *xpathDensity* function, this might not be the case and different *xpaths* can be produced. If the highest scoring *xpath* is consistent, we mark it as reliable. In case different *xpaths* are produced at different runs, we check the compatibility of results produced by each *xpath* on all the pages and we pick the *xpath* with highest overlap with the pertinent dictionary.

4 Experiments

We perform experiments on a publicly available WI dataset [7], which covers 8 concepts, including *Autos, Books, Cameras, Jobs, Movies, NBA Players, Restaurants, and Universities*, with 10 websites per concept and around 2000 pages per website. The dataset is accompanied by ground truth values for 3 to 5 common attributes per concept. Our method depends on the availability of suitable dictionaries for each concept-attribute. For the concepts *Autos, Cameras* and *Jobs* our dictionary generation step could not generate dictionaries using *LD*, as we were not able to find a suitable class on *LD* to represent the concept. We tested the method on the concept-attributs: *Book* (*title, author, isbn, publisher, publish-date*), *Film* (*title, director, genre, rating*), *NBA player* (*name, team*),*Restaurant* (*name, address, phone, cuisine*), *University* (*name, phone, website*). To generate the dictionaries for the above concept-attributes we explore *LD* and we manually create queries to retrieve possible values for all attributes of interest[4]. We executed the queries over different SPARQL endpoints and federated the results. We apply the noise filtering procedure (Section 3): we classify results of each query by data type, and only keep results classed as the majority type. For example, when creating the

[4] An example of SPARQL query to retrieve film titles could be "SELECT DISTINCT ?title WHERE{ ?film a ⟨`http://schema.org/Movie`⟩; ⟨`http://dbpedia.org/property/name`⟩?title . FILTER (langMatches(lang(?title), 'EN')).}".

dictionary for *Book ISBN*, the majority of results are classified as number (e.g. 978-0-671-87743-9), while some are classified as short text (e.g. "0143017861(penguin group canada)"). We only keep in the dictionary the one classified as numbers. Once we created dictionaries for all attributes of each concept, we check for values present in more than one attribute dictionary; e.g. for the concept *Book*, "Allen Robert" is present both in the *author* and *publisher* dictionary, or "A Fictional Guide to Scotland" is both in *title* and *publisher*. We consider those as noise and remove them from all the dictionaries for the concept. The resulting dictionaries have varying sizes, ranging from less than 200 seeds for some attributes, to more than 70,000 seeds for others (an indication of the number of seeds for each dictionary can be found in Table 1).

Table 1. Results of our method in terms of Precision (P), Recall (R) and F measure (F). Results are reported as the average on all websites for each concept (or domain) and attribute.

Domain	attribute	P	R	F	Domain	attribute	P	R	F	Average		
book	author	0.99	0.86	0.88	nbaplayer	name	0.78	0.77	0.77	**P**	**R**	**F**
book	publisher	0.59	0.55	0.56	nbaplayer	team	1.00	0.93	0.95	0.78	0.74	0.75
book	title	0.90	0.90	0.90	restaurant	cuisine	0.39	0.37	0.38			
film	director	0.77	0.71	0.74	restaurant	name	0.88	0.85	0.86			
film	genre	0.66	0.64	0.65	university	name	1.00	1.00	1.00			
film	title	0.77	0.76	0.77	university	website	1.00	0.94	0.97			

Results in terms of Precision, Recall and F-measure are reported in table 1, for each concept and attribute. We excluded the attributes: *Book* (*isbn, publish-date*), *Film* (*rating*), *NBA player* (*name, team*), *Restaurant* (*address, phone*), *University* (*phone*), as the generated dictionaries had no coverage on all the websites of the dataset (some were too small to be representative). The *coverage* indicates the percentage of true answers (obtained from the groundtruth) which are contained in our generated dictionaries. In some cases, none of the true answers is covered by the specific dictionary, which means that the training data will not contain any true positive at all, but only false positive (i.e. dictionary seeds matching nodes of the pages other that the node of interest). This is a limitation of our current method, partially overcome by our reliability strategy (Section 3) which checks the compatibility of results with the dictionary (in terms of datatype). This will discard highly incorrect patterns, although it will not avoid the selection of incorrect patterns that extract results of compatible types. Our method achieves an average (micro-average) F measure of 0.78 (with P = 0.81 and R = 0.77) considering all results, while when excluding websites where the dictionaries have no coverage (marked as bold in Table 1) the F measure rises to F 0.85 (with P = 0.87 and R = 0.84).

Table 2 reports the average of results by concept (both including and excluding websites with no coverage) and the overall macro-average. It also compare results with available figures from [7] and [6]. When comparing to [6], figures excluding no coverage websites should be considered, as they discarded them as well for computing final figures. Our method (F=0.85) outperforms [6] (F=0.80), and achieves same performance as [7] (which is a supervised method).

Table 2. Precision (P), Recall (R) and F measure (F) by concept, including and excluding websites with no coverage. Comparison with state of the art methods.

	All results			Covered websites			SoA	
CONCEPT	P	R	F	P	R	F	F[7]	F[6]
book	0.83	0.77	0.78	0.88	0.83	0.84	0.87	0.78
film	0.73	0.70	0.72	0.79	0.76	0.78	0.79	0.76
NBA player	0.88	0.84	0.86	1.00	0.96	0.97	0.82	0.87
restaurant	0.62	0.60	0.61	0.69	0.67	0.68	0.96	0.69
university	1.00	0.97	0.98	1.00	0.97	0.98	0.83	0.91
	0.81	0.78	0.79	0.87	0.84	0.85	0.85	0.80

One assumption of our work is that the method works as long as the concept is fairly covered on LD. Table 2 shows that results for the concept *Restaurant* are worst than average. This is compatible with the fact that dictionaries for this concept are less rich than others. For example we could collect only 629 restaurant names, against 27,720 book titles or 72,354 film titles. Given the continuous and exponential growth of *LD*, it is reasonable to believe that the availability and coverage of resources will get better and better.

5 Conclusions and Future Work

We propose a novel method for Wrapper Induction, where the training data is automatically generated using *Linked Data* and obtained results comparable with a supervised solution. Two strong assumptions are taken. First, that LD contains resources to generate training data; this is not always the case, but it is reasonable to believe that coverage will improve fast. Second, our method assumes that the values to extract are fully and exactly contained in a page node. Although this a common feature amongst websites for observed attributes, it is not always the case, e.g. in the used dataset, the film website *iheartmovies* includes the film title and release year in a single HTML node, therefore our method will not find a match, although it does not return a wrong pattern (none is returned). Ongoing work is aimed at addressing these limitations.

References

1. Arasu, A., Garcia-Molina, H.: Extracting structured data from web pages. In: ACM SIGMOD/PODS 2003, pp. 337–348. ACM (2003), `http://dl.acm.org/citation.cfm?id=872799`
2. Blanco, R., Halpin, H., Herzig, D., Mika, P.: Entity search evaluation over structured web data. In: SIGIR 2011, pp. 65–71 (2011), `http://www.aifb.kit.edu/images/d/d9/EOS-SIGIR2011.pdf`
3. Carlson, A., Schafer, C.: Bootstrapping information extraction from semi-structured web pages. In: Daelemans, W., Goethals, B., Morik, K. (eds.) ECML PKDD 2008, Part I. LNCS (LNAI), vol. 5211, pp. 195–210. Springer, Heidelberg (2008)

4. Crescenzi, V., Mecca, G.: Automatic information extraction from large websites. Journal of the ACM 51(5), 731–779 (2004), http://portal.acm.org/citation.cfm?doid=1017460.1017462
5. Dalvi, N., Kumar, R., Soliman, M.: Automatic wrappers for large scale web extraction. In: VLDB 2011, vol. 4(4), pp. 219–230 (2011), http://dl.acm.org/citation.cfm?id=1938547
6. Gentile, A.L., Zhang, Z., Augenstein, I., Ciravegna, F.: Unsupervised wrapper induction using linked data. In: K-CAP 2013, pp. 41–48. ACM (2013), http://doi.acm.org/10.1145/2479832.2479845
7. Hao, Q., Cai, R., Pang, Y., Zhang, L.: From One Tree to a Forest: a Unified Solution for Structured Web Data Extraction. In: SIGIR 2011, pp. 775–784 (2011), http://research.microsoft.com/pubs/152207/StructedDataExtraction_SIGIR2011.pdf
8. Kobilarov, G., Bizer, C., Auer, S., Lehmann, J.: DBpedia-A Linked Data Hub and Data Source for Web and Enterprise Applications. In: WWW 2009, pp. 1–3 (2009), http://jens-lehmann.org/files/2009/dbpedia_www_developers.pdf
9. Kushmerick, N.: Wrapper Induction for information Extraction. In: IJCAI 1997, pp. 729–735 (1997), http://www.icst.pku.edu.cn/course/mining/11-12spring/%E5%8F%82%E8%80%83%E6%96%87%E7%8C%AE/10-01WrapperInductionforInformationExtraction.pdf
10. Mulwad, V., Finin, T., Syed, Z., Joshi, A.: Using linked data to interpret tables. In: COLD 2010, pp. 1–12 (2010)
11. Muslea, I., Minton, S., Knoblock, C.: Active Learning with Strong and Weak Views: A Case Study on Wrapper Induction. In: IJCAI 2003, pp. 415–420 (2003), http://www.isi.edu/integration/papers/muslea03-ijcai.pdf
12. Muslea, I., Minton, S., Knoblock, C.: Hierarchical wrapper induction for semistructured information sources. Auton. Agents and Multi-Agent Syst., 1–28 (2001), http://www.springerlink.com/index/XMG5W31380116467.pdf
13. Soderland, S.: Learning information extraction rules for semi-structured and free text. Mach. Learn. 34(1-3), 233–272 (1999), http://dx.doi.org/10.1023/A:1007562322031
14. Wong, T., Lam, W.: Learning to adapt web information extraction knowledge and discovering new attributes via a Bayesian approach. IEEE Knowledge and Data Engineering 22(4), 523–536 (2010), http://ieeexplore.ieee.org/xpls/abs_all.jsp?arnumber=4906994

Paraphrase and Textual Entailment Generation

Zuzana Nevěřilová

NLP Centre, Faculty of Informatics,
Masaryk University, Botanická 68a,
602 00 Brno, Czech Republic
xpopelk@fi.muni.cz

Abstract. One particular information can be conveyed by many different sentences. This variety concerns the choice of vocabulary and style as well as the level of detail (from laconism or succinctness to total verbosity). Although verbosity in written texts is considered bad style, generated verbosity can help natural language processing (NLP) systems to fill in the implicit knowledge.

The paper presents a rule-based system for paraphrasing and textual entailment generation in Czech. The inner representation of the input text is transformed syntactically or lexically in order to produce two type of new sentences: paraphrases (sentences with similar meaning) and entailments (sentences that humans will infer from the input text). The transformations make use of several language resources as well as a natural language generation (NLG) subsystem.

The paraphrases and entailments are annotated by one or more annotators. So far, we annotated 3,321 paraphrases and textual entailments, from which 1,563 were judged correct (47.1 %), 1,238 (37.3 %) were judged incorrect entailments, and 520 (15.6 %) were judged non-sense.

Paraphrasing and textual entailment can be put into effect in chatbots, text summarization or question answering systems. The results can encourage application-driven creation of new language resources or improvement of the current ones.

Keywords: textual entailment, paraphrase, natural language generation.

1 Introduction

In human communication a lot of information is not mentioned, e.g. [1] observes that "[i]n human communication meaning is not conveyed by the text alone, but crucially relies on the inferential combination of the text with a context". The non-mentioned (implicit) information deserves attention since it is considerable part of communication in natural languages. [2, p. 149] estimates the ratio explicit:implicit information is up to 1:8.22 which means that the vast majority of information is to be inferred. Computer programs that simulate natural language understanding have to have this ability. This problem (also known as missing common sense knowledge) is well known and studied in artificial intelligence, cognitive science and linguistics.

When people explain something, they proceed in two ways: they can express the same thing in other words or they can explicitely voice the implicit knowledge. The former phenomenon is called *paraphrase*, the latter makes part of *textual entailment*.

P. Sojka et al. (Eds.): TSD 2014, LNAI 8655, pp. 293–300, 2014.

Textual entailment can be defined as a "relationship between a coherent text T and a language expression H, which is considered as a hypothesis. T *entails* H if the meaning of H, as interpreted in the context of T, can be deduced from the meaning of T" [3]. Entailment is often marked by the arrow symbol: $T \rightarrow H$.

Paraphrases are sentences with the same or almost the same meaning. Paraphrases can be seen as mutual entailments: if $T \rightarrow H$ and $H \rightarrow T$ then T and H are paraphrases.

In this work, we present a software that produces both entailments and paraphrases from one or more input sentences. By turning the implicit knowledge into explicit one, the system simulates natural language understanding. Paraphrasing and entailment generation can be used in question answering systems, text summarization, plagiarism detection, or chatbots.

The paper is organized as follows: in Section 2 we present works related to textual entailment recognition and generation, Section 3 describes the methods we use. Section 4 discusses the correctness of the generated sentences, in Section 5 we conclude our work and outline future tasks.

2 Related Work

2.1 Paraphrasing and Textual Entailment Datasets

Authors [4] distinguish paraphrase and textual entailment generation systems along another dimension: whether they recognize, generate or extract paraphrases or entailments. From this viewpoint, recognizing textual entailment (RTE) is the most studied topic. Eight workshops on RTE took place from 2004 to 2013, at first as the Pascal RTE challenges, then as tracks on Text Analysis Conference (TAC), and recently as a track on SemEval challenge[1]. All datasets from Pascal RTE challenges are available, so future RTE systems can undergo an evaluation using these benchmark datasets.

Independently on these RTE challenges, The Boeing-Princeton-ISI (BPI) Textual Entailment Test Suite[2] was developed. According to [5], it is syntactically simpler than Pascal RTE challenges but semantically more challenging. BPI focuses more on the knowledge than on linguistic requirements. The authors also classified the types of knowledge needed to successfully decide whether T entails H; moreover examples of these types are provided together with information about availability of the knowledge in Princeton WordNet (PWN) [6].

Microsoft Research Paraphrase Corpus (MSR) is the most widely used benchmark dataset for paraphrase recognition. From more than 5,000 pairs of sentences about two thirds are annotated as correct paraphrases.

Evaluation datasets do not exist only for English but also for Italian [7] and German [8]. The problem for non-English resources is the smaller diversity and size of language resources. For English language resources, the standardization and their aggregate use

[1] The up-to-date overview can be found in the ACLWiki `http://aclweb.org/aclwiki/index.php?title=Recognizing_Textual_Entailment`

[2] `http://www.cs.utexas.edu/~pclark/bpi-test-suite/`

is indoubtely the most developed. From the classification of knowledge types[3] made by [5], it is clear that the entailments and paraphrases can be recognized, generated or extracted only from big and manifold language resources.

2.2 Paraphrasing and Textual Entailment Applications

Authors [9] describe different RTE methods, from bag-of-words or vector space models to logic-based representations, syntactic interpretations, similarity measures on symbolic meaning representation, and decoding techniques. In the latter, rule-based transformations based on replacing synonyms, hyponyms by hypernyms, and application of paraphrasing patterns result in new sentences.

Some representants of existing RTE systems are:

- VENSES (Venice Semantic Evaluation System)[4]—a cross-platform system for RTE based on two subsystems: GETARUN (a system for text understanding) and the semantic evaluator initialy created for summary and question evaluation.
- EXCITEMENT Open Platform[5] is an open source platform for RTE. The system separates linguistics processing and entailments. It is pre-trained in three languages and it is further trainable. The software has an online demo. BIUTEE[6] (Bar Ilan University Textual Entailment Engine) was formerly a separate software, currently it is part of the EXCITEMENT project.
- EDITS (Edit Distance Textual Entailment Suite)[7] is a RTE software based on edit distance. It consists of three main modules: edit distance algorithm, cost scheme for edit operations, and a set of rules expressing either entailment or contradiction. EDITS works either in Italian or in English.
- Nutcracker[8] is a RTE system based on first order logic and theorem prover.

3 Methods

In this section, we describe our new system. Since it relies on different tools and language resources, we first describe them in short. Afterwards, we describe some paraphrasing and textual entailment generation tools. Each output sentence keeps information on the transformation type—we call this information a *signature*—which is helful in further analysis of the paraphrases and entailments.

[3] http://www.cs.utexas.edu/~pclark/bpi-test-suite/bpi-rte-knowledge-types.txt
[4] http://project.cgm.unive.it/venses.html
[5] http://hltfbk.github.io/Excitement-Open-Platform/
[6] http://u.cs.biu.ac.il/~nlp/downloads/biutee/protected-biutee.html
[7] http://edits.fbk.eu/
[8] http://svn.ask.it.usyd.edu.au/trac/candc/wiki/nutcracker

Table 1. Example LOSOP with the pronoun *jej* (*him*) replaced by the corresponding coreferent (*Jan Novák* or *John Doe*)

1: Jan Novák šel na procházku do temného lesa.		John Doe went for a walk in a dark forest.	
id: 1 *John Doe* word: Jan Novák lemma: Jan Novák tag: SG, NOM part: subject head: Jan constraint: +person	id: 3 *go* word: šel lemma: jít tag: SG, MASC, PAST, POSITIVE, 3RD PERS part: predicate	id: 5 *for a walk* word: procházku lemma: procházka tag: SG, ACC part: object constraint: -person ili: ENG20-00271999 prep: na	id: 7 *in the dark forest* word: temného lesa lemma: temný les tag: SG, GEN part: object head: les constraint: -person ili: ENG20-07926765 prep: do
2: Po setmění jej zachvátil šílený strach.		After the dusk a terrible panic seized him.	
id: 2 *after the dusk* word: setmění lemma: setmění tag: SG, LOC part: object head: setmění constraint: -person prep: po	id: 3 *John Doe* word: Jana Nováka lemma: Jan Novák tag: SG, ACC part: object constraint: +person dep: 4 coref: 1.1	id: 4 *seize* word: zachvátil lemma: zachvátit tag: SG, MASC, PAST, POSITIVE, 3RD PERS part: predicate	id: 5 *a terrible panic* word: šílený strach lemma: šílený strach tag: SG, NOM part: subject constraint: -person ili: ENG20-07058289 ENG20-07058791 …

3.1 Preprocessing

In our approach, we make use of several language resources that exist for Czech. The software is based on morphological analyser `majka` [10], tagger `desamb` [11], syntactic parser `SET` [12], and partial anaphora resolution tool Aara (yet unpublished). The input text is processed by these tools and then converted to its inner representation: we call it a list of set of phrases (LOSOP). An example LOSOP is shown in Table 1: the nodes are phrases (and not tokens as in a usual parse tree). Each phrase is annotated both syntactically (see the properties: *word*, *lemma*, *tag*[9], sentence *part*, *head*) and semantically (see the properties *constraint* and *ili*). We classify each subject or object according to the shallow ontology Sholva [13] (see the property *constraint*), and Czech WordNet [14] (see the property *ili*). So far, we only use two Sholva classes: `+person` and `-person`. Since we do not apply word sense disambiguation, each phrase is linked to all possible inter-lingual indices (ILI) of the Czech WordNet.

[9] We use the following abbreviations for the grammar categories: SG – singular, MASC – masculine, NOM – nominative, GEN – genitive, ACC – accusative, LOC – locative.

3.2 Synonym and Hypernym Replacement

Synonym replacement is one of the basic paraphrasing strategies. For synonym and hypernym replacement, we use Czech WordNet. The algorithm extracts maximum subphrases that exist as a literal in Czech WordNet. The subphrase must contain the head of the original phrase (e.g. *former minister of education* is a *minister* but not *education*). All adjective modifiers transform in order to preserve the grammatical agreement. For example, we can replace *auto* (neuter) by *vůz* (masculine, both meanint *the car*) and similarly *Janovo auto* to *Janův vůz* (*John's car*).

Hyponym replacement is very similiar to synonym replacement. Phrases are replaced by their hypernym, only if the predicate is positive (i.e. the head verb is not negative). Thus, *Peter came in his new convertible* can be transformed into *Peter came in his new car* but *Peter did not come in his new convertible* is not transformed.

As far as we do not employ word sense disambiguation, the result can be ambiguous as well. For example, the word *strach* can refer to `fear:1`, `worry:2`, or `panic:1` in PWN. The likelihood of a correct replacement is estimated upon the language model that computes the likelihood upon n-grams for $n = \{2, 3, 4, 5\}$ (further description can be found in [15]).

3.3 Verb Frame Equivalence

For entailments, we use the verb valency lexicon VerbaLex [16]. From VerbaLex, we extracted all synsets with phrase slots (i.e. no frames with idioms or subordinate clauses). We obtained 152,127 transformation rules that express verb frame equivalence (e.g. *to come* means the same as *to arrive*). In addition, we added 71 manual rules concerning not only equivalence but also preconditions and effects. An example rule can be seen in Figure 1. The rule can produce entailments such as *terrible panic seized* $X \rightarrow X$ *had fear*.

Verbs are more polysemous than nouns. Similarly to synonym and hypernym replacement, we do not disambiguate verbs, however, the verb frame syntactic structure is less ambiguous than the verb alone. We again rank the output sentences according to the language model mentioned in Section 3.2.

```
effect: zachvátit-bát se
 1:[type="predicate" lemma="zachvátit"] 2:[type="subject"
lemma="strach|panika|hrůza"] 3:[type="object" case="ACC"
constraint="+person"]
-> 1: [type="predicate" lemma="bát"] 3:[type="subject"]
5:[type="reflexive" lemma="se"]
```

Fig. 1. Verb frame inference rule that sets relation between *zachvátit* (*seize*) and (*strach*/*panika*/*hrůza* (*fear*/*panic*/*horror*), and *bát se* (*to have fear*)

3.4 Predicator Chains

[5] formulated 19 types of knowledge needed for successfull textual entailment. Our system covers at least three of them. However, three more could be covered by verb frame inference. The problem is the lack of language resources, although the phenomena of *predicator chains* is well-known and described by linguists.

Predicators are verb phrases in its functional relation to the clause. [17, p.32] distinguishes predicators on elementary and mutation predicators. Elementary predicators either describe a state or an elementary (impartible) process, mutation predicators indicate a change of state. Mutation predicators are semantically compound: a mutation predicator expresses a transition from state a to state b. State a is not explicitely mentioned but it is supposed to exist before the change takes place. For example the verb *zblednout* (to turn pale) implies that the subject was not pale before. Examples of each type of predicators is shown below:

- *process predicator* describes an action, e.g. *Matka suší prádlo* (Mother dries the laundry.)
- *mutation predicator* describes a process or state change, e.g. *Prádlo schne* (The laundry is drying.)
- *state predicators* describes a state, e.g. the predicator "to be" in sentences such as *Prádlo je suché* (The laundry is dry.)

The predicators of all three types form typical chains describing the states and their changes (here *to dry–to be drying–to be dry*). Computational linguists and cognitive scientists approach this phenomena in a complex manner by describing semantic frames (e.g. the FrameNet Project) or prototype theory. Nevertheless, this—purely linguistic—approach would cover many cases since predicator chains are often morphological (e.g. *to break–break–to be broken*).

Because the lack of a large language resource we manually crafted 34 rules (from those 71 mentioned in Section 3.3) that cover chains such as *to die–to be dead, set fire–burn–to be burnt, to sell–to be sold*. As we show in Section 4, the verb frame inference produces reasonable entailments.

4 Evaluation

The generated paraphrases and entailment were evaluated according to human judgment. For this purpose, we designed and implemented an annotation game presented in [15].

So far, we have collected 3,321 (non-unique) H–T pairs. From these pairs, 1,563 were judged correct (47.1 %), 1,238 (37.3%) were judged incorrect entailments, 520 (15.6 %) were on average judged non-sense. The game allows repeated annotations but the results show that players are not much motivated to annotate some previous text. Only 456 pairs were annotated more than once.

The overview of individual paraphrase and textual entailment generation methods is in Table 2. The method *no change* means that the input sentence was only analysed and generated from the corresponding LOSOP. *Other methods* comprise phrase reordering and a possibility to enter a correct entailment manually. It can be seen that synonym replacement and verb frame equivalence are the less successful methods. Both methods work with ambiguous input and both produce a lot of noise. In future, we will concentrate on selecting the correct paraphrase rules. Currently, we implemented a ranking algorithm: the correlation between past annotations and particular signatures suggests the correctness of a paraphrase. For example, even though *kolo* can mean *wheel*, *round*, *lap*, or *bicycle*, the correlation between signatures *kolo*→ * and annotations shows that the dominating sense of *kolo* is *a bicycle*. For future paraphrases, the transformations of *kolo* in this sense will be preferred.

Table 2. Number of generated sentences per method

method	correct	incorrect	non-sense	% of correct	% of incorrect	total
no change	316	8	25	90.54	2.29	349
hypernym replacement	22	6	19	46.81	12.77	47
synonym replacement	434	854	173	29.71	58.45	1,461
anaphora resolution	276	127	95	55.42	25.5	498
verb frame equivalence	153	142	156	33.92	31.49	451
predicator chains	41	10	4	74.55	18.18	55
other methods	321	91	48	69.78	19.7	460
total	1,563	1,238	520	47.06	37.28	3,321

5 Conclusion and Future Work

This paper presents the generation phase of a bigger project. Although it is developed for Czech, the described techniques can be used for other languages as well. The only conditions are a variety and considerable size of the language resources. The system relies entirely on other NLP tools such as morphological analyser, tagger and syntactic parser. Their accuracy also affects the system's performance. Since the tools are in continuous development, our system's accuracy can increase in future.

There are many questions concerning the topic, for example: how one sentence can be generated from several input sentences? Currently, the system produces one sentence from one of the input sentences but it does not use wider context.

In future, we want to focus our research on two different directions: (1) error analysis (e.g. what annotators consider incorrect?) and (2) coverage estimation. It is clear that humans can produce paraphrases and entailments that are not covered by our system at all. In future, we want to identify more precisely those other types of paraphrases and entailments. We plan to extract paraphrases from texts describing the same topic.

This work can also encourage further development in language resources as it identifies clearly what types of language knowledge are missing.

Acknowledgments. This work has been partly supported by the Ministry of Education of CR within the LINDAT-Clarin project LM2010013 and by the Czech Science Foundation under the project P401/10/0792.

References

1. Gutt, E.A.: Implicit information in literary translation: A relevance-theoretic perspective. Target: International Journal of Translation Studies 8(2), 239–256 (1996)
2. Graesser, A.: Prose Comprehension Beyond the Word. Springer (1981)
3. Akhmatova, E.: Textual entailment resolution via atomic propositions. In: Proceedings of the PASCAL Challenges Workshop on Recognising Textual Entailment (April 2005)
4. Androutsopoulos, I., Malakasiotis, P.: A survey of paraphrasing and textual entailment methods. CoRR abs/0912.3747 (2009)
5. Clark, P., Fellbaum, C., Hobbs, J.R.: The boeing-princeton-ISI (BPI) textual entailment test suite (December 2006), `http://www.cs.utexas.edu/~pclark/bpi-test-suite/` (accessed online April 14, 2014)
6. Fellbaum, C.: WordNet: An Electronic Lexical Database (Language, Speech, and Communication). The MIT Press (May 1998)
7. Pakray, P., Neogi, S., Bandyopadhyay, S., Gelbukh, A.: Recognizing textual entailment in non-English text via automatic translation into English. In: Batyrshin, I., Mendoza, M.G. (eds.) MICAI 2012, Part II. LNCS, vol. 7630, pp. 26–35. Springer, Heidelberg (2013)
8. Zeller, B., Padó, S.: A search task dataset for German textual entailment. In: Proceedings of the 10th International Conference on Computational Semantics (IWCS), Potsdam (2013)
9. Dagan, I., Roth, D., Zanzotto, F.M.: Tutorial notes. In: 5th Annual Meeting of the Association of Computational Linguistics. The Association of Computational Linguistics, Prague (2007)
10. Šmerk, P.: Towards Computational Morphological Analysis of Czech. Dissertation, Masaryk University in Brno (2010)
11. Šmerk, P.: K morfologické desambiguaci češtiny (Towards morphological disambiguation of Czech). Thesis proposal, Masaryk University (2008)
12. Kovář, V., Horák, A., Jakubíček, M.: Syntactic analysis using finite patterns: A new parsing system for Czech. In: Human Language Technology. Challenges for Computer Science and Linguistics, Poznań, Poland, November 6-8, Revised Selected Papers, pp. 161–171 (2011)
13. Grác, M.: Rapid Development of Language Resources. Dissertation, Masaryk University in Brno (2013)
14. Pala, K., Smrž, P.: Building Czech WordNet. Romanian Journal of Information Science and Technology 2004(7), 79–88 (2004)
15. Nevěřilová, Z.: Annotation game for textual entailment evaluation. In: Gelbukh, A. (ed.) CICLing 2014, Part I. LNCS, vol. 8403, pp. 340–350. Springer, Heidelberg (2014)
16. Hlaváčková, D., Horák, A.: VerbaLex – new comprehensive lexicon of verb valencies for Czech. In: Proceedings of the Slovko Conference (2005)
17. Grepl, M., Karlík, P.: Skladba spisovné češtiny. Edice Učebnice pro vysoké školy. Státní naklad (1986)

Clustering in a News Corpus

Richard Elling Moe

Department of Information Science and Media Studies,
University of Bergen

Abstract. We adapt the Suffix Tree Clustering method for application within a corpus of Norwegian news articles. Specifically, suffixes are replaced with n-grams and we propose a new measure for cluster similarity as well as a scoring-function for base-clusters. These modifications lead to substantial improvements in effectiveness and efficiency compared to the original algorithm.

1 Background

This investigation came about as part of a project with the long term goal to model the flow of news in Norwegian online newspapers over time and to visualize the concentration of coverage related to various topics. An essential question is then how much overlap and recirculation there is in news production.

Since 2006 the Norwegian Newspaper Corpus [9] has downloaded the front pages of the 8 largest online newspapers and stored them in HTML format. From this, a sample corpus consisting of the daily 10 top stories from December 7 to 18, 2009 had been extracted and prepared for experimentation [6]. A total of 960 articles had been manually coded based on categories that are used by media scholars to classify news, cf [1] and [3]. Each article received a tag, consisting of five categories, characterizing the content of the article. For example

International – Economy – FinanceCrisis – Debt – Dubai

The data had been further pre-processed by reducing words to their ground form and keeping only certain kinds of words: nouns, verbs, adjectives and adverbs. This was achieved by marking up the text with syntactic information using the Oslo-Bergen tagger [10] and subsequently filter the document to leave only the desired words in the desired form.

The ability to cluster documents on the basis of having similar content would be instrumental to our goal of detecting reuse and overlap. Therefore we have explored the application of a clustering technique to our news corpus.

Zamir and Etzioni [12] demonstrate that documents can be clustered by applying the Suffix Tree method to short excerpts from them, referred to as *snippets*. This is an attractive feature in our context since such snippets may be readily available for news articles in the form of front-page matter such as headlines, captions and ingresses. For this reason we chose to adapt their use of Suffix Tree Clustering and also because they report it to outperform a number of other algorithms. More recently, Eissen et al. [2] present a more nuanced picture. They point out that the technique has some weaknesses

P. Sojka et al. (Eds.): TSD 2014, LNAI 8655, pp. 301–307, 2014.

but maintains that these have little impact when applied to shorter texts and therefore represent no great problem in our specific context.

The current investigation is a continuation of our previous reports [8] on the initial charting of territory and [7] exploring the potential for improving the Suffix Tree Clustering in general terms. Now the focus is on the Norwegian Newspaper Corpus and the adaptation of the technique to that specific context in order to further improve its performance.

2 Suffix Tree Clustering

The backbone of Suffix Tree Clustering is the data structure known as a *compact trie* [11]. A trie is a tree for storing sequences of tokens. Each arc is labelled with a token such that a path from the root represents the sequence of labels along the path. This simple structure effectively represents sequences and their subsequences as paths whereas the branching captures shared initial sequences. Note that a path does not necessarily represent a *stored* sequence. A stored sequence will have an end-of-sequence mark attached to its final node. The trie structure can be refined for the purpose of saving space. The idea is that sections of a path containing no end-of-sequence marks and no branching can be collapsed into a single arc. The *compact* trie thus allows arcs to be labeled with sequences of tokens. Figure 1 shows the compact trie for the sequences $aa, ab, abab, abc, babb, bc, cbba$, and $cbbc$, with the S_i as end-of-sequence marks.

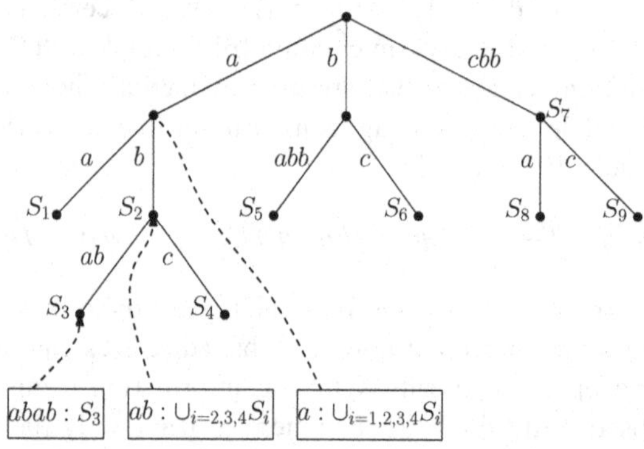

Fig. 1. A compact trie and base-clusters

The suffix tree employed by Zamir and Etzioni is a compact trie storing all the suffixes of a set of given *phrases*, i.e. the snippets. That is, the arcs are labeled with sequences of *words*. Furthermore, the end-of-sequence mark is now the set of documents that the phrase occurs in. (In practice, the set of document ID's.)

The suffix tree forms the basis for constructing clusters of documents. Each node in the tree corresponds to a *base-cluster*. A base-cluster σ:S is basically the set S of documents associated with the subtree rooted in the node. The *label* σ is composed of the labels along the path from the root to the node in question. Three examples of base-clusters are illustrated in Figure 1.

The base-clusters will be further processed to form the final clusters. That is, they are merged on grounds of being similar. Specifically, two base-clusters $\sigma{:}S$ and $\sigma'{:}S'$ are similar if and only if $\frac{|S \cap S'|}{|S|} > 0.5$ and $\frac{|S \cap S'|}{|S'|} > 0.5$.

Now consider the *similarity-graph* where the base-clusters are nodes and there is an edge between nodes if and only if they are similar. The final clusters correspond to the connected components of the similarity-graph. That is, a cluster is the union of the document-sets found in the base-clusters of a connected component. Originally the cluster is not given a designated label of its own. However, we will find use for such a label so we add one by collecting the words from the base-cluster labels and sort them by their frequencies therein.

Clearly, the construction of final clusters requires every base-cluster to be checked for similarity with every other base-cluster. This is a bottleneck in the process but Zamir and Etzioni circumvent the problem by restricting the merging to just a selection of the base-clusters. For this purpose they introduce a *score* and form the final clusters from from only the 500 highest scoring base-clusters. We refer to this limit as λ, i.e. in [12] we have $\lambda = 500$.

The score of a base-cluster $\sigma{:}B$ is defined to be the number $|B| \times f(\sigma)$ where the function f returns the *effective length* of σ. The effective length of a phrase is the number of words it contains that are neither too frequent, too rare nor appear in a given *stop-list* of irrelevant words. Specifically, 'too frequent' means appearing in more than a percentage $\theta_{\max}$ of the (total collection of) documents whereas a word is too rare if it appears in less than $\theta_{\min}$ documents. Originally these thresholds are set to $\theta_{\min}$ =3 and $\theta_{\max}$ =0.4. Furthermore, f penalizes single-word phrases, is linear for phrases that are two to six words long and becomes constant for longer phrases. See [12] and [5].

3 Modifications

In the present context we can make use of the front page matter, i.e. headline, ingress and caption, should there be a photo. So, for each news article its snippet will be the collection of such phrases. Given a snippet containing multiple phrases, each of them will be inserted into the trie separately, i.e. a suffix-tree would hold all the suffixes of each phrase in the snippet.

Our first modification is to abandon the confinement to suffixes and instead fill the compact trie with all n-grams of snippets for a suitable n. One reason is that the use of suffixes shifts emphasis towards the end of the phrase in the sense that a word will appear more times in the suffix-tree than the words that precede it. For example, the suffixes of the phrase 'one two three four' are 'one two three four', 'two three four', 'three four' and 'four' so the suffix-tree contains four occurrences of the word 'four' and three of 'three' whereas 'two' occurs twice and 'one' only once. In contrast, the corresponding 2-grams are 'one two', 'two three' and 'three four'. Now the words 'one' and 'four' appear once while 'two' and 'three' appears twice. Generally, the words on the rims of the phrase will have some fewer occurrences in the trie, but the heavy bias toward the end is gone. Furthermore, except from some uninteresting special cases, the number of words contained in the n-grams of a phrase is strictly fewer than the number of words in the corresponding suffixes. With fewer words to process we expect the algorithm to work faster but there is also the concern that the information held by the data then becomes impoverished. In an attempt to strike a balance, we choose n so as to maximize the number of words inserted into the trie. This is achieved by expanding a phrase of length k into its $\lceil k/2 \rceil$-grams.

Secondly, we disregard clusters where the label consists of a single word only. The presence of very frequent words may cause texts to gravitate towards each other when clusters are formed. Even if the use of a stop-list can help reduce the impact of some very common and irrelevant words we can not blacklist every common word there is. There will inevitably be clusters cemented by the co-occurrences of a single common word. Such clusters are often large and inaccurate. A one-word cluster is not necessarily a bad cluster but it seems reasonable to assume that this is the case more often than not. Then the net effect of removing them would be positive.

Thirdly, we believe the original scoring is somewhat arbitrary and sensitive to the kind of text it is applied to. In the case at hand there is a high proportion of articles that should make up a cluster of its own, being the only texts on their topics. This is a natural characteristic for a corpus such as ours because of petty local news that are only reported once and do not spread nationwide. Unfortunately, the original scoring favors bigger base-clusters and so singleton clusters are never passed on to further processing. Experimentation with different scoring-functions revealed a significant potential for improvement relative to our specific data. Here we use the scoring obtained from the original one by tweaking the θ_{min} and θ_{max} thresholds, to 6 and 0.5 respectively, and reversing the order.

Finally, we will apply a more sophisticated similarity measure. Originally, the similarity of two base-clusters is determined solely on the basis of the amount of overlap between the document-sets they are composed of. It seems likely that the decision would benefit from taking into account additional cluster characteristics such as word frequencies and label overlap.

Notation: We write $\hat{\sigma}$ to denote the set of words occurring in a label σ.

We define a new similarity measure by making the additional requirement that the labels should have a certain amount $\theta_{\cap}$ of overlap and that the average frequency of the words they contain is below a certain limit θ_{freq}. Specifically, base-clusters $\sigma{:}S$ and $\sigma'{:}S'$ are similar iff they satisfy the original measure in conjunction with

$$|\hat{\sigma} \cap \hat{\sigma}'| \geq \theta_{\cap} \quad \text{and} \quad \frac{\sum\limits_{w \in \hat{\sigma} \cup \hat{\sigma}'} cf(w)}{|\hat{\sigma} \cup \hat{\sigma}'|} \leq \theta_{\text{freq}}$$

where $cf(w)$ denotes the *corpus frequency* of the word w, i.e. the total number of times w occurs in our documents. In our experiments we set $\theta_{\cap}$ =2 and $\theta_{\text{freq}} = 4$.

4 Ground Truth, Precision and Recall

The manually tagged portion of our corpus can serve as ground truth for evaluation in terms of precision and recall. Since the tags represent a human judgement as to what the document is about we think it is fair to assume that a high degree of overlap in tags will indicate overlap in content.

A *ground truth cluster* consists of all documents having identical tags, and only those documents. Thus, ground truth clusters are identifiable by tags.

Precision/recall studies rely on a notion of *relevance*. Here, the basic idea is that a good cluster contains only documents with the same tags. That is, a cluster is considered relevant if it matches a subset of some ground truth cluster. The question is, should we require a *perfect* match?

Consider a cluster containing three articles, two of which are tagged

International – Politics – Climate – Obama – Copenhagen

and one with the tag

International – Politics – Climate – Draft – Copenhagen

These articles are all about the 2009 Copenhagen Climate Change Conference, and the cluster would appear to be good. However, there can be no matching ground truth cluster because of the discrepancy of one word in the tags. Is it reasonable to deem this cluster irrelevant? This is largely a matter of the intended use and human opinion so the question has no definite answer. However, by incorporating a degree of perfection we get the flexibility that might allow for the cluster to be considered relevant. A cluster C *matches ground truth with discrepancy* $5-d$ if and only if d is the number of categories common to all tags in C. Intuitively, discrepancy 0 means a perfect match, i.e. $C \subseteq G$ for some ground truth cluster G, while 5 means that there is no category that appears in all tags and we can hardly claim a match at all.

Assuming that C is the set of clusters generated by the algorithm and R the set of relevant clusters, precision would measure the proportion $\frac{|C \cap R|}{|C|}$ of relevant clusters among the clusters generated by the algorithm. Recall measures the extent to which the algorithm will recreate the set of relevant clusters, i.e. $\frac{|C \cap R|}{|R|}$.

The computation of recall-values presents us with a serious challenge. Clearly, checking to what extent the relevant clusters has been generated involves checking the $2^{|G|}$ subsets of each ground truth cluster G. If ground truth clusters are large this job becomes too massive. We escape the problem by introducing a limit λ_{rec} on the size of relevant clusters to be considered. That is, if $|G|$ exceeds this limit a random selection G' of size λ_{rec} is extracted from G and only the subsets of G' are considered for recall. Inspection of our sample corpus reveals that only 3 ground clusters have more than 9 elements. We set $\lambda_{\text{rec}}= 9$ for our experiments.

As described above, the algorithm sets the limit λ on the number of base-clusters that proceed to be merged into final clusters. This poses a serious threat to recall. Specifically, only 500 of the original 25,378 base-clusters are retained. When the initial data for generating clusters is cut short by such a large amount we can not expect the algorithm to be able to fully recreate the ground truth. In fact, our data contains more than 4,500 relevant clusters, while Zamir and Etzioni's original setup of the algorithm produces a total of only 372 clusters. With our corpus this has a devastating effect, causing great harm to recall. For these reasons we prioritized precision over recall when modifying the algorithm.

5 Evaluation

Tests have been carried out to evaluate our modifications of the algorithm. Our benchmark is the performances of the original algorithm shown in Table 1.

Table 1. Original algorithm with $\lambda = 500$.

Discrepancy	Precision	Recall
$d = 0$	0.726	0.053
$d \leq 1$	0.742	0.114
$d \leq 2$	0.753	0.252

Initial experiments have shown that each of our three modification will improve performance. Together they make a considerable difference. Table 2 a) shows precision and recall for the modified algorithm, as well as the average change in performance compared to the benchmark.

Table 2. Modified algorithm

	a) $\lambda = 500$		b)$\lambda = 25,000$	
Discrepancy	Precision	Recall	Precision	Recall
$d = 0$	0.984	0.052	0.954	0.200
$d \leq 1$	0.984	0.052	0.965	0.510
$d \leq 2$	0.984	0.052	0.969	0.608
Average change	+33%	−63%	+30%	+217%

As expected the modified algorithm is more efficient, running 24% faster than the benchmark run.

We have already noted that recall suffers as a result of discarding base-clusters. Indeed, increasing the mass by merging more base-clusters will boost recall. This can be observed in Table 2 b) but, unfortunately, it comes with a punishing cost in running-time. Clearly, the higher number of base-clusters floods the bottleneck of merging them.

6 Conclusion

We have modified the Suffix Tree Clustering technique with some success. Because of our focus on a particular data set we can not claim external validity for our results. Beyond the adaptation for the Norwegian Newspaper Corpus, our contribution is merely to demonstrate an interesting potential for improvement and also to point out directions for further work.

There are several issues that could be pursued. First, we believe that there are other varieties of similarity measures that deserve to be explored. Secondly, we observed that scoring can be sensitive to the kind of text it is applied to. We believe scoring could make more sophisticated use of cluster characteristics, such as labels, size and word-frequencies. Finally, we see that the potential for good recall values is severely hampered by the computational bottleneck of merging base-clusters. A research challenge lies in finding faster algorithms or alternative ways of forming clusters from base-clusters.

References

1. Allern, S.: Newsvalue: On marketing and journalism in ten norwegian newspapers. IJ Forlaget (Publisher) (2001) (in Norwegian)
2. Zu Eissen, S.M., Stein, B., Potthast, M.: The Suffix Tree Document Model Revisited. In: Tochtermann, M. (ed.) Proceedings of the I-KNOW 2005, Graz 5th International Conference on Knowledge Management, pp. 596–603 (2005); Journal of Universal Computer Science
3. Elgesem, D., Moe, H., Sjøvaag, H., Stavelin, E.: The national public service broadcaster's (NRK) news on the internet in 2009. Report to the Norwegian Media Authority, Department of information science and media studies, University of Bergen (2010) (in Norwegian)
4. Erdal, J.: Where does the news come from? On the flow of news between newspapers, broadcasters and the internet (in Norwegian). Official Norwegian Reports NOU2010:14, appendix 1 (2010)
5. Gulla, J.A., Borch, H.O., Ingvaldsen, J.E.: Contextualized Clustering in Exploratory Web Search. In: do Prado, H.A., Ferneda, E. (eds.) Emerging Technologies of Text Mining: Techniques and Applications, pp. 184–207. IGI Global (2007)
6. Losnegaard, G.: Automatic extraction of news text from online newspapers. Project report, Department of information science and media studies, University of Bergen (2012)
7. Moe, R.: Improvements to Suffix Tree Clustering. In: de Rijke, M., Kenter, T., de Vries, A.P., Zhai, C., de Jong, F., Radinsky, K., Hofmann, K. (eds.) ECIR 2014. LNCS, vol. 8416, pp. 662–667. Springer, Heidelberg (2014)
8. Moe, R., Elgesem, D.: Compact trie clustering for overlap detection in news. In: Proceedings of the Norwegian Informatics Conference (NIK 2013) (2013)
9. Norwegian Newspaper Corpus, `http://avis.uib.no/om-aviskorpuset/english`
10. Oslo-Bergen Tagger, `http://tekstlab.uio.no/obt-ny/english/index.html`
11. Smyth, B.: Computing Patterns in Strings. Addison Wesley (2003)
12. Zamir, O., Etzioni, O.: Web Document Clustering: A Feasibility Demonstration. In: Proceedings of the 21st Annual International ACM SIGIR Conference on Research and Development in Information Retrieval, pp. 46–54. ACM, New York (1998)

Partial Grammar Checking for Czech Using the SET Parser

Vojtěch Kovář

NLP Centre, Faculty of Informatics, Masaryk University,
Botanická 68a, 602 00 Brno, Czech Republic
xkovar3@fi.muni.cz

Abstract. Checking people's writing for correctness is one of the prominent language technology applications. In the Czech language, punctuation errors and mistakes in subject-predicate agreement belong to the most severe and most frequent errors people make, as there are complex and non-intuitive rules for both of these phenomena. At the same time, they include numerous syntactic, semantic and pragmatic aspects which makes them very difficult to be formalized for automatic checking. In this paper, we present an automatic method for fixing errors in commas and subject-predicate agreement, using pattern-matching rule-based syntactic analysis provided by the SET parsing system. We explain the method and present first evaluation of the overall accuracy.

Keywords: parser, SET, Czech, grammar checking, punctuation detection, syntactic analysis.

1 Introduction

Reliable checking people's writing for correctness is one of the important goals in natural language processing. Spelling checkers became a common part of our lives, but checking more complex language phenomena still presents a challenge. Although there are "grammar checkers" available in software packages like Microsoft Office or as stand-alone programs, they can address only a restricted range of grammar error types, and they are far from being able to find all the errors, wisely following the "minimum number of false alerts" philosophy.

In the Czech language, punctuation errors and mistakes in subject-predicate agreement belong to the most severe and most frequent errors people make, as there are complex and non-intuitive rules for both writing punctuation and correct usage of subject-predicate agreement.[1] At the same time, these rules include numerous syntactic, semantic and pragmatic aspects which makes them very difficult to be formalized for automatic checking.

Punctuation detection and fixing errors in the Czech grammar is often used as a textbook example of how automatic syntactic analysis can be exploited for a prominent

[1] In Czech, subject-predicate agreement is difficult mainly because of homophonic verb endings (i/y) and differences between standard and colloquial language. E.g. "psi štěkali" ("dogs barked") is correct, "psi štěkaly" is wrong (but it reads the same), "děvčata šla" ("girls went") is correct, but very frequent colloquial form "děvčata šly" is wrong.

P. Sojka et al. (Eds.): TSD 2014, LNAI 8655, pp. 308–314, 2014.

practical application. However, in real life, the full parsing is rarely used (and if, the results are not convincing [1]), and the current methods use rather various types of common error patterns or light-weight modifications of the full syntactic formalisms.

In this paper, we introduce case studies of new methods for punctuation correction, and detection of subject-predicate agreement violations in Czech. Both of the studies exploit syntactic parser SET [2].

2 Related Work

There are two commercial systems for grammar checking of Czech: The Grammar checker built into the Microsoft Office, developed by the Institute of the Czech language [3], and the Grammaticon checker created by the Lingea company [4]. Not much has been published about the principles these are based on; most of the available materials are Czech-only and have rather advertising character. According to available information, both tools are trying to describe negative (wrong) constructions and minimize number of false alerts, i.e. prefer precision over recall significantly (frequent false alerts bother users and make them stop using the tool). The available tests of these tools [5,6] (available only in Czech) indicate that the tools are able to fix 25-35 percent of errors, with the number of false alerts around 6-30 percent.

The Czech parsing community also contributed to the grammar checking problem. Holan et al. [1] proposed using automatic dependency parsing, however, authors conclude that the results have only a prototype character and much work is still needed to achieve practically usable product. Jakubíček and Horák [7] reported on using the Synt parser [8], together with a specialized grammar for Czech to detect punctuation in sentences. They report over 80 percent precision and recall in punctuation *detection* which means that the system fills in the commas into the text without commas (rather than into a text with errors). 80 percent in detection roughly means that every fifth comma is missing and every fifth is wrong. It is not completely clear how the system would behave on real erroneous texts and it is not possible to re-test, as the tool is not available at the moment.

3 The SET Parser

The SET parsing system,[2] firstly introduced in [2], was designed according to the principles of agile and rapid software development [9,10] that we adopt in our solutions, too. Namely, design simplicity and practical usability was the highest priority that was taken into account in all phases of development, rather than accuracy compared to the data annotated according to linguistic theories.

The core of the SET system is formed by a pattern matching engine and a variant of maximum spanning tree algorithm. The tool is open source and its distribution contains several sets of pattern matching rules ("grammars"), the default one being the grammar for parsing general Czech. The rule syntax is illustrated in Figure 1, and explained more

[2] SET is an abbreviation of "syntactic engineering tool".

```
TMPL: verb  ...  $AND  ...  verb   MARK 0 2 4 <coord>  HEAD 2
    $AND(word): , a ani nebo
```

Fig. 1. Example of a SET rule, describing coordination of two verbs using one of the Czech conjunctions *a*, *ani*, *nebo* (and, neither, or), or a comma, with any gap between the verbs and the conjunction. If the rule is matched and selected, the relevant tokens are to be marked as a coordination in the tree, with the conjunction being the head of this constituent.

in detail in [2], or on the SET project page.[3] The primary output of the system, *hybrid tree*, combines dependency and constituent structure features (in form of special phrasal tokens inserted into a dependency tree), and allows conversion into pure dependency or pure constituent structure formalisms.

4 Punctuation Detection

We have designed a specialized SET grammar for punctuation detection, together with an added special output function which prints a comma before each word marked by a special phrasal token (we used <c>, as illustrated in the examples). The grammar contains 10 rules for analysis of the most important patterns where a missing punctuation should be added, that are used for building a reduced tree where the only important information are the tokens marked with <c>. The rules are dealing with following phenomena:

- commas between coordination members (2 rules)
- relative clause boundaries (6 rules)
- 1 particular type of apposition (1 rule)
- 1 rule is negative and specifies where the comma should not be written before relative pronoun (which is normally a clause boundary)

This approach is deliberately approximative, and follows the more straightforward pattern matching idea of Grammaticon and Grammar checker, rather than the full syntactic analysis introduced by Jakubíček and Horák [7]. However, it is one of our future goals to combine the added functionality with the full power of the standard SET grammar and compare the results with the shallow approach.

Examples of a punctuation rule, a reduced syntactic tree for a sentence with missing punctuation, and the resulting sentence with completed punctuation, are given in Figures 2 and 3. As we can see, the "syntactic tree" on the SET output contains practically no syntactic information, except the <c> guidelines for completing the sentence punctuation – rather than that, the SET parser is used as an economical pattern matching engine.

Evaluation of the functionality was performed using the Desam corpus [12], using the same methodology as Jakubíček and Horák [7] – deleting all commas from the input sentences and comparing the original texts with the output of the parser.

[3] `nlp.fi.muni.cz/projects/set`

```
TMPL: $NEG $PREP $REL  MARK 1 <c>  HEAD 1
    $REL(tag): k3.*y[RQ] k6.*y[RQ]
    $PREP(tag): k7
    $NEG(tag not): k7 k3.*y[RQ] k6.*y[RQ] k8
    $NEG(word not): a * " tak přitom
```

Fig. 2. One of the punctuation detection rules in SET, matching preposition (`k7`) and relative pronoun (`k3.*y[RQ]`) or adverb (`k6.*y[RQ]`), not preceded by preposition or conjunction (`k8`) or relative pronoun/adverb and few other selected words (the `tag not` and `word not` lines express negative condition – token must not match any of the listed items). Ajka morphological tagset is used [11].

```
Input: Neví na jaký úřad má jít.
```

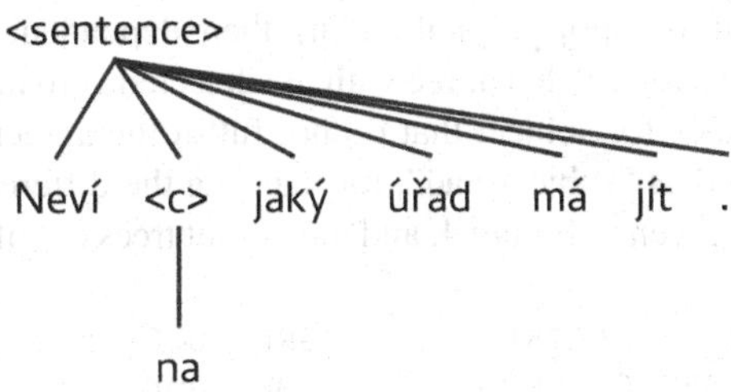


```
Output: Neví, na jaký úřad má jít.
```

Fig. 3. Illustration of SET punctuation analysis – reduced tree and the output sentence with completed punctuation. The rule from Figure 2 was matched. Sentence: *"Neví na jaký úřad má jít."* (missing comma before *"na"* – *"(He) does not know what bureau to go in."*).

The results are summarized in Table 1. We have distinguished a sample of first 500 sentences from the corpus, and the whole corpus of 50,000 sentences; also, we worked with both automatic and correct manual morphological tagging. We can see that the results are very similar for all the testing sets, and it can be concluded that errors in automatic tagging do not influence punctuation detection significantly.

The system shows very high, nearly 95 percent precision, which is very good as it minimizes the number of false alerts. Recall is rather low which means that the system is able to find only about 50 percent of errors. Speed of the analysis was in all cases rather high – 313 sentences per second, on a single Intel Xeon 2.66 GHz core.

We have performed a manual investigation of the differences between the parser output and the correct punctuation, on first 150 sentences of the testing data. This insight showed that many of the parser errors are actually not errors – in Czech, in some places the comma is not necessary but writing it is not a mistake. Out of the missed commas, 21.4 percent were not necessary according to the Czech writing rules (most frequent real errors were in coordinations). From the false positives, 50 percent were actually correctly placed commas. If we extrapolate these percentages to the whole Desam testing set, we get the numbers as in the Extrapolation row.

Table 1. Results of punctuation detection within the SET system

Testing set	Precision (%)	Recall (%)	F-measure (%)
Desam 500 (manual tagging)	94.7	47.3	63.1
Desam full (manual tagging)	94.1	45.0	60.9
Desam 500 (automatic tagging)	95.3	45.4	61.5
Extrapolation	97.1	56.8	71.6

5 Subject-Predicate Agreement

Unlike the previous case study, detecting errors in subject-predicate agreement in Czech sentences uses the standard full SET grammar. The rules detecting subjects of clauses (labelling them as "subject" and adding their dependency on the verb) were differentiated to correct subjects that agree with the detected verb in gender and number, and the salient candidates for subject that do not fulfill the agreement condition. The latter ones were labelled as "subject-bad", for marking the difference. Example of the respective SET rules is given in Figure 4, and the output trees are illustrated in Figure 5.

```
TMPL: $MAINVERB $...* $LIKESUBJ   AGREE 0 2 gn
         MARK 2  DEP 0   PROB 602  LABEL subject
TMPL: $MAINVERB $...* $LIKESUBJ
         MARK 2  DEP 0   PROB 601  LABEL subject-bad

   $MAINVERB(tag): k5.*mF ...
   $LIKESUBJ(tag) $LIKESUBJ(tag): k1.*c1 k3.*c1.*xP ...
```

Fig. 4. One of the SET subject rules, and its twin detecting bad subject-predicate agreement. The main difference is the AGREE action associated with the first rule, which enforces agreement in gender (g) and number (n). MAINVERB and LIKESUBJ are common variable definitions for both rules.

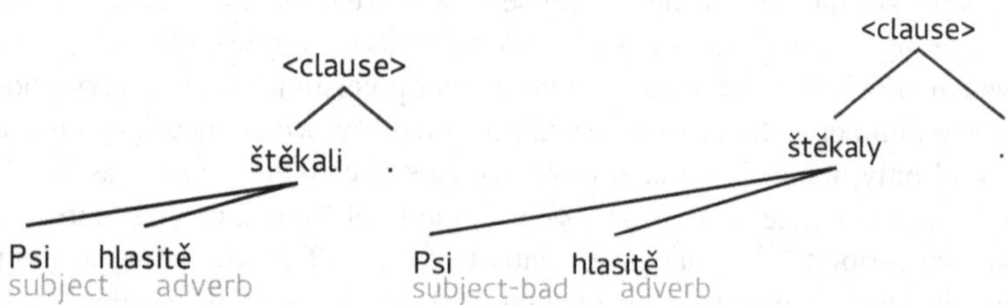

Fig. 5. SET output tree for correct and incorrect version of sentence *"Psi hlasitě štěkali."* (*"Dogs loudly barked."*).

Again, the current rules within the SET grammar cover the most frequent patterns. There are more complicated cases where the subject consists of a complex coordination, error in which would not be detected by our solution, in certain cases. However, according to the YAGNI principle[4] which is part of rapid application development philosophy, we first implement and test the straightforward approach, then identify the real drawbacks and then plan how to fix them, rather than devising a complete solution at the beginning and suppose that we are able to anticipate possible problems. Correctness of the YAGNI principle showed very early in this case.

As there is no large available database of frequent Czech subject-predicate agreement errors, we have decided to use a small set of sentences from a Czech primary school dictation, where frequent errors were manually identified and classified [13]. The set contained 26 sentences with 11 subject-predicate errors. Although the testing set is small, from Table 2 we can clearly see that there is a problem in automatic morphological tagging. The difference in recall between the manual and automatic version is immense, and the reason is that the subjects in the erroneous clauses were tagged as non-subjects, e.g. as accusative instead of nominative (there is very frequent nominative-accusative homonymy in Czech), and therefore they were not recognized as subjects by the parser. This is probably caused by the fact that the tagger (Desamb [14]), as it is usual for taggers, was trained on correct texts and the non-agreement between subject and predicate is so rare in these texts, that it chooses rather another option. Actually, most of the tagging errors resulted in syntactically correct Czech sentences, sometimes even semantically correct, although not suitable in the given context. This is a complex problem that will require a new approach to Czech tagging.

Table 2. Results of subject-predicate agreement checking within the SET system

# sentences	26
# errors	11
# errors spotted (automatic tagging)	2 (18%)
# false alerts	0
# errors spotted after tagging correction	7 (64%)

Another problem are sentences with unvoiced subject (usually present in the previous sentence) – this was in 3 of the 11 sentences. Solution to this problem requires quality anaphora resolution, and we did not attempt to solve it within this case study.

Notable is the 100 percent precision that we have obtained in case of both manual and automatic tagging – there was no false alert.

6 Conclusions

Our system for **punctuation detection**, using as few as 10 rules, outperforms the general reported results for Grammaticon and Czech grammar checker, in terms of both precision and recall – number of false alerts below 3% is very good compared to them, and the recall is better as well. Jakubíček and Horák [7] reported better recall but lower precision; and we are confident that the precision is more important here, due

[4] `en.wikipedia.org/wiki/You_aren't_gonna_need_it`

to the bothering character of false alerts, and any tool with precision lower than 90–95 percent is not suitable for practical usage. Thanks to its results, our tool is ready to be built into a grammar checking application.

The **subject-predicate agreement** case study revealed a serious problem in automatic tagging of erroneous Czech texts. Nevertheless, thanks to the 100 percent precision, the system can be immediately employed in a grammar checker as well.

Because of the specific Czech writing rules, the proposed grammar modifications cannot be used for other languages without further changes. However, the approach proved very expressive – it is able to produce good results with very few rules, so it should be straightforward to adapt it for other languages.

Acknowledgement. This work has been partly supported by the Ministry of Education of the Czech Republic within the LINDAT-Clarin project LM2010013.

References

1. Holan, T., Kuboň, V., Plátek, M.: A prototype of a grammar checker for Czech. In: Proceedings of the 5th Conference on Applied Natural Language Processing, pp. 147–154. Association for Computational Linguistics (1997)
2. Kovář, V., Horák, A., Jakubíček, M.: Syntactic analysis using finite patterns: A new parsing system for Czech. In: Vetulani, Z. (ed.) LTC 2009. LNCS, vol. 6562, pp. 161–171. Springer, Heidelberg (2011)
3. Oliva, K., Petkevič, V.: Microsoft s.r.o.: Czech grammar checker (2005), `http://office.microsoft.com/word`
4. Lingea s.r.o.: Grammaticon (2003), `http://www.lingea.cz/grammaticon.htm`
5. Pala, K.: Pište dopisy konečně bez chyb – Česká gramatickÝ korektor pro Microsoft Office. Computer, 13–14 (2005)
6. Behún, D.: Kontrola české gramatiky pro MS Office - konec korektorů v Čechách (2005), `http://interval.cz/clanky/kontrola-ceske-gramatiky-pro-ms-office-konec-korektoru-v-cechach`
7. Jakubíček, M., Horák, A.: Punctuation detection with full syntactic parsing. Research in Computing Science, Special issue: Natural Language Processing and its Applications 46, 335–343 (2010)
8. Horák, A.: Computer Processing of Czech Syntax and Semantics. Librix.eu, Brno (2008)
9. Martin, J.: Rapid application development. Macmillan (1991)
10. Gabriel, R.P.: Lisp: Good news, bad news, how to win big. AI Expert 6, 30–39 (1991)
11. Sedláček, R., Smrž, P.: A new Czech morphological analyser ajka. In: Matoušek, V., Mautner, P., Mouček, R., Tauser, K. (eds.) TSD 2001. LNCS (LNAI), vol. 2166, pp. 100–107. Springer, Heidelberg (2001)
12. Pala, K., Rychlý, P., Smrž, P.: DESAM — annotated corpus for Czech. In: Jeffery, K. (ed.) SOFSEM 1997. LNCS, vol. 1338, pp. 523–530. Springer, Heidelberg (1997)
13. Trifanová, B.: Analýza chyb v diktátech žáků po absolvování 1. stupně ZŠ. Bachelor thesis, Masaryk University (2014), `http://is.muni.cz/th/382965/ff_b`
14. Šmerk, P.: Unsupervised learning of rules for morphological disambiguation. In: Sojka, P., Kopeček, I., Pala, K. (eds.) TSD 2004. LNCS (LNAI), vol. 3206, pp. 211–216. Springer, Heidelberg (2004)

Russian Learner Translator Corpus:
Design, Research Potential and Applications

Andrey Kutuzov[1] and Maria Kunilovskaya[2]

[1] National Research University Higher School of Economics, Moscow, Russia
[2] Tyumen State University, Tyumen, Russia
{akutuzov72,mkunilovskaya}@gmail.com

Abstract. The project we present – Russian Learner Translator Corpus (RusLTC) is a multiple learner translator corpus which stores Russian students' translations out of English and into it. The project is being developed by a cross-functional team of translator trainers and computational linguists in Russia. Translations are collected from several Russian universities; all translations are made as part of routine and exam assignments or as submissions for translation contests by students majoring in translation. As of March 2014 RusLTC contains the total of nearly 1.2 million word tokens, 258 source texts, and 1,795 translations. The paper gives a brief overview of the related research, describes the corpus structure and corpus-building technologies used; it also covers the query tool features and our error annotation solutions. In the final part we make a summary of the RusLTC-based research, its current practical applications and suggest research prospects and possibilities.

Keywords: corpus building, learner corpora, multiple translation corpora, query tool, linguistic mark-up, mistakes annotation, corpus-driven translator education.

1 Introduction

This paper aims to provide a detailed description of an English-Russian learner translator corpus, which brings together traditional parallel corpus-building methods and a specific type of annotation used in translator training, a combination never attempted before.

Russian Learner Translator Corpus (RusLTC) is a parallel corpus of translation trainees' target texts aligned with their sources at sentence level. It includes translations into English and out of English. The learners' mother tongue is Russian. The project sets out to create a representative and reliable resource to be used in translation studies research and to inform translation pedagogy. The corpus is available under Creative Commons license at `http://rus-ltc.org` website. The corpus is enriched with meta data searchable via the query interface on different external and linguistic factors that can affect students' performance.

RusLTC is a large corpus which is not designed with a specific research purpose in mind, but rather for a broad research agenda in translation studies, including

1. exploring variation and choice in translation, when different translations of the same source are compared;

P. Sojka et al. (Eds.): TSD 2014, LNAI 8655, pp. 315–323, 2014.

2. comparing learner translator output to native data which can bring to conclusions about non-nativeness and 'translationese'; the translator inter-language and consequences of the constraints of translation as communicative activity as opposed to free speech;
3. exploring interdependence between the translation characteristics and various meta data (direction and conditions of translation, source text genre);
4. analysis of concordances of multiple translations (comparing several translations to sources) that can help to develop and test hypotheses about error-prone linguistic items ("problem areas");
5. computer-aided error analysis of the most common translation errors to draw conclusions about the weaker components in the current translator population competences; this strand of research can be extended to include the study into the didactics of translation quality assessment (TQA);
6. apart from learner corpora and translation studies research, the results of which can be applied in the curriculum and materials design, there are numerous ways how the corpus can be directly used as a teaching and learning aid.

The present paper opens with a brief overview of the related research in Section 2; Section 3 describes the corpus design, its content and types of annotations as well as query interface; the next part of the paper (Section 4) outlines the RusLTC-based research and its classroom use. In Section 5 we provide the outlook for further research and developments.

2 Current Learner Translator Corpora Projects and Research

The use of learner translator corpora in research and translator education seems to have been first reported by Robert Spence [13], who compiled a Corpus of Student L1–L2 Translations. It was followed by a learner translator corpus within the PELCRA project by Raf Uzar and Jacek Waliński [15] and the Student Translation Archive by Lynne Bowker and Peter Bennison [2]. These corpora vary in the number of languages they involve, directionality of translation and design technologies used, but they are similar in being unavailable and relatively small in size.

A new excitement to the field was added by two learner corpora projects which were online. The first one – ENTRAD – was introduced by Celia Florén in 2006 [6]. The corpus is a bank of about 45 English sources translated into Spanish by trainees whose mother tongue is marked as mostly Spanish or French. The corpus is text-level aligned and provides no multiple concordances. The query can be narrowed down by the meta data, including the translator's age, gender and mother tongue. The error annotation is based on a colour code and graphical marks and is not machine-readable.

Another notable achievement in multiple LTC development is multilingual MeLLANGE LTC, which is well-documented and easily available online. It provides extensive searchable meta data and proper error-tagging based on prior linguistic annotation. It comes with an elaborate but user-friendly query interface, which allows to retrieve contexts containing specific error category. MeLLANGE seems to be the only LTC which provides one-to-many concordances and reference translation [4]. Sara Castagnoli is also the author of Multiple Italian Student Translation Corpus (MISTiC) which stands out for being compiled for specific research purposes. Her PhD thesis reported in [3] sets a benchmark for multiple learner corpora research in translation studies and demonstrates its potential.

In connection with the present English-Russian project we should mention a similar project undertaken at the Department of Applied Linguistics, Ulyanovsk State Technical University (Russia) [12]. Sosnina's RuTLC is reported to be 1 mln tokens in size. It stores translations of technical texts by part-time translator trainees, who carried out translations as their final paper of 10000 words each. Alignment is not reported. The corpus bears HTML-based error-annotation which allows automatic analysis.

The recent developments in the field include the project LTC-UPF presented by Anna Espunya, Andrea Wurm's KOPTE[1] and NEST developed by Anne-Line Graedler. The first two corpora are enriched with state-of-the-art linguistic annotation; both corpora provide proper error annotation. The related publications describe the error typologies and discuss technical problems which are very familiar [5]. While LTC-UPF is relatively small in size (10 sources, 194 targets), KOPTE is considerably larger (77+ sources, 971+ targets) and has an informative site. The third corpus mentioned above is still fledging; the interface is not yet in place, no linguistic or error annotation is reported. The corpus size is limited to 18 sources in Norwegian, which have been rendered into the translator's foreign language (English) [7].

The overview above makes it possible to define RusLTC as **the third learner translator corpus available** online after ENTRAD and MeLLANGE LTC; it is a large corpus, which provides **sentence-level aligned multiple concordances** and searchable meta data. It is **versatile** in terms of source text genres and is balanced as to **directionality** of translation in English-Russian pair. Besides, it can be used both for a "traditional" corpus research and for error-analysis.

3 Corpus Design

3.1 Data Source

We collect translations from 10 Russian universities offering professional translator training. All translations are made as part of routine and exam assignments or as submissions for translation contests. There are, however, translations from trainees who study translation as a supplementary course or study translation part-time. We also include translations made as internship tasks by students majoring in translation and as graduation translation projects by part-time students. All these trainee populations are described in the searchable corpus meta data.

As of March 14, 2014 RusLTC contains the total of nearly 1.2 mln word tokens, 258 source texts (41 Russian sources and their 589 translation; 217 English sources and their 1,206 translation). The number of translations to a single source varies from 1 to more than 60. All translations and meta data are anonymised.

[1] `http://fr46.uni-saarland.de/index.php?id=3702`

3.2 Corpus Alignment

The parallel nature of the corpus poses some difficulties and does not allow to re-use many popular corpus engines. Thus, we had to design a lot ourselves.

The corpus source texts are aligned with their translations at sentence level. The alignment is done with *hunalign* library [16] and then manually corrected with a TMX-editor *Okapi Olifant*[2]. Sentences that were added by the translator are included in the preceding unit of alignment; untranslated segments of the original are left out, which means that the corpus has no segments with blank translation. When there are several translations of one and the same text, we build several pairs of source and target texts with one and the same source and align them pair-wise. Then redundancy is removed (see below).

Aligned bitexts are stored in **TMX** (Translation Memory eXchange) format, with source and translation segments identified by corresponding XML attributes. The principal unit here is the sentence, however, links to original files are also preserved which allows the interface to retrieve the whole text at need.

Originally TMX format is intended to store translations of one and the same sentence to several languages ('translation unit variants'). We use it differently, at the same time retaining full compatibility. In our TMX base translation unit variants are translations of one and the same sentence made by different students, with different corresponding meta data. Technically this is achieved by our home-grown script which takes the output of *hunalign* and searches for pairs with identical source sentences. Then it joins them into one translation unit, thus avoiding redundancy.

3.3 Meta Data

All translated texts in the corpus are equipped with meta data falling into three major groups: personal, on the source text and translation situation. Personal data include:

1. Trainee's gender
2. Trainee's experience (year of study/education programme or contest)
3. Trainee's affiliation

Translation situation data include:

1. Grade for the translation
2. Conditions of production (routine/exam; home/classroom)
3. Year of production

It should be noted that translations were made under very different conditions (with different amount of stress, under tight or no time limit, as huge projects that had taken weeks to complete and as mini-tasks). No restrictions on use of reference materials was ever reported. The researcher might want to be very specific about the sample he or she is querying. For example, home routine translation are usually done after the source and draft translations have been discussed in class and the translator is relatively at ease as

[2] `https://code.google.com/p/okapi-olifant/`

far as time factor is concerned, while class exam translations are made under significant time constraint (usually 400–450 words in two hours).

Source text data are limited to its genre. We approach genre as a category assigned on the basis of external criteria such as intended audience, purpose, and activity type. RusLTC includes translations of sources classified into the following eleven genres: academic, informational, essay, interview, tech, fiction, educational, speech, letters, advertisement, review.

Meta data are stored as separate plain text files (headers). Up-to-date automatically updated statistics over the whole corpus can be accessed online at `http://dev.rus-ltc.org/statistics`.

3.4 Linguistic Mark-Up

All the texts in the corpus are tokenized and POS-tagged with the help of Freeling library [11]. We are now in the process of designing a query tool which will make proper use of this mark-up. Experimental version which allows to search for particular part-of-speech is available at `http://dev.rus-ltc.org/search`.

There is also a small, but rapidly growing, subcorpus equipped with error annotation. The process of error annotation is powered by customized *brat* text annotation framework [14] and is based on the pre-defined error classification which is a 3-level hierarchy of 31 types. This scheme is based on the fundamental distinction between content-related and language-related errors. Each of the mentioned main categories includes sub-categories such as (wrong) referent, cohesion and pragmatics and lexis, morphology, syntax, spelling and punctuation with further subdivisions respectively. We tried to make our types mutually exclusive, but have found out that in many cases double annotation is possible. In this case the annotators were instructed to evaluate the damage to content transfer and, if present, to tag the mistake as content-related rather than language-related. Apart from this major type of mistakes description, the annotators can use two extra mistakes characteristics, such as weight and technology. The former refers to the seriousness of mistakes and includes a scheme of three points (critical, major and minor), and the latter offers a non-exhaustive variety of possible reasons behind a mistake such as "too literal", "lack of SL knowledge", "lack of background information", etc. It is also important that the error-annotation interface offers an option of a Note by annotator in which he or she can explain the mistake. The note is revealed by hovering with the mouse over error-tags. Because of length limits we do not elaborate much on mistakes classification here, but it is available in full on the corpus site[3].

As of March 2014, we have annotated 241 translation (198 translations into Russian and 43 translations into English), but as the annotation tool is in everyday classroom use, this data is rapidly growing. So far this part of the Corpus can be queried only with standard *brat* means and is available at `http://dev.rus-ltc.org/brat/#/rusltc/`.

[3] `http://rus-ltc.org/classification.html`

3.5 User Interface and Query Tool

Java-based query tool at `http://rus-ltc.org` that we developed supports lexical search for both sources and targets and returns all occurrences of the query item in respective texts along with their targets/sources aligned at sentence level. The search for the query actually happens in the above-mentioned TMX file which allows to return aligned pairs. There is also the possibility to export search results in CSV format and to view full texts and corresponding meta data. Unfortunately, this tool does not support neither search for all morphological variants of the query nor search for words belonging to a particular part of speech. Also, one can search for particular mistakes only in *brat* interface (see subsection 3.4).

As mentioned above, we are also developing a new morphology-enriched query tool in Python which will eventually replace the current one.

Beta status of query tools is mitigated by the fact that the whole Corpus is available under Creative Commons Attribution-ShareAlike license. It can be downloaded both in TMX format or as a compressed set of plain text files and their headers with meta data.

4 Use in Translation Studies and Translator Training

4.1 RusLTC-Based Translation Studies Research

Though our corpus is relatively young (it is under development since 2011), its data have been used in two translation studies research projects. The first one deals with gender asymmetry in translated texts. This work, based on the statistical analysis of the translations in the corpus, has shown that translations made by males and females reproduce the same gender asymmetry (in terms of lexical variety calculated as type to token ratio) that is known for Russian originals. The author has found, though, that texts translated by females tend to contain longer sentences than those translated by males, which doesn't reflect similar statistics for originals [10].

Another RusLTC-based research was to study one of the techniques in translation which is usually referred to as sentence-splitting [9]. It offers detailed analysis of typologically justified types of syntactic splitting in English-Russian translation. More importantly, it describes translation mistakes associated with splitting. Most of these mistakes result in damage to target text coherence and erratic discourse structure. Careless splitting most often leads to loss or misinterpretation of semantic relations between propositions, issues with anaphora resolution and greater communicative value acquired by upgraded sentences. These findings are used to draw students' attention to the textual quality of their production and are implemented in editing exercises aimed at detecting incoherence at semantic and pragmatic levels and editing them.

As part of the corpus involves translation error-tagging, which is claimed to be essentially subjective, there was an attempt to measure inter-rater reliability of our error-annotation system and define most subjective and relatively objective areas in marking translations [8]. The measure was based on annotations provided by three different evaluators who have been trained to use the error typology. It was found out that the experts agree about location of the target text span containing a mistake on average in 60% of cases and demonstrate different target language rigour. The percent agreement on the type of mistake (in terms of main categories in our hierarchy) is 80,5%, with slightly more agreement on content errors than on language errors. But disappointingly little agreement (34,8%) was seen on the weight of these mistakes and on good decisions offered by trainees. Besides, the statistical data showed that while awarding grades for translations the evaluators tend to agree more on poor translations, rather than on good ones. We have taken steps to improve our typology and reduce disagreement, although it is possible for different teachers to use customized and compatible versions of the classification.

4.2 Classroom Use

Apart from these research applications RusLTC is used as a source of data for awareness-raising and revision/editing exercises aimed at prevention of most frequent translation mistakes and introducing trainees to other corpora usefulness in translation practice. The usefulness of a learner corpus-driven approach in foreign language and translator training has been convincingly shown in a number of works ([1] among others). We have developed Python scripts to generate mistakes statistics visualised in a bar chart for any sample of texts, which helps to detect and address most common errors and underlying weaker translator competences for each group or a translator and to register their progress. Language mistakes most often signal lack of transfer skills when a student produces a literal translation, or they can be blamed on the lack in TL competence when a student is unfamiliar with the language use in a particular sphere. In both cases checking the translator's variants with national corpora helps. Editing these mistakes is aimed, therefore, not only at raising trainees' awareness, but also at the study and use of query syntax of the national and special corpora of the working languages. Content mistakes are usually provoked by lack of background information. It can be remedied by extending the scope of world knowledge through information search and raising awareness of cultural differences. Another common reason behind content errors is text analysis and text comprehension problems. We put special stress on these categories of mistakes which, despite having lower frequency ranks, are most dramatic in terms of harm done to the overall quality of translation. Editing these mistakes involves the study and practice in information-mining technology. We feel particularly bad about logic errors which result in incoherent unreadable target texts with vague or non-existent message, and, consequently, defeat the very purpose of translation. Editing these mistakes often requires the analysis of significant chunks of the source and target texts and builds up translators' textual competence.

5 Conclusion and Future Work

With multiple learner translator corpora remaining a fairly new and scarce resource in translation studies, RusLTC seems a valuable contribution to the scene. In the very least it can be utilised as a significant well-organised and documented source of raw data on translational behaviour, including variation and problem-solving techniques. This corpus is a large bidirectional corpus that provides one-to-many concordances of source/target sentences, containing the query item, aligned with their respective translation unit variants in the other language. There are a lot of challenging tasks lying ahead, both technological and linguistic, as well as methodological. We are committed to improve the interface to integrate the functions that are now isolated. Translator training demands error-tag query and easy-to-use statistics. It would be practicable to provide the user with an opportunity to create subcorpora using meta data filters. We are looking for technical solutions to produce parallel concordances of error-annotated subcorpus.

As for the corpus application we are working on a more efficient technology to create tailor-made exercises to address individual translator learning issues that could be used as self-study materials. The corpus will also be used for applied research into translation quality assessment and to measure changes in translation quality and strategies over time. RusLTC-driven research can also help to describe "problem areas" in English-Russian and Russian-English translation, which can be accounted for in the curriculum design.

Acknowledgements. The study was implemented in the framework of the Basic Research Program at the National Research University Higher School of Economics (HSE) in 2014.

References

1. Bernardini, S., Castagnoli, S.: Corpora for translator education and translation practice. In: Rodrigo, E. (ed.) Topics in Language Resources for Translation and Localisation. Benjamins translation library: EST subseries, vol. 79, pp. 39–57. John Benjamins Publishing Company (2008)
2. Bowker, L., Bennison, P.: Student translation archive: design, development and application. In: Zanettin, F., Bernardini, S., Stewart, D. (eds.) Corpora in Translator Education, pp. 103–117. Saint Jerome Publishing (2003)
3. Castagnoli, S.: Variation and regularities in translation: insights from multiple translation corpora. In: UCCTS 2010 - Using Corpora in Contrastive and Translation Studies (2010)
4. Castagnoli, S., Kunz, K., Kübler, N., Volanschi, A.: Designing a learner translator corpus for training purposes (2006)
5. Espunya, A.: Investigating lexical difficulties of learners in the error-annotated upf learner translation corpus. In: Granger, S., Gilquin, G., Meunier, F. (eds.) Twenty Years of Learner Corpus Research: Looking back, Moving ahead. Corpora and Language in Use – Proceedings. Presses Universitaires de Louvain (2013)
6. Florén, C., Sanz, R.: The application of a parallel corpus (english-spanish) to the teaching of translation (entrad project). In: Muñoz Calvo, M., Buesa-Gómez, C., Ruiz-Moneva, M.A. (eds.) New Trends in Translation and Cultural Identity, pp. 433–443. Cambridge Scholars Publishing (2008)
7. Graedler, A.L.: Nest – a corpus in the brooding box. In: Huber, M., Mukherjee, J. (eds.) Corpus Linguistics and Variation in English: Focus on Non-Native Englishes. Studies in Variation, Contacts and Change in English, University of Giessen (2013)
8. Ilyushchenya, T., Kunilovskaya, M.: Inter-rater reliability in student translation evaluation). In: Proceedings of International Conference on Translation Studies Ecology of Translation: Interdisciplinary Research and Perspectives, Tyumen, Russia, pp. 105–115 (2013) (in Russian)

9. Kunilovskaya, M., Morgoun, N.: Gains and pitfalls of sentence-splitting in translation. In: Perm National Research Polytechnic University Herald, pp. 152–166. Linguistic and Pedagogy, Perm National Research Polytechnic University (2013)
10. Kutuzov, A.: Is there a difference between male and female translations (based on the rusltc data). In: Proceedings of International Conference on Translatology, Problems of Translation and Methods of Teaching Translation, vol. 1, pp. 97–104. Nizhny Novgorod, Russia (2012) (in Russian)
11. Padró, L., Stanilovsky, E.: Freeling 3.0: Towards wider multilinguality. In: Calzolari, N., Choukri, K., Declerck, T., Doğan, M.U., Maegaard, B., Mariani, J., Odijk, J., Piperidis, S. (eds.) Proceedings of the Eight International Conference on Language Resources and Evaluation (LREC 2012). European Language Resources Association (ELRA), Istanbul (2012)
12. Sosnina, E.: Russian translation learner corpus: The first insights. In: The Proceedings of the 6 International Scientific Conference Interactive Systems: Problems of Human-computer Interaction (2005)
13. Spence, R.: A corpus of student l1-l2 translations. In: Granger, S., Hung, J. (eds.) Proceedings of the International Symposium on Computer Learner Corpora, Second Language Acquisition and Foreign Language Teaching, pp. 110–112. The Chinese University of Hong Kong (1998)
14. Stenetorp, P., Pyysalo, S., Topic, G., Ohta, T., Ananiadou, S., Tsujii, J.: Brat: a web-based tool for nlp-assisted text annotation. In: EACL, pp. 102–107 (2012)
15. Uzar, R., Waliski, J.: Analysing the fluency of translators. International Journal of Corpus Linguistics 6(1), 155–166 (2001-12-01T00:00:00)
16. Varga, D., Németh, L., Halácsy, P., Kornai, A., Trón, V., Nagy, V.: Parallel corpora for medium density languages. In: Recent Advances in Natural Language Processing (RANLP 2005), pp. 590–596 (2005)

Development of a Semantic and Syntactic Model of Natural Language by Means of Non-negative Matrix and Tensor Factorization

Anatoly Anisimov, Oleksandr Marchenko, Volodymyr Taranukha, and Taras Vozniuk

Faculty of Cybernetics, Taras Shevchenko National University of Kyiv, Ukraine
ava@unicyb.kiev.ua, rozenkrans@yandex.ua, taranukha@ukr.net, taarraas@gmail.com

Abstract. A method for developing a structural model of natural language syntax and semantics is proposed. Syntactic and semantic relations between parts of a sentence are presented in the form of a recursive structure called a control space. Numerical characteristics of these data are stored in multidimensional arrays. After factorization, the arrays serve as the basis for the development of procedures for analyses of natural language semantics and syntax.

Keywords: Information Extraction, WordNet, Wikipedia, Knowledge Representation, Ontologies.

1 Introduction

Recently, the non-negative tensor factorization (NTF) method has become a widely used technology in such fields as information retrieval, image processing, machine learning and natural language processing. The approach is most promising for detection and analysis of linkages and relations in the data where objects of N different types are presented. In computational linguistics, the N-dimensional tensor is implemented as a multiway array of data obtained from the frequency analysis of large text corpora. Factorization of N-dimensional tensor with decomposition rank k generates N matrices. Such matrices consist of k columns that represent mapping of each individual dimension of the tensor on k factor-dimensions of latent semantic space. It is a unique tool to model and explore correlations of linguistic variables in an array of N-dimensional data.

The NTF method is looked upon as a promising technique for solving problems of computational linguistics [1,2,3,4]. Two works are of particular interest [1,2]. The authors describe models for the tensor representation of frequency of various types of syntactic word combinations in sentences, such as 3-dimensional combinations "Subject – Verb – Object", or 4-dimensional combinations of "Subject – Verb – Direct_Object – Indirect_Object" and other syntactic combinations no longer than the dimension of tensor N. Each dimension in tensor is responsible for a certain part of a sentence, i.e. Subject, Predicate, Direct Object, etc.

The N-dimensional tensors contain estimates for the frequency of word combinations sets in text corpora. The model takes into account syntactic positions of words.

P. Sojka et al. (Eds.): TSD 2014, LNAI 8655, pp. 324–335, 2014.

After large text corpora are processed and sufficient amounts of data are accumulated in the tensor, an N-way array is formed. It contains commutational properties of lexical items in the sentences of natural language. For the words presented in the tensor, the properties include: syntactic relations the word tends to be engaged into, other words in the tensor these relations point to, and frequencies of the corresponding relations. Moreover, these relations are multi-dimensional rather than binary, with N being the maximum number of possible dimensions. Then non-negative factorization for the obtained tensor is performed, as it significantly transforms the presentation model. Originally, a multi-dimension tensor is sparse and huge in its volume. Each of the N axes of the syntactic space contains tens of thousands or hundreds of thousands of points representing words. After the tensor has been factorized, its data are represented as N matrices consisting of k columns (where k is much smaller than the number of points in any of the tensor's N dimensions). Parameter k is a degree of factorization, the number of dimensions of the latent semantic space, and the number of attribute dimensions in it. In addition to a much more compact data representation, the probability of every possible word combination can be easily estimated in different syntactic sentence structures. This can be done by calculating the sum of the products of the components for N k-dimensional vectors corresponding to the words chosen from the matrices corresponding, in turn, to their syntactic positions.

For example, one needs to test how feasible is the sentence "The monkey eats banana", in the matrix SUBJECT one finds the k-dimensional vector s, which corresponds to the Noun "monkey", then in the matrix VERB the k-dimensional vector v is found, which corresponds to the Verb "eats". After that in the matrix DIRECT_OBJECT one finds the k-dimensional vector *do* which corresponds to the Noun "banana". The result is calculated as the sum of the products of the corresponding components for three vectors ($N = 3$):

$$x_{svdo} = \sum_{i=1}^{k} s_i v_i do_i,$$

where s_i is the i-th element of vector s, v_i is the i-th element of vector v, and do_i is the i-th element of vector *do*.

Where the sum exceeds a certain threshold level, a conclusion is drawn that such a sequence of words in the sentence is plausible. The calculation for the combination of ("Banana", "eats", "monkey") shows the impossibility of such an option.

This model allows for successful automatic extraction of special linguistic structures from the corpus, such as *selectional preferences* [1] and *Verb SubCategorization Frames* [2], which combine data on the syntactic and semantic properties of relations between verbs and their noun arguments in sentences.

This model is promising and powerful, but lacks flexibility and represents the syntax of natural language in a limited way. The number of dimensions in the tensor restricts the maximum length of sentences and phrases described by this model. Each axis corresponds to a particular syntactic position. Van de Cruys describes the three-dimensional tensor for modeling a syntactic combination: Subject – Verb – Object [1]. Subsequently Van de Cruys and colleagues describe tensors of 9 and 12 dimensions to simulate up to twenty different types of syntactic relations and connections [2]. The mere increase in the tensor dimension number, however, does not seem to be a very convincing way of improving the model and handling more types of extended arity syntactic relations. It is quite reasonable, therefore, to look for other universal representation models for syntactical structures of natural language sentences.

The control spaces [5] have been chosen from among numerous time-tested classical formal models of language syntax representation due to the fact that in this model an arbitrary complex structure is described using recursion through superposition of two basic syntactic relationships - syntagmatic and predicative. The proposed lexical and syntactic tensor model consists of a 3-dimensional tensor for predicative relations and a matrix for syntagmatic relations. The use of control spaces appears to be an efficient means to reduce arbitrary n-ary syntactic relation to the superposition of binary and ternary relations.

Understanding natural language requires knowledge of the language per se (vocabulary, morphology, syntax), and knowledge of the extralinguistic world. The tensor models include data on communicative properties only of the words from the texts already processed and only within the sentences and phrases in which these words are used. This paper proposes to use the hierarchical lexical database WordNet to generalize descriptions of communicative properties of words using the implicit mechanisms of inheritance by taxonomy tree branches. Assuming a word A belongs to a synset S and has a certain property P, there is a high probability that the other words from S will also have the property P. Also, some words of the children synsets of S will almost certainly have P and words of the parent synsets of S are also likely to have P. These assumptions underpin the implementation of the generalization mechanism that describes communicative semantic and syntactic properties of words applying the principle of taxonomic inheritance.

The training set contains texts from The Wall Street Journal (WSJ) corpus, along with the English Wikipedia and the Simple English Wikipedia articles. The latter two contain the definitions and basic information about concepts, which enhances semantics in the model.

2 Control Space of Natural Language Syntactic Structures

The basic syntactic structures are typically described in classical grammar patterns. Rather subtle relations of government among words are expressed in the linguistic models of constituent systems mostly developed by Chomsky [6] and in the models of dependency trees proposed by Tesnière [7]. These models only approximately describe the actual communication properties of syntactic structures.

Attempts to build models more suitable for machine processing that are able to generalize properties of dependency trees and constituent systems led to the creation of more convenient representations.

In work [5] the author proposed to use space for representation, which is independent of the order of text entries. The space expressing all predicative and syntagmatic relations contained in syntactic structures is named *the control space*.

Let us consider the algorithmic model of natural language sentence in terms of control spaces. A sentence is regarded as a dynamic recursive computational process developing in the control space that connects syntactically grouped parts of the sentence with informational channels. The structure of the control space reflects the relations of syntagmatic and predicative language constructions.

Apart from having the referential function, language expresses relations that objects enter into, where the Verb establishes relations between the objects involved in the scheme of this Verb; the Adjective specifies the connection of the object to itself. The syntactic model should indicate which parts of the sentence are linked through relations and of what type the relations are. Predicative relation expresses the relation between syntactic objects through the concept that implies an action and is usually expressed by the predicate, a verb. A syntagma is a combination of two syntactic objects, one of which specifies the other, so the model must fully cover these types of relations. Moreover, in the broadest sense, syntagmas should form syntactic groups. An adequate model of the syntactic structure should also reflect the basic property of being recursive [5].

In control space formalism the conventional linguistic approach is intentionally violated. The Verb is not considered as the principal member of the sentence. In the control space model it is more convenient to define the syntactic relations of *generation* and *transmission of relations*, which ensures a more accurate description of the government connections.

When two objects A and B enter into relation C, we distinguish between an object (say A) that brings about relation C, and the object to which the relation is transferred, which is B. Thus, two types of directed links are differentiated: from *the relation generator object* to *the relation* and from *the relation* to *the subordinate*. The first type of connection is the α-connection (*generation connection*). The second is the β-connection (*propagation connection*). Objects A, B and C are placed in relevant locations in the control space and thus the formal representation of the relation C that connects A and B acquires the form: $A - \alpha \rightarrow C - \beta \rightarrow B$.

Verbs define relations between objects, in the typical pattern of the simple sentence: "Noun – Verb – Noun" the α-connection is directed from the first Noun to the Verb, and the β-connection is directed from the Verb to the second Noun. Let us consider an example: ***"Jim bought a ball"***. The object ***"Jim"*** generates the relation ***"bought"*** and directs it to the object ***"ball"***. Therefore the α-β-structure of the sentence has the form: $Jim - \alpha \rightarrow bought - \beta \rightarrow a\ ball.$

In the phrase: ***"Tall Jim"***, the ***"Jim"*** generates the unary relation ***tall*** and transmits the relation onto itself: $Jim - \alpha \rightarrow Tall - \beta \rightarrow Jim.$

Applying similar reasoning to the phrase ***Tall Jim really loves football*** we obtain the following structure:

$$(Jim - \alpha \rightarrow tall - \beta \rightarrow Jim) - \alpha \rightarrow (loves - \alpha \rightarrow really - \beta \rightarrow loves) - \beta \rightarrow football.$$

The sentences have two types of α-β-links: a strictly linear relation and a closed cyclic dependency. The first is called a linear structure, and the second is a definition. The first corresponds to the predicative language constructs, the second to the syntagmatic ones.

For control spaces the formal model oriented to forming complex structures of a required type is constructed as follows.

A base set of objects U is given. Each object is associated with a certain type. The number of types is finite. The types can be expressed as numbers from the interval 0, N.

An ambiguity arises when mapping objects to types where type function φ maps U into the set of all subsets generated by numbers from the interval $0, N$. The constructions are either objects of U or obtained from other constructions by substituting the latter in terms of linear or defining dependency. The construction types are calculated after the following rules:

1. If in a linear dependence an i-type object A is α-connected with a j-type object B, and the j-type object B is β-connected to a k-type object C, then the type of construction is $f(i, j, k)$.
2. If an i-type object A is α-connected with a j-type object B in the definition structure, and B is β-connected to A, then the entire construction is attributed to $d(i, j) = i$-type.

Since the set of base types is finite, the functions f and d can be specified by tables. Rule 2 allows us to easily calculate any type of complex construction which will coincide with the type of one of the basic constructions defined by functions f or d.

The type of construction is an incorrect one if we can not calculate it. The entirety of the correct constructions composes the control spaces of set U.

Regarding syntactic structures, the definition reads as follows: the basic objects are words and collocations that represent parts of speech (Nouns, Adjectives, Verbs, Particles, etc.) with the appropriate morphological features, as well as compound relations and correlators, designed to connect subordinate sentences with the principal ones. The type of the word contains its complete grammatical description. For example, the type of the word ***book*** is ("Noun", "inanimate", "singular", "Nominative Case"). The notion of type can be extended with some semantic attributes. The ambiguity of the type definition lies in the ambiguity of of some words taken out of context. For example, the word ***book*** belongs to both the Nouns and Verbs classes. The function f specifies the types of simple sentences and complex sentences, depending on the construction of the upper level. Function d specifies the conditions of matching defined and defining objects. For example, the definitions of the Noun can be defined by Adjectives, Prepositions or a subordinate clause, the Verb can be defined by Adverbs, Gerunds or a subordinate clause, Verbs not forming a cyclic α-β-link with Nouns, etc. Thus, the functions f and d are used as a filter to identify the constructions allowed. In the definition constructions the role of the subordinate part is set to a comment or a clarification of the main part. So the type value for the whole syntagmatic structure is set to the generator value as the main object.

The control space of arbitrary sentences can be converted into a dependency tree and a CFG parse tree [5]. Therefore, the structure of the control space can be regarded as a generalization of both dependency trees and CFG parse trees. So, control spaces can express the syntactic structure of arbitrary complexity and arity as a set of binary and ternary relations. This allows for an accurate recording of all data on the semantic and syntactic relations with a single matrix D and one three-dimensional tensor F.

A special empty word is used instead of the missing object in 3-dimensional tensor for intransitive verbs. The construction ***"The boy runs"*** is written in the tensor F as a triplet (boy, runs, Ø). For ditransitive verbs, the parallel reduction procedure is used: construction ***"John gave Mary a toy"*** turns into $(John, gave, Mary) + (John, gave, toy)$ [5].

3 Building a Lexical-Syntactical Model of Natural Language

In order to construct the semantic-syntactic model of natural language, the method for automatic filling the three-dimensional tensor F and the matrix D was designed. It is used in the syntactic analysis and post-processing of sentences from large corpora. The method requires the following steps:

- Sentences from a text corpus are taken and parsed by the Stanford Parser module, which generates the syntactic structures of sentences in the form of dependency trees and parse trees for CF phrase structure grammar [8,9];
- The program examines the dependency tree and the CFG parse tree of the current sentence. It constructs the control space of the syntactic structure, analyzing relations between corresponding words to identify predicate combinations of length 3 (e.g., Subject-Verb-Object, etc.) and syntagmatic combinations of length 2 (Noun-Adjective, Verb-Adverb, etc.);
- Having assembled the control space of this sentence for every triad of points (i, j, k) connected with the linear predicative sequence of α-β-links, tensor F receives the value for the cell FI, J, K: $FI, J, K = FI, J, K + 1$. The coordinates I, J, K of the tensor cell correspond to pairs (w_i, A_i), (w_j, A_j) and (w_k, A_k), where w means words that are lexical values of the corresponding points (i, j, k), and A is a coded description of the characteristics of these words (part of speech, gender, number of lexical units, etc.).
- Similarly, in the control space of the syntactic structure of the current sentence for each pair of points (i, j) interconnected with the cyclic syntagmatic α-β-link, matrix D receives the value for the cell DI, J: $DI, J = DI, J + 1$. The coordinates I, J correspond to pairs (w_i, A_i) and (w_j, A_j), where w stands for words representing lexical values of the corresponding points (i, j), and A is a coded description of these words.

After processing large amounts of text, matrix D and three-dimensional tensor F accumulate sufficient lexical and syntactic communicative information to efficiently implement the lexical and syntactic model of natural language.

An extremely large dimension and sparsity of matrix D and tensor F demand for non-negative matrix and tensor factorization in order to store the data in a more economical way. Matrix D is factorized using Lee and Seung Non-negative Matrix Factorization algorithm [10] that decomposes matrix $D(N \times M)$ as a product of two matrices $W(N \times k) \times H(k \times M)$, where $k << N, M$. Tensor F is factorized using the non-negative three-dimensional tensor factorization parallel algorithm PARAFAC [11]. The factorization yield corresponding matrices X, Y and Z.

4 Properties of a Lexical Model of Natural Language

After matrix D and tensor F factorization, the system forms a strong knowledge base which contains information about the syntactic framework of natural language sentences. The description of semantic relations between the words is integrated into the structures. Apart from the description of general syntax that defines the structure of the sentences in a general abstract form, the base also contains semantic restrictions that determine which words can form a syntactic connection of a certain type. To determine whether two words a and b form a cyclic syntagmatic relation, one has to take vector-row W_a from matrix W corresponding to word a, and vector-column matrix H_b from matrix H which corresponds to word b, and calculate the scalar product of vectors (W_a, H_b^T). If the product is greater than a certain threshold T, then this relation is defined. In order to determine whether the three words a, b and c enter into predicative relation $(a \rightarrow b \rightarrow c)$, it is necessary to take vector X_a corresponding to word a, vector Y_b corresponding to word b, and vector Z_c corresponding to word c and to calculate the value:

$$S_{abc} = \sum_{i=1}^{k} X_{a}i * Y_{b}i * Z_{c}i$$

If S_{abc} value is greater than a threshold, then this relation is defined. If not, it is considered undefined.

These matrices implicitly define a set of defined language clauses, the set being specified with the input text corpus. The vectors of words from the derived matrices implicitly describe their "structural behavior". They define in which syntactic relation these words may join and which words they have joined. With the resulting matrix, one may parse sentences and generate the control space of their syntactic structures, using the ascending algorithms such as CYK [12]. The control space is built where possible.

5 Implementation

As the initial training text corpus, sets of articles from the English Wikipedia, the Simple English Wikipedia and the WSJ corpus are used. The texts are processed sequentially with the parser and with the program that constructs the control space of syntactic structures. First, the sentences are analyzed with the Stanford Parser yielding a CFG parse trees (for phrase structure grammar) and a dependency trees. Also, an algorithm has been developed to construct control spaces by converting a dependency tree and a parse tree into the control space of a sentence. The algorithm is a recursive traversal from left to right of the sentence tree which creates points of the control space in each node of the CFG parse tree and performs conversion of corresponding relations of the dependency tree into α-β-connections of control space (either predicative or syntagmatic connections). Each point of the space is assigned a specific lexical value (a word or phrase) and characteristics (part of speech, gender, number, etc.). At the outset every word is an isolated point in the control space. When points A and B are connected to form a new point S in the space, representing the α-β-relationship between A and B, this new point gains its own lexical value. This value can be inherited from the main element of the pair (A, B), e.g., the phrase *cold water* consists of a pair *(cold, water)* that has a Noun as the main word. Consequently, the new point will inherit value from *water*. Also, the merger of two points may result in their lexical value forming a fixed collocation. For example, the combined value of point A *(Weierstrass)* and B *(theorem)* is the *Weierstrass theorem*, which is the lexical value of the new generated point C. Fixed collocations are obtained based on Wikipedia articles with a corresponding titles.

After the control space has been built, for each cyclic α-β-syntagmatic link the value dI, J is increased by 1 in cyclic links matrix D (where I is the index of the first word, J is the index of the second word): $dI, J = dI, J + 1$. For each of the triplets in linear relations $A - \alpha - B - \beta - C$ the three-dimensional tensor F cell fI, J, K is increased by 1 (where I is the index of word A, J is the index of word B, and K is the index of word C): $fI, J, K = fI, J, K + 1$.

800,000 articles from the English Wikipedia and the Simple Wikipedia have been processed, along with the WSJ corpus. As the WSJ corpus is annotated manually and contains correct syntactic structures, a high number of quality syntactic structures control spaces are received.

The processing yielded the large matrix D for cyclic links (numbering approximately 2.3 million words $\times$ 2.3 million words, with up to 57 million non-zero elements) and the large three-dimensional tensor F for linear predicative connections (consisting of approximately 2.3 million words $\times$ 52 thousand words $\times$ 2.3 million words, with about 78 million non-zero elements). These arrays were factorized by the non-negative matrix factorization algorithm [10] and the non-negative tensor factorization parallel algorithm PARAFAC [11].

Factorized data sets allow for efficient computing of probability for cyclic syntagmatic relations between any two words using the scalar product of two corresponding vectors. To form linear predicative relations between any three words the probability can be efficiently and easily calculated.

To investigate the applicability of this model for practical NLP tasks a parser for the English language based on the obtained arrays of lexical-syntactic combinability has been implemented. This parser, based on the Cocke-Younger-Kasami algorithm, directly constructs the control space of a sentence.

The model describes only the relations among those words which actually occur in the corpus sentences and have been processed accordingly. When a pair of words A and B make a cyclic syntagmatic link and has value in the array, the pair A_1 and B_1 (where A_1 is synonymous with A and B_1 is synonymous with B) will not have the link if A_1 and B_1 are absent in the data. The same holds for linearly predicative relations. The matter can be easily dealt with using synonym dictionaries. In the system we developed the WordNet is used to this end. We assume that if between A and B a relation exists, it also exists between an arbitrary pair of A_i and B_i, where A_i is any word from the synset that contains A, while B_i is any word from the synset that contains B. However, the question of homonymy arises when one word corresponds to several synsets in the WordNet. Every time a sentence is parsed, the point at issue is how to determine whether a pair or triplet of synsets is correct.

On the one hand, there are several standard approaches to solving this classic problem of ambiguous words (WSD). On the other hand, the two matrices W and H resulting from the non-negative matrix factorization of D can be considered powerful tools for determining the degree of semantic similarity between words according to the methods of latent semantic analysis [13].

So, to determine the presence of cyclic syntagmatic α-β-connections and to solve the problem of ambiguous words the following steps are carried out:

A: Take vector W_a corresponding to word a from term matrix W, vector column H_b which corresponds to the word b from matrix H, and calculate the scalar product of the vectors (W_a, H_b^T). If the value $(W_a, H_b^T) > T$, then this link is **defined**. T is the threshold. The optimal value of T is found experimentally. If it fails:

B: Take synsets for words a and b from the WordNet. The set of synsets $\{A_i\}$ refers to word a, and the set of synsets $\{B_i\}$ refers to word b. Check the pairs of the words formed from the elements of $\{A_i\}$ and $\{B_i\}$. If there is word $\acute{a}_k$ from $A_k \in \{A_i\}$ and word $\acute{b}_j$ from $B_j \in \{B_i\}$ such that scalar product of vectors $(W_{\acute{a}_k}, H_{\acute{b}_j}^T) > T$, then this link between a and b is **defined**. If not:

C: The set $\{A_i\}$ is expanded with synsets linked with nodes from $\{A_i\}$ with hyponym and hypernym relations in the WordNet. The set $\{B_i\}$ is expanded in the same way. Check the pairs of words formed from elements of $\{A_i\}_{exp}$ and $\{B_i\}_{exp}$ (excluding the pairs already checked on step B). If there is a word $\acute{a}_k$ from the synset $A_k \in \{A_i\}_{exp}$ and a word $\acute{b}_j$ from the synset $B_j \in \{B_i\}_{exp}$ such that the scalar product of vectors $(W_{\acute{a}_k}, H_{\acute{b}_j}^T) > T$, then the link between a and b is **defined**. If it fails: expand $\{A_i\}_{exp}$ and $\{B_i\}_{exp}$ recursively 2 or 3 times and repeat step (C).

If it is always $(W_{\acute{a}_k}, H_{\acute{b}_j}^T) < T$, then the link **does not exist**.

When expanding $\{A_i\}$ and $\{B_i\}$, one should avoid adding synsets from the list of the concepts with the most general meanings from the top of the WordNet hierarchy. If $\{A_i\}_{exp}$ and $\{B_i\}_{exp}$ are extended with such concepts, the semantic similarity between $\acute{a}_k$ and $\acute{b}_j$ quickly deteriorates. Inheritance of properties through hyponymy/hypernymy is not correct for such synsets.

For the linear predicative α-β-link this algorithm works in the same way.

The taxonomic hierarchy of the WordNet lexical database together with the mechanism of inheritance allows us to generalize this representation model of syntactic and semantic relations of natural language. This turns the constructed system into a versatile tool for syntactic and semantic analysis of natural language texts.

6 Experiments

To form a robust syntactical and semantic relations base, it is crucial to have a huge corpus of correctly tagged texts. Usage of the WSJ corpus has a significant effect on the quality assurance of the resulting model. To construct tagged texts from the English Wikipedia and the Simple English Wikipedia, the Stanford parser is used. The accuracy of parse trees is about 87%, while the accuracy of dependency trees is close to 84%. As some of the trees are incorrect, it is natural that they yield some inaccurate descriptions of the syntactic structures control spaces. The algorithm for converting CFG parse trees and dependency trees into control spaces of syntactic structures shows no errors on correct trees.

The development of the system for parsing and control spaces generation for natural language sentences based on created syntactic and semantic relation databases was followed by experiments. The accuracy was measured by computing control spaces of the syntactic structures. To generate test samples, 1,500 sentences were taken from

Table 1. Precision estimation of cyclic α-β-syntagmatic links and linear predicative α-β-links on sentences from the Simple English Wikipedia, the English Wikipedia and the WSJ corpus

Cyclic α-β-syntagmatic links	Simple Wikipedia	Wikipedia	WSJ corpus
Case A	95,17%	91,23%	93,71%
Case B	91,29%	89,91%	91,05%
Case C	89,17%	83,06%	85,07%
Linear predicative α-β-links	**Simple Wikipedia**	**Wikipedia**	**WSJ corpus**
Case A	96,17%	92,24%	94,37%
Case B	93,21%	90,01%	91,33%
Case C	91,03%	87,79%	89,79%

the Simple Wikipedia articles; 1,500 sentences - from the Wikipedia articles (using the texts not included in the 800,000 articles processed for constructing matrix D and tensor F).

The syntax trees of the sets of texts from the Wikipedia and the Simple Wikipedia that were processed with the Stanford parser were automatically transformed into control spaces. The obtained control spaces were manually verified and corrected by experts. This annotated text corpus was formed for the purpose of checking the quality of parsing and generating syntactic structure control spaces for the Simple Wikipedia and the English Wikipedia texts.

The system for parsing and control spaces generation constructs control spaces of syntactic structures for sentences from the annotated corpus. Subsequently, the obtained control spaces were compared with the corresponding correct control spaces from the annotated test corpus.

Each cyclic syntagmatic α-β-link and each linear predicative α-β-link that were found was automatically tested. The test was carried out with due regard for the algorithmic case in which a particular syntactic relation was found. Case **A** describes the identification of the direct link between words through the scalar product of their vectors; case **B** describes the usage of synonyms to compute the probability of the link. Case **C** describes the usage of the hyponym and hypernym WordNet connections for these words to find the probability of the link. The test was performed only for the sentences that had been successfully processed with the complete building of the syntactic structure control spaces (94.1% from 1,500 sentences from the Simple English Wikipedia and 83.4% from 1,500 sentences from English Wikipedia were successfully processed in the test set). Also, the test was performed on the WSJ corpus using cross-validation (when checking the quality of the system on 1 part of the corpus out of 10, the corresponding data obtained from the above mentioned part were temporarily excluded from the base of the model). The test on the WSJ corpus was performed automatically and 92.7% of sentences from the WSJ corpus obtained complete parse. The results are summarized in Table 1.

It should be noted that the precision estimates of the linear predicative α-β-links are higher than the precision estimates of the cyclic syntagmatic α-β-links. It seems natural to consider the relative positional stability for relations of type *Subject-Verb-Object* structure in the sentences. A certain small percentage of errors occurs even in the simplest case **A**. It indicates that errors must be present in the training set of control spaces of sentences that served as the base for constructing the cyclic links matrix D and the three-dimensional linear predicative relations tensor F. The model can be improved by checking and correcting the training set. The best estimates correspond to sentences from the Simple Wikipedia, which is quite understandable due to the simple and clear syntactic structure of its sentences. The English Wikipedia sentences are much more complicated, leaving more room for different interpretations of grammatical structures. Hence the precision of processing the WSJ corpus sentences is higher than that for the English Wikipedia sentences. It indicates that the high quality training data from the WSJ corpus allows for improving the model to a great extent.

7 Conclusions

The recursiveness of syntactic structures control spaces allows us to describe sentence structures of arbitrary complexity, length and depth. This enables the development of a semantic-syntactic model based on a single three-dimensional tensor and a single matrix instead of increasing the number of dimensions of connectivity arrays for lexical items. To investigate the applicability of this model for practical NLP tasks a system for analysis and constructing syntactic structure control spaces has been developed on the basis of factorized arrays. It shows high quality and accuracy, thus proving the correctness and efficiency of the developed model.

References

1. Van de Cruys, T.: A Non-negative Tensor Factorization Model for Selectional Preference Induction. Journal of Natural Language Engineering 16(4), 417–437 (2010)
2. Van de Cruys, T., Rimell, L., Poibeau, T., Korhonen, A.: Multi-way Tensor Factorization for Unsupervised Lexical Acquisition. In: Proceedings of COLING 2012, pp. 2703–2720 (2012)
3. Cohen, S.B., Collins, M.: Tensor Decomposition for Fast Parsing with Latent-Variable PCFGs. In: NIPS 2012, pp. 2528–2536 (2012)
4. Wei, P., Tao, L.: On the equivalence between nonnegative tensor factorization and tensorial probabilistic latent semantic analysis. Applied Intelligence, Springer Journals 35(2), 285–295 (2011)
5. Anisimov, A.V.: Control space of syntactic structures of natural language. Cybernetics and System Analysis 93, 11–17 (1990)
6. Chomsky, N.: Syntactic Structures, 117 p. Mouton & Co. (1957)
7. Tesnière, L.: Èlèments de syntaxe structurale. Klincksieck, Paris (1959)
8. Klein, D., Manning, C.D.: Accurate Unlexicalized Parsing. In: Proceedings of ACL 2003, pp. 423–430 (2003)
9. de Marneffe, M.-C., MacCartney, B., Manning, C.D.: Generating Typed Dependency Parses from Phrase Structure Parses. In: Proceedings of LREC (2006), `http://nlp.stanford.edu/pubs/LREC06_dependencies.pdf`
10. Lee, D.D., Seung, H.S.: Algorithms for Non-Negative Matrix Factorization. In: NIPS (2000), `http://hebb.mit.edu/people/seung/papers/nmfconverge.pdf`

11. Cichocki, A., Zdunek, R., Phan, A.-H., Amari, S.-I.: Nonnegative Matrix and Tensor Factorizations: Applications to Exploratory Multi-way Data Analysis and Blind Source Separation. J. Wiley & Sons, Chichester (2009)
12. Kasami, T.: An efficient recognition and syntax-analysis algorithm for context-free languages. Scientific report AFCRL-65-758. Air Force Cambridge Research Lab, Bedford, MA (1965)
13. Deerwester, S., Dumais, S.T., Furnas, G.W., Landauer, T.K., Harshman, R.: Indexing by Latent Semantic Analysis. Journal of the American Society for Information Science 41(6), 391–407 (1990)

Partial Measure of Semantic Relatedness Based on the Local Feature Selection

Maciej Piasecki and Michał Wendelberger

Institute of Informatics, Wrocław University of Technology, Poland
maciej.piasecki@pwr.edu.pl

Abstract. A corpus-based Measure of Semantic Relatedness can be calculated for every pair of words occurring in the corpus, but it can produce erroneous results for many word pairs due to accidental associations derived on the basis of several context features. We propose a novel idea of a *partial measure* that assigns relatedness values only to word pairs well enough supported by corpus data. Three simple implementations of this idea are presented and evaluated on large corpora and wordnets for two languages. Partial Measures of Semantic Relatedness are shown to perform better in tasks focused on wordnet development than a state-of-the-art 'full' Measure of Semantic Relatedness. A comparison of the partial measure with a globally filtered measure is also presented.

1 Introduction

Measures of Semantic Relatedness (henceforth, MSRs) built within Distributional Semantics are one of ways to mine for lexical semantic knowledge out in large text corpora. The extracted knowledge can be utilised in wordnet development, cf [10]. MSR assigns a numerical value to a word pair specifying their semantic relatedness on the basis of their corpus distribution and contextual *features* characterising word occurrences, e.g. a feature is the number of co-occurrences with a word or a lexico-syntactic structure in a specified text context. MSRs construction can be tuned to different needs [1]. For instance, in wordnet development, we expect that words linked by lexico-semantic relations are assigned higher values by MSR.

MSR can be used as knowledge source for wordnet expansion algorithms, e.g. [14,13], but it can also be directly consulted by lexicographers. In both cases, the basic information we want to obtain from MSR is a list of k words that are most semantically related to the given word and such a list is called here a *k-list*. It can be thousands words long for a word, but only the top part is useful in both applications. K-lists can include many errors of two main kinds: words associated by plenty of other relations than those used in wordnets and words linked completely accidentally. The latter errors are especially characteristic for less frequent words. For instance an MSR built on British National Corpus (BNC)[1] (Sec. 3) produces for *frost* (relatedness values in brackets): ***rain*** (0.151), *forsbrand* (0.15), *gale* (0.146), *hawksworth* (0.124), *fitzwilliam* (0.122), ***sleet*** (0.114) ***snow*** (0.11) *rime* (0.1), ***dew*** (0.1), *buda* (0.1), *lawley* (0.1), ***fog*** (0.096), ... While links of non-wordnet relations can be helpful in wordnet expansion

[1] http://www.natcorp.ox.ac.uk

P. Sojka et al. (Eds.): TSD 2014, LNAI 8655, pp. 336–343, 2014.

– e.g. to characterise semantic fields – the accidental associations are problematic for both lexicographers and algorithms (except those that express idiosyncratic properties).

A simple but commonly used method is a threshold-based filtering of words and features during MSR construction, e.g. MSR computation only for those words that are frequent enough in the corpus or have weight value high enough, e.g. [9,15,11,1]. Such global filtering is a radical solution and results in information loss. In relation to particular words, the global frequency threshold can be too high or too low, e.g. some words occurring around 100 times in the 1 billion corpus can still have good description. A large wordnet mostly consists of less frequent words and wordnet expansion is done mostly for words that are less frequent even in a huge corpus.

The strength of association between two words depends on features shared by them. Some words can be mistakenly associated due to accidental features, e.g. an MSR associated *blue spruce* with *antler* because both were described by adjectives *branchy* and *towerring*, and only a few features more. For such words we should decide whether they are semantically related on the basis of the collected information. Our idea is to recognise word pairs for which we have enough amount of the corpus-based information characterising their relationship and compute MSR only for them. The goal is to create a method which will identify words that can be compared and for which the MSR value can be calculated. In sum, we aim at *a partial MSR* that assigns values only to those word pairs for which enough information was collected from the corpus.

2 Method

Two words should be assigned a higher value by an MSR only if they share *a large enough number* of good *quality* features. However, if they do not, we cannot assess their semantic association, as the corpus data are always partial. Thus, when the data supporting a word pair association are too limited, MSR should abstain and does not assign a value to it.

Good quality features must provide enough information to support semantic association of the compared words. The key issues are: how to measure feature quality and how many features must be shared? Statistical association measures or information theoretic measures are commonly used to weigh descriptions provided by features for individual words. For instance, Pointwise Mutual Information (PMI) was often applied and reported to express good performance, e.g. [9,13,2]. However, PMI overestimates some features, especially for less frequent words and there is no universal threshold for PMI values that guarantees appropriate feature selection, i.e. it is clear that only positive PMI values should be used but there are no further thresholds. Most statistical association measures can be misinterpreted in the case of infrequent words. Thus, instead of filtering procedures working on the global scale, we propose a general scheme of *a partial MSR computation* that can be instantiated with different specific solutions. We identify *globally unimportant* features as those that do not discriminate among different words, e.g. e% of features with the highest entropy in the matrix, and *locally unimportant* features that according to the assumed *measures of feature importance* cannot be treated as an important part of the word description.

Let: **M** be a coincidence matrix of words and features, C_E – a set of globally unimportant features, σ – a matrix raw similarity function, x, y – words, R_x, R_y – unweighted row vectors (of frequencies), and W_x, W_y – weighted row vectors.

A partial MSR for x and y is calculated in two steps. First, locally important features for x and y are identified. Next, the partial MSR value is calculated only if the features shared between x and y fulfill the specified conditions.

1. For each $i \in x, y$ the set of locally important features LF_i is defined as:
   (a) LF_i = all non-zero features from R_i minus features from C_E
   (b) $LF_i = f_{imp}(LF_i, R_i, W_i)$
2. Similarity computation:
   (a) If $f_{part}(LF_x, R_x, W_x, LF_y, R_y, W_y)$ equals **true**
   then $\text{MSR}(x, y) = \sigma(W_x, W_y)$ else **unknown**.

The functions: f_{imp} for filtering out locally unimportant features and f_{part} for deciding about sufficient amount of information for comparison, are parameters.

Simple frequency-based partial MSR is based on a simple heuristics originating from the manual inspection of a sample of k-lists (later excluded from tests). Single co-occurrences of a feature and a word are often accidental, so f_{imp} returns only features j such that $R_i j > 1$.

Next, we observed that two 'good' features are mostly enough to make the MSR value meaningful, so f_{part} =

1. If $|LF_x \cap LF_y| > 2$ then return **true**.
2. If $\exists c.(LF_x \cap LF_y| = \{c\}$ and
   – $\exists c' \in LF_x.(c' \neq c \wedge R_y c' > 0)$
   – or $\exists c'' \in LF_y.(c'' \neq s \wedge R_x c'' > 0)$ then return **true**, else **false**.

So, according to the simple heuristics MSR value is calculated only for words sharing at least two features co-occurring with each of them at least twice. These criteria are applied in parallel to the weighting of features and similarity computation. In all experiments we used PMI weighting and cosine similarity.

PMI-based partial MSR – is based on the assumption that features (W_i) are weighted by popular and well-performing PMI. On the basis of the previous experience, we selected Lin's version of PMI [9] and the cosine similarity measure, cf [13,2]. Lin used co-occurrences of words and features as events:

$$\text{PMI}(w, f) = \log(\frac{c(w, f) \sum_{w', f'} c(w', f')}{\sum_{f'} c(w, f') \sum_{w'} c(w', f)}$$

The aim was to exchange frequency criteria to PMI-based criteria, but zero seemed to be too weak and higher values are not theoretically motivated. On the basis of the manual inspection of the same date sample, we set the PMI threshold for locally important features to $\geq$ 1.0 (the global threshold was kept on 0). In addition, we wanted to favour features grouping words into semantic classes or shared among words of the same class. Broad semantic classes were identified on the basis of the wordnet hypernymy structure. Starting from the top synsets hyponymic subtrees were iteratively divided into smaller ones. In each step the largest hypernymy subtree was selected and divided into two subclasses. The direct and indirect hypernyms of the selected subtree were included into both created subclasses. The process ended when the predefined number of subclasses was reached.

Globally unimportant features were defined as 1% with the highest entropy in relation to words and 5% of the highest entropy to semantic classes. Class-base coincidence matrix was constructed on the basis of the word-based matrix. Occurrences of the polysemous words were added to all their classes.
The f_{imp} function was re-defined on the basis of the PMI threshold:
f_{imp} filters out all features j such that $W_i j \leq 1$
and used in the function f_{part}, which stayed unchanged.

Hypernymy-based partial MSR – explores more the developed semantic noun classification. The criterion for locally important features was not changed, because semantic classes of words outside wordnet are not known. However, the associations between features and semantic classes can be discovered on the basis of a large wordnet and a large corpus. The function f_{part} was updated to promote words linked by features of the same class[2]:

1. If $\exists A \in Classes.$
   $|\{f \;\; f \in (LF_x \cap LF_y) \wedge desc(f, A)\}| \geq k$ then return **true**.
2. If $|LF_x \cap LF_y| > n$ then **true** else **false**.

As the first condition is to strict alone, MSR value is also calculated for words sharing at least n locally important features, but we intend to have $n \geq k$.

3 Evaluation

Several approaches to the evaluation of MSRs were proposed, e.g. a comparison to human decisions [16], solving tests similar to TOEFL [6,4,12], and comparison with the wordnet-based similarity [9,15]. However, for the purposes of wordnet development, the most important is to have possibly many instances of wordnet relations in the top of the k-list, in our test:

- for each test word x, a set of all words connected to it in the wordnet is generated,
- and the set is compared with the k-list of x generated by the MSR.

Cut-off *precision* for k-lists is defined as $|L_M(k)|/|L_W|$ and *recall* as $|L_M(k) \cap L_W|/|L_W|$, where $L_M(k)$ is a set of word pairs $\langle x, y \rangle$ such that y belongs to k-list of x, and L_W is a set of all pairs extracted from the wordnet for the test words. The wordnet set includes: synonyms, direct and indirect hypo/hypernyms (up to 3 links), cousins (up to 2 hypernymic and hyponymic links), mero/holonyms and the words linked directly by lexical relation. In the case of polysemous words all synsets were considered.

In addition, we also applied the second evaluation method based on testing MSR ability to predict wordnet-based semantic similarity [15]. Correlation between k-lists

[2] In practice, due to the polysemy of the words in the original matrix most features are associated with many semantic classes.

Table 1. Tests on British National Corpus: k – No of positions for the cut-off precision P and recall R, *Cor.* – the correlation with the similarity measure on WordNet 3.1; *Cov* – ratio between the No pairs extracted by a MSR and the full MSR, *Hits* – No of word pairs in the k-best lists of the MSR.

Func.	k	P [%]	R [%]	F1 [%]	Cov. [%]	Hits	Cor.
Full	**5**	11.42	0.03	0.06	100.00	9 926	0.003459
	10	10.87	0.05	0.11	100.00	18 884	0.000797
	20	10.40	0.10	0.20	100.00	36 102	0.000164
	50	9.77	0.24	0.47	100.00	84 499	0.000018
Full globally filtered	**5**	14.37	0.04	0.07	96.52	12 054	—
	10	13.58	0.07	0.13	96.60	22 786	—
	20	12.69	0.12	0.25	96.71	42 596	—
	50	11.54	0.28	0.55	96.96	96 791	—
Simple	**5**	**31.09**	**0.07**	**0.15**	70.04	18 931	**0.003901**
	10	**29.07**	**0.14**	**0.27**	68.42	34 553	**0.000899**
	20	**27.03**	**0.24**	**0.48**	66.07	61 978	**0.000186**
	50	**24.09**	**0.50**	**0.98**	61.51	128 175	**0.000021**
PMI-based	**5**	24.32	0.06	0.12	**93.26**	19 719	0.003741
	10	23.23	0.11	0.22	**92.05**	37 155	0.000867
	20	22.20	0.21	0.41	**90.03**	**69 355**	0.000179
	50	20.36	0.45	0.88	**85.80**	**151 065**	0.000020
Hypernymy-based	**5**	24.33	0.06	0.12	93.25	**19 725**	0.003766
	10	23.27	0.11	0.22	92.00	**37 186**	0.000874
	20	22.23	0.21	0.41	89.87	**69 335**	0.000180
	50	20.41	0.45	0.88	85.38	150 722	0.000020

(treated as ranking lists) produced by an MSR and a wordnet-based similarity measure was analysed. Following [15], we used the Jiang and Conrath similarity measure [5] (JC measure) to generate wordnet-based k-lists for test words. For ranking list comparison, neighbour set comparison technique proposed by [8] and adapted in [15] for this task was applied:

$$Cor(S, S') = \frac{\sum_{w \in S \cap S'} (w, S)(w, S')}{\sum_{i=1}^{k} i^2}$$

where S, S' are two k-lists (ranking lists) and (w, S) returns $k - ranking(w, S)$.

Experiments were performed on the two world largest wordnets: Princeton WordNet 3.1 (PWN) [3] and plWordNet 2.1 – the Polish wordnet (plWN) [10], which is not translated from PWN and has also a slightly different character. For the experiments on English, we used BNC, which has been often used for building MSRs. For Polish, we have built a joint corpus of 1.8 billion tokens by merging together freely available corpora and larger texts acquired from Internet. BNC was processed by MiniPar dependency parser [7], and the Polish joint corpus with morpho-syntactic constraints from [13]. As a result English, and Polish words were described by lexico-syntactic relations used as features in the matrices.

Table 2. Tests on the Polish joint corpus (the same labels as in Tab. 2)

Func.	k	P	R	F1	Cov. [%]	Hits	Cor.
Full	**5**	24.80	0.11	0.22	100.00	62 239	–
	10	22.85	0.21	0.41	100.00	114 473	–
	20	20.96	0.38	0.75	100.00	209 401	–
	50	18.49	0.82	1.56	100.00	450 468	–
Full globally filtered	**5**	29.26	0.13	0.26	95.13	69 843	0.01124
	10	26.96	0.24	0.49	95.30	128 696	0.002585
	20	24.62	0.45	0.88	95.56	235 049	0.000540
	50	21.34	0.97	1.85	97.88	508 680	0.00006
Simple	**5**	**40.08**	**0.18**	**0.35**	79.08	79 535	0.011253
	10	**36.45**	**0.32**	**0.63**	78.51	143 353	0.002588
	20	**32.72**	**0.57**	**1.11**	77.60	253 646	0.000541
	50	**27.55**	**1.15**	**2.21**	77.15	517 699	0.00006
PMI-based	**5**	36.38	0.16	0.32	**90.92**	**82 990**	**0.011254**
	10	33.59	0.30	0.59	**90.29**	**151 934**	**0.002589**
	20	30.73	0.54	1.06	**89.31**	**274 222**	**0.000541**
	50	26.56	1.13	2.17	**89.04**	**576 006**	**0.00006**
Hypernymy-based	**5**	36.39	0.16	0.33	90.89	82 987	**0.011254**
	10	33.61	0.30	0.59	90.26	151 936	**0.002589**
	20	30.74	0.54	1.06	89.26	274 185	**0.000541**
	50	26.58	1.13	2.17	88.92	575 810	**0.00006**

The baseline *full MSRs* were built for all one-word nouns from both corpora. All possible features were collected initially, but first 1% features with the highest entropy. So, the English matrix was initially of the size: 39,411 nouns and 124,830 features, and 58,781 nouns and 643,894 features for Polish. Next, the influence of the *global filtering* was tested. In addition to the entropy threshold, minimal frequency for words was set to 5, for features to 20, and all features with less than 20 non-zero cells after PMI weighting were removed from the matrix. Concerning parameter values, *simple partial MSR* (PMSR) was used as described earlier. For *PMI-based PMSR*, we set f_{imp}: 200 semantic classes, minimal PMI value 1, top entropy features vs words 1%, top entropy features vs classes 5%, and f_{part} unchanged. In hypernymy-based PMSR, f_{part} was instantiated to $k = 2$ and $n = 2$. In all cases global feature frequency threshold was 5, entropy threshold was 1%, and cosine similarity was used.

The evaluation results are presented in Tab. 1 for English and in Tab. 2 for Polish. All partial MSRs have higher precision than the 'full' MSRs and the globally filtered MRSs, i.e. mostly non-related words were filtered out. Proper words are sometimes eliminated too, but, surprisingly, partial MSRs also have better recall. This is caused by lifting up proper words from lower ranking positions to the *k* top words. The difference between all partial MSRs and both types of the 'full' MSRs is statistically significant, with the confidence level higher than $1 - 0.005$ (measured by t-test applied to average results on 50 random samples) in both types of evaluation. Only some differences between partial MSRs are statistically significant with the confidence $1 - 0.05$.

Coverage – a ratio of the number of word pairs generated by partial and full MSR, is higher in the case of PMI-based and hypernymy-based PMSRs. They generate more word pairs. Some of them can be proper, but not covered by wordnets. Both PMSRs have the highest numbers of hits, i.e. they produce longer k-lists on average that cover more wordnet relation instances.

The following examples illustrate changes introduced to k-lists by a simple PMRS. For the English word *frontier* the full MSR produces: ***border*** (0.108), *euphrates* (0.104), *mindanao* (0.0838) *findhorn* (0.0791), *memnon* (0.0755), *negus* (0.0709), ***province*** (0.0686), *demerit* (0.0681) *eurythmics* (0.0661), *weinstock* (0.0661) *vase-painting* (0.0655), ***ambassador*** (0.0639) *pattaya* (0.0638) *non-russian* (0.0626), ***siberia*** (0.0617), *reik* (0.0615), ***territory*** (0.0603), ***coast*** (0.0602), *tunney* (0.0596), *quad-fx* (0.0584), *gondwana* (0.0554), *descendent* (0.0553), *beowulf* (0.0551), ***national*** (0.0548), ***sovereignty*** (0.0548). The simple partial MSR filters out all associations from this list except the ones in bold.

The MSR built for Polish provides better description for many words, as the joint corpus is more than 18 times bigger than BNC. However, less frequent words can cause accidental associations, e.g. for the Polish noun *kapsuła* 'capsule' the full MSR generates: ***właz*** 'hatch' (0.1011), ***statek kosmiczny*** 'spacecraft' (0.0816), *konserwatornia* '≈restoration workshop' (0.0803), *izostazja* 'isostasy' (0.078), ***luk*** 'hatchway' (0.0766), *odcumowanie* 'unmooring' (0.075), ***kabina*** 'cabin' (0.071), ***wahadłowiec*** 'shuttle' (0.0696), ***moduł*** 'module' (0.0687), *smażalnik* '≈machine for frying' (0.0685), *odżelaziacz* '≈iron remover' (0.0685), ***zasobnik*** 'tray' (0.0677), *sferolit* 'spherulite' (0.0668), *łuszczynka* 'siliqua' (0.0661), *ser ementalski* 'Emmental cheese' (0.066), ***prom*** 'space shuttle' (0.0653), ...

4 Conclusions

A novel idea of the partial MSR was proposed that assigns semantic relatedness values only to word pairs that are enough well described by the corpus-based information. A partial MSR seems to be a better knowledge source for wordnet development and lexicography than a 'full' MSR. Simple but effective implementations of partial MSRs were tested on the two world largest wordnets, and two corpora: a medium and huge one and two very different languages. All evaluated variants are language and resource independent. An obvious drawback of a partial MSR is that it does not assign values to some word pairs. However, a full MSR often assigns unreliable values that are not well supported by corpus data.

Acknowledgments. Partially financed by the Polish Ministry of Science and Higher Education as a Investment in CLARIN-PL Research Infrastructure.

References

1. Baroni, M., Lenci, A.: Distributional memory: A general framework for corpus-based semantics. Computational Linguistics 36(4), 637–721 (2010)

2. Bullinaria, J.A., Levy, J.P.: Extracting semantic representations from word co-occurrence statistics: stop-lists, stemming, and SVD. Behav. Res. Methods 44(3), 890–907 (2012)
3. Fellbaum, C. (ed.): WordNet – An Electronic Lexical Database. The MIT Press (1998)
4. Freitag, D., Blume, M., Byrnes, J., Chow, E., Kapadia, S., Rohwer, R., Wang, Z.: New experiments in distributional representations of synonymy. In: Proc. of the 9th Conf. on Computational Natural Language Learning, pp. 25–32. ACL, Ann Arbor (2005)
5. Jiang, J.J., Conrath, D.W.: Semantic similarity based on corpus statistics and lexical taxonomy. In: Proceedings of the International Conference on Research in Computational Linguistics (ROCLING X), Taiwan (1997)
6. Landauer, T.K., Dumais, S.T.: A solution to Plato's problem: The Latent Semantic Analysis theory of acquisition. Psychological Review 104(2), 211–240 (1997)
7. Lin, D.: Principle-based parsing without overgeneration. In: Proceedings of the 31st Annual Meeting of the Association for Computational Linguistics (1993)
8. Lin, D.: Using syntactic dependency as local context to resolve word sense ambiguity. In: Proc. of the 35th ACL and 8th EACL, pp. 64–71. ACL, Madrid (1997)
9. Lin, D.: Automatic retrieval and clustering of similar words. In: Proc. of the 35th ACL and 17th Inter. Conf. on Computational Linguistics, pp. 768–774. ACL (1998)
10. Maziarz, M., Piasecki, M., Rudnicka, E., Szpakowicz, S.: Beyond the transfer-and-merge wordnet construction: plWordNet and a comparison with WordNet. In: Proc. of the Inter. Conf. Recent Advances in Natural Language Processing, RANLP 2013. INCOMA Ltd. and ACL, Hissar, Bulgaria (2013)
11. Navigli, R., Velardi, P., Faralli, S.: A graph-based algorithm for inducing lexical taxonomies from scratch. In: Proceedings of IJCAI (2011)
12. Piasecki, M., Szpakowicz, S., Broda, B.: Extended similarity test for the evaluation of semantic similarity functions. In: Vetulani, Z. (ed.) Proce. of the 3rd Language and Technology Conference, Poznań, pp. 104–108 (2007)
13. Piasecki, M., Szpakowicz, S., Broda, B.: A Wordnet from the Ground Up. Oficyna Wydawnicza Politechniki Wrocławskiej (2009), http://www.plwordnet.pwr.wroc.pl/main/content/files/publications/A_Wordnet_from_the_Ground_Up.pdf
14. Snow, R., Jurafsky, D., Ng, A.Y.: Semantic taxonomy induction from heterogenous evidence. In: Proc. of the Joint Conf. of the International Committee on Computational Linguistics and ACL, pp. 801–808 (2006)
15. Weeds, J., Weir, D.: Co-occurrence retrieval: A flexible framework for lexical distributional similarity. Computational Linguistics 31(4), 439–475 (2005)
16. Zesch, T., Gurevych, I.: Automatically creating datasets for measures of semantic relatedness. In: Proceedings of the Workshop on Linguistic Distances, pp. 16–24. Association for Computational Linguistics, Sydney (2006)

A Method for Parallel Non-negative Sparse Large Matrix Factorization

Anatoly Anisimov, Oleksandr Marchenko, Emil Nasirov, and Stepan Palamarchuk

Faculty of Cybernetics, Taras Shevchenko National University of Kyiv, Ukraine

Abstract. This paper proposes parallel methods of non-negative sparse large matrix factorization. The described methods are tested and compared on large matrices processing.

Keywords: computational linguistics, parallel computations, non-negative matrix factorization.

1 Introduction

Non-negative matrix and tensor factorization are very popular techniques in computational linguistics. With the help of non-negative matrix and tensor factorization within the paradigm of latent semantic analysis [1] computational linguists solve applied problems such as classification, clustering of texts and terms [2,3], construction of measures of semantic similarity [4,5], automatic extraction of linguistic structures and relations (Selectional Preferences) and Verb Sub-Categorization Frames), etc. [6]

This work describes the construction of a model for parallel non-negative factorization of a sparse large matrix. Such a model can be used in large NLP systems not limited to narrow domains.

The problem of non-negative factorization for a sparse large matrix emerged in the development of a measure of semantic similarity between words with Latent Semantic Analysis usage. To cover a wide range of topics a great amount of articles from the English Wikipedia was processed to construct the similarity measure. Lexical analysis of the various Wikipedia articles was performed to calculate the frequency of using words and collocations. As a result, a large matrix Terms $\times$ Articles was constructed. It contains frequency estimations of using terms in texts. The precise size of the matrix equals to 2,437,234 terms $\times$ 4,475,180 articles of the English Wikipedia. The frequency threshold T=3 was set to remove the noise. The resulting matrix contains 156,236,043 non-zero elements. To factorize a sparse matrix of such size it is necessary to develop a specific model for parallelizing matrix computations. The model has been implemented using distributed and parallel computing on the GPU. Recently a plenty number of powerful parallel models for Non-Negative Matrix Factorization (NMF) have been developed [7,8,9,10]. However none of the developed applications for them is an acceptable solution for the defined task. Some of them do not satisfy the requirements of the matrix dimensions [7,8,9]. The model presented in work [10] performs NMF for sparse matrices of required dimensions in an acceptable time, but it requires excessively large computational resources and it is not always affordable.

P. Sojka et al. (Eds.): TSD 2014, LNAI 8655, pp. 344–352, 2014.

2 NMF Algorithm

Non-negative matrix factorization of matrix V of size $[n; m]$ is a process of calculating two matrices W and H of size $[n; k]$ and $[k; m]$ respectively, such that $V \approx WH$.

$$F(W,H) = \sqrt{\sum_{i=1}^{n}\sum_{j=1}^{m}(V_{i,j} - (WH)_{i,j})^2} \tag{1}$$

The goal of the algorithm is to minimize the cost function quantifying the approximation quality. There are a lot of different cost functions. In this paper the root-mean-square distance between V and WH is used 1.

In [11], the authors proposed a simple iterative algorithm to approximate the matrices. It consists of two consequent updates of matrices W and H given by and .

$$(H_t)_{ij} = (H_{t-1})_{ij}\frac{(W_{t-1}^T V)_{ij}}{(W_{t-1}^T W_{t-1} H_{t-1})_{ij}} \tag{2}$$

$$(W_t)_{ij} = (W_{t-1})_{ij}\frac{(V H_t^T)_{ij}}{(W_{t-1} H_t H_t^T)_{ij}} \tag{3}$$

In 2 and 3, H_t and W_t are the matrices obtained on iteration t. H_0 and W_0 are initialized with random values from [0; 1) range.

The algorithm continues until either a stationary point is reached or a certain number of iterations is performed.

3 Model Analysis

The goal is to solve the NMF problem for different k values and compare results for all of them. Table 1 shows memory requirements for storing W and H for different k. On each iteration the algorithm described in Section 2 requires twice as much memory as required for matrices storage. This does not include memory required for V. Due to such excessive memory requirements of the algorithm it is difficult to execute it on a single machine, without dumping data to the hard drive. Two variants of the algorithm implementation are described below: local (with intensive hard drive usage) and distributed (with intensive network usage).

Table 1. Memory requirements for storing of W and H for different k, based on 32-bit float

k	100	200	300
W	0.98Gb	1.95Gb	2.92Gb
H	1.79Gb	3.58Gb	5.37Gb
total	2.76Gb	5.53Gb	8.29Gb

4 A GPU Version of the Algorithm

To simplify explanations the substitution ($H' = H^T$) and transformation of 2 and 3 result in:

$$(H'_t)_{ij} = (H'_{t-1})_{ij} \frac{(V^T W_{t-1})_{ij}}{(H'_{t-1} W^T_{t-1} W_{t-1})_{ij}} \quad (4a)$$

$$(W_t)_{ij} = (W_{t-1})_{ij} \frac{(V H'_t)_{ij}}{(W_{t-1} H'^T_t H'_t)_{ij}} \quad (4b)$$

It allows for treating both formulas in the same way, by simply substituting either H', W and V^T or W, H' and V instead of A, B and S into 5.

$$A_{ij} = A_{ij} \frac{(SB)_{ij}}{(AB^T B)_{ij}} \quad (5)$$

From this point, only evaluation of 5 with a configuration W, H' and V is discussed, since other configuration can be obtained in the same way.

Formula 5 can be calculated as a series of four steps as in 6.

$$C = SB \quad (6a)$$

$$K = B^T B \quad (6b)$$

$$D = AK \quad (6c)$$

$$A_{ij} = A_{ij} \frac{C_{ij}}{D_{ij}} \quad (6d)$$

This order of computation 5 requires a minimal number of calculations. The steps have computational complexity of $O(k * (nnz(S) + n))$, $O(k^2 m)$, $O(k^2 n)$ and $O(kn)$ correspondingly, where $nnz(S)$ is a number of non-zero cells in matrix S. The first three steps are natively supported by CUDA cuSPARSE [12] and cuBLAS [13] libraries (or other similar libraries for AMD). The fourth step requires custom GPU kernel implementation, but at the same time it is a relatively cheap operation and thus it can be performed on CPU.

Also these matrices are too large to be stored in the memory of GPU, thus operations should be performed by parts in a manner that reduces amount of excessive memory copying.

So for 6a matrices can be written as $S = (S'_1|S'_2|...|S'_t)^T$ and $B = (B_1|B_2|...|B_r)$ and each cell of C calculated as shown in 7. Since B is larger than S in terms of memory usage multiplications should be grouped by pieces of B (to upload them only once). Also it is rational to minimize r and keep t reasonably small, otherwise most of GPU cores will be idle. C is matrix of size m; k for H' and n; k for W.

$$C = \begin{pmatrix} S'_1 B_1 & ... & S'_1 B_r \\ ... & & ... \\ S'_t B_1 & ... & S'_t B_r \end{pmatrix} \quad (7)$$

For 6b it is preferable to write matrices as $B = (B'_1|...|B'_t)^T$ and $K = B'^T_1 B'_1 + ... + B'^T_t B'_t$, because it doesn't require any redundant matrix uploads to GPU. K is matrix of size k; k.

For 6c matrix K should be kept in memory and A should be multiplied by blocks of rows. D is matrix of size m; k for H' and n; k for W.

There is no need to store D in memory if 6d is applied on the piece of matrix A that was used to obtain a piece of matrix D.

The complexity of operation is $O(nm)$ and is straightforward to implement with CUDA toolkit.

5 Distributed Algorithm

The next step to improve the performance is to use a distributed grid of PCs of same configuration. There are several distribution models. In the case of 2 nodes in the grid, there are next three distribution models:

1. **W and H' are separately calculated on different nodes.** Both nodes work in one of the two modes alternatively. They either support the other node (supplying data to the other node) or lead (calculating by using the data received form supporting node). In this distribution model, on each iteration it is necessary to transmit over the network amount of data equal to $sizeof(W) + sizeof(H)$, where $sizeof(X)$ is the amount of memory required to store matrix X. Also lead node will be mostly idle, because 6a is the most resource-demanding step out of all the 4.
2. **W and H' are split in chunks of rows and evenly distributed between nodes.** Where $H' = (H'_1|H'_2)$ and $W = (W_1|W_2)$, the first node responsible for H'_1, W_1 and the second node for H'_2, W_2. In this model each node behaves as supporting and leading node at the same time. Nodes need to transmit amount of data equal to $1.5 * (sizeof(W) + sizeof(H))$ over the network.
3. **W and H' are split in chunks of columns and evenly distributed between the nodes.** Where $H' = (H_1|H_2)^T$ and $W = (W'_1|W'_2)^T$, the first node is responsible for H_1, W'_1 and the second for H_2, W'_2. In this model, similarly to the previous one, each node works in both modes at the same time. Nodes need to transmit the amount of data equal $sizeof(W) + sizeof(H)$ over the network because it is possible to calculate pieces of $H^T * H, W^T * W$ on each node separately and there is no need to transmit H'_2 to H'_1 and W_2 to W_1 as in second model.

In each of the above models nodes also need to transmit one or several matrices of size k; k, but their total size is neglectable comparing to the size of W and H. For metrics calculation for both the first and the third model it is necessary to transmit the amount of data equal to $\frac{(sizeof(W)+sizeof(H))*K}{2}$, where K is the number of nodes in the grid, the second model requires $(K - 1)$ times more transmitted data.

The last model is the most preferable, because it is better in both network and GPU utilization, thus it is used in the implementation.

Also it should be noticed that in case of the grid expansion, the total amount of the data transmitted over the network rises polynomially, but per node it will be limited by $2 * (sizeof(W) + sizeof(H))$.

Since V is a sparse matrix, it may contain an unevenly distributed amount of non-zero cells and this may badly impact the performance of the distributed algorithm. To optimize distribution of work between the nodes it is reasonable to rearrange the rows and columns of V in a way that equalizes amount of non-zero cells in each large cell of matrix V.

The third model is implemented to perform NMF of input matrix and used on a grid of four nodes, so this case will be described.

Where matrices W, H' and V are partitioned according to the selected model 3:

$$W = (W_1'|W_2'|W_3'|W_4')^T, H' = (H_1|H_2|H_3|H_4)^T, V = \begin{matrix} V_{11} \ldots V_{14} \\ \ldots \quad \ldots \\ V_{41} \ldots V_{44} \end{matrix}.$$

The algorithm consists of three main phases: initialization, iterations and metrics calculation. At initialization phase W, H and V are distributed between all the 4 nodes. Node i gets W_i', H_i and V_{ki}, V_{ik}, $k = \overline{1, .., 4}$, this phase is represented by the scheme in Figure 1 (left).

The iteration phase consists of two similar steps, one for calculation of H' and the other for W. Each of them is subdivided into 3 smaller sub-steps as was described further.

At the first sub-step each node calculates $k \times k$ matrix $W_i' * W_i'^T$ and sends it to the aggregator. The aggregator sums all received pieces into one matrix K_w and sends the aggregated result to all the nodes. This sub-step is represented by the scheme in Figure 1 (left).

At the next sub-step each node calculates its own $(V_{1i}^T|V_{2i}^T|V_{3i}^T|V_{4i}^T)^T * W_i'$. The resulting matrix has the same size as H'. Finally each node divides its matrix according to the initial partitioning of matrix H and transmits these pieces to the corresponding nodes. This sub-step is represented on Figure 1 (right).

At the third sub-step the nodes calculate matrix $H_i * K_w$ and perform an in-place update of matrix H_i. This sub-step does not require any network communication.

These three sub-steps are intended for calculating matrix H'. After updating H', the same sub-steps should be made for W. Specifically next products should be calculated $H_i * H_i^T$, $(V_{i1}|V_{i2}|V_{i3}|V_{i4}) * H_i$ and $W_i' * K_h$.

At the metrics calculation phase each node transmits its piece of matrix H' to all other nodes. After receiving a piece of matrix H' each node calculates the corresponding part of the metrics. This phase is also represented by in Figure 1 (right).

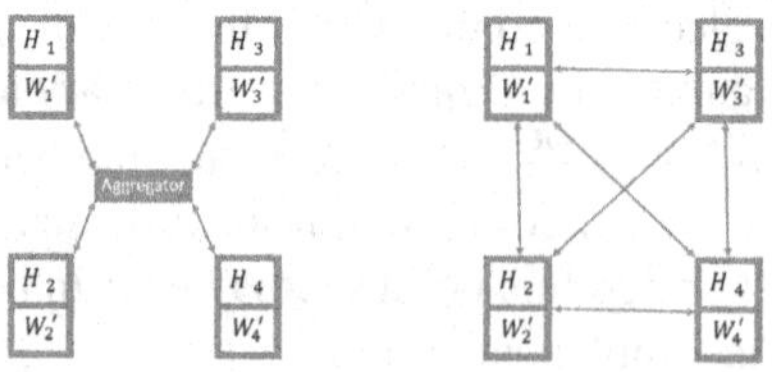

Fig. 1. Initial partitioning in the model with 4 nodes. (left) The iteration phase of the distributed model with 4 nodes. (right)

6 Results of Analysis

The previously described distributed algorithm with GPU usage has been implemented. The local GPU algorithm that dumps and reads data from the local hard drive has also been implemented to compare the performance of the models.

Table 2. Performance of local and distributed implementations for iteration and metrics calculation

	Iteration		Metrics calculation	
	Local	Distributed	Local	Distributed
Data reads	34.44Gb	6.22Gb	13.66Gb	6.22Gb
Data writes	16.58Gb	6.22Gb	0	6.22Gb
Time (computation)	58s	15s	45865s	11371s
Time (data IO)	729s	287s	192s	280s

Both implementations are executed with the same input matrix.

We used next hardware configuration for tests: Intel Core i7 CPU, NVIDIA GeForce GTX560 1Gb, 8Gb of RAM (available 6 Gb), 1Gbit LAN and SATA III hard drive.

Table 2 shows the time and resources required for each version of the algorithm to perform the iteration. The data for distributed model are per node, so the total data IO (read & write together) across all 4 nodes is 49.76Gb. Table 2 shows comparison of metrics calculation. The data in both tables are obtained for $k = 300$.

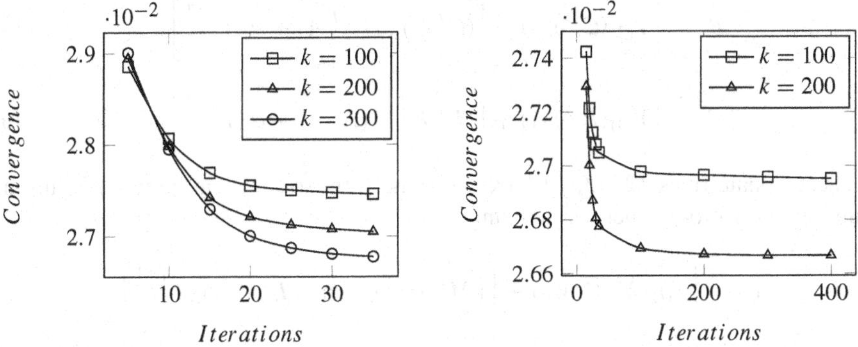

Fig. 2. Convergence of NMF with different k. Metrics value is calculated at each 5th iteration. (left) Convergence of NMF with $k = 200$ and $k = 300$. Metrics value is calculated at each 100th iteration. (right)

The experiments show that the process of matrices calculation converges after approximately 100 iterations. Therefore, the calculation of the non-negative factorization for the given sparse large matrix with the proposed model takes approximately 9.6 hours, for the distributed implementation and almost 21 hours for the local.

7 Parallelization of Non-negative Tensor Factorization

Distributed approach can be easily used for non-negative tensor factorization. To compute the non-negative component matrices A, B, C usually constrained optimization approach is applied as by minimizing a suitable cost function. Typically, the following global cost function is minimizing (with respect the component matrices):

$$D_F(\underline{Y}||[A,B,C]) = \|\underline{Y} - [A,B,C]\|_F^2 + \alpha_A\|A\|_F^2 + \alpha_B\|B\|_F^2 + \alpha_C\|C\|_F^2 \quad (8)$$

, where $\alpha_A, \alpha_B, \alpha_C$ are non-negative regularization parameters.

The most popular approach is to apply the ALS technique [14]. In this approach the gradient of the cost function for each component matrix is computed individually.

$$\nabla_A D_F = -Y_{(1)}(C \odot B) + A(C^T C) \circledast (B^T B) + \alpha_A I \quad (9)$$

$$\nabla_B D_F = -Y_{(2)}(C \odot A) + A(C^T C) \circledast (A^T A) + \alpha_B I \quad (10)$$

$$\nabla_C D_F = -Y_{(3)}(B \odot A) + A(B^T B) \circledast (A^T A) + \alpha_C I \quad (11)$$

ALS update rules for the NTF are obtained by equating the gradient components to zero:

$$A \leftarrow \left[Y_{(1)}(C \odot B) + \left[(C^T C) \circledast (B^T B) + \alpha_A I\right]^{-1}\right]_+ \quad (12)$$

$$B \leftarrow \left[Y_{(2)}(C \odot A) + \left[(C^T C) \circledast (A^T A) + \alpha_B I\right]^{-1}\right]_+ \quad (13)$$

$$C \leftarrow \left[Y_{(3)}(B \odot A) + \left[(B^T B) \circledast (A^T A) + \alpha_C I\right]^{-1}\right]_+ \quad (14)$$

These update rules 12, 13, 14 have the same form so they can be rewrited in one common rule 15 for further parallelizing:

$$M_i \leftarrow \left[Y_{(i)}(M_j \odot M_k) + \left[\left(M_j^T M_j\right) \circledast (M_k^T M_k) + \alpha_{M_i} I\right]^{-1}\right]_+ \quad (15)$$

Such update rule can be computed distributively in a similar way, like it was described for matrices factorization.

For the grid of two nodes M_j and M_k' should be split in chunks of columns and evenly distributed between the nodes. Where $M_k' = (M_{k_1}|M_{k_2})^T$ and $M_j = (M_{j_1}'|M_{j_2}')^T$, the first node is responsible for M_{k_1}, M_{j_1}' and the second for M_{k_2}, M_{j_2}'.

In such distributed model every node has to calculate matrix $(M_{j_t}^T M_{j_t})$ of size $k \times k$ and send it to aggregator. Aggregator after receiving data from all nodes merges blocks into one matrix K_j and sends corresponding parts to the nodes. The same step should be done for calculation of matrix $(M_k^T M_k)$.

On the next step nodes have to calculate $S_t = \left[\left(M_j^T M_j\right)_t \circledast (M_k^T M_k)_t + \alpha_{M_i} I\right]_t^{-1}$ and send then to the aggregator which after receiving of all parts needs merge them, calculate S^{-1}, divide result on blocks and send them to corresponding nodes. After that nodes can calculate parts of the result matrix as $(M_j \odot M_k)_t + (S^{-1})_t$.

Therefore, the distributed model described in Section 5 can be easily transformed for non-negative factorization of large tensors.

8 Conclusion

We have combined the GPU-based and distributed algorithms, and also paid special attention to memory usage, which allows larger input matrices to be factorized. The experiments showed the constructed model is effective. It can be used to perform the tasks of industrial scale to factorize sparse matrices of large dimension with an acceptable time using available computing resources.

Proposed distributed model can be easily modified to speed up non-negative factorization of large tensors.

References

1. Deerwester, S., Dumais, S.T., Furnas, G.W., Landauer, T.K., Harshman, R.: Indexing by latent semantic analysis. Journal of the American Society for Information Science, 391–407 (1990)
2. Xu, W., Liu, X., Gong, Y.: Document Clustering Based on Non-negative Matrix Factorization. In: Proceedings of the 26th Annual International ACM SIGIR Conference on Research and Development in Informaion Retrieval, SIGIR 2003, pp. 267–273. ACM, New York (2003)
3. Shahnaz, F., Berry, M.W., Pauca, V.P., Plemmons, R.J.: Document Clustering Using Non-negative Matrix Factorization. Inf. Process. Manage., 373–386 (2006)
4. Landauer, T.K., Foltz, P.W., Laham, D.: An introduction to latent semantic analysis Discourse processes, pp. 259–284. Ablex Publishing Co (1998)
5. Mihalcea, R., Corley, C., Strapparava, C.: Corpus-based and knowledge-based measures of text semantic similarity. In: AAAI 2006, pp. 775–780. AAAI Press, Menlo Park (2006)
6. Van de Cruys, T.: A non-negative tensor factorization model for selectional preference induction. Natural Language Engineering, 417–437 (2010)
7. Brett, W.: Bader and Tamara G. Kolda: MATLAB Tensor Toolbox Version 2.5 (2012), `http://www.sandia.gov/tgkolda/TensorToolbox/`
8. Kanjani, K.: Parallel Non Negative Matrix Factorization for Document Clustering Texas A & M University (2007)
9. Kysenko, V., Rupp, K., Marchenko, O., Selberherr, S., Anisimov, A.: GPU-Accelerated Non-negative Matrix Factorization for Text Mining. In: Bouma, G., Ittoo, A., Métais, E., Wortmann, H. (eds.) NLDB 2012. LNCS, vol. 7337, pp. 158–163. Springer, Heidelberg (2012)
10. Liu, C., Yang, H.-C., Fan, J., He, L.-W., Wang, Y.-M.: Distributed Nonnegative Matrix Factorization for Web-scale Dyadic Data Analysis on Mapreduce. In: Proceedings of the 19th International Conference on World Wide Web, WWW 2010, pp. 681–690. ACM, Raleigh (2010)

11. Lee, D.D., Seung, H.S.: Algorithms for Non-negative Matrix Factorization In NIPS, pp. 556–562. MIT Press (2000)
12. Naumov, M., Chien, L.S., Vandermersch, P., Kapasi, U.: CUDA CUSPARSE Library NVIDIA, San Jose, CA (2010)
13. NVIDIA: CUBLAS Library User Guide (2013), http://docs.nvidia.com/cublas/index.html
14. Cickocki, A., Zdunek, R., Phan, A.H., Amari, S.-I.: Non-negative matrix and tensor factorizations: applications to exploratory multiway data analysis and blind source separation Fabulous, Singapore, pp. 237–240 (2009)

Using Graph Transformation Algorithms to Generate Natural Language Equivalents of Icons Expressing Medical Concepts

Pascal Vaillant and Jean-Baptiste Lamy

Université Paris 13, Sorbonne Paris Cité, LIMICS, (UMRS 1142),
74 rue Marcel Cachin, 93017, Bobigny cedex, France
INSERM, U1142, LIMICS, 75006, Paris, France
Sorbonne Universités, UPMC Univ Paris 06, UMRS 1142, LIMICS, 75006, Paris, France
vaillant@univ-paris13.fr

Abstract. A graphical language addresses the need to communicate medical information in a synthetic way. Medical concepts are expressed by icons conveying fast visual information about patients' current state or about the known effects of drugs. In order to increase the visual language's acceptance and usability, a natural language generation interface is currently developed. In this context, this paper describes the use of an informatics method – graph transformation – to prepare data consisting of concepts in an OWL-DL ontology for use in a natural language generation component. The OWL concept may be considered as a star-shaped graph with a central node. The method transforms it into a graph representing the deep semantic structure of a natural language phrase. This work may be of future use in other contexts where ontology concepts have to be mapped to half-formalized natural language expressions.

Keywords: Graph grammars, Natural Language Generation, Health and Medicine, Iconic Language.

1 Introduction

This work takes place in the field of medical knowledge visualization. It is part of an ongoing project aiming at developing and promoting the use of new interfaces for accessing medical information systems, including a graphical representation language, VCM [1], whereby medical concepts are expressed by icons, and a multi-lingual interface. The graphical language is used to provide complex information in a form adapted to synthetic visual perception.

To be accepted and used more widely within different medical information systems, icons needs to be made as easy to learn and to use as possible. In this view, it is necessary to provide users with easily accessible natural language expressions of the meaning conveyed by any icon, e.g. in the form of a pop-up balloon appearing when the mouse cursor is hovering over the icon.

The icons express meanings which result from the combination of a finite set of elementary meaning components. As there are hundreds of thousands of potential combinations, and since the language design is still expanding, icons are dynamically

P. Sojka et al. (Eds.): TSD 2014, LNAI 8655, pp. 353–362, 2014.

generated from the elementary components. Thus, the natural language utterances also have to be automatically generated. To this purpose, we develop a natural language generation module, which outputs phrases in two languages.

The graphical language VCM is built against an ontology of medical concepts, defined with the OWL-DL representation formalism [2]. Every icon expresses a concept in the VCM ontology. So, the primary input data is an OWL concept, which corresponds to the medical concept to be expressed. This one concept is used in one process to generate an image object (not discussed here), and in another process to generate a natural language phrase. In the application context, more specifically, it is generally wished that the phrase should be a noun phrase (NP); but this should be a mere parameter of the generation process, and should it be desirable that the output be e.g. a sentence (S), it would be possible as well.

So, the stake of the work described here is the automatic generation of natural language expressions of concepts defined within a formal ontology. This problem has already raised interest in the semantic web community [3], and has given way to approaches allowing to precisely verbalize the set of logical restrictions and specifications which define concepts in a logical description language like OWL-DL [4]. In the medical field, work has been done towards automatic generation of case descriptions in natural language from an RDF input [5].

The present work adopts another approach. It applies a method based on the principle of graph transformation (specifically, graph grammar) to the problem of preparing data into a form suitable for natural language generation. There are reasons to think that this approach is suited to the nature of the problem, and, moreover, that it has a potential to generalization.

2 Background: Context of Use, Input Data

The minimal "visual utterances" of the graphical language are icons. Those icons actually have a well-defined internal composition, built against a standardized visual grammar. The elements which make up an icon are graphical primitives, each of which contributes to the overall meaning of the visual sign. An icon may for instance (Figure 1.a) display: a central pictogram representing a liver; embedded in a square colored in orange (conventionally meaning "risk"); with a shape modifier on the left side made of a small graphical symbol representing a virus; and another shape modifier, located in the top right corner, showing a blue square (conventionally meaning "monitoring"). As an example, such a combination of graphical primitives (central pictogram, shape, color, side modifier, superscript modifier), with their respective meanings (iconic or conventional) and their assembly rules, constitute a sign conveying the concept "viral hepatitis risk monitoring".

Hence, the primary starting point for the text generation process is the same as the one for the icon generation process: a standardized code, made up of 7 positional fields, each of which corresponds to a possible graphical primitive (cf. caption of Figure 1).

The first step of the process actually is a parsing step. It consists in projecting the 7-fields code, that essentially is the specification of a syntagm in a graphical grammar, onto an ontology of concepts. To this end, a dictionary defines a mapping between

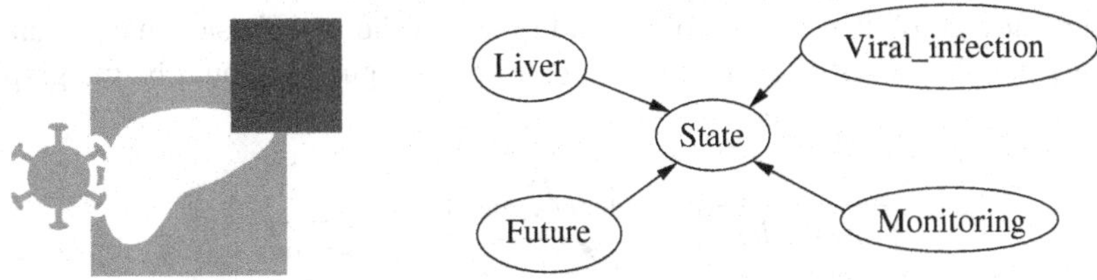

Fig. 1. a. Icon with internal code: *risk–virus–liver–monitoring–null–null–null* (Viral hepatitis risk monitoring). b. Graph giving the logical definition of the concept.

graphical elements and concepts. More precisely, every graphical element considered within the context of a type of graphical relation between part and whole (e.g. "blue square *as superscript modifier*"), maps to one (or more) ontology entries that define a specific property constraining the most generic concept of the ontology, namely that of "medical state". For instance, one of the rules specifies that when a visual sign *has as central color* the *color: orange*, then the state thus represented is linked to the *temporality: future*. Similarly, if the visual sign *has as side modifier* the *element: virus*, then the state is linked to a *viral infection*. Again, if the sign *has as central pictogram* the *pictogram: liver*, then the state is linked to the *organ: liver* or to the *function: hepatic*.

The result of the parsing is a concept of a medical state, specified by a number of restrictive properties, which may be represented in the form of a graph. For example, the icon on Figure 1.a is a visual expression of the graph in Figure 1.b. Possible ambiguities of some graphical elements (e.g. the "liver" pictogram being used at times to represent the *organ: liver*, and at other times to represent the *function: hepatic*), are removed at that stage by a reasoner that filters valid OWL concepts. For instance, in the case exhibited above, it filters out the function, since a virus may infect an organ, but not (directly) a function.

The format of the data in Figure 1.b is not fit to be fed in to an automatic text generation process. Natural language generation takes as input data something that should be close to a deep semantic representation of some natural language fragment, that is a semantic graph (we will avoid the use of the term "conceptual graph" coined by John Sowa [6], since it has a more specific formal definition; moreover, it may mislead the reader into confusing "concepts" of conceptual graphs with "concepts" of OWL-DL). As a matter of fact, the graph which represents the concept in the ontology is: (1) of a regular shape (star-shaped); (2) non-ambiguous (within the reference ontology).

On the contrary, a semantic graph should represent the semantic structure underlying a given linguistic phrase. Hence, it possesses the properties expected from that level of representation, namely, it is (1) of an irregular shape (not necessarily star-shaped, or linear); (2) made up of ambiguous, i.e. multivocal, units. The node labels in the semantic graphs are multivocal as much in their relation to the reference ontology (a "semanteme" may match more than one concept, and a concept more than one semantemes) as in their relation to the surface linguistic forms (a semanteme may be expressed by different lexical units depending on its syntactic context, e.g. "eye" or "ocular").

A semantic graph that would correspond to one of the most basic, "naive", among the many possible ways of expressing the concept of Figure 1, would be the graph in Figure 2.

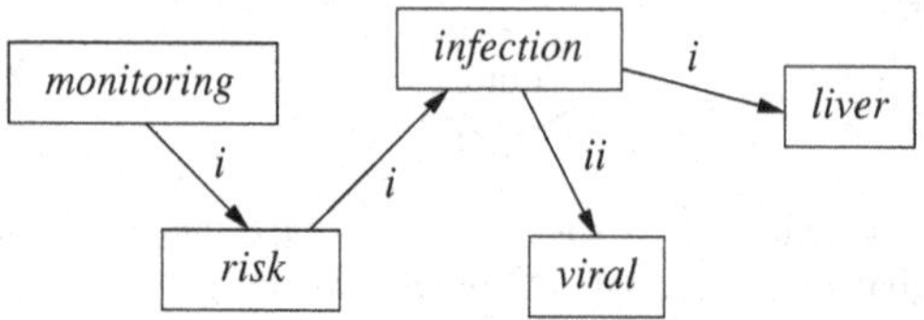

Fig. 2. Semantic graph corresponding to the English noun phrase "monitoring of the risk of viral infection of the liver". The nodes of the graph are semantemes, the edges are weakly typed semantic relations, *à la* Tesnière (the lower case roman number reflects the order of "centrality" of the actant relative to its predicate).

Of course, in this particular case, there are more elegant ways to express the concept, which at the same time are closer to the actual use by health professionals (here, "viral hepatitis risk monitoring"). But we deliberately take the naive phrase as example, because the goal of the natural language generation process described here precisely *is not* to provide the most frequent term: it is to provide a verbalization of what concept exactly is covered by the icon, with a view to help users of health information systems better understand the logic of the graphic language. The idea explored in this work is that from the same exact input code, two different functions will generate an image on one side, and a natural language string (or a sorted set of alternative natural language strings) on another.

In a medical classification like e.g. the one used by ICD-10, the base concept will be referred to by the term "viral hepatitis", not as "viral infection of the liver". But our graphical language is designed with the ambition to be able to express medical concepts by combining primary elements, not to reflect an exact mapping with a specific medical classification. Consequences of this are: that it is possible to build icons for concepts which are not relevant nor frequent; that some icons might be more specific, in what they actually express, than a most common medical term; and that other icons might be more generic (for instance, a "myocardial infarction" is represented by an icon, the exact meaning of which is "blocked blood vessel in the heart"). We view this independance from medical terminology as a feature of the language. There are many reasons for this: first, there are different medical terminologies, and our visual language must be able to be used in conjunction with any of them; second, the ontology approach, with a few discrete atomic axes, permits to express a much greater number of possible combinations of medical concepts than the terminology approach, which sets in advance a finite set of possibilities (this rationale also is behind the GRAIL language, defined within the GALEN project [7]). Third, there is no 100% agreement between different experts on exact mappings of even widespread medical concepts – one more evidence that there may be no bijective mapping between any given classification and even a subset of a graphical language.

The initial step of the text generation process hence consists in transforming the primary input data (the OWL concept, expressed as a graph, like in Figure 1.b) into a deep semantic structure (the semantic graph, like in Figure 2). So, it is a graph transformation process.

3 Method: Graph Transformation

The problem with the preparation of data lies in the fact that some specific configurations of properties of the initial medical concept are jointly expressed by set words or phrases from the human language (English in the example given here). For example, the fact that the medical state affects the liver (top left part of the graph in Figure 1.b), and that it is connected with a viral pathology (top right part), is expressed in English (among other possibilities) by the phrase "viral infection of the liver" (the present paper concentrates on the graph transformation process, and hence does not address the issue of generating multiple possible expressions, by using different roots – like 'hepat-' instead of 'liver' – or by using different linguistic mechanisms – like morphological derivation instead of syntax).

Another property of the initial concept, e.g. in this case the fact that it refers to a future possibility, is spontaneously expressed as a noun phrase headed by 'risk', and taking as a syntactic argument the already built phrase ("risk of viral infection of the liver"), as in Figure 2. This underlying graph structure is different from the one directly drawn from concept properties (Figure 1.b), which, if linearized as natural language, would rather yield some text like "There is a risk of a state. The state affects the liver. The state is related to a viral infection."

Transformations of that type are systematic. Changing parameters in the entry graph would yield structurally identical phrases like "parasitic infection of the liver" or "viral infection of the respiratory tract".

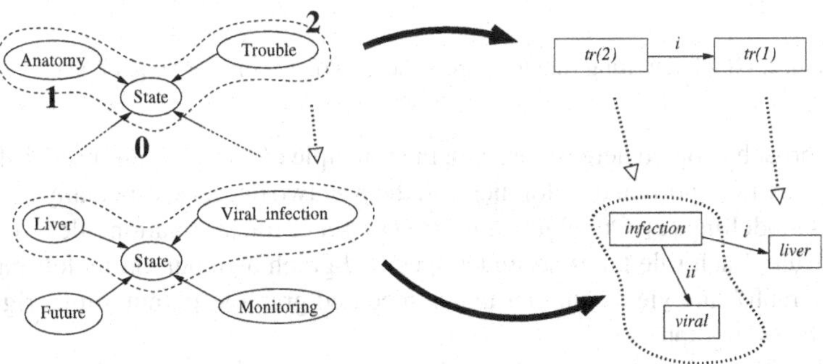

Fig. 3. Rewriting rule to express the generic pattern for "trouble of an organ". The arab numerals as function arguments on the right side of the figure refer to boldface indices on the left side.

To this end, we need graph rewriting rules allowing to specify the systematic transformation of a sub-structure of the input graph (corresponding to the pattern: medical state affecting an organ Y; medical state connected to a pathology X) into a sub-structure of the output semantic graph (corresponding to the under-specified phrase: "*<pathology*

Y> of *<organ X>*"). Such a transformation does not imply preserving either the number or the "perimeter" of the nodes in the input graph when transferring their meaning into the output graph. For instance, the node "Viral_infection" (Figure 1.b), a unique individual entity in the ontology, should be translated by two different semantemes of the English language: "viral" and "infection" (Figure 3).

Similarly, we need a rule able to express the systematic transformation of the subgraph in the input graph expressing the property "Future" into a a subgraph in the output graph corresponding to the lexical unit "risk of . . . " (Figure 4).

There is a difference between the examples in Figure 3 and Figure 4: in the first case, the output semantic subgraph is complete, or saturated (it could be an output graph in itself); whereas in the second case, the output subgraph is awaiting completion by being grafted to another, saturated, semantic graph. The node marked with a star in the right side of Figure 4 may be called *substitution node* (in analogy with the technical term used in the frame of the Tree-Adjoining Grammars to refer to a comparable operation on phrase-structure trees): it is a non-instanciated node that has to be *substituted* for by another graph, given as a function argument.

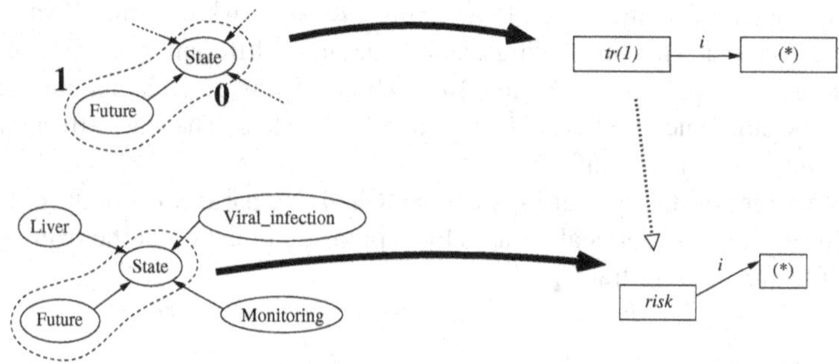

Fig. 4. Graph rewriting rule to express the generic pattern for "risk of a state"

The approach adopted here is based on the principle of "graph grammars" [8], which has given way to important developments in the past two decades, especially in the field of process modelling [9]. Our algorithm defines a set of transformation rules – or graph-rewriting rule. Each rule takes an under-specified graph as input, on its left-hand side (*filter* subgraph), and yields another under-specified graph as output, on its right-hand side (*product* subgraph).

Our algorithm may be classified in the category of *graph grammars* proper, not simply *graph-rewriting systems* (following the distinction drawn by Blostein [10]), because it makes a difference between terminal graphs and non-terminal graphs, analogous to the similar difference that phrase-structure grammars (PSG) make between terminal strings (made up of terminal symbols only) and non-terminal strings in a linear language.

In the present case, a terminal graph is a graph that contains only nodes of the type semanteme, and has no more node of the type concept. Semantemes and concepts belong to two different XML/RDF namespaces.

4 Method: Implementation

The generic rewriting system is implemented as a module in the *python* programming language. It relies on four specific mechanisms: (1) an operation of unification of graph topological structures, along with unification of node and edge labels; (2) a translation function, mapping the set of input node labels onto the set of output node labels; (3) a co-indexing mechanism to manage glueing the incident edges (left loose after removing a node of the input graph) to a node in the rewritten graph; (4) a substitution mechanism, defined at unsaturated nodes, to manage glueing the neighboring (saturated) nodes to edges pertaining to the rewritten graph.

(1) The detection of matching sites for a filter graph (left-hand side graph of a rewriting rule) implies: (a) detecting an *isomorphism* between part of the complete input graph and the filter graph, and (b) identifying *subtype-to-supertype* ("is a") relations between (more specific) node labels in the input graph and (more generic) node labels in the filter graph. Such "is a" relations depend on the concept type hierarchy defined within the graphical language ontology. They allow e.g. to recognize that the subgraph circled by a dotted line, in the bottom right corner of Figure 5, is a specific instance of the generic filter graph displayer in the top right part of the same figure (by making sure that a "viral infection" is a sort of "infection", which is a sort of "trouble"; and that the "liver" is a sort of "organ of the digestive system", which is a sort of "anatomy").

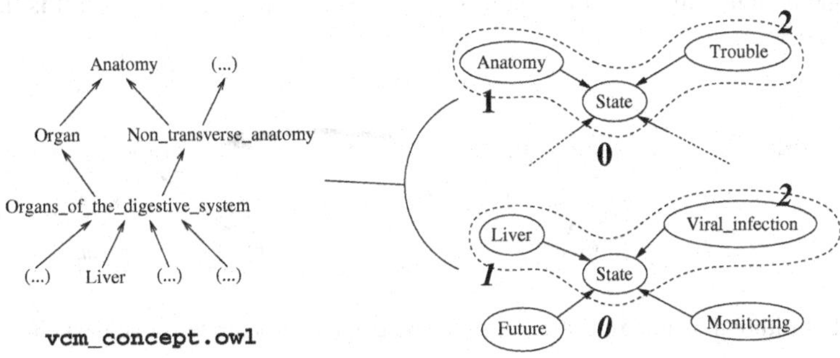

Fig. 5. Detecting subtype-to-supertype relations when matching a filter graph

(2) The generation of the product subgraph, when a rule is being applied to a matching site, relies on a *translation* function (noted *tr* in Figures 3 and 4), which maps every element in the OWL concept ontology onto a small semantic graph (generally, but not necessarily, made up of one single semanteme node). In fact, since the concepts of the filter graph are under-specified, it is not possible to specify in advance, for every rule, the exact type of the nodes in the product subgraph.

(3) When the filter subgraph of a rule finds a matching site on a bigger input graph, the result of the rewriting operation is a new graph where the subgraph found at the matching site is replaced by the product subgraph of the rule. The "glueing" of that product subgraph with the remaining parts of the input graph relies on a *co-indexing* mechanism between product graph and filter graph. Co-indices are attributes present on

both filter-side nodes and product-side nodes, that get numeric values; when a filter-side node and a product-side node share the same co-index, it means that they should match the same node in the input graph. The actual integer number used as value for a co-index in the definition of a rule may be arbitrary: its only purpose is to be shared by the left-hand side and the right-hand side. If there are more than one co-index, different integer values mean that the relevant nodes should match distinct nodes in the input graph. Hence, co-indices allow to spot the nodes in the input graph where loose incident relations of the product subgraph have to be "glued".

(4) Some product subgraphs are made up of a set of fully determined semanteme-nodes, that express all the concepts of the input graph which were captured when matching the filter subgraph (Figure 3). Other, oppositely, have a loose edge – to put it another way, they include an edge between a node which is already fully determined in the product subgraph, and a node which has to be determined somewhere else (Figure 4). Such product subgraphs contain a *substitution* node. After the application of the rule, the substitution node must be unified with a saturated node from the remaining of the graph, to build the whole rewritten graph (Figure 6). Substitution is compulsory.

Remark: Points (3) and (4) actually are implemented by the same underlying computer function operating on graphs, and taking two arguments: the "graft" and the "trunk". This function attempts to find co-indexed nodes on both sides sharing the same value, and it "glues" the two graphs on those nodes. For every such "co-indexed site", one of the sides must be filled and the other side blank. (3) is implemented when the trunk node is blank and the graft node is filled; (4) is implemented when it is the other way around.

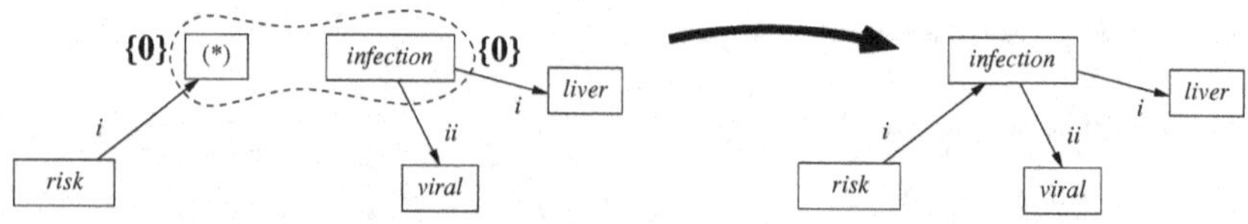

Fig. 6. Substitution of an unspecified node in one graph by a saturated node in another graph

The algorithm works by iterated rewritings:

At step 0, the set of rewritings R_0 is initialized to G; G being the original input graph, representing the OWL concept.

At every step $n+1$, R_{n+1} is augmented with the set of rewritings yielded by applying matching rules to elements of R_n, when those results do not already belong to R_n:

$$R_{n+1} \leftarrow R_n \cup \{R_k(g) \mid g \in R_n\},$$

where $R_k(g)$ denotes the result of rewriting a graph g (present in R_n) by one of the applicable rewriting rules, k.

When the set R_n ceases to grow between two iterations, the loop is exited, and R_n is filtered so that only the "terminal" graphs are kept (the graphs where all the nodes are semantemes, and no more concepts).

In our system, the generic processing mechanisms are separated from the description of specific rewriting rules, like it is common practice in the field of formal grammars (it is an instance of the more general principle that data should be treated separately from processes). The former are implemented by functions in the *python* programming language, taking graph-rewriting rule identifiers as en input parameter; the latter are stored in XML documents following an *ad hoc* document schema.

5 Conclusion and Perspective

The next step in the present work is the development of a complete text generation module, based on the generation of phrase structure trees by derivation of elementary trees in a TAG lexicalized grammar [11].

The graphical language is built on minimal segments of expression called *icons*, a description of which has been given above (Section 2). Those icons may be combined together, following a constrained visual syntax, to compose more complex iconic utterances: on bidimensional surfaces, structured in predefined fields, they form synthetic visualization grids displaying information about the complete set of contraindications or side effects of a drug, or the clinical condition of a patient.

A future extension of the natural language generation work will be taking into account that visual syntax, to be able to translate complex graphical utterances in texts in the chosen target natural language. It is envisioned that future developments shall include other output languages, so that the visual language approach actually allows embedding in multi-lingual systems for displaying medical information.

We believe that the method presented here has a potential for generalization. It can be used in other cases where generation of natural language equivalents of OWL concepts may be desirable as a tool to help ontology users; and, more generally, when the pre-linguistic input for natural language generation is expressed in a knowledge representation formalism translatable in the form of graphs. This might be of use in other application fields, like automatic explanation generation in health information systems, or help in decision making.

References

1. Lamy, J.-B., Duclos, C., Bar-Hen, A., Ouvrard, P., Venot, A.: An iconic language for the graphical representation of medical concepts. BMC Med. Inform. Decis. Mak. 8 (16) (2008)
2. Welty, C., McGuinness, D.L., Smith, M.K.: OWL web ontology language guide. W3C Recommandation. W3C (2004), `http://www.w3.org/TR/owl-guide/`
3. Wilcock, G.: Talking OWLs: Towards an ontology verbalizer. In: Proc. ISWC Workshop on Human Language Technology for the Semantic Web and Web Services, pp. 109–112 (2003)
4. Hewlett, D., Kalyanpur, A., Kolovski, V., Halaschek-Wiener, C.: Effective NL paraphrasing of ontologies on the semantic web. In: Proc. ISWC Workshop on End User Semantic Web Interaction, vol. 172. CEUR-WS.org (2005)
5. Bontcheva, K., Wilks, Y.: Automatic report generation from ontologies: the MIAKT approach. In: Meziane, F., Métais, E. (eds.) NLDB 2004. LNCS, vol. 3136, pp. 324–335. Springer, Heidelberg (2004)

6. Sowa, J.: Conceptual structures: information processing in mind and machine. Addison Wesley, New York (1984)
7. Rector, A.L., Bechhofer, S., Goble, C.A., Horrocks, I., Nowlan, W.A., Solomon, W.D.: The GRAIL concept modelling language for medical terminology. Artif. Intell. Med. 9(2), 139–171 (1997)
8. Ehrig, H., Habel, A., Kreowski, H.J.: Introduction to Graph Grammars with Applications to Semantic Networks. Comput. Math. Appl. 23(6-9), 557–572 (1992)
9. Schürr, A., Winter, A.J., Zündorf, A.: Graph grammar engineering with PROGRES. In: Botella, P., Schäfer, W. (eds.) ESEC 1995. LNCS, vol. 989, pp. 219–234. Springer, Heidelberg (1995)
10. Blostein, D., Fahmy, H., Grbavec, A.: Issues in the practical use of graph rewriting. In: Cuny, J., Engels, G., Ehrig, H., Rozenberg, G. (eds.) Graph Grammars 1994. LNCS, vol. 1073, pp. 38–55. Springer, Heidelberg (1996)
11. Schabes, Y., Abeillé, A., Joshi, A.K.: Parsing strategies with 'lexicalized' grammars: application to Tree Adjoining Grammars. In: COLING 1988: Proc. 12th International Conference on Computational Linguistics, Budapest, August 22-27, pp. 578–583 (1988)

Speech

"**Speech**: the expression of or the ability to express thoughts and feelings by articulate sounds: *he was born deaf and without the power of speech.*"

NODE (The New Oxford Dictionary of English), Oxford, OUP, 1998, page 1788, meaning 1.

GMM Classification of Text-to-Speech Synthesis: Identification of Original Speaker's Voice*

Jiří Přibil[1,2], Anna Přibilová[3], and Jindřich Matoušek[1]

[1] University of West Bohemia, Faculty of Applied Sciences, Dept. of Cybernetics, Univerzitní 8, 306 14 Plzeň, Czech Republic
jmatouse@kky.zcu.cz

[2] SAS, Institute of Measurement Science, Dúbravská cesta 9, SK-841 04 Bratislava, Slovakia
Jiri.Pribil@savba.sk

[3] Slovak University of Technology, Faculty of Electrical Engineering & Information Technology, Institute of Electronics and Photonics, Ilkovičova 3, SK-812 19 Bratislava, Slovakia
Anna.Pribilova@stuba.sk

Abstract. This paper describes two experiments. The first one deals with evaluation of synthetic speech quality by reverse identification of original speakers whose voices had been used for several Czech text-to-speech (TTS) systems. The second experiment was aimed at evaluation of the influence of voice transformation on the original speaker recognition. The paper further describes an analysis of the influence of initial settings for creation and training of the Gaussian mixture models (GMM), and the influence of different types of used speech features (spectral and/or supra-segmental) on correctness of GMM identification. The stability of the identification process with respect to the duration of the tested sentence (number of the processed frames) was analysed, too.

Keywords: quality of synthetic speech, text-to-speech system, GMM classification, statistical analysis.

1 Introduction

The text-to-speech system (TTS) usually represents the output part of the whole voice communication system with a human-machine interface. The quality, and first of all, the intelligibility of the produced synthetic speech is a basic condition for its usability. Furthermore, it enables setting of a suitable strategy for the dialogue management. Higher quality and naturalness of synthetic speech can be achieved by various methods of speech synthesis, structures of TTS systems, used types of speech inventories, approaches to prosody generation, etc. Several subjective and objective methods are used to verify the quality of produced synthetic speech [1]. The most often used subjective method for giving the feedback information about users' opinion is the listening test. On the other hand, the objective method based on automatic speech recognition system yielding the final evaluation in the form of a recognition score can

* The work has been supported by the Technology Agency of the Czech Republic, project No. TA01011264, the Grant Agency of the Slovak Academy of Sciences (VEGA 2/0013/14), and the Ministry of Education of the Slovak Republic (KEGA 022STU-4/2014).

P. Sojka et al. (Eds.): TSD 2014, LNAI 8655, pp. 365–373, 2014.

be used [2]. These recognition systems are often based on neural networks [3], hidden Markov models [4], [5], or Gaussian mixture models (GMM) [6]. The main advantage of these statistical evaluation methods is that they work automatically without human interaction and the obtained results can be numerically judged.

We investigate whether the quality of synthetic speech produced by a TTS system can be evaluated by a reverse identification of the original speaker and whether the re-identification score depends on the used method of speech modelling and synthetic speech production. To verify this hypothesis, the one-level GMM recognizer for identification of the original male and female speakers from the synthetic speech produced by various Czech TTS systems was developed.

Motivation of this work was to analyse further the influence of initial settings in the GMM creation and training phases (number of used mixture components) and different types of used speech features (spectral and/or supra-segmental) on correctness of GMM identification. The GMMs are created and trained on the original speech of the male and female Czech speakers and tested on the speech produced by the Czech TTS systems with several speech synthesis methods. In addition, the stability of the identification process with respect to the duration of the tested sentence (number of the processed frames) is analysed in the paper.

2 Method

The Gaussian mixture models can be defined as a linear combination of multiple Gaussian probability distribution functions (GPDFs) of the input data vector [6]

$$f(x) = \sum_{k=1}^{N_{gmix}} \alpha_k P_k(x), \tag{1}$$

where $P_k(x)$ is the GPDF, α_k is a weighting parameter, and N_{gmix} is the number of these functions. For GMM creation, it is necessary to determine the covariance matrix, the vector of mean values, and the weighting parameters from the input training data. Using the expectation-maximization (EM) iteration algorithm, the maximum likelihood function of GMM is found [6]. The performance of the EM algorithm is controlled by the N_{gmix} parameter representing the number of applied mixtures of GPDFs in each of the GMM models. In standard use of the GMM classifier, the resulting score of the model is given by the maximum overall probability for the given class

$$i^* = \arg\max_{1 \leq n \leq N} score(T, n), \tag{2}$$

where the $score(T, n)$ represents the probability value of the GMM classifier for the models trained for the current n-th class in the evaluation process, and T is the input vector of the features obtained from the tested sentence.

For our purpose we need to quantify and compare differences between probability values of the obtained scores; therefore, these values are normalized and the additional

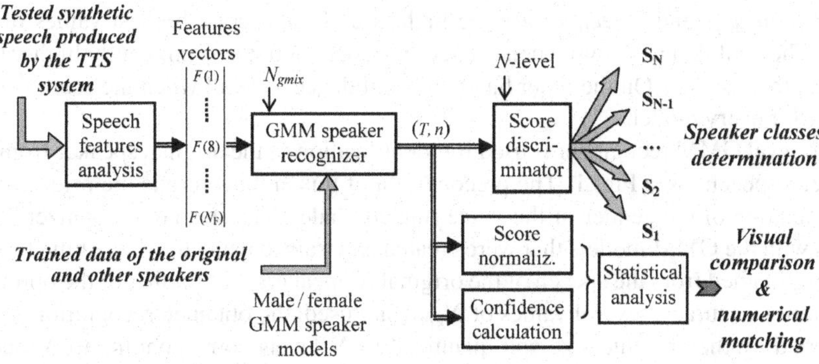

Fig. 1. Block diagram of the GMM recognizer for identification of the original speaker from the synthetic speech produced by the TTS system

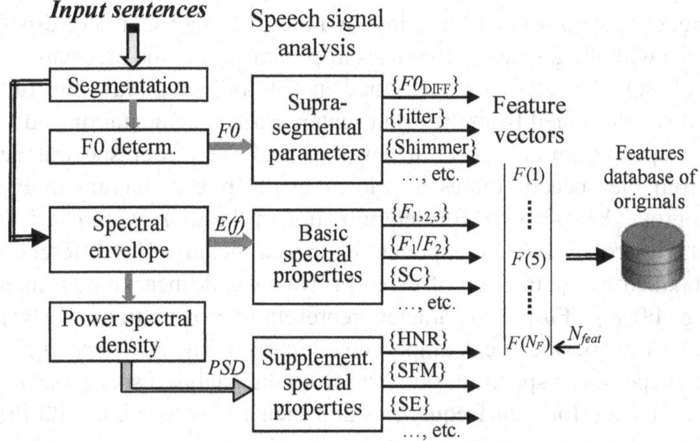

Fig. 2. Block diagram of the feature database creation from the spectral properties and supra-segmental parameters of the original speech

Table 1. Basic specification of tested synthetic speech produced by TTS systems

Type	Synthesis method	TTS name (specification)	Voice
$TTS1_{M/F}$	cepstral/harmonic	PC VOX (diphone speech inventory)	Kubec / Ellen
$TTS2_{M/F}$	PSOLA	Epos (triphone speech inventory)	Machac / Violka
$TTS3_{M/F}$	unit selection	ARTIC	Jan / Radka

parameters are calculated for a subsequent statistical analysis. The next evaluated parameter is based on the maximum confidence used for selection of features. The confidence measure (CM) gives information how distinctive is the assessment of the given classifier [7]

$$\mathrm{CM} = 1 - \frac{score_{\max 2}}{score_{\max 1}}, \tag{3}$$

where $score_{\max 1}$ and $score_{\max 2}$ are the highest and the second highest values of the score. The confidence is high when the score for one model is significantly higher than for the other models. On the other hand, the confidence is small when the score is very similar for every model.

One-level GMM recognizer is used for identification of the original speaker from the synthetic speech—see Fig. 1. The precondition of this architecture is the prior correct determination of the gender of the voice (male/female). The speaker recognizer block works with the GMM models that were created and trained using the data of the feature vectors obtained from the speech of the original N speakers. For finding of the optimum recognition accuracy, several values of N_{gmix} are used, the obtained recognition scores are sorted by the absolute size and quantized to N levels corresponding to N output classes in the score discriminator block. In the classification phase, we obtain the scores using the input feature vectors from the tested sentences synthesized by various TTS systems. It means that the highest obtained score represents the synthesized sentences with the values of the speech features that are most similar to those obtained from the original sentences used for GMM training; and the minimum score corresponds to the tested sentence with the greatest differences in comparison to the originals.

The speech signal analysis is performed in the following way: the fundamental frequency F0 is determined from the input sentence after segmentation and weighting. In the next step, the smooth spectral envelope and the power spectral density are computed from the speech frames as shown in the block diagram in Fig. 2. The virtual F0 contour (VF0) is used for determination of the *supra-segmental parameters* describing the microintonation component of speech melody. The differential contour $F0_{\mathrm{DIFF}}$ is obtained by subtraction of mean F0 values and linear trends (including the zero crossings $F0_{\mathrm{ZCR}}$). Further parameters represent microvariations of F0 (jitter) and the variability of the peak-to-peak amplitude (shimmer). The *basic spectral properties* describe the shape of the spectrum obtained from the analysed speech segment. They include the first three formant frequencies and their ratios together with the spectral centroid (SC) and the spectral decrease (tilt). The *supplementary spectral features* are determined from the smoothed magnitude or power spectrum envelope: the spectral flatness measure (SFM), the spectral entropy (SE), and the harmonics-to-noise ratio (HNR) providing an indication of the overall periodicity of the speech signal. Obtained values in the form of the feature vectors with the length N_{feat} are subsequently stored in a database containing the features of the original speakers in dependence on the voice type (male/female) for further processing.

3 Material, Experiments, and Results

The speech material for GMM creation and training consists of the short sentences with duration from 0.5 to 2.5 seconds, resampled at 16 kHz, representing the original speech in Czech language uttered by five male and five female speakers (already used in another research [8])—typically 50 sentences per speaker. We have also additional sentences originated from the speaker whose voice was used for building of the speech corpus (male/female) of the tested TTS systems with basic parameters given in Table 1. So we finally have the speech material consisting of 6+6 speakers for testing in each of our identification experiments (designated as Orig1-$6_{M/F}$) where the voice number indicates the original speech material for the tested TTS system ($Orig1_M$ is the source speech material for synthesis of the voice $TTS1_M$ and so on). As regards the synthetic speech (TTS1-$3_{M/F}$), the database consists of testing sets including 25 short sentences produced by the TTS systems using different types of speech modelling (cepstral [9], harmonic [10], PSOLA [11], unit selection [12], [13]).

The main experiment was focused on identification of the original speaker from the synthetic speech produced by the Czech TTS systems. The second experiment was aimed at evaluation of the influence of voice transformation on the original speaker recognition. The speech material used here was the synthetic speech produced by the TTS system PCVOX—implemented in the special aids for blind and partially sighted people [14], [15]. Four synthetic voices were compared: the basic male voice (synthesis from the original speaker $TTS1_M$—see Table 1) and the transformed voices of a young male (Tr-young), a female (Tr-female) and a child (Tr-child) [8]. In addition, our research was aimed at investigation of:

- influence of the number of used mixtures (from 2 to 8) on GMM evaluation,
- influence of the used feature set (P1-P3),
- stability of the identification process depending on the tested sentence duration.

The input data vector for GMM training and classification contains the supra-segmental parameters {VF0, $F0_{DIFF}$, $F0_{ZCR}$, jitter, and shimmer}, the basic spectral features determined from the spectral envelopes {$F_{1,2,3}$, F_1/F_2, SC, and tilt}, and the supplementary spectral parameters {HNR, SFM, SE}. In the case of the spectral features, the basic statistical parameters—mean values and standard deviations (std)—were used as the representative values in the feature vectors for GMM evaluation. For implementation of the supra-segmental parameters of speech, the statistical types—median values, range of values, std, and/or relative maximum and minimum were used in the feature vectors. The length of the input feature vector $N_{feat} = 16$ was experimentally chosen in correspondence with the obtained results of our previous research [16]. The three tested feature sets were: P1 consisting of the basic spectral features together with the supra-segmental parameters, P2 consisting of the supplementary spectral features and the supra-segmental parameters, and P3 being a mix of the basic and the supplementary spectral features with the prosodic parameters.

As regards the GMM classifier, the simple diagonal covariance matrix of mixture models was applied in this identification experiment. The basic functions from the Ian T. Nabney "Netlab" pattern analysis toolbox [17] were used for creation of the GMM models, data training, and classification.

The obtained results of the GMM identification are presented in a graphical form (for visual comparison) and also as the values for numerical matching separately with respect to the TTS voice gender. The used order of tables and figures corresponds to the course and evaluation of the performed experiments. If not otherwise stated, the presented graphs and tables were determined with the following parameter setting: $N_{gmix} = 5$, feature set P2.

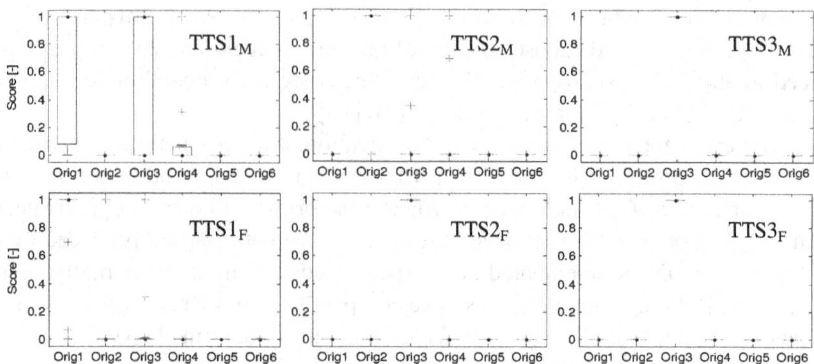

Fig. 3. The boxplot of the basic statistical parameters of the normalised GMM score: for male (upper set) and female (bottom set) TTS voices

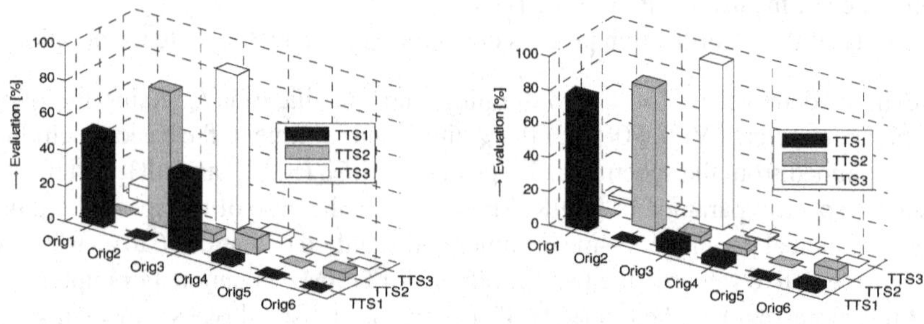

Fig. 4. Confusion matrices of original speaker identification for male (left) and female (right) TTS voices of six originals, three TTS synthesis systems

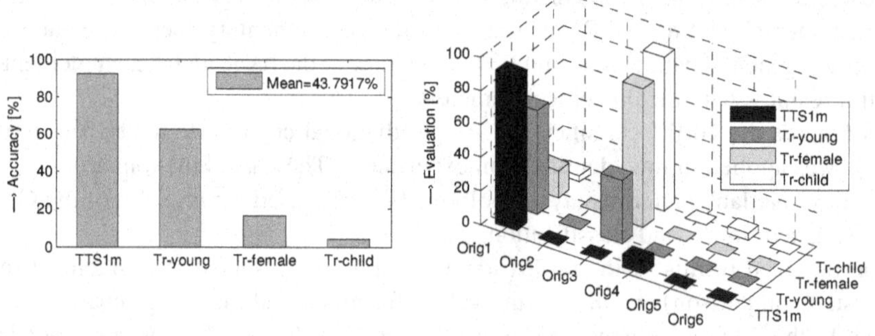

Fig. 5. Results of the second identification experiment of TTS synthesis with the male voice and its conversion to young male, female, and childish voice: a bar graph of the mean recognition accuracy (left), a detailed confusion matrix (right)

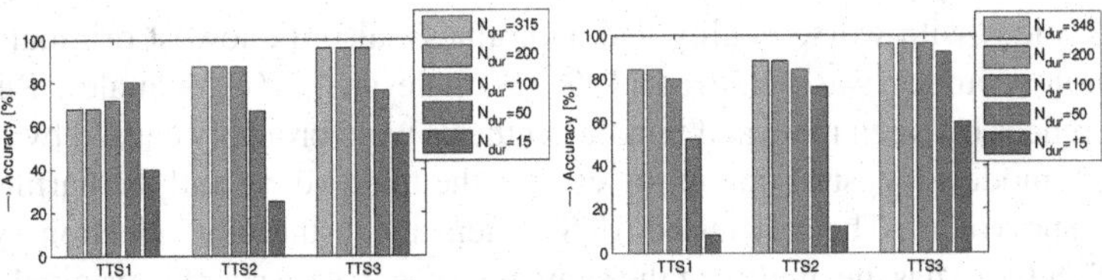

Fig. 6. Influence of the tested sentence lengths Ndur in [frames] on the original speaker GMM identification accuracy: for male (left) and female (right) TTS voices

Table 2. Basic statistical parameters of the CM values calculated from the GMM score for the male and female TTS voices

TTS voice	Min*)	Mean	Std
$TTS1_{M/F}$	0.311 / 0.807	0.937 / 0.948	0.1522 / 0.0455
$TTS2_{M/F}$	0.647 / 0.943	0.958 / 0.996	0.0985 / 0.0127
$TTS3_{M/F}$	0.996 / 0.980	0.999 / 0.999	0.0040 / 0.0007

*) Maximum is equal to 1 in all cases.

Table 3. Mean original speaker GMM identification accuracy in [%] in dependence on the number of used mixtures

TTS voice	$N_{gmix}=2$	$N_{gmix}=3$	$N_{gmix}=4$	$N_{gmix}=5$	$N_{gmix}=6$	$N_{gmix}=7$	$N_{gmix}=8$
$TTS1_{M/F}$	42/70	42/72	50/76	52/78	54/80	50/82	54/80
$TTS2_{M/F}$	72/84	72/86	72/88	76/84	72/86	75/92	72/89
$TTS3_{M/F}$	87/96	85/96	88 /97	88/98	88/99	89/99	88/99

Table 4. Summary results of original speaker GMM identification accuracy in [%] for different types of used feature vectors

TTS voice/ feature vector type	P1	P2	P3
$TTS1_{M/F}$	48.0/76.3	52.0/78.2	68.0/82.5
$TTS2_{M/F}$	75.0/84.0	75.5/84.1	87.5/95.5
$TTS3_{M/F}$	100/92.8	87.5/98.2	100/100

4 Discussion and Conclusion

The performed experiments have shown that there exists a principal influence of a chosen type of parameters in a feature vector on the stability and accuracy of the GMM identification. The best results are produced by the feature set P3 consisting of a mix of spectral and prosodic features, the worst results correspond to the set P2 (see results in Table 4) when only the supplementary spectral and supra-segmental features were used. Therefore, the detailed comparison for this worst case was performed next. For the original speaker with the worst identification from the synthetic speech the confidence measure has the lowest value of the minimum and the highest value of the standard deviation—see boxplots of the basic statistical parameters of the normalised GMM score in Fig. 3 and CM values in Table 2. In the case of the TTS system PCVOX producing voices with worse quality and naturalness, also the lowest original speaker identification accuracy was achieved (49% for male and 77% for male). Relatively great differences between male and female synthesis were probably caused by different used speech models (cepstral one in the case of the male voice, and the harmonic one for the female voice). The last tested TTS system using the unit selection synthesis method ($TTS3_{M/F}$) has the quality of the synthetic speech very near the original voice as shown by the best results of identification accuracy (96% for male and 95% for female voices)—see the corresponding 2D representation of confusion matrices in Fig. 4. From the second recognition experiment follows that the obtained score values correspond to the degree of the voices transformation: the highest score corresponds to the basic male voice and the lowest one to the transformation to the childish voice (see the bar-graph and the 2D confusion matrix in Fig. 5). Contrary to our expectations, the number of used mixtures has not great significance (see values in Table 3), so the setting of $N_{gmix} = 5$ was chosen for next processing and comparison. Finally, it can be said that the results obtained in this way are in good correspondence with the predicted working hypothesis. The last part of our experiment showed that the length limitation of the processed speech signal practically does not play essential role (see Fig. 6) because our GMM original speaker identifier was developed for testing of continuous speech (i.e. sentences—not isolated words).

Increase of the original speaker identification accuracy can be expected if the full covariance matrix is used for GMM model creation, training, and employment in the classification process, so in near future we will compare approaches using the diagonal and the full covariance matrices.

References

1. Blauert, J., Jekosch, U.: A Layer Model of Sound Quality. Journal of the Audio Engineering Society 60, 4–12 (2012)
2. Kondo, K.: Subjective Quality Measurement of Speech: Its Evaluation, Estimation and Applications. Springer (2012)
3. Zelinka, J., Trmal, J., Müller, L.: On Context-Dependent Neural Networks and Speaker Adaptation. In: Proc. IEEE Conf. Signal Processing 2012, Beijing, China, pp. 515–518 (2012)
4. Pražák, A., Psutka, J.V., Psutka, J., Loose, Z.: Towards Live Subtitling of TV Ice-Hockey Commentary. In: Proc. SIGMAP 2013, Reykjavík, Iceland, pp. 151–155 (2013)
5. Jeong, Y.: Joint Speaker and Environment Adaptation Using TensorVoice for Robust Speech Recognition. Speech Communication 58, 1–10 (2014)
6. Reynolds, D.A., Rose, R.C.: Robust Text-Independent Speaker Identification Using Gaussian Mixture Speaker Models. IEEE Trans. on Speech and Audio Processing 3, 72–83 (1995)
7. Vondra, M., Vích, R.: Evaluation of Speech Emotion Classification Based on GMM and Data Fusion. In: Esposito, A., Vích, R. (eds.) Cross-Modal Analysis. LNCS (LNAI), vol. 5641, pp. 98–105. Springer, Heidelberg (2009)
8. Přibilová, A., Přibil, J.: Non-Linear Frequency Scale Mapping for Voice Conversion in Text-to-Speech System with Cepstral Description. Speech Commun 48(12), 1691–1703 (2006)
9. Vích, R., Přibil, J., Smékal, Z.: New Cepstral Zero-Pole Vocal Tract Models for TTS Synthesis. In: Proc. IEEE Region 8 EUROCON 2001, vol. 2, pp. 458–462 (2001)

10. Přibilová, A., Přibil, J.: Harmonic Model for Female Voice Emotional Synthesis. In: Fierrez, J., Ortega-Garcia, J., Esposito, A., Drygajlo, A., Faundez-Zanuy, M. (eds.) BioID MultiComm2009. LNCS, vol. 5707, pp. 41–48. Springer, Heidelberg (2009)
11. Horák, P.: Czech Pitch Contour Modeling Using Linear Prediction. In: Sojka, P., Horák, A., Kopeček, I., Pala, K. (eds.) TSD 2008. LNCS (LNAI), vol. 5246, pp. 333–339. Springer, Heidelberg (2008)
12. Tihelka, D., Kala, J., Matoušek, J.: Enhancements of Viterbi Search for Fast Unit Selection Synthesis. In: Proc. INTERSPEECH 2010, Makuhari, Japan, pp. 174–177 (2010)
13. Romportl, J., Matoušek, J.: Formal Prosodic Structures and Their Application in NLP. In: Matoušek, V., Mautner, P., Pavelka, T. (eds.) TSD 2005. LNCS (LNAI), vol. 3658, pp. 371–378. Springer, Heidelberg (2005)
14. Přibil, J., Přibilová, A.: Czech TTS Engine for BraillePen Device Based on Pocket PC Platform. In: Proc. Conf. Electronic Speech Signal Processing (ESSP 2005), pp. 402–408 (2005)
15. Personal Computer Voices: PCVOX. Spektra v.d.n., `http://www.pcvox.cz/pcvox/pcvox-index.html` (accessed February 5, 2014)
16. Přibil, J., Přibilová, A.: Evaluation of Influence of Spectral and Prosodic Features on GMM Classification of Czech and Slovak Emotional Speech. EURASIP Journal on Audio, Speech, and Music Processing 2013(8), 1–22 (2013)
17. Nabney, I.T.: Netlab Pattern Analysis Toolbox, `http://www.mathworks.com/matlabcentral/fileexchange/2654-netlab` (retrieved October 2, 2013)

Phonation and Articulation Analysis of Spanish Vowels for Automatic Detection of Parkinson's Disease

Juan Rafael Orozco-Arroyave[1,2], Elkyn Alexander Belalcázar-Bolaños[1], Julián David Arias-Londoño[1], Jesús Francisco Vargas-Bonilla[1], Tino Haderlein[2], and Elmar Nöth[2,3]

[1] Universidad de Antioquia, Medellín, Colombia
[2] Friedrich-Alexander-Universität, Erlangen-Nürnberg, Germany
[3] King Abdulaziz University, Jeddah, Saudi Arabia

Abstract. Parkinson's disease (PD) is a chronic neurodegenerative disorder of the nervous central system and it can affect the communication skills of the patients. There is an interest in the research community to develop computer aided tools for the analysis of the speech of people with PD for detection and monitoring.

In this paper, three new acoustic measures for the simultaneous analysis of the phonation and articulation of patients with PD are presented. These new measures along with other classical articulation and perturbation features are objectively evaluated with a discriminant criterion. According to the results, the speech of people with PD can be detected with an accuracy of 81% when phonation and articulation features are combined.

Keywords: Parkinson's disease, phonation, articulation, acoustics, nonlinear dynamics.

1 Introduction

Parkinson's disease (PD) is the second most prevalent neurodegenerative disorder after the Alzheimer's disease [1] and about 1% of the people older than 65 suffer from this disease. About 90% of people with PD have disordered speech and such disorders are associated to motor impairments such as rigidity, bradykinesia, hypokinesia and tremor. Perceptually, speech and voice of people with PD are characterized by reduced loudness, monopitch, monoloudness, reduced stress, breathy, hoarse voice quality, imprecise articulation, among others [2]. All these symptoms are grouped and called hypokinetic dysarthria [3].

Voice problems are typically one of the first symptoms of PD, and while the disease is progressing, other speech problems appear affecting different speech characteristics, such as prosody, articulation, and fluency [2]. There are works focused on the automatic classification of speech of people with PD and people without any speech disorder or neurological disease, also called healthy controls (HC) [4,5,6]; however, the real contribution of the features considered on those works remains unclear.

In order to provide an informative feedback during the speech therapy of people with PD, there exists the need for analyzing the speech of people with PD considering

P. Sojka et al. (Eds.): TSD 2014, LNAI 8655, pp. 374–381, 2014.

different characteristics. The Royal College of Physicians in the United Kingdom states that it is required to tackle the speech therapy of the patients considering respiratory exercises whose direct impact is in the loudness and phonation [7]. In [4] the authors analyze different acoustic measures to detect disordered speech in people with PD. The analysis includes different versions of jitter and shimmer along with several noise measures. A correlation analysis to decide which feature may be included in the classification stage is performed; however, there is not presented an analysis of the discriminant capability of each feature. Therefore, it is not clear which of the considered measures is more relevant for the classification process.

Recent works claim that it is required to consider phonation, articulation and prosody to fulfill a complete analysis of speech in patients with PD. In [8] the authors present an acoustic analysis of speech of people with PD considering phonation, articulation, and prosody. Phonation is evaluated with sustained phonation of vowels in order to have the vibration of the vocal folds for a long period of time. Articulation analysis is considered in two tasks: with a diadochokinetic (DDK) exercise which consists in the rapid repetition of the syllables /pa/-/ta/-/ka/, and using sustained phonation of vowels for the analysis of their first ($F1$) and the second ($F2$) spectral formant, respectively. In [5], the authors use acoustic and prosodic features along with features derived from a two-mass model of the vocal folds to evaluate the speech of 88 patients with PD and 88 HC. The accuracy rate reported in the paper is 90.5% when the prosodic features are considered. Articulation capability of people with PD is evaluated in [9], the authors calculate the vowel formants $F1$ and $F2$ and show that triangular vowel space area (tVSA) is lower in speech of people with PD than in HC.

This paper presents an analysis of several perturbation and articulation features of speech of people with PD and HC. The discriminant capability of the features is evaluated in different experiments considering a total of 300 recordings of the five Spanish vowels, 150 uttered by patients with PD and 150 by HC.

The paper is organized as follows: Section 2 includes details of the speech corpus, the characterization process and the experiments carried out. In Section 3, the results and the discussion are presented. Finally, the conclusions derived from this work are included in Section 4.

2 Experimental Setup

2.1 Corpus of Speakers

The speech corpus contains three repetitions of the five Spanish vowels uttered by 50 patients with PD and 50 HC (150 recordings per vowel on each group, PD and HC). The recordings were sampled at 44.1 KHz with a resolution of 16 bits. All of the speakers are balanced by gender and age, i.e. the age of the 25 male patients ranges from 33 to 77 years (mean 62.2 ± 11.2), and the age of the 25 female patients ranges from 44 to 75 years (mean 60.1 ± 7.8). For the case of HC, the age of the 25 men ranges from 31 to 86 years (mean 61.2 ± 11.3) and the age of the 25 women ranges from 43 to 76 years (mean 60.7 ± 7.7). All of the patients were diagnosed by neurologist experts; the values of their evaluation according to the UPDRS-III [10] and Hoehn & Yahr [11] scales are 36.7 ± 18.7 and 2.3 ± 0.8, respectively. None of the people in the HC group has a history of symptoms related to Parkinson's disease or any other kind of neurological disorder. Further details of this database can be found in [12].

2.2 Characterization

Acoustic Measures: Articulation. The acoustic feature set includes measures that are calculated in the F1-F2 plane including a new set of features that are introduced in this paper. The classical acoustic features considered here are: The first two formants of each Spanish vowel calculated using a standard Linear Predictive Coding (LPC) filter with 12 coefficients: ($F1_a$, $F2_a$, $F1_e$, $F2_e$, etc.), the pitch value of each vowel: ($F0_a$, $F0_e$, $F0_i$, $F0_o$, $F0_u$), the vowel articulation index (VAI) [9] and the triangular vowel space area ($tVSA$) [9]. $F0$ value and its variability in time and amplitude measured through jitter and shimmer respectively, are features traditionally included in the analysis of speech of people with PD. With the aim of considering phonation and articulation information, in this paper we propose several measures: (1) the *vocal prism* whose base is the $tVSA$ and whose altitude is given by the variability of the pitch value estimated on the vowels /a/, /i/ and /u/, (2) the *vocal pentagon*, whose vertexes are the values of the F1 and F2 for the five Spanish vowels, and (3) the *vocal polyhedron* whose base is formed by the vocal pentagon and whose edges are given by the pitch variability obtained from the five Spanish vowels.

From this set of measures, different vocal features are derived: the volume of the vocal prism (*vPrism*), the area of the vocal pentagon (*aPenta*), the volume of the vocal polyhedron (*vPoly*) and the set of five measures that correspond to the coordinates of the centroid of each vertex in the *vocal pentagon*; they are named $CentPentaFk_j$, where $k \in \{1, 2\}$ indicates the axes (F1 horizontal and F2 vertical) and $j \in \{a, e, i, o, u\}$. Figure 1 is included to give a graphical idea of the information that can be provided by the features proposed here. Note that the behavior of $tVSA$ and $aPenta$ is similar, e.g. the phonations of people with PD have a smaller area in the F1-F2 plane than HC, indicating that the articulation capability of the patients is reduced compared to HC. Note also that *aPenta* includes information of the five Spanish vowels while *tVSA* only can provide information about three.

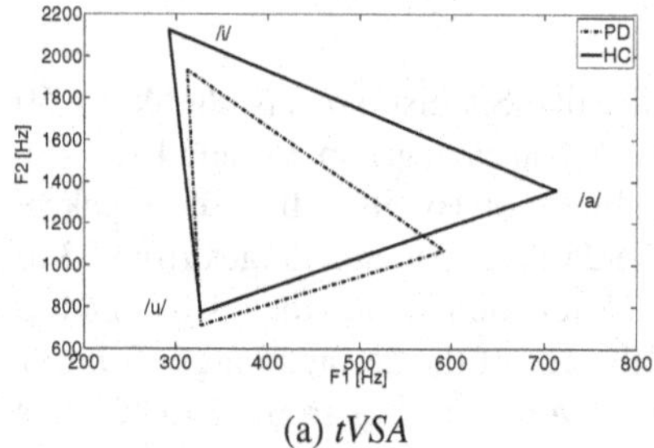

(a) *tVSA*

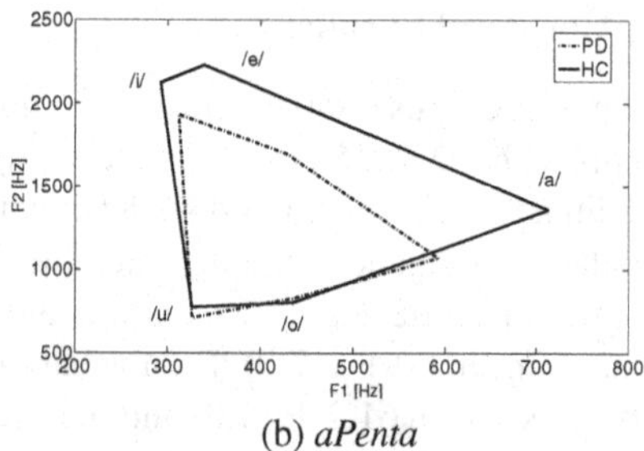

(b) *aPenta*

Fig. 1. Triangular vowel space area and vocal pentagon for PD and HC

vPrism is the other measure introduced in this paper, its base is the vocal triangle and its altitude is the pitch variability; this measure can give information about articulation and phonation simultaneously. The measure is illustrated in Figure 2.

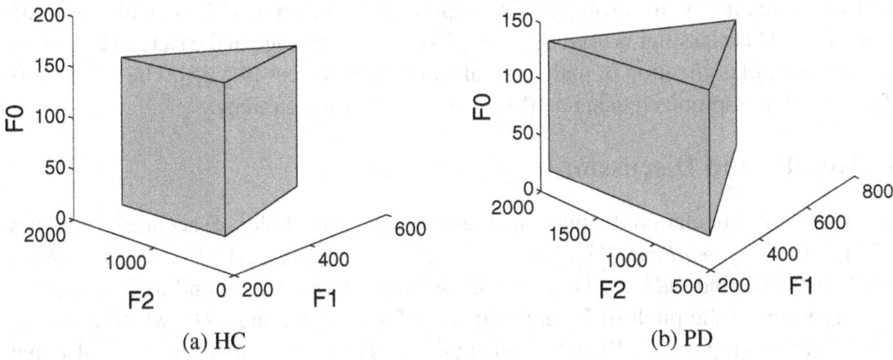

Fig. 2. Vocal prism

Perturbation Measures: Phonation. Temporal and amplitude variation of the pitch, also known as jitter (*Jitt*) and shimmer (*Shimm*) respectively, are included along with another perturbation measure which is called Correlation Dimension *CD* [13].

Jitt and *Shimm* are measures commonly used to evaluate speech of people with PD; however they assume near periodicity in the speech signal, thus they could not be reliable in cases where the severity of the disease disables the estimation of the pitch [14]. In this work, the CD is included in the set of perturbation features due to its capability to describe periodicity in a speech signal. According to [14], the vibrations of the vocal folds can be represented as trajectories in the state space with time evolution, and such trajectories can give qualitative information about the phenomenon, thus periodic vibrations produce closed trajectories, whereas aperiodic vibrations produce irregular trajectories. With CD it is possible to quantify how aperiodic a speech signal is, so this measure can be considered as a perturbation feature, with the advantage of being able to give information even in cases where the estimation of other perturbation measures is not possible, i.e. when the disease is advanced [13]. See [15], for more details about the estimation of the CD.

2.3 Experiments

The recordings are windowed each 40 ms with an overlap of 20 ms. The features are calculated for each frame, forming a feature vector for each measure. Afterward, four statistics are calculated for each feature: mean value (m), standard deviation (*std*), kurtosis (k) and skewness (*sk*).

The discrimination analysis is performed in two stages. The first is the linear Bayesian classification of recordings from PD and HC speakers. With the aim of obtain a sub set with the most discriminative measures, after the first classification stage, only the features with the minimum accuracy of 61% (after several tests with different numbers this was the accuracy that offered the best compromise between classification and number of features), are included in the second classification stage which consists of the classification with a soft-margin support vector machine (SVM) with Gaussian kernel [16]. This classifier is used because of its extensive usage in the state of the art for the automatic classification of pathological and healthy voices [4,5,15]. The parameters of the SVM are optimized using a 10-fold cross validation strategy.

3 Results and Discussion

The set with articulation features that exceed the threshold of accuracy includes $stdF1_o$, $stdF0_u$, $centPentaF2_u$, *stdVprism*, $mF2_u$, $stdF0_a$, *mVAI*, $stdF1_i$, $stdF1_a$, $stdF0_i$, $stdF0_e$ and $stdF1_u$. Four of the twelve features correspond to measures of the variability of the pitch (different statistics of *Jitt*, *Shim*, and *CD*), which confirms previous results reported in [9] evaluating utterances of German speakers. Note also that two of the measures introduced in this paper are in the selected group: $centPentaF2_u$ and *stdVprism*. These features can give information about phonation and articulation processes simultaneously, and have similar performance to classical acoustic features, such as vocal formants and *VAI*.

In addition, the perturbation features are tested on each Spanish vowel and the same criteria (accuracy 61%) is applied to decide whether a feature is included in the feature set of the second stage of classification.

The second stage of classification is performed with a soft-margin SVM with margin parameter C and Gaussian kernel with parameter γ, both optimized in a grid-search following a cross-validation strategy. Articulation and phonation features are combined in the same representation space, and the performance of the system is improved from 74.0% to 81.3% with a total of 73 features (phonation + articulation). Figure 3 shows the accuracy obtained when more features are added to the classification process incrementally.

Table 1. Classification results with SVM

	Accuracy(%)	Sensitivity(%)	Specificity(%)	AUC
Articulation	79.3 ± 5.8	80.2 ± 10	79.3 ± 12.4	0.76
Phonation	74.0 ± 9.6	73.1 ± 9.9	73.8 ± 14.1	0.74
Union	81.3 ± 5.5	82.5 ± 9.9	80.9 ± 4.8	0.85

The results obtained with each set of features (phonation and articulation), as well as with the union of all features, are presented in Table 1. The performance is presented in terms of accuracy, sensitivity, specificity and the area under the ROC curve (AUC).

The performance of each feature set can be observed in Figure 4 in a more compact way. Note the improvement in the accuracy when phonation and articulation features are combined.

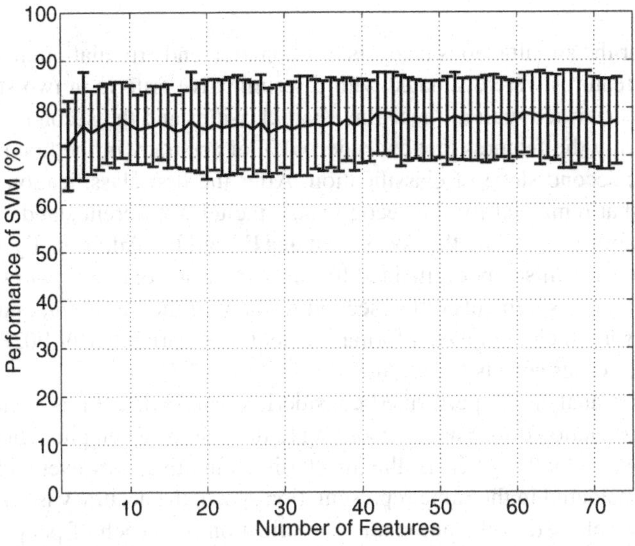

Fig. 3. Performance with the phonation and articulation features

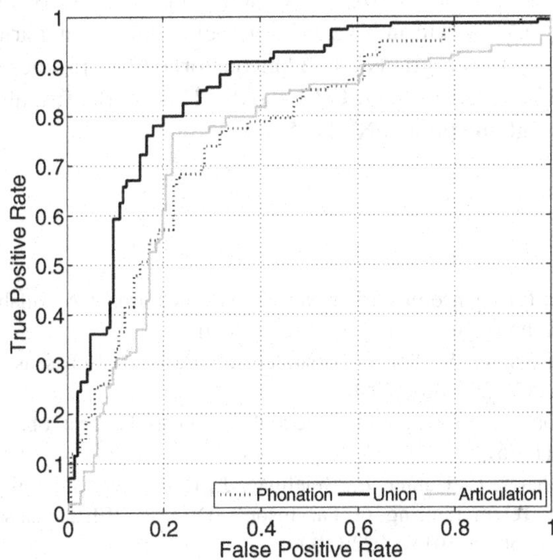

Fig. 4. ROC curves obtained with phonation and articulation features and their union

4 Conclusions

New features for the simultaneous analysis of phonation and articulation in speech are presented. The features were evaluated with discriminative criteria in two stages. The first stage was performed with a linear Bayesian classifier and the second with a soft-margin SVM. Only the features that exceeded 61% of accuracy in the first stage were included for the second stage of classification. After the first classification stage, the set of features that remained for the second stage includes different standard features such as the variability of F0 and the vocal formants F1 and F2. Additionally, two of the features introduced in this paper remained for the second stage; one is $centPentaF2_u$ which is based on measurements of the second formant of the vowel /u/, and the other one is $stdVprism$ which compiles information about the variability of F0, F1, and F2 measured on the corner vowels /a/, /i/, /u/.

The phonation analysis is performed considering standard features such as jitter, shimmer, and correlation dimension. According to the results, when phonation features are evaluated individually, jitter is the most discriminative; however, when these measures are considered in the same representation space, the accuracy is improved.

The accuracy obtained in the automatic classification of speech of people with PD and HC was improved from 74.0% to 81.3% when articulation and phonation features are combined.

Acknowledgments. Juan Rafael Orozco Arroyave is under grants of COLCIENCIAS through the program "Convocatoria N$^{\underline{o}}$ 528, generación del bicentenario 2011". Elkyn Belalcazar Bolaños is under grants of "Convocatoria 617, jovenes investigadores e innovadores 2013" funded by COLCIENCIAS. This work was also financed by COLCIENCIAS through the project N$^{\underline{o}}$ 111556933858.

References

1. de Rijk, M., et al.: Prevalence of Parkinson's Disease in Europe: A collaborative study of population-based cohorts. Neurology 54, 21–23 (2000)
2. Ramig, L., Fox, C., Sapir, S.: Speech Treatment for Parkinson's Disease. Expert Review Neurotherapeutics 8(2), 297–309 (2008)
3. Darley, F., Aronson, A., Brown, J.: Differential Diagnosis Patterns of Dysarthria. Motor Speech Disorders (1975)
4. Little, M.A., McSharry, P., Hunter, E., Spielman, J., Ramig, L.: Suitability of Dysphonia Measurements for Telemonitoring of Parkinson's Disease. IEEE Transactions on Biomedical Engineering 56(4), 1015–1022 (2009)
5. Bocklet, T., Nöth, E., Stemmer, G., Ruzickova, H., Rusz, J.: Detection of Persons with Parkinson's Disease by Acoustic, Vocal and Prosodic Analysis. In: Proceedings of the IEEE Workshop on Automatic Speech Recognition and Understanding (ASRU), pp. 478–483 (2011)
6. Bayestehtashk, A., Asgari, M., Shafran, I., Mcnames, J.: Fully Automated Assessment of the Severity of Parkinson's Disease from Speech. Computer Speech & Language, 1–14 (to appear, 2014)

7. National Collaborating Centre for Chronic Conditions: Parkinson's Disease: National Clinical Guideline for Diagnosis and Management in Primary and Secondary Care. Royal College of Physicians, London (2006)
8. Rusz, J., Cmejla, R., Ruzickova, H., Ruzicka, E.: Quantitative Acoustic Measurements for Characterization of Speech and Voice Disorders in Early Untreated Parkinson's Disease. The Journal of the Acoustical Society of America 129(1), 350–367 (2011)
9. Skodda, S., Visser, W., Schlegel, U.: Vowel Articulation in Parkinson's Disease. Journal of Voice 25(4), 467–472 (2011)
10. Stebbing, G., Goetz, C.: Factor Structure of the Unified Parkinson's Disease Rating Scale: Motor Examination Section. Movement Disorders 13, 633–636 (1998)
11. Hoehn, M.M., Yahr, M.D.: Parkinsonism: Onset, Progression, and Mortality. Neurology 17, 427–442 (1967)
12. Orozco-Arroyave, J., Arias-Londoño, J., Vargas-Bonilla, J., González-Rátiva, M., Nöth, E.: New Spanish Speech Corpus Database for the Analysis of People Suffering from Parkinson's Disease. In: Proceedings of the 9th Language Resources and Evaluation Conference, LREC (to appear, 2014)
13. Lee, V., Zhou, X., Rahn, D., Wang, E., Jiang, J.: Perturbation and Nonlinear Dynamic Analysis of Acoustic Phonatory Signal in Parkinsonian Patients Receiving Deep Brain Stimulation. Journal of Communication Disorders 41(6), 485–500 (2008)
14. Jiang, J., Zhang, Y., McGilligan, C.: Chaos in Voice: From Modeling to Measurement. Journal of Voice 20(1), 2–17 (2006)
15. Orozco-Arroyave, J.R., Vargas-Bonilla, J.F., Arias-Londoño, J.D., Murillo-Rendón, S., Castellanos-Domínguez, G., Garcés, J.: Nonlinear Dynamics for Hypernasality Detection in Spanish Vowels and Words. Cognitive Computation 5(4), 448–457 (2013)
16. Schölkopf, B., Smola, A.: Learning With Kernel. The MIT Press (2002)

Speaker Identification by Combining Various Vocal Tract and Vocal Source Features

Yuta Kawakami[1], Longbiao Wang[1], Atsuhiko Kai[2], and Seiichi Nakagawa[3]

[1] Nagaoka University of Technology, Japan
[2] Shizuoka University, Japan
[3] Toyohashi University of Technology, Japan
{wang@vos, s123118@stn}.nagaokaut.ac.jp

Abstract. Previously, we proposed a speaker recognition system using a combination of MFCC-based vocal tract feature and phase information which includes rich vocal source information. In this paper, we investigate the efficiency of combination of various vocal tract features (MFCC and LPCC) and vocal source features (phase and LPC residual) for normal-duration and short-duration utterance. The Japanese Newspaper Article Sentence (JNAS) database was used to evaluate our proposed method. The combination of various vocal tract and vocal source features achieved remarkable improvement than the conventional MFCC-based vocal tract feature for both normal-duration and short-duration utterances.

Keywords: Speaker identification, phase, LPC residual, LPCC, GMM.

1 Introduction

For the speaker identification task, many feature parameters had been used [1]. Mel-Frequency Cepstral Coefficients (MFCCs) [2] are basic feature for general speech processing. Linear Predictive Coding (LPC) [3,4] based features like the LPC Cepstral Coefficients (LPCC) [5] are also used. Line Spectral Frequencies (LSFs) [5] are coefficients in frequency domain, which are equivalent of LPC coefficients, this method is used for speech coding. Perceptual Linear Prediction (PLP) coefficients [6] consider psychophysics by using some human auditory-based filters, this also uses LPC method. These methods perform good also for speaker identification. Wang et al. used MFCC-based Gaussian Mixture Model (GMM) and LPCC-based Hidden Markov Model (HMM) for the distant speaker recognition, which worked well [7,8]. However, these feature parameters contain much vocal tract characteristics than that of vocal source. Vocal source characteristics are considered to be important for the speaker identification.

To catch vocal source characteristics, Markov et al. proposed a GMM-based speaker identification system that integrates pitch and the LPC residual with the LPC-derived cepstral coefficients [9]. Their experimental results show that using pitch information is most effective when the correlation between pitch and the cepstral coefficients is taken into consideration. Zheng et al. proposed Wavelet Octave Coefficients of Residues (WOCOR) which are based on LPC residual. They reported the improvement of speaker recognition performance by combining WOCOR with MFCC [10]. Recently, group

P. Sojka et al. (Eds.): TSD 2014, LNAI 8655, pp. 382–389, 2014.

delay-based phase information has been used [11]- [13]. Group delay is defined as the negative derivative of the phase of the Fourier transform of a signal. Hedge et al. reported the improvement of the speaker recognition performance by combining MFCC with group delay [11]. However, the group delay based phase also contains power spectrum information, therefore, the complementary nature of the power spectrum-based MFCC and group delay phase was not sufficient enough.

Previously, Wang et al. proposed phase related features which is directly extracted from the Fourier transform of the speech wave for speaker recognition [14]- [21]. The phase information is valid for speaker identification, because it captures rich vocal source information. The combination of MFCC and phase information outperformed than the MFCC because the complementary nature of the power spectrum-based MFCC (vocal tract information) and phase spectrum-based feature (vocal source information) was used. However, the sufficient performance could not be achieved especially for short-duration utterance. That seems to be improved by combining various vocal tract and vocal source features which have complementary speaker information. In this paper, we investigate the efficiency of combination of various vocal tract features (MFCC and LPCC) and vocal source features (phase and LPC residual) for normal-duration and short-duration utterance.

The rest of this paper is organized as follows: Section 2 presents feature extraction method for the phase information and the LPC residual based feature. Section 3 describes combining method for two features. Section 4 discusses experimental setup and speaker identification results. Section 5 gives our conclusions.

2 Feature Extraction Method

In this section, we introduce two vocal source-based feature extraction methods, phase related feature and LPC residual-based feature.

2.1 Phase Related Features

The spectrum $S(\omega, t)$ of a signal is obtained by DFT of an input speech signal sequence

$$\begin{aligned} S(\omega, t) &= X(\omega, t) + jY(\omega, t) \\ &= \sqrt{X^2(\omega, t) + Y^2(\omega, t)} \times e^{j\theta(\omega, t)}. \end{aligned} \tag{1}$$

However, the phase changes, depending on the clipping position of the input speech even at the same frequency ω. To overcome this problem, the phase of a certain basis frequency ω is kept constant, and the phases of other frequencies are estimated relative to this. For example, by setting the phase of basis frequency ω to 0, we obtain

$$S'(\omega, t) = \sqrt{X^2(\omega, t) + Y^2(\omega, t)} \times e^{j\theta(\omega, t)} \times e^{-j\theta(\omega, t)}, \tag{2}$$

whereas for the other frequency $\omega' = 2\pi f'$, the spectrum becomes

$$\begin{aligned} S'(\omega', t) &= \sqrt{X^2(\omega', t) + Y^2(\omega', t)} \times e^{j\theta(\omega', t)} \times e^{-j\frac{\omega'}{\omega}\theta(\omega, t)} \\ &= \tilde{X}(\omega', t) + j\tilde{Y}(\omega', t). \end{aligned} \tag{3}$$

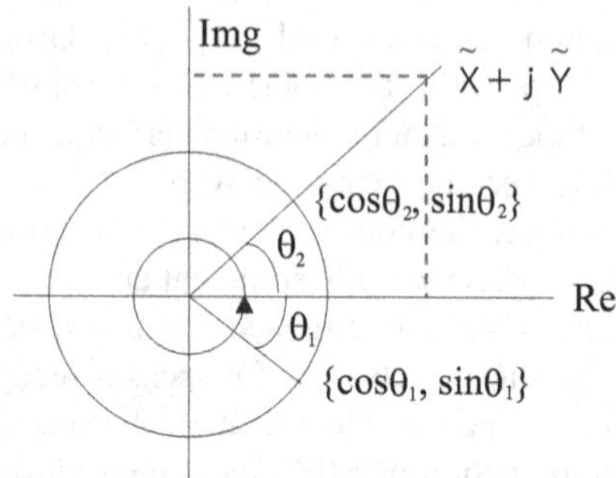

Fig. 1. Modified phase information

In this way, the phase can be normalized. Then, the real and imaginary parts of (3) become

$$\tilde{X}(\omega', t) = \sqrt{X^2(\omega', t) + Y^2(\omega', t)} \times \cos\left(\theta(\omega', t) - \frac{\omega'}{\omega}\theta(\omega, t)\right) \tag{4}$$

$$\tilde{Y}(\omega', t) = \sqrt{X^2(\omega', t) + Y^2(\omega', t)} \times \sin\left(\theta(\omega', t) - \frac{\omega'}{\omega}\theta(\omega, t)\right), \tag{5}$$

and the phase information is normalized as follows:

$$\tilde{\theta}(\omega', t) = \theta(\omega', t) - \frac{\omega'}{\omega}\theta(\omega, t). \tag{6}$$

In the experiments described in this paper, the basis frequency ω is set to $2\pi \times 1000Hz$. In a previous study, to reduce the number of feature parameters, we used phase information in a sub-band frequency range only. However, a problem arose with this method when comparing two phase values. For example, for two values $\pi - \tilde{\theta}_1$ and $\tilde{\theta}_2 = -\pi + \tilde{\theta}_1$, the difference is $2\pi - 2\tilde{\theta}_1$. If $\tilde{\theta}_1 \approx 0$, then the difference $\approx 2\pi$, despite the two phases being very similar to each other. Therefore, we modified the phase into coordinates on a unit circle [19], like fig. 1, that is,

$$\tilde{\theta} \rightarrow \{\cos\tilde{\theta}, \sin\tilde{\theta}\}. \tag{7}$$

In addition, we used the pseudo pitch synchronize method when clipping input speech. This method searches peak positions of the signal, and the positions are used as the center positions of the clipping window. Fig. 2 shows the overview of the method. We had confirmed the improvement of the speaker identification performance even in noisy environments by using pseudo pitch synchronization [20,21].

2.2 LPC Residual Based Features

Linear Predictive Coding (LPC) is a basic method to get vocal tract characteristics, and its cepstram coefficients (LPCC) are generally used as the feature parameters [5,8].

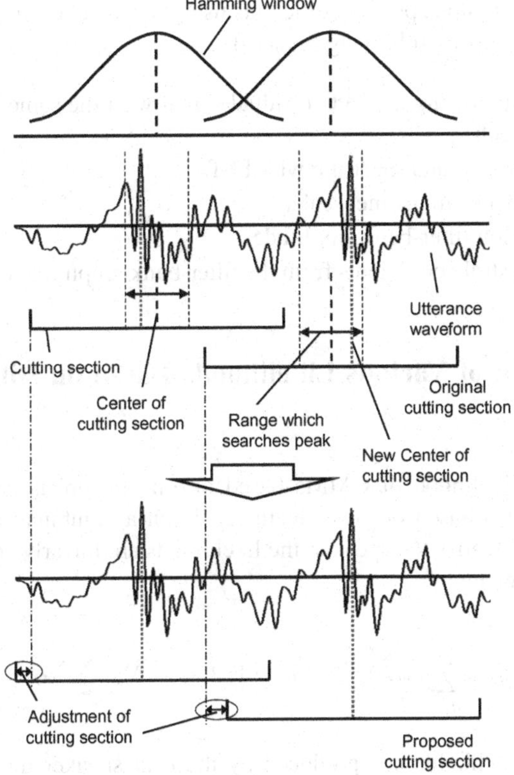

Fig. 2. Overview of the pseudo pitch synchronize method

From LPC coefficients, the source signal is approximated as $\hat{s}(n)$ by a weighted sum of past samples

$$\hat{s}(n) = \sum_{k=1}^{p} a_k s(n-k), \tag{8}$$

where p is the order of prediction, and a_k are LPC coefficients. The LPC residual signal, or prediction error $e(n)$ is calculated as the difference between the source signal and the predicted signal

$$\begin{aligned} e(n) &= s(n) - \hat{s}(n) \\ &= s(n) - \sum_{k=1}^{p} a_k s(n-k). \end{aligned} \tag{9}$$

The LPCC contains vocal tract characteristics, in contrast, the LPC residual is interpreted as the vocal source information [9,10]. In this work, the obtained LPC residual signal is transformed into cepstral coefficients using the standard mel-frequency filter-bank analysis technique by following steps [9].

1. Framing and windowing the LPC residual signal with the same rate and length as the original speech.
2. Obtaining the magnitude spectrum with FFT.
3. Forming filter banks in the mel scale.
4. Computing the log filter-bank amplitudes.
5. Calculating cepstral coefficients from the filter-bank amplitude using DCT.

3 Combination of Various Likelihoods Based on Different Features

In this paper, the likelihood of GMMs based on one feature is combined with the likelihoods of GMMs based on other features. When a combination of the multiple methods is used to identify the speaker, the likelihoods are linearly coupled to produce a new score L^n_{comb} given by

$$L^n_{comb} = \sum_{i=1}^{I} \alpha_i L^n_{feat_i}, \quad n = 1, 2, \cdots, N, \ \sum_i \alpha_i = 1, \tag{10}$$

where $L^n_{feat_i}$ are the likelihoods produced by the n-th speaker model based on i-th feature. N is the number of speakers registered, I is the number of feature and α_i donates the weighting coefficients for i-th feature, which are determined empirically. The speaker (or speaker model) with maximum likelihood is judged to be the target speaker.

4 Experiments

4.1 Experimental Setup

We conducted speaker identification experiments for the JNAS (Japanese Newspaper Article Sentence) database [22] which contains 135 males and 135 females, about 100 clean utterances per person. The input speech was sampled at 16 kHz. Each utterance had about 6 seconds on average.

Speakers were modeled by GMMs with 128 mixtures from scratch. Each speaker models were trained by 5 utterances for 4 features (MFCC, LPCC, phase, LPC residual). The feature extraction conditions are shown in Table 1. We used the rest of database for the test, the number of test utterance was 23,160 (about 85 utterance per person). GMM likelihoods for multiple features were coupled as combination score. In this work, for the test data, we also used short utterance by cutting whole utterances into 2, 1 and 0.5 seconds, in addition to the whole one.

Table 1. Feature extract conditions

Feature	Vocal tract features		Vocal source features	
	MFCC	LPCC	Phase	LPC residual
LPC order		14		14
Frame length	25 ms	25 ms	12.5ms	25 ms
Frame shift	10 ms	10 ms	5 ms	10 ms
Dimensions	25 12 MFCCs, 12 Δs and a Δ power	25 12 LPCCs, 12 Δs and a Δ power	24 $\sin\tilde{\theta}$, $\cos\tilde{\theta}$ of the first 12 $\tilde{\theta}$ s of the phase spectrum (60-750 Hz range)	25 12 MFCCs, 12 Δs and a Δ power of the residual signal

4.2 Experimental Results

Speaker identification rates by single features are shown in Table 2, and combination results are indicated in Table 3. Comparing with the vocal tract features, the LPCC performed better than the MFCC, and with the vocal source features, the LPC residual was better than the phase, for any length. Nevertheless, in the combinations of two features, the rates by "MFCC+Phase" exceeded that of "LPCC+LPC residual". This means MFCC and Phase information has better complementarity than other combinations.

By combining various vocal tract and vocal source feature (combination of 4 features), the best identification rates were obtained. The results verify that performance of "MFCC+phase" or "LPCC+LPC residual" is not sufficient. For shorter utterances,

Table 2. Speaker identification rates by single feature (%)

feature	whole	2 sec	1 sec	0.5 sec
MFCC	95.1	89.8	81.6	66.2
LPCC	95.3	90.1	82.3	68.0
Phase	90.8	79.3	64.0	46.2
LPC residual	94.5	87.2	78.5	63.7

Table 3. Speaker identification rates by multiple features (%), The combination coefficients are fine tuned for each test

feature	whole	2 sec	1 sec	0.5 sec
MFCC	95.1	89.8	81.6	66.2
MFCC+Phase	98.4	96.0	91.1	79.8
LPCC+LPC residual	96.3	92.7	86.7	74.5
Vocal tract feature (MFCC+LPCC)	96.5	93.0	87.1	75.5
all features	**98.4**	**96.3**	**92.2**	**82.4**

identification rates were degraded. However, combination of all features achieved 48.0 % relative error reduction (66.2 % to 82.4 %) from the MFCC only, for 0.5 seconds utterances. This means we can get complementary information from the utterances by each feature, and the combination of them is effective for the short-utterance speaker identification.

5 Conclusions and Future Work

In this paper, we confirmed the efficiency of the vocal source information for speaker identification. Then the combination of MFCC and Phase performed the best in two features combinations. By combining 4 features, the identification rates were improved and the best performance was obtained. The results indicate that various vocal tract and vocal source features have complementarity for speaker recognition. In addition, the combination method was effective for short-utterance speaker identification. However, for short utterances, the improvement of the identification rates might be insufficient. Hereafter, we address this problem by improving the feature extraction method.

References

1. Kinnunen, T., Li, H.: An overview of text-independent speaker recognition: From features to supervectors. Speech Communication 52(1), 12–40 (2010)
2. Davis, S., Santa, B., Mermelstein, P.: Comparison of parametric representations for monosyllabic word recognition in continuously spoken sentences. IEEE Trans. on Acoustics, Speech and Signal Processing 28(4), 357–366 (1980)
3. Makhoul, J., Bolt, B.: Linear prediction: A tutorial review. Proc. of IEEE 63(4), 561–580 (1975)
4. Mammone, R.J., Zhang, X., Ramachandran, R.P.: Robust speaker recognition: A feature-based approach. IEEE Signal Processing Magazine 13, 58–71 (1996)
5. Huang, X., Acero, A., Hon, H.W.: Spoken Language Processing: A Guide to Theory, Algorithm, and System Development. Prentice-Hall, New Jersey (2001)
6. Hermansky, H.: Perceptual linear predictive (PLP) analysis of speech. The Journal of the Acoustical Society of America 87(4), 1738–1752
7. Wang, L., Kitaoka, N., Nakagawa, S.: Robust Distant Speaker Recognition Based on Position Dependent Cepstral Mean Normalization. In: Proceedings of the 9th European Conference on Speech Communication and Technology (Interspeech 2005-Eurospeech), pp. 1977–1980 (2005)
8. Wang, L., Kitaoka, N., Nakagawa, S.: Robust distant speaker recognition based on position dependent CMN by combining speaker-specific GMM with speaker-adapted HMM. Speech Communication 49, 501–513 (2007)
9. Markov, K.P., Nakagawa, S.: Integrating pitch and LPC-residual information with LPC-cepstrum for text-independent speaker recognition. Jour. ASJ (E) 20(4), 281–291 (1999)
10. Zheng, N., Lee, T., Ching, P.C.: Integration of complementary acoustic features for speaker recognition. IEEE Signal Processing Letters 14(3), 181–184 (2007)
11. Hedge, R.M., Murthy, H.A., Rao, G.V.R.: Application of the modified group delay function to speaker identification and discrimination. In: Proc. ICASSP 2004, vol. 1, pp. 517–520 (2004)

12. Padmanabhan, R., Parthasarathi, S., Murthy, H.: Robustness of phase based features for speaker recognition. In: Proc. Interspeech, pp. 2355–2358 (2009)
13. Kua, J., Epps, J., Ambikairajah, E., Choi, E.: LS regularization of group delay features for speaker recognition. In: Proc. Interspeech, pp. 2887–2890 (2009)
14. Nakagawa, S., Asakawa, K., Wang, L.: Speaker recognition by combining MFCC and phase information. In: Proc. InterSpeech, pp. 2005–2008 (2007)
15. Wang, L., Ohtsuka, S., Nakagawa, S.: High improvement of speaker identification and verification by combining MFCC and phase information. In: Proc. ICASSP, pp. 4529–4532 (2009)
16. Wang, L., Minami, K., Yamamoto, K., Nakagawa, S.: Speaker identification by combining MFCC and phase information in noisy environments. In: Proc. ICASSP, pp. 4502–4505 (2010)
17. Wang, L., Minami, K., Yamamoto, K., Nakagawa, S.: Speaker recognition by combining MFCC and phase information in noisy conditions. IEICE Transactions on Information and Systems E93-Dd(9), 2397–2406 (2010)
18. Hirano, Y., Wang, L., Kai, A., Nakagawa, S.: On the Use of Phase Information-based Joint Factor Analysis for Speaker Verification under Channel Mismatch Condition. In: Proc. of APSIPA ASC 2012, 4 pages (2012)
19. Nakagawa, S., Wang, L., Ohtsuka, S.: Speaker Identification and Verification by Combining MFCC and Phase Information. IEEE Trans. on Audio, Speech, and Language Processing 20(4), 1085–1095 (2012)
20. Shimada, K., Yamamoto, K., Nakagawa, S.: Speaker identification using pseudo pitch/synchronized phase information in voiced sound. In: Proc. APSIPA ASC 2011, pp. 1–6 (2011)
21. Kawakami, Y., Wang, L., Nakagawa, S.: Speaker Identification Using Pseudo Pitch Synchronized Phase Information in Noisy Environments. In: Proc. APSIPA ASC 2012, 5 pages (2013)
22. Itou, K., Yamamoto, M., Takeda, K., Takezawa, T., Matsuoka, T., Kobayashi, T., Shikano, K., Itahashi, S.: JNAS:Japanese speech coupus for large vocabulary continuous speech recognition research. J. Acoust. Soc. Jpn. (E) 20(13), 199–206 (1999)

Inter-Annotator Agreement on Spontaneous Czech Language

Limits of Automatic Speech Recognition Accuracy

Tomáš Valenta, Luboš Šmídl, Jan Švec, and Daniel Soutner

University of West Bohemia, Faculty of Applied Sciences, Department of Cybernetics,
Univerzitní 22, 306 14 Plzeň, Czech Republic
{valentat,smidl,honzas,dsoutner}@kky.zcu.cz

Abstract. The goal of this article is to show that for some tasks in automatic speech recognition (ASR), especially for recognition of spontaneous telephony speech, the reference annotation differs substantially among human annotators and thus sets the upper bound of the ASR accuracy. In this paper, we focus on the evaluation of the inter-annotator agreement (IAA) and ASR accuracy in the context of imperfect IAA. We evaluated it using a part of our Czech Switchboard-like spontaneous speech corpus called Toll-free calls. This data set was annotated by three different annotators rendering three parallel transcriptions. The results give us additional insights for understanding the ASR accuracy.

Keywords: automatic speech recognition, inter-annotator agreement, accuracy.

1 Introduction

Automatic speech recognition accuracy differs significantly among various domains and tasks. On particular tasks in certain domains, the recognition accuracy almost achieves 100 %, whereas in others, it is about 60 % or less, see Fig. 1 [1].

A similar trend can be observed for human transcriptions when applying the inter-annotator agreement (IAA). It can be almost perfect in domains like dictation of written text, whereas human transcriptions of spontaneous spoken language may differ substantially. The estimate of human transcription error in Fig. 1 is quite optimistic and does not take into account the domain.

The inter-annotator agreement of classification tasks such as assigning labels from few classes (e.g. parts-of-speech tagging, named entity recognition etc.) is commonly evaluated. On text, the IAA is evaluated in transcription and translation of historical texts [10] or psychoanalysis, and sometimes in speech synthesis [12]. Surprisingly, the IAA is not commonly evaluated in the area of speech recognition. The transcription(s) available are viewed as a gold-standard no matter how accurate they are.

For the evaluation of the IAA in classification tasks assigning few labels only, statistics compensating the chance agreement are often used, e.g. Cohen's κ [2]. The chance agreement in ASR tasks is, however, inversely proportional to the vocabulary size, hence it has virtually no effect on the value of the IAA.

P. Sojka et al. (Eds.): TSD 2014, LNAI 8655, pp. 390–397, 2014.

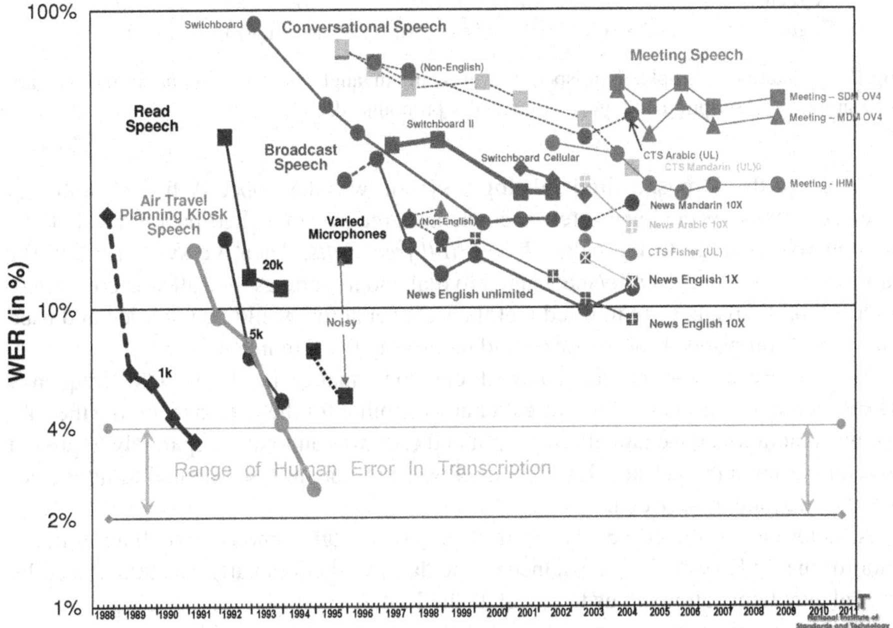

Fig. 1. National Institute of Standards and Technology speech-to-text benchmark history — May 2009 [1]. Word error rate used as opposed to the accuracy used in the rest of the article.

The goal of this article is to show that on some tasks, human transcription accuracy (i.e. IAA) above 90 % is almost unachievable. By implication, this sets the upper bound for the automatic speech recognition accuracy far below 100 %.

To demonstrate a lower IAA and its impact on the recognition accuracy, a Czech Switchboard-like corpus was chosen. It contains recordings of telephone communication of two people. The callers usually know each other very well, they use lots of non-standard or local words and they speak colloquially. This reduces the recognition performance significantly as well as the ability to recognize and understand the conversation by other people.

In Section 2, the evaluation corpus is described. All results, including the human and automatic speech recognizer, were computed on evaluation sets of the corpus. In Section 3, the inter-annotator agreement methods are described and the results obtained are presented. In Section 4, the IAA results are compared to an automatic speech recognition system. Finally, Section 5 summarizes the results and draws some conclusions.

2 The Corpus

The evaluation corpus is called *Toll-free Calls*. It is similar to the standard Switchboard corpus [3]. It consists of many hours of spontaneous spoken Czech language used in phone calls.

Czech: `<ehm_ANO> <unintelligible> (nějak(ňák)) zvláštně`
English: `<erm_YES> <unintelligible> (some(sum)) weirdly`

Fig. 2. Annotation example. Non-speech events are in angle brackets, round brackets mark orthographic transcription and what was actually pronounced.

To record the corpus, a simple dialogue system was developed. When the call was received, it was rejected and a few seconds later the system called back the caller. As nothing was charged to the caller, hence *Toll-free Calls*. Then the system asked the caller for a phone number he/she wanted to dial and as soon as the call was connected, both call participants were notified that their conversation would be recorded and used for research purposes. Both parties could talk for up to 15 minutes.

The audio was recorded in a standard telephone quality, i.e. 16bit PCM, frequency 8 kHz, stereo — one channel for the caller and the other for the person receiving the call. For the annotation, the channels were split and each was annotated separately to protect the call participants' privacy. Each channel was subsequently segmented to utterances which were then transcribed.

A major part of the corpus was transcribed by a single annotator to obtain training data for the ASR system. The balance of the data (evaluation part) was transcribed by several annotators allowing us to calculate the IAA.

The transcriptions were made by well-experienced annotators with focus on acoustic modelling which means that the transcriptions were not orthographic and several non-speech event tags, exact pronunciations etc. were used in the annotation, see Fig. 2 for an example.

2.1 Evaluation Sets

The part of the corpus used for evaluation was annotated by several annotators. Each sentence was annotated three times by different annotators. Two evaluation sets, *Set1* and *Set2*, were selected so that the number of sentences covered by three same annotators was maximal. The sets contain 2,003 and 2,838 sentences respectively. Set1 was annotated by annotators A, B and C, Set2 was annotated by annotators A, B and D.

Prior to the evaluation, the text was pre-processed in the following way:

- non-speech events were removed;
- converted to lower-case;
- punctuation removed (replaced by spaces);
- exact pronunciations removed (orthographic transcription used if available, see Fig. 2);
- multiwords (e.g. good_morning) were split;
- white-space normalized;
- commonly confused words (frequency greater than two) were considered as equivalent.

In spoken Czech, some words pronounced similarly (*sem*, *jsem*; *byli*, *byly*), orthographically vs. colloquially (*mladý*, *mladej*) or just typing errors were fixed this way.

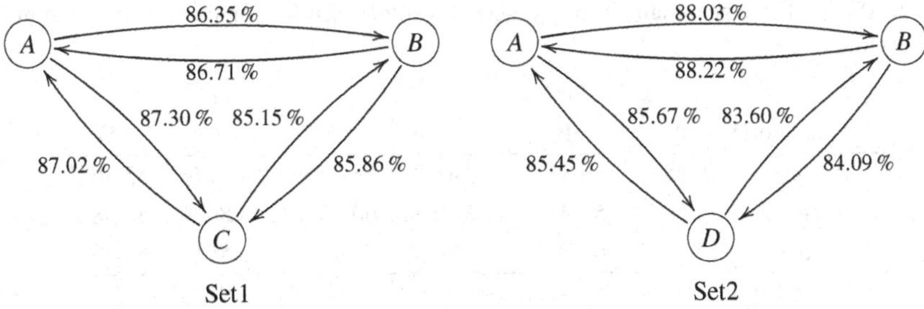

Fig. 3. Inter-annotator agreement on evaluation sets. Annotator at the origin of the arrow was used as a reference and at the tip as a (recognition) hypothesis.

Table 1. Average inter-annotator agreement on evaluation sets

Set1	A	B
B	86.53 %	
C	87.16 %	85.51 %

Set2	A	B
B	88.13 %	
D	85.56 %	83.85 %

3 Inter-Annotator Agreement

The inter-annotator agreement was calculated in the same way as recognition accuracy is calculated. First, the annotations were aligned so that the Levenshtein distance [7] was minimal. Penalties for substitution, insertion and deletion were set to be the same as in HTK toolkit, i.e. $p_S = 10$, $p_I = 7$ and $p_D = 7$ respectively [15]. Accuracy is then defined by (1):

$$Acc = \frac{N - S - I - D}{N}, \tag{1}$$

where N is the number of words in the reference, S is the number of substitutions, I is the number of insertions, and D is the number of deletions.

Being penalties $p_S \neq p_I + p_D$ and annotations (one used as a reference and one as a recognition hypothesis) having different length, the accuracy depends on which annotation is used as reference, as shown in Fig. 3. In Table 1, weighted average, with respect to the number of words in the reference, is taken from the numbers to obtain a simple measure. It gives the same result as if the reference consisted of concatenation $A\|B$ and the hypothesis of $B\|A$.

Averaging the numbers in Table 1 we get

$$IAA_1 = 86.40\,\% \qquad \text{and} \qquad IAA_2 = 85.84\,\% \tag{2}$$

for evaluation sets Set1 and Set2 respectively. If we wished to get an overall estimate of inter-annotator agreement on Toll-free Calls corpus, we can take a weighted average, with respect to the number of sentences, of the two numbers and we get

$$IAA = 86.01\,\%. \tag{3}$$

Table 2. ASR performance: plain accuracy, accuracy after rescoring, oracle accuracy and multi-oracle accuracy

	Set1			Set2		
Annotator	A	B	C	A	B	D
Accuracy	50.35 %	49.43 %	48.41 %	55.70 %	54.17 %	53.46 %
Rescore-Acc	54.50 %	53.98 %	52.69 %	60.45 %	58.89 %	57.98 %
Ora-Acc	67.73 %	66.66 %	65.56 %	72.60 %	70.57 %	69.69 %
Mul-Ora-Acc		52.41 %			58.07 %	

4 Evaluating an ASR System

The recognition engine follows a standard design [9]. The acoustic models in our system are based on the hidden Markov models (HMM). Standard 3-state left-to-right models with a mixture of 16 Gaussians in each state are used (5000 states totally). The speech data were parametrized as 12-dimensional PLP [5] cepstral features including their delta and delta-delta derivatives. Cepstral mean subtraction was applied per speaker. We used a real-time large vocabulary continuous speech recognizer (LVCSR) to achieve a high degree of interactivity. The LVCSR system [11] uses lexical trees and Viterbi search using 3-gram language models with Witten-Bell discounting [14]. The ASR recognition vocabulary contains 123,038 words and the OOV rate on evaluation data is below 1 %.

In Figure 4 and Table 2 recognition accuracy is calculated against each annotation taken as a reference. Average IAA is put on arcs connecting the annotators.

4.1 Rescoring

We took the word lattices from the first recognition pass and extracted n best hypotheses (with $n = 1,000$) for each utterance; this n-best list was the base for our experiments with language models. The second pass of the recognition was provided with a more complex language model to discover if the results could be improved. After several

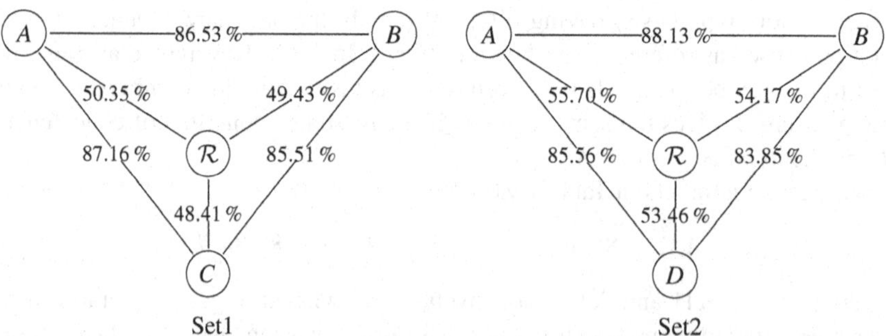

Fig. 4. Recognition accuracy compared to the average inter-annotator agreement, $\mathcal{R}$ stands for the recognizer

experiments we rescored the n-best list with recurrent neural network language model (RNN LM) [8] in combination with background LM which is a standard 5-gram smoothed with Kneser-Ney discounting [6].

The choice of RNN was motivated by its ability to capture longer context than n-gram models and because of better performance on smaller corpus, which is our case (training data size is 5.4 M words). The vocabulary of RNN LM was shortlisted to 40 k most frequent words, the size of the hidden layer was chosen $h = 400$ and the model was trained together with 8 M maximum entropy features.

Running the second recognition pass described above, the recognition accuracy can be increased slightly, see Table 2.

The upper bound of rescoring performance is given by the oracle accuracy. It is the maximum accuracy of any recognition hypothesis from the n-best list. See Table 2.

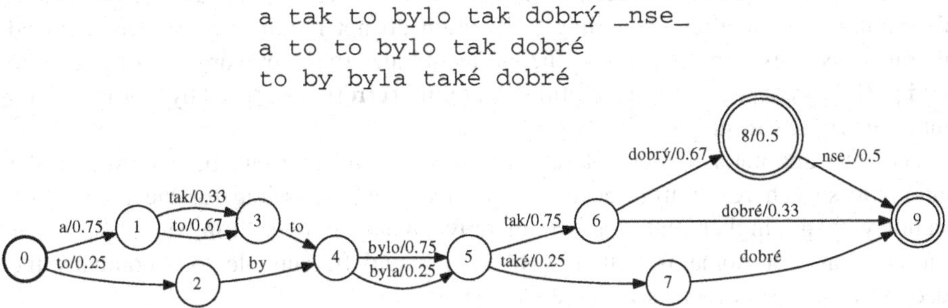

Fig. 5. Multi-oracle accuracy lattice made from three transcriptions. _nse_ means an optional non-speech event.

Having three parallel annotations, we can reverse the oracle accuracy calculation. We take the best hypothesis from the recognizer (not rescored) and try to match it as closely as possible against a lattice created from all annotations for the utterance to get a multi-oracle accuracy. For an example of this type of lattice (generalized confusion network), see Fig. 5. The results are shown in Table 2.

5 Conclusion

In the article we defined the inter-annotator agreement calculation on a speech corpus. We evaluated it on spontaneous speech corpus called Toll-free calls, where it shows to be surprisingly low.

Spoken Czech language differs from its written form significantly, a lot of colloquialisms is used. In spontaneous and expressive [4] communication, lots of non-verbal sounds are also used that are hard to transcribe. Moreover, written form of some words is dependent on the context (gender of the subject) which cannot be told from the speech segments transcribed by the annotators. Most of these phenomenons were, however, mitigated by the text preprocessing and the rest is the very matter of the inter-annotator disagreement, which sets the upper bound to the recognition accuracy.

Unquestionably, the ASR results are worse than human transcription, although it should be noted that the ASR processes the audio in real time in a single pass. In contrast, a human annotator works about 8 times slower than real time, and also has the opportunity to play back the recording repeatedly. The annotator also "recognizes" the spoken data in real time. However transcribing the text with all of the non-speech tags is considerably time-demanding. Undertaking a second recognition pass takes about twice as long as real time and can increase the accuracy. Yet it still fails to meet the IAA.

The oracle accuracy, a measure of how we can increase the accuracy by improving the language model or by rescoring, comes quite close to the IAA. On the other hand, the language modelling of such data is a difficult task, because a lot of non-verbal sounds are used in the communication, the utterances are very short. The number of words used is relatively small, although theoretically unlimited.

The presented results demonstrate that achieving 100% recognition accuracy in spontaneous or expressive speech is impossible and often not necessary. For example, natural language understanding tasks often do not require perfectly recognized utterance (best hypothesis), more efficient techniques (e.g. keyword spotting) can be used [13]. Moreover, most transcription/recognition errors are caused by words with no important sense (hesitation, fillings etc.).

The inter-annotator agreement sets the upper bound that can be achieved in the automatic speech recognition accuracy. In theory, it is possible for the recognition accuracy to go higher, but that would only signal an over-fitting to a particular transcription. For domains with low IAA, it would be suitable to accompany the recognition accuracy reports with the IAA value.

Acknowledgement. Access to computing and storage facilities owned by parties and projects contributing to the National Grid Infrastructure MetaCentrum, provided under the programme "Projects of Large Infrastructure for Research, Development, and Innovations" (LM2010005), is greatly appreciated.

This work was supported by the European Regional Development Fund (ERDF), project "New Technologies for Information Society" (NTIS), European Centre of Excellence, ED1.1.00/02.0090. Ministry of Education, Youth and Sports of the Czech Republic project No. LM2010013. The work has been supported by the grant of the University of West Bohemia, project No. SGS-2013-032.

References

1. Ajot, J., Fiscus, J.: Speech-To-Text (STT) and Speaker Attributed STT (SASTT) Results. NIST Rich Transcription Evaluation Workshop (2009)
2. Cohen, J.: A Coefficient of Agreement for Nominal Scales. Educational and Psychological Measurement 20(1), 37–46 (1960)
3. Godfrey, J., Holliman, E., McDaniel, J.: SWITCHBOARD: telephone speech corpus for research and development. In: Proceedings ICASSP 1992: 1992 IEEE International Conference on Acoustics, Speech, and Signal Processing, vol. 1, pp. 517–520. IEEE (1992)
4. Grůber, M., Tihelka, D.: Expressive speech synthesis for Czech limited domain dialogue system — Basic experiments. In: Proceedings of the IEEE 10th International Conference on Signal Processing, pp. 561–564. IEEE (October 2010)

5. Heřmanský, H.: Perceptual linear predictive (PLP) analysis of speech. The Journal of the Acoustical Society of America 87(4), 1738 (1990)
6. Kneser, R., Ney, H.: Improved backing-off for M-gram language modeling. In: Proceedings of Acoustics, Speech, and Signal Processing, ICASSP 1995, vol. 1, pp. 181–184 (1995)
7. Levenshtein, V.I.: Binary Codes Capable of Correcting Deletions, Insertions and Reversals. Soviet Physics Doklady 10, 707 (1966)
8. Mikolov, T., Kombrink, S., Deoras, A., Burget, L., Černocký, J.: RNNLM — Recurrent Neural Network Language Modeling Toolkit. In: Proceedings of ASRU 2011, pp. 1–4 (2011)
9. Müller, L., Psutka, J.V., Šmídl, L.: Design of speech recognition engine. In: Sojka, P., Kopeček, I., Pala, K. (eds.) TSD 2000. LNCS (LNAI), vol. 1902, pp. 259–264. Springer, Heidelberg (2000)
10. Munyaradzi, N.: Transcription of the Bleek and Lloyd Collection using the Bossa Volunteer Thinking Framework. Ph.D. thesis, Department of Computer Science, University of Cape Town, Cape Town (2013)
11. Pražák, A., Psutka, J.V., Hoidekr, J., Kanis, J., Müller, L., Psutka, J.: Automatic online subtitling of the Czech parliament meetings. In: Sojka, P., Kopeček, I., Pala, K. (eds.) TSD 2006. LNCS (LNAI), vol. 4188, pp. 501–508. Springer, Heidelberg (2006)
12. Tihelka, D., Romportl, J.: Statistical evaluation of reliability of large scale listening tests. In: Proceedings of 9th International Conference on Signal Processing, pp. 631–636. IEEE (October 2008)
13. Švec, J., Šmídl, L., Ircing, P.: Hierarchical Discriminative Model for Spoken Language Understanding. In: Proceedings of IEEE International Conference on Acoustics, Speech, and Signal Processing 2013, pp. 8322–8326. IEEE, Vancouver (2013)
14. Witten, I.H., Bell, T.C.: The zero-frequency problem: Estimating the probabilities of novel events in adaptive text compression. IEEE Transactions on Information Theory 37(4), 1085–1094 (1991)
15. Young, S.J., Evermann, G., Gales, M.J.F., Hain, T., Kershaw, D., Moore, G., Odell, J., Ollason, D., Povey, D., Valtchev, V., Woodland, P.C.: The HTK Book, version 3.4. Cambridge University Engineering Department, Cambridge (2006)

Minimum Text Corpus Selection for Limited Domain Speech Synthesis*

Markéta Jůzová and Daniel Tihelka

University of West Bohemia, Univerzitní 8, Plzeň, Czech Republic
{juzova,dtihelka}@kky.zcu.cz
http://www.kky.zcu.cz

Abstract. This paper concerns limited domain TTS system based on the concatenative method, and presents an algorithm capable to extract the minimal domain-oriented text corpus from the real data of the given domain, while still reaching the maximum coverage of the domain. The proposed approach ensures that the least amount of texts are extracted, containing the most common phrases and (possibly) all the words from the domain. At the same time, it ensures that appropriate phrase overlapping is kept, allowing to find smooth concatenation in the overlapped regions to reach high quality synthesized speech. In addition, several recommendations allowing a speaker to record the corpus more fluently and comfortably are presented and discussed. The corpus building is tested and evaluated on several domains differing in size and nature, and the authors present the results of the algorithm and demonstrate the advantages of using the domain oriented corpus for speech synthesis.

Keywords: limited domain speech synthesis, concatenative speech synthesis, text corpus, speech units, text chunks, unit concatenation.

1 Introduction

The limited domain speech synthesis (LDS) may at first seem to be a trivial task, when compared to the general-purpose text-to-speech (TTS) synthesis system. However, to reach a high-level of naturalness within a LDS system is not as simple as it looks like, mainly due to the fact that any non-natural artefact occurrence following longer natural-sounding speech is perceived very negatively [1]. Therefore, we must be very careful in the process of limited-domain (LD) speech corpus preparation, since a concatenation algorithm does not have to have many speech unit candidates available to ensure smooth segments concatenation, as it is the case of "classic" general-purpose unit selection TTS system [2].

More specifically, when speech corpus for a general TTS system[1] is being prepared, we focus on short speech units (e.g. diphones) to make the speech corpus rich in, trying

* This work was supported by the European Regional Development Fund (ERDF), project "New Technologies for Information Society" (NTIS), European Centre of Excellence, CZ.1.05/1.1.00/02.0090, the Technology Agency of the Czech Republic, project No. TA01030476 and SGS-2013-032.

[1] We will limit ourself entirely to a system embedding unit selection method, since it firstly still provides more natural output than the HMM-based synthesis [3], and secondly its comparison with the LD synthesis operation is much more straightforward.

P. Sojka et al. (Eds.): TSD 2014, LNAI 8655, pp. 398–407, 2014.

to select such a text material (from various domains) which contains all possible units (regarding their phonetic and prosodic context) with the number of occurrences as large as possible or manageable [4,5]. Only in such a way we maximize the chance that an appropriate sequence of (short) units will be selected and concatenated (at least unless units' prosodic synonymy and homonymy is taken into an account [6], which is out of focus of the present paper), to realize speech without audible unnatural artefacts.

The nature of limited domain, however, attracts with the advantage of the possibility to have much smaller, and thus much cheaper speech corpus, while reaching higher level of speech naturalness than with a general TTS system [7], especially if the required domain is away from the domains of texts used in the TTS system. Naturally, it means that much longer units, like words or whole phrases, must be used in this case. On the other hand, the design of the corpus must be carried out much more carefully, since there is no "units redundancy" which we can rely on in the case of general TTS. Inappropriate corpus design can easily lead either to significantly larger corpus than would be necessary, or to the need to concatenate recorded phrases in inappropriate places, e.g. in pauses or phrase breaks. It is the score of the present paper to show how to design a minimum text corpus suitable for a limited domain speech synthesis, while preserving sufficient units redundancy for the smooth concatenation of (longer) units the system is built on.

2 Limited Domain Text Representation

There are many forms of LD text representation, depending on the size of the domain and on the variability of texts within it. The most trivial example is a strictly-limited domain characterized by several (tens, hundreds) fixed sentence structures with variable items (let us call them *slots*), and sets of words to be placed in. The example can be an automatic system informing about departures and arrivals of trains, the fragment of which is shown in the following snippet:

`Jak vám mohu pomoci?`	*How can I help you?*
`V kolik hodin chcete jet ze stanice [Rokycany] do stanice [Strakonice]?`	*What time do you want to go from station [Rokycany] to station [Strakonice]?*
`V kolik hodin chcete jet ze stanice [Chrást u Plzně] do stanice [Domažlice]?`	*What time do you want to go from station [Chrást u Plzně] to station [Domažlice]?*
`Cesta trvá [3] hodiny.`	*The journey takes [3] hours.*

where [boxed items] are variable parts in otherwise fixed text frames.

However, in general, the domain, although still limited somehow, is wider with texts showing a larger degree of variability. The example of such domain can be weather forecast, transcripts of real ATC communications (*Air Traffic Control*, provides information and advisory services to planes using a defined phraseology), or ATIS (*Airport Terminal Information System*, broadcasting informations to planes before landing), as the last two are real examples of domains we currently build the LDS system for. Few representative examples of the latest are as follows:

WIND [*CALM*].
[*CROSS*] *WIND*.
[*GUSTING*] [*CROSS*] *WIND*.
WIND [*DATA NOT AVAILABLE*].
WIND [*120*] *DEGREES* [*3*] *KNOTS BETWEEN* [*060*] *AND* [*150*] *DEGREES*.
WIND [*230*] *DEGREES* [*9*] *KNOTS*.
WIND [*240*] *DEGREES* [*13*] *GUSTING* [*23*] *KNOTS*.
EXPECT [*TAIL*] *WIND*.
EXPECT WIND [*VARIABLE*].
WIND [*VARIABLE*].
WIND [*VARIABLE*] [*1*] *KNOT*.
WIND SPEED [*180*] *DEGREES* [*4*] [*METRES PER SECOND*].
WIND SPEED [*25*] [*KILOMETRES PER HOUR*].
WIND SPEED [*10*] [*METRES PER SECOND*] [*GUSTING*].
WIND SPEED [*5*] [*METRES PER SECOND*] [*MAXIMUM*].
WIND SPEED [*12*] [*METRES PER SECOND*] [*MINIMUM*].
WIND SPEED [*6*] [*METRES PER SECOND*].
WIND SPEED [*VARIABLE*] [*8*] [*METRES PER SECOND*].

where it can be seen that the structure of phrases is much more variable with significantly shorter fixed text frames, if there are even any such at all.

Thus, in the real situations, we (or the LDS system creators in general) are facing the task of representative phrases selection, given a (large) set of "raw" texts covering the given domain. In other words, having a set of texts which may appear in the domain, the task is to select the set of segments (chunks) to be recorded which is both the smallest in size and rich in variability (i.e. covers the domain as the whole or allows a high-quality generation of the missing pieces). In principle, the following are rough ways of how texts to be recorded can be extracted from the given texts:

1. record all available sentences
   pros: no concatenation is necessary, because all possible sentences are in the corpus (theoretically), the recorded sentences are only replayed back with the highest level of naturalness
   cons: huge (= expensive) speech corpus and storage resources consumption
2. split the given text to a disjunctive set of text chunks and to record each chunk individually
   pros: the smallest possible speech corpus
   cons: chaining in the pauses between words – result will not sound fluently, speech artefacts may appear at the concatenation points
3. split the given text to a disjunctive set of text chunks, but enhance each chunk with its left and right context of the appropriate length, and to record each extended chunk individually (discussed in Section 2.1)
   pros: small speech corpus, the possibility to find the optimal concatenation point in the contexts which naturally overlap
   cons: not the smallest possible speech corpus

Here, the third way seems to be a suitable compromise, and that is the approach we adopted. The question now is how to extract the chunks, so further follows the algorithm we have developed for this task.

2.1 The Role of the Context

Let us suppose now, that we want the corpus of the minimum size. To ensure this, there must be minimum duplicity in the recorded texts. Thus, having for example the following chunks in the corpus:

```
WIND SPEED ...
... 4 ..., ... 5 ..., ... 6 ...
... METRES PER SECOND ...
... MINIMUM.
```

we want to synthesized the sentence `WIND SPEED 6 METRES PER SECOND MINIMUM`. Due to no possible overlap, the output sentence would have to be composed from the chunks:

WIND SPEED	6	METRES PER SECOND	MINIMUM.

Since the individual chunks start and end with pause, they have to be recorded in this way, short pause will naturally appear at each concatenation of two chunks. Even if the pause is minimized, there is still a possibility of speech style twist, since people have a natural tendency to pronounce isolated words in a different way than they would be pronounced within a phrase [8].

The presence of context word (or any arbitrary, yet natural text chunk) helps avoiding this, of course at the cost of a bit larger speech corpus. The situation is illustrated on Figure 1 and in details described in [9].

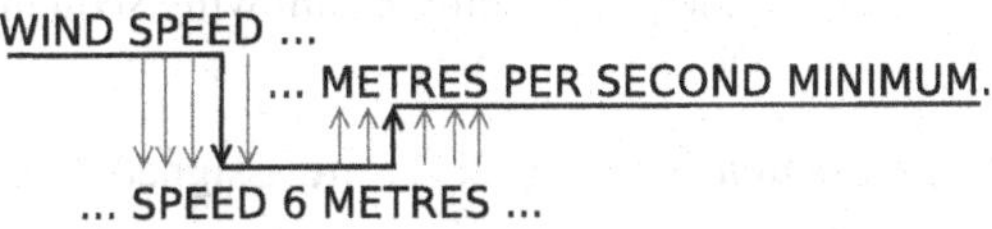

Fig. 1. The concatenation of text chunks in their contexts. The arrow mark all possible concatenation point from which the best (highlighted) can be dynamically selected during synthesis.

3 Text Selection Algorithm

To formalize further description, let $S = \{S_1, S_2, \dots\}$ be a set of all sentences from the domain we have in disposal, each consisting of a sequence of words $S_s = w_1, w_2, \dots$. Let us define $W = \{w_1, w_2, \dots\}$ to be the set of all unique words in the whole set S, and $W(s)$ to be the set of all unique words in a s-th sentence S_s. Similarly, let W^{II} and $W^{II}(s)$ be defined for all unique bigrams in S and S_s respectively.

Now, for each word we can simply count the number of occurrences $C(w)$, $w \in W$, as well as the number of occurrences for each bigram $C^{II}(w)$, $w \in W^{II}$. Let further

$$C = \sum_{w \in W} C(w) \tag{1}$$

$$C^{II} = \sum_{w \in W^{II}} C^{II}(w) \tag{2}$$

denote the total number of unigrams (i.e. words) and bigrams in S, and

$$C(s) = \sum_{w \in W(s)} C(w) \tag{3}$$

$$C^{II}(s) = \sum_{w \in W^{II}(s)} C^{II}(w) \tag{4}$$

denote the total number of unigrams and bigrams in a particular sentence S_s.[2]

The task of the selection algorithm described further is to select new set of sentences S^* containing (all) text chunks which appeared in the original set S, but minimum in size and preserving a context through which individual corresponding chunks can be concatenated together, as discussed in Section 2.1.

For the algorithm we also define a constant $R \in (0, 1)$, managing the ratio between unigrams (value closer to 0) and bigrams (value closer to 1), which affects the selection algorithm as described in Section 3.1. It is also possible to define the maximum number of text chunks M which will be selected by the algorithm and recorded later on. Note however, that too small value of M will lead to S^* where $W^* \subset W$ (i.e. some words of the domain will not exist in the selected chunks), but the choice of the suitable value of M depends on many factors like the size and character of the domain, budget limitations, speaker availability, and so on.

The proposed chunks selection algorithm for the speech corpus building is a simple loop in the interval $l = 0, 1, \ldots, M$, in which the following sequence of operations is carried out:

Sentence evaluation – all sentences $s = 1, 2, \ldots$ are assigned with score computed as

$$\sigma_s = R\frac{C^{II}(s)}{C^{II}} + (1 - R)\frac{C(s)}{C}\,. \tag{5}$$

The reason of adding counts of different bigrams and unigrams is to cover more data with the output text corpus, the example is illustrated in Section 3.1.

Choice of the best sentence – that sentence with the highest σ is selected:

$$s^* = \arg\max_s \sigma(s) \tag{6}$$

The sentence S_{s^*} will contribute the most to the coverage of the limited domain, so it is added to the set of text chunks S^* to be recorded.

Resetting of counters – we does not have to consider the words from the selected sentence (and thus the bigrams as well) any more, since they are from now

[2] Note that although $C(w)$ and $C(s)$ can be interchanged, we will strictly use indexing s for sentences and w for words or bigrams in case of $C^{II}(w)$.

included in the LD text corpus and their repetition will only increase the size of the corpus without adding any real benefits (plus it violates our requirement for disjunctive set S^*). Therefore, we can set the $C(w) = 0, \forall w \in W(s^*)$ and $C^{II}(w) = 0, \forall w \in W^{II}(s^*)$.

Cutting of phrase to chunks – to prevent the consideration of (now) needless bigrams in the selection procedure, we split the sentences in S containing bigrams in $W^{II}(s^*)$ according to the following scheme:

1. examine all $s = 1, 2, \ldots, s \neq s^*$
2. for the given s examine all bigrams from $w = W^{II}(s)$ in the order of their occurrence in S_s
3. if $w \in W^{II}(s^*)$,
   - split the sentence into two in such a way that the left part S_s^l will end with the first word of the bigram and the right part S_s^r will start with the second bigram's word
   - clear S_s^l or S_s^r if it contains one phone only (that from w)

   otherwise keep S_s unchanged
4. remove the original S_s and add the splits into the set (if they are not empty) $S = (S \setminus \{S_s) \cup \{S_1^l, S_2^r\}$ and start from 1 with the updated set

Each sentence can, of course, be split to more than two chunks, depending on the bigrams in the selected sentence S_{s^*}. In the following example, S_{s^*} contains text chunks `WIND SPEED` and `METRES PER SECOND` which are also contained in another sentence in the set:

S_{s^*}:	`WIND SPEED 5 METRES PER SECOND GUSTING.`
S_s, $s \neq s^*$:	`WIND SPEED 12 METRES PER SECOND MINIMUM.`
S^l from 4th step:	`... SPEED 12 METRES ...`
S^r from 6th step:	`... SECOND MINIMUM.`

Naturally, the better the domain is described (i.e. more texts from the domain we have in disposal), the better the algorithm will work, and of course the lower is the chance that some of the words or texts chunks are missing in the speech corpus. Sometimes, however, it is not simply possible, for example where names (both first and especially surnames) appear in the domain. Before the algorithm start, nevertheless, we can extend the set S with "cloned" sentences, where additional names are added. Similarly, we can extend the set with numbers, hours and any other slot types for which we need to ensure full (or higher) coverage.

3.1 The Impact of R Value

The choice of value R has a decisive influence on the sentence selection. To illustrate it, we created an artificial set of sentences $S = \{S_1, S_2, S_3, S_4\}$

S_1: A B C D E
S_1: A A A D
S_1: A A A D
S_1: A B A A E D

Table 1. The counts of unigram and bigram occurrences per token $C(w)$ and $C^{II}(w)$, and their sum as defined by Equations (1) and (2)

w	A	B	C	D	E
$C(w)$	10	4	3	3	2
C	22				

w	AA	AB	AD	AE	BA	BC	CA	CD	DE	ED
$C^{II}(w)$	4	4	1	1	1	3	1	1	1	1
C^{II}	18									

Table 2. The illustration of the impact of R-value on the sentence choice. Bold values are the highest values of Equation (5) denoting the selected sentence s for the given R.

			R										
s	$C(s)$	$C^{II}(s)$	1.0	0.9	0.8	0.7	0.6	0.5	0.4	0.3	0.2	0.1	0.0
1	22	9	0.50	0.55	0.60	0.65	0.70	**0.75**	**0.80**	**0.85**	**0.90**	**0.95**	**1.00**
2	13	5	0.28	0.31	0.34	0.37	0.40	0.43	0.47	0.50	0.53	0.56	0.59
3	19	11	0.61	0.64	0.66	0.69	**0.71**	0.74	0.76	0.79	0.81	0.84	0.86
4	17	12	**0.67**	**0.68**	**0.69**	**0.7**	0.71	0.72	0.73	0.74	0.75	0.76	0.77

with the values of unigram/bigram counts shown in Table 1, and with the values of per-sentence unigram, bigram and $\sigma(s)$ (see Equation (5)) shown in Table 2.

When the value of R is shifted towards bigrams ($\rightarrow$ 1), the algorithm prefers sentences with the higher number of higher frequent bigrams (AA, AB), while it prefers unigrams (words) when the value ($\rightarrow$ 0). The value near 0.5 leads to the balance of choice between both the most common unigrams and bigrams.

If the value M is defined as $infinity$, meaning that we want to generate all chunks from the set S (i.e. 100% coverage of the domain), any value of R has no influence on the resulting S^* set, only the order of chunks selection is different.

4 Helping Speaker in the Recording

For the recording of the corpus, we have used our own tool described in [4]. Contrary to the recording for general-purpose TTS, where whole phrases and/or sentences are presented to the speaker, the text from the domains were smashed to chunks which can be difficult for a speaker to record them in the required prosody style. The fact is that a chunk is required to be pronounced in such a way to fit seamlessly into the phrase (or phrases) where it belongs to.

To help the speaker keep the style while recording, we decided to present the chunks in relation with the original phrase:

```
WIND SPEED 2 KILOMETRES PER HOUR.
... 3 KILOMETRES ...
... 4 KILOMETRES ...
... 9 KILOMETRES ...
```

In the recording tool, the whole text is recorded into a single prompt (which is cut later). The speaker was instructed to read the first line as a whole, and to read the following chunks in the same style as the chunk within the first line, with short pause between chunks. Since the human memory is capable to remember recent events better than further, and since the speaker is usually professional or semi–professional, there is good chance that the chunks will be recorded in very similar (or even undistinguishable) style to keep prosody and co-articulation. In the case that there are more chunks, new recording prompts are generated, like:

```
# WIND SPEED 2 KILOMETRES PER HOUR.
... 2 KILOMETRES ...
... 7 KILOMETRES ...
... 5 KILOMETRES ...
... SPEED 1 KILOMETRE PER HOUR.
```

The speaker was instructed not to read the '#'–starting line (although he/she could), it is intended to make the speaker recall the style in which the preceding prompt was read (naturally, chunks from the given phrase follow during the recording). And still, if the chunks' style diverges slightly (which cannot be entirely avoided), there is the context, in which the optimal concatenation point is searched for to ensure as smooth transition as possible.

5 Results and Evaluation

We tested the algorithm on several domains. The results of using this approach are described in the following subsections, for all experiments we set $R = 0.5$. To quantify the behaviour of the algorithm somehow, we compare the size of the original set S and the generated set S^* from the viewpoint of:

- the number of items in the sets, where the items are sentences in S and chunks in S^*,
- the average number of words per set item (average sentence/chunk length),
- the total number of words in the sets (i.e. C from Equation (1)),
- the number of different words in the sets (i.e. cardinality $\|W\|$),

while the set S can be considered either in *full* form (with "classic" texts) or in *reduced* form where particular slot values can be replaced by appropriate place–holders, e.g. numbers by `$NUM` and so on. The size of S^* is always presented in its "full" form.

5.1 ATIS Limited Domain

For this domain, we had thousands of "real" sentences, automatically downloaded every 2 hours from the public web page where actual ATIS informations are published. The downloading started at autumn 2013.

Looking at the corpus sizes in the Table 3, it can be seen that S^* is much lower than full S. The selected chunks occupy only 0.15% of all the possible sentences, while covering 100% of the full set.

Table 3. The comparison of original sentences set size to the size of selected chunks

	S, reduced	S, full	S^*
items in sets	1,177	510,311	3,368
number of words C	7,408	8,952,311	13,507
number of different words	728	737	737
average sentence length	6.3	17	4.0

5.2 ATC Communication Messages

The domain of *Air Traffic Control* has a rather strictly defined phraseology, nevertheless it is both quite large and not strictly abided by pilots and controllers. In the input, moreover, we have transcripts of real communications instead of phrases defined by the phraseology, so there is much higher variability in the form of a many particular phrases (inserted or missing "filler" words, change in the word order, etc.).

Due to the more vague nature of the domain, we, therefore, did not required full coverage in the selection, but we have choose $M = 1,000, 2,000, \ldots, 5,000$ of selected chunks. Therefore, in Table 4 we further present the perceptual coverage of words (comparing W items from full S and S^*) showing how big the intersection of the two sets is, and the number of words $\notin S^*$ as well as the highest $C(w)$ for $w \notin S^*$.

Table 4. The comparison of original sentences set size to the size of selected chunks

	S, full	S^*				
number of sentences	45,551	1,000	2,000	3,000	4,000	5,000
number of words C	391,429	6,919	12,276	15,455	19,567	23,431
number of different words $\|\|W\|\|$	2,012	938	1197	1347	1513	1617
coverage of $\|\|W\|\|$	100%	46.6%	59.5%	66.9%	75.1%	80.4%
number of missing words	–	1,074	815	665	499	395
$\max(C(w))$ for missing words	–	18	8	4	3	2

It can be seen from Table 4, that the coverage reached is rather high and all the most frequent words are included in S^*. Although there is still a significant number of word and bigram tokens missing in the selected text, and thus they will have to be generated by a general TTS, their frequency is very low.

6 Conclusion

The paper presents the algorithm for building a limited–domain text corpus, which is as small as possible possible, while still allowing natural and smooth transition between the recorded chunks by keeping sufficient context in which an optimal (not being a pause) concatenation point can be found. It is demonstrated, that even with the context, the number of prompts to be recorded is only a fragment of the original domain. What remains to be said is, that all the presented corpora have already been recorded, tested and evaluated, and the dominance of the LD synthesis using them has been proved, as discussed in [9].

Special thanks are due to the National Grid Infrastructure MetaCentrum providing access to computing and storage facilities under the program LM2010005 "Projects of Large Infrastructure for Research, Development, and Innovations".

References

1. Brenton, H., Gillies, M., Ballin, D., Chatting, D.: The uncanny valley: does it exist. In: 19th British HCI Group Annual Conference: Workshop on Human-animated Character Interaction (2005)
2. Tihelka, D., Kala, J., Matoušek, J.: Enhancements of Viterbi search for fast unit selection synthesis. In: INTERSPEECH 2010, Proceedings of 11th Annual Conference of the International Speech Communication Association, pp. 174–177 (2010)
3. Grůber, M., Hanzlíček, Z.: Czech expressive speech synthesis in limited domain: Comparison of unit selection and HMM-based approaches. In: Sojka, P., Horák, A., Kopeček, I., Pala, K. (eds.) TSD 2012. LNCS, vol. 7499, pp. 656–664. Springer, Heidelberg (2012)
4. Matoušek, J., Tihelka, D., Romportl, J.: Building of a speech corpus optimised for unit selection tts synthesis. In: LREC 2008, Proceedings of 6th International Conference on Language Resources and Evaluation. ELRA (2008)
5. Matoušek, J., Romportl, J.: On building phonetically and prosodically rich speech corpus for text-to-speech synthesis. In: Proc. of the Second IASTED Int. Conf. on Computational intelligence, pp. 442–447. ACTA Press, San Francisco (2006)
6. Tihelka, D.: Towards automatic measure of similarity for use in unit selection. In: 9th Int. Conf. on Signal Processing, ICSP 2008, Beijing, China, pp. 637–642 (2008)
7. Black, A.W., Zen, H., Tokuda, K.: Statistical parametric speech synthesis. In: Proc. ICASSP 2007, pp. 1229–1232 (2007)
8. Labov, W.: The Social Stratification of English in New York City. Center for Applied Linguistics, Washington, DC (1966)
9. Jůzová, M., Tihelka, D.: Tuning limited domain speech synthesis using general tts system. Accepted at Text, Speech and Dialogue 2014 (2014)

Tuning Limited Domain Speech Synthesis Using General Text-to-Speech System*

Markéta Jůzová and Daniel Tihelka

University of West Bohemia, Univerzitní 8, Plzeň, Czech Republic
{juzova,dtihelka}@kky.zcu.cz
http://www.kky.zcu.cz

Abstract. The subject of the present paper is the building of a limited domain speech synthesis system, where longer units, like words and phrases, can naturally be concatenated together. However, instead of building a single-purpose domain-oriented engine working with longer units, we show that a general-purpose TTS system can be used as a good emulation tool to ensure that a real domain-oriented engine will work correctly. Since the current general speech synthesis system embedding unit selection method concatenates short speech units (diphones), the selection algorithm has been modified to pretend the concatenation of words or even the whole phrases, while still concatenating diphones internally. The behaviour of the system is tested on two limited domains and its output is compared to the output of general (unmodified) version of the same TTS system. The results show clear encouragement for the build of the "real" domain-oriented engine.

Keywords: limited domain, concatenative speech synthesis, speech units, units concatenation, target cost, concatenation cost.

1 Introduction

Limited domain speech synthesis (LDTS, [1,2,3,4]), i.e. the domain-oriented synthesis embedding unit selection method [5,6], attracts with the advantage of using longer units for concatenations instead of short units, e.g. diphones. The usage of longer units brings a lower number of concatenations in the synthesized sentence, which positively affects the probability of the speech artefacts occurrence, as well as it lowers the computational complexity due to the much lower number of candidates of speech units to search optimal path through. While it may seem for the first time that the build of LDTS system is fairly easy task, the opposite is true – the concatenation of longer units, being natural-sounding by their nature, can cause a phenomenon called "Uncanny Valley", describing the level of aversion to unnatural artefacts in otherwise natural surrounding.

To ensure the high-quality LDTS, the limited domain (LD) text corpus has to be prepared carefully. It is recommended that the corpus should cover well the given

* This work was supported by the European Regional Development Fund (ERDF), project "New Technologies for Information Society" (NTIS), European Centre of Excellence, CZ.1.05/1.1.00/02.0090, the Technology Agency of the Czech Republic, project No. TA01030476 and SGS-2013-032.

P. Sojka et al. (Eds.): TSD 2014, LNAI 8655, pp. 408–415, 2014.

domain [1,8], i.e. to contain the most (or ideally all) common words and phrases in contexts. The presence of contexts brings the advantage of finding the optimal concatenation point in the overlap, which is much better compared to concatenations in pauses between words. The context enlarges the corpus a bit, but it is still much smaller (and cheaper) than the corpus for a general-purpose TTS system. The general text corpus should, on the contrary, meet the requirement of a good coverage of diphones in different phonetic and prosodic contexts [7], since they are required when synthesizing a text unknown beforehand.

1.1 Our Goal

Based on the assumption about the ability of LD system to generate higher–quality speech, we aimed to create such a system (and verify if it is truth) and use it for applications, in which it would fit better than a general synthesizer. Since we have only specialized LD text corpus at disposal, but not any real LDTS system, we decided to use our general TTS system ARTIC [6] working on diphones, and twist it in such a way that it works exactly as LDTS system should work. That means, the emulation tool created in this way and presented in this paper is a general TTS system containing several modifications to ensure the pretence of the concatenation of longer units. The outputs of the designed system are compared with the outputs of the general (not modified) system to find out, if the LD synthesis achieves higher level of naturalness and thus if LDTS system is worth building and using instead of the general one.

1.2 Concatenative Method Analysis

In both versions of speech synthesizer, general and LDTS emulator, the optimal sequence of speech unit candidates is searched for the given input text. Generally, after decomposition of the text into speech units, candidates of these units stored in the speech units database are used to build the graph being evaluated by *target cost* (nodes) and *concatenation cost* (edges):

- *Target cost* $C^t(t_i, u_i)$ quantifies the difference between the speech unit candidate u_i and the requirements for the unit t_i; this regards the phonetic context, sentence type, position features etc. [10]; $C^t = 0$ for candidates originating in the same surroundings as required to be placed into.
- *Concatenation cost* $C^c(u_{i-1}, u_i)$ indicates how much the join of candidates would fit together; $C^c = 0$ for candidates neighbouring in the source corpus.

The *cumulation cost* is defined as

$$C(t_1^N, u_1^N) = \sum_{i=1}^{N} C^t(t_i, u_i) + C^c(\#, u_1) + \sum_{i=1}^{N} C^c(u_{i-1}, u_i) + C^c(u_N, \#)^1 \quad (1)$$

Having evaluated the graph, the Viterbi algorithm [11] is used for the optimal path search, choosing the path with the minimum cumulation cost.

[1] # = pause.

2 Phrase Segments Searching and Dividing into Chunks

The process of text-to-units decomposition differs significantly between the two versions of the synthesizer. In the LDTS case, we used the approach described in [4], where the longest phrase segments were searched for. For the given input sentence, this method first creates a set of segments by detracting words from the end and from the beginning of the sentence, as shown below:

`Dnes bude zataženo a zima.`	*Today it will be cloudy and cold.*
`Dnes bude zataženo a ...`	*Today it will be cloudy and ...*
`... bude zataženo a zima.`	*... it will be cloudy and cold*
`Dnes bude zataženo ...`	*Today it will be cloudy ...*
`... bude zataženo a ...`	*... it will be cloudy and ...*
`... zataženo a zima.`	*... cloudy and cold.*
`Dnes bude ...`	*Today it will be ...*
`... bude zataženo ...`	*... it will be cloudy ...*
`... zataženo a ...`	*... cloudy and ...*
`a zima.`	*... and cold.*
`Dnes ...`	*Today ...*
`... bude ...`	*... it will be ...*
`... zataženo ...`	*... cloudy ...*
`... a ...`	*... and ...*
`... zima.`	*... cold.*

It is evident from the example that the term "phrase segment" refers to any word sequence (sometimes even containing only one word).

After decomposing the given sentence into segments, the longest segments contained in the LD speech corpus must be found. Let's say, the corpus contains only one sentence and one phrase:

`Dnes bude slunečno a větrno.`	*Today it will be sunny and windy.*
`... bude zataženo a ...`	*... it will be cloudy and ...*

For the given sentence

`Dnes bude zataženo a zima.`	*Today it will be cloudy and cold.*

the following longest segments are found in the corpus:

`Dnes bude ...`	*Today it will be ...*
`... bude zataženo a ...`	*... it will be cloudy and ...*

The fact that the phrases overlap with the word "bude" is intentional (the corpus has been designed to ensure this, as described in [8]), and the advantage of the overlap is discussed further in Section 3. Unfortunately, they do not cover the whole given sentence – there is the last word

`... zima.`	*... cold.*

not contained in the corpus. The reason for this may be that this word does not belong to the domain, or due to a huge size of the domain the word is missing in the corpus. Note that this is a model example, in which we intentionally do not include the word "zima" in the corpus to show the behaviour of the system in such a case. In reality, we try to prevent this situation as far as possible when building the specialized text corpus by the algorithm introduced in [8].

Before the next step we transform the segments into three types of chunks:

- type 0 – chunks appearing in found phrase segments only once
  `Dnes ...` *Today ...*
  `... zataženo a ...` *...cloudy and ...*
- type 1 – chunks representing the overlapping regions of segments, in which the optimal concatenation point is searched, see Section 3
  `... bude ...` *... it will be ...*
- type 2 – chunks corresponding to the part not found in the corpus
  `... zima.` *... cold.*

3 Synthesis of Individual Sentence's Chunks

The "real" LD synthesizer would just take type 0 chunks, find the optimal concatenation point to concatenate them together and, if needed, the general TTS system will be used for the synthesis of chunks of type 2 using short speech units.

Since we use the general TTS system as the underlying platform, it decomposes the synthesized text into diphones[2]. To ensure that the general system emulates the LDTS way of work, we modified the unit selection algorithm [11] to pre–prune the candidates graph according to the type of synthesized chunk.

Type 0

The real LDTS system would take the chunk of type 0 as an atomic unit, and will as a whole concatenate it with the following chunk of type 1 or 2. Therefore, we have to ensure that only unit candidates from the phrase representing the chunk will remain in the selection algorithm, as drawn in Figure 1. Although we cannot avoid the evaluation of the costs, they are both $C^t(t_i, u_i) = C^c(u_{i-1}, u_i) = 0, \forall i$ and there is only one possible path through the graph.

Type 1

In the case of overlapping segments, the real LDTS system must choose the optimal concatenation point within the segment. To achieve this, the emulator prunes out all unit candidates except those from the two overlapping segments, as illustrated in Figure 2. During the costs evaluation, $C^t(t_i, u_i) = 0, \forall i$ due to the same reasons as for type 0, and $C^c(u_{i-1}, u_i) = 0$ for all i except the point in segments transition.

Type 2

The chunks marked as type 2 are not contained in the LD corpus at all. Therefore, all candidates for the given units can be used to find the optimal sequence in the same way

[2] Let's note that all examples are demonstrated on letters for their better understandability; the fact that the real TTS works with diphones does not affect the described algorithms in any way, one must just put diphone in place of letter in the examples.

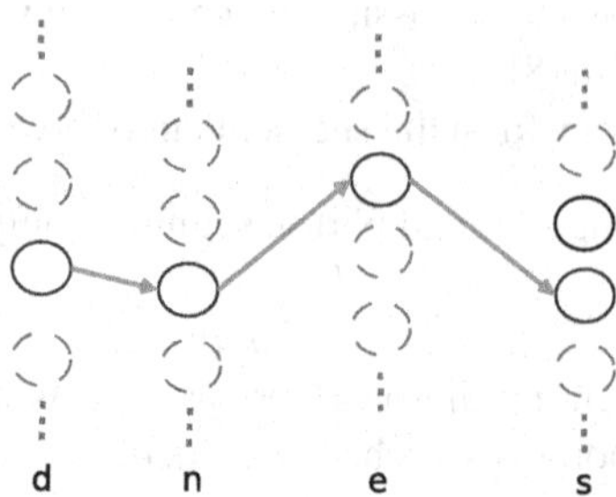

Fig. 1. Graph structure for units in the chunk of type 0 (`dnes` = *today*). The dashed–lined circles represent the unit candidates originated from different places in the LD corpus, which must not be considered. The arrows represent the only possible path.

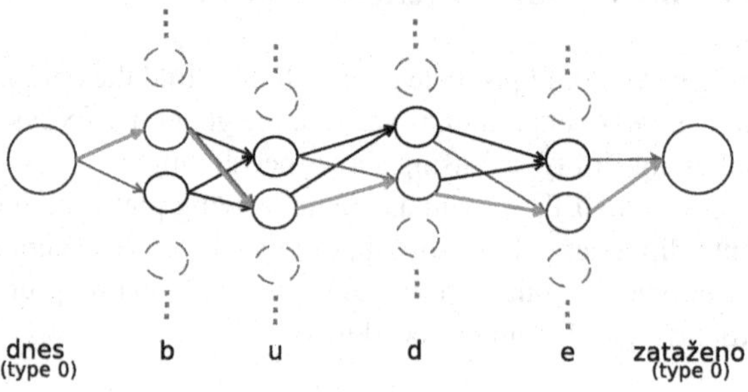

Fig. 2. Graph structure for a chunk of type 1 (`dnes` = *today*, `bude` = *it will be*, `zataženo` = *cloudy*). The edges connecting the nodes from the first segment to the nodes from the second segment represent the possible concatenation points in the chunk, the highlighted ones represent some of the optimal paths through the graph.

as it works in the general-purpose TTS; see Figure 3 for the illustration. Both costs are important (getting generally non-zero values), because there are many candidates available for every unit in the chunk.

4 Evaluation

Although the LDTS emulator builds much less extensive graph of candidates for Viterbi searching, which significantly reduces the computational complexity of the synthesizer, there are still unnecessary cost evaluations which will not occur with larger units (all the zero costs in the chunks of type 0 and 1). Nevertheless, rough estimation of performance gain has been concluded in Section 4.1.

However, the reduction of computational cost is only one of the reasons why to prefer domain-oriented system to that general–purpose. The main motivation was to verify whenever speech generated from much smaller speech corpus using LDTS system can

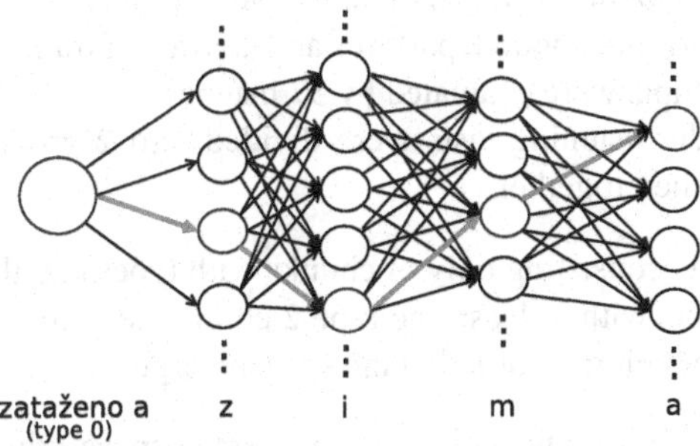

Fig. 3. Graph structure for a chunk of type 2 (`zataženo a` = *cloudy and*, `zima` = *cold*). The highlighted edges represent an example of the optimal path in the graph.

outperform the naturalness of speech generated from large speech corpus [7] by our well tuned general TTS system; see Section 4.2.

4.1 The Evaluating of the Computational Complexity

In the case of general synthesis, there are tens or hundreds of candidates for every diphone in the corpus [11], leading to huge amount of $C^t(t_i, u_i)$ and $C^c(u_{i-1}, u_i)$ evaluations. It is illustrated in Table 1 that there is only a fraction of evaluations for the same synthesized phrase in the emulator, even when the unnecessary evaluations are considered as well. On the other hand, the task of text–to-units decomposition is more complex in case of LDTS system (and not very optimized in this experimental version). Therefore, the performance gain of LDTS version is bit lower than it would be estimated from the ratio of costs evaluations, but still an order of magnitude above the general–purpose TTS system. This is summarized in t_1/t_2 column in Table 1.

Table 1. The evaluating of the computational complexity in the number of costs evaluations. t_1/t_2 is the ratio of times required to synthesize the same set of LD texts.

		LDTS emulator		general TTS		
number	chunks' types	C^t	C^c	C^t	C^c	t_1/t_2
1	only type 0	38	36	35 824	18 538 254	1/96
2	only types 0 and 1	62	99	41 612	20 230 985	1/35
3	one short chunk of type 2	505	767	44 393	30 397 914	1/23
4	types 0, 1, 2	8 683	1 484 595	42 508	30 613 890	1/16

4.2 The Evaluating of the Quality of Synthesized Sentences

To obtain the comparison of the quality of LDTS-generated speech with the quality achieved by the general TTS system, we carried out listening tests on two limited domains for which we have speech corpus recorded. The first one is the telephone automaton informing about results of exams on our university, the second is the automatic system informing about departures and arrivals of trains and ticket prices [9]. All the generated prompts were evaluated by 57 listeners.

The prompts for the listening tests were divided into 2 groups, which were first evaluated separately, then together:

- 1st group: prompts consisting only of chunks with types 0 and 1
- 2nd group: prompts with at least one type 2 chunk, i.e. with at least one word out of the domain, not being recorded in the speech corpus

The same number of prompts in both groups were purposefully chosen to represent, in authors' view, "the worst case" — in reality, there would be more phrases from the 1st group during the usual LDTS system operation, supposing that the domain is covered well with the specialized corpus [8].

For the evaluating, we used *CCR tests* with a simple 3-point scale:

- 1 (LD) - the output of the emulated LDTS system sounds better,
- 0 (S) - the outputs are about the same (sound either good or bad),
- -1 (G) - the output of general TTS system sounds better.

The quality is then defined as

$$q = \frac{1 \cdot LD \; + \; (-1) \cdot G}{LD + S + G} \,, \tag{2}$$

where LD, S, G are the numbers of listeners' answers in the given category. The positive value of q means that the output of the LD synthesizer sounds better.

To ensure the validity of results, we also carried out the *sign test* with the null hypothesis "H_0: *the outputs of the both synthesizers are of the same quality.*" compared against the alternative hypothesis "H_1: *the output of one synthesizer sounds better,*" the p-values were determined for the significance level $\alpha = 0.05$.

All the results are shown in the Table 2. It is evident, that the quality of LDTS is evaluated higher than the quality of the general synthesis.

Table 2. The evaluating of the quality. The numbers of listeners' answers, the value q counted using the (2) and the $p-value$ of *sign test*.

	1st group	2nd group	all sentences
LD system is better (LD)	280	159	439
general TTS is better (G)	105	184	289
systems are of the same quality (S)	71	56	127
q	0.384	-0.063	0.175
p-value for $\alpha = 0.05$	< 0.0001	0.0975	< 0.0001
conclusion	LD system better	the same quality	LD system better

5 Conclusion

Summing up the results, it was confirmed that the LDTS system achieves better results, both in terms of quality and performance. Although this is rather expected result, most importantly it confirms that the development of LDTS system still makes a lot of sense for applications with requirements for the highest possible quality of generated speech.

On the other hand, it must be mentioned that the outputs of a domain-oriented LDTS system can be better only if the domain is covered well with the speech corpus used together with the LDTS system. The results show that if the synthesized text contains unknown word, the quality is, naturally, lowered to the level achieved with the general–purpose TTS system. It is the fact that the 100% coverage of given domain can not usually be met due to constraints put on the corpus size, recording time or development budget. There still may be chunks not coverable by the corpus at all (e.g. surnames, street names, etc.), or there may appear new texts required to be synthesized which had originally not been considered and recorded. Therefore, the corpus preparation must be taken very carefully, and it generally wise to extend the LD corpus with an appropriate amount of out-of-domain texts to ensure the coverage of rare (as related to the domain, not rare in general) units which will have to be used for type 2 chunks.

References

1. Black, A.W.: Perfect Synthesis for all of the people all of the time. In: IEEE Workshop on Speech Synthesis, Santa Monica, USA (2002)
2. Black, A.W., Lenzo, K.A.: Limited Domain Synthesis. In: ICSLP 2000 (2000)
3. Yi, J., Glass, J.: Natural-sounding speech synthesis using variable-length units. In: Proceedings of ICSLP, Sydney, Australia, pp. 1167–1170 (1998)
4. Donovan, R.E., Franz, M., Sorensen, J.S., Roukos, S.: Phrase splicing and variable substitution using the ibm trainable speech synthesis system. In: Proc. of the Acoustics, Speech, and Signal Processing 1999, pp. 373–376. IEEE Computer Society, Washington, DC (1999)
5. Matoušek, J., Romportl, J., Tihelka, D., Tychtl, Z.: Recent Improvements on ARTIC: Czech text-to-speech system. In: INTERSPEECH 2004 – ICSLP, Proc. of the 8th Int. Conf. on Spoken Language Processing, Korea, pp. 1933–1936 (2004)
6. Matoušek, J., Tihelka, D., Romportl, J.: Current state of Czech text-to-speech system ARTIC. In: Sojka, P., Kopeček, I., Pala, K., et al. (eds.) TSD 2006. LNCS (LNAI), vol. 4188, pp. 439–446. Springer, Heidelberg (2006)
7. Matoušek, J., Tihelka, D., Romportl, J.: Building of a Speech Corpus Optimised for Unit Selection TTS Synthesis. In: LREC 2008, Proc. of 6th Int. Conf. on Language Resources and Evaluation. ELRA (2008)
8. Jůzová, M., Tihelka, D.: Minimum Text Corpus Selection for Limited Domain Speech Synthesis. In: Sojka, P., et al. (eds.) Proc. of Text, Speech and Dialogue (2014)
9. Švec, J., Šmídl, L.: Prototype of Czech Spoken Dialog System with Mixed Initiative for Railway Information Service. In: Sojka, P., Horák, A., Kopeček, I., Pala, K., et al. (eds.) TSD 2010. LNCS (LNAI), vol. 6231, pp. 568–575. Springer, Heidelberg (2010)
10. Ramportl, J., Tihelka, D.: Exploring Automatic Similarity Measures for Unit Selection Tuning. In: Proc. of Int. Conf. Interspeech 2009, pp. 736–739 (2009)
11. Tihelka, D., Kala, J., Matoušek, J.: Enhancements of Viterbi Search for Fast Unit Selection Synthesis. In: Proc. of Int. Conf. Interspeech 2010, pp. 174–177 (2010)

Study on Phrases Used for Semi-automatic Text-Based Speakers' Names Extraction in the Czech Radio Broadcasts News

Michaela Kuchařová[1], Svatava Škodová[2], Ladislav Šeps[1], and Marek Boháč[1]

[1] Institute of Information Technology and Electronics, Technical University of Liberec, Studentskáa 2, 461 17, Liberec, Czech Republic
{michaela.kucharova1,ladislav.seps,marek.bohac}@tul.cz

[2] Department of the Czech Language and Literature, Technical University of Liberec, Studentská 2, 461 17, Liberec, Czech Republic
svatava.skodova@tul.cz

Abstract. In this paper we introduce a methodology leading to the extension of speakers' database used in the process of automatic transcription of spoken documents stored in the largest Czech Radio audio archive. We address the issue of the conversion of spoken speech to written texts – the automatic detection of speakers and their names. We work with a subset of the archive that consists of 8,020 hours of broadcasting news and 58,914,179 words within the years 1968–2011. We observed the occurrence of thousands of speakers' names during the period and therefore it is necessary to use their automatic or semi-automatic identification. Another investigated issue leading to the extension of speakers' database is the co-occurrence of a speaker's name in a specific phrase in the text transcription linked with the speaker's change in the audio recording.

Keywords: audio archive processing, spoken formal speech, radio broadcast news, automatic speech recognition, text-based speakers' names extraction.

1 Introduction

The research described in this paper comes from the needs of a project which aims at making the Czech Radio audio archive publicly accessible and searchable [1]. This task demands many speech recognition technologies: i) a Large-Vocabulary Continuous Speech Recogniser (LVCSR) able to recognise Czech and Slovak (and to automatically determine which language is being spoken), ii) a document segmentation module able to split the document into speaker-homogenous segments, iii) a speaker identification tool and iv) a post-processing module, which makes the recognised text comfortably readable. The third mentioned module – the speaker identification tool – must be provided with a database of speakers. This database can be viewed at two levels – firstly, as a list of important speakers and persons (speakers who appear or are mentioned often enough) with some basic information (e.g. gender, language) and secondly, as the training data and corresponding model linked to one of the enlisted speakers.

As the oldest recordings are 90 years old and the amount of the recordings approaches 100,000 hours, the number of speakers occurring in the broadcasts is

P. Sojka et al. (Eds.): TSD 2014, LNAI 8655, pp. 416–423, 2014.

enormous. Hence we need an unsupervised way to perform the following tasks: i) to find names in the recognised text (this part covers the problem of recognition of surnames originating in common words, i.e., *Jan Medvěd* (bear) is recognised as *Jan medvěd*), ii) to decide whether the person is a speaker or he/she is only mentioned by the actual speaker (e.g. *"reporter X.Y. is speaking from the country Z"* vs. *"president X.Y. visited the country Z"*) and iii) to estimate if there is a recording segment pronounced by the found speaker (e.g. *"our reporter X.Y. is reporting from..."*).

To improve the automatic transcripts we plan to use a speaker adaptation technique [2] that can improve the recognition accuracy by approx. 4%. For this technique we need to identify speakers and we have to prepare acoustic data for each speaker.

2 Technical Background – Systems and Programs

For our research, we needed several systems and programs described in this part. First, an automatic speech recognition system, its description is given in Sec. 2.1; next, a program NanoTrans [3] to correct several transcripts that we use for training and testing. Above that, we had to prepare several scripts to find names, phrases and to count the number of their occurrences. All these scripts were especially designed to work with a vast amount of data – the transcripts of the recordings are stored in 18,239 files.

2.1 Automatic Speech Recognition System

The transcriptions of recordings of spoken documents stored in the Czech Radio audio archive were prepared by the standard LVCSR system [1]. This system was developed at the Technical University of Liberec and processes 16 kHz audio data, parameterises them into 39 MFCC features, and then applies global or floating CMS. The acoustic model uses a context-dependent triphone HMMs to represent 41 Czech phonemes and seven types of noises. It has been trained on 320 hours of speech (microphone and broadcasting speech). The decoder works in real time with a vocabulary whose size is about 500,000 lexical words. The language model is based on bigrams and is smoothed by the Kneser-Ney method.

2.2 Document Segmentation and Speaker Identification Modules

Document segmentation is a part of diarisation. It follows a standard framework consisting of three parts – voice activity detection, speaker turn detection and speaker clustering. You can see more in [2]. This module splits transcript into segments based on distinct speakers detected locally in a given recording.

The speaker identification module assigns the speaker (his/her ID or name) to the locally identified segments. Input to this module are speaker models that are trained from an acoustic data labelled by speaker ID. For our purpose we used the first mentioned module, and we attempted to find adequate acoustic data for the second one.

2.3 Phrase Patterns Used for the Proper Names Search

Experiments performed previously [4] indicated that the limiting factor for crawling through the transcriptions is the actual read speed of a hard drive, not the speed of a CPU. The more complex nature of present experiments described in this paper prevented us from using simple text-based searches because more complex pattern-matching techniques were required. The regular expressions proved to be too slow for the given task (as one experiment took more than 3 hours). To improve the search speed, we implemented our own simple pattern-matching algorithm. To make it as fast as possible, we limited the smallest searchable unit to the whole word, instead of a character used in the regular expressions. Additionally, by restricting patterns to start with a word, we reduced the time of one experiment below 15 minutes.

Table 1. Examples of the phrase patterns and retrieved phrases

Phrase patterns	Examples of retrieved phrases
více ? * * * * $ (m/f) *more ? * * * * $*	**více** už ale prozradí XY (m/f) **více** už o tom naše zpravodajka XY (f)
telefonuje * * **zpravodaj** (m) **telefonuje** * * **zpravodajka** (f) *is phoning * * rapporteur*	**telefonuje** z Moskvy **zpravodaj** (m) **telefonuje** náš pařížský **zpravodaj** (m) **telefonuje** naše moskevská **zpravodajka** (f)
sleduje * * **zpravodaj** (m) **sleduje** * * **zpravodajka** (f) *monitors * * rapporteur*	**sleduje** náš **zpravodaj** (m) **sleduje** stálý londýnský **zpravodaj** (m) **sleduje** naše spolupracovnice **zpravodajka** (f)
pokračuje ? * * * * $ (m/f) *continues ? * * * * $*	**pokračuje** jablonecká zpravodajka XY (f) **pokračuje** redaktor XY (m)
potvrdil to ? * * * * $ (m) **potvrdila to** ? * * * * $ (f) *confirmed by ? * * * * $*	**potvrdil to** našemu zpravodajovi hlavní inženýr XY (m) **potvrdila to** její tisková mluvčí XY (f)

These patterns have been assembled on the basis of a previous experimental probe that we have performed to delimit key words and phrases preceding proper names of speakers. Each pattern consists of one or two key words (usually a verb of speaking [5], [6] and a designation of the speaker's role, i.e., *moderator*, *reporter*, *correspondent*, etc., and several defined wildcards. As the basis of the patterns we used valence patterns with the verbs of speaking [7], to deconcretize them we use combinations of the following wildcards: "?" for any word in the phrase pattern stream; "*" for zero or one word; "$" for a word starting with capital character. Example of the phrase pattern would be *"popsala českému rozhlasu ? * * * * $"* (*"described to Czech Radio? * * * * $"*). The number of asterisks in each wildcard was determined according to the results of the tested hypothesis. Overall, we used a maximum of six wildcards in one phrase.

Table 1 illustrates some examples of the patterns and their specific occurrences within the texts. The column Phrase Patterns shows examples of the patterns in Czech and their English translation; the disproportion of phrases in Czech and English is caused by the highly inflective character of Czech, which specifies the information in various lexemes or morphemes. The examples of retrieved phrases are not translated to English as far as their meaning is not significant to the text; but they are used to provide illustration of the words variability represented by wildcards. All real names in the illustrational phrases were replaced by the signs XY for space considerations. We present only several phrase patterns with a limited amount of illustrative phrases. The symbols (f/m) in the Tables 1 and 2 are used to determine masculine or feminine gender reflected in a phrase.

3 Methodology

In Part 3, we describe in detail the complete scheme of the speakers' database extension in the following steps: i) finding all speakers´ names (name entities which are parts of speaker indicating phrase); ii) counting the speaker occurrences in the documents of interest (potentially the whole archive or some sub-period). The speakers of interest are chosen and added to the database (if not already included); iii) finding the utterances of important speakers, checking it and using it for the training of the speakers' models.

As long as the method is based on the automatic process and its successive manual correction, it is not necessary to evaluate the error rate of the method.

For this project, we need well-built acoustic and language models for automatic recognition. To improve the models, we use the specially written program NanoTrans for semi-manually rewriting the recordings from broadcasting; labelling the speakers and creating the speakers' database.

The term semi-manual rewriting is used for a combination of automatic and manual work: the automatically-generated transcriptions are corrected by people. We have 375 recordings (1 recording = 1 daily news broadcast) that store 160 hours of Czech Radio (from which there are approximately 152 hours of Czech speech, the rest being mainly Slovak, with small amounts of both Russian and English speech). These were recorded through the time period 1968–2005.

By rewriting these recordings we get a basic speaker database and labelled acoustic data to train some models to recognise speakers. We cannot use the complete set of data from rewritten recordings to train an acoustic model for speakers because there were a lot of data from telephone, some parts were spoken over music, or the recordings were of very poor quality. Also, in the entire basic speaker database not all of the speakers are relevant and some are missing. For the time period selected for the transcription (1923–2014), there are hundreds of relevant speakers and not all of them are present in our semi-manually rewritten recordings.

Therefore we decided to find all possible speakers in the recordings, choose the most frequent ones, and then train the acoustic model for them.

To search the speakers' names, we have selected 196 introductory phrases that frequently co-occur with speaker names. Inasmuch as our system lacks the lemmatisation, it was necessary to supply all indispensable morphological variations of the key words in the phrases, i.e., morphological forms of selected verbs for the present and past tense, singular and plural forms of nomina agentis. Examples of the wildcards and their realisations are seen in Table 1.

As stated in Sec 2.1, we used phrase patterns for speakers´ name search, or the co-occurrence of a speaker´s name in a specific phrase in the text transcription connected with the speaker´s change in the audio recording.

Table 2. Speakers change

	Before phrase		Number of occurrences	After phrase	
	Average distance	Standard deviation		Average distance	Standard deviation
telefonuje * * zpravodaj (m) telefonuje * * zpravodajka (f) *is phoning * * rapporteur*	7.7 8.9	45.1 58.8	1877 416	32.2 46.0	27.1 42.9
telefonoval o tom (m) telefonovala o tom (f) *phoned about*	7.9 6.9	40.8 60.2	78 11	28.7 35.9	22.0 29.0
z ? * * se přihlásil * * zpravodaj (m) z ? * * se přihlásila * * zpravodajka (f) *from ? * * has entered * * rapporteur*	13.9 11.4	41.9 27.9	708 45	26.6 26.2	30.2 14.9
jak dodává * * reportérka (f) *how adds * * rapporteur*	20.6	58.4	198	59.3	40.6
jak dodává mluvčí (m/f) *how adds speaker*	15.7	48.2	43	43.0	27.6
blíže k tomu náš zpravodaj (m) *in more details on the topic our rapporteur*	7.8	25.0	153	21.7	14.1

As far as the amount of data in the period between 1923 and 1966 is negligible from the statistical point of view, we analysed the data starting in 1966 for our purposes.

We assumed that the broadcast news follow typical text patterns using stereotypical phrases when introducing each speaker who will be next to speak. We proceeded as follows: i) we searched the data; seeking the typical phrases connected with the new speaker's occurrence, ii) we created patterns combining particular words with variable positions that hypothetically combine with the occurrence of a speaker's name, iii) we found all the occurrences of these phrases with names in all transcriptions, iv) we chose names (actually they were proper names, we had to filter them) and added the important speakers' names to the speaker database.

Among the steps leading to the extension of the database we included the findings of acoustic data for the frequent speaker (we assume that we will find the data mostly for the reporters). We took a phrase, found it in all transcripts and counted the distance between this phrase and the change of the speaker. We assume that if the change of speaker is close, there are some phrases that indicate the name of the changed speaker.

Table 2 shows the distance (count in words) from the phrase to the change of the speaker. In the first column you can see the average distance from phrase to change of a speaker before the phrase, the fourth column shows the distance between the phrase and the change of a speaker after the phrase. The second and fifth columns give the standard deviation. In the third column is the number of occurrences. You can also see that there are differences between occurrences of male and female speakers.

From the results shown in Table 2 we decided to choose acoustic data for the speakers in several steps: i) find a phrase with a speaker's name, ii) count the distance between the phrase and the following speaker and decide if it is close enough, iii) extract the speaker

Table 3. Examples of patterns occurrence

	1966-9	1970-4	1975-9	1980-4	1985-9	1990-4	1995-9	2000-4	2005-9
u mikrofonu ? * * * * $	112	247	645	823	1010	369	431	178	201
u mikrofonu $	15	17	121	304	331	204	186	69	49
at the microphone									
sleduje * * zpravodajka	0	0	0	0	0	1	64	183	50
sleduje * * zpravodaj	0	1	0	6	22	40	194	379	246
sledovala * * zpravodajka	0	0	1	0	0	6	60	142	48
sledoval * * zpravodaj	6	1	13	25	36	48	146	164	105
sledovala $	0	0	7	18	17	97	280	175	10
sledoval $	6	4	24	47	217	297	542	176	44
monitors (monitored)									
více ? * * * * $	445	493	639	704	787	1377	3726	3191	1704
více * * zpravodajka	0	1	0	0	0	1	79	191	7
více * * zpravodaj	0	0	3	9	11	27	138	271	116
more									

name, iv) automatically split the audio recording and the transcript, and v) manually check to make sure the speaker was correctly identified.

Table 3 presents the examples of patterns for three key words incorporated in the appropriate phrases and the normalised number of their occurrences through the years in the time spacing of five years.

Figure 1 illustrates selected nomina agentis (the names of agent) occurrence in the time span from 1966 to 2009 in the time spacing of five years. From the total amount of 17 nomina agentis closely related to the speakers appearing in the radio broadcasting we have chosen the following ones: spolupracovník – associate (m), spolupracovnice – associate (f), zpravodaj – rapporteur (m), zpravodajka – rapporteur (f), redaktor – editor (m), redaktorka – editor (f). The graph shows not only the popularity of the names used across the period but also the male/female speaker occurrences in broadcasting.

4 Speakers

Based on the phrase patterns, we have found 60,172 potential speakers' names. There were not only names of actual speakers but also proper names (e.g. Brno, Radiožurnál - the names of Czech cities and the name of broadcasting stations). We downloaded the list of first names and surnames from Ministry of the Interior of the Czech Republic. By comparing these lists with the found names, we get a resulting list of 20,313 speakers' names.

In this list were speakers about whom they were talking on the radio. We need only the list of speakers that speak on the radio. We know that almost all editors are among the important speakers and because they should be named in the broadcasting, we found how many times the name of the speaker occurred in the transcripts. There were several

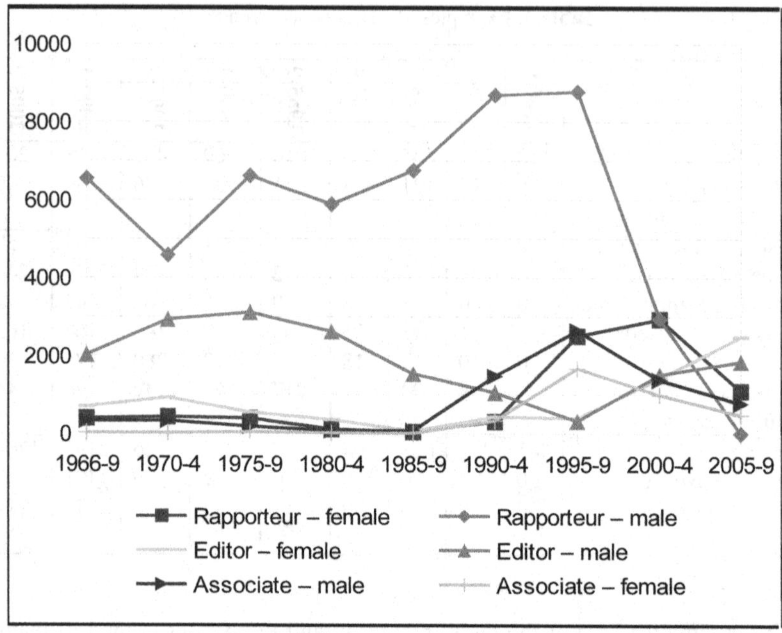

Fig. 1. Selected nomina agentis occurrence

speaker names that occurred more than 500times (15 speakers' names). We decided to add all speaker names that occurred 10 or more times to the speaker database (1,478 speaker names).

5 Conclusions and Future Work

In this paper we investigated a methodology for automatic extension of a speaker database based on automatic transcription. We want to add the most frequent speakers to our speaker database; we assume that these speakers are publicly important persons (presidents, politicians, famous artists, etc.) and reporters or frequent guests of the radio – they are contemporary speakers. With the introduced tools and methodology we extended our speaker database by 1,478 most frequent speakers.

We prepared a methodology showing how to extract acoustic data necessary to train models for speaker recognition. In our future work we will apply it. The extraction is not fully automated, there will be necessary manual work (mostly with the decision and confirmation that the data are chosen for the correct speaker). Hence we will try to lower the demands on human work (by automatic data clustering).

It has been observed that the performance of speaker recognition depends on the number of potential speakers. Therefore, we would like to investigate the benefit of restricting the potential set of speakers to the respective time period in which they could be active.

Acknowledgments. This work was supported by project no. DF11P01OVV013 provided by The Czech Ministry of culture in research program NAKI.

References

1. Nouza, J., et al.: Making Czech Historical Radio Archive Accessible and Searchable for Wide Public. Journal of Multimedia 7(2012), 159–169 (2012)
2. Cerva, P., Silovsky, J., Zdansky, J., Nouza, J., Seps, L.: Speaker-adaptive speech recognition using speaker diarization for improved transcription of large spoken archives. Speech Communication 55(10), 1033–1046 (2013)
3. Seps, L.: NanoTrans – Editor for orthographic and phonetic transcriptions. In: 36th International Conference on Tel. and Signal Processing (TSP), pp. 479–483 (2013)
4. Kuchařová, M., Škodová, S., Šeps, L., Lábus, V., Nouza, J., Boháč, M.: On the quantitative and qualitative speech changes of the Czech radio broadcasts news within years 1969-2005. In: Habernal, I. (ed.) TSD 2013. LNCS (LNAI), vol. 8082, pp. 360–368. Springer, Heidelberg (2013)
5. Soltys, O.: Verba dicendi a metajazyková informace. Ústav pro jazyk český, Praha (1983)
6. Hirschova, M.: Česká verba dicendi v performativním užití: Příspěvek ke zkoumání komunikativních funkcí výpovědi. FF UPOL, Olomouc (1988)
7. Lopatkova, M., Zabokrtsky, Z., Kettnerova, V.: Valenční slovník českých sloves. Karolinum, Praha (2008)

Development of a Large Spontaneous Speech Database of Agglutinative Hungarian Language

Tilda Neuberger, Dorottya Gyarmathy, Tekla Etelka Gráczi, Viktória Horváth, Mária Gósy, and András Beke

Research Institute for Linguistics of the Hungarian Academy of Sciences
Departement of Phonetics, Benczúr 33, 1068 Budapest, Hungary
{neuberger.tilda,gyarmathy.dorottya,graczi.tekla, horvath.viktoria,gosy.maria,beke.andras}@nytud.mta.hu

Abstract. In this paper, a large Hungarian spoken language database is introduced. This phonetically-based multi-purpose database contains various types of spontaneous and read speech from 333 monolingual speakers (about 50 minutes of speech sample per speaker). This study presents the background and motivation of the development of the BEA Hungarian database, describes its protocol and the transcription procedure, and also presents existing and proposed research using this database. Due to its recording protocol and the transcription it provides a challenging material for various comparisons of segmental structures of speech also across languages.

Keywords: database, spontaneous speech, multi-level annotation.

1 Introduction

Nowadays the application of corpus-based and statistical approaches in various fields of speech research is a challenging task. Linguistic analyses have become increasingly data-driven, creating a need for reliable and large spoken language databases. In our study, we aim to introduce the Hungarian database named BEA that provides a useful material for various segmental-level comparisons of speech also across languages. Hungarian, unlike English and other Germanic languages, is an agglutinating language with diverse inflectional characteristics and a very rich morphology. This language is characterized by a relatively free word order. There are a few spoken language databases for highly agglutinating languages, for example Turkish [1], Finnish [2]. Language modeling of agglutinating languages needs to be different than modeling of languages like English [3]. There are corpora of various sizes, different numbers of speakers and diverse levels of transcription. TIMIT Acoustic-Phonetic Continuous Speech Corpus was created for training speaker-independent speech recognizers. This database consists of sentence reading from 630 American English speakers; includes time-aligned orthographic, phonetic and word transcriptions [4]. The Verbmobil database (of 885 speakers) was developed also in the 90's with speech technological purposes [5]. The spoken part of the British National Corpus (100 million words) [6] consists of informal dialogues that were collected in different contexts, ranging from formal business or government meetings to radio shows. The London–Lund Corpus contains 100 texts

P. Sojka et al. (Eds.): TSD 2014, LNAI 8655, pp. 424–431, 2014.

of spoken British English. The basic prosodic features, simultaneous talk, contextual comment (laughs, coughs, telephone rings, etc.) were marked in the annotation [7]. The Switchboard corpus [8] includes 2,400 telephone dialogues of 543 American English speakers. It was developed mostly for the applications in speaker identification and speech recognition. There are also some corpora of audio and transcripts of conversational speech, such as HCRC map task corpus [9] or Buckeye corpus [10], and natural meetings, such as ICSI (International Computer Science Institute) Meeting Corpus [11] or AMI (Augmented Multi-party Interaction) Meeting Corpus [12]. Although the earliest databases had consisted of written and spoken English texts, new corpora were developed also in other languages in the past decades (e.g, the German Kiel Corpus [13], Danish spoken corpus [14]. The CSJ (Corpus of Spontaneous Japanese) is one of the largest databases; it contains 661 hours of speech by 1,395 speakers including 7.2 million words [15]. EUROM1 [16] and BABEL [17] are multilingual databases, containing samples of various languages giving possibility to compare the phonetic structures of these languages using similar materials and recording protocols in all languages. Recordings of spoken Hungarian were first compiled at the beginning of the twentieth century; unfortunately, this material was destroyed. Various types of dialectical speech materials were recorded in the 1940s; these recordings were archived in the late nineties and are available for studying at the Research Institute for Linguistics of the Hungarian Academy of Sciences, RIL. The Budapest Sociolinguistic Interview contains tape recorded interviews with 250 speakers (2–3 hours each) made in the late eighties [18]. The Hungarian telephone speech database (MTBA) is a speech corpus containing read speech recorded via phone by 500 subjects. It was designed to support research and developments in the fields of speech technology [19]. The HuComTech Multimodal Database contains audio-visual recordings (about 60 hours) of 121 young adult speakers that represent North-East Hungary [20]. The developing of the largest Hungarian spontaneous speech database, BEA (the abbreviation stands for the letters of the original name of the database: BEszélt nyelvi Adatbázis 'Speech Database 'Speech Database') started at the Phonetics Department of RIL in 2007. This database involves a great number of speakers who speak relatively long, contains various styles of speech materials, and has various levels of transcriptions.

2 Database Specification

At the moment of writing this paper, the total recorded material of BEA comprises 333 recordings, meaning 300 hours of speech material (approximately 4,500,000 words). The shortest recording lasts 24 minutes and 27 seconds, the duration of the longest is 2 hours, 24 minutes and 47 seconds; the average length is 51 minutes (SD: 15.8). The majority of them appear between 40 and 60 minutes. Speech materials from 184 female and 149 male speakers are available at the moment. For each recording, the following data are documented: the participant's age, schooling, job, height, weight, whether s/he is a smoker. The youngest participant is 19 years old, while the oldest one is 90 years old. The mean age of speakers is 39 years (SD: 18.8). The majority of the participants are in their twenties and thirties.

The database contains various types of speech materials: mainly spontaneous speech, but it also includes sentence repetitions and read texts; which provides an opportunity for comparison among speech styles. The protocol consists of six modules: 1. sentence repetition (25 phonetically-rich sentences), 2. spontaneous narrative about the subject's life, family, job, and hobbies, 3. opinion about a topic of current interest, 4. directed spontaneous speech; summary of content of two heard text, 5. three-party conversation, and 5. reading of sentences and text (for further details see [21]). In 95% of all recordings, the interviewer was the same young woman. Recordings are invariably made in the same room, under identical technical conditions: in the sound-proof booth of the Phonetics Department, specially designed for the purpose. The size of the room is $3.4 \times 2.1 \times 3.0$ m. The walls of the room are provided with a sound-absorbing layer in order to avoid reverberation. The degree of sound damping as compared to the outside environment is 35 dB at 50 Hz, and ≥ 65 dB above 250 Hz. The recording microphone is AT4040. Recording is made digitally, direct to the computer, with GoldWave sound editing software, with sampling at 44.1 kHz (storage: 16 bits, 86 kbytes/s, mono).

3 Transcription, Segmentation and Labeling

The BEA database has three types of transcription. The first was done in MS Word, the second in Transcriber, the third is being done in Praat. This chapter introduces the first two in nutshell, as they have been introduced already in details [21], and the third one is to be described in details first time in this study.

1. The primary transcription in MS Word (.doc format) is based on the orthography but without punctuation. The participants are uniformly abbreviated in these transcriptions as A (subject), T1 (interviewer and first conversational partner), T2 (second conversational partner). The proper names are capitalized, and some phenomena are marked: disfluencies (bold), hesitations, hummings, and other non-verbal noises like laughter (exclamation mark), as well as speaking simultaneously (parentheses), perceived pauses (❑) (see Fig. 1). 47% of the recordings were transcribed in this format.

T2 ([1]![2] **oo**[3]r dono[4] [I don't know][5]) → [1]simultaneous speech, [2]breath, [3]lengthening, [4]causal word form [5]intended form/expression

A (s**oo**)

A long ago the old people **eer**[6]![7] so! said that! th**aa**t! **probem**[8] [problem] tha [that] **tho**[9] [though] any degrees you should learn a manual occupation → [6]hesitation, [7]breath/noise... → [8]slip-of-the-tongue, [9]unfinished word

A (my son!)

T1 (yes yes)

T2 (mhm[10]) → [10]humming

A and this this vie**ww** **eer** in the past forty years

A (disappeared)

T2 (did disappear)

Fig. 1. Sample fragment of conversation in primary transcription (translated into English, with transcription marks)

2. Transcription made in Transcriber (`http://trans.sourceforge.net`) is basically time-aligned pause-to-pause labelling, also follows the rules of Hungarian orthography but without punctuation. The label boundaries are set at approximately the middle of at least 200 ms long pauses. Even longer pauses are marked separately. Noises from the speakers or the environment, non-speech events (like laughter, cough), hummings, hesitations, unfinished words, simultaneous utterances, unintelligible speech, and other phenomena are marked with special abbreviations and codes. The speakers are identified using the same characters as in the Word-transcriptions. 51% of the recordings were transcribed using this software.

3. The third type of transcription is done in Praat (`http://praat.org`) at several levels (Fig. 2). This transcription is being done at present. Criteria and rules were developed in the second half of 2013, and the transcription procedure started by trained annotators. The transcription includes 9 levels, where the first three levels include the interviewer's speech, the second three levels include the subject's talk, while the last three levels (only in part 5: conversation) are devoted to what the second conversational partner said. The first level of each speaker includes pause-to-pause labels in that speech samples are transcribed in orthography without punctuation. The second level of each speaker means word-level segmentation. The third levels are speech sound labels. Some specific simplifications are defined. Neither hyphens, nor capitals are used as opposed to the orthographical requirements since both of them have special functions. Silent ('SIL') and filled (e.g., 'M') pauses are marked in separate labels in each row of the speaker whose speech sample they belong to. When the speaker is not speaking but listening to the other(s), their lines are marked by 'PAUSE'. Unintelligible or noisy speech segments are labelled in each row as unusable parts. Simultaneous speech samples are not transcribed, but marked in each row just as overlapping speech ('E'). These transcriptions also include non-speech events (e.g. humming, laughter, sigh), disfluency phenomena, speech errors, word fragments. Slip-of-the-tongue phenomena are written as pronounced at the pause-to-pause level, and the intended word is added in square brackets. The word level includes the intended word and the sound level includes the pronounced speech sounds.

In the sound-level annotation the segment label set is phonetic and has several rules that concern specific problematic realizations. Here we give some examples. In cases where a silent pause is followed by a voiceless closure phase (of p, t, c, k, ts, tS), the boundary of the pause is consensually placed 30 ms before the first closure release. The cases, where a vowel is followed by a consonant at the phonological level that does not appear in the pronunciation but influences the realization of the vowel, are marked in the vowel label. Irregular and breathy voice and aspiration are marked also in the sound label ('Y', 'W', 'H', respectively). Further special cases are also appropriately discussed in the instructions for the transcribers.

4 Research Based on BEA

In this chapter the usefulness of BEA will be evaluated from a point of view of speech science. Research has been initiated in the following areas of phonetics, psycholinguistics and speech technology: the segmental structure of speech, coarticulation, suprasegmental features of speech, fluency of speech, temporal factors, disfluencies, non-verbal

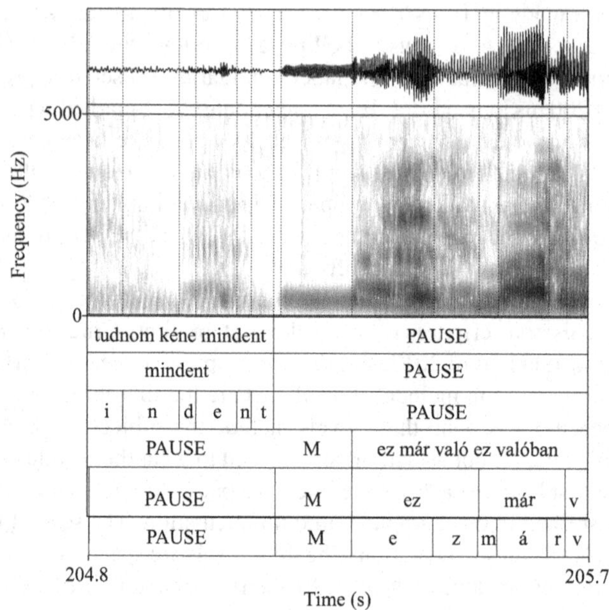

Fig. 2. A sample of turn taking in Praat annotations (M = hesitation)

vocalizations, automatic classification of various speech phenomena, speech detection, overlapping speech detection, speaker diarization.

A collection of studies has been published in Hungarian focusing on recent investigations where this database was used [22]. The durations and formant frequencies of the Hungarian vowels (more than 10,000 tokens) were measured in spontaneous speech of BEA [23,24]. One of the questions of the various investigations was how different vowels can be discriminated in spite of the large overlaps of the formant frequencies. The results showed that the accuracy of J48 classifier was higher depending on the horizontal tongue movements (87%) than depending on the vertical ones (69.7%). Fricative phoneme realizations (total duration of the consonants, duration of the voiced part, mean HNR, COG and other features) and the frequency of neutralization or weakening of the voicing oppositions were analyzed using both spontaneous and read speech samples of eight speakers of BEA [25]. Multilayer Perceptron neural network method was used for automatic classification. In a recent study [26] an attempt was made to define various units of spontaneous narratives and capture objective acoustic-phonetic properties of boundary marking. The results showed that (i) the majority of the speakers organize their narratives in similar temporal structures, (ii) thematic units can be identified in terms of certain prosodic criteria, and (iii) there are statistically valid correlations between factors like the duration of phrases, the word count of phrases, the rate of articulation of phrases, and pausing characteristics. Several investigations focused on the examination of speech planning and self-monitoring mechanisms by analyzing disfluency phenomena. An analysis [27] of the frequency and phonetic characteristics of anticipations and perseverations (in spontaneous speech samples by twenty-seven speakers) revealed that higher-organized units could drift away from their planned position to a relatively longer distance in time than lower-organized units while the latter tended to do so more frequently than the former. Temporal patterns confirmed that the speech production mechanism controls pre-planning more successfully. False starts and false words were investigated in a large amount of spontaneous speech samples (16-hour speech material consists of narratives of 70 adults) [28]. The results confirmed that false starts occur more frequently than false words, which indicates the appropriate functioning of the covert self-monitoring. The duration of the editing phase is affected by its structure the same way in both types of the analyzed disfluencies; and depends on the type of the word, and on the relation between the reparandum and the repair. A PhD thesis addressed the topic of speaker diarization for 100 spontaneous conversations of BEA [29]. The presented speaker diarization system was based on unsupervised learning method which could be easily adapted to another speech corpus. The best result (DER: 28.71%) was yielded using BIC-base method where the penalty value was 1, the features were MFCC(2,5–3,5) and the system contained the VAD and overlapping detection algorithm as well. Spontaneous conversations (also in the BEA database) frequently contain various non-verbal vocalizations such as laughter. The sound sequence of laughter may acoustically resemble to speech sounds; F0, formant structure, and RMS amplitude of laughter seem to be rather speech-like. There was an attempt to develop an accurate and efficient method to differentiate laughter from other speech events [30]. The results showed that the GMM-SVM system trained on acoustic parameters, MFCC and PLP could be a particularly good method for solving this problem.

5 Conclusion and Future Work

There are many research possibilities provided by BEA. It records the contemporary state of spoken Hungarian, providing the foundation for later comparative studies of linguistic change. In a number of areas like phonetics, laboratory phonology, speech technology, psycholinguistics, applied speech research, pragmatics, spontaneous speech grammatics, socio-phonetics, speaker identification (forensic phonetics) or speech-based medical diagnostics, most examinations can only be done on the basis of a large amount of speech material meeting the criteria of database technology. Although the BEA corpus was created to study phonetic aspects of speech, it should be useful to scientists interested in many other (linguistic) aspects of spontaneous speech.

The database is available for any researcher by contacting the developers. The files are not uploaded to the internet in order to warrant the speakers' privacy rights. However, we are planning the elaboration of an open access infrastructure, which provides an access to the corpus (both recordings and annotation) with privacy and security conditions. The corpus will be made available to the scientific community when transcription is completed.

Acknowledgments. This work was supported by OTKA 108762.

References

1. Mengusoglu, E., Deroo, O.: Turkish LVCSR: Database preparation and language modeling for an agglutinative language. In: IEEE International Conference on Acoustics Speech And Signal Processing, vol. 6, pp. 4018–4018. IEEE (1999, 2001)
2. Seppänen, T., Toivanen, J., Väyrynen, E.: MediaTeam speech corpus: a first large Finnish emotional speech database. In: Proceedings of the Proceedings of XV International Conference of Phonetic Science, pp. 2469–2472 (2003)
3. Mihajlik, P., Fegyó, T., Tüske, Z., Ircing, P.: A morphographemic approach for the recognition of spontaneous speech in agglutinative languages - like Hungarian. In: Proc. Interspeech 2007, Antwerp, Belgium, pp. 1497–1500 (2007)
4. Keating, P., Byrd, D., Flemming, E., Todaka, Y.: Phonetic analyses of word and segment variation using the TIMIT corpus of American English. Speech Communication 14(2), 131–142 (1994)
5. Bael, C.V., Boves, L., van den Heuvel, D., Strik, H.: Automatic phonetic transcription of large speech corpora. Journal of Computer Speech and Language 21(4), 652–668 (2007)
6. Aston, G., Burnard, L.: The BNC Handbook. Exploring the British National Corpus with SARA. Oxford University Press (1998)
7. Svartvik, J. (ed.): The London Corpus of Spoken English: Description and Research. Lund Studies in English, 82. Lund University Press, Lund (1990)
8. Godfrey, J.J., Holliman, E.C., Daniel, J.: SWITCHBOARD: telephone speech corpus for research and development. In: Acoustics, Speech, and Signal Processing, ICASSP 1992, vol. 1, pp. 517–520 (1992)
9. Anderson, A.H., Bader, M., Bard, E.G., Boyle, E., Doherty, G., Garrod, S.,... Weinert, R.: The HCRC map task corpus. Language and Speech 34(4), 351–366 (1991)
10. Pitt, M.A., Johnson, K., Hume, E., Kiesling, S., Raymond, W.: The Buckeye corpus of conversational speech: labeling conventions and a test of transcriber reliability. Speech Communication 45, 89–95 (2005)
11. Janin, A., Baron, D., Edwards, J., Ellis, D., Gelbart, D., Morgan, N., ... Wooters, C.: The ICSI meeting corpus. In: Proceedings of the 2003 IEEE International Conference on Acoustics, Speech, and Signal Processing, ICASSP 2003, vol. 1, pp. 364–367 (2003)
12. Carletta, J.E., et al.: The AMI meeting corpus: A pre-announcement. In: Renals, S., Bengio, S. (eds.) MLMI 2005. LNCS, vol. 3869, pp. 28–39. Springer, Heidelberg (2006)
13. Kohler, K.J., Pätzold, M., Simpson, A.P.: From the acoustic data collection to a labelled speech data bank of spoken Standard German. Arbeitsberichte des Instituts für Phonetik und digitale Sprachverarbeitung der Universität Kiel (AIPUK) 32, 1–29 (1997)
14. Grønnum, N.: A Danish phonetically annotated spontaneous speech corpus (DanPASS). Speech Communication 51(7), 594–603 (2009)
15. Maekawa, K.: Corpus of Spontaneous Japanese: Its design and evaluation. In: ISCA IEEE Workshop on Spontaneous Speech Processing and Recognition (2003)
16. Chan, D., et al.: EUROM: a spoken language resource for the EU. In: Proceedings of the 4th European Conference on Speech Communication and Speech Tecnology, Eurospeech 1995, Madrid, vol. 1, pp. 867–880 (1995)
17. Roach, P., Arnfield, S., Barry, W.J., Baltova, J., Boldea, M., Fourcin, A., ... Vicsi, K.: BABEL: an eastern european multi-language database. In: ICSLP (1996)
18. Váradi, T.: A Budapesti Szociolingvisztikai Interjú. In: Kiefer F, Siptár P. (ed.). A magyar nyelv kézikönyve Akadémiai Kiadó, Budapest, pp. 339–359 (2003)
19. Vicsi, K., Tóth, L., Kocsor, Gordos, G., Csirik, J.: MTBA – magyar nyelvű telefonbeszéd-adatbázis. Híradástechnika 8, 35–39 (2002)

20. Papay, K.: Designing a Hungarian multimodal database – speech recording and annotation. In: Esposito, A., Esposito, A.M., Martone, R., Müller, V.C., Scarpetta, G. (eds.) COST 2102 Int. Training School 2010. LNCS, vol. 6456, pp. 403–411. Springer, Heidelberg (2011)
21. Gósy, M.: BEA – A multifunctional Hungarian spoken language database. The Phonetician 105(106), 50–61 (2012)
22. Gósy, M. (ed.): Beszéd, adatbázis, kutatások. Akadémiai Kiadó, Budapest (2012)
23. Gráczi, T.E., Horváth, V.: A magánhangzók realizációja spontán beszédben. In: Beszédkutatás 2010, pp. 5–16 (2010)
24. Beke, A., Gósy, M.: Characteristic and spectral features used in automatic prediction of vowel duration in spontaneous speech. In: Institute of Electrical Electronics Engineers (eds.): CogInfoCom 2012: 3rd International Conference on Cognitive Infocommunications, pp. 65–71 (2012)
25. Gráczi, T.E., Beke, A.: Fricatives in spontaneous speech. In: ExAPP 2013, Copenhagen, March 20-22 (2013)
26. Beke, A., Gósy, M., Horváth, V.: Temporal variability in spontaneous Hungarian speech. In: Proceedings of 6th Language Technology Conference: Human Language Technologies as a Challenge for Computer Science and Linguistics, Poznan, December 7-9, pp. 219–223 (2013)
27. Gósy, M., Gyarmathy, D., Horváth, V.: Improper activation and monitoring failures in speech planning. Govor / Speech 29(1), 3–22 (2012)
28. Gyarmathy, D., Neuberger, T.: Self-monitoring strategies: the factor of age. In: Presentation at the 19th International Congress of Linguists, Geneva, July 21-27 (2012)
29. Beke, A.: Automatic speaker diarization in Hungarian spontaneous conversations. PhD thesis. ELTE, Budapest (2013)
30. Neuberger, T., Beke, A.: Automatic laughter detection in spontaneous speech using GMM-SVM method. In: Habernal, I. (ed.) TSD 2013. LNCS (LNAI), vol. 8082, pp. 113–120. Springer, Heidelberg (2013)

Unit Selection Cost Function Exploration Using an A* Based Text-to-Speech System

David Guennec and Damien Lolive

IRISA, University of Rennes 1, Lannion, France
{david.guennec,damien.lolive}@irisa.fr

Abstract. Speech synthesis systems usually use the Viterbi algorithm as a basis for unit selection, while it is not the only possible choice. In this paper, we study a speech synthesis system relying on the A^* algorithm, which is a general pathfinding strategy developing a graph rather than a lattice. Using state of the art techniques, we propose and analyze different selection strategies and evaluate them using a subjective evaluation on the N-best paths returned. The best strategy achieves a MOS score of 3.29 ($\pm$0.18). More interesting, the proposed system enables an in-depth analysis of unit selection.

Keywords: Corpus-based Unit Selection TTS, Evaluation, Cost function accuracy, Concatenation cost, Cost weighting.

1 Introduction

Currently, two main techniques dominate the field of Text-to-Speech synthesis. One of them is the parametric approach for which HTS [1] is the main system. The second one is the historical concatenative approach extended by [2] and widely implemented, for example in [3,4,5,6,7]. Even if the majority of research works concern the HTS framework, it is the corpus-based approach which is mainly used in commercial systems. In both cases, the current main research problem is expressivity control during synthesis [8]. Then, to enable the exploration of expressive corpus-based synthesis, we have to explore the in-depth behavior of the system and especially cost functions.

In corpus-based synthesis, the fundamental question is how the most appropriate sequence of units in an annotated speech corpus can be efficiently selected? This most appropriate units sequence is generally found by minimizing two criteria: the number of concatenations and the risk of generating hearable concatenation artefacts. Numerous criteria have been proposed in [9] but generally the cost function to optimize follows the form [3]:

$$U^* = \underset{U}{argmin} \left(\sum_{n=1}^{card(U)} C_t(u_n) + \sum_{n=2}^{card(U)} C_c(u_{n-1}, u_n) \right) \tag{1}$$

where U^* is the best unit sequence according to the cost function and u_n the candidate unit in the candidate sequence U intending to match the nth target unit. The sub-cost $C_t(u_n)$ (target cost) represents the distance between candidate and corresponding target

P. Sojka et al. (Eds.): TSD 2014, LNAI 8655, pp. 432–440, 2014.

unit. The second sub-cost (concatenation cost), $C_c(u_{n-1}, u_n)$, is the distance between the current candidate u_n and the previous one u_{n-1} in the path, used to minimize artifacts in concatenation areas.

Since unit selection has to process millions of candidate units, pruning is needed to give a result in an acceptable time or even in real time for most industrial applications. In most cases, the Viterbi algorithm [10] is used to find the best unit sequence (for example in [4,11,7]), but it is not the only possible one. An attempt to use a genetic algorithm in [12] can also be mentioned. Practically, the Viterbi algorithm is optimal and achieves a complexity of $\mathcal{O}(N * K^2)$, K being the maximum of candidate units for a given target phoneme. Since unit selection can be formulated as a path finding problem, other graph exploration algorithms can also be applied. In particular, some algorithms like A^* are better-suited for heuristic introduction to speed up the unit selection step.

In this paper, we describe in Section 2 a new unit selection based TTS system relying on the A^* algorithm, built to analyze cost functions behavior and exploration strategies. Then the general architecture of the system is proposed in Section 3. A description of the corpus used follows in Section 4 and the experimental setup is depicted in Section 5. Finally, the results are presented in Section 6. The paper is viewed as a preliminary study aiming at demonstrating the feasibility of using a A* algorithm to drive a unit selection process. The second objective of the paper is to establish a reference set of preselection filters and cost function for a comparison with other algorithms.

2 Unit Selection as a Pathfinding Problem

In this section we first introduce the A^* algorithm used to find the best unit sequence according to the cost function. Then, some implementation details are given.

2.1 A^* Algorithm

Contrary to the Viterbi algorithm, which computes a lattice containing all the candidate nodes (or at least M nodes for each time instant), A^* algorithm develops a graph. At each time instant, it explores the best node of the graph using a cost function that depends on both the path from the source node and the estimated cost to the target.

Originally introduced in 1968 by [13], the algorithm basically operates by searching for a path in a directed graph, whose nodes only have a finite number of successors, between a start node and a target node. To avoid arbitrary choice of the start node, we introduce a dedicated start node s which has the first candidate units as successors. We make a similar choice by introducing a unique target node t. 614bibThe algorithm uses a cost function of the form $f(n) = g(n) + h(n)$ with $g(n)$ being the cost of the sub-path between s and current node n and $h(n)$ being the estimated (heuristic) cost between n and t.

At each step, A^* takes the most promising node according to $f(n)$ and expends its successors (computing $f(n)$ by the way) until t is reached. $h(n)$ is a heuristic that enables to speed up the algorithm by privileging the nodes that seem to be on an optimal path over those which have a better $g(n)$ cost but may lead to greater costs in the future [14].

Considering a unique target node, one of the main advantages of A^* is that the algorithm delivers an optimal solution if the heuristic is admissible, ie. if $h(n) \leq h^*(n)$, where $h^*(n)$ is the real minimum value of the distance to the target node. In particular, note that the algorithm is optimal in the trivial case $h(n) = 0$, ie. if there is no heuristic, and turns out to be equivalent to Dijkstra's algorithm.

2.2 Adaptation to the Unit Selection Problem

Considering the algorithm presented above, the main functions that need to be adapted to our problem are (1) the cost function computation and (2) the successor function. In this work, we only consider the $g(n)$ part of the cost function, thus putting $h(n)$ to 0 which insures algorithm optimality. The cost function is detailed in Section 3.2.

Concerning the successor function, it needs to consider domain-based knowledge. During the search process, each phone of the target sequence is considered as the start of a potential unit for developing the graph. Moreover, for a particular unit, all the candidates within a window of width $w = 5$ are explored. Window width has been chosen arbitrarily to give satisfying search speed.

Furthermore, to improve algorithmic performance, the OPEN list is implemented as a binary heap sorted according to the cost function and a joined hash table to get quick membership queries. In addition, all the graph nodes are not computed, only those expanded during the successors search are really created.

In order to explore cost functions behavior, we modified the algorithm to be able to get the N-best paths, and also to get the N-best paths between a minimum and a maximum cost.

3 System Architecture

The system is interfaced with the ROOTS toolkit as described in [15]. For the synthesis, feature extraction starts from ROOTS files (generated from textual source using automatic tools) containing the target sequence with the needed annotations. Then the unit selection step is done using the A^* algorithm presented in Section 2. This step is parameterized by a cost function and user parameters (for example, requesting the best path or the N-best paths). The selection process makes queries to the corpus through the pre-selection filters as explained in Section 3.1. Finally, signal generation is performed by mixing each two units on an horizon of 2 pitch periods, using Hann windows.

3.1 Pre-selection Filters

As the problem of searching for variable-sized units in a corpus is computationally expensive, hash tables and pre-selection filters are implemented to speed up the unit selection process [11]. To achieve this, a key containing discrete information (mostly binary) is created for each speech segment (phoneme or non speech sound) in the corpus. That enables A^* to take or reject the unit quickly by just comparing the values in the key with target values. The key contains phonetic, linguistic and prosodic information. Binary masks are used to get access only to the desired information during runtime.

The pre-selection filters, using data included in segments keys, are integrated to the hash functions used to access the units in the corpus in order to reduce the number of candidates added to the OPEN list. The set of filters used for these experiments is the following, for each speech segment constituting the unit:

1. Is the segment a non speech sound?
2. Is it in the onset of the syllable?
3. Is it in the coda of the syllable?
4. Is it in the last syllable of its breath group?
5. Is the current syllable in word end?
6. Is the current syllable in word beginning?

In the case of a non speech sound, the only feature that matters is the first one, the others being all set to false. if no unit corresponding to the current set of filters is found, the pre-selection filters are relaxed one by one, starting from the end of the list. The priority order of the filters is the one given above. One drawback is that we can explore candidates far from the target features we want, thus risking to produce artefacts but this backtracking mechanism insures to actually find a unit and so insures the production of a solution.

3.2 The Target and Concatenation Costs

The cost of a unit is generally divided into a target cost and a concatenation cost as described in equation (1). Here, only the concatenation cost is included in the function, the target cost part being achieved by the filters. Weights are introduced here to balance the magnitude differences of the sub-costs.

Concatenation Cost. Here, we consider three sub-costs, well rated in the state of the art, for concatenation quality evaluation which are MFCC, amplitude and f0 distances, all weighted to give equal importance to each sub-cost on a phoneme basis. Basic rules addressing duration where first included and then dropped, first because they did not show a real improvement and also because generated speech seems generally well enough. Nevertheless, the inclusion of real duration or intonation models would be very interesting. Hence, we have the following concatenation cost:

$$\begin{aligned} C_c(u,v) = {} & W_{mfcc}(u,v)C_{mfcc}(u,v) + \\ & W_{amp}(u,v)C_{amp}(u,v) + W_{F0}(u,v)C_{F0}(u,v) \end{aligned}$$

where u and v are the two units under comparison, $W_{mfcc}(u,v)$ is the normalization coefficient for the MFCC cost $C_{mfcc}(u,v)$, $W_{amp}(u,v)$ is the normalization coefficient for the amplitude cost $C_{amp}(u,v)$ and $W_{F0}(u,v)$ is the normalization coefficient for the F0 cost $C_{F0}(u,v)$.

Weights tuning is a completely distinct problem, and many methods have been experimented. [16] includes a good review of the most common techniques. Actually, subjective methods remain widely used among TTS systems, in the absence of objective methods strongly correlated to perception. Here, we consider that the weights are linked to the phoneme to be concatenated and are computed using means and standard deviations for each phoneme of the corpus.

In addition, given that in particular context-dependent and voiced phonemes are more susceptible to produce concatenation artefacts than others because of higher spectrum, pitch and energy variability, we decided to evaluate the relevance of a feature addressing this particular issue. Different additions to cost functions have been experimented [9] [17], but the one we viewed as the most promising was only used, to our knowledge, for corpus reduction. This feature, "vocalic sandwich" cost [18] [19] is a sequence of phonemes following the regular expression $C(W^*VW^*)^+C$ where V is the vowels set, C a set including all the consonants considered to feature good splicing properties and W contains the remaining. In our implementation, we put a penalty to the units not respecting the sandwich expression.

Target Cost. In our implementation, the target cost is not directly incorporated in the cost function. Indeed, we consider that there is no need to integrate the nodes failing to show a certain fitting to the target sequence in the graph. As other works showed, the nodes achieving a good target cost are generally equally satisfying. Hence, features used for preselection also stand as binary target sub-costs. This means there is no target cost mark in our implementation, units are processed by pass or reject preselection filters. As the values we use are binary, their integration into the preselection filters is easy. Thus, the units that satisfy a given level of filters are considered equivalent regarding the target cost.

4 Corpus Analysis

For all the experiments, we used a French expressive corpus built from an audiobook spoken by a single male speaker. The corpus is automatically annotated using the global process described in [15] and is represented using the ROOTS toolkit. Speech signal is sampled at 44.1kHz (1 channel) and the mean F0 value for the speaker considering voiced segments is as low as 87Hz.

Since it is an audiobook, the content of the corpus and expressivity are not controlled, which may be a problem. For instance, as the voice has not been recorded for TTS purposes, prosody is sometimes exaggerated. Nevertheless, such a corpus has some interesting properties like homogeneous speech or good signal quality.

Concerning the size of the full corpus, it consists of 3 339 utterances, totalizing 419 742 speech segments (388 251 phonemes & 31 491 non speech sounds). The overall length of the corpus is 10h 45'. Diphoneme covering (and other non speech sounds) is 78%. The missing diphones are most of the time unused in French or even impossible.

The full corpus is then split into three sub-corpora: learning (3 139 utterances), testing (100 utterances) and validation (100 utterances). Testing and validation corpora are manual checked and represent about 25' each.

5 Experiments

5.1 Experimental Setup

The experiments we have conducted intend to:

1. Prove that a TTS system using A* to drive a unit selection process is viable.
2. Demonstrate the ability of the system to extract the N-best paths easily.
3. Assess the overall performance of the system.
4. Establish a reference cost function for further experiments.
5. Verify the stability of the cost functions presented above: the main idea is to explore the variability & ranking accuracy for the 100 best paths found by our system.

Figure 1 shows the evolution of the global cost for the 100 best paths found. The target sentence is "Car ce n'est pas le chagrin qui la fit partir" (Because it is not grief that caused her to leave). Due to the differences between selected units among the paths, costs are reported to each phoneme on the sequence (x axis). At first glance, we observe little variability among the paths. Most changes seem to occur on the first/last units (Non Speech Sounds here actually). Due to lack of space we cannot show figures for individual sub-cost illustrating the compensations existing between them. The mean number of units passing the filters is approximately 200, with a mean size of 3.8 phones for selected units, which is satisfying. We noted that the relaxation of filters was quite rare, which could signify other filters can be added to refine our target cost.

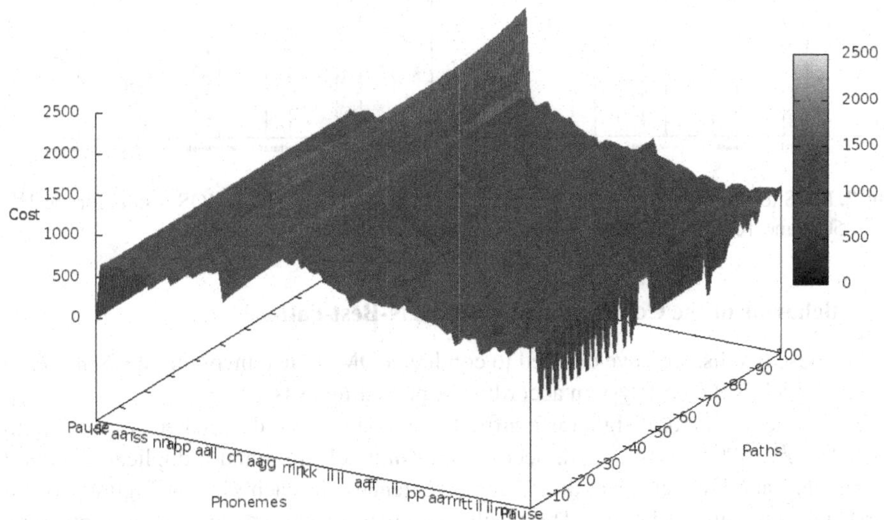

Fig. 1. Global cost evolution for 100-best paths (French sentence "Car ce n'est pas le chagrin qui la fit partir")

Three listening tests were also accomplished. For each, we have followed ITU-T P.800 recommendation, with 8 expert testers, questions/answers being those described in the recommendation.

6 Results and Discussion

6.1 Overall Quality

The first subjective test is a MOS test intending to rate the global performance of two variants of the system: (1) *Base* with all weights set to 1, (2) *Smoothing*: adjusted weights but no sandwich, (3) *Sandwich*: weights set to 1 + sandwich cost and finally (4) *All*: adjusted weights + sandwich costs. Each listener evaluates 20 samples for each system.

The results of the MOS test are shown on Figure 2 on the left. The *Smoothing* function gives the best overall quality with a MOS score of 3.29 ($\pm$0.18). This result is better than *Base*'s result (2.93 $\pm$0.17). Surprisingly, *Base* achieves a reasonably good performance considering the distribution of the sub-costs. Indeed, an unweighted amplitude cost tends to upper bound the values of other sub-costs. Then, we can think that the amplitude cost has a significant impact (in well) on listeners' decisions. More analysis on the individual sub-costs, particularly correlation tests, should be performed to complete this result. On the contrary, the use of sandwich penalties seems inconclusive (3.05 ($\pm$0.18)), even when combined with weighted sub-costs(2.98 ($\pm$0.20)). We believe that trying to find a vocalic sandwich path may result in worse fitted units on the prosodic plan than those selected only over spectral considerations.

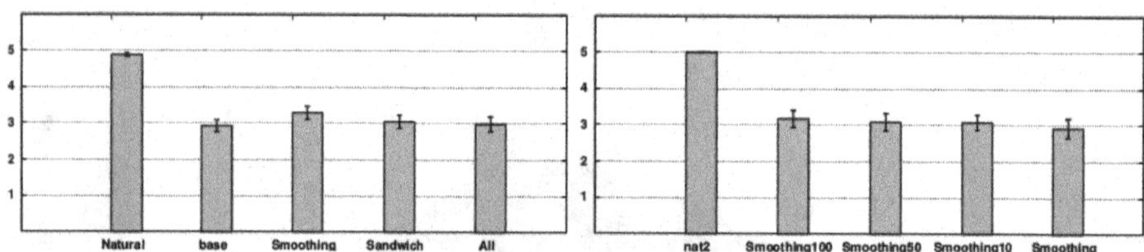

Fig. 2. MOS test between the 4 variants of the cost function (left) and DMOS results for the 1st, 10th, 50th and 100th paths of the *Smoothing* cost function (right)

6.2 Behavior of the Cost Function with the N-Best Paths

Given these results, we have decided to conduct a DMOS test involving the *Smoothing* system which is the best system according to preceding tests.

Each time, the natural signal is confronted to a synthesized signal corresponding to the 1st, 10th, 50th or 100th path according to unit selection and a duplicated natural reference. Each listener hears the 12 same sentences for each system. Figure 2 (right part) shows results of the test. The results for all paths are approximately 3, meaning slightly annoying degradation. Confidence intervals are thiner but no preference can be observed. No correlation between functions rankings and the testers choices are observed and minimal variability is spotted. These results raise many interrogations. First, the concatenation costs or the distance used may not be accurate. Secondly, as other works also pointed out low variability over the first paths, it would be more relevant to lookup further, up to the 1 000 first paths for example. These questions should be considered in order to enhance the ranking function.

7 Conclusion

In this paper, we have presented a new speech synthesis system with an original unit selection algorithm implemented as an A^* rather than the usual Viterbi. The system, designed as an experimental platform to explore the behavior of concatenative speech synthesis in depth, implements state of the art mechanisms to perform the unit selection. In particular, using A^* makes it very easy to explore the N-best paths in a unit selection problem while a good heuristic can drastically improve the time necessary to sort the problem out without sacrificing the result's optimality. Several cost functions have been evaluated for this new system, showing that vocalic sandwiches, very efficient for corpus reduction, do not enhance speech quality when used as part of the selection process. Furthermore, a reference cost function, *Smoothing*, has been established for further experiments. Further works will focus on 3 aspects: (1) Filters refinement, (2) cost function improvement and finally (3) comparing our A* with Viterbi considering efficiency, speed and overall quality using the same set of filters, cost function and corpora.

References

1. Yamagishi, J., Ling, Z., King, S.: Robustness of HMM-based Speech Synthesis. In: Proc. of the International Conference on Speech Prosody, pp. 581–584 (2008)
2. Sagisaka, Y.: Speech synthesis by rule using an optimal selection of non-uniform synthesis units. In: Proc. of IEEE ICASSP, pp. 679–682 (1988)
3. Black, A.W., Taylor, P.: CHATR: a generic speech synthesis system. In: Proc. of the 15th Conference on Computational Linguistics, vol. 2, pp. 983–986 (1994)
4. Hunt, A.J., Black, A.W.: Unit selection in a concatenative speech synthesis system using a large speech database. In: Proc. of IEEE ICASSP, vol. 1, pp. 373–376 (1996)
5. Taylor, P., Black, A., Caley, R.: The architecture of the Festival speech synthesis system. In: Proc. of the ESCA Workshop in Speech Synthesis, pp. 147–151 (1998)
6. Breen, A.P., Jackson, P.: Non-uniform unit selection and the similarity metric within BT's Laureate TTS system. In: Proc. of the ESCA Workshop on Speech Synthesis, pp. 373–376 (1998)
7. Clark, R.A., Richmond, K., King, S.: Multisyn: Open-domain unit selection for the Festival speech synthesis system. Speech Communication 49(4), 317–330 (2007)
8. Rebordao, A.R.F., Shaikh, M., Hirose, K., Minematsu, N.: How to Improve TTS Systems for Emotional Expressivity. In: Proc. of Interspeech, pp. 524–527 (2009)
9. Yi, J.R.W.: Natural-sounding speech synthesis using variable-length units. Master thesis, Massachusetts Institute of Technology (1998)
10. Viterbi, A.J.: Error bounds for convolutional codes and an asymptotically optimum decoding algorithm. IEEE Tr. on Information Theory, 260–269 (1967)
11. Conkie, A., Beutnagel, M.C., Syrdal, A.K., Philip, E.: Preselection of candidate units in a unit selection-based text-to-speech synthesis system. In: ICSLP, vol. 3, pp. 314–317 (2000)
12. Kumar, R.: A genetic algorithm for unit selection based speech synthesis. In: Proc. of INTERSPEECH (2004)
13. Hart, P.E., Nilsson, N.J., Raphael, B.: A Formal Basis for the Heuristic Determination of Minimum Cost Paths. IEEE Transactions on System Science and Cybernetics 4, 100–107 (1968)
14. Nilsson, N.J.: Principles of artificial intelligence, pp. 61–88. Springer-Verlag editions (1982)

15. Boeffard, O., Charonnat, L., Le Maguer, S., Lolive, D., Vidal, G.: Towards fully automatic annotation of audiobooks for TTS. In: Proc. of LREC (2012)
16. Alías, F., Formiga, L., Llorá, X.: Efficient and reliable perceptual weight tuning for unit-selection text-to-speech synthesis based on active interactive genetic algorithms: A proof-of-concept. Speech Communication 53(5), 786–800 (2011)
17. Donovan, R.E.: A new distance measure for costing spectral discontinuities in concatenative speech synthesizers. In: SSW4 (2001)
18. Cadic, D., Boidin, C., D'Alessandro, C.: Vocalic sandwich, a unit designed for unit selection TTS. In: Tenth Annual Conference of the International Speech Communication Association, pp. 2079–2082 (2009)
19. Cadic, D., Boidin, C., D'Alessandro, C.: Towards optimal TTS corpora. In: Proceedings of the Seventh International Conference on Language Resources and Evaluation, Valetta, Malta, pp. 99–104 (2010)

LIUM and CRIM ASR System Combination for the REPERE Evaluation Campaign

Anthony Rousseau[1], Gilles Boulianne[2], Paul Deléglise[1], Yannick Estève[1], Vishwa Gupta[2], and Sylvain Meignier[1]

[1] LIUM , University of Le Mans, France
http://www-lium.univ-lemans.fr
[2] Centre de Recherche Informatique de Montréal (CRIM), Québec, Canada
http://www.crim.ca

Abstract. This paper describes the ASR system proposed by the SODA consortium to participate in the ASR task of the French REPERE evaluation campaign. The official test REPERE corpus is composed of TV shows. The entire ASR system was produced by combining two ASR systems built by two members of the consortium. Each ASR system has some specificities: one uses an i-vector-based speaker adaptation of deep neural networks for acoustic modeling, while the other one rescores word-lattices with continuous space language models. The entire ASR system won the REPERE evaluation campaign on the ASR task. On the REPERE test corpus, this composite ASR system reaches a word error rate of 13.5 %.

1 Introduction

REPERE is an evaluation project in the field of people recognition in television documents [2], funded by the DGA (French defence procurement agency) and ending in 2014. Several evaluation tasks were organized, including an evaluation of automatic speech recognition systems on French TV shows.

This paper describes the ASR system proposed by the SODA consortium, including CRIM and LIUM institutions. This system, which combines CRIM's and LIUM's individual ASR systems, won the evaluation task.

Both systems are built on the Kaldi project [14], but each one has some specificities. For instance, CRIM has developed for its system an i-vector-based speaker adaptation of deep neural networks for acoustic modeling [8], while LIUM system has developed a tool to rescore word-lattices by using continuous space language models [15].

In addition to the speaker adaptation approach and the linguistic rescoring of word-lattices, main differences between the two ASR systems are vocabulary, tokenization, training data, and acoustic features. The combination of the two systems provides a very significant reduction of word error rate.

2 ASR System

As seen above, the ASR system which participated in the ASR task of the REPERE evaluation campaign is a composite ASR system. The combination of the two single

P. Sojka et al. (Eds.): TSD 2014, LNAI 8655, pp. 441–448, 2014.

ASR systems which are involved in the composite system is made by merging word-lattices. In order to make this merging easier, both ASR systems use the same speech segmentation.

2.1 Speaker Segmentation

To segment the audio recordings and to cluster speech segments by speaker, we used the *LIUM_SpkDiarization* speaker diarization toolkit [12]. This speaker diarization system is composed of an acoustic Bayesian Information Criterion (BIC)-based segmentation followed by a BIC-based hierarchical clustering. Each cluster represents a speaker and is modeled with a full covariance Gaussian. A Viterbi decoding re-segments the signal using GMMs with 8 diagonal components learned by EM-ML, for each cluster. Segmentation, clustering and decoding are performed with 12 MFCC+E, computed with a 10ms frame rate. Gender and bandwidth are detected before transcribing the signal with the two ASR systems.

2.2 LIUM ASR System

The LIUM ASR system built for the REPERE evaluation campaign is based on the Kaldi Speech Recognition Toolkit, which uses finite state transducers (FSTs) for decoding (the general approach is described in [13]). A first step is performed with the Kaldi decoder by using a bigram language model and classical GMM/HMM models to compute a fMLLR matrix transformation. Another step is performed by using the same language model and deep neural network acoustic models. This pass generates word-lattices: an in-house tool, derived from a rescoring tool from the CMU Sphinx project, is used to rescore word-lattices with a 5-gram continuous space language model [15].

In this section we will first present the training data used to estimate LIUM's models, then describe how the system was built using this toolkit.

Training Data. The training set used to build LIUM's system consists of 145,781 speech segments from several sources: the radiophonic broadcast ESTER [3] and ESTER2 [4] corpora, which accounts for about 100 hours of speech each; the TV broadcast ETAPE corpus [5], accounting for about 30 hours of speech; the TV broadcast REPERE train corpus, accounting for about 35 hours of speech and other LIUM radio and TV broadcast data for about 300 hours of speech, which have been segmented using the speaker diarization system described above. The training dictionary has 107.603 phonetized entries. Table 1 summarizes the characteristics of each dataset.

For language modeling, the training data is composed of the manual transcriptions from the training corpus used to estimate the acoustic models, of articles extracted from of TV websites, of articles extracted from Google News, of the French Gigaword corpus, of articles from newspaper 'Le Monde'. All of these data were collected before January 2013.

Table 2 presents of the number of words in each corpus in the training corpus used to estimate language models.

Table 1. Characteristics of the training data for acoustic modeling

Sources	Speech	Segments
ESTER	100h	12,902
ESTER2	100h	15,162
ETAPE	30h	8,378
REPERE	35h	10,269
LIUM v8	300h	99,070
Total	565h	145,781

Table 2. Characteristics of the training data for language models

Sources	Number of words
Manual transcriptions from the training corpora used to train the acoustic models	8M
Articles from TV websites (≤2012)	5M
Google News (≤2012)	204M
French Gigaword (≤2012)	1015M
Newspapers (≤2012)	366M
Subtitles of TV Newspaper (≤2012)	11M
Total	1609M

Acoustic Modeling. The GMM-HMM (Gaussian Mixture Model – Hidden Markov Model) models are trained on 13-dimension PLP features with first and second derivatives by frame. By concatenating the four previous frames and the four next frames, this corresponds to $39 \times 9 = 351$ features projected to 40 dimensions with linear discriminant analysis (LDA) and maximum likelihood linear transform (MLLT). Speaker adaptive training (SAT) is performed using feature-space maximum likelihood linear regression (fMLLR) transforms. Using these features, the models are trained on the full 565 hours set, with 12,000 tied triphone states and 450,000 Gaussians. On top of these models, we train a deep neural network (DNN) based on the same fMLLR transforms as the GMM-HMM models and on state-level minimum Bayes risk (sMBR) [10] as discriminative criterion. Again we use the full 565 hours set as the training material. The resulting network is composed of 7 layers for a total of 42.5 millions parameters and each of the 6 hidden layers has 2,048 neurons. The output dimension is 9,866 units and the input dimension is 440, which corresponds to an 11 frames window with 40 LDA parameters each.

Weights for the network are initialized using 6 restricted Boltzmann machines (RBMs) stacked as a deep belief network (DBN). The first RBM (Gaussian-Bernoulli) is trained with a learning rate of 0.01 and the 5 following RBMs (Bernoulli-Bernoulli) are trained with a rate of 0.4. The learning rate for the DNN training is 0.00001. The segments and frames are processed randomly during the network training with stochastic gradient descent in order to minimize cross-entropy between the training data and network output. When these training steps are done, the last step of training is processed, by applying the minimum Bayes risk criterion, as indicated above. To speed up the learning process, we used a general-purpose graphics processing unit (GPGPU) and the CUDA toolkit for computations.

Language Modeling. The vocabulary used in the LIUM ASR system has 160K words. The bigram language model used during the decoding with Kaldi is trained on the data presented in section 1.1 by using the SRILM toolkit [16]. No cut-off was applied and the modified Kneser-Ney discounting is applied.

To rescore word-lattices generated by Kaldi, trigram and quadgram LMs are trained with the same toolkit. A 5G continuous-space language model (CSLM) is also estimated for the final lattice rescoring. No cut-off is applied and the same discounting method as for the bigram language model is applied.

2.3 CRIM ASR System

This system is also based on the Kaldi toolkit, with the addition of DNN speaker adaptation based on i-vectors [8].

Training Data. CRIM training data to estimate acoustic models contains the same ESTER, ESTER 2, ETAPE and REPERE corpora as LIUM's, for a total of 335 hours of audio: this number is higher than the number of hours used by LIUM from these corpora because LIUM put aside about 50 hours from ESTER 2.

In addition to these 335 hours, CRIM had 178 hours of internally transcribed audio from French TV broadcasts in Quebec. Overall, CRIM had 513 hours of transcribed audio for training. In all the training audio, speaker segments were manually labeled in order to facilitate speaker-adapted training.

Acoustic Modeling. For training the deep neural network (DNN) using back propagation, 3 hours of the training audio were set aside for validation. CRIM uses TRAP (TempoRAl Pattern) features [6] extracted from filter-bank as input to the neural net. To compute TRAP features, 23-dimensional filterbank features are normalized to zero mean per speaker. Then 31 frames of these 23-dimensional filterbank features (15 frames on each side of current frame) are spliced together to form a 713-dimensional feature vector. This 713-dimensional feature vector is transformed using a Hamming window (to emphasize the center), passed through a discrete cosine transform and the dimensionality is reduced to 368. This 368-dimensional feature vector is globally normalized to have zero mean and unit variance.

The i-vector extractor is trained from the same data used for training the DNN, using speaker labels from the transcriptions. At test time, for each speaker identified by the automatic segmentation, one i-vector of dimension 100 is extracted. The TRAP features are then augmented with the 100-dimensional i-vector corresponding to the current speaker. This 468-dimensional feature vector is then input to the 7-layer DNN, as illustrated in Figure 1. The feature vector is advanced by one frame every time (note that the i-vector part stays fixed for a given speaker).

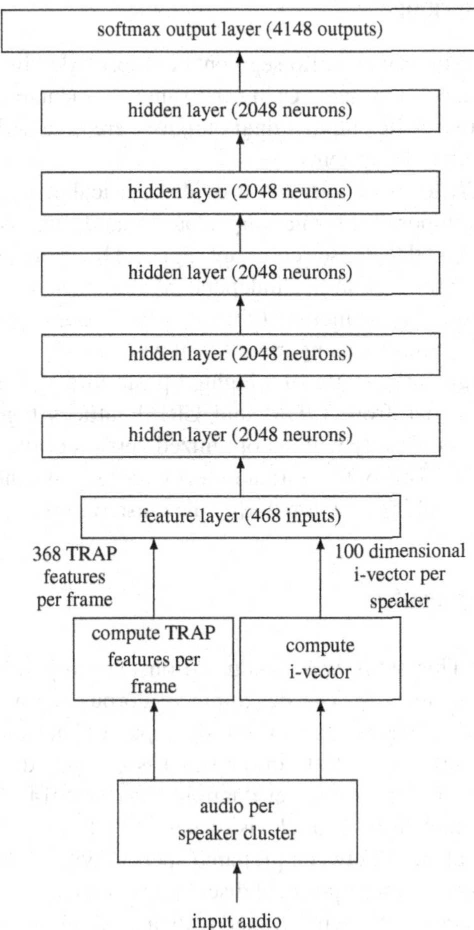

Fig. 1. Deep neural network architecture used for speaker adaptation of acoustic models in the CRIM ASR system

Language Modeling. Sources for CRIM include broadcast news transcriptions from EPAC and ESTER campaigns, transcripts from ETAPE, 350,000 sentences selected from French Gigaword database, and Google 4-grams (closely following [7]). Entropy-based pruning was applied to reduce language model size to 1.8M trigrams for search and 20M quadgrams for rescoring word lattices. Perplexities on the REPERE development text are 162 for the search trigram and 134 for the rescoring quadgram, with an out-of-vocabulary rate of 0.65%. The initial vocabulary was selected by taking words with the highest frequency count weighted inversely with source size until a vocabulary size of 100,000 words was obtained. To this, words from REPERE training transcripts were added, as well as proper names found in ETAPE, EPAC, ESTER sources, and also French departments, Paris metro stations, and French acronyms taken from the Web. The final vocabulary was 144,000 words.

2.4 Word-Lattice Merging

CRIM and LIUM used the same audio segmentation, provided by the *LIUM_SpkDiarization* speaker diarization system. Using the same segmentation makes easier the merging between the two ASR outputs: final outputs were obtained by merging word-lattices provided by both ASR systems.

Both LIUM and CRIM ASR systems provide classical word-lattices with usual information: words, temporal information, acoustic and linguistic scores. Before merging lattices, for each edge, these scores are replaced by its *a posteriori* probability. Posteriors are computed for each lattice independently, then weighted by $\frac{1}{n}$, where n is the number of word-lattices to be merged (here, $n = 2$). In our experiments, we did not find significant improvements by using more tuned weights.

For each speech segment, the use of weighted posteriors allows to merge starting (respectively ending) nodes from LIUM and CRIM lattices together into a single lattice in order to process directly with an optimized version of the consensus network confusion algorithm [11]. This optimization reduces very significantly the computation time by managing temporal information during the clustering steps.

3 Experimental Results

Experimental Data. This study was conducted on two corpora from the REPERE French evaluation campaign [9]. The development corpus (dev) is composed of 28 TV shows. This corpus corresponds to the test corpus of the first evaluation which took place in January 2013. The test corpus (test) is composed of 62 TV shows. It corresponds to the test set of the second evaluation (January 2014). Shows are recorded from the two digital French terrestrial television stations BFM and LCP.

These corpora are balanced between prepared speech, with 23 broadcast news, and more spontaneous speech, from 67 political discussions or street interviews. Only a part of the recordings are annotated, giving respectively a total duration of 3 hours for dev corpus and 10 hours for the test corpus.

Results and Discussion. A first evaluation on the ASR task was organized last year in 2013, in which the LIUM ASR system ranked first, on similar but different test data. This system appears in Table 3 under the name *old 2013 LIUM* system: it can be used to measure improvements achieved since last year.

The *old 2013 LIUM* ASR system was based on the CMU Sphinx toolkit, with some improvements, for instance the use of hybrid MLP/HMM acoustic models. A variant of this system is described in [1].

The main difference between the new ASR system developed by LIUM and the old one comes from the use of DNN acoustic models and the use of the *finite state machine* paradigm. These functionalities are both offered by the Kaldi toolkit. Notice that the linguistic rescoring tool is the same one in both LIUM ASR systems. With the same language models and the same training data for the estimation of acoustic models, the word error rate (WER) of the LIUM ASR system is reduced of 2.6 points (14%) to 16.0%.

The CRIM system achieves a word error rate of 16.3%. When the linguistic rescoring tool of LIUM system is applied to CRIM word-lattices (called CRIM+CSLM in Table 3), the WER is 1 point smaller than the WER of the LIUM system. This can be explained by better acoustic models provided by the DNN adaptation approach proposed by CRIM.

Combining the single-best hypothesis of each system with ROVER (and by using confidence measures) fails to provide an improvement (line ROVER in Table 3).

In contrast, merging word lattices achieves a large reduction in error over both individual systems (line CRIM $\oplus$ LIUM in Table 3), bringing the WER down by about 2 points (13.1% relative) when applied to LIUM and initial CRIM systems.

Notice that when applied to LIUM and CRIM+CSLM, the WER is reduced by 1.5 point (10% relative). The same training data were used to train the CSLMs of the CRIM+CSLM and the LIUM systems: this may explain this smaller improvement provided by the merging process.

Table 3. Word error rates on REPERE test corpus (TV shows)

ASR system	WER
old 2013 LIUM	18.6%
LIUM	16.0%
CRIM	16.3%
ROVER(CRIM,LIUM)	16.3%
CRIM $\oplus$ LIUM	13.9%
CRIM+CSLM	15.0%
CRIM+CSLM $\oplus$ LIUM	13.5%

4 Conclusion

Both LIUM and CRIM ASR systems are based on the Kaldi toolkit. Each one has noticeable specificities: the CRIM system uses a DNN speaker adaptation approach, while the LIUM system uses a 5g CSLM to rescore word-lattices. The word-lattice merging used in this work in order to build a composite ASR system permits to get significant improvements in terms of word error rate, and is very simple to use: no constraint about vocabulary, tokenization, nature of acoustic models. Only classical word-lattices are necessary, with acoustic and linguistic scores in order to compute posteriors. Merging LIUM and CRIM ASR systems was easy, and these systems were sufficiently accurate and complementary to get such performances.

The *old 2013 LIUM* ASR system, which won the two last evaluation campaigns on French language in 2012 and 2013, achieves a WER of 18.6% on the test data of the 2014 REPERE campaign. From this starting point, the composite system presented in this paper reduces the WER down to 13.5% on the test set (WER reduction of 27% relative), a significant advance in state-of-the-art French ASR.

References

1. Bougares, F., Deléglise, P., Estève, Y., Rouvier, M.: LIUM ASR system for Etape French evaluation campaign: experiments on system combination using open-source recognizers. In: Habernal, I. (ed.) TSD 2013. LNCS, vol. 8082, pp. 319–326. Springer, Heidelberg (2013)
2. Galibert, O., Kahn, J.: The first official REPERE evaluation. In: First Workshop on Speech, Language and Audio in Multimedia (SLAM), Marseille, France, pp. 43–48 (2013)
3. Galliano, S., Geoffrois, E., Gravier, G.F., Bonastre, J., Mostefa, D., Choukri, K.: Corpus description of the Ester evaluation campaign for the rich transcription of French broadcast news. In: 5th International Conference on Language Resources and Evaluation (LREC), pp. 315–320 (2006)
4. Galliano, S., Gravier, G., Chaubard, L.: The Ester 2 evaluation campaign for the rich transcription of french radio broadcasts. In: Interspeech (2009)
5. Gravier, G., Adda, G., Paulsson, N., Carré, M., Giraudel, A., Galibert, O.: The ETAPE corpus for the evaluation of speech-based TV content processing in the French language. In: Eighth International Conference on Language Resources and Evaluation (LREC), Istanbul, Turkey, pp. 114–118 (2012)
6. Grézl, F.: TRAP-based Probabilistic Features for Automatic Speech Recognition. Ph.D. thesis, dept. Computer Graphics & Multimedia, Brno University of Technology (2007)
7. Gupta, V., Boulianne, G., Osterrath, F., Ouellet, P.: CRIM's french speech transcription system for ETAPE 2011. In: WOSSPA (2013)
8. Gupta, V., Kenny, P., Ouellet, P., Stafylakis, T.: I-vector-based speaker adaptation of deep neural networks for french broadcast audio transcription. In: ICASSP, Florence, Italy (2014)
9. Kahn, J., Galibert, O., Quintard, L., Carre, M., Giraudel, A., Joly, P.: A presentation of the REPERE challenge. In: International Workshop on Content-Based Multimedia Indexing (CBMI), pp. 1–6 (2012)
10. Kingsbury, B.: Lattice-based optimization of sequence classification criteria for neural-network acoustic modeling. In: ICASSP, pp. 3761–3764 (2009)
11. Mangu, L., Brill, E., Stolcke, A.: Finding consensus in speech recognition: word error minimization and other applications of confusion networks. Computer Speech & Language 14(4), 373–400 (2000)
12. Meignier, S., Merlin, T.: LIUM SpkDiarization: an open source toolkit for diarization. In: CMU SPUD Workshop, Dallas, Texas, USA (2010)
13. Mohri, M., Pereira, F., Riley, M.: Speech recognition with weighted finite-state transducers. In: Springer Handbook of Speech Processing, pp. 559–584. Springer, Heidelberg (2008)
14. Povey, D., Ghoshal, A., Boulianne, G., Burget, L., Glembek, O., Goel, N., Hannemann, M., Motlíček, P., Qian, Y., Schwarz, P., Silovský, J., Stemmer, G., Veselý, K.: The Kaldi Speech Recognition Toolkit. In: ASRU Workshop, pp. 1–4 (2011)
15. Schwenk, H.: CSLM – a modular open-source continuous space language modeling toolkit. In: Interspeech, Lyon, France, pp. 1198–1202 (2013)
16. Stolcke, A.: SRILM – an extensible language modeling toolkit. In: Interspeech, pp. 901–904 (2002)

Anti-Models: An Alternative Way to Discriminative Training

Jan Vaněk* and Josef Psutka

University of West Bohemia in Pilsen, Univerzitní 22, 306 14 Pilsen
Faculty of Applied Sciences, Department of Cybernetics
vanekyj@kky.zcu.cz

Abstract. Traditional discriminative training methods modify Hidden Markov Model (HMM) parameters obtained via a Maximum Likelihood (ML) criterion based estimator. In this paper, anti-models are introduced instead. The anti-models are used in tandem with ML models to incorporate a discriminative information from training data set and modify the HMM output likelihood in a discriminative way. Traditional discriminative training methods are prone to over-fitting and require an extra stabilization. Also, convergence is not ensured and usually "a proper" number of iterations is done. In the proposed anti-models concept, two parts, positive model and anti-model, are trained via ML criterion. Therefore, the convergence and the stability are ensured.

Keywords: ASR, HMM, Acoustic Modeling, Discriminative Training, Anti-Models, MMI, MCE, MPE.

1 Introduction

Discriminative training (DT) techniques have been shown to outperform the ML-based training in automatic speech recognition (ASR). But they require a proper tuning and use a number of heuristics [3]. Moreover, they usually do not converge and maximization of a training criterion may not lead to maximum recognition accuracy with unseen data. ASR systems used acoustic models with a reduced complexity in past due to limited computing power. DT techniques gain is better with the less-complex models trained from the same amount of training data [7]. The more-complex models are more sensitive to a DT setup and require a finer tuning to get a significant gain over the ML models. Nowadays, a multi-core computer architecture has enough computing power to run any single diagonal-covariance acoustic model in real-time and, with a GPU acceleration, even multiple models [1] or full-covariance models [2]. Therefore, the more and more-complex HMMs (even with full-covariance matrices) are used and it is obvious that some simpler and more robust DT technique could be helpful.

In this paper, a concept of anti-models is introduced. The idea is simple: Instead of modification of the Gaussian parameters to fit the model into an unnatural (non-Gaussian) shape using negative statistic or gradient methods, we directly construct

* This research was supported by the Grant Agency of the Czech Republic, project No. GAČR GBP103/12/G084.

P. Sojka et al. (Eds.): TSD 2014, LNAI 8655, pp. 449–456, 2014.

the anti-model, model of the data that belongs to the others HMM states. The idea is general and can be adopted to all major DT criteria, e.g. Maximal Mutual Information (MMI), boosted-MMI, Minimum Classification Error (MCE), and Minimum Phone Error (MPE). To keep a clarity and to fit into limited paper size, the MMI case is shown in detail in this paper. Nevertheless, the derivation for the other criteria is analogous.

This paper is organized as follows. A brief overview of DT techniques is given in Section 2. The concept of anti-models is introduced in Section 3. The derivation of the anti-models with the MMI criterion is described in Section 4. Experiments and results are presented and discussed in Section 5.

2 Discriminative Training

In principle, the ML based training is a machine learning method from positive examples only. In contrast to ML, discriminative approaches take into account an information about class competition during the training. This extra information may improve results, but it brings an extra computation burden also. A short review of the most frequently used discriminative criteria follows.

2.1 Maximum Mutual Information – MMI

In the MMI case, a training algorithm seeks to maximize the posterior probability of the correct utterance given the used models [8]:

$$\mathcal{F}_{MMI}(\lambda) = \sum_{r=1}^{R} \log \frac{P_\lambda(O_r|s_r)^\kappa P(s_r)^\kappa}{\sum_S P_\lambda(O_r|s)^\kappa P(s)^\kappa}, \tag{1}$$

where λ represents the acoustic model parameters, O_r is the training utterance feature set, s_r is the correct transcription for the r'th utterance, κ is the acoustic scale which is used to amply confusions and herewith increases the test-set performance. $P(s)$ is a language model part. Optimization of the criterion (1) requires to generate lattices or many-hypotheses recognition run with appropriate language model. The lattices generation is highly time consuming. Furthermore, these methods require good correspondence between training and testing dictionary and language model. If the correspondence is weak, e.g. there are many words which are only in the test dictionary then the results of these methods are not good. In this case, we can employ Frame-Discriminative training (MMI-FD), which is independent on the used dictionary and language model [6]. In addition, this approach is much faster.

Optimization of the MMI objective function uses Extended Baum-Welch update equations [5] and it requires two sets of statistics. The first set, corresponding to the numerator (num) of the equation (1), is the correct transcription. The second one corresponds to the denominator (den) and it is a recognition/lattice model containing all possible words. An accumulation of statistics is done by the forward-backward algorithm on reference transcriptions (numerator) as well as generated lattices (denominator). The Gaussian means and variances are updated as follows [6]:

$$\hat{\mu}_{jm} = \frac{\Theta_{jm}^{num}(O) - \Theta_{jm}^{den}(O) + D_{jm}\mu'_{jm}}{\gamma_{jm}^{num} - \gamma_{jm}^{den} + D_{jm}} \tag{2}$$

$$\hat{\sigma}_{jm}^2 = \frac{\Theta_{jm}^{num}(O^2) - \Theta_{jm}^{den}(O^2) + D_{jm}(\sigma_{jm}'^2 + \mu_{jm}'^2)}{\gamma_{jm}^{num} - \gamma_{jm}^{den} + D_{jm}} - \mu_{jm}^2, \tag{3}$$

where j and m are HMM-state and Gaussian indexes, respectively, γ_{jm} is the accumulated occupancy of the Gaussian, $\Theta_{jm}(O)$ and $\Theta_{jm}(O^2)$ are a posterior probability weighted by the first and the second order accumulated statistics, respectively. Gaussian-specific stabilization constants D_{jm} are set to maximum of (i) double of the smallest value which ensures positive estimated variances, and (ii) value $E\gamma_{jm}^{den}$, where constant E determines the stability/learning-rate and it is a compromise between stability and number of iteration which is needed for well-trained models [9]. To bring a stability to HMM states with low-data, I-smoothing was introduced in [10] that uses a prior from the ML model or the MMI model from the previous iteration.

2.2 Boosted-MMI

The Boosted-MMI [14] is a modification of the MMI method. The denominator lattice trajectories are weighted by an error in this method. The error is defined in the same way like in MPE or MWE (see below). Therefore, the Boosted-MMI is a combination between MMI and MPE/MWE.

2.3 Minimum Classification Error – MCE

Another popular criterion is MCE [11,12,13]. The MCE criterion directly minimizes an error in a recognized word sequence. In contrast, MMI maximizes a probability of the correct sequence against others. The non-smooth classification error is smoothed via a sigmoid function to allow gradient-based optimization. It operates with n-best lists or with lattices that are more suitable for large vocabulary continuous speech recognition.

2.4 Minimum Word Error – MWE

The MWE criterion is defined formally similar to (1), but with a word error incorporation:

$$\mathcal{F}_{MWE}(\lambda) = \sum_{r=1}^{R} \log \frac{\sum_S P_\lambda(O_r|s_r)^\kappa P(s_r)^\kappa \text{RawAccuracy}(s)}{\sum_S P_\lambda(O_r|s)^\kappa P(s)^\kappa}, \tag{4}$$

where RawAccuracy(s) is the error rate of the word sequence s, κ is the acoustic model weight. This criterion is a weighted mean of the correct recognized words out of all other possible word sequences. Maximization of the MWE criterion improve a number of correctly recognized words in the most probable word sequences.

2.5 Minimum Phone Error – MPE

The MWE shows poor ability to generalize the training performance to unseen data. Therefore, more robust phone-based criteria were introduced. The maximization is aimed to the phone-level here [7]. Formal description of the MPE criterion is identical to (4). Only the error part RawAccuracy(s) is different in MPE. It express a relative count of correct recognized phones. An approximation to the error is used in practice for efficiency reasons [15].

2.6 Minimum Phone Frame Error – MPFE

Another approximation to the phone error was introduced in [16]. The MPE described above does not penalize deletion errors sufficiently. In addition, dynamic range of phone RawAccuracy(s) is typically quite narrow, which makes MPE occupancies considerably lower than MMI occupancies [10]. This may lead to an MPE robustness problem when training data are not abundant. MPFE uses phone-lattices and modified criterion to overcome the MPE shortcomings.

3 Concept of Anti-Models

The traditional DT techniques have two main shortcomings: Low stability and proneness to over-fitting. An implementation of the DT training that successfully copes with the shortcomings often seems to be more art than science. In contrast, the ML training is stable and there is a considerable experience how to prevent the over-fitting [17,18,19,20,21,22]. The concept of the anti-models is based on advantages of the ML training and in addition of the HMM state-concurrency information. Other-states data that have high likelihood in the model of the particular state are modeled separately by the anti-model. Thus, each state has standard ML model and the anti-model. The final state observation probability $p(o_t)$ of the feature vector $p(o_t)$ is calculated as follows

$$p(o_t) = \frac{p_{ML}(o_t)^2}{p_{ML}(o_t) + w_A p_A(o_t)}, \tag{5}$$

where $p_{ML}(o_t)$ is the output probability of the ML model and $p_A(o_t)$ is probability of the anti-model. w_A is the weight of the anti-model that is proportional to an amount of the wrongly modeled data. An illustration of the anti-model concept is in Figure 1. The figure shows that the anti-model modifies only the part of the distribution, where the wrongly modeled data were observed. In contrast, MMI modifies the entire distribution even in the part where no data were observed and therefore, it may be less optimal for unseen data. The figure also shows that the anti-model concept is able to model even non-Gaussian distributions. A log-domain is usually used in practice for HMM observation probabilities evaluation. The log form of the equation (5) is

$$\log(p(o_t)) = 2\log(p_{ML}(o_t)) - \log(p_{ML}(o_t) + w_A p_A(o_t)). \tag{6}$$

The sum of two probabilities evaluated in the log-domain can be calculated in a more robust and faster way:

$$\begin{aligned} \log(e^{lp_1} + e^{lp_2}) &= \max(lp_1, lp_2) + \log(1 + e^{(\min(lp_1,lp_2)-\max(lp_1,lp_2))}) \\ &= \max(lp_1, lp_2) + f(\min(lp_1, lp_2) - \max(lp_1, lp_2)), \end{aligned} \tag{7}$$

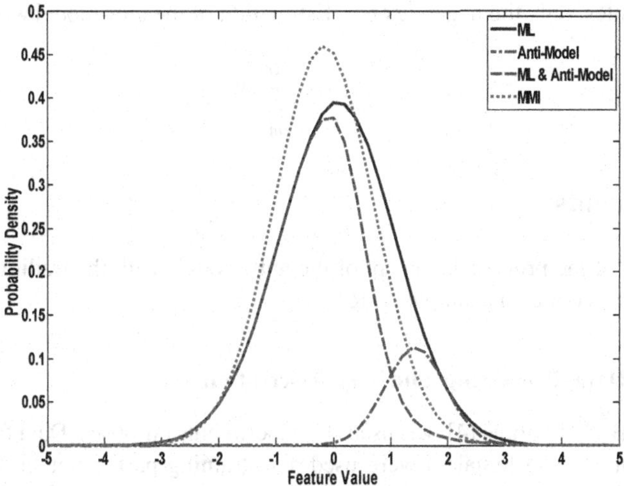

Fig. 1. Illustration of the anti-model concept. Probability density functions are compared for ML, MMI, anti-model, and ML & anti-model as the application of equation (6).

where lp_1 and lp_1 are two probabilities evaluated in the log-domain and the function $f(\min(lp_1, lp_2) - \max(lp_1, lp_2))$ is a smooth function that fast limits to zero for larger values. The function can be effectively approximated via Taylor series decomposition or calculated via lower-accuracy hardware implemented GPU instructions. Note, that this approach used to be implemented in the HMM evaluation already, thus it is simple to reuse it for the anti-models also. The resulting anti-models based HMM has about twice as many parameters to estimate. However, the more reliable ML training is used to the estimation. Note that the ML model and the anti-model of the state do not need to have the equal number of components and a proper model complexity can be chosen for both parts.

4 MMI-Based Anti-Models

Derivation of the anti-model parameters estimation for MMI is straight forward and simple. The equations are the same like for ML training. Only the denominator statistics of the MMI criterion is used:

$$\hat{\mu}_{Ajm} = \frac{\Theta_{jm}^{den}(O)}{\gamma_{jm}^{den}} \tag{8}$$

$$\hat{\sigma}_{Ajm}^2 = \frac{\Theta_{jm}^{den}(O^2)}{\gamma_{jm}^{den}} - \mu_{jm}^2. \tag{9}$$

No additional stabilization nor the I-smoothing is needed. There is only one difference, the denominator statistic for the anti-model does not contain the state's own data – the positive examples, only the negative ones. Estimation of the anti-model weight w_{Ajm} is also simple

$$\hat{w}_{Ajm} = \frac{\gamma_{jm}^{den}}{\gamma_{jm}^{num}}. \quad (10)$$

5 Experiments

A comparison of the proposed concept of the anti-models with the traditional ML and MMI criteria was done on a simple ASR task.

5.1 Speech Data, Processing, and Test Description

A part of UWB_S01 corpus [4] was used for experiments purposes. Data from first 100 speakers (57 males, 43 females) were used as a training part. Another 100 speakers (64 males, 36 females) make a test part. The digitization of an analogue signal was provided at 22.05 kHz sample rate and 16-bit resolution format. In order to extract features, Mel-frequency cepstral coefficients (MFCCs) were utilized, 15 dimensional feature vectors were extracted each 10 ms utilizing a 32 ms hamming window, including the energy coefficient. Then, Cepstral Mean Normalization (CMN) was applied, and Δ, Δ^2 coefficients were added. A 3 state HMM based on triphones with 425 states in total and 8 component GMM with diagonal covariances in each of the states was trained via ML criterion. In the case of anti-models, 8 component GMMs were trained for both the parts, ML and anti-model. The anti-models as well as the MMI models are based on MMI-FD criterion. In the case of MMI, three variants that differ in setup of the stabilization constant E were trained (see 2.1). Two variants with the fixed E equal 1 and 2 were done and one variant with dynamic E equal $\sqrt{\text{iteration}}$ has been added. I-smoothing was employed for all the MMI variants with $\tau = 100$.

To test the performance a simple 476-words zero-gram language model (LM) with no OOV was used. The simple LM was used to boost sensitivity to the acoustic part of the system. The zero-gram variant of the recognizer published in [23] was used and Word Error Rate (WER) was evaluated.

5.2 Results

The proposed concept of the anti-models were compared to the ML baseline and the MMI discriminative training. Both, MMI and anti-models, was initialized by the ML baseline model. The results are shown in Figure 2. It is clear that MMI as well as the anti-models outperform ML training. MMI produces very good models just after a single iteration on this task. However, the stability is a problem and after a few additional iterations WER increases. In contrast, the anti-models needs at least two iterations to proper results, but it keeps the achieved level.

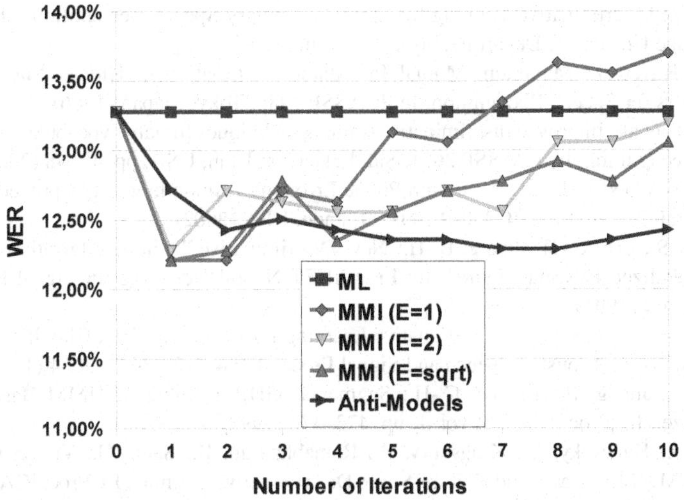

Fig. 2. Results of the ASR experiment. WER was evaluated for ML, three variants of MMI, and for the anti-models.

6 Conclusion

The alternative way to discriminative training, the concept of the anti-models, was introduced in this paper. Advantages and disadvantages were discussed. The main advantage is the stability and robustness that is brought by ML training. The main disadvantage is a more complex final HMM that requires a higher computation power to evaluate. The concept was tested on the ASR experiment. The results confirm equivalent WERs but the better stability in comparison to MMI.

References

1. Vanek, J., Trmal, J., Psutka, J.V., Psutka, J.: Optimized Acoustic Likelihoods Computation for NVIDIA and ATI/AMD Graphics Processors. IEEE Transactions on Audio, Speech and Language Processing 20(6), 1818–1828 (2012)
2. Vanek, J., Trmal, J., Psutka, J., Psutka, J.: Full Covariance Gaussian Mixture Models Evaluation on GPU. In: Proc. IEEE ISSPIT, Vietnam, Ho Chi Minh City (2012)
3. Heigold, G., Schlüter, R., Ney, H., Wiesler, S.: Discriminative Training for Automatic Speech Recognition: Modeling, Criteria, Optimization, Implementation, and Performance. IEEE Signal Processing Magazine 29, 58–69 (2012), doi:10.1109/MSP.2012.2197232
4. Radová, V., Psutka, J.: UWB_S01 Corpus – A Czech Read-Speech Corpus. In: Proc. of the ICSLP 2000, Beijing, China, pp. 732–735 (2000)
5. Normandin, Y., Morgera, D.: An Improved MMIE Training Algorithm for Speaker-Independent Small Vocabulary, Continuous Speech Recognition. In: Proc. of the IEEE, ICASSP 1991, Toronto, Canada, pp. 537–540 (1991)
6. Kapadia, S.: Discriminative Training of Hidden Markov Models. Ph.D. thesis, Cambridge University, Department of Engineering (1998)

7. Povey, D.: Discriminative Training for Large Vocabulary Speech Recognition. Ph.D. thesis, Cambridge University, Department of Engineering (2003)
8. Bahl, L.R., et al.: Maximum Mutual Information Estimation of Hidden Markov Model Parameters for Speech Recognition. In: ICASSP 1986, Tokyo, Japan (1986)
9. Povey, D., et al.: Improved discriminative training techniques for large vocabulary continuous speech recognition. In: ICASSP 2001, Salt Lake City, Utah, USA, pp. 45–48 (2001)
10. Povey, D., Woodland, P.C.: Minimum Phone Error and I-Smoothing for Improved Discriminative Training. In: Proc. ICASSP 2002, Orlando, USA (2002)
11. Katagiri, S., Lee, C.-H., Juang, B.-H.: New Discriminative Training Algorithms Based on the Generalized Descent Method. In: Proc. IEEE Neural Networks for Signal Processing, pp. 299–308 (1991)
12. Juang, B.-H., Katagiri, S.: Discriminative Learning for Minimum Error Classification. IEEE Transactions on Acoustic, Speech and Signal Processing 40(12), 3043–3054 (1992)
13. Chou, W., Juang, B.-H., Lee, C.-H.: Segmental GDP Training of HMM Based Speech Recognizer. In: Proc. ICASSP, vol. 1, pp. 473–476 (1992)
14. Povey, D., Kanevsky, D., Kingsbury, B., Ramabhadran, B., Saon, G., Visweswariah, K.: Boosted MMI for Model and Feature-Space Discriminative Training. In: Proc. ICASSP 2008, Las Vegas, USA, pp. 4057–4060 (2008)
15. Povey, D., Woodland, P.C., Gales, M.J.F.: Discriminative MAP for Acoustic Model Adaptation. In: Proc. ICASSP 2003, Hong Kong, pp. pp. I–312 (2003)
16. Zheng, J., Stolcke, A.: Improved Discriminative Training. Using Phone Lattices. In: Proc. Eurospeech 2005, Lisbon, Portugal (2005)
17. Bell, P.: Full Covariance Modelling for Speech Recognition. Ph.D. Thesis, The University of Edinburgh
18. Lee, Y., Lee, K.Y., Lee, J.: The Estimating Optimal Number of Gaussian Mixtures Based on Incremental k-means for Speaker Identification. International Journal of Information Technology 12(7), 13–21 (2006)
19. Figueiredo, M., Leitão, J., Jain, A.: On Fitting Mixture Models. In: Hancock, E.R., Pelillo, M. (eds.) EMMCVPR 1999. LNCS, vol. 1654, pp. 54–69. Springer, Heidelberg (1999)
20. Mclachlan, G.J., Peel, D.: On a Resampling Approach to Choosing the Number of Components in Normal Mixture Models. Computing Science and Statistics 28, 260–266 (1997)
21. Paclík, P., Novovičová, J.: Number of Components and Initialization in Gaussian Mixture Model for Pattern Recognition. In: Proc. Artificial Neural Nets and Genetic Algorithms, pp. 406–409. Springer, Wien (2001)
22. Vaněk, J., Machlica, L., Psutka, J., Psutka, J.: Covariance Matrix Enhancement Approach to Train Robust Gaussian Mixture Models of Speech Data. In: Železný, M., Habernal, I., Ronzhin, A. (eds.) SPECOM 2013. LNCS, vol. 8113, pp. 92–99. Springer, Heidelberg (2013)
23. Pražák, A., Psutka, J.V., Hoidekr, J., Kanis, J., Müller, L., Psutka, J.?: Automatic online subtitling of the Czech parliament meetings. In: Sojka, P., Kopeček, I., Pala, K., et al. (eds.) TSD 2006. LNCS (LNAI), vol. 4188, pp. 501–508. Springer, Heidelberg (2006)

Modelling F_0 Dynamics in Unit Selection Based Speech Synthesis*

Daniel Tihelka, Jindřich Matoušek, and Zdeněk Hanzlíček

University of West Bohemia, Faculty of Applied Sciences, Dept. of Cybernetics
Univerzitni 8, 306 14 Plzeň, Czech Republic
{dtihelka, jmatouse, zhanzlic}@kky.zcu.cz

Abstract. In the common unit selection implementations, F_0 continuity is measured as one of concatenation cost features with the expectation that smooth units transition (regarding speech melody) is ensured when the difference of F_0 is low enough. This measure generally uses a static F_0 value computed at the units boundary. In the present paper we show, however, that the use of static F_0 values is not enough for smooth speech units concatenation, and that a dynamic nature of the F_0 contour must be taken into account. Two schemes of dynamic F_0 handling are presented, and speech generated by both schemes is compared by means of listening tests on specially selected phrases which are known to carry unnatural artefacts. Advantages and disadvantages of the individual schemes are also discussed.

Keywords: text-to-speech synthesis, unit selection, concatenation cost, fundamental frequency F_0.

1 Introduction

There have been many papers describing concatenation cost features in unit selection speech synthesis, [20,21,15,4,16,1,13,14] to name a few. While most of them aim at determining the sources of spectral discontinuities, with results often in contradiction, in [5] was shown that a large number of audible discontinuities tend to appear at joins with incoherent F_0 values in the wider area around concatenation points.

There is a general agreement across unit selection researches that the incorporation of F_0 continuity measure at the units boundaries is an essential condition of smooth concatenation achievement. Usually, however, the authors limit this feature to simple "static" F_0 difference in Hz (or log(Hz) in some cases) [3,2,12], i.e. $d = \left|f^e(i) - f^b(i+1)\right|$, where $f^e(i)$ denotes the F_0 value assigned to the end of the ith unit, and $f^b(i+1)$ value assigned to the beginning of the $(i+1)^{\text{th}}$ unit. The manner of F_0 computation may differ (and it usually does) for individual approaches, but it basically is an average through several epoch periods to eliminate the F_0 fast changes

* The research has been supported by the European Regional Development Fund (ERDF), project "New Technologies for Information Society" (NTIS), European Centre of Excellence, ED1.1.00/02.0090, and by the Technology Agency of the Czech Republic, project No. TA01011264.

P. Sojka et al. (Eds.): TSD 2014, LNAI 8655, pp. 457–464, 2014.

in microprosody. Nevertheless, whatever the F_0 computing scheme, it must be ensured that $f^e(i) = f^b(i+1)$, and thus $d = 0$, for the two units following each other in the speech corpus.

In this paper, the preliminary experiments taking into account wider F_0 context are described and discussed. In our work we extend [8], but instead of evaluating specially designed phrases with a single concatenation point in the middle of vowels [5], we will employ a real TTS system on which the results are obtained.

2 The Ways of F_0 Dynamics Modelling

First, let us describe what the baseline implementation of concatenation cost computation looks like in our TTS system ARTIC [11,6]. When concatenating two units (diphones in our case) i and $i+1$, the concatenation cost $C^c(i, i+1)$ is computed as

$$C^c(i, i+1) = \frac{C^c_S(i, i+1) + C^c_E(i, i+1) + C^c_F(i, i+1)}{3} \tag{1}$$

where $C^c_S(i, i+1)$ is the Euclidean distance of 12 MFCC coefficients expecting to reflect spectral smoothness of the concatenated units[1], $C^c_E(i, i+1)$ is the absolute difference of energy, and $C^c_F(i, i+1)$ reflects the "static" F_0 difference at the units boundaries, computed according to Equation (7) as $C^c_F(i, i+1) = \left|\delta\left(f^e(i), f^b(i+1)\right)\right|$.

All the features are computed in the pitch-synchronous way, meaning that each pitch-mark (see [7] for the definition) has been assigned the value of energy, F_0 (being `NaN` for unvoiced pitch-marks), and the vector of MFCC coefficients. All the values are z-score normalized to align their ranges. Then, each unit boundary obtained by HTK-alignment process [10,9] is tied with the set of features computed for the pitch-mark being the closest to the given boundary.

While both energy and MFCC are computed from a window of fixed length, centred around the pitch-mark the resulting value is assigned to, the computation of F_0 is slightly more complicated. For a sequence of voiced pitch-marks $p(k), k = 1, 2, \ldots, K$, each pitch-mark has assigned mean F_0 value $f(k)$:

$$f(k) = \frac{\sum_{l=x}^{y-1} \frac{1}{p(l+1)-p(l)}}{y - x} \tag{2}$$

$$x = k - w\left\lfloor\frac{k}{K} + 0.5\right\rfloor \tag{3}$$

$$y = k + w\left\lfloor\frac{K-k}{K} + 0.5\right\rfloor \tag{4}$$

where w is the fixed number of epochs through which the F_0 is computed. As illustrated on Figure 1, it enables the use of a fixed number of epochs regardless of whether the F_0 value is computed at the beginning, middle, or at the end of the voiced pitch-marks sequence.

[1] The use of MFCC is revised currently, since our evidence suggests that it does not seem to be an appropriate feature for such measure. Therefore, although it is used in this experiment, it may become invalid in foreseeable future.

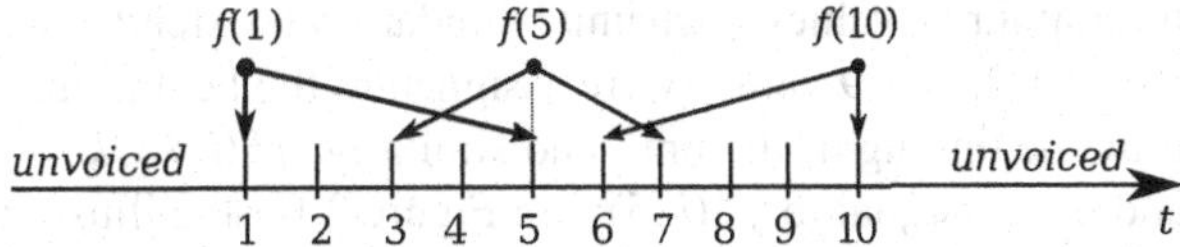

Fig. 1. The illustration of pitch-synchronous F0 values computation. $K = 10$, $w = 4$, vertical lines represent pitch-marks.

In this paper we experiment only with values of $C_F^c(i, i+1)$ in Equation (1). The remaining features, as well as the manner of target cost computation, stay untouched.

2.1 Delta Coefficients

Natural consideration of F0 dynamics employment is to use "classic" *delta* coefficients. However, they are not very suitable for unit selection synthesis, since they reflect the dynamics only in a relatively near point around the concatenation point. Although such dynamics are usually used together with spectral and other features in HMM synthesis (where their use is legitimate since HMM works as generator on model states), their use for longer cross-unit F0 fluency policing is not very effective.

To illustrate it, let us take HTK [22] toolkit as an example. For k^{th} feature value, its dynamic coefficients are computed as:

$$D(k) = \frac{\sum_{i=1}^{I} i\left(F(k+i) - F(k-i)\right)}{2\sum_{i=1}^{I} i^2} \quad (5)$$

where I is the configurable length of window through the dynamics are computed and F is the value of the feature (F0 in out case). It is obvious that even with $I > 1$, the largest portion of the delta value is taken from the difference to $k-1$ and $k+1$ point. And the same situation is for the *acceleration* (*delta-delta*) coefficients, which are computed by Equation (5), except using the computed D in place of F.

Contrary to this, we aimed at involving the wider tendency of F0 behaviour, since it is natural supra-segmental feature expressing the communication function of a phrase crossing several adjacent phones. On the other hand, the considered context must not be too wide (e.g. crossing several diphones) since the feature would not reflect what it is intended to (i.e. the smoothness of join), but it would instead describe something like supra-segmental prosody tendencies, while a local audible unnatural artefact could still happen.

2.2 Contour Comparison

Quite encouraging results were reported in [8], in which the vector of 8 F0 values extracted from the vicinity of carefully designed concatenation point is able to detect audible discontinuities with accuracy about 90%. Therefore, we were curious whether the scheme is able to provide a similar result when employed in the real TTS system.

To compare the contour F_0 values, each unit boundary was extended with 9 F_0 values. That is, $f^b(i,k), k = 1.2, \ldots, 9$ values were assigned to the beginning of ith unit with $f^b(i,5)$ equal to the beginning of the unit, and similarly, $f^e(i,k), k = 1.2, \ldots, 9$ are assigned to the end of unit equal to $f^e(i,5)$; see Figure 2 for the illustration. Thus, the requirement $f^e(i,k) = f^b(i+1,k)$ from Section 1 is still valid $\forall i, k$ for adjacent units, while the context 9 pitch-marks long is compared in the concatenation cost.

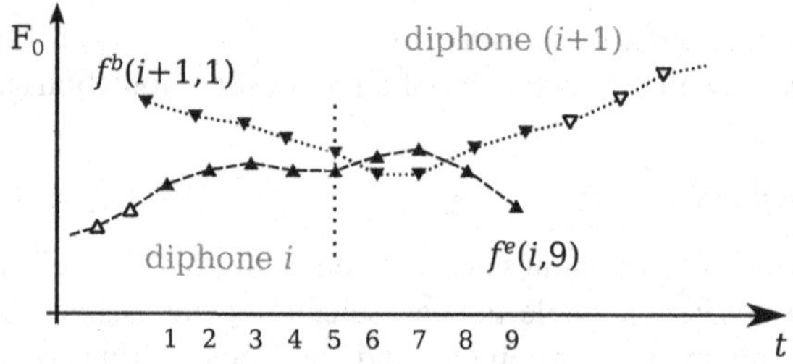

Fig. 2. The illustration of F_0 contour computation on a join of voiced diphones boundary. The dashed line connects values $f^e(i)$, dotted connects values $f^b(i+1)$, black triangles represent the $k = 1.2, \ldots, 9$ pairs of the F_0 values used in Equation (6). Concatenation point is dotted vertical line.

The value of F_0-related sub-cost is computed here using Euclidean distance between the corresponding f values as:

$$C^c_F(i, i+1) = \sqrt{\sum_{k=1}^{9} \delta\Big(f^e(i,k), f^b(i+1,k)\Big)^2} \tag{6}$$

where $\delta(a,b)$ is function defined as:

$$\delta(a,b) = \begin{cases} a - b, & a \neq \texttt{NaN},\ b \neq \texttt{NaN} \\ 0, & a = b = \texttt{NaN} \\ 6, & \text{otherwise} \end{cases} \tag{7}$$

with the value 6 chosen as large enough, since the difference of z-score normalized values f will exceed it for large F_0 differences only (exactly it is in case $a \leq -3,\ b \geq 3$, each having 0.1% likelihood). However, the particular value does not matter a great deal.

2.3 Slope

The main disadvantage of F_0 contour comparison scheme is its higher computation cost — there are 9 floating points multiplications followed by square root evaluation. Considering the number of evaluations which are carried out during the concatenation cost computing (may approach 250 millions, as described in [18]), such a scheme will have a significant negative impact on the performance of the TTS system, which would be notable especially on lower-resource devices, e.g. smart-phones [19].

This is the reason why we have experimented with another scheme of F_0 dynamics embedding – the comparison of the F_0 slope of the concatenated units; in [8] it was reported as only slightly worse than the use of contour. Firstly, we have computed the slope of F_0 using the linear regression of all the F_0 values measured on a voiced *phone* (such covered by voiced pitch-marks in more than 70% of length[2]); let us mark it as $S(j)$, where j is the index of phone within a phrase. Then, the sequence of phones is converted to *diphones*, so two values $S(j), S(j+1)$ are assigned to diphones as $S(j) = S^e(i-1) = S^b(i), S(j+1) = S^e(i) = S^b(i+1)$; the whole scheme is illustrated on Figure 3. The value of F_0-related sub-cost is then computed as:

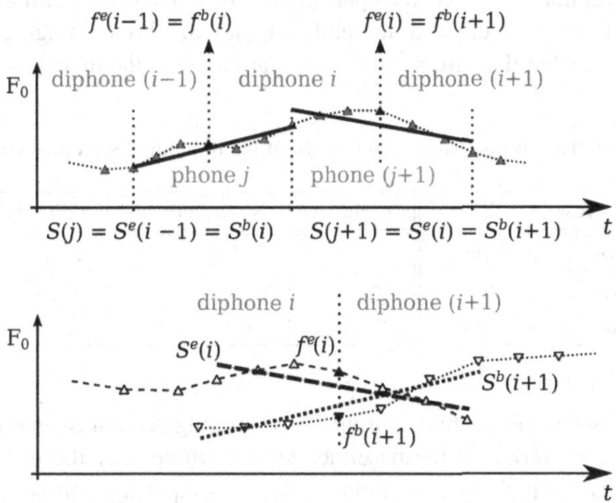

Fig. 3. The illustration of F_0 slope (line) computation and its phone-to-diphone distribution in the upper part of the Figure. In the lower part, the dashed line represents F_0 contour ($f^e(i) = \blacktriangle$) and slope $S^e(i)$, dotted line illustrates contour ($f^b(i+1) = \blacktriangledown$) and slope $S^b(i+1)$ used in Equation (8). Concatenation point is dotted vertical line.

$$C_F^c(i,i+1) = \begin{cases} 3\left|f^e(i) - f^b(i+1)\right| + 2\left|S^e(i) - S^b(i+1)\right|, & f^e(i) \neq \text{NaN}, \\ & f^b(i+1) \neq \text{NaN} \\ 0, & f^e(i) = \\ & f^b(i+1) = \text{NaN} \\ 15 = 9 + 6, & \text{otherwise} \end{cases} \tag{8}$$

[2] This is slight difference from [8] when the slope was computed only through 4 pitch-periods around the concatenation point

where the first part is the difference of z-score F_0 values at the diphone boundaries computed exactly as in the baseline system, and the second part is the difference of F_0 slopes. The weights were chosen to slightly prefer the static F_0 value to the slope.

3 Evaluation

To evaluate the effect of F_0 dynamics modelling, we have designed listening tests with 20 phrases taken from the set in which the largest number of unnatural artefacts were evaluated in our internal research. There were 14 people involved in the test for which 3-point scale CCR (comparison category rating) form was used. The pairs compared were *baseline* × *slope* and *contour* × *slope*, presented in the randomized ordering. The *baseline* × *contour* test had been carried out earlier during the research to clarify results from [8], but lower number of listeners participated in it. Therefore, although we present the results of this test as well, and the tendency they display is in agreement with the overall results, note that they are not fully comparable with the main tests.

Table 1. The comparison of preference of the individual system versions

Test (A × B)	Prefer A	No preference	Prefer B
contour × *slope*	38.8%	36.9%	24.3%
baseline × *slope*	16.5%	43.5%	40.1%
baseline × *contour*	9.0%	34.0%	57.0%

It can clearly be seen that while *contour*-incorporating version is generally preferred, both versions are preferred to the baseline system, where only the difference of unit boundaries-related "static" F_0 is computed. When comparing *contour* to *slope*, there is slight preference for the use of *contour*; we expect that the reason is more precise F_0 contour comparison. On the other hand, this computation scheme is much more demanding, as mentioned in Section 2.3. It may seem that the use *contour*-based dynamics are evidently more preferred to the *baseline* that the *slope* is, but note again that this test has not been carried out by the same number of listeners, although on the same set of phrases.

4 Conclusion

The results presented are in general agreement with the results of [8], so it may be concluded that the use of dynamic F_0 features as a part of the concatenation cost has noticeable effect on the quality of speech synthesis. What remains to be found is the most effective, both in terms quality improvement and computation speed, scheme of the features comparison. We plan experiments, where, for example, fewer F_0 points will be compared in the *contour* scheme, or where the Euclidean distance in Equation (6) will be replaced by the mean of $\left|\delta\left(f^e(i,k), f^b(i,k)\right)\right|$ absolute differences. We also need to check the slope computed exactly as described in [8].

Moreover, as a part of listening tests stimuli, we plan to use phrases where clear F_0 artefact is found, when generated by baseline system. Currently, although the evaluated phrases do contain unnatural artefacts, they may be of any type. Due to the rather small range of listening test and the fact that only phrases used for evaluation have been synthesized, we did not also carry out the evaluation of results reliability, as described in [17]. We plan to do so in the near future.

Special thanks are due to National Grid Infrastructure MetaCentrum, providing the access to computing and storage facilities under the program LM2010005 “Projects of Large Infrastructure for Research, Development, and Innovations”.

References

1. Bellegarda, J.R.: A novel discontinuity metric for unit selection text-to-speech synthesis. In: Proc. of 5th Speech Synthesis Workshop (SSW5), Pittsburgh, PA, USA, pp. 133–138 (2004)
2. Conkie, A., Syrdal, A.K.: Using F0 to constrain the unit selection Viterbi network. In: Proc. of Acoustics, Speech, and Signal Processing ICASSP, pp. 5376–5379. IEEE (2011)
3. Hunt, A.J., Black, A.W.: Unit selection in a concatenative speech synthesis system using a large speech database. In: Proc. of Acoustics, Speech, and Signal Processing ICASSP 1996, vol. 1, pp. 373–376. IEEE (1996)
4. Klabbers, E., Veldhuis, R.N.J.: Reducing audible spectral discontinuities. IEEE Transactions on Speech and Audio Processing 9(1), 39–51 (2001), http://dblp.uni-trier.de/db/journals/taslp/taslp9.html#KlabbersV01
5. Legát, M., Matoušek, J.: Design of the test stimuli for the evaluation of concatenation cost functions. In: Matoušek, V., Mautner, P. (eds.) TSD 2009. LNCS, vol. 5729, pp. 339–346. Springer, Heidelberg (2009)
6. Legát, M., Matoušek, J.: Collection and analysis of data for evaluation of concatenation cost functions. In: Sojka, P., Horák, A., Kopeček, I., Pala, K. (eds.) TSD 2010. LNCS, vol. 6231, pp. 345–352. Springer, Heidelberg (2010)
7. Legát, M., Matoušek, J., Tihelka, D.: On the detection of pitch marks using a robust multi-phase algorithm. Speech Communication, 552–566 (2011), http://www.kky.zcu.cz/en/publications/LegatM_2011_Onthedetectionof
8. Legát, M., Matoušek, J.: Pitch contours as predictors of audible concatenation artifacts. In: Proc. of World Congress on Engineering and Computer Science 2011, San Francisco, USA, pp. 525–529 (2011)
9. Matoušek, J., Romportl, J.: Automatic pitch-synchronous phonetic segmentation. In: INTERSPEECH 2008, Proc. of 9th Annual Conference of International Speech Communication Association, Brisbane, Australia, pp. 1626–1629 (2008)
10. Matoušek, J., Tihelka, D., Psutka, J.V.: Experiments with automatic segmentation for Czech speech synthesis. In: Matoušek, V., Mautner, P. (eds.) TSD 2003. LNCS (LNAI), vol. 2807, pp. 287–294. Springer, Heidelberg (2003), http://dx.doi.org/10.1007/978-3-540-39398-6_41
11. Matoušek, J., Tihelka, D., Romportl, J.: Current state of Czech text-to-speech system ARTIC. In: Sojka, P., Kopeček, I., Pala, K. (eds.) TSD 2006. LNCS (LNAI), vol. 4188, pp. 439–446. Springer, Heidelberg (2006), http://dx.doi.org/10.1007/11846406_55
12. Narendra, N.P., Rao, K.S.: Syllable specific unit selection cost functions for text-to-speech synthesis. ACM Transactions on Speech and Language Processing 9(3), 5:1–5:24 (2012), http://doi.acm.org/10.1145/2382434.2382435

13. Pantazis, Y., Stylianou, Y.: On the detection of discontinuities in concatenative speech synthesis. In: Stylianou, Y., Faundez-Zanuy, M., Esposito, A. (eds.) COST 277. LNCS, vol. 4391, pp. 89–100. Springer, Heidelberg (2007), http://dx.doi.org/10.1007/978-3-540-71505-4_6
14. Přibil, J., Přibilová, A.: Evaluation of influence of spectral and prosodic features on GMM classification of Czech and Slovak emotional speech. EURASIP Journal on Audio, Speech, and Music Processing 33(3), 1–22 (2013), http://dx.doi.org/10.1186/1687-4722-2013-8
15. Stylianou, Y., Syrdal, A.K.: Perceptual and objective detection of discontinuities in concatenative speech synthesis. In: Proc. IEEE Acoustics, Speech, and Signal Processing (ICASSP), pp. 837–840 (2001)
16. Syrdal, A.K., Conkie, A.D.: Data-driven perceptually based join costs. In: Proc. of 5th Speech Synthesis Workshop (SSW5), Pittsburgh, PA, USA, pp. 49–54 (2004)
17. Tihelka, D., Grůber, M., Hanzlíček, Z.: Robust methodology for TTS enhancement evaluation. In: Habernal, I. (ed.) TSD 2013. LNCS, vol. 8082, pp. 442–449. Springer, Heidelberg (2013), http://dx.doi.org/10.1007/978-3-642-40585-3_56
18. Tihelka, D., Kala, J., Matoušek, J.: Enhancements of Viterbi search for fast unit selection synthesis. In: INTERSPEECH 2010, Proc. of 11th Annual Conference of the International Speech Communication Association, pp. 174–177 (2010), http://www.isca-speech.org/archive/interspeech_2010/i10_0174.html
19. Tihelka, D., Stanislav, P.: ARTIC for assistive technologies: Transformation to resource-limited hardware. In: Proc. of World Congress on Engineering and Computer Science 2011, San Francisco, USA, pp. 581–584 (2011)
20. Vepa, J., King, S.: Kalman–filter based join cost for unit–selection speech synthesis. In: Proc. EUROSPEECH 2003 – INTERSPEECH 2003, Proc. of 8th European Conference on Speech Communication and Technology, pp. 293–296. ISCA (2003)
21. Vepa, J., King, S.: Join cost for unit selection speech synthesis. Ph.D. thesis, The University of Edinburgh, College of Science and Engineering, School of Informatics (2004), https://www.era.lib.ed.ac.uk/handle/1842/1452
22. Young, S., Kershaw, D., Odell, J., Ollason, D., Valtchev, V., Woodland, P.: The HTK Book Version 3.4. Cambridge University Press (2006)

Audio-Video Speaker Diarization for Unsupervised Speaker and Face Model Creation

Pavel Campr[1], Marie Kunešová[1], Jan Vaněk[1], Jan Čech[2], and Josef Psutka[1]

[1] University of West Bohemia, Faculty of Applied Sciences, Dept. of Cybernetics
Univerzitni 8, 306 14 Plzen, Czech Republic
{campr,mkunes,vanekyj,psutka}@kky.zcu.cz
[2] Czech Technical University in Prague, Faculty of Electrical Engineering,
Department of Cybernetics, Center for Machine Perception
Technicka 2, 166 27 Praha 6, Czech Republic
cechj@cmp.felk.cvut.cz

Abstract. Our goal is to create speaker models in audio domain and face models in video domain from a set of videos in an unsupervised manner. Such models can be used later for speaker identification in audio domain (answering the question "Who was speaking and when") and/or for face recognition ("Who was seen and when") for given videos that contain speaking persons. The proposed system is based on an audio-video diarization system that tries to resolve the disadvantages of the individual modalities. Experiments on broadcasts of Czech parliament meetings show that the proposed combination of individual audio and video diarization systems yields an improvement of the diarization error rate (DER).

Keywords: audio-video speaker diarization, audio speaker recognition, face recognition.

1 Introduction

With the increasing amount of multimedia data it is necessary to develop techniques that detect the presence of people and find out their identities. Such information can be used for indexing and searching purposes, for enhancement of automatic speech recognition systems, for building audio or video identification systems or for building audio or video corpora.

Our main goal is to create speaker models in the audio domain and face models in the video domain, in an unsupervised manner, so that the models can be used later for audio-video person identification. Contrary to most other existing systems [3,4] our method produces a slightly larger number of speaker-model candidates than the real number of speakers. The reason is that it is more important that no two different speakers are assigned the same identity than that each speaker is only assigned one. The former error has a negative impact on the performance of the whole system and can never be corrected, while the latter can be discovered and resolved in later processing by automatic or manual verification.

The paper is organized as follows. Section 2 describes the diarization system in audio-only domain, Section 3 in video-only domain. Combined audio-video diarization is described in Section 4 and evaluated in Section 5.

P. Sojka et al. (Eds.): TSD 2014, LNAI 8655, pp. 465–472, 2014.

2 Audio Speaker Diarization

For audio diarization we use an approach based on Gaussian Mixture Models (GMMs). The system is largely based on the ones proposed in [5] and [6]. However, there are some differences.

Most notably, as this particular application does not require the diarization to be done online, we perform additional offline clustering after all audio files have been processed, to identify and merge speaker-model candidates corresponding to the same speakers, both within a single file and between different ones. Unlike the authors of [5], we also use energy-based speech activity detection as opposed to model-based.

At the beginning, the system starts with only two GMMs, one for each gender, which are trained in advance. Afterwards, for every speech segment, the system decides if the segment corresponds to an already known speaker, or a new one. In the case of a new speaker, a new model is created by copying one of the gender dependent models. Otherwise, one of the existing models is selected. The assigned model is then adapted using the data from the segment.

The system for audio speaker diarization consists of several modules:

1. Feature extraction and voice activity detection
2. Speech segmentation
3. Speaker identification and novelty detection
4. Online GMM learning
5. Offline clustering

2.1 Feature Extraction and Voice Activity Detection

For feature extraction, we used the LFCC with 25 filters in range from 50 Hz to 8 kHz based on 25ms FFT window with 10ms shift. 20 cepstral coefficients were computed without the energy coefficient. No cepstral normalization was performed.

This module also performs an energy-based voice activity detection (VAD), with every frame being labeled as *speech* or *silence* based on a threshold.

2.2 Speech Segmentation

Using the information obtained from the VAD and parameters such as the minimum and maximum segment length and the maximum pause length in a segment, the speech is divided into short segments. Of each segment, only the frames labeled as *speech* are used for the novelty detection and GMM learning modules. In our experiments, this has lead to both reduced computation time and improved performance.

2.3 Speaker Identification and Novelty Detection

For each speech segment the system uses a maximum-likelihood classification to determine both the speaker's gender (using the gender dependent models) and their most likely identity out of the existing speaker-model candidates. Afterwards, a likelihood ratio test is used to decide whether the segment belongs to the chosen identity, or represents an entirely new speaker.

The likelihood ratio is as follows:

$$L(X) = \frac{P_{sp}}{P_{gen}}, \tag{1}$$

where X is a speech segment and P_{sp} and P_{gen} are the likelihoods of the winning speaker and the appropriate gender dependent model, respectively.

If $L(X) \geq \theta$, the segment X belongs to the old speaker. Otherwise it belongs to an entirely new speaker.

The optimal value of decision threshold θ was found experimentally, and chosen in such a way that the system produces a slightly larger number of speaker models than the real number of speakers. The reason for this is that it is more important to us in this stage that no two different speakers are assigned the same identity than that each speaker is only assigned one. The former error has a negative impact on the performance of the whole system and can never be corrected, while the latter can be discovered and resolved in later stages, either during clustering or once the results from both audio and video are combined.

2.4 Online GMM Learning

For GMM adaptation we use an online variant of the Expectation-Maximization algorithm, as described in [7], with values of the parameters as proposed by [5].

2.5 Audio Clustering

After we have processed all audio recordings, we perform clustering. The main purpose is to find the labels corresponding to the same real speakers between different recordings, although the system also resolves most cases of multiple labels being assigned to the same real speaker within a single recording.

For every pair of speaker models λ_i and λ_j, we calculate the value of the following expression, which is similar to the Cross-Likelihood Ratio [8]:

$$L(i, j) = \frac{1}{N_i} \cdot \log \left(\frac{p(X_i|\lambda_j)}{\max(p(X_i|\lambda_m), p(X_i|\lambda_f))} \right), \tag{2}$$

where X_i represents all the speech data assigned to the speaker model λ_i, N_i is the number of frames in X_i, and λ_m and λ_f are the gender dependent models.

If both $L(i, j)$ and $L(j, i)$ exceed a certain threshold, we consider λ_i and λ_j to be the same speaker. In order for more than two models to be merged, the condition must be fulfilled for every two of them.

3 Video Speaker Diarization

The goal of this module is to detect and track faces in a video, to extract features from each face image and to perform clustering. The result is a set of clustered face tracks, each cluster representing one identity.

The task of face detection, tracking, and identification has been widely studied. Existing solutions can solve this task with high accuracy [1,2].

3.1 Face Tracking and Feature Extraction

In this paper we use a facial landmarks detector based on the Deformable Part Models [1]. In addition to the detected position of the face, this face detector provides a set of facial landmarks: nose, mouth and canthi corners. Such landmarks are used for the construction of normalized face images from which the face features are computed. The feature computation is based on Local Binary Patterns (LBP).

During the face tracking process, normalized face images are computed for the whole face track. For all normalized face images in the face track, the distances from all the previous normalized images in the same face track are computed. If at least one distance is lower than a threshold θ_1, a similar face appearance was already seen and the image is ignored. As a result, each of the N_T face tracks is represented by a set of key face images and corresponding features λ^V.

3.2 Clustering

Clustering is performed after the processing of all video recordings. The purpose is to find the labels corresponding to the face of the same person between different face tracks, in all videos. The whole process is visualised in Figure 1.

For every pair of face tracks i and j, we compute their distances $D_T(i, j)$, based on the features λ_i^V and λ_j^V. The distance is computed as a min-min distance between the sets of features λ_i^V and λ_j^V. If their minimal distance is lower than a certain threshold θ_2, then both face tracks are considered to be of the same identity with the same label. In other words, if there are similar faces in the first face track i and the second track j, then the identity of both is considered to be the same.

As a result we have N_C clusters, each representing the face of one person. The threshold θ_2 was experimentally set so that we have multiple clusters with the same

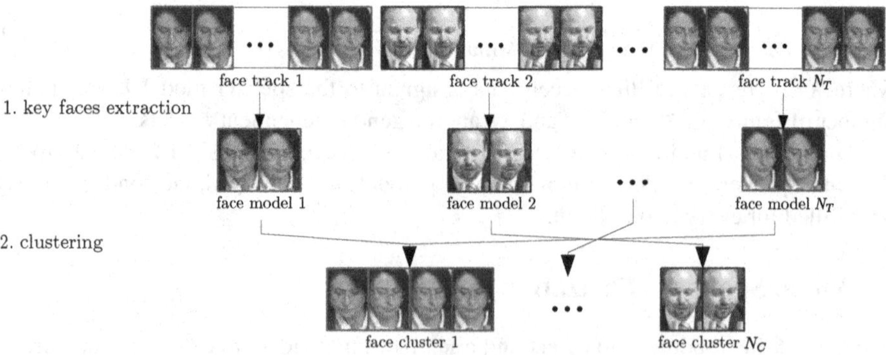

Fig. 1. Video diarization process

identity, but there is no cluster that represents multiple identities. The reason is that the additional merging of clusters is performed in later stage using audio modality.

4 Audio-Video Speaker Diarization

The combined audio-video speaker diarization system tries to resolve the disadvantages of particular modalities. The audio-only diarization requires longer intervals (usually several seconds) to produce a certain decision about the speaker's identity. For the faces in the video domain, the decision can be made in one frame only, but the appearance of one speaker's face can change in time greatly. When compared to audio, the speech style of one speaker changes only a little during the time.

The proposed system is built with the assumption that the speaker's face is present in the video most of the time during his speech. The generated audio and video speakers' models can be used later for any videos where this condition is not fulfilled.

4.1 Audio-Video Speaker Models Clustering

Only the best matching segments (models) were clustered in the previous stages of the processing where the audio and the video modalities were treated independently. Therefore, the number of obtained models was too high number. Further clustering cannot be done accurately according to individual modalities. However, fused audio-video merging is able to reduce the number of models with acceptable error rate. The clustering of the models was done in the following way:

1. The audio similarity matrix was based on the symmetric Kullback-Leibler divergence:

$$L_{KLD}(i,j) = \frac{1}{2}\left[\mathcal{L}(X_i|\lambda_i) + \mathcal{L}(X_j|\lambda_j) - \mathcal{L}(X_i|\lambda_j) - \mathcal{L}(X_j|\lambda_i)\right], \quad (3)$$

   where

$$\mathcal{L}(X_i|\lambda_j) = \frac{1}{N_i}\sum_i \log(p(X_i|\lambda_j)). \quad (4)$$

2. The video similarity matrix was based on a transformation of $D(i,j)$ values produced by the face-model distance function. The transformation was chosen to match the audio L_{KLD} similarity:

$$L_V(i,j) = \frac{D(i,j)-\alpha}{\beta}, \quad (5)$$

   where α and β are parameters tuned on a small development data set.
3. The final fused similarity is defined as

$$L_{AV}(i,j) = L_{KLD}(i,j) + L_V(i,j). \quad (6)$$

   The evaluated $L_{AV}i, j$ values were compared with a threshold θ and the model pairs with a value higher than the threshold were clustered.

5 Experiments

For testing purposes we used recordings from Czech parliament meetings broadcasted by the Czech Television. A sample image from the broadcasts can be seen in Figure 2. We had 8 recordings with a total of 30 hours of labeled audio.

To evaluate the system we compared the audio-only and audio-video clustering results.

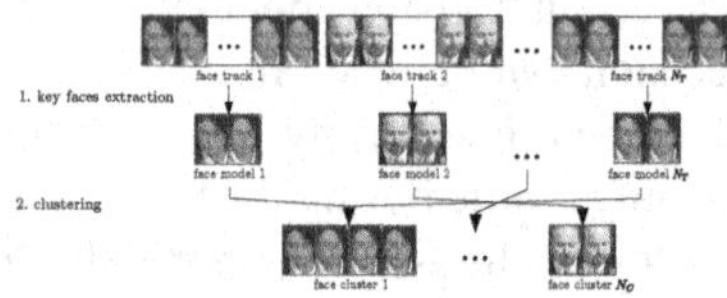

Fig. 2. Sample image from the Czech parliament meetings broadcasts. The two rectangles denote the detected faces.

5.1 Audio Clustering

Gender dependent models were trained using 30 seconds of speech from each of 16 women and 70 men.

In the experiments, we used GMMs with only 8 components and the minimum and maximum segment lengths were set to 1 and 5 seconds respectively.

In order for the diarization results to better correspond to the reference labels, we relabeled any short pauses between two consecutive speech segments which belong to the same speaker as speech as well. The optimal maximum length of such pauses was determined from the reference labels to be 3 seconds.

To evaluate the performance of the system, we used the diarization error rate (DER), which is the sum of three values: the rate of missed speech (i.e. the speech frames that were incorrectly labeled as silence), false alarm (FA, silence incorrectly labeled as speech) and speaker error (SE, the speech labeled as a wrong speaker). It is measured as the fraction of time that is not assigned correctly to a speaker or to non-speech [9].

Additionally, we also computed the speaker error rate and the DER when tolerating the use of multiple models for each real speaker (SE_{MM} and DER_{MM} in the table). These values essentially represent the ideal error rates we would obtain by performing additional oracle clustering.

Average values from all recordings, obtained both before and after the clustering, are shown in Table 1. The values given were obtained without any forgiveness collar around reference segment boundaries.

The audio clustering has lead to a significant decrease of the speaker error, though it does not reach the ideal value represented by SE_{MM}. The slight increase of SE_{MM} suggests that we have also clustered a small number models which did not truly belong to the same speaker, while the decrease of DER_{MM} was caused by the decreased miss rate.

Table 1. Audio diarization performance (%)

	miss	FA	SE	DER	SE_{MM}	DER_{MM}
before clustering	2.58	1.09	7.33	11.0	1.68	5.35
after clustering	2.25	1.11	4.15	7.51	1.85	5.22

5.2 Audio-Video Clustering

The table 2 presents the results of audio-video clustering method. The previous results of audio-only clustering are compared to the audio-video clustering.

Table 2. Audio-video diarization performance

	DER (%)	number of speakers
audio clustering	7.51	381
audio-video clustering	7.18	292
truth	-	86

The results show that the DER decreased from 7.51% to 7.18% when the video modality is used. The number of resulting clusters decreased from 381 to 292. The merged clusters were either short or contained some noise and the audio-only clustering system was unable to merge these ambiguous clusters. Because the DER is measured as the fraction of time and we clustered mostly short segments, the decrease of DER is 4.5% relatively, although the decrease of number of clusters is 23.4% relatively.

As the face is not present in the video all the time, the proposed algorithm is not able to merge clusters where only audio modality is used and the segments are too short or ambiguous.

6 Conclusion

We addressed the problem of audio-visual speaker diarization. After the description of the baseline audio-only diarization system we presented our proposed method for association of individual models from the audio and video modalities, all in an unsupervised manner. The method was evaluated on 30 hours of video. The diarization error rate (DER) decreased from 7.51% to 7.18% and the number of clusters decreased from 381 to 292, where the real number of speakers was 86. These results show that the short utterances which the audio-only diarization is unable to associate to a correct identity can be associated more reliably with the addition of video modality and face recognition.

Acknowledgments. This research[1] was supported by the Grant Agency of the Czech Republic, project No. GAČR GBP103/12/G084.

[1] The access to computing and storage facilities owned by parties and projects contributing to the National Grid Infrastructure MetaCentrum, provided under the programme "Projects of Large Infrastructure for Research, Development, and Innovations" (LM2010005) is highly appreciated.

References

1. Uřičář, M., Franc, V., Hlaváč, V.: Detector of Facial Landmarks Learned by the Structured Output SVM. In: VISAPP 2012: Proceedings of the 7th International Conference on Computer Vision Theory and Applications, pp. 547–556 (2012)
2. Sonnenburg, S., Franc, V.: COFFIN: A Computational Framework for Linear SVMs. Technical Report, Center for Machine Perception, Czech Technical University, Prague, Czech Republic (2009)
3. Bendris, M., Charlet, D., Chollet, G.: People indexing in TV-content using lip-activity and unsupervised audio-visual identity verification. In: 9th International Workshop on Content-Based Multimedia Indexing (CBMI), pp. 139–144 (2011)
4. El Khoury, E., Sénac, C., Joly, P.: Audiovisual diarization of people in video content. Multimedia Tools and Applications (2012)
5. Markov, K., Nakamura, S.: Never-Ending Learning System for Online Speaker Diarization. In: IEEE Workshop on Automatic Speech Recognition & Understanding, ASRU 2007, pp. 699–704 (2007)
6. Geiger, J., Wallhoff, F., Rigoll, G.: GMM-UBM based open-set online speaker diarization. In: INTERSPEECH 2010, pp. 2330–2333 (2010)
7. Sato, M., Ishii, S.: On-line EM algorithm for the Normalized Gaussian Network. Neural Computation 12, 407–432 (2000)
8. Reynolds, D., Singer, E., Carlson, B., O'Leary, J., McLaughlin, J., Zissman, M.: Blind clustering of speech utterances based on speaker and language characteristics. In: Proceedings of the 5th International Conference on Spoken Language Processing, vol. 7, pp. 3193–3196 (1998)
9. National Institute of Standards and Technology, `http://www.itl.nist.gov`

Improving a Long Audio Aligner through Phone-Relatedness Matrices for English, Spanish and Basque

Aitor Álvarez, Pablo Ruiz, and Haritz Arzelus

Human Speech and Language Technologies, Vicomtech-IK4, San Sebastián, Spain
{aalvarez,pruiz,harzelus}@vicomtech.org

Abstract. A multilingual long audio alignment system is presented in the automatic subtitling domain, supporting English, Spanish and Basque. Pre-recorded contents are recognized at phoneme level through language-dependent triphone-based decoders. In addition, the transcripts are phonetically translated using grapheme-to-phoneme transcriptors. An optimized version of Hirschberg's algorithm performs an alignment between both phoneme sequences to find matches. The correctly aligned phonemes and their time-codes obtained in the recognition step are used as the reference to obtain near-perfectly aligned subtitles. The performance of the alignment algorithm is evaluated using different non-binary scoring matrices based on phone confusion-pairs from each decoder, on phonological similarity and on human perception errors. This system is an evolution of our previous successful system for long audio alignment.

Keywords: Long audio alignment, automatic subtitling, phonological similarity matrices, perceptual confusion matrices.

1 Introduction

Subtitling is one of the most important means to make audiovisual content accessible. To promote accessibility, current European audiovisual law is forcing TV channels to subtitle a huge proportion of their contents. To address this increased demand, broadcasters and subtitlers are seeking alternatives more productive than manual subtitling. Speech recognition technologies have proved useful in this respect. One efficient approach, when the script for the content exists, is speech-text alignment, which relies on aligning audio with its script to automatically recover time stamps. Forced-alignment is challenging with long signals, because of the widely-used Viterbi algorithm, which forms very large lattices during decoding, requiring a lot of memory.

In this work, the system presented in [1] for long audio alignment in an automatic subtitling scenario has been improved and extended. Phone-decoder accuracy was improved using context-dependent acoustic models, besides implementing an adaptation of the generic language models to the script of the contents to subtitle. The system was also extended to Basque, its original languages being English and Spanish, and additional linguistic resources were created for the Spanish aligner.

The paper is structured as follows. Section 2 looks at related work in long audio alignment and in phone-relatedness measures. Section 3 describes our speech-text alignment system, and Section 4 presents the phoneme similarity matrices created.

P. Sojka et al. (Eds.): TSD 2014, LNAI 8655, pp. 473–480, 2014.

Section 5 discusses the evaluation method and results. Section 6 presents conclusions and suggestions for further work.

2 Related Work

The reference for many of the related studies is the work done in [2], where the forced alignment was turned into a recursive and iteratively adapted speech recognition process. They used dynamic programming to align the hypothesis text and the reference transcript at word level. Subsequent works proposed improvements of this system, to deal with scenarios in which transcripts are not exact. In [3] a Driven Decoding Algorithm (DDA) was proposed to simultaneously align and correct the imperfect transcripts. At a new generated assumption of the speech recognizer in the lattice, DDA aligned it with the approximated transcript and a new matching score was computed and integrated with the language model for linguistic rescoring. An efficient, and simpler, long audio alignment approach was presented in [4]. They developed a system based on Hirschberg's dynamic programming algorithm [5] to align the phone decoder output with the transcription at phoneme level. They used a binary matrix to score alignment operations, with a cost of one for insertions, deletions and substitutions, and a cost of 0 for matches. Inspired on [4], for our experiments in [1] we created several scoring matrices, based on criteria like phonological similarity, phone-decoder confusion and phone confusion in human perception.

Concerning literature relevant for the creation of our scoring matrices, our phonological similarity metric is based on [6], where Kondrak constructed a metric that outperformed previously available ones, evaluating it with cognate alignment tasks. The metric was also successfully employed in spoken document retrieval in [7]. Regarding phone confusion in human perception, our American English matrices rely on perceptual error data reported in [8], who used a phoneset that closely corresponds to our phone-decoder's phoneset. Our Spanish data are based on the corpus of misperceptions developed by [9], which provides data covering our entire phoneset.

3 Long Speech-Text Alignment System

The goal of any speech-text alignment system is to obtain a perfect timing synchronization between the source audio and related text recovering the time codes for each word in the transcript. Our multilingual long speech-text alignment system is trained to align long audios and related transcripts for English, Spanish and Basque. For each language, a language-dependent phone decoder was developed, in addition to a grapheme-to-phoneme transcriptor. The aim of the alignment algorithm is to find matches between the phones recognized by the phone-decoder and the reference phoneme transcription. Only the time-codes of the correctly aligned phones will be used as reference times for further synchronization.

However, all the phonemes are not always correctly aligned during alignment; substitutions, deletions and insertions may occur. In fact, using the evaluation contents presented in Section 5, only 34% of the phonemes were correctly aligned for English, while 57% and 48% of the phonemes were matched for Spanish and Basque respectively in the best-performing configuration. These time-codes at phoneme level are then used to estimate the start time of each word and thus of each subtitle. The promising results presented in this paper prove that the time-codes recovered by the aligner are good enough to generate near-perfectly aligned subtitles.

3.1 Context-Dependent Phone Decoders

The phone-decoders have been improved from the last version of the system presented in [1], in which monophone models were employed. For this study, cross-word triphone models were built for each language to deal with coarticulation effects. With the aim of reducing linguistic variability, the language model consisted of an interpolation of the generic language model and a specific model created for each transcript. The interpolated models were bigram triphone models. The triphone-based phone decoders were trained using the HTK[1] tool. The parametrization of the signal consisted of 18 Mel-Frequency Cepstral Coefficients plus the energy and their delta and delta-delta coefficients, using 16-bit PCM audios sampled at 16 KHz.

The English triphone-based decoder system was built using the TIMIT database [11], which is composed by 5 hours and 23 minutes of clean speech data. Texts totaling 369 million words, gathered from digital newspapers, were used to train the generic language model. The Phone Error Rate (PER) of this decoder was 24.71%.

The Spanish triphone-based decoder system was based on 20 hours of clean-speech from three databases; Albayzin [12], Multext [13], and records of broadcast news contents from the SAVAS corpus [14]. The generic language model was trained with texts crawled from national newspapers, toting up 45 million words. The PER of the Spanish decoder was 31.79%.

The Basque triphone-based decoder system was generated using 36 hours of clean speech records of broadcast news contents. The generic language model was built using texts crawled from national newspaper, totaling 91 million words. The PER of the Basque decoder was 20.92%.

For all three languages, the corpora were split between training and test sets containing 70% and 30% of the data, respectively.

3.2 Grapheme-to-Phoneme Transcriptors

The grapheme-to-phoneme (G2P) transcriptors used for English and Spanish were the same used in the previous work [1]. The Spanish G2P was ruled based and inspired on the tool provided by Lopez[2]. The English transcriptor was inferred from the Carnegie Mellon Pronouncing Dictionary[3] using Phonetisaurus[4] tool. The Basque G2P transcriptor was based on manually created heuristic rules. The phonesets for all the languages are available on our project's website[5].

[1] http://htk.eng.cam.ac.uk/

[2] http://www.aucel.com/pln/

[3] http://svn.code.sf.net/p/cmusphinx/code/trunk/cmudict/

[4] http://code.google.com/p/phonetisaurus/

[5] http://sites.google.com/site/similaritymatrices/

3.3 Algorithm for Alignment of Phoneme Sequences

Our alignment algorithm is a slightly modified version of the well-known divide-and-conquer Hirschberg's algorithm. These modifications were established once their effectiveness in the alignment process was tested.

Given the two phoneme sequences $X = \{x_1, \ldots, x_n\}$ and $Y = \{y_1, \ldots, y_m\}$ to be aligned, the algorithm forces them to be recursively divided at indexes x_{mid} and y_{mid} respectively. Hirschberg defined x_{mid} as *round(length(x)/2)*. Nevertheless, following the procedure several candidates can arise for y_{mid}. In our algorithm, y_{mid} always corresponds to the candidate-index closest to the middle of Y. The other modification relies on forcing a substitution operation, even if the phonemes do not match, when the recursive algorithm only has sequences of one symbol left to align.

Four edit-operations are allowed in the alignment algorithm: matches, substitutions, deletions and insertions. The scores for matches and substitutions are defined by the scoring matrices (See Section 4), while deletions and insertions incur a gap penalty. Since each matrix-type tested has a different range of values, the gap penalties are also different for each matrix-type. In our binary matrix, the gap penalty was 2. For all other matrices, the penalty was a quarter of the matrix' maximum value, following one of the practices for gap penalties referenced in [6].

4 Phoneme-Relatedness Scoring Matrices

The phoneme-relatedness matrices provide information to the aligner about how likely it is for an alignment between two phonemes to be correct. The matrices favour aligning similar phonemes, by giving such alignments higher scores than to alignments between less similar phonemes. The matrices give the lowest scores to alignments between highly dissimilar phones, which are unlikely to be correct.

We created different scoring matrices for each language, applying different phoneme-relatedness criteria. The first scoring matrix is decoder-dependent, based on errors made by the phone decoder. The second matrix is decoder-independent, and based on phonological similarity, assessed by comparing largely articulatory features. The final matrix is also decoder-independent, and relies on phoneme confusion in human perception. Samples for all types of matrices are available on our project's website.

4.1 Matrices Based on Phone-Decoding Errors

The matrices were created based on HTK's HResults logs, when aligning the phone-decoding output and the G2P transcription for sequences of approx. 200,000 phonemes in English, 1,000,000 in Spanish and 2,000,000 in Basque. For each phone in the phoneset, the matrices contain the percentages of misrecognitions and correct recognitions by the decoder, normalized to a 1–1000 integer range. For instance, if 4% of the occurences of /ɲ/ were misrecognized as /n/, the matrix shows a score of 40 for the [ɲ,n] phoneme pair. In order to prevent substitutions between phonemes never mistaken by the decoder, a score of −500 was entered in the matrix for such phoneme-pairs. This score corresponds to $1/2 \times (0 - \max(\{\text{Score Range}\}))$.

4.2 Matrices Based on Phonological Similarity

Our phonological similarity scores are based on the metric devised by Kondrak in [6], as part of the ALINE cognate alignment system[6]. Phonemes are described with Ladefoged's [14] multivalued features, and a *salience* factor weights each feature according to its impact for phoneme similarity. The features, values and saliences employed for each language are available on our project's website.

$$\sigma_{\text{sub}}(p,q) = (C_{\text{sub}} - \delta(p,q) - V(p) - V(q))/100$$
$$\text{where } V(p) = \begin{cases} 0 & \text{if } p \text{ is a consonant or } p = q \\ C_{\text{vwl}} & \text{otherwise} \end{cases}$$
$$\delta(p,q) = \sum_{f \in R} \mathit{diff}(p,q,f) \times \mathit{salience}(f)$$
$$\sigma_{skip}(p) = \mathit{ceiling}(|C_{\text{sub}}/400|)$$

Fig. 1. Similarity function, based on Kondrak (2002)

Fig. 1 shows equations with our scoring function. $\sigma_{\text{sub}}(p, q)$ returns the similarity score for phonemes p and q, $C_{\text{sub}}/100$ being the maximum possible similarity score. C_{vwl} represents the relative weight of consonants and vowels. Values for C_{sub} and C_{vwl} are set heuristically as described in [15]. The function diff(p, q, f) yields the similarity score between phonemes p and q for feature f, and the feature-set R is configurable. Last, $\sigma_{\text{skip}}(p)$ returns the penalty for insertions and deletions used in the aligner. We defined heuristically a C_{sub} value of 3,500 (i.e. a maximum similarity score of 35), and a gap penalty of 9 for alignment, which corresponds to ceiling($|C_{\text{sub}}/400|$).

Kondrak's original function was designed for cognate alignment. We modified the function, for coherence with our audio aligner, and to adapt it to audio alignment tasks, achieving better results with the modified version than with the original. Details about the modifications are discussed in [1] and in the project's website.

4.3 Matrices Based on Perceptual Errors

The English matrices were based on human perceptual error data from [8]. They performed a phoneme identification study with native speakers of American English, asking them to identify the initial or final phoneme of 645 syllables of types CV (ConsonantVowel) and VC, at signal-to-noise ratios (SNR) of 0, 8 and 16. The noise type was multi-speaker babble. Participants chose a response among several possibilities presented to them visually. The phoneme-set in the study covers all of our decoder's phoneset except schwa. We only used the SNR 16 data, since a matrix based exclusively on this subset of the data yielded better alignment results than when considering data at other SNR for building the matrix.

The Spanish matrix was based on an extended version, provided by the authors directly, of the corpus of human misperceptions in noise developed in [9]. The

6 ALINE is available at
`http://webdocs.cs.ualberta.ca/~kondrak/#Resources`

methodology involved presenting 69 native speakers of Spanish with over 20,000 single-word stimuli, under different masking-noise conditions, and asking the speakers to write the word they had heard. Only stimuli for which certain agreement thresholds were reached among participants' responses were kept for the final misperception corpus, which consists of 3,294 stimuli and their associated responses. The study is thus a free-response error-elicitation task, not a closed-response task like [8]. However, we chose [9] as our data source, since, unlike other Spanish perception studies, it provides data for all phonemes in our decoder's phoneset. For coherence with our English data, we based our matrices on the 1,838 stimuli where multi-speaker babble was used as the masker. SNR in these stimuli ranged between −8 and +1. For computing our confusion matrix, we compared the corpus' stimulus and responses in cases where the response involved a single-phoneme error. We recorded the percentage of matches and mismatches between each stimulus and each response in the stimulus' response-set (a maximum of 15 responses were available per stimulus). Match and mismatch percentages were normalized to a 1–1000 range. For phoneme pairs where no confusion had taken place, a score of −500 (i.e. 1/2 × *(0 – max({Score Range})* was entered in the matrix. The matrix was based on 6,807 stimulus-response pairings.

Perceptual-relatedness matrices were not created for Basque, since we are not aware of appropriate data that could be exploited for their creation.

5 Evaluation and Results

The English test-set totaled 21,310 phonemes, 4,732 words and 471 subtitles, and contained non-clean speech from television audios. Its reference subtitles contained some stretches where transcription was imperfect, with subtitles missing for some parts of the audio. The Spanish test-set consisted of 47,480 phonemes, 8,774 words and 1,249 subtitles, and was composed of clean speech from documentaries. The Basque test-set totaled 26,712 phonemes, 4,331 words and 726 subtitles, containing a concatenation of a documentary and a film, and included noisy-speech.

Long audio alignment accuracies using different phone-relatedness matrices for English, Spanish and Basque are presented in Table 1. The results present the percentage or words and subtitles correctly aligned within the specified deviation range from the reference. The real time-codes at word level were obtained applying a forced-alignment algorithm for each subtitle in the reference material, which was composed of time-coded subtitles manually created by professional subtitlers. For subtitle-level evaluation, the deviation of the first and last words of the subtitles were measured.

The results show the effectiveness of our long audio alignment system, even with contents containing noisy-speech and imperfect transcriptions. Besides, the improvements using non-binary matrices are clearly proved comparing to the accuracies obtained with the binary matrix. Considering that a maximum deviation of 1 second is not long enough for listeners to have difficulties associating the subtitle and the audio, near-perfectly aligned subtitles were obtained for all three languages. In fact, alignment accuracies of 91.30%, 96.72% and 95.18% were obtained for English, Spanish and Basque respectively at this maximum deviation time.

Regarding non-binary matrices performance, the PDE matrices achieve the most accurate alignment results for English and Basque. It was expectable since these

Table 1. Alignment accuracy at word and subtitle level. **PDE**: Phone-decoder-error based matrix, **PHS**: Phonological similarity, **PCE**: Perceptual error matrix.

	Word-level deviation (seconds)						Subtitle-level deviation (seconds)					
	Matrix	**0**	**≤0.1**	**≤0.5**	**≤1.0**	**≤2.0**	**0**	**≤0.1**	**≤0.5**	**≤1.0**	**≤2.0**	**Matrix**
English	Binary	0.25	8.13	28.12	40.36	56.14	0.42	4.46	38.64	83.65	100	Binary
	PDE	**1.02**	**29.88**	**60.17**	**72.94**	**84.52**	**0.85**	**14.86**	**54.35**	**91.30**	100	PDE
	PHS	0.87	25.41	56.10	69.55	79.73	0.64	11.25	53.50	88.54	100	PHS
	PCE	0.72	26.66	57.26	70.31	82.57	0.64	14.65	53.29	90.02	100	PCE
Spanish	Binary	2.47	47.70	69.11	75.55	80.21	0.48	21.06	63.49	92.47	100	Binary
	PDE	**5.55**	77.42	**92.21**	**94.39**	95.93	1.12	40.19	**80.22**	96.64	100	PDE
	PHS	5.44	**77.45**	92.17	94.31	**95.97**	**1.20**	**40.67**	80.14	96.64	100	PHS
	PCE	5.22	74.83	92.03	94.48	96.35	1.28	38.83	78.78	**96.72**	100	PCE
Basque	Binary	1.55	34.98	56.63	61.06	65.25	0.83	24.24	64.05	92.29	100	Binary
	PDE	2.34	48.91	76.21	80.91	85.05	**1.65**	**44.63**	**75.76**	**95.18**	100	PDE
	PHS	**2.49**	**49.97**	**77.00**	**82.10**	**86.45**	1.38	35.54	74.24	95.04	100	PHS

matrices were based on each phone-decoder phone confusion-pairs. However, the improvements with the PDE matrix comparing to improvements with the other non-binary matrices are not relevant. For Spanish, the PCE matrix obtained the best results, although the PDE and PHS matrices achieved very similar accuracies.

6 Conclusions and Further Work

The adequate performance of our multilingual long audio alignment system in the automatic subtitling scenario was presented in this work. We established the effectiveness of a customized version of the well-known Hirschberg algorithm, and proved that using several scoring matrices based on different phoneme-relatedness criteria obtains well-performed alignments.

Since the current system works with triphone-based phone decoders, ongoing work is focused on the development of context-dependent phoneme scoring matrices. The goal behind this approach will be to improve the alignment process considering not only phones, but also biphones and triphones, to deal with coarticulation effects.

References

1. Álvarez, A., Arzelus, H., Ruiz, P.: Long audio alignment for automatic subtitling using different phone-relatedness measures. In: IEEE International Conference on Acoustics, Speech and Signal Processing, ICASSP, Florence, Italy (2014)
2. Moreno, P.J., Joerg, C., Van Thong, J.-M., Glickman, O.: A recursive algorithm for the forced alignment of very long audio segments. In: Proceedings of the 5th International Conference on Spoken Language Processing, ICSLP, Sydney, Australia (1998)
3. Lecouteux, B., Linàres, G., Nocéra, P., Bonastre, J.: Imperfect transcript driven speech recognition. In: Proceedings of INTERSPEECH, pp. 1626–1629 (2006)
4. Bordel, G., Nieto, S., Peñagarikano, M., Rodríguez-Fuentes, L.J., Varona, A.: A simple and efficient method to align very long speech signals to acoustically imperfect transcriptions. In: Proceedings of INTERSPEECH, Portland, Oregon (2012)

5. Hirschberg, D.S.: A linear space algorithm for computing maximal common subsequences. Communications of the ACM 18(6), 341–343 (1975)
6. Kondrak, G.: Algorithms for Language Reconstruction. PhD Thesis. University of Toronto (2002)
7. Comas, P.: Factoid Question Answering for Spoken Documents. PhD Thesis. Universitat Politècnica de Catalunya (2012)
8. Cutler, A., Weber, A., Smits, R., Cooper, N.: Patterns of English phoneme confusions by native and non-native listeners. Journal of the Acoustical Society of America 116(6), 3668–3678 (2004)
9. García Lecumberri, M.L., Toth, A.M., Tang, Y., Cooke, M.: Elicitation and analysis of a corpus of robust noise-induced word misperceptions in Spanish. In: Proceedings of INTERSPEECH, pp. 2807–2811 (2013)
10. Garafolo, J.S.L., Fisher, W., Fiscus, J., Pallett, D., Dahlgren, N., Zue, V.: TIMIT Acoustic-Phonetic Continuous Speech Corpus. Linguistic Data Consortium, Philadelphia (1993)
11. Díaz, J.E., Peinado, A., Rubio, A., Segarra, E., Prieto, N., Casacuberta, F.: Albayzín: a task-oriented Spanish speech corpus. In: Proceedings of LREC, Granada, Spain (1998)
12. Campione, E., Véronis, J.: A multilingual prosodic database. In: Proceedings of the 5th International Conference on Spoken Language Processing, ICSLP, Sydney, Australia (1998)
13. Del Pozo, A., Aliprandi, C., Álvarez, A., Mendes, C., Neto, J.P., Paulo, S., Piccinini, N., Rafaelli, M.: SAVAS: Collecting, Annotating and Sharing Audiovisual Language Resources for Automatic Subtitling. In: Proceedings of LREC, Reykjavik, Iceland (2014)
14. Ladefoged, P.: A Course in Phonetics. Harcourt Brace Jovanovich, New York (1995)
15. Ruiz, P., Álvarez, A., Arzelus, H.: Phoneme similarity matrices to improve long audio alignment for automatic subtitling. In: Proceedings of LREC, Reykjavik, Iceland (2014)

Initial Experiments on Automatic Correction of Prosodic Annotation of Large Speech Corpora*

Zdeněk Hanzlíček and Martin Grůber

NTIS - New Technology for the Information Society,
Faculty of Applied Sciences, University of West Bohemia,
Univerzitní 22, 306 14 Plzeň, Czech Republic
{zhanzlic,gruber}@ntis.zcu.cz
http://www.ntis.zcu.cz/en

Abstract. Most modern speech synthesis systems utilize large speech corpora to learn new voices. These speech corpora usually contain several hours of speech spoken by talented speakers who are able to record such an amount of speech data in a sufficient quality. An appropriate phonetic and prosodic annotation of the recorded utterances is necessary for a high quality of synthesized speech. For many languages, the pitch shape within the last prosodic word of a phrase is characteristic for particular types of sentences and phrase structure of compound/complex sentences. However in the real data, this formal convention can be breached and a different pitch shape than expected can be present. This can be a source of prosody inconsistency in synthesized speech. This article presents some experiments on automatic detection of prosodic mismatch in recorded utterances. A simple classifier based on GMM was proposed for this task. Experiments were performed on 5 large speech corpora. The classification results were successfully verified by listening tests.

Keywords: speech corpora, prosodic annotation, prosodeme.

1 Introduction

Most modern speech synthesis systems [1,2] utilize large speech corpora to learn new voices. These speech corpora usually contain several hours of speech spoken by talented speakers who are able to record such an amount of speech data in a sufficient quality. An appropriate phonetic and prosodic annotation of the recorded utterances is necessary for a high quality of synthesized speech [3]. Generally, the knowledge of presence of various prosodic events in speech data and their detailed description can be useful for many other applications as well.

In connection with using the large speech corpora, the automatic phonetic and prosodic annotation of speech [4,5] became an important task. This article presents some initial experiments on automatic detection of prosodic mismatch in recorded utterances.

* This work was supported by the Technology Agency of the Czech Republic, project No. TA01011264 and by the European Regional Development Fund (ERDF), project "New Technologies for Information Society" (NTIS), European Centre of Excellence, ED1.1.00/02.0090.

P. Sojka et al. (Eds.): TSD 2014, LNAI 8655, pp. 481–488, 2014.

For many languages, the pitch shape within the last prosodic word of a phrase (corresponding to functionally involved prosodeme[1]) is characteristic for particular types of sentences and for the phrase structure of compound/complex sentences. However, in the real speech data, this formal convention can be breached and a different type of prosodeme than expected can be present. Using a speech corpus with bad prosodeme labels can be source of prosody inconsistency in synthesized speech. Prosodemes whose type does not correspond to the given sentence structure should be revealed and corrected or removed from the corpus. This should improve the overall quality of resulting synthetic speech.

This paper is organized as follows, Section 2 explains the prosody model used in this work. Procedure for prosodeme classification is proposed in Section 3. Section 4 describes performed experiments and their results. Finally, Section 5 concludes this paper and outlines the future work.

2 Prosody Model and Prosodemes

Within this paper, the formal prosody model proposed by Romportl [6] is used. According to this model, an utterance can be divided into prosodic clauses separated by short pauses. Each prosodic clause includes one or more prosodic phrases, which contain certain continuous intonation scheme. A prosodic phrase consists of two prosodemes: null prosodeme and functionally involved prosodeme which is usually related to the last prosodic word in the phrase.

For the Czech language[2], the following basic classes of functionally involved prosodemes are distinguished (for detailed prosodme categorization see [6]):

P1 – prosodemes terminating satisfactorily (specific for declarative sentences)
P2 – prosodemes terminating unsatisfactorily (specific for questions)
P3 – prosodemes non-terminating (specific for non-terminal phrases in compound/ complex sentences)

This paper is focused on compound/complex sentences. We assume that the last phrase in these sentences ends with prosodeme P1 and all previous phrases end with prosodeme P3. In the case of neutral speech (no emphasis, expression etc.), prosodemes P1-1 and P3-1 are expected. Typical examples of prosodemes P1-1 and P3-1 are depicted on Figures 1 and 2.

A typical feature for prosodeme P1-1 is a pitch decrease within its last syllable. For prosodeme P3-1, a pitch increase within the last syllable is specific. In some cases, the pitch increase/decrease can be realized as a value contrast between pitch of last and previous syllable.

Beside the pitch shape, spectral, duration and energy features can be characteristic for particular prosodemes. However, their impact seems to be not so relevant for prosody perception or the dependence is more complex.

In real speech data, a different prosodeme than expected could be present. This problem appears even in utterances spoken by a professional speaker. A typical example

[1] Prosodemes are described in Section 2.

[2] A different/modified set of prosodemes can be specific for other languages.

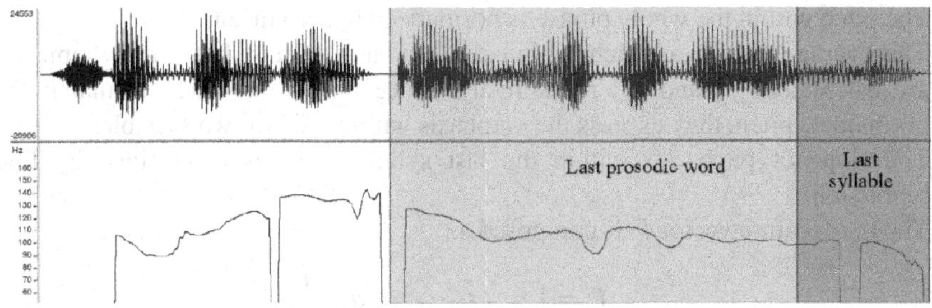

Fig. 1. Example of a phrase terminated with prosodeme P1-1 (waveform and pitch)

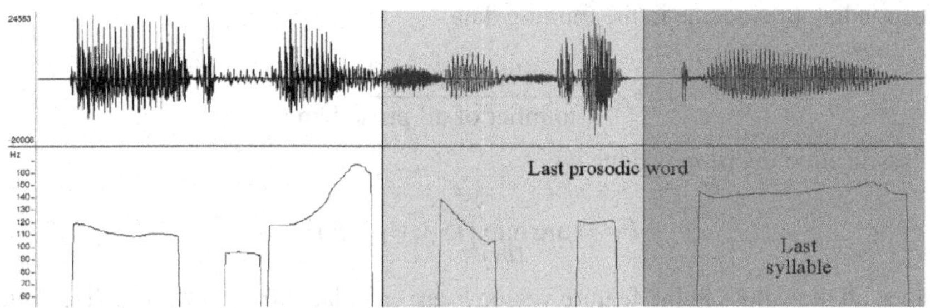

Fig. 2. Example of a phrase terminated with prosodeme P3-1 (waveform and pitch)

is the compound sentence that can be split into several independent sentences. Within the compound sentence, all phrases (except the last one) should be terminated with the prosodeme P3-1. However, the independent sentences are naturally terminated by the prosodeme P1-1.

Badly annotated corpus can be a source of various troubles in some applications. In speech synthesis (specifically, in unit selection method), prosodeme labels are used for selecting sequence of optimal speech units for building resulting speech [7]. Using units from an inappropriate prosodeme or mixing units from different types of prosodemes can cause a decrease in the overall speech quality – prosody of synthesized speech does not correspond to the type or the structure of the sentence, some unnatural pitch fluctuation can occur, etc.

3 Prosodeme Classification

Gaussian mixture models (GMMs) are widely used in various speech classification tasks, such as speaker identification [8], emotion recognition [9], etc. Since the usage of GMMs is straightforward and the performance is usually satisfactory, we decided to use them in our experiments as a baseline.

First, each prosodic phrase is represented by a feature vector F and a default prosodeme type P_X. Pitch is extracted from audio files by using the RAPT algorithm [10] implemented in the SPTK toolkit [11]. The feature vector is computed from the extracted pitch as follows

1. The pitch within the whole phrase is normalized to zero mean.
2. The average values of pitch within the last and the last but one syllabic core are calculated: $\bar{f}_1$ and $\bar{f}_2$, respectively. Since both values are calculated from normalized pitch, they express the emphasis within the last two syllables.
3. The slope of pitch df_1 within the last syllabic core is determined by linear regression.
4. The final feature vector F is composed as

$$F = \left[\bar{f}_1, (\bar{f}_1 - \bar{f}_2), df_1\right]^T$$

For each prosodeme P_X, a simple Gaussian mixture model $\mathcal{G}_{P_X}(F)$ is trained. Moreover, the weigh $\mathcal{W}_{P_X}$ for particular models is given as a relative number of corresponding prosodeme in the training data.

$$\mathcal{W}_{P_X} = \frac{\text{number of prosodeme } P_X}{\text{number of all prosodemes}}$$

Classification decision is done by

$$P_Y = \arg\max_{\{P_X\}}\left[\mathcal{W}_{P_X}\mathcal{G}_{P_X}(F)\right]$$

Since all the values in the feature vector F are calculated from the normalized pitch, they seems to be (partly) speaker-independent. Thus, it would be possible to train one speaker-independent set of classifiers; however, to capture the speaker specific features more precisely, individual classifiers were used for particular speakers in our initial experiments.

The training and the classification are performed on whole speech corpora. This approach is based on the assumption that most prosodemes are correct and the minority of incorrect prosodemes should not influence the training of the classifiers since as outliers they are not taken into account. In the case of large amount of incorrect prosodemes, the classifiers would be probably poorly trained and would be inapplicable.

4 Experiments and Results

4.1 Experimental Data

For our experiments, we used 5 large speech corpora recorded for the purposes of speech synthesis [12]: 3 male voices (denoted as M_{AJ}, M_{JS}, M_{TJ}) and 2 female voices (denoted as F_{MR}, F_{KI}). Each corpus contains about 10,000 utterances[3]. With the exception of M_{TJ}, all corpora contain the same sentences[4]; corpus M_{TJ} was partly different.

Since a proper phonetic segmentation [4] (including pauses) was available for all corpora, splitting particular utterances into prosodic clauses was straightforward.

[3] Particular corpora contained larger number of utterances, but only declarative sentences were selected for our experiments. Thus, the accurate number of utterances selected from particular corpora corresponds to the number of prosodemes P1-1.

[4] The numbers of utterances slightly differ because some defective utterances were discarded.

Table 1. Number of prosodemes in particular corpora

prosodeme	M_{AJ}	M_{JS}	M_{TJ}	F_{MR}	F_{KI}
P1-1	10,001	9,896	9,896	9,897	9,878
P3-1	13,051	10,545	17,479	8,581	3,562

However, dividing clauses into phrases is a more complex task because a sophisticated text analysis is necessary – a simple detection of conjunctions and punctuation marks is not sufficient. From that reason, we decided to perform our initial experiments only on the last phrase in each clause where the prosodeme occurrence was guaranteed.

A simple prosodic profiles of particular corpora are presented in Table. 1. Different values illustrate various speaking styles of particular speakers. Since only last phrases in particular prosodic clauses were taken into account, the number of prosodeme P3-1 corresponds to the number of pauses in speech.

4.2 Classification Results

An independent set of classifiers was trained for each speaker. In all cases, GMMs for particular prosodemes contained 5 mixtures. The classification results are presented in Table 2. Some speakers (namely M_{AJ}, M_{JS} and F_{KI}) have obviously and extraordinarily consistent speaking style because only a few individual prosodeme were classified as of a different type. The other speakers (M_{TJ} and F_{MR}) apparently often separated compound sentences into independent declarative phrases. This has 2 consequences:

1. Prosodemes P3-1 were classified as P1-1 because they actually correspond to that prosodeme type.
2. Prosodemes P1-1 were classified as P3-1 because training data for the P3-1 classifier contained a lot of P1-1 samples; therefore, the classifier was poorly trained.

Table 2. Classification of prosodemes in particular corpora (total numbers and percentages)

speaker	default prosodeme	classification			
		P1-1		P3-1	
M_{AJ}	P1-1	9,982	99.81 %	19	0.19 %
	P3-1	57	0.44 %	12,994	99.56 %
M_{JS}	P1-1	9,887	99.91 %	9	0.09 %
	P3-1	8	0.07 %	10,537	99.93 %
M_{TJ}	P1-1	9,075	91.70 %	821	8.30 %
	P3-1	3,516	20.12 %	12,994	79.88 %
F_{KI}	P1-1	9,884	99.87 %	13	0.13 %
	P3-1	25	0.76 %	3,537	99.24 %
F_{MR}	P1-1	9,523	96.41 %	355	3.59 %
	P3-1	1,198	13.96 %	7,383	86.04 %

Besides the GMM-based classification described in Section 3, some comparative experiments with support vector machines with various kernels [13] were also performed. The results were very similar and are therefore not presented in this paper. A more detailed classifier comparison is planned to be performed in our future work.

4.3 Listening Tests

The functionality of the proposed GMM-based classifiers was evaluated by listening tests. 10 participants took part in this listening test, most of them were speech processing experts who understood the theoretical background of the problem and had some former experience with listening tests.

The test contained 20 utterances for each speaker:

a) 5 phrases terminated with prosodeme P1-1,
b) 5 phrases terminated with prosodeme P3-1,
c) 10 phrases where the default prosodeme P3-1 was classified as P1-1; hereinafter, this prosodeme is denoted PX-Y.

The phrases a) and b) were selected to be prosodically unambiguous for the required prosodeme. Phrases c) were selected randomly. However, all the utterances were semantically neutral, i.e. the type of the phrase could not be determined from the text content.

Utterances of speakers M_{JS} and F_{KI} were not included in the test because of lack of phrases of type c). Thus, the test contained only utterances of 2 male speakers and one female speaker (M_{AJ}, M_{TJ} and F_{MR}, respectively).

The test results presented in Table 3 show that all the listeners were able to distinguish the prosodemes P1-1 and P3-1. Moreover, in most cases, they identified the prosodeme PX-Y as P1-1. That is in agreement with the results of the classifiers.

Table 3. Results of listening tests

speaker	prosodeme	listeners' decision [%]		
		P1-1	P3-1	undecided
M_{AJ}	P1-1	**100.0**	0.0	0.0
	P3-1	0.0	**94.0**	6.0
	PX-Y	**96.0**	0.0	4.0
M_{TJ}	P1-1	**98.0**	0.0	2.0
	P3-1	0.0	**100.0**	0.0
	PX-Y	**90.0**	1.0	9.0
F_{MR}	P1-1	**100.0**	0.0	0.0
	P3-1	2.0	**96.0**	2.0
	PX-Y	**100.0**	0.0	0.0
all	P1-1	**99.3**	0.0	0.7
	P3-1	0.7	**96.7**	2.7
	PX-Y	**95.3**	0.3	4.3

5 Conclusion

This paper presented some initial experiments on the detection of errors in the prosodic annotation of large speech corpora. The annotation errors are identified by simple GMM-based classifiers. Experiments performed on 5 large speech corpora revealed various numbers of suspicious prosodemes whose default prosodeme label did not match its new classification.

Listening test confirmed that the decision of the classifiers was correct in most cases and the new prosodeme label was correct. In the remaining cases, closer examination revealed two secondary causes of different classification: problems with the pitch extraction and problems with the default phonetic segmentation.

5.1 Future Work

In our future work, the experiments on classification of other types of prosodemes will be performed. By including the null prosodeme model, possibly incorrect phrase boundaries (shifted segmentation) could be also detected.

Given the promising initial results, the classifiers are planned to be used for an automatic correction of particular corpora. We expect that using corpora with corrected prosodeme labels for training a new voice in a TTS system should improve the overall quality of synthesized speech, especially its prosodic features.

Another aim is to develop speaker-independent classifiers that could be used for speech data from non-professional speakers whose speech prosody is not consistent enough to train new independent classifiers or the amount of speech data is low.

Last but not least, we intend to perform a more thorough comparison with other types of classifiers, e.g. support vector machines [13].

References

1. Hunt, A., Black, A.W.: Unit selection in a concatenative speech synthesis system using a large speech database. In: Proceedings of ICASSP 1996, Atlanta, Georgia, pp. 373–376 (1996)
2. Zen, H., Tokuda, K., Black, A.W.: Statistical parametric speech synthesis. Speech Communication 51, 1039–1064 (2009)
3. Ross, K., Ostendorf, M.: Prediction of abstract prosodic labels for speech synthesis. Computer Speech and Language 10, 155–185 (1996)
4. Toledano, D., Gómez, L., Grande, L.: Automatic Phonetic Segmentation. IEEE Transactions on Speech and Audio Processing 11(6), 617–625 (2003)
5. Wightman, C., Ostendorf, M.: Automatic labeling of prosodic patterns. IEEE Transactions on Speech and Audio Processing 2(4), 469–481 (1994)
6. Romportl, J., Matoušek, J., Tihelka, D.: Advanced Prosody Modelling. In: Sojka, P., Kopeček, I., Pala, K. (eds.) TSD 2004. LNCS (LNAI), vol. 3206, pp. 441–447. Springer, Heidelberg (2004)
7. Tihelka, D., Matoušek, J.: Unit Selection and its Relation to Symbolic Prosody: A New Approach. In: Proceedings of Interspeech 2006, Pittsburgh, Pennsylvania, USA, pp. 2042–2045 (2006)

8. Reynolds, D., Rose, R.: Robust text-independent speaker identification using Gaussian mixture speaker models. IEEE Transactions Speech Audio Processing 3(1), 72–83 (1995)
9. Přibil, J., Přibilová, A.: Evaluation of influence of spectral and prosodic features on GMM classification of Czech and Slovak emotional speech. EURASIP Journal on Audio, Speech, and Music Processing 8, 1–22 (2013)
10. Talkin, D.: A Robust Algorithm for Pitch Tracking (RAPT). In: Kleijn, W.B., Paliwal, K.K. (eds.) Speech Coding and Synthesis, ch. 14, pp. 495–518. Elsevier Science (1995)
11. Speech Signal Processing Toolkit (SPTK), http://sp-tk.sourceforge.net
12. Matoušek, J., Tihelka, D., Romportl, J.: Building of a Speech Corpus Optimised for Unit Selection TTS Synthesis. In: Proc. of LREC 2008, Marrakech, Morocco (2008)
13. Vapnik, V.: Statistical Learning Theory. Wiley, Chichester (1998)

Automatic Speech Recognition Texts Clustering*

Svetlana Popova[1,2], Ivan Khodyrev[2], Irina Ponomareva[3], and Tatiana Krivosheeva[3]

[1] Saint-Petersburg State University, Saint-Petersburg, Russia
[2] ITMO University, Saint-Petersburg, Russia
svp@list.ru, kivan.mih@gmail.com
[3] Speech Technology Center, Saint-Petersburg, Russia
{ponomareva,krivosheeva}@speechpro.com

Abstract. Abstract. This paper deals with the clustering task for Russian texts obtained using automatic speech recognition (ASR). The input for processing are recognition result for phone call recordings and manual text transcripts for these calls. We present a comparative analysis of clustering results for recognition texts and manual text transcripts, make an evaluation of how recognition quality affects clustering and explore approaches to increasing clustering quality by using stop words and Latent Semantic Indexing (LSI).

Keywords: clustering, speech-to-text, recognition result clustering, Latent Semantic Indexing, information retrieval, stop words.

1 Introduction

The development of Internet communication and multimedia raises the issue of searching and structuring data not only in textual form, but also in the form of sound recordings, graphics and video. Spoken content retrieval is becoming increasingly important and this fact motivates extensive research on techniques and technologies in this area [1]. There are two approaches to solving this task. The first one involves transforming speech into text and further processing of text data. The second one is based on analyzing the acoustic signal itself without preliminary transformation of speech into text [2]. This paper uses the first approach.

Our aim is to group together documents that are thematically close, and to classify documents on different topics into different groups. We explore possible approaches to clustering Russian data produced by a speech-to-text system.

Our results demonstrate an extremely small influence of recognition word error rate (WER) of about 20–35% for Russian database. For clustering algorithms we also found a substantial improvement in clustering quality if high-frequency words, excepting context words (keywords) are removed.

In natural language processing it is common to use methods for detecting and using latent features of documents, for instance, Latent Semantic Indexing (LSI) [3]. We explored the possibility of improving clustering quality by means of LSI for a database of near-spontaneous speech with a large overlap of high frequency words between

* This work was partially financially supported by the Government of Russian Federation, Grant 074-U01.

P. Sojka et al. (Eds.): TSD 2014, LNAI 8655, pp. 489–498, 2014.

documents. We demonstrate that using LSI leads to an improvement in the results of the EM (Expectation Maximization [4]) clustering algorithm but does not make much difference for k-means [5]. This fact shows that using LSI does help to detect latent features, however the overlap between clusters is retained in the semantic space built using LSI.

2 The Goal of the Research

The goal of our research is to explore and evaluate different approaches to the task of clustering Russian texts produced by speech-to-text conversion (namely, k-means [5] and the EM (Expectation Maximization) [4] algorithms), as well as to estimate the influence of recognition quality (100% versus 80–65%) on clustering results. We used the implementation of these algorithms provided by the WEKA library (`http://www.cs.waikato.ac.nz/ml/weka/`).

Our choice of algorithms was determined by the fact that k-means and EM are both classical iterative optimization algorithms whose results depend on their initialization. EM is less sensitive to mutual cluster overlapping than k-means, while k-means can work more efficiently if clusters are separated well. Next task was to evaluate the possibility of improving the results of each algorithm using LSI and a domain dependent list of stop words.

We speculated that LSI would detect latent features in the documents, which would make it possible to better identify thematically close documents. The stop list was expected to help remove high-frequency words that did not define the topic and could be treated as noise.

3 Experimental Data

A speech dataset was collected for the purposes of the research. The dataset consists of spontaneous speech recordings (8kHz sample rate) recorded in different analogue and digital telephone channels. The recordings were provided by Speech Technology Center Ltd (www.speechpro.ru) and contain customer telephone calls to several large Russian contact centers. All test recordings have manual text transcripts. In order to be able to evaluate clustering quality, the test dataset was manually labeled with the most frequent call topics. Each text document that contained a transcript of a call was analyzed by three experts. In difficult cases the decision of attributing the text to a certain topic was made by vote. The experts could attribute the text to one or several thematic categories out of the list they were provided with. Only the recordings that were attributed to a single topic by the majority of experts were later used for the test dataset.

As a result, we obtained a dataset of manually prepared text transcripts of the speech recordings, which were divided into five thematic clusters.

We used the speaker-independent continuous speech recognition system for Russian developed by Speech Technology Center Ltd [6,7] which is based on a CD-DNN-HMM acoustic model [8]. The ASR system included interpolation of a general language model (LM) trained on a 6GB text corpus of news articles (300k words, 5 million n-grams) with a thematic language model trained on a set of text transcripts of customer calls to the contact center (70MB of training data, the training and test datasets did not overlap). We used Good-Turing smoothing (cutoff=1 for all orders of n-grams). Recognition accuracy on the test dataset under these conditions was 80–65%. The recognition results were used to create a second experimental dataset, which corresponded to the same sound files as the first one but contained texts that were produced by using an automatic (rather than manual) speech-to-text transformation with recognition accuracy below 100%.

Table 1 contains a description of both datasets. Each text in the dataset is the recognition result for one short phone call, which is a text of small length. This means that we are faced with the short text clustering task and it becomes difficult to gather enough statistics to improve text processing [9,10].

Table 1. Text datasets

	Manual transcripts	Recognition results
Cluster number	5	
Cluster sizes	44, 24, 28, 55, 35	
Cluster topics	Municipal issues, Military service issues, Political issues, Family and maternity issues, Transport issues	
Word count	12641	11784
Dataset lexicon size	3819	3519

4 Text Pre-processing and Stop Words

Text pre-processing included lowering the case of all characters, removing punctuation marks and deleting words that were found in fewer than three documents (collection size 186 documents). We expected that this would remove words specific to a particular speaker, as well as incorrectly recognized words. The experiments confirmed that using this threshold improves clustering quality.

Stop words. Manual text analysis showed that most texts contain the same high-frequency words and phrases, which carry no information about the clusters topic. For instance, all texts contain greetings, goodbyes and thanks, as well as common function words. Although the clusters are well segmented thematically (they do not belong to similar topics), such uninformative words can introduce a lot of noise in the texts that need to be clustered, and lead to high cluster overlapping. Table 2 shows the top 20 most frequent words for each cluster. Words that carry information about the cluster topic are given in bold.

These words are high-frequency expressions that are not informative and do not refer to the contents of the document. We created a frequency lexicon using both the manual transcripts and the recognition results, after which an expert selected

Table 2. Demonstrates the high rate of cluster overlap for common words

Russian	English
Cluster: Municipal issues	
в, и, не, у, я, нас, за, на, по, а, мы, это, вот, что, с, спасибо, вопрос, нам, город, меня	*v "in", i "and", ne "not", u "by", ya "I", nas "us", za "for", na "on", po "along", a "and", my "we", eto "this", vot "so", chto "that", s "with", spasibo "thanks", vopros "question", nam "to us", gorod "city", menya "me"*
Cluster: Military service issues	
в, и, я, с, не, вот, на, что, по, спасибо, это, город, как, **пенсии**, вопрос, у, а, меня, так	*v "in", i "and", ya "I", s "with", ne "not", vot "so", na "on", chto "that", po "along", spasibo "thanks", eto "this", gorod "city", kak "how", **pensii "pensions"**, vopros "question", u "by", a "and", menya "me", tak "this way"*
Cluster: Political issues	
в, и, не, на, вот, по, вопрос, у, бы, как, это, меня, а, за, так, что, я, спасибо, город	*v "in", i "and", ne "not", na "on", vot "so", po "along", vopros "question", u "by", by (particle), kak "how", eto "this", menya "me", a "and", za "for", tak "this way", chto "that", ya "I", spasibo "thanks", gorod "city"*
Cluster: Family and maternity issues	
я, в, и, не, на, у, что, меня, вот, как, спасибо, вопрос, здравствуйте, это, **детей**, по, бы, нас, почему, до	*ya "I", v "in", i "and", ne "not", na "on", u "by", chto "that", menya "me", vot "so", kak "how", spasibo "thanks", vopros "question", zdravstvuyte "hello", eto "this", **detey "children"**, po "along", by (particle), nas "us", pochemu "why", do "until"*
Transport issues	
И, в, у, не, нас, на, я, это, вот, с, ни, вопрос, нет, что, нам, **дороги**, мы, меня, как, по	*i "and", v "in", u "by", ne "not", nas "us", na "on", ya "I", eto "this", vot "so", s "with", ni "not", vopros "question", net "no", chto "that", nam "to us", **dorogi "roads"**, my "we", menya "me", kak "how", po "along"*

high-frequency common function words from the lexicon and added them to the stop word list. Removing stop words was expected to facilitate thematic clustering. We should note that the list of high-frequency words was analyzed by an expert, and some words from it were not added to the stop list. It is important because the call transcripts contain few context words that reflect the topic. Deleting context words (such as "pensija" (pension), "deti" (children), "dorogi" (roads)) deteriorates clustering quality, which cancels out the effect of stop word removal. For this reason, adding all high-frequency words to the stop list does not improve clustering quality (experiment result), and the list needs to be edited by an expert.

5 The Algorithms

We use two algorithms that are well-known in the field of information retrieval: k-means [5] and EM [4], in the implementation provided by the WEKA library. To define the feature space we used the dataset lexicon obtained after text pre-processing. Each document was represented as a vector in the obtained feature space. The weight of the feature (word) in the document was estimated using *tf-idf* [11]. Both clustering algorithms were given input information about the number of clusters, which equaled 5. For k-means, we used the value 1−(cosine of the angle between document vectors [11]) to calculate the distance between documents. For EM, we set the minimum allowable standard deviation 1.0E-6 and the maximum iteration number as 100.

6 Evaluation

We used the classic clustering quality estimate based on combining information about the cluster Precision and Recall [12,13], F-measure, we will sign it as FM:

$$FM = \sum_i \frac{G_i}{|D|} \max_j F_{ij}, \quad \text{where } F_{ij} = \frac{2 \cdot P_{ij} \cdot R_{ij}}{P_{ij} + R_{ij}},$$
$$P_{ij} = \frac{|G_i \cap C_j|}{|G_i|}, \quad R_{ij} = \frac{|G_i \cap C_j|}{|C_j|}, \tag{1}$$

$G = \{G_i\}_{i=\overline{1,m}}$ is an obtained set of clusters, $C = \{C_j\}_{j=\overline{1,n}}$ is set of classes, defined by experts, D – number of documents in the collection.

7 Experiments and Results

All experiments were performed on two versions of the same data collection: one contained manual transcripts, the other contained recognition results (with 80–65% recognition quality, see the description of the data in Section "Experimental Data" and in Table 1).

7.1 Experiments Group 1

For these experiments, all documents were pre-processed without removing stop words. Then the documents were clustered using k-means and EM algorithm. We also tested the possibility of improving clustering quality by using LSI, due to mapping the feature space to a lower dimensional semantic space.

7.2 Experiments Group 2

In this group of experiments, all documents were pre-processed and stop words were removed. Then the documents were clustered using k-means and EM algorithm. In these experiments we also tested the possibility of improving clustering quality by using LSI.

7.3 Results

For the Group 1 experiments we obtained the following results. Table 3 shows clustering results for k-means and EM for the experimental conditions without removing stop words. Since both algorithms do not have a single constant result and depend on the initial solution, we performed 100 tests for each algorithm. For each test, we estimated clustering quality using the *FM* (1). We then chose the highest (max) and lowest (min) score for each algorithm, and calculated the average score for all 100 experiments (avg). Table 3 demonstrates two conditions: when LSI was not used, and when LSI was used (for the case of the optimal choice of semantic space dimensions).

Table 3. Clustering results for k-means and EM on the manual transcript dataset and recognition results dataset with and without LSI. Stop words were not removed

	Clustering result for text transcripts			Clustering result for ASR results		
	Without LSI					
	avg	max	min	avg	max	min
k-means	0.36	0.46	0.29	0.35	0.47	0.29
EM	0.35	0.43	0.29	0.36	0.48	0.29
	With LSI (for the case of the optimal choice of semantic space dimensions)					
K-means	0.42	0.49	0.34	0.40	0.42	0.37
EM	0.41	0.42	0.36	0.40	0.47	0.35

Figure 1 shows how mapping the initial feature space onto a lower-dimensional semantic space (LSI) influences clustering results. We selected space dimensions of 2 to 49. First diagram shows the dependency of the average clustering result (avg) for both algorithms on the dimension of the semantic space. Next diagram shows the same dependency for the best clustering results (max). Average FM=0.36 and maximum FM=0.46 were chosen as baseline because these values reflect the algorithms results without stop word deletion and LSI. Both diagrams show clustering result both for manual transcripts and for recognized text.

For Group 2 experiments we obtained the following results. Table 4 demonstrates clustering results for k-means and EM when stop words were removed, with and without LSI. As in Table 3, we show the estimate for the best (max), worst (min) and average (avg) clustering result over 100 tests.

Figure 2 shows how using LSI and mapping the initial feature space onto a lower-dimensional semantic space influences clustering results when stop words are removed from the texts (2 to 49 dimensions). First diagram shows the dependency of the average

Table 4. Clustering results for k-means and EM on the manual transcript dataset and recognition results dataset with and without LSI. Stop words were removed

	Clustering result for text transcripts			Clustering result for ASR results		
	Without LSI					
	avg	max	Min	avg	max	min
k-means	0.44	**0.57**	0.34	0.42	**0.58**	0.35
EM	0.36	0.45	0.30	0.37	0.48	0.29
	With LSI (for the case of the optimal choice of semantic space dimensions)					
K-means	0.41	0.51	0.32	0.40	0.49	0.34
EM	**0.47**	0.56	**0.38**	**0.46**	0.53	**0.39**

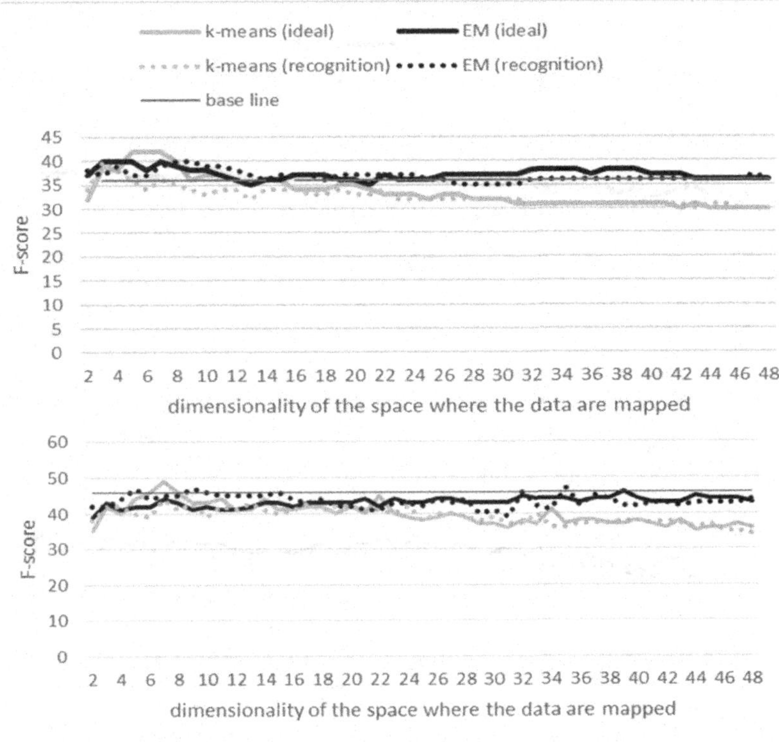

Fig. 1. Dependency of clustering quality on LSI semantic space dimension. First: dependency of the average score for 100 tests (avg). Next: best score (max). Ideal – result for manual text transcripts, recognition – result for ASR results. Stop words were not removed

clustering result (avg) for both algorithms on the dimension of the semantic space. Next diagram shows the same dependency for the best clustering results (max).

8 Discussion

Our results show that recognition quality has no strong influence on the k-means and EM clustering results when comparing manual text transcripts and recognition results with 80–65% accuracy. This shows that a decrease in recognition quality does not have a strong influence on clustering if the accuracy is about 80–65% (it should be noted that most data in the collection have recognition accuracy close to 80%).

Both algorithms show a similar result if we do not use stop word deletion and LSI. Using LSI improves average clustering quality (avg) when stop words are not removed. However, for EM this improvement is practically stable on 2 to 49 – dimensional semantic space, while for k-means the improvement is only observed when the dimension is lower than 15; when the dimension increases further, clustering quality begins to decline. This is probably due to the assumption that as semantic space

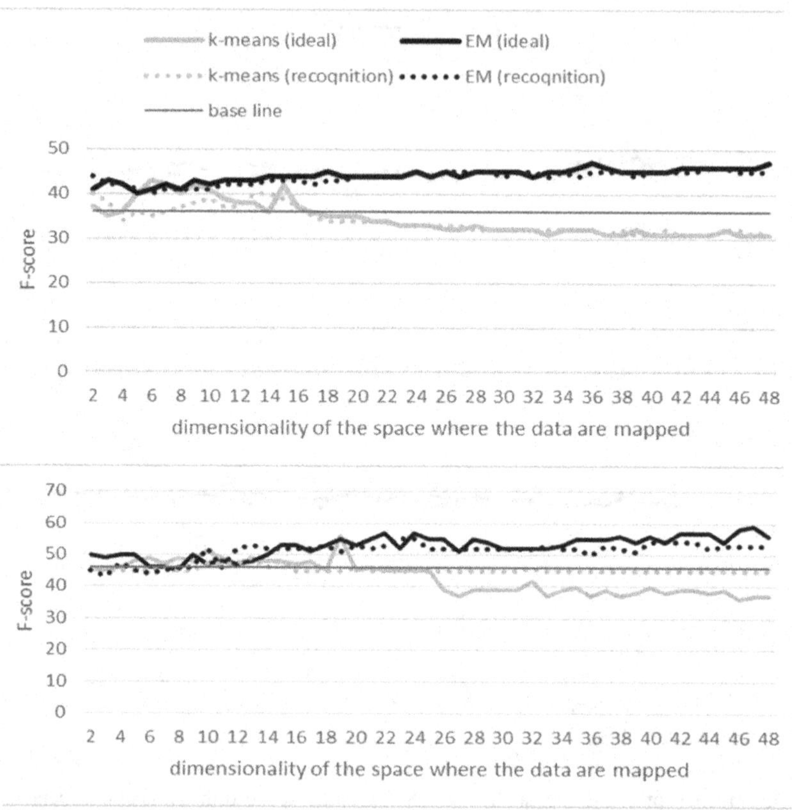

Fig. 2. Dependency of clustering quality on LSI semantic space dimension. First: dependency of the average score for 100 tests (avg). Next: best score (max). Ideal – result for manual text transcripts, recognition – result for ASR results. Stop words were removed

dimension increases, clusters begin to overlap more strongly. That has a negative effect on k-means, which is more sensitive to cluster overlap than EM. K-means provides the highest average score under the optimal semantic space dimension (avg FM=0.42 for manual transcripts), however this value only slightly improves upon the best average EM result (FM=0.41).

If we compare the best results (max) obtained with LSI, we observe a situation similar to the average results (avg), although the difference in maximum results is more pronounced than in average results.

Removing stop words considerably improves k-means results, which illustrates better cluster segmentation when a stop word list is used. EM results remain virtually unchanged compared to baseline if we use stop words.

If we use both the stop word list and LSI we observe a steady increase in EM results when semantic space dimension is increased (from 2 to 49). In case of k-means, clustering results with LSI are worse compared to those without LSI, and the result deteriorates steadily when semantic space dimension is increased. This is probably due to the assumption that clusters overlap more in the semantic space than in the feature space before its dimension decreases.

When we use both the stop word list and LSI, EM average result (avg) exceeds the average result of k-means. The results for the best scores (max) when using both stop words and LSI behave similarly to the average scores (avg). It should be noted that the highest (max) clustering quality is reached when using k-means, stop word deletion and no LSI (FM=0.58), which improves the EM result under the same conditions but with LSI by 0.02.

To sum up, on average, the best results are demonstrated by EM when stop word deletion and LSI are used.

9 Conclusion

Our research demonstrates that average clustering results of k-means and EM on manual transcripts and recognition results with 80-65% accuracy do not show a large difference. Using a stop word list of high-frequency common function words (without context words) leads to a substantial increase in k-means clustering quality if LSI is not used. Removing stop words improves EM results if we use feature space mapping onto a semantic space by means of LSI. In the latter case we observe stable improvements in quality as the semantic space dimension increases from 2 to 49. The following conclusions can be drawn from experiments. Using a list of stop words improves cluster segmentation, which influences the performance of k-means and LSI. The use of the latter improves the performance of the EM algorithm.

On average, the best results are achieved when using a domain dependent stop list, LSI and the EM algorithm.

References

1. Larson, M., Jones, G.J.F.: Spoken content retrieval: A survey of techniques and technologies. Foundations and Trends in Information Retrieval 5(4-5), 235–422 (2012) ISSN 1554-0669
2. Park, A., Glass, J.R.: Unsupervised pattern discovery in speech. IEEE Trans. Acoustics, Speech and Language Processing 8(1), 186–197 (2008)
3. Deerwester, S., et al.: Improving Information Retrieval with Latent Semantic Indexing. In: Proceedings of the 51st Annual Meeting of the American Society for Information Science, vol. 25, pp. 36–40 (1988)
4. Dempster, A.P., Laird, N.M., Rubin, D.B.: Maximum Likelihood from Incomplete Data via the EM Algorithm. Journal of the Royal Statistical Society, Series B (1977)
5. MacQueen, J.B.: Some Methods for classification and Analysis of Multivariate Observations. In: Proceedings of 5th Berkeley Symposium on Mathematical Statistics and Probability, vol. 1, pp. 281–297. University of California Press, Berkeley (1967)
6. Chernykh, G., Korenevsky, M., Levin, K., Ponomareva, I., Tomashenko, N.: Cross-Validation State Control in Acoustic Model Training of Automatic Speech Recognition System. Scientific and Technical Journal Priborostroenie 57(2), 23–28 (2014)
7. Kudashev, O., Kozlov, A.: The Diarization System for an Unknown Number of Speakers. In: Železný, M., Habernal, I., Ronzhin, A. (eds.) SPECOM 2013. LNCS, vol. 8113, pp. 340–344. Springer, Heidelberg (2013)

8. Dahl, G.E., Yu, D., Deng, L., Acero, A.: Context-Dependent Pre-Trained Deep Neural Networks for Large-Vocabulary Speech Recognition. IEEE Trans. Audio, Speech and Language Proc. 20(1), 30–42 (2012)
9. Pinto, D.: Analysis of narrow-domain short texts clustering. In: Research report for Diploma de Estudios Avanzados (DEA). Department of Information Systems and Computation, UPV (2007)
10. Pinto, D., Rosso, P., Jimenez, H.: A Self-Enriching Methodology for Clustering Narrow Domain Short Texts. Comput. J. 54(7), 1148–1165 (2011)
11. Manning, C., Raghavan, P., Schutze, H.: Introduction to Information Retrieval. Cambridge University Press (2009)
12. Eissen, S.M.z., Stein, B.: Analysis of Clustering Algorithms for Web-based Search. In: Karagiannis, D., Reimer, U. (eds.) PAKM 2002. LNCS (LNAI), vol. 2569, pp. 168–178. Springer, Heidelberg (2002)
13. Stein, B., zu Eissen, S.M., Wibbrock, F.: On Cluster Validity and the Information Need of Users. In: Hanza, M.H. (ed.) 3rd IASTED Int. Conference on Artificial Intelligence and Applications (AIA 2003), Benalmadena, Spain, pp. 216–221. ACTA Press, IASTED (2003) ISBN 0-88986-390-3

Impact of Irregular Pronunciation on Phonetic Segmentation of Nijmegen Corpus of Casual Czech

Petr Mizera[1], Petr Pollak[1], Alice Kolman[2], and Mirjam Ernestus[3]

[1] Faculty of Electrical Engineering, Czech Technical University in Prague
{mizerpet,pollak}@fel.cvut.cz

[2] Radboud University Nijmegen & Christian University of Applied Sciences CHE
akolman@che.nl

[3] Radboud University Nijmegen & Max Planck Institute for Psycholinguistics
mirjam.ernestus@mpi.nl

Abstract. This paper describes the pilot study of phonetic segmentation applied to Nijmegen Corpus of Casual Czech (NCCCz). This corpus contains informal speech of strong spontaneous nature which influences the character of produced speech at various levels. This work is the part of wider research related to the analysis of pronunciation reduction in such informal speech. We present the analysis of the accuracy of phonetic segmentation when canonical or reduced pronunciation is used. The achieved accuracy of realized phonetic segmentation provides information about general accuracy of proper acoustic modelling which is supposed to be applied in spontaneous speech recognition. As a byproduct of presented spontaneous speech segmentation, this paper also describes the created lexicon with canonical pronunciations of words in NCCCz, a tool supporting pronunciation check of lexicon items, and finally also a minidatabase of selected utterances from NCCCz manually labelled on phonetic level suitable for evaluation purposes.

Keywords: spontaneous speech, casual speech, pronunciation reduction, phonetic segmentation, NCCCz.

1 Introduction

In the past decades, speech technology applications have started being focused on the processing of spontaneous and informal speech which can be seen e.g. in automated transcription of various informal recordings from meetings or transcription of TV or broadcast programs for on-line subtitling or for archiving purposes. Due to this fact, researchers have become increasingly interested in the characteristics of spontaneous and casual speech in the most important world languages such as German, Dutch or English [1,2,3] and first steps have been taken in this field also for Czech [4,5].

The current speech recognition systems usually work very precisely for standard speech and we can find many works describing such systems for all world languages including Czech. However, the accuracy of speech recognition of stron- gly spontaneous speech is significantly lower and the amount of published works describing the analysis or recognition of informal speech is also smaller. In this paper we present the first

P. Sojka et al. (Eds.): TSD 2014, LNAI 8655, pp. 499–506, 2014.

pilot study of phonetic segmentation accuracy applied to Nijmegen Corpus of Casual Czech (NCCCz), which was created to bring a missing corpus of Czech containing high-quality recordings from naturally occurring interaction which are suitable for detailed analysis of spontaneous speech in Czech [6].

The paper is organized as follows: firstly, the brief description of NCCCz is presented, secondly, the creation procedure of the lexicon with regular canonical pronunciations for NCCCz data is mentioned (together with the description of the tool supporting this step), and finally the results of the first analyses of the phonetic segmentation accuracy applied to casual Czech speech are presented.

2 The Nijmegen Corpus of Casual Czech

For the development of standard recognition systems, corpora of read speech or speech produced during formal interviews are used, e.g. SPEECON and SpeechDat database [7,8,9]. The Nijmegen Corpus of Casual Czech (NCCCz) used in this work contains more than 30 hours of high-quality recordings of casual conversations among 10 triplets of male and 10 triplets of female friends. One speaker from each triplet always acted as a confederate who asked two friends of the same gender (henceforth the naive speakers) to participate in recordings of natural conversations. The recording procedure was controlled by an experimenter.

Each session was recorded in a soundproof booth and in the first part of the recording, the confederate pretended to have received an important phone call that had to be answered immediately and the two naive speakers were left alone without information about whether they were already being recorded. Depending on the liveliness of the conversations between the two naive speakers, the confederate returned to the booth. Then the second part of the recording started, which consisted of free conversation among the three speakers. Various topics including school, relationships, common hobbies, and stories about all sorts of encounters were addressed. In the third part of the recordings, the experimenter entered the room with a list of questions on political and social issues and the speakers were asked to discuss at least four issues from the list and negotiate a common opinion for each question. The speakers were engaged in conversations approximately for 90 minutes and the recordings obtained by the above mentioned procedure contain very informal spontaneous data.

All speakers were recorded simultaneously on separate audio channels using cardioid microphones avoiding possible cross-talks in particular channels for each speaker. The whole corpus has been annotated at orthographic level using standard non-reduced transcription joined by additional marks for non-speech events. Corpus is freely available on demand as it is described in more details in [6].

3 The NCCCz Lexicon

For the purpose of further studies and developments, the orthographic transcription of records had to be completed by the pronunciation lexicon. It always represents an important component which has significant impact on the accuracy of target ASR system. It is especially even more important in the case of spontaneous or casual speech

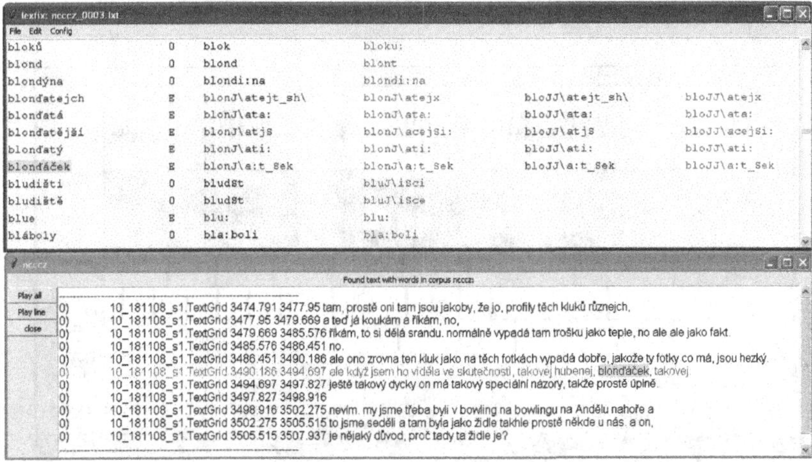

Fig. 1. Illustrative example of the work with the LexFix tool

recognition in which the process of coarticulation, assimilation and reduction often appears. Therefore the pronunciation lexicon should contain as many pronunciation variants as possible to capture this variability in informal spontaneous speech.

Due to very informal speaking style yielding many rare and non-standard words, the lexicon of regular canonical pronunciation was created manually in cooperation with Czech native speakers with background in phonetics. The development of proper pronunciation lexicon can be described in the following steps: the first version was created using the rules of conversion from Czech orthographic transcription to canonical pronunciation [10]. The automatically generated pronunciation contained large amount of incorrect pronunciations mainly for foreign or the above mentioned non-standard words, but also due to poor voicing assimilation, phone softening, etc.

The correction (editing) of the pronunciation lexicon was supported by the extraction of the information about the context of given word form in the corpus and possibly also by listening to unclear words. For this purpose, we modified the *LexFix* tool for lexicon editing which provides the linking with both the orthographic transcription and audio signal to enable listening of recorded utterance.

The tool was created generally to support the determination of correct pronunciation of particular word forms. At the same time it was extended by other functions which simplify the work with a huge corpus. The possibility to search for the word in a huge corpus and display the neighbouring context can help determine the correct pronunciation. Typically, for foreign, rare, or generally non-standard words it may be difficult to decide about the pronunciation without listening, so it could be necessary to play the particular sentence with the given word in found context. The illustrative example of the work with the *LexFix* tool is at Fig. 1.

Finally, the created pronunciation lexicon for NCCCz contains approx 30 000 word forms. During these checks, reduced pronunciations (e.g. "nějaký" vs. "ňáký") were not marked so the current lexicon contains only canonical pronunciation with a small amount of pronunciation variants. The reduction of the pronunciation is supposed to be

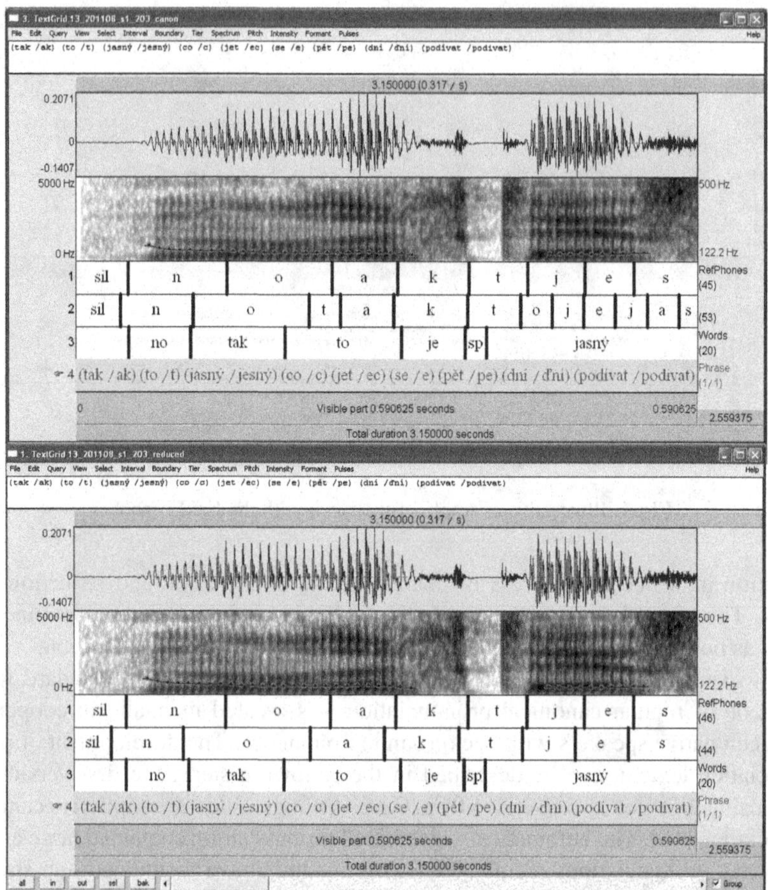

Fig. 2. Illustrative example of phonetic segmentation results with the canonical (top picture) and reduced (bottom picture) pronunciation

solved automatically at further steps of this this wider research. The process of looking for the reduction rules is also supposed to be supported by the NCCCz data together with the possible listening of particular occurrences of lexicon items in the corpus.

4 Phonetic Segmentation

Automatic phonetic segmentation can be implemented in various ways. HMM-based automatic phonetic segmentation, which is well-known as a forced alignment is widely used technique. However, other approaches for the phoneme localization using Bayesian changepoint detector, or artificial neural networks are also used by some authors [11,12,13].

The HMM-based forced alignment is a well known algorithm, looking for the maximum likelihood path through a composed acoustic model for an utterance with known contents. Phone boundaries are then determined by the occupacy of HMM states representing particular phones over the found optimum path. The selection of proper pronunciations, ideally those which have been really realized in the given utterance, plays a significant role in the segmentation accuracy. However, it has not often been fulfilled in case of casual speech.

The phonetic contents of each word used in the above mentioned algorithm is typically taken from the lexicon. Therefore, one important purpose is to analyze the precision of phonetic segmentation in three basic cases, i.e. using three variants of pronunciation generation for the HMM-based forced alignment:

- using the lexicon with *canonical pronunciation*,
- using *actually realized reduced pronunciation* in each utterance which was transcribed manually,
- using the lexicon with *more pronunciation variants* containing several levels of pronunciation reduction.

This study should demonstrate the general impact of proper pronunciation selection on the basis of casual speech phonetic segmentation accuracy as an objective criterion (as it is shown illustratively in Fig. 2) which is supposed further to improve also the accuracy of casual speech recognition.

5 Experimental Setup

HMM-based forced alignment was implemented by the open-source Kaldi Speech Recognition Toolkit [14] in a rather standard setup. As speech features we used common Mel-Frequency Cepstral Coefficients with the additional zeroth cepstral coefficient, completed by their delta and delta-delta features (MFFC_0_D_A). Cepstral mean normalization (CMN) was also applied to minimize the small mismatch between training and processed data. Short-time analysis used the frame length of 25 ms and frame shift of 10 ms. The GMM-HMM based acoustic model (AM) was based on triphones and trained on utterances from the Czech SPEECON database. The procedure of AM training was inspired by an example for the Wall Street Journal and the TIMIT database (KALDI recipes s4 and s5). The set used for the training of our AM contained utterances recorded in rather clean office enviroment and the amount was about 52 hours of read speech.

Finally, we used the *gmm-align* tool for forced alignment realization. As this tool produces the output of state-level alignments of utterances which are represented by transiton-ids, we have created a simple tool for the conversion to phoneme-level alignments of utterances. The output of this tool is in HTK MLF format, containing also the information on time boundaries of particular phones.

The experiments were carried out using selected utterances from the NCCCz corpus having the lengths of about 15 – 20 words. For this pilot experiment we have selected utterances of speakers with rather standard level of reduced pronunciation and we have selected data without further disturbance such as high background noise, high frequency of non-speech acoustic events or overlapping speech. This evaluation subset contains 19 utterances from 8 speakers which were now manually segmented at the phonetic level.

Eventually, in this evaluation subset we suppose to have at least 100 utterances from all speakers. The amount of data in this evaluation subset is summarized in Table 1 (values for the target state are estimated).

Table 1. The NCCCz evaluation subset statistics

	Current state				*Target state*			
Sex	minutes	speakers	sentences	phones	*minutes*	*speakers*	*sentences*	*phones*
Male	1.09	6	16	923	*3.33*	*30*	*50*	*3000*
Female	0.20	2	3	172	*3.33*	*30*	*50*	*3000*
Total	1.29	8	19	1095	*6.66*	*60*	*100*	*6000*

6 Results and Discussion

This section presents the results of above discussed phonetic segmentation of casual Czech speech which were perfomed with various level of prounciation reduction. The accuracy was quantified using the following criteria: *Shift of the Phone Beginning* (*SPB*), *Shift of the Phone End* (*SPE*), *Change of the Phone Length* (*CPL*), and *Phone Error Rate* (*PER*), more details can be found in [12].

The overall results are presented in Table 2. The mean values and standard deviation of *SPB*, *SPE*, *CPL*, and *PER* were computed across all phones and we can observe the improvement of global accuracy of automatic phonetic segmentation (for all analyzed criteria) when reduced pronunciation is used, i.e. both mean values and standard deviations decreased significantly. Significant improvement of accuracy on an average of 3.2 ms (across all criteria SPB, SPE, CLP) was observed when reduced pronunciation was used instead of canonical one, slightly smaller improvement about 1.9 ms when the lexicon contained more pronunciations variants. Secondly, the results for phone categories are in Fig. 3, here we can see in more detail the achieved segmentation accuracy for phones of similar character [15]. The mean values and standard deviation for VOW, FRI, VOWNM, FRIAFF in particular, had significantly improved for the criteria SPB and SPE.

Although presented results were obtained by experiments performed on small evaluation subset of manually segmented utterances, the results had already proved the contribution of the information about pronunciation reduction. Currently, further manual segmentation of the evaluation test utterances is being processed thus the

Table 2. The overall results of phonetic segmentation for all phones

	canonical	*reduced*	*variants*
SPB [ms]	-5.24 ± 47.59	-1.05 ± 18.96	-3.39 ± 39.53
SPE [ms]	-4.16 ± 41.39	-1.64 ± 22.49	-1.42 ± 38.53
CPL [ms]	-3.17 ± 20.75	-0.07 ± 21.65	-2.07 ± 19.12
PER [%]	21.04	1.18	14.94

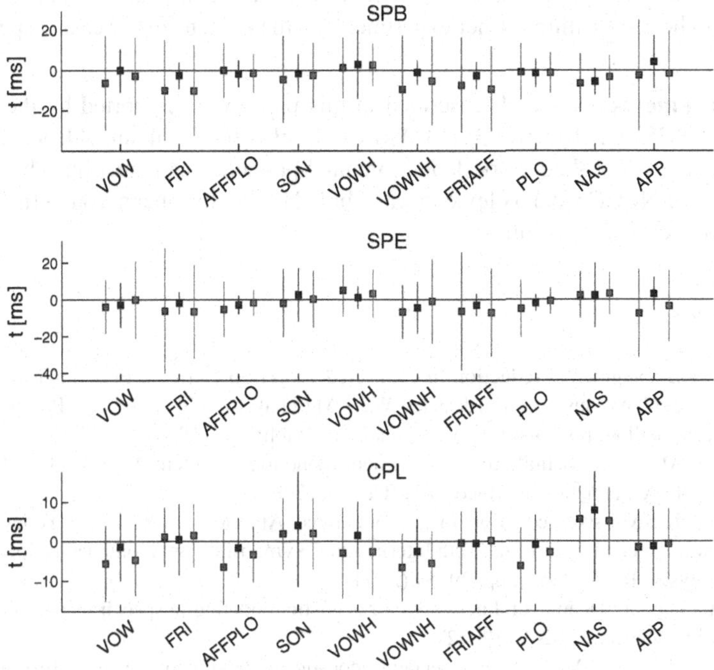

Fig. 3. The results of phonetic segmentation for particular phone groups (red – canonical pronunciation, blue – reduced, magenta – more pronunciation variants)

processing we assume that these results will be precised during the presentation at the workshop.

7 Conclusions

This study is the first step in further detailed research on pronunciation reduction in NCCCz which is also supposed to be used for better modelling of spontaneous speech for recognition purposes. The realized pilot analysis of HMM-based phonetic segmentation accuracy with regard to the usage of canonical or reduced pronunciations is the main contribution of this paper. The experiments done with the speech from Nijmegen Corpus of Casual Czech (NCCCz) with very strong spontaneous nature demonstrated the significant impact of pronunciation reduction on the proper acoustic modelling of spontaneous speech (applied currently on phonetic segmentation). Further contribution of this paper is in the basic information about NCCCz corpus and its lexicon containing typical casual words. The created tool LexFix also represents the important contribution of this work because it generally supports lexicon editing with possible checks of a word context in the corpus, just as does selected listening. Finally, the created evaluation database with utterances from NCCCz completed by manual labels at phonetic level is the last important contribution. It was used in the experimental part described in this paper but it is suitable for evaluation purposes in general and it is supposed to be used within further experiments with spontaneous or casual speech.

Acknowledgments. Research described in this paper was supported by the internal CTU grant SGS14/191/OHK3/3T/13. We would also like to thank Zdeněk Patc and Helena Pollaková for their work done in manual phonetic segmentation. The work on the creation of NCCCz and its lexicon was funded by the European Young Investigator Award given to the fourth author.

References

1. Kohler, K.J.: Segmental reduction in connected speech in German: Phonological facts and phonetic explanations. In: Hardcastle, W.J., Marchal, A. (eds.) Speech Production and Speech Modelling, pp. 69–92. Kluwer Academic Publishers (1990)
2. Ernestus, M.: Voice assimilation and segment reduction in Dutch: A corpus-based study of the phonology-phonetics interface. LOT, Utrecht (2000)
3. Johnson, K.: Massive reduction in conversational American English. In: Yoneyama, K., Maekawa, K. (eds.) Proc. of the 10th International Symposium on Spontaneous Speech: Data and Analysis, Tokyo, Japan, pp. 29–54 (2004)
4. Hinton, G., et al.: Deep neural networks for acoustic modeling in speech recognition. Signal Processing Magazine, 82–97 (2012)
5. Vaněk, J., Psutka, J.V.: Gender-dependent acoustic models fusion developed for automatic subtitling of parliament meetings broadcasted by the Czech TV. In: Sojka, P., Horák, A., Kopeček, I., Pala, K. (eds.) TSD 2010. LNCS (LNAI), vol. 6231, pp. 431–438. Springer, Heidelberg (2010)
6. Ernestus, M., Kočková-Amortová, L., Pollak, P.: The Nijmegen Corpus of Casual Czech. In: LREC 2014, Reykjavik, Iceland, May 26-31 (2014)
7. Pollak, P., Černocký, J.: Czech SPEECON adult database. Technical report (November 2003), `http://www.speechdat.org/speecon`
8. Pollak, P., Černocký, J., et al.: Speechdat(E) – Eastern European telephone speech databases. In: Proc of XLDB, Athens, Greece (2000)
9. Siemund, R., Höge, H., Kunzmann, S., Marasek, K.: SPEECON – Speech data for consumer devices. In: Proc. of the LREC 2000, Athens, Greece (2000)
10. Hanzl, V., Pollak, P.: Tool for Czech Pronunciation Generation Combining Fixed Rules with Pronunciation Lexicon and Lexicon Management Tool. In: Proc. of LREC 2002, Las Palmas de Gran Canaria, Spain, pp. 1264–1269 (2002)
11. Cmejla, R., et al.: Bayesian changepoint detection for the automatic assessment of fluency and articulatory disorders. Speech Communication, 178–189 (2013)
12. Mizera, P., Pollak, P.: Accuracy of HMM-based phonetic segmentation using monophone or triphone acoustic model. In: Proc. of Applied Electronics, Pilsen, Czech Republic (2013)
13. Schwarz, P.: Phoneme recognition based on long temporal context. PhD Thesis, Brno University of Technology (2009)
14. Povey, D., Ghoshal, A., et al.: The Kaldi Speech Recognition Toolkit. In: Proc. of ASRU, Hawaii, USA (2011)
15. Pollak, P., Volin, J., Skarnitzl, R.: Phone Segmentation Tool with Integrated Pronunciation Lexicon and Czech Phonetically Labelled Reference Database. In: Proc of LREC, Marrakech, Morocco (2008)

Parametric Speech Coding Framework for Voice Conversion Based on Mixed Excitation Model

Michał Lenarczyk

Institute of Computer Science, Polish Academy of Sciences,
ul. Jana Kazimierza 5, 01-248 Warsaw, Poland

Abstract. Adaptation of mixed-excitation linear predictive (MELP) model for application in voice conversion is presented. The adapted model features only numerical parameters which can be used for phonetic space transformation from source to target speaker using methods of machine learning. The validity of the model was demonstrated by applying transformation to both the pitch and the spectral envelope of voice.

1 Introduction

Voice conversion is a technique of transforming speaker individuality in speech signal. The aim is to change voice of an original (source) speaker to sound like another (target) speaker, while preserving the semantic content of the utterance. A voice conversion system follows the general structure of Fig. 1.

Signal processing techniques are applied in analysis and synthesis stages, to transform speech signal to and from a parametric representation that is suitable for transformation, which is most commonly done using machine learning techniques. The analysis and synthesis can thus be viewed as a pre- and post-processing for representing data in a concise form, such that transformation function can be estimated given speech data of the source and target speakers. One approach to learning the source-target transformation originally introduced in the classic work of Abe et al. [1] is to use parallel corpora (composed of the same set of utterances from both speakers) and extract numerical features from time aligned source and target speech signals. In this case, there are a number of general purpose methods that can be applied for transformation. They include, among others: frequency warping functions [2], maximum likelihood modeling including gaussian mixture models (GMM) [3,4], hidden Markov models (HMM) [5,6], support vector regression [9], or artificial neural networks (ANN) [8,7]. On the other hand, the representation of speech (coding) is typically defined differently by different authors, making it difficult to evaluate fitness and compare particular transformation methods in the task of interest.

Fig. 1. General structure of a voice conversion system

P. Sojka et al. (Eds.): TSD 2014, LNAI 8655, pp. 507–514, 2014.

Considering the chain of Fig. 1 with the transformation function removed, i.e. replaced by identity function (channel), we recognize it as a prototype vocoder. Rather than building a VC system from scratch, use of one of established voice coding methods is therefore possible. In this paper, adaptation of the mixed excitation linear predictive (MELP) coding is presented that lends itself for use in selective transformation of voice features.

There are a number of speech features that define speaker individuality. The instantaneous configuration of formants in signal defines the phoneme uttered but the overall characteristics of the frequency envelope of speech over entire acoustic space is the individual voice, or *timbre*. Pitch, or perceived fundamental frequency in a signal, is the primary distinguishing feature of male and female voices. Accentuation and prosodic features such as the modulation of pitch and amplitude in time, duration of phonemes, etc. also count as characteristic properties of speakers. It is generally agreed that timbre and pitch are the most important and they are the focus of this work.

2 Parametric Modeling of Speech

The mixed excitation linear predictive (MELP) coder, originally introduced in [10], is an example of a highly (yet not fully) parametric coding scheme. The basis of this coder is the linear predictive coding (LPC), which models the speech production as a source-filter process. It captures the general shape of speech spectrum (envelope), which corresponds roughly to the effect of vocal tract and lips, in the form of a filter defined by a small number of coefficients. The model residual is believed to represent phonation, be it voiced (harmonic) or unvoiced (noisy). MELP models the excitation as a combination of harmonic pulse train and white noise, mixed in appropriate proportion.

The initial motivation for MELP was low bitrate coding. As such, it features quantization and several other nonparametric aspects that are an obstacle for manipulation, such as switched voicing state, switched pitch aperiodicity, noise thresholds, and unfavourable spectral envelope representation. This section briefly describes the structure of the vocoder and details changes introduced into the adapted model. As a baseline, the 2.4 kb/s MELP specification accepted as the U.S. Federal Standard is taken [12].

Modification of parameters can be done with any method. Preliminary experiments conducted with static and neural network-learned transformations are described in Section 3.

2.1 Structure of MELP

The foundation of MELP synthesis is a mixed source composed of white noise and a harmonic component represented by a sequence of pulses at defined pitch intervals, followed by an all-pole synthesis filter. The two basic types of excitation are combined in a harmonic/noise proportion that corresponds to one of three allowed states of voicing: fully voiced (0.8/0.2), weakly voiced (0.5/0.5) and unvoiced (0.0/1.0), with interpolation between the two voiced states. On top of this basic structure, a number of additional features have been added to enhance speech quality while keeping the bit rate low. They include:

1. switched aperiodicity that adds jitter to the harmonic excitation, applied in the weakly voiced case,
2. encoded amplitudes of initial harmonics to better represent the characteristics of the periodic excitation,
3. two conditioning filters for shaping pulse and noise spectra (boost low frequency part of harmonic spectrum and high frequency part of noise spectrum to approximate the behaviour observed in speech) while assuring spectral flatness of the excitation mixture,
4. subband voicing analysis allowing finer shaping of the excitation's harmonic to noise ratio with a staircase approximation,
5. perceptual weighting filter in the postprocessing phase for formant sharpening (adaptive spectral enhancement).

2.2 Proposed Vocoder Framework

The vocoder elaborated by the author is proposed as a flexible and fully parametric framework for application in voice conversion that builds on the mixed-excitation model. The structure preserves essential parts of the original MELP coder and modifications were designed to make the speech representation suitable for learning and transformation.

The framework has been adapted to process speech signal sampled with 16 kHz as opposed to 8 kHz used in telephone-band speech coding. This adds an additional octave that is important in fricatives. The frame size is set to 10 ms (160 samples) as opposed to 22.5 ms in MELP, and analysis is performed in 30 ms window (480 samples), as opposed to 25 ms.

The synthesis section (Fig. 2-b) preserves the mixed excitation composed of white noise and periodic pulse with period given by the pitch lag parameter. Envelopes for both source types are generated by dividing the overall gain into harmonic and noise parts according to voicing level parameter, and interpolating with a filter based on a Hamming-windowed sinc function truncated at ± 320. The mixer shapes the periodic and noise signals using the filters $H_P(z) = 1 + az^{-1}$ and $H_S(z) = 1 - bz^{-1}$, respectively, with b fixed at 1 and a calculated according to $a = \frac{G_S^2}{G_P^2}$ where G_P, G_S denote pulse and noise gain. This is equivalent to the formulations in [10] where a different gain scheme is used. The process of generating the excitation can be performed in subbands. The synthesis filter is lattice IIR of order 18. The parameters for synthesis are estimated in the analysis section (Fig. 2-a) which differs from baseline MELP both in representation and methods used. The following paragraphs detail the differences.

Preprocessing: a first order preemphasis is applied during preprocessing to compensate the spectral tilt. This preconditioning is sufficient to accurately model the entire spectrum and thus adaptive spectral enhancement is not employed.

Spectral envelope representation: spectral envelope is represented by filter estimated using standard autocorrelation method of LPC and the coefficients are found in the Levinson-Durbin procedure. The LP filter order is increased from 10 to 18 to account for the wider frequency band.

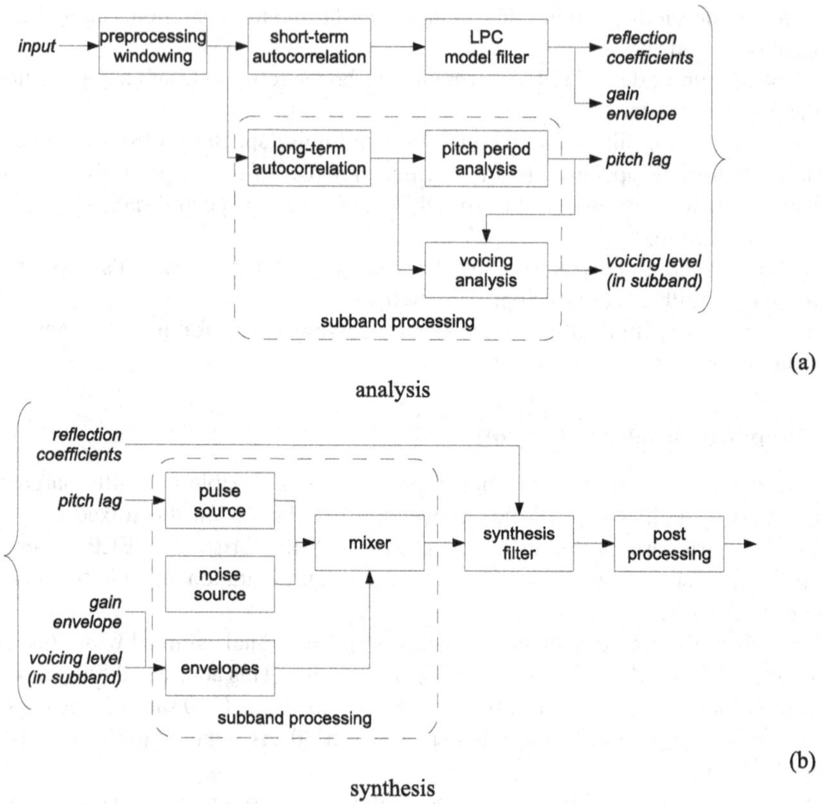

Fig. 2. Structure of the parametric framework

Representation of the envelope is of primary importance. The proposed framework abandons line spectrum frequencies (LSF) representation, as used in most modern speech coders, in favor of reflection coefficients of a lattice filter. LSF have favourable quantization and interpolation properties, but require strict ordering to preserve filter stability which may be hard to assure when transforming using general purpose methods of machine learning. On the other hand, stability can be easily assured by bounding the reflection coefficients to the range $k_i \in (-1, +1)$. In experiments conducted with neural network based transformation, the requirement for restricted range was naturally met by selecting a bipolar activation function. If other methods of transformation are to be used, log area ratios (LAR) given by $\log \frac{1-k_i}{1+k_i} \in \mathbb{R}$ may be a preferable representation. Interpolation in reflection coefficient domain gives poor results and is thus avoided by reducing the frame length to 10 ms.

Correction of harmonic amplitudes: in the standard MELP [12], Fourier magnitudes of initial harmonics of the LPC residual of voiced speech were added because it is known [13] that LPC fails to model harmonic spectra accurately. Because harmonic frequencies change when pitch is altered, and the residual values are not predictable from the model, they are a non-parametric feature and are thus omitted. Study on relative importance of parameters in bitrate-reduced version of MELP [11] revealed Fourier magnitudes to be of secondary importance for quality.

Pitch: the pitch period is evaluated by searching the position of the dominant peak of autocorrelation function in a defined interval. To increase reliability, the original and $2\times$ upsampled and interpolated autocorrelation vectors are upper halfwave rectified and multiplied to produce a periodicity index, which peaks at double the fundamental period. It was found to be a good estimator with a half sample resolution. Explicit checks are additionally performed to avoid selecting halved or doubled pitch period. Aperiodicity (pitch jitter) is not modeled.

Voicing estimation: the voicing level is estimated as a continuous fraction of the harmonic component in the overall mixture. This is in contrast with MELP where the level of voicing is nominal with three possible levels (fully voiced, jittery voiced and unvoiced). The associated ratios of harmonic to total energy are fixed at 0.8, 0.5 and 0, respectively, with interpolation between the voiced states. Such a coding produces a degenerate distribution which is hard to model using machine learning. The proposed method avoids quantization of voicing which reduces the amount of logic and allows gradual transformation towards breathy or whispered speech.

The evaluation of this proportion is based on a weighted sum of peak levels of autocorrelation at the pitch lag and its multiples. The value is always clipped to the range $0, +1$. Typically, it is close to 1 for fully voiced and slightly above 0 for unvoiced speech, i.e. there remains some residual energy of the other kind. This is not a problem : in voiced phonation, the additional noise adds naturalness and in unvoiced speech the weak harmonic component is dominated by noise. Moreover, in unvoiced segments the pitch estimate tends to fluctuate, making the impulse train less regular and easier to vanish in noise. Fig. 3 illustrates the method with a sample speech utterance.

Gain envelopes: the gain is estimated every 10 ms from residual energy over the window. In the synthesis stage, the vocing level is used to divide the energy into harmonic/stochastic parts, and smooth envelopes are generated at the sample rate by interpolating in log domain using Hamming windowed sinc function filter.

2.3 Subband Analysis

For a more realistic modeling of phonemes that are both noisy and harmonic, but with a different ratio depending on frequency region, subband processing was proposed in [10] and is followed in this work. For this mode of analysis, four frequency bands are defined and the level of voicing is estimated from band filtered speech. When generating excitation, each voicing level is used to generate a band limited mixture with a different proportion of harmonic and noisy components. The subband excitations are finally added to form a total excitation that preserves spectral flatness. As shown on Fig. 4, higher frequency bands tend to be more noisy when the overall voicing is high.

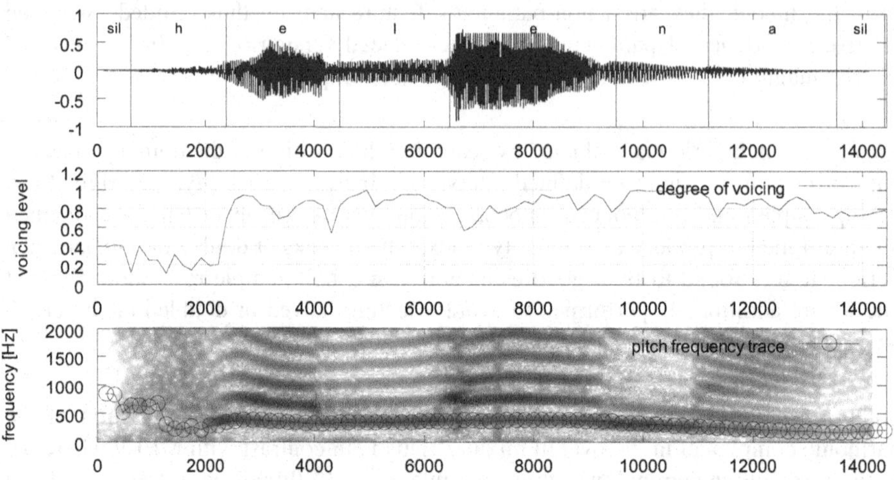

Fig. 3. Pitch tracking and voicing level detection for the word "Helena" uttered by a boy. From top to bottom: original speech waveform, voicing level estimate, spectrogram with pitch frequency estimate overlaid.

3 Transcoding Experiments

To evaluate the quality of the coding method and its applicability for conversion of voice characteristics, transcoding was applied to a small subset of samples from the *Corpora* Polish speech corpus [14]. Informal listening tests were conducted in three conditions: no transformation, pitch and voicing scaling and spectral envelope transformation. This

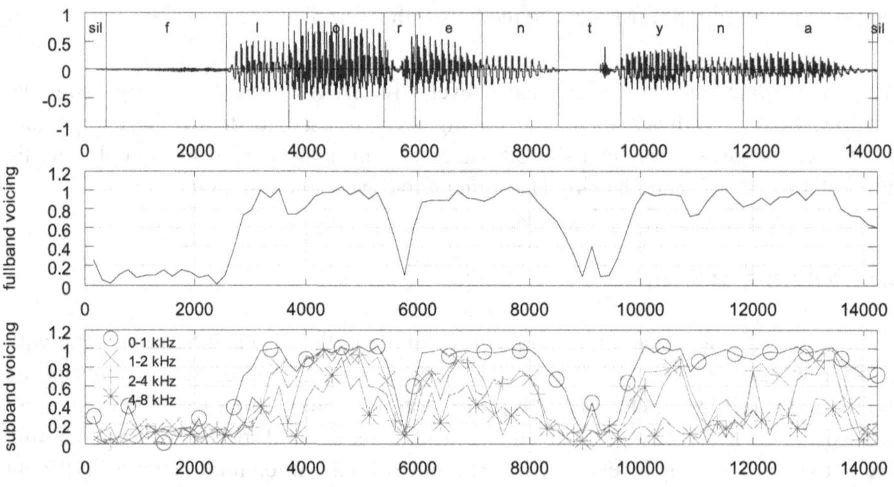

Fig. 4. Male utterance "Florentyna" (top) and associated voicing traces from full band (middle) and subband (bottom) analysis. Four subbands are used: 0–1 kHz; 1–2 kHz; 2–4 kHz; 4–8 kHz.

evaluation is a preliminary proof of validity of the framework and formal evaluation is planned. The obtained samples can be accessed from the author's website [15].

In the first scenario, the parameters were extracted and used directly for speech reconstruction. This provides a baseline estimate for the level of quality achievable through this coding scheme. Two versions were tested : with a single level of voicing evaluated over the full frequency band, and with voicing calculated over four subbands (0–1 kHz, 1–2 kHz, 2–4 kHz and 4–8 kHz). In both cases, different kind of speech degradation is audible. Fullband processing yields a buzzy quality of voice especially in male speech. Subband approach offers a voice quality that is very close to the source but not transparent, and suffers from occasional tonal thumps that are attibuted to less reliable estimation of voicing, which results in noise bursts (cf. Fig. 4). This effect is especially pronounced at plosives. Better subband pitch tracking may help reduce the problem.

The second scenario is concerned with pitch modification, which was done by rescaling the frequency estimate by a predefined ratio. Informal listening indicates that a considerable amount of scaling (by a factor of two, or one octave, up or down) can be applied without introducing additional distortion compared to unscaled pitch. Similar modification of voicing was also performed and demonstrates the ability to generate breathy and whispered speech.

In the last scenario, the frequency envelope of speech was transformed toward a desired target. To achieve this, the reflection coefficients of the model filter were transformed with the use of a neural network. The network had a layered structure with one hidden layer of 40 units and input and output layers each of 18 units which is the size of the reflection coefficients vector. A bipolar activation function was selected and the network was trained using backpropagation algorithm with a learning constant of 0.0001. A parallel set of over 10000 input-output vectors was created from 251 short utterances (names, numbers, control words) of the speech corpus. In repeated experiments, the learning process consistently led to a 10 to 20-fold reduction in error within initial 100 iterations, around 10% reduction within the next 100 iterations, and no appreciable further gain. The error contribution per vector stabilised at around 0.5.

As a result of transformation of the vocal tract parameters, the speech underwent definite qualitative transformation. The most obvious artifact in the resynthesised samples is a muffling of voice. This muffling is presumably due to variability in pronounciation of phonemes across utterances and the fact that alignment was only based on phonetic annotation and not on numerical procedures like the commonly used dynamic time warping (DTW). Another possible cause is the attraction of filter polynomial roots toward the center of the Z-plane ($z = 0$) as a result of averaging, which causes loss of formant sharpness.

4 Concluding Remarks

A parametric speech coding framework has been proposed allowing selective transformations of voice properties. Modification of pitch, voicing and frequency envelope was demonstrated. With direct transcoding the obtained quality was good but not transparent, which puts a limit on quality with conversion applied. Solution to some issues (e.g.

subband mode thumps) seems possible. On the other hand, the framework has the potential for the emerging field of real time voice conversion. Further work is directed towards this goal and formal listening tests are planned.

Acknowledgement. Study was supported by research fellowship within "Information technologies: research and their interdisciplinary applications" agreement number POKL.04.01.01-00-051/10-00.

References

1. Abe, M., Nakamura, S., Shikano, K., Kuwabara, H.: Voice conversion through vector quantization. In: Proc. Int. Conf. Acoustics, Speech and Signal Processing, vol. 1, pp. 655–658 (1988)
2. Hanzlíček, Z., Matoušek, J.: On using warping function for LSFs transformation in a voice conversion system. In: Proc. Int. Conf. Signal Processing, pp. 2725–2728 (2008)
3. Stylianou, Y., Cappé, O., Moulines, E.: Continuous probabilistic transform for voice conversion. IEEE Transactions on Speech and Audio Processing 6(2), 131–142 (1998)
4. Ye, H., Young, S.: Quality-enhanced voice morphing using maximum likelihood transformations. IEEE Transactions on Audio, Speech and Language Processing 14(4), 1301–1312 (2006)
5. Arslan, L.M., Talkin, D.: Speaker transformation using sentence HMM based alignments and detailed prosody modification. In: Proc. Int. Conf. Acoustics, Speech and Signal Processing, vol. 1, pp. 289–292 (1998)
6. Rentzos, D., Vaseghi, S., Yan, Q., Ho, C.: Parametric Formant Modelling and Transformation in Voice Conversion. International Journal of Speech Technology 8(3), 227–245 (2005)
7. Guido, R.C., Sasso Vieira, L., Barbon, J.S., Sanchez, F.L., Maciel, C.D., Everthon, S.F., Pereira, C.J.: A Neural-wavelet Architecture for Voice Conversion. Neurocomputing 71(1-3), 174–180 (2007)
8. Orphanidou, C., Moroz, I.M., Roberts, S.J.: Wavelet-based voice morphing. WSEAS Journal on Systems 10, 3297–3302 (2004)
9. Laskar, R., Talukdar, F., Bhattacharjee, R., Das, S.: Voice Conversion by Mapping the Spectral and Prosodic Features Using Support Vector Machine. Applications of Soft Computing 28 (2009)
10. McCree, A., Barnwell, T.P.: A new mixed excitation LPC vocoder. In: Proc. Int. Conf. Acoustics, Speech and Signal Processing, vol. 1, pp. 593–596 (1991)
11. McCree, A., De Martin, J.C.: A 1.6 kb/s MELP coder for wireless communications. In: Proc. IEEE Workshop on Speech Coding for Telecommunications, pp. 23–24 (1997)
12. Supplee, L.M., Cohn, R.P., Collura, J.S., McCree, A.V.: MELP: The new Federal Standard at 2400 bps. In: Proc. Int. Conf. Acoustics, Speech and Signal Processing, vol. 2, pp. 1591–1594 (1997)
13. Makhoul, J.: Linear prediction: A tutorial review. Proc. IEEE 63(4), 561–580 (1975)
14. Grocholewski, S.: CORPORA - speech database for Polish diphones. In: Proc. EUROSPEECH (1997)
15. Author's web page, http://phd.ipipan.waw.pl/~m.lenarczyk/TSD2014.html

Captioning of Live TV Commentaries from the Olympic Games in Sochi: Some Interesting Insights*

Josef V. Psutka, Aleš Pražák, Josef Psutka, and Vlasta Radová

Department of Cybernetics, West Bohemia University, Pilsen, Czech Republic
{psutka_j,aprazak,psutka,radova}@kky.zcu.cz

Abstract. In this paper, we describe our effort and some interesting insights obtained during captioning more than 70 hours of live TV broadcasts from the Olympic Games in Sochi. The closed captioning was prepared for ČT Sport, the sport channel of the public service broadcaster in the Czech Republic. We will briefly discuss our solution for distributed captioning architecture on live TV programs using re-speaking approach as well as several modifications of existing live captioning application (especially LVCSR system), but also the way of re-speaking of a real TV commentary for individual sports. We will show that a re-speaker after hard training can achieve such accuracy (more than 98 %) and readability of captions which clearly outperform accuracy of captions created by automatic recognition of TV soundtrack.

Keywords: live captioning, speech recognition, re-speaking.

1 Introduction

One very interesting application area of an automatic speech recognition is the captioning of live TV programs [2,3,6,9]. There are basically two approaches to this task – automatic recognition of a TV soundtrack or human re-speaker who listens to the TV commentary and re-speaks it to the LVCSR system. Both methods have their advantages/disadvantages and are suitable for different types of TV genres and programs. The subtitling from a TV soundtrack is cheaper, but it is more prone to errors that can lead to the creation of obscure subtitles. The choice of a suitable method may also be influenced by the level of background noise, the possibility of overlapping speakers, a manner of their speech, expected frequency of OOV words (even if the vocabulary and language model are tuned to a given genre), etc.

Our research group has extensive experience with both methods of subtitling. During the past years we have subtitled for the Czech TV hundreds of hours of TV broadcasts of meetings from both chambers of the Parliament of the Czech republic (by processing of the TV soundtrack) [8], as well as live programs, such as political debates, entertainment shows and also sports events (using a re-speaker) [12].

* This paper was supported by the Technology Agency of the Czech Republic, project No. TA01011264.

P. Sojka et al. (Eds.): TSD 2014, LNAI 8655, pp. 515–522, 2014.

The theme of this article was inspired by recent subtitling more than 70 hours of live broadcasts from the Olympic Games in Sochi. We subtitled selected broadcasts of ice-hockey matches, speed skating, figure skating, biathlon, alpine skiing, ski jumping, cross-country skiing and also opening and closing ceremonies for ČT4 Sport channel. Everything was done using re-speakers who employed our in-house LVCSR system equipped with specific vocabularies and language models specially prepared for each sport discipline.

At the end of the Olympic competitions, we wanted to answer the question whether it would be possible to provide subtitles for live broadcasts of some sports directly by the processing of accompanying soundtrack, i.e. by recognition of TV commentaries. It is evident that such method of subtitling is cheaper and if it achieved a satisfactory accuracy and intelligibility of generated subtitles, it could be used for some sports during future broadcasts. For this reason we performed a series of experiments where we compared the results of subtitling real broadcasts from Sochi (with our re-speakers) and those that would be obtained if we used directly the automatic recognition of TV commentaries. The results of these experiments, including a brief description of our LVCSR system are discussed in the next sections of this article.

2 Acoustic Modeling

2.1 Acoustic Model for Recognition of Original Soundtrack

For recognition of speech of TV commentators directly from the soundtrack we had to develop a special acoustic model. The acoustic data was collected over several years, especially from TV broadcasts of the Ice-hockey World Championships, the FIFA World Cups and also the last Winter Olympic Games in Vancouver. All these sports events were broadcasted by the Czech television and the total amount of data was more than 100 hours of speech.

The PLP parameterization was used in the front-end module. The sampling frequency was 22 kHz and we used 27 band pass filters and 16 cepstral coefficients with delta and delta-delta features. Due to a very intense background noise, many noise reduction techniques were tested and compared (more details can be found in [4]). The best recognition results were achieved using J-Rasta techniques [1].

The basic speech unit in all our experiments was a three-state HMM with a continuous output probability density function assigned to each state. Phonetic decision trees were utilized to tie states of the triphones. Various experiments were done to ascertain the best recognition results depending on the number of clustered states and also on the number of mixtures per state. The best setting from the recognition point of view was 32 mixtures of multivariate Gaussians for each of 3018 states (see [5] for methodology).

2.2 Acoustic Model for Re-speaking

The main objective of a re-speaker is to take heed to the original speakers and re-speak their dialogues. Unlike the direct recognition of the original soundtrack, a re-speaker can achieve higher transcription accuracy and create easily readable subtitles.

An equally important task is simplifying of subtitles, if appropriate, to achieve greater intelligibility. There are many people among the target audience (for example elderly people), who have limited reading speed capability and subtitles with more than 180 words per minute are frustrating for them [7]. This simplification of subtitles or their re-formulation are in case of direct soundtrack recognition yet unsolvable problem. Similar problems appear in case of incoherent speech or overlapping speakers. In these cases, the re-speaker is expected to simplify and rephrase the original speech by clear and grammatically correct sentences with the same semantic meaning, so viewers are capable to keep up.

Each re-speaker has his/her own personal acoustic model. The front-end consisted of PLP features (19 filters, 100 frames per second) with delta and delta-delta coefficients followed by a cepstral mean subtraction. The total dimension of feature vectors was 36. Similar as in a direct recognition system the basic speech unit is a clustered three-state HMM with a continuous output probability density function assigned to each of 4,922 states. The number of mixtures depends on the specific speaker and it is about 22 mixtures of Gaussians per state. Of course, it is assumed that the speech of a re-speaker will be uttered clearly in a quiet environment.

In case of misrecognition, a re-speaker is able to erase the subtitle by a keyboard command and re-speaks it again. Other keyboard commands are used for punctuation marks and original speaker coloring (details can be found in [12]). The job of a re-speaker is quite difficult and requires special long-term training [10]. Unfortunately, not all re-speaker candidates possess the necessary skills. In our experience, a skilled re-speaker can handle maximally one to two hours of subtitling without a pause. A more detailed description of the training of re-speakers can be found in [12].

3 Vocabularies and Language Models for Individual Sports

Although the application of automatic speech recognition technology is a better solution for live captioning that surpasses the manual transcription using the keyboard, stenotype or velotype, it still causes some issues that result from the natural limitations of the recognition system. For example, it may be expected that even with very large vocabulary (we use LVCSR with more than one million words), some words needed during live captioning will always be out-of-vocabulary. The re-speaker can use his hands to write the new word or even to add it to the recognition system directly during re-speaking. However, if the vocabulary is too sparse, this may have considerable impact on quality and delay of final captions.

Since each sport has its specific terms and expressions that are commonly used during TV commentary of the sport, the best way is to use the transcriptions of the commentary of the given sport for vocabulary preparation and language model training. In addition, the transcriptions contain the names of sportsmen, their nationalities and teams involved in the match or competition. Even when these names are added to the recognition system they cannot embrace all the names of sportsmen in the matches or competitions that will be captioned in the future. This problem leads to a class-based language model, where its classes should be filled before each live captioning.

We did manual transcriptions of TV commentaries of nine sports participating in the Olympic Games in Sochi. The commentaries of the last Olympic Games in Vancouver were used. To make the corpus independent of specific Olympic Games and their participants, each name of player, competitor, team, nationality or sport place was labeled. The names which did not relate to the transcribed match or competition were not labeled, because the commentators may use them freely (for example legendary sportsmen – "Bjoerndalen", "Plyushchenko", "Jagr" etc.). Different labels were used for the names of players and competitors, for the names of competing teams or countries and participant nationalities and for the designations of sport places. In addition, each label was supplemented by a number representing one of 6 (excluding vocative) grammatical cases of the expression.

Later on, during language model training, first two labels were automatically divided into basic and possessive forms based on the word ending. Finally, taking into account the above mentioned labels instead of the individual words, six class-based language models, each with 30 classes (12 classes for players and competitors, 12 classes for teams and nationalities and 6 classes for sport places) were trained. These language models comprise also some punctuation marks (comma, colon, full stop and question mark) indicated by the re-speaker during re-speaking. Counts of tokens and labels of each type in the training corpus for individual sports are reported in Table 1.

Table 1. Statistics of the training corpus

	# of tokens	**# of players & competitors**	**# of teams & nationalities**	**# of sport places**
Alpine skiing	119,191	4,413	1,293	554
Biathlon	68,960	4,085	1,400	269
Cross-country skiing, ski jumping, Nordic combined	227,680	11,113	4,362	1,141
Figure skating	86,395	2,170	876	201
Ice-hockey	922,694	89,870	24,790	1,047
Speed skating, short track	126,233	4,953	2,590	530

Even complete transcriptions of TV commentaries from the last Olympic Games in Vancouver are not sufficient for robust language modeling with large vocabularies. That is why we used other language models based on the data from the sport domain (9M tokens from TV news transcriptions, 88M tokens from newspapers and 94M tokens from internet news) and mixed them with the class-based language models for individual sports. The weights of individual language models were set to minimize the perplexity on the transcriptions referred in Table 1 by the cross-validation technique.

While each Olympic sport has its own fans among deaf and hard-of-hearing, the opening and closing ceremonies of the Olympic Games are watched by the most viewers. At the same time, it is the most challenging task for live captioning, because the scope of the commentary is very wide and the problem of sparse vocabulary is even more significant. Unfortunately, we did not receive any detailed information about the ceremonies ahead of their captioning. While we covered the names of legendary sportsmen and sportsmen participating in the Olympic Games by using class-based language models created for individual sports, names of artists, songs and work of arts were covered only partially.

To prepare the best language model for captioning of opening and closing ceremonies of the Olympic Games, we used other data from non-sport domains, some texts concerning Russian history and the transcriptions of ceremonies from several Olympic Games as well (see [11] for more detailed information on the selection of appropriate data). Anyway, the largest part of live captioning was made by the re-speaker who has to overcome the imperfectness of the recognition system with the vocabulary containing over 1.1M words. For example, in the case of repeated out-of-vocabulary word he can simply (on-line) add such new word to the recognition system by typing it.

4 Experimental Results

The Olympic Games were broadcasted on two channels of the Czech TV. Although we captioned only some live transmissions chosen by the Czech Television, some days we had to employ all six skilled re-speakers who alternated every two hours. Totally we captioned 74 hours of live transmissions during two weeks. The live captioning was performed by specially trained re-speakers from their homes connected through the internet and ISDN network to the Czech Television [12].

As the described class-based language models specifically trained for each sport were used, the re-speaker had to fill the classes before each live captioning. We enhanced our captioning software for a class management. Before each live captioning session the re-speaker adds the names and surnames of sportsmen participating in the match or competition based on official start lists on the web pages. Since the pronunciation of foreign name depends on its original language, the re-speaker enters a pseudotranscription (a transcription using Czech letters) for each name, which is processed by common rule-based grapheme-to-phoneme conversion. We do not know the exact pronunciation that will be used by a TV commentator, so we make the best guess based on our knowledge of foreign languages and if the pronunciation then varies, the re-speaker translates the name to the right form demanded by the recognition system. Twelve language model classes are filled with names and surnames automatically declined (based on indicated sex) to 6 basic grammatical cases and 6 grammatical cases of their possessive forms. Both a surname only and a combination of name and surname are generated. Since the declension of foreign names is a very complex problem, the re-speaker may check and correct all generated items. These items are then added to the language model classes based on their grammatical case and form. The same process is used for the names of teams, nationalities and sport places. All items may be prepared in advanced, so the re-speaker only chooses predefined selection, items are added to the language model of the recognition system (including all class n-grams) and the re-speaker immediately starts captioning.

The live captioning through re-speaking of Winter Olympic sports has some specifics compared to the live captioning of political debates or TV shows. Due to the delay of live captions (about 5 seconds on average) some information may be irrelevant when the captions are displayed to the viewer, for example designations of jumps during figure skating or passes between ice-hockey players. In addition, superfluous captions may distract viewer's attention from the sport itself. Thus an additional aim of the re-speaker is to filter the information contained in sport commentaries and to deliver to the viewer all the important and interesting information in the form that do not bother nor distract the viewer. In Table 2, you can see the average number of words per minute in original and re-spoken commentaries. The re-speaking factor shows the proportion of re-spoken words against the original comment.

Table 2. Re-speaking factor

	Source of speech		**Re-speaking factor [%]**
	Original	**Re-spoken**	
	words per minute		
Alpine skiing	107	89	82.9
Biathlon	114	88	77.6
Ceremonies	81	56	69.2
Cross-country skiing, ski jumping, Nordic combined	116	91	77.8
Figure skating	50	42	84.5
Ice-hockey	91	57	62.7
Speed skating, short track	113	86	76.0

As can be seen in Table 2, there are not big differences between re-speaking factors of individual Olympic sports with the exception of ice-hockey and the opening and closing ceremony. The problem with subtitling ice-hockey by re-speaking is (already mentioned above) 5 second delay between the actual commentary and the subtitle displayed to the viewer. Some in-game situations may be irrelevant with such delay. Furthermore, redundant subtitles during the play (e.g. a possession of the puck is visible) excessively occupy the viewer. Based on the experience of deaf and hard-of-hearing viewers, re-speakers concentrate mainly on rich commentary of interesting situations not covered by the video content.

During captioning of the opening and closing ceremonies very difficult tasks had to be solved by re-speakers, because they added to the recognition system many OOV words. These often refilling OOV words, then induced a significant simplification of the re-spoken commentaries due to a lack of time.

One hour of each Olympic sport was manually transcribed both the original as well as the re-speaker soundtrack to evaluate the type of suitable captioning method. The Table 3 shows OOV, perplexity and word error rate (WER) for recognition from the original TV soundtrack and also the WER for recognition from the re-spoken soundtrack. The OOV is not shown in the second case, because it is almost zero due to adding of all OOV words to the recognition system by hand. All recognition experiments were performed with the corresponding class-based language model.

A re-speaker evaluation was performed according to the evaluation scheme proposed especially for captioning through re-speaking introduced in [12]. The first level evaluation is represented by a standard word error rate, while the semantic accuracy represents the third level of the evaluation – the overall error rate, including recognition,

Table 3. Experimental results

	Source of speech				
	Original soundtrack			Re-spoken	
	OOV [%]	Perplexity	WER [%]	WER [%]	Semantic Acc [%]
Alpine skiing	0.99	579	32.02	1.34	93.06
Biathlon	0.66	514	34.53	1.23	91.20
Ceremonies	3.58	910	45.77	2.83	83.50
Cross-country skiing, ski jumping, Nordic combined	0.90	602	32.93	1.61	92.70
Figure skating	0.87	395	35.45	1.30	97.84
Ice-hockey	0.93	545	25.97	1.03	92.77
Speed skating, short track	0.28	315	27.66	1.14	86.29

syntactic and semantic errors, in other words a percentual ability to express original ideas in the text form (by means of the recognition system).

5 Conclusion

In this article we described our efforts to build a system for captioning broadcasts of the Olympics Games in Sochi. We discussed not only the LVCSR system, which is equipped with special class-based language models, but also two principal methods for captioning live TV programs, especially sports events. The specially prepared class-based language models reduce significantly the perplexity and OOV and contribute to the robust recognition of key information such as names of players, nationalities, etc. It is evident, according to the obtained results, that we cannot use the recognition of the original soundtrack due to a very high WER. This is caused partly by a very noisy background (cheering, music, whistles, drums, etc.) in the original commentary and partly by a relatively higher OOV (opening ceremony) even if we used 1.1M words in vocabulary.

Such difficulties disappear when we use re-speaking method for captioning. In this method, an experienced and highly trained speaker re-speaks (in a quiet environment) and optionally simplifies the original commentary. Moreover, a perplexity of the task is considerably reduced, because the re-speaker rephrases original often incoherent commentary to more comprehensible sentences. The WER decreased rapidly from 33.5% (direct recognition) to 1.5% (re-speaking) on average.

References

1. Koehler, J., Morgan, N., Hermansky, H., Hirsch, H.G., Tong, G.: Integrating RASTA-PLP into speech recognition. In: IEEE International Conference on Acoustics, Speech and Signal Processing, ICASSP 1994, vol. 1, pp. 421–424 (1994)
2. Evans, M.J.: Speech Recognition in Assisted and Live Subtitling for Television. R&D White Paper WHP 065. BBC Research & Development (2003)

3. Marks, M.: A distributed live subtitling system. R&D White Paper WHP 070, BBC Research & Development (2003)
4. Psutka, J., Psutka, J.V., Ircing, P., Hoidekr, J.: Recognition of spontaneously pronounced TV ice-hockey commentary. In: ISCA & IEEE Workshop on Spontaneous Speech Processing and Recognition, pp. 169–172 (2003)
5. Psutka, J.V.: Robust PLP-Based Parameterization for ASR Systems. In: SPECOM, International Conference on Speech and Computer, pp. 509–515 (2007)
6. Ortega, A., Garcia, J.E., Miguel, A., Lleida, E.: Real-Time Live Broadcast News Subtitling System for Spanish. In: 10th Annual Conference of the International Speech Communication Association, pp. 2095–2098. Causal Productions (2009)
7. Romero-Fresco, P.: More haste less speed: Edited versus verbatim respoken subtitles. Vigo International Journal of Applied Linguistics, University of Vigo, Number 6, 109–133 (2009)
8. Trmal, J., Pražák, A., Loose, Z., Psutka, J.: Online TV Captioning of Czech Parliamentary Sessions. In: Sojka, P., Horák, A., Kopeček, I., Pala, K. (eds.) TSD 2010. LNCS (LNAI), vol. 6231, pp. 416–422. Springer, Heidelberg (2010)
9. Bordel, G., Nieto, S., Penagarikano, M., Rodriguez-Fuentes, L.J., Varona, A.: Automatic Subtitling of the Basque Parliament Plenary Sessions Videos. In: 12th Annual Conference of the International Speech Communication Association, pp. 1613–1616. Causal Productions (2011)
10. Pražák, A., Loose, Z., Psutka, J., Radová, V.: Four-phase Re-speaker Training System. In: SIGMAP, International Conference on Signal Processing and Multimedia Applications, pp. 217–220 (2011)
11. Švec, J., Hoidekr, J., Soutner, D., Vavruška, J.: Web text data mining for building large scale language modelling corpus. In: Habernal, I., Matoušek, V. (eds.) TSD 2011. LNCS (LNAI), vol. 6836, pp. 356–363. Springer, Heidelberg (2011)
12. Pražák, A., Loose, Z., Trmal, J., Psutka, J.V., Psutka, J.: Novel Approach to Live Captioning Through Re-speaking: Tailoring Speech Recognition to Re-speaker's Needs. In: 13th Annual Conference of the International Speech Communication Association, pp. 1370–1373. Curran Associates, Inc., Red Hook (2012)

Language Resources and Evaluation for the Support of the Greek Language in the MARY Text-to-Speech

Pepi Stavropoulou[1,2], Dimitrios Tsonos[1], and Georgios Kouroupetroglou[1]

[1] National and Kapodistrian University of Athens,
Department of Informatics and Telecommunications, Greece
{pepis,dtsonos,koupe}@di.uoa.gr
[2] University of Ioannina, Department of Philology, Greece

Abstract. The paper outlines the process of creating a new voice in the MARY Text-to-Speech Platform, evaluating and proposing extensions on the existing tools and methodology. It particularly focuses on the development of the phoneme set, the Grapheme to Phone (GtP) conversion module and the subsequent process for generating a corpus for building the new voice. The work presented in this paper was carried out as part of the process for the support of the Greek Language in the MARY TtS system, however the outlined methodology should be applicable for other languages as well.

Keywords: MaryTTS, Greek Language, Grapheme to Phone, Diphone Database.

1 Introduction

Among the most important factors for determining the success of a Text to Speech system is the quality of the lexicon, the Grapheme to Phone (GtP) module, and most importantly the quality of the speech database. The latter is conditioned on the set of phonemes used for defining the pronunciations and consequently the set of diphones for the TtS language, as well as the corpora and methods used for selecting the final set of utterances to be recorded by the voice talent. The final set should ideally provide maximum coverage of the possible diphones in a language and also accommodate important linguistic and prosodic events, such as question or negation contours.

The ultimate goal is the generation of a corpus that can effectively and adequately support the development of a more "context-aware" voice, which could be used in contexts such as natural language spoken dialogue systems, companion robots and so forth. Most generic TtS systems are trained on neutral, read speech databases, which both differ in style and lack pragmatic, "context-sensitive" events often occurring in dialogue [1]. Dialogue, on the other hand, poses specific requirements on the corpus used for developing the TtS voice. At minimum there should be an adequate representation of different types of questions, list structures, and deaccented materials as a result of early, non-default focus position [2,3].

As a first step we used the existing tools and processes for building a new language in the MARY TtS platform [4]. More specifically, in this paper we describe and evaluate the tools and processes for building the speech database, so as to ultimately extend them in the line of thought described above.

P. Sojka et al. (Eds.): TSD 2014, LNAI 8655, pp. 523–528, 2014.

2 The MARY TtS – New Voice Support Process

The MARY TtS system follows the Client-Server (CS) model. Server side executes the: text preprocessing/normalization, natural language processing, calculation of acoustic parameters and synthesis. The client sends the server the requests including the text to be processed and the parameters for the text handling by the server side. The system is: multi-threaded, due to CS implementation, flexible (modular architecture) and transparent and understandable as possible using XML-based, state-of-the-art technologies such as DOM and XSLT [4]. MARY TtS also includes several tools, in order to easily add support for a new language and build Unit Selection and HMM-based synthesis voices [5,6]. The tools can be grouped into: a) New Language Support Tools (NLST) and b) Voice Creation Tools (VCT). Using NLST we are able to create a new language support, providing the minimum NLP component for MARY TtS, and a text corpus in order to support the next stage using VCT, for the implementation of Unit Selection or HMM-based voice.

Greek language currently is not supported in MARY TtS. Following are the basic, necessary steps for baseline support of a new language in the MARY framework [5]:

- Define the set of allophones for the new language.
- Build the pronunciation lexicon, train Letter to Sound rules (i.e. GtP module) based on handcrafted transcriptions and define a list of functional words for the development of a primitive POS tagger.
- Prepare recording script, using Wikipedia corpus in combination with provided tools.
- Record the prompts, for the speech corpus creation based on Wikipedia corpus.
- Run Voice Import tools to build a Unit Selection or HMM-based voice.

3 Allophone Set Definition

The first step for developing a new language is the definition of the language's allophone set. The set is then used for the development of the grapheme to phone module and the generation of the diphone database. In general, the set of allophones should provide an adequate representation of the phonemic structure of the language, while at the same time the number of phones is kept manageable. Standard principles for deriving the set of allophones are the minimal pair distinction principle (cf. e.g. [7]) and phonetic similarity [8]. The latter ensures that same half phone units within diphones join well together.

Accordingly, the set defined for the Greek language was based on the state of the art descriptions of the language's phonemic inventory [9]. The selection of allophones was further conditioned on their systematic – and hence predictable – behavior which also allowed for a consistent, standardized representation in the lexicon. In this line of thought, within word freely alternating allophones such as [mb] and [b] were denoted by a single abstract phone. The choice between the two pronunciations is often speaker dependent subject to characteristics such as speaker's age, origin and dialect. On a later step, this abstract pronunciation could be specified to meet individual's speakers idiolect. On the other hand, mutually exclusive allophones such as [k] and [c] had distinct representations. Approximants [ts] and [tz] were represented as a single phoneme, contrary to [ks] and [ps], which correspond to two distinct phones and are thus not represented as a single unit in the phones set. Finally, contrary to previous representations for the Greek Language [10], no distinction was made between stressed and unstressed vowels in the inventory (i.e. no distinct allophones). We thus reduced the size of the inventory and represented stress as an abstract separate linguistic entity instead, affecting the complete syllable and consequent unit selection. Table 1 illustrates the final set of allophones used.

Table 1. Greek Allophone Set

Symbol	Example Word	Symbol	Example Word
a	anemos "wind"	G	Gala "milk"
e	eTnos "nation"	J	jenos "origin"
i	isos "maybe"	x	xara "joy"
o	oli "all"	C	Ceri "hand"
u	urios "favourable"	m	moni "alone"
p	poli "city"	M	Mazo "resemble"
b	bala "ball"	n	neos "new"
t	telos "end"	N	NoTo "feel"
d	dino "dress"	r	roi "flow"
k	kozmos "world"	R	tReno "train"
c	cima "wave"	l	lemoni "lemon"
g	gol "goal"	L	Lono "melt"
q	qiNa "bad luck"	ts	tsiGaro "cigarette"
f	filos "friend"	dz	dzami "glass"
v	velos "arrow"	W	laWo "glow"
T	Telo "want"	V	meVa "mint"
D	Dino "give"	Y	aYaLa "hug"
s	stelno "send"	Q	aQizo "touch"
z	zoi "life"	-	-

4 Grapheme to Phone Conversion

The Grapheme to Phone (GtP) module generates the pronunciation of words based on graphemes, the letters they are comprised of. Typically the word is first looked up in the pronunciation lexicon and if not found, its pronunciation is automatically generated by the GtP module. In the Open Mary platform the grapheme to phone rules are automatically learnt from phonetically transcribed data. The technique is based on CART decision trees that use questions on each grapheme's context (i.e. preceding and subsequent letters/graphemes). In our case, questions based on a context of two graphemes both subsequent and following turned out to yield the best results.

A set of 4900 words was initially used for learning the GtP rules. 3242 words and their corresponding pronunciation were automatically extracted from an annotated, transcribed set of 600 newspaper sentences, which were developed as part of the RHETOR project [11]. The rest of the words were manually added, to account for grapheme sequences that did not occur in the original set and corresponded to various phonological rules affecting pronunciation. Furthermore, transcriptions were enriched to include syllable structure and stress patterns. Syllabification adhered to the maximum onset principle. Following is a typical entry from the lexicon: πληροφορική | p l i - r o - f o - r i - 'ci

The current model had a 5,2% word error rate on a test set consisting of 500 frequently occurring Greek words. Word frequencies were based on statistics of use from the Greek Wikipedia repository. To improve the model's performance we reduced the "minLeafData" option to 35 from 100 which was the default value. The "minLeafData" value determines the minimum number of instances that have to occur in at least two subsets induced by split, i.e. at a leaf node. 35 proved to be a safe threshold, given that Greek has a rather regular orthography. In general, though, in languages such as e.g. Greek, German or Spanish, where the grapheme to phone relationship is highly structured and the orthography regular, rule based approaches can be just as – or even more – effective as data driven techniques.

5 Recording Script Preparation – Diphone and Intonation Coverage

The Greek Wikipedia repository [12] and the basic NLP module of MARY TtS were used for the recording script preparation. The size of Wikipedia xml file was 824.5MB. The file was processed by the sentence cleanup procedure, and stripped-off from any annotation/xml tag, keeping only the text content. Next, an automated cleanup procedure of Wikipedia content was executed, in order to exclude sentences with unknown words or strange symbols and to extract only reliable sentences with the optimum diphone coverage.

Of course, the automated procedure cannot completely remove all unreliable sentences, thus a manual selection is mandatory. We excluded sentences, which are missed by the automated procedure, containing e.g. duplicate sentence entries, any special/unknown characters, non-Greek words/characters, difficult to be read aloud by the voice talent during voice recording.

The tools for the recording script preparation provide statistical results at the end of each process. Table 2 presents the initial and final corpus diphone coverage. Initial corpus consists of 13,876 sentences, covering a total number of 1,990,345 diphones and 834 different diphones. It should be noted that not all diphones are truly possible given language specific phonotactic rules. The final manually selected set is comprised of 1,243 sentences and drastically reduced diphone coverage (524 different diphones).

With regards to the coverage of different sentence types (i.e. questions, negation, list structures), affirmative declaratives comprised the vast majority of the selected sentences, while wh-questions, polar questions and negation comprised a mere 3.18%, 1.62% and 7.78% of the total sentences respectively. Accordingly, diphone coverage in these contexts was significantly limited, which is expected to have a negative effect on the final output of the synthesizer, especially in the case of unit selection synthesis

Table 2. Initial diphone coverage distribution and corresponding number of sentences for the Wikipedia Corpus before and after manual selection respectively

	Initial Coverage	Manual Selection
Number of Sentences	13,876	1,243
Diphone Coverage	834	524

whereas there is often limited signal processing modification. Furthermore, at this point the simple intonation coverage algorithm makes no distinction between different types of questions or differences in polarity. Nevertheless, in many languages, e.g. English and Greek among others, wh-questions and polar questions have distinct phonological melodies; it is a fundamental distinction that should therefore be modeled. We recorded the final sentence set (approximately 90 minutes of recordings, 16 bit – 44.1 KHz sampling rate) in order to create a speech corpus for the later voice import procedure. Voice import tools (for a detailed description see [13]) were executed, changing the sampling rate to 16 KHz, with their default configurations and a Unit Selection Voice and a HMM-based Voice were created.

6 Conclusions

In conclusion, diphone coverage seems to be rather low, especially after manual selection. The number of distinct diphones drops to 524 which is almost half the number achieved in e.g. [14]. In their study [14], the construction of a database for a Greek TtS system is presented, which achieves coverage for 813 diphones. The initial corpus included 300 most frequent diphones. They inserted new sentences containing the missing diphones, achieving maximum coverage. In our case, since manual selection is a post process, it is not accounted for by the greedy algorithm used for the initial utterance selection.

Furthermore, only the basic intonation coverage is achieved. Wikipedia does not seem to be the optimal resource for ensuring a sufficient coverage for context-aware dialogue based applications. It is notable that most instances of questions occurred in the comments section rather than the main article. In addition, the simple "punctuation based" features used in the greedy algorithm should be enriched with additional features (e.g. wh-words, negation particles) to identify different types of sentences corresponding to different prosodic realizations. The selection process would be more complete with the use of additional features such as utterance position, position relative to the operator (wh-word, negation particle) and so forth, always taking into account the issues of data sparsity and inventory size.

Accordingly, we need to explore other types of databases containing actual transcribed dialogues (examples are e.g. the Switchboard corpus for English or the Corpus of Greek Text [15] for Greek). Nevertheless, human-human dialogues may not be as restricted and still lack events that are particular to the "human-machine dialogues" genre (e.g. lists). Therefore, we may need to resort to a hybrid approach where the "normal", real life text material is supplemented with hand crafted material, designed to ensure sufficient coverage.

Acknowledgments. This research has been co-financed by the European Union (European Social Fund – ESF) and Greek national funds through the Operational Program "Education and Lifelong Learning" of the National Strategic Reference Framework (NSRF) under the Research Funding Project: "THALIS-University of Macedonia-KAIKOS: Audio and Tactile Access to Knowledge for Individuals with Visual Impairments", MIS 380442.

References

1. Syrdal, A., Kim, Y.-J.: Dialog speech acts and prosody: Considerations for TTS. In: Proc. of the Speech Prosody, Brazil (2008)
2. Huang, X., Acero, A., Hon, H.W.: Spoken Language Processing: A Guide to Theory, Algorithm and System Development. Prentice Hall PTR (2001)
3. Stavropoulou, P., Spiliotopoulos, D., Kouroupetroglou, G.: Where Greek Text to Speech Fails. In: Proc. of the 11th International Conference on Greek Linguistics, Rhodes (September 2013)
4. Schröder, M., Trouvain, J.: The German Text-to-Speech Synthesis System MARY: A Tool for Research, Development and Teaching. International Journal of Speech Technology 6(4), 365–377 (2003)
5. Pammi, S., Charfuelan, M., Schröder, M.: Multilingual Voice Creation Toolkit for the MARY TTS Platform. In: LREC 2010, Malta (2010)
6. Schröder, M., Charfuelan, M., Pammi, S., Steiner, I.: Open source voice creation toolkit for the MARY TTS Platform. In: Proc. Interspeech, Florence, Italy (2011)
7. Ladefoged, P., Johnson, K.: A Course in Phonetics. Wadsworth, Cengage Learning Inc., Boston (2010)
8. Taylor, P.: Text to Speech Synthesis. Cambridge University Press, Cambridge (2009)
9. Arvaniti, A.: Greek Phonetics: The State of the Art. Journal of Greek Linguistics 8, 97–208 (2007)
10. Fotinea, S.-E., Tambouratzis, G.: A Methodology for Creating a Segment Inventory for Greek Time Domain Speech Synthesis. International Journal of Speech Technology 8(2), 161–172 (2005)
11. Fourli-Kartsouni, F., Slavakis, K., Kouroupetroglou, G., Theodoridis, S.: A Bayesian Network Approach to Semantic Labelling of Text Formatting in XML Corpora of Documents. In: Stephanidis, C. (ed.) Universal Access in HCI, Part III, HCII 2007. LNCS, vol. 4556, pp. 299–308. Springer, Heidelberg (2007)
12. Wikipedia, http://dumps.wikimedia.org/elwiki/latest/elwiki-latest-pages-articles.xml.bz2 (accessed May 2013)
13. Voice Import Tools Tutorial: How to build a new Voice with Voice Import Tools, http://mary.opendfki.de/wiki/VoiceImportToolsTutorial
14. Fotinea, S.-E., Tambouratzis, G., Carayannis, G.: Constructing a segment database for greek time domain speech synthesis. In: Proc. of EUROSPEECH 2001 Scandinavia, 7th European Conference on Speech Communication and Technology, 2nd INTERSPEECH Event, Aalborg, Denmark, September 3-7 (2001)
15. Corpus of Greek Text, http://www.sek.edu.gr

Intelligibility Assessment of the De-Identified Speech Obtained Using Phoneme Recognition and Speech Synthesis Systems

Tadej Justin, France Mihelič, and Simon Dobrišek

Faculty of Electrical Engineering, University of Ljubljana,
1000 Ljubljana, Tržaška 25, Slovenia
{tadej.justin, france.mihelic,simon.dobrisek}@fe.uni-lj.si
http://luks.fe.uni-lj.si

Abstract. The paper presents and evaluates a speaker de-identification technique using speech recognition and two speech synthesis techniques. The phoneme recognition system is built using HMM-based acoustical models of context-dependent diphone speech units, and two different speech synthesis systems (diphone TD-PSOLA-based and HMM-based) are employed for re-synthesizing the recognized sequences of speech units. Since the acoustical models of the two speech synthesis systems are assumed to be completely independent of the input speaker's voice, the highest level of input speaker de-identification is ensured. The proposed de-identification system is considered to be language dependent, but is, however, vocabulary and speaker independent since it is based mainly on acoustical modelling of the selected diphone speech units. Due to the relatively simple computing methods, the whole de-identification procedure runs in real-time.

The speech outputs are compared and assessed by testing the intelligibility of the re-synthesized speech from different points of view. The assessment results show interesting variabilities of the evaluators' transcriptions depending on the input speaker, the synthesis method applied and the evaluators capabilities. But in spite of the relatively high phoneme recognition error rate (approx. 19%), the re-synthesized speech is in many cases still fully intelligible.

Keywords: Voice de-identification, phoneme recognition, speech synthesis, diphone speech units, HMM modelling, intelligibility evaluation.

1 Introduction

De-identification can be defined as the process of concealing the identities of individuals captured in a given set of data (e.g., video, audio or text) for the purpose of protecting their privacy [1]. De-identification techniques should not only be capable of preventing humans from recognizing subjects in the multi-media content, they should also be able to conceal identities from automated recognition systems, such as face recognition [2], speaker verification and others.

Many different techniques to obtain the speaker's de-identified voice have already been reported. We can divide them into techniques using a voice-degradation approach,

P. Sojka et al. (Eds.): TSD 2014, LNAI 8655, pp. 529–536, 2014.

a voice-conversion approach or a voice-morphing approach [3,4,5]. These techniques can be used in real applications, but each with their own limitations, which have to be considered. The voice obtained with the voice degradation technique often runs as an on-line process, but the resulting output speech is more or less unnatural. The aim in the voice conversion approach is to assess the mapping transformation from the input speaker to the target speaker. For such an operation the access to the input speaker and the target speaker audio recordings and/or acoustic models must to be provided in advance. With the goal to obtain an on-line speaker de-identification system, we propose and investigate the intelligibility of a language-dependent method, which is lexical independent, but still uses the possibility of a speech synthesis system. Despite the fact that it is not possible to obtain robust automatic word recognition only from a phoneme-recognition system (a phonetic typewriter) the evaluation of automatically recognised phonemes shows that errors made by the recognition system are often realized as substitutions between phonetically similar phonemes. By listening to an utterance with such substitution errors, the listener with a suitable linguistic knowledge can still recognise most of the uttered words and understand the meaning. If we assume that the speech synthesis system uses its own target speaker's voice, which is different from the input speaker, then the output voice can be seen as a de-identified input voice.

The system is described more precisely in the next section. Since the intelligibility of the speech generated using the presented approach is questionable, our experiment was focused on a subjective assessment of the intelligibility of the de-identified speech. We describe and comment on the evaluation procedure in Sections 3 and 4. We summarize and further discuss our experimental results in the conclusion, where we also propose further steps for improvements.

2 System Description

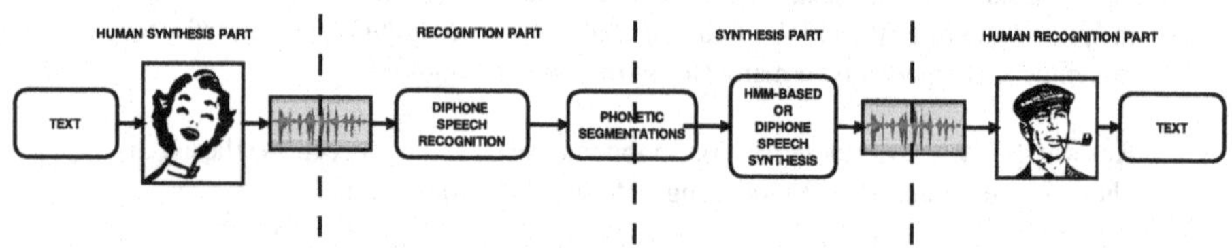

Fig. 1. Speaker de-identification system evaluation scheme

The experiments were conducted with two different configuration set-ups (Figure 1). In each set-up we used our own implementation of the Slovene phoneme-recognition system [6]. The system was developed with the use of Hidden Markov Models (HMMs) and bigram phoneme language modelling. In our implementation, the basic acoustic units were presented as context-dependent diphones, called bi-diphones. These diphone units are defined as speech segments containing mostly a transition between the two neighbouring phonemes and were in the beginning introduced in the field of speech synthesis [7]. Bi-diphone recognition provides better results in comparison with the classic triphone HMM-based recognition on the same evaluation test corpora that was used for our present evaluation study [8].

The Slovene speech synthesis systems used in the experimental set-ups have also been developed in our institution. Each of the two speech synthesis systems was trained with a different speech database; therefore, the synthesised voice from one or other system has different target-speaker characteristics. In the first set-up the de-identified voice was obtained with the use of a diphone speech synthesis system [9]. In parallel, the second set-up represents the use of a HMM-based speech synthesis system [10] to achieve the same goal. A second system was developed with the use of the HTS toolkit, version 2.2 [11], where contextual quin-phones were used for the base units. It is well known and also proved in previous works, such as [12], that the synthetic speech synthesised with the HMM-based approach is more natural in comparison with the speech obtained with the diphone synthesis system. Nevertheless, the comparison of intelligibility of such speech systems still remains an open issue, especially in our case where diphone speech units were also the base units in our recognition system.

The output of a bi-diphone recognition system represents a string of Slovene allophones with their associated estimations of duration and provides the input to the speech synthesis systems. Pitch estimation and loudness can also be obtained from the recognizer, but since the output speech signal has to present the de-identified voice, this was discarded from subsequent processing. Speech synthesis systems could also not take into account any additional prosodic modelling based on the word units, which are usually obtained in text-to-speech systems.

3 System Evaluation Setup

The proposed system was tested on the Slovene speech database GOPOLIS [13], which contains the speech signals (read speech) and their transcriptions from 50 (25 male and 25 female) speakers. The word-recognition system using this database was primarily developed in our previous work [14], with the aim being to build an automatic speech dialogue system for querying flight information. Using the standard protocol, the training set contained recordings from first 18 male and 18 female speakers and a test set of the remaining 7 male and 7 female speakers' speech recordings. The training part of the database was used to train our bi-diphone speech recognition system.

For the voice de-identification system's evaluation we randomly picked 28 test sentences from the test set with the following limitations:

- Only two sentences from the same speaker.
- Each sentence had to be between 5 and 8 words long.

With such limitations we ensured that, on one hand, the synthesised sentences are not too short and, in addition, are not too simple to understand. And on the other hand, the sentences were not allowed to be too long and consequently easy to forget, since the evaluator's task was to remember and transcribe the recognised words from the artificial speech utterance.

Using both limitations we appropriately distributed the test sentences between all the different test speakers. The final evaluation set consisted of $2 \cdot$ (7 different male) and

$2 \cdot$ (7 different female) input sentences, resulting in 56 (28 diphone synthesis, 28 HMM synthesis) different synthesised utterances.

All the evaluation tests were made with our own web-based evaluation system. Even though the application can be accessed from every computer with an internet connection and an Adobe–Flash–enabled web browser, the evaluation process took place in a controlled environment at the Faculty of Electrical Engineering the University of Ljubljana. During the evaluation process all the evaluators wore headphones. The evaluation test was successfully completed by 26 evaluators. All the evaluators were 3rd–year university program students from the University of Ljubljana, Faculty of Electrical Engineering. The evaluators had a limited, basic knowledge of speech technologies.

Before the evaluation task was started a brief explanation of the developed system and the evaluation process was introduced to the evaluators. Between the introduction also the semantic-domain (flight-service information queries) of the evaluation utterances was explained. Each evaluator transcribed 7 sentences synthesised with the diphone speech synthesis system and 7 sentences synthesised with the HMM–based synthesis system. The evaluators were divided into two groups. The first group with 13 evaluators evaluated the first randomly picked sentence from each speaker in the test dataset. The second group evaluated the sentences that were not evaluated by the first group of evaluators. With such an evaluation set-up we ensured that each evaluator listened to each sentence only once and also that all the evaluator's transcriptions belong to different input speakers. With such an evaluation process we obtained $(13 + 13) \cdot 14 = 364$ transcriptions.

The evaluated system for speaker de-identification can produce two kinds of errors. The first one is related to the bi-diphone speech recognition system, and it can be measured as the Phoneme Error Rate (PER). The second one can be presented as the output system error. This error results as a combination of influences from errors made by the speech recognition system, the performance of the synthesis system and the evaluator's capabilities. It can be measured from the analysis of the evaluator's transcriptions in relation to the reference sentence's transcriptions as the Word Error Rate (WER).

The error rate (ER) is defined from the accuracy (A)

$$ER = 1 - A \quad , \tag{1}$$

where the accuracy is determined as in [15] from the number of correctly recognized units (N_{cor}), the number of insertion errors (I) and the number of reference units (N_{ref})

$$A = \frac{N_{cor} - I}{N_{ref}} \quad .$$

4 Experimental Results

The evaluation analysis results can provide some interesting conclusions about the human ability to recognize words based on the acoustic representation of phoneme

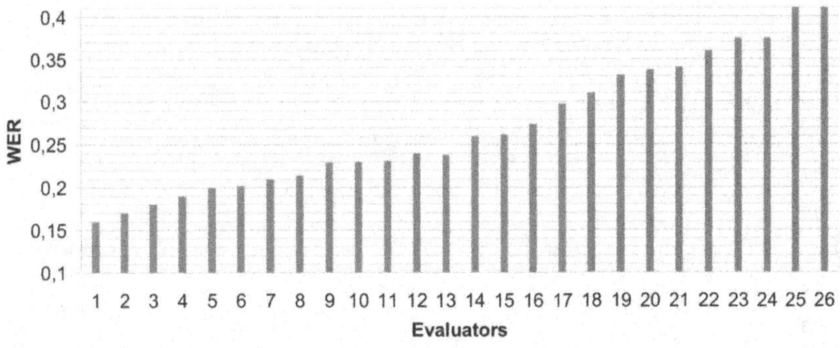

Fig. 2. Evaluators average word error rate (WER) for all the listening tests

recognition. Such a process of human perception is comparable to the process used in automatic systems for speech recognition, where we commonly realize it with the help of a phonetic word lexicon, sentence syntax and semantic language modelling.

Figure 2 presents the variance of 26 evaluators transcription WER capabilities. We can conclude that the results are strongly dependent on the the evaluator's identity. A detailed analysis shows that there are many transcriptions from the same evaluator's identity that are fully recognized (WER = 0) or with a very low WER, although - in terms of PER - not even one input sentence was completely correctly recognized by the bi-diphone speech recognition system. We can also observe this phenomenon in Figure 3.

Figure 3 also shows the trends of the average transcriptions WER for input sentences depending on the recognition PER. As expected we can observe positive linear regressions. From there we can also notice the difference between the average WER depending on the speech synthesis system. In the present voice de-indentification system the diphone speech synthesis system is more intelligible than the HMM-based speech synthesis system. Significant differences in the WER can also be confirmed from Table 1, displaying the average WER for different types of speech synthesis.

Table 1. Average word error rate (WER) for different types of speech synthesis

Number of listening tests	WER HMM	WER Dif
182	0.33	0.21

If we compare the obtained WER results with the WER from the words-recognition system presented in our previous work [14] describing a spoken dialogue system for air-flight queries, where the same acoustic models were used, we obtained - at first sight - a surprising paradox. The average WER 21 % obtained with the diphone speech synthesis

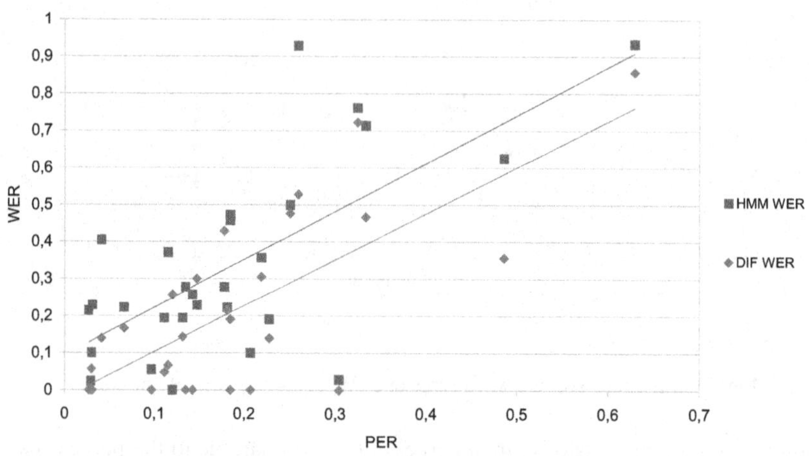

Fig. 3. Average word error rate WER per utterance for HMM synthesis and Diphone synthesis depending on the recognition phoneme error rate (PER)

Table 2. Correlations between (WER) and different types of phoneme recognition errors: deletions (D), substitutions (S), insertions (I) and phoneme error rate PER

	D	S	I	D + S + I	PER
WER HMM	0.71	0.40	0.34	0.71	0.69
WER Dif	0.58	0.53	0.40	0.76	0.73

is significantly higher than the WER 8% obtained from the dialogue system, thus the machine speech recognition system outperforms the human recognition abilities? In this case this apparent phenomenon can be explained by the fact that the word-recognition system used a relatively small word lexicon (829 words) and a syntax model with a very low perplexity (5.7) [13].

Based on the evaluation results we can roughly estimate the value of the PER from input sentence where the evaluators could still understand the de-identified synthesised utterance. Certainly, we do not need a 0% PER. In fact, in some occurrences the evaluators achieved the correct transcription (WER = 0), although the recognition PER was near to 50%. However the WER and PER are strongly correlated. If we further investigate the phoneme recognition errors' influences on the WER (Table 2), we can speculate that the phoneme deletions error have the major influence on de-identified voice intelligibility, especially for the HMM based synthesis (0.71 estimated correlation) and insertions are the least critical. It also seems that the error-measure defined as $D + S + I$ is more descriptive in our case than the usually obtained ER

Table 3. Average evaluators word error rate (WER) for different types of speech synthesis and average phoneme error rate (PER) for all test utterances, depending on the speaker's gender

gender	WER HMM	WER Dif	PER
female	0.44	0.29	0.23
male	0.23	0.13	0.14

measure, defined in (1). On the other hand, the PER and consequently the WER results are most likely also dependent on input the speaker's voice. For instance, we can see (Table 3) that the PER and, consequently, the WER are speaker–gender dependent.

5 Conclusion

In this article we propose an approach for a speaker-de-identification system and its evaluation with the use of subjective listening tests for intelligibility measuring. The evaluation results showed major differences between the transcriptions of the individual evaluators. Such results are not desirable, but also not critical for the evaluation of our proposed system. From the results it is evident that every result obtained from the subjective evaluations is strongly related to the evaluator's motivation in participating in the evaluation process. On the other hand it should be noted that in some cases due to a low recognition phoneme error rate, some sentences were not understandable to all the evaluators at all. This fact suggests the questionable usage of such a system in real applications.

Future improvements can be expected with the use of speaker-adaptation techniques in a speech recognition system, with the inclusion of additional application-dependent word spotting or even the replacement of the phoneme speech recognition system with a word recognition system. A possible solution to achieve better results could also be the use of pre-defined user training. Since it is an online system, the instant system de-indentified voice output could be used for training as an acoustic feedback.

Acknowledgments. The work presented in this paper was supported in parts by the national research program P2-0250(C) Metrology and Biometric Systems, and the European Union's Seventh Framework Programme (FP7-SEC-2011.20.6) under grant agreement number 285582 (RESPECT). The authors additionally appreciate the support of COST Actions IC1106 and IC1206.

References

1. Ribarić, S., et al.: De-identification for privacy protection in mutlimedia content. COST Action MOU (2013)
2. Poh, N., Štruc, V., Pavešić, N., et al.: An evaluation of video-to-video face verification. IEEE Transactions on Information Forensics and Security 5(4), 781–801 (2010)
3. Stylianou, Y.: Voice Transformation: A survey. In: ICASSP 1999, pp. 3585–3588 (1999) ISSN 1520-6149

4. Pfitzinger, H.R.: Unsupervised Speech Morphing between Utterances of any Speakers. In: Cassidy, S., Cox, F., Mannell, R., Palethorpe, S. (eds.) Proceedings of the 10th Australian International Conference on Speech Science & Technology, pp. 545–550 (2004)
5. Qin, J., Toth, A.R., Schultz, T., Black, A.W.: Speaker de-identification via voice transformation. In: IEEE Workshop on Automatic Speech Recognition & Understanding, ASRU 2009, pp. 529–533 (2009) ISBN 978-1-4244-5478-5
6. Dobrišek, S., Mihelič, F., Pavešić, N.: Acoustical modelling of phone transitions: biphones and diphones - what are the differences? In: Olaszy, G., Nemeth, G., Erdohegyi, K. (eds.) Proceedings of Eurospeech 1999, vol. 3, pp. 1307–1310 (1999)
7. O'Shaughnessy, D., Barbeau, L., Bernardi, D., Archambault, D.: Diphone speech synthesis. Speech Communication 7(1), 55–65 (1988)
8. Dobrišek, S.: Analysis and Recognition of Phones in Speech Signals, PhD Thesis, University of Ljubljana (2001)
9. Žganec Gros, J., Pavešić, N., Mihelič, F.: Text-to-Speech synthesis: A complete system for the Slovenian language, vol. 5(1), pp. 11–19. CIT (1997) ISSN 1330-1136.
10. Pobar, M., Justin, T., Žibert, J., Mihelič, F., Ipšić, I.: A Comparison of Two Approaches to Bilingual HMM-Based Speech Synthesis. In: Habernal, I., Matousek, V. (eds.) TSD 2013. LNCS (LNAI), vol. 8082, pp. 44–51. Springer, Heidelberg (2013)
11. Zen, H., Nose, T., Yamagishi, J., et al.: The hmm-based speech synthesis system (hts) version 2.0. In: Proc. of Sixth ISCA Workshop on Speech Synthesis, pp. 294–299 (2007)
12. Vesnicer, B., Mihelič, F.: Evaluation of the Slovenian HMM-based speech synthesis system. In: Sojka, P., Kopeček, I., Pala, K. (eds.) TSD 2004. LNCS (LNAI), vol. 3206, pp. 513–520. Springer, Heidelberg (2004)
13. Mihelič, F., Žganec Gros, J., Dobrišek, S., Žibert, J., Pavešić, N.: Spoken language resources at LUKS of the University of Ljubljana. Int. J. Speech Technol. 6(3), 221–232 (2003)
14. Ipšić, I., Mihelič, F., Dobrišek, S., Gros, J., Pavešić, N.: A Slovenian spoken dialog system for air flight inquiries. In: Olaszy, G., Nemeth, G., Erdohegyi, K. (eds.) Proceedings of Eurospeech 1999, vol. 6, pp. 2659–2662 (1999)
15. Young, S.J., Evermann, G., Gales, M.J.F., et al.: The HTK Book, version 3.4.1. Cambridge University Engineering Department, Cambridge (2009)

Dialogue

"**Dialogue**: a discussion between two or more people or groups, especially one directed towards exploration of a particular subject or resolution of a problem: *interfaith dialogue*."

NODE (The New Oxford Dictionary of English), Oxford, OUP, 1998, page 509.

Referring Expression Generation: Taking Speakers' Preferences into Account

Thiago Castro Ferreira and Ivandré Paraboni

School of Arts, Sciences and Humanities, University of São Paulo (EACH / USP)
Av. Arlindo Bettio, 1000 - São Paulo, Brazil
{thiago.castro.ferreira,ivandre}@usp.br

Abstract. We describe a classification-based approach to referring expression generation (REG) making use of standard context-related features, and an extension that adds speaker-related features. Results show that taking speakers' preferences into account outperforms the standard REG model in four test corpora of definite descriptions.

Keywords: Natural Language Generation, Referring Expressions.

1 Introduction

In Natural Language Generation (NLG), Referring Expression Generation (REG) is the computational task of uniquely describing a given target object r within its context C.[1] For instance, given a visual context as in Figure 1, we may compute descriptions of a particular entity r as 'the cone next to a box', 'the cone closest to the centre' etc., among other possibilities.

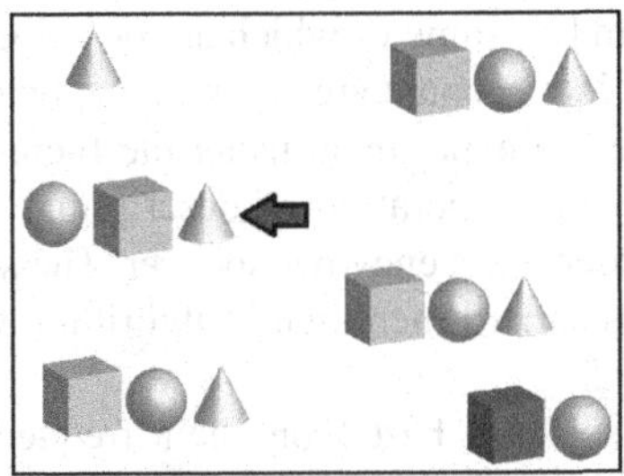

Fig. 1. A simple visual context, with a target object pointed by the arrow.

The input to the REG task is a set of context objects C (including the intended referent r) and their possible semantic properties represented as attribute-value pairs, as in (*type-cone*) or (*colour-grey*). The goal is to produce a list L of properties of r so

[1] We presently focus on content selection. For surface realisation issues, see, e.g., [11].

P. Sojka et al. (Eds.): TSD 2014, LNAI 8655, pp. 539–546, 2014.

as to distinguish r from all other objects (or distractors) in C [2]. Moreover, as in many NLG tasks, it is generally assumed that L should reflect the human choice of contents as closely as possible.

REG implementations range from purely algorithmic proposals (e.g., [2,10]), to the more recent use of machine learning techniques [5]. Regardless of implementation, however, the output description tends to be the same (i.e., fixed) for each input context, and this is so despite evidence that speakers may differ from each other in their choices of reference strategy. In other words, many existing REG algorithms do not pay regard to the fact that different people may refer to the same object in different ways.

Differences across speakers are observed in many aspects of language production. For instance, speakers may vary the choices of referential attributes, or the degree of reference specification. Regarding attribute choice, the work in [13], for instance, describes an experiment in which 15% of speakers always made use of relational descriptions (e.g., 'the cone *next to* a box'), whereas 31% never used them (e.g., favouring atomic properties as in 'the *middle* one ').

As for reference specification, the experiment in [12], for instance, shows that 12% of the speakers always chose a minimal description (e.g., 'the cone next to the box'), whereas 16% always overspecified (e.g., 'the cone next to the *grey* box', in a context in which colour was redundant).

The reasons why reference strategies may vary across speakers are beyond the scope of this work. In what follows, we simplify and assume that these differences represent linguistic preferences at some level, and we focus instead on the use of machine learning techniques that take advantage of speaker-related information, an issue that few approaches to REG have taken into account to date.

2 Background

The NLG literature presents a wide range of approaches to REG that are mainly based on the Incremental algorithm [2], some of which are described in [9]. Studies that take speaker-related information into account are summarized below.

The work in [6] describes an experiment using the Incremental algorithm and one of its extensions applied to the generation of descriptions in two dialogue corpora. Attributes are selected in order of recency per speaker. The speaker-dependent version is shown to outperform the standard Incremental algorithm with a fixed preference list for attribute selection.

In [1,4], attributes are selected based on their frequency per speaker. In both studies, the use of speaker-related information is once again shown to outperform the Incremental algorithm with a fixed preference list. The work in [4] presents also a speaker-dependent REG strategy that computes all possible descriptions of a given target, and selects either the most frequent or most recent form for each speaker. In both cases, using speaker-related information outperforms a standard algorithm that selects the most frequent form in the training data.

The more recent availability of larger sets of referring expressions and their accompanying contexts (or REG corpora) has enabled a number of corpus-based approaches to REG as well. Among these, of special interest to our present work are *GRE3D3* and *GRE3D7* [13,15], *Stars* [12] and *Stars2* [7] corpora. Each of these corpora were produced by a relatively large number of human participants, making them more likely to show variation across speakers. *GRE3D3* [13] and *GRE3D7* [15] concern descriptions of objects in simple 3D scenes, which may be either atomic (e.g., 'the red box') or involve a spatial relation (e.g., 'the red box on top of the cube'). *GRE3D3* contains 630 descriptions produced by 63 speakers, and *GRE3D7* contains 4480 descriptions produced by 287 speakers.

Stars [12] and *Stars2* [7] concern descriptions in 2D scenes involving up to three objects each (namely, a target, first and second landmark objects, as in 'the cube next to the cone below the red sphere'). Previous Figure 1 illustrates a scene from the *Stars2* corpus. *Stars* contains 704 descriptions produced by 64 speakers, and *Stars2* [7] contains 1216 descriptions produced by 76 speakers.

The availability of corpus knowledge allows the use of machine learning techniques in REG. For instance, the work in [14] applies decision-tree induction to predict reference strategies. In one of their experiments, the REG model includes information that identifies each speaker, which generally outperforms alternatives that did not take speaker information into account.

The work in [5] applies support vector machine (SVM) classifiers to the REG task. Results show that the SVM approach outperformed a standard implementation of the Incremental algorithm on both *GRE3D3* and *GRE3D7* corpora. Our present work follows the SVM approach in [5], which is presently expanded in a number of ways: by considering additional context features, additional corpora for training and evaluation purposes, and a set of speaker-related features.

3 Current Work

As in [5], we make use of a classification-based approach to REG built on support vector machines with radial basis function kernel. Two kinds of classifier are considered: binary classifiers for each individual atomic attribute prediction (e.g., colour, size etc.), and multi-class classifiers for relational attribute prediction (e.g., the choice between left(x), above(x) etc. where x is a landmark object.)

The classifiers for each corpus may be related to a target, first or second landmark object mentioned in each description. For instance, 'the box next to the sphere below the cone' conveys a reference to a target (box), a first (sphere) and second (cone) landmarks. Given our goal to evaluate REG algorithms in four corpora of referring expressions (namely, *GRE3D3*, *GRE3D7*, *Stars* and *Stars2*, cf. previous section), the number of classifiers is slightly different for each domain. For instance, attributes representing vertical and horizontal screen position are only relevant to the 2D scenes in *Stars* and *Stars2* data, but not to *GRE3D3* and *GRE3D7* 3D scenes. Furthermore, *Stars* and *Stars2* descriptions may refer to up to two landmark objects, whereas for *GRE3D3* and *GRE3D7* descriptions there is a one landmark maximum.

GRE3D3 and *GRE3D7* descriptions required 8 binary classifier for individual attribute prediction, and one multi-class classifier for the relation attribute prediction. Possible values for the relation attribute in this case are *no relation*, *right-of*, *left-of*, *next-to*, *on-top-of* and, in the case of *GRE3D3*, also *in-front-of*.

Stars descriptions required 12 binary classifier for individual attribute prediction, and *Stars2* descriptions required 15. In both cases, two multi-class classifiers were trained: one for the relation attribute between the target and the first landmark, and another for the relation between the first and second landmarks. Possible values for the relation attributes are *no relation*, *right-of*, *left-of*, *next-to*, *below*, *above* and, in the case of *Stars2*, also *in-front-of*.

From the set of binary and multi-class classifiers for each individual attribute, a description is built as follows. First, atomic target attributes are considered. For each positive class prediction, the corresponding attribute will be included in the target description. Next, relational attributes are considered. If no relational attribute is predicted, the algorithm terminates by returning an atomic description L of the target. On the other hand, if L involves a relation then the related landmark object is included in the output description and the algorithm is called once again to describe it recursively.

The algorithm does not explicitly test for uniqueness of the output description under generation, that is, every attribute that corresponds to a positive class is always included in L, which in many cases will become overspecified. This, as we shall see in the results described in Section 4, turns out to produce descriptions that resemble those produced by human speakers in the test data.

As most existing REG algorithms, the current model - which is solely based on context-related features - will always produce the same (i.e., fixed) output description for a given target input. This model, hereby called $-SP$, will be used as a standard, speaker-independent REG approach. The context features used by this model are summarized in Table 1. In the *Stars* corpus, the features related to the size of objects (TG_Size, LM_Size, Num_TG_Size, Num_LM_Size, TG_LM_Same_Size) are not used.

Table 1. Context-related features, taken from [14]

Feature	Description
TG_Size	target size
LM_Size	landmark size
Relation_Type	type of relation between target and landmark
Num_TG_Size	number of objects of same size as the target
Num_LM_Size	number of objects of same size as landmark
TG_LM_Same_Size	target and landmark share size
Num_TG_Col	number of objects of same colour as target
Num_LM_Col	number of objects of same colour as landmark
TG_LM_Same_Col	target and landmark share colour
Num_TG_Type	number of objects of same type as target
Num_LM_Type	number of objects of same type as landmark
TG_LM_Same_Type	target and landmark share type

In addition to that, we will also consider an expanded version of the model containing a number of speaker-related features. This speaker-dependent model will be called $+SP$, and conveys, besides the information in $-SP$, personal and attribute usage information as seen in the training data. These features are summarized in Table 2.

Table 2. Speaker-related Features

Feature	Description
Speaker_ID	speaker's unique identifier
Speaker_Gender	speaker's gender
Speaker_AgeGroup	speaker's age group
Speaker_Frequency	speaker's attribute frequency vector

All classifiers were trained, validated and tested using 10-fold cross validation, except for the case of the *Stars* corpus, in which we used 6-fold cross validation. The use of only six folds in the case of the *Stars* corpus reflects the number of descriptions available for each speaker in that corpus, which is kept balanced within each fold.

In order to optimize the C and γ parameter values of the SVM radial basis function kernel, grid search was performed. In every step of cross validation, one fold was left out as the test fold, and grid search was performed within the remaining folds, guided by the cross validation method as the performance metric on the training data. In the search, multiple SVM classifiers were attempted using different value combinations for C (1, 10, 100 and 1000) and γ (0.1, 0.01, 0.001, 0.0001). The winning parameter combinations were applied to train the SVM classifiers in all the remaining folds so as to make predictions in the test fold.

As in [5], we used Python *Scikit-learn* software for the actual SVM implementation. For the multi-class prediction of the relation attributes, we followed a 'one-against-one' approach [8].

4 Evaluation

Our work was evaluated against four corpora of definite descriptions: *GRE3D3* [13], *GRE3D7* [15], *Stars* [12] and *Stars2* [7]. In all cases, our goal was to compare the two models - with ($+SP$) and without ($-SP$) speaker-related features - as described in the previous section.

F-Score and Area under the ROC Curve (AUC) values for *GRE3D3/7* and *Stars/Stars2* individual classifiers are presented in Tables 3 and 4. Generally speaking, classifiers that took speakers' preferences into account outperformed those that did not in all corpora. The exceptions (*vpos* and *hpos* in *Stars* and *lm2_size* in *Stars2*) were due to data sparsity.

Each set of classifiers makes a REG algorithm as described in Section 3. Evaluation of the REG task was carried out by comparing the descriptions produced by each algorithm with the reference description found in each corpus. The overall precision of each algorithm was computed by measuring Accuracy (i.e., the number of exact matches between System and Reference description pairs). The degree of overlap between each System-Reference description pair was measured by Dice [3] scores. Results are summarized in Table 5.

We applied Wilcoxon's Signed-rank test over Dice scores, and the Chi-squared test over accuracy scores. Differences between the two algorithms are significant as summarized in Table 6.

Table 3. GRE3D3 and GRE3D7 classifier results

	GRE3D3				GRE3D7			
	-SP		+SP		-SP		+SP	
Classifier	F_1	AUC	F_1	AUC	F_1	AUC	F_1	AUC
tg_type	1.00	1.00	1.00	1.00	1.00	1.00	1.00	1.00
tg_colour	0.88	0.16	0.95	0.94	0.99	0.35	0.99	0.85
tg_size	0.88	0.73	0.94	0.98	0.74	0.67	0.88	0.92
tg_location	0.00	0.15	0.30	0.74	0.00	0.23	0.00	0.75
lm_type	1.00	1.00	1.00	1.00	1.00	1.00	1.00	1.00
lm_colour	0.82	0.26	0.91	0.89	0.93	0.34	0.93	0.79
lm_size	0.73	0.58	0.77	0.90	0.66	0.52	0.84	0.87
lm_location	0.75	0.70	0.59	0.92	0.00	0.32	0.00	0.80
relation	0.15	0.39	0.87	0.96	0.00	0.50	0.75	0.91

Table 4. Stars and Stars2 classifier results

	Stars				Stars2			
	-SP		+SP		-SP		+SP	
Classifier	F_1	AUC	F_1	AUC	F_1	AUC	F_1	AUC
tg_type	1.00	1.00	1.00	1.00	0.99	0.27	1.00	0.86
tg_colour	0.00	0.27	0.69	0.87	0.71	0.80	0.80	0.94
tg_size	-	-	-	-	0.02	0.79	0.07	0.96
tg_hpos	0.00	0.23	0.47	0.79	0.00	0.00	0.00	0.00
tg_vpos	0.00	0.23	0.47	0.78	0.00	0.00	0.00	0.00
lm_type	0.99	0.31	1.00	0.75	0.99	0.15	0.99	0.43
lm_colour	0.58	0.43	0.79	0.81	0.84	0.84	0.88	0.95
lm_size	-	-	-	-	0.78	0.92	0.85	0.96
lm_hpos	0.00	0.00	0.00	0.00	0.00	0.00	0.00	0.00
lm_vpos	0.00	0.00	0.00	0.00	0.00	0.00	0.00	0.00
lm2_type	0.99	0.14	0.98	0.19	1.00	1.00	1.00	1.00
lm2_colour	0.20	0.30	0.68	0.79	0.00	0.36	0.35	0.88
lm2_size	-	-	-	-	0.00	0.00	0.00	0.00
lm2_hpos	0.00	0.00	0.00	0.00	0.00	0.00	0.00	0.00
lm2_vpos	0.00	0.00	0.00	0.00	0.00	0.00	0.00	0.00
relation	0.98	0.22	0.98	0.66	0.90	0.85	0.95	0.97
lm_relation	0.00	0.34	0.73	0.89	0.00	0.40	0.43	0.70

Table 5. REG results with (+SP) and without (-SP) speaker-related features

	GRE3D3		GRE3D7		Stars		Stars2	
Alg.	Dice	Acc.	Dice	Acc.	Dice	Acc.	Dice	Acc.
-SP	0.78	0.46	0.88	0.61	0.72	0.18	0.66	0.30
+SP	0.92	0.74	0.94	0.77	0.73	0.29	0.76	0.36

Table 6. -SP and +SP results comparison

	Dice		Accuracy	
Corpus	W	p	χ^2	p
GRE3D3	1974.0	< .0001	245.75	< .0001
GRE3D7	261574.0	< .0001	475.27	< .0001
Stars	29927.0	< .8165	24.28	< .0001
Stars2	118278.0	< .0001	19.27	< .0001

Results show that taking speakers' preferences into account significantly increases both Dice and accuracy scores in all corpora, with the only exception of Dice scores on *Stars* data. This generally confirms our main research hypothesis.

The present results for *GRE3D3* are superior to those in [14], with reported 0.58 accuracy when using unpruned trees with 10-fold cross-validation, and also in [16], with reported Dice=0.85 and accuracy=0.60 for a 'longest first' selection strategy. Our results for *GRE3D7* are also superior to those in [14], with reported 0.67 accuracy when using pruned trees with 10-fold cross-validation.

Regarding *Stars* data, the present results are also superior to those obtained in [12], with reported Dice=0.61 and accuracy=0.11 when combining the Incremental algorithm with a decision-tree model of reference underspecification.

5 Final Remarks

This paper discussed the generation of referring expressions using SVM classifiers with and without speaker-related features. Results in four REG corpora suggest that the model that takes speakers' preferences into account outperforms the model that does not. Moreover, present results on *GRE3D3*, *GRE3D7* and *Stars* data are all superior to previous work in the field.

Speakers however do not stick to a single reference strategy in all situations, and may in fact produce different descriptions on different occasions even when the context remains unchanged. As future work, we intend to model this kind of non-determinism to account for both between-speakers and within-speakers variation in REG.

Acknowledgments. The authors acknowledge support by CAPES and FAPESP.

References

1. Bohnet, B.: The fingerprint of human referring expressions and their surface realization with graph transducers. In: INLG 2008, Stroudsburg, USA, pp. 207–210 (2008)
2. Dale, R., Reiter, E.: Computational interpretations of the Gricean maxims in the generation of referring expressions. Cognitive Science 19(2), 233–263 (1995)
3. Dice, L.R.: Measures of the amount of ecologic association between species. Ecology 26(3), 297–302 (1945)

4. Fabbrizio, G.D., Stent, A.J., Bangalore, S.: Trainable speaker-based referring expression generation. In: Proceedings of the Twelfth Conference on Computational Natural Language Learning, CoNLL 2008, Stroudsburg, PA, USA, pp. 151–158 (2008), http://dl.acm.org/citation.cfm?id=1596324.1596350
5. Ferreira, T.C., Paraboni, I.: Classification-based referring expression generation. In: Gelbukh, A. (ed.) CICLing 2014, Part I. LNCS, vol. 8403, pp. 481–491. Springer, Heidelberg (2014)
6. Gupta, S., Stent, A.J.: Automatic evaluation of referring expression generation using corpora. In: Proceedings of the 1st Workshop on Using Corpora in Natural Language Generation (UCNLG), Birmingham, pp. 1–6 (2005)
7. Iacovelli, D., Galindo, M.R., Paraboni, I.: Lausanne: A framework for collaborative online NLP experiments. In: 11th International Conference on Computational Processing of Portuguese, PROPOR 2014 (to appear, 2014)
8. Knerr, S., Personnaz, L., Dreyfus, G.: Single-layer learning revisited: A stepwise procedure for building and training a neural network. In: Soulié, F., Hérault, J. (eds.) Neurocomputing. NATO ASI Series, vol. 68, pp. 41–50. Springer (1990)
9. Krahmer, E., van Deemter, K.: Computational generation of referring expressions: A survey. Computational Linguistics 38(1), 173–218 (2012)
10. de Lucena, D.J., Pereira, D.B., Paraboni, I.: From semantic properties to surface text: The generation of domain object descriptions. Inteligencia Artificial. Revista Iberoamericana de Inteligencia Artificial 14(45), 48–58 (2010)
11. Pereira, D.B., Paraboni, I.: Statistical surface realisation of portuguese referring expressions. In: Nordström, B., Ranta, A. (eds.) GoTAL 2008. LNCS (LNAI), vol. 5221, pp. 383–392. Springer, Heidelberg (2008)
12. Teixeira, C.V.M., Paraboni, I., da Silva, A.S.R., Yamasaki, A.K.: Generating relational descriptions involving mutual disambiguation. In: Gelbukh, A. (ed.) CICLing 2014, Part I. LNCS, vol. 8403, pp. 492–502. Springer, Heidelberg (2014)
13. Viethen, J., Dale, R.: The use of spatial relations in referring expression generation. In: INLG 2008, Stroudsburg, USA, pp. 59–67 (2008)
14. Viethen, J., Dale, R.: Speaker-dependent variation in content selection for referring expression generation. In: Proceedings of the Australasian Language Technology Association Workshop 2010, Melbourne, Australia, pp. 81–89 (December 2010)
15. Viethen, J., Dale, R.: GRE3D7: A corpus of distinguishing descriptions for objects in visual scenes. In: Proceedings of the UCNLG+Eval: Language Generation and Evaluation Workshop, Edinburgh, Scotland, pp. 12–22 (July 2011)
16. Viethen, J., Mitchell, M., Krahmer, E.: Graphs and spatial relations in the generation of referring expressions. In: EACL 2013, Sofia, pp. 72–81 (2013)

Visualization of Intelligibility Measured by Language-Independent Features

Tino Haderlein[1,2], Catherine Middag[3], Andreas Maier[1], Jean-Pierre Martens[3], Michael Döllinger[2], and Elmar Nöth[1]

[1] Universität Erlangen-Nürnberg, Lehrstuhl für Mustererkennung (Informatik 5), Martensstraße 3, 91058 Erlangen, Germany
Tino.Haderlein@cs.fau.de
http://www5.cs.fau.de

[2] Klinikum der Universität Erlangen-Nürnberg, Phoniatrische und pädaudiologische Abteilung, Bohlenplatz 21, 91054 Erlangen, Germany

[3] Universiteit Gent, Vakgroep voor Elektronica en Informatiesystemen (ELIS), Sint-Pietersnieuwstraat 41, 9000 Gent, Belgium

Abstract. Automatic intelligibility assessment using automatic speech recognition is usually language-specific. In this study, a language-independent approach based on alignment-free phonological and phonemic features is proposed. It utilizes models that are trained with Flemish speech, and it is applied to assess dysphonic German speakers. In order to visualize the results, two techniques were tested: a plain selection of most relevant features emerging from Ensemble Linear Regression involving feature selection, and a Sammon transform of all the features to a 3-D space. The test data comprised recordings of 73 hoarse persons (48.3 ± 16.8 years) who read the German version of the text "The North Wind and the Sun". The reference evaluation was obtained by five speech therapists and physicians who rated intelligibility according to a 5-point Likert scale. In the 3-D visualization, the different levels of intelligibility were clearly separated. This could be the basis for an objective support for diagnostics in voice and speech rehabilitation.

1 Introduction

Evaluation of voice distortions is still mostly performed perception-based. This, however, is too inconsistent among single raters to establish a standardized and unified classification. Perception experiments are applied to spontaneous speech, read-out standard sentences, or standard texts. In contrast, already used methods of automatic analysis rely mostly on sustained vowels [9]. The advantage of speech recordings, however, is that they contain phonation onsets, variation of F_0, and pauses [14]. Furthermore, they allow us to evaluate speech-related criteria, such as intelligibility [4]. Intelligibility has been identified as one of the most important aspects of voice and speech assessment [1,13,17]. Experimental tools on intelligibility assessment usually employ an automatic speech recognition (ASR) system to compare the patient's utterance with the target text. However, ASR encompasses acoustic models for representing the basic sounds of a language. Therefore, it can only be used to assess speech of that language. Recently, ASR-free methods were developed [2,11]. They

P. Sojka et al. (Eds.): TSD 2014, LNAI 8655, pp. 547–554, 2014.

embed acoustic models that can track the phonological properties of the utterance as a function of time. Such a tracking does not rely on what was actually said anymore. It has been demonstrated that these ASR-free techniques are able to assess intelligibility of different voice pathology groups even if the members of the group speak a language that was not included in the training of the phonological models [11].

The results of such an automatic evaluation are basically a sequence of numbers. This will be of no help for the technically uneducated medical personnel. Therefore, the goal of our work is to provide a graphical visualization of a small number of features which are extracted from a high number of "technical" features by some automatic dimension reduction method. The basis of the distance measure between different speakers are the phonological and phonemic features in this study. In order to provide 3-D visualization, two approaches were applied. One is to select the three most relevant features that emerge from an Ensemble Linear Regression involving feature selection. The other is to apply Sammon mapping [15] to the full feature vectors. It allows the graphical representation of abstract data, unveiling underlying structures and configurations. This method itself is not new, but it has never been applied to this concrete problem. In earlier studies, it has been successfully used for the visualization of different levels of voice quality, even from different recording conditions [5,7]. The test set of this study is composed of speech recordings of chronically hoarse speakers. The following questions will be addressed: Can different levels of intelligibility of hoarse speakers be visualized by phonological and phonemic features? Are there also phonological and phonemic properties that can visibly separate the two subgroups of functional and organic dysphonia?

Section 2 describes the test data, Sect. 3 gives an overview on the features obtained from the data, and Sect. 4 reviews the dimensionality reduction methods applied for 3-D visualization. The results are presented and discussed in Sect. 5.

2 Test Data and Subjective Evaluation

73 German persons with chronic hoarseness participated in this study. Patients suffering from cancer were excluded. The most common pathologies were grouped into functional (n=45) and organic dysphonia (n=24; see Table 1). Functional dysphonia is usually caused by too few or too much speaking effort due to psychogenic reasons or vocal misuse. Organic dysphonia has its origin in anatomical changes, such as vocal fold polyps, edemas or pareses. The remaining four speakers suffered from chronic laryngitis. Each person read the phonetically balanced text "Der Nordwind und die Sonne" ("The North Wind and the Sun", [6]), which is frequently used in medical speech evaluation in German-speaking countries. It contains 108 words (71 distinct) with 172 syllables. Additional remarks by the speaker were removed manually from the recordings. The data were recorded with a sampling frequency of 16 kHz and 16 bit amplitude resolution using an AKG C 420 microphone (AKG Acoustics, Vienna, Austria).

Five voice professionals estimated the patients' intelligibility on a five-point Likert scale while listening to a play-back of the recordings. They marked one of the grades "very high", "rather high", "medium", "rather low", or "very low", which were converted to integer values from 1 (very high) to 5 for computation. An averaged mark, expressed as a floating point value, was calculated for each patient as the mean of the single scores. These marks served as ground truth in our experiments. The inter-rater agreement, computed as the mean correlation of one rater against the average of the four others, is given in Table 1.

Table 1. Number of speakers, age statistics, perceptual evaluation results (intelligibility on a 5-point scale), and inter-rater correlation r for the patient groups

	no. of speakers			age				intelligibility scores				
persons	all	men	women	μ	σ	min	max	μ	σ	min	max	r
total group	73	24	49	48.3	16.8	19	85	2.51	1.02	1.00	5.00	0.82
functional	45	13	32	47.1	16.3	20	85	2.27	1.00	1.00	5.00	0.83
organic	24	9	15	52.2	15.6	25	79	3.06	0.91	1.60	4.80	0.75

3 Features Computed from the Speech Data

The pre-processing stage returns a spectro-temporal representation of the acoustic signal. From this representation, speaker features are extracted which constitute a compact characterization of the speech of the tested person. An intelligibility score is finally computed on the basis of these speaker features.

During the pre-processing stage, a stream of Mel-frequency cepstral coefficients (MFCC) is extracted from the recording. For each 25 ms speech frame (frame shift: 10 ms), 12 MFCCs and an energy value are returned. Based on the stream of MFCCs, two text-independent feature extraction method, focusing on phonological and phonemic aspects, have been explored.

Alignment-free Phonological Features (ALF-PLFs): First described in [12], these features follow from a tracking of the temporal evolutions of the individual outputs of an artificial neural network that was trained (see [10,12] for more details) to generate 14 phonological properties per frame. These properties describe:

- vocal source: voicing
- manner of articulation: silence, consonant-nasality, vowel-nasality, turbulence (referring to fricative and plosive sounds)
- place of consonant articulation: labial, labio-dental, alveolar, velar, glottal, palatal
- vowel features: vowel height, vowel place, vowel rounding

Every phonological property is analyzed by two sub-networks. One of them determines whether the property is relevant at a given time (e.g. it is not relevant to investigate vowel place during utterance of a consonant); the other one determines whether the characteristic (e.g. “labial”) is actually present or not.

The hypothesis is that temporal fluctuations in the network outputs can reveal articulatory deficiencies, regardless of the exact phonetic content of the text that was read, at least as long as this text is sufficiently rich in phonetic content. The temporal analysis of each network output generates a set of parameters, such as the mean and standard deviation, the percentage of the time the output is high (above 0.66), intermediate or low (below 0.33), respectively, the mean height of the peaks (maxima), and the mean time it takes to make a transition from low to high. The overall number of output features is currently 504, and it is acknowledged that several of them may carry similar information. These speaker features are computed without knowledge of the text that was read. Hence, we expect them to be text-independent.

Alignment-free phonemic features (ALF-PMFs): The features, introduced in [10], are based on the hypothesis that intelligibility degradation is correlated with problems in realizing a certain *combination* of phonological classes that is needed for the production of a certain phone. Therefore, the ALF-PMFs follow from a plain analysis of posterior phone probabilities which are themselves retrieved from the phonological properties by means of a neural network. Considering all frames for which the maximal posterior probability is assigned to a particular phone, one computes the mean and standard deviation of that probability, the mean of the peaks (maxima) and the valleys (minima) found in the temporal evolution of that probability. In addition, the percentage of the time a frame is assigned to the phone, and the mean probability of this phone over all frames are computed. Clearly, these features are computed without any knowledge of the text that was read and can therefore be expected to be text-independent. The number of ALF-PMFs is equal to 495.

4 Reduction of Dimension for Visualization

The aim of the dimensionality reduction is to construct a 3-D space that can be visualized using 3-D graphics.

The first method consists of creating a lot of linear regression models on different randomly chosen subsets of the training data and to allow each model to select three features (principle of Ensemble Linear Regression). The three features that were selected most frequently are then retained for visualization.

The second method is Sammon mapping. It performs a non-linear transformation preserving data topology which is represented by a matrix of inter-utterance distances. The distance metric can be chosen without any mathematical restrictions, such as linearity etc. This is the great advantage of the Sammon transform against other dimension reduction operations, such as PCA or LDA [8]. In the pilot experiments, the Euclidean distance between the feature vectors of two respective speakers was used as the distance metric. In the future, we plan to extend the feature set by other feature types which call for another distance metric, e.g. a metric that is more suitable for comparing distances between HMM parameters, as in [5]. The heart of Sammon's method is its special error function E, yielding a stress factor between the actual configuration of n points in an m-dimensional target domain and the original data in a d-dimensional space ($m < d$):

$$E = \frac{1}{\sum_{p=1}^{n-1} \sum_{q=p+1}^{n} \delta_{pq}} \sum_{p=1}^{n-1} \sum_{q=p+1}^{n} \frac{(\delta_{pq} - \nu_{pq})^2}{\delta_{pq}} \tag{1}$$

δ_{pq} denotes the Euclidean distance between feature vectors with number p and q, v_{pq} is the distance between points $s(p)$ and $s(q)$ in the Sammon map. E is within [0,1], where $E = 0$ means a lossless projection from d- to m-dimensional space. Utterances forming clusters in original space will tend to cluster also in destination space. The same holds for utterances being far apart from each other. In order to achieve the final map, we apply standard steepest descent to (1).

5 Results and Discussion

The three most relevant features retained from the set of ALF-PLF and ALF-PMF features for the representation of intelligibility were:

- *turbulence_presence_meanmax* represents the mean of the peaks in the turbulence feature pattern over the file. It is obvious that intelligibility decreases with rising turbulence in the voice.
- *alveolar_relevance_meanmin* represents the value of the valleys (minima) in the relevance of the property *alveolar*. These valleys are supposed to occur where vowels are spoken. In distorted voices, there is not always a clear distinction between vowels and consonants, and the valleys are less deep than in normal voices. Obviously, this lack of distinction degrades intelligibility.
- *duration* is the overall duration of the read-out text. Since all persons read the same text, the duration is inversely proportional to the speaking rate. It is intuitive that slower speakers are more intelligible. However, a lower speaking rate may also point towards higher speaking effort caused by anatomical problems in the articulatory organs. This has been shown for totally laryngectomized persons with a substitute voice after removal of the larynx [3].

For the separation of functional and organic dysphonia, two features were found to be most relevant:

- *consonantnasality_presence_meanmin* describes the minima of the presence of nasality.
- *h_meanneg* is a phonemic feature describing the mean low evidence that a /h/ has been uttered.

Figure 1 shows a visualization of the most relevant features for each speaker. In the upper left figure, the shading of gray denotes the human reference evaluation of intelligibility. For visualization purposes, the *duration* values were normalized (divided by 5,000), and the *turbulence_presence_max* axis shows the value of $(1 - \mathit{turbulence\_presence\_max}) * 30$ in order to achieve about the same range for all three dimensions. The axes were rotated, so that the ability of the features to depict different levels of intelligibility was best visible.

On the right-hand side, Fig. 1 also shows a 3-D visualization computed by the Sammon transform using all available ALF-PLF and ALF-PMF features. The normalization of the feature values was done during the computation of the distance matrix. Again, in the upper image the levels of intelligibility are clearly visible. A transform directly into 2-D did not show such a good result, however.

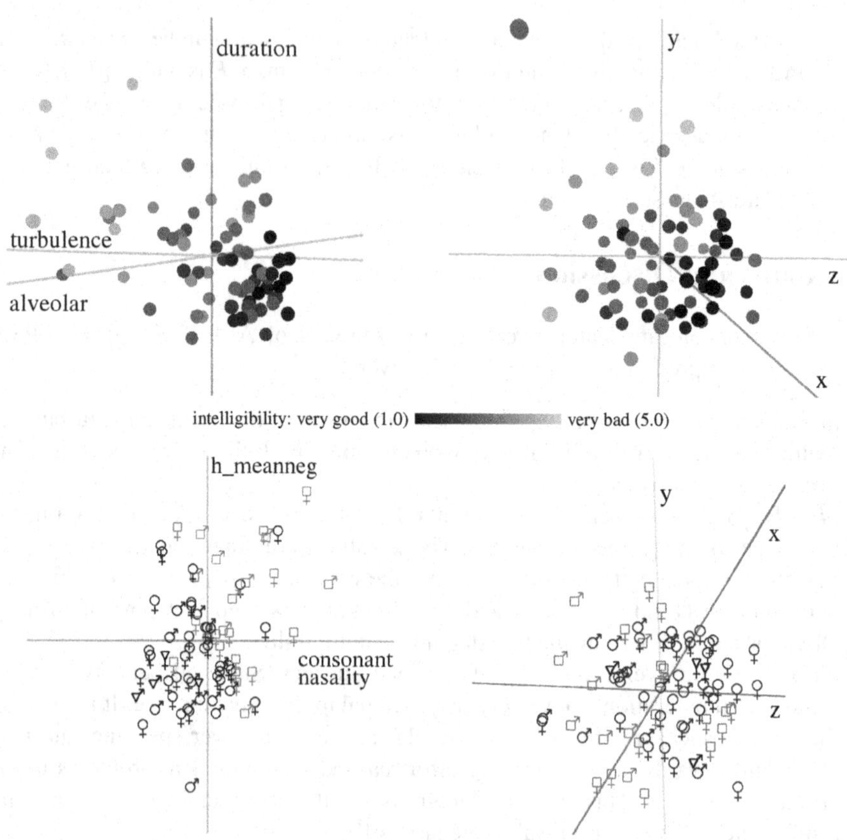

Fig. 1. Visualization of 73 hoarse speakers using the most relevant features only *(left side)* and the Sammon transform on all features *(right side)*; the axes are rotated so that the relevant information is best visible. *Top row:* Human-rated intelligibility is denoted by the shades of gray of the points. *Bottom row:* Functional (○) and organic dysphonia (□), and the four remaining speakers with laryngitis (▽).

In the bottom panels of Fig. 1, the subgroups of hoarseness in the test data are depicted. Both the two most relevant features and all features together resulted in visualizations that show some tendencies for the separation of different diagnoses. The Sammon map shows a tendency to group men to the left and women to the right which is very likely an indicator for F_0. Women with functional hoarseness seem to concentrate in one octant. However, which voice properties are arranged in which direction by the Sammon transform, is dependent on the data and not known in advance. This phenomenon was already reported in [16] where such a map was suggested to have an unlimited number of axes. Most of them represent complex properties of the data and are thus difficult to describe. In the case of the two most relevant features, rather organic dysphonia tends to have higher feature values. The reason could be higher breathiness in these voices. This is confirmed by preliminary findings by the speech therapists who rated the breathiness on a 4-point scale. For the speakers with functional dysphonia, the average rating was much lower (0.97) than for the persons with organic dysphonia (1.64).

Another research question was the language-independence of the features. In [11], intelligibility models were trained with Flemish and German pathologic speech and tested on the same language, respectively. Here, the test persons spoke German, and the phonological models had been trained with Flemish speech. Additionally, all the test speakers show a similar type of dysphonia, and the training was done with normal speakers. When Support Vector Regression models for intelligibility were trained for the test speakers, the average root mean square error between the computed score and the reference intelligibility was 0.74, corresponding to a Pearson's correlation of r=0.70. This confirms the suitability of the features for evaluation of intelligibility. The basic approach is described in detail in [10].

The results obtained in this study seem to support the following conclusions: Phonologic and phonemic features can be used to display levels of intelligibility. They are even suitable for language-independent analysis. The current feature set can also serve as a basis for the separation of different types of hoarseness automatically. With the integration of more features, the method might be a helpful objective support in the field of voice rehabilitation in the future.

Acknowledgments This work was partially funded by the Else Kröner-Fresenius-Stiftung (Bad Homburg v. d. H., Germany) under grant 2011_A167 and supported by "Kom op tegen Kanker", the campaign of the Vlaamse Liga tegen Kanker VZW and of The Netherlands Cancer Institute/Antoni van Leeuwenhoek Hospital (Amsterdam). The responsibility for the contents of this study lies with the authors.

References

1. Bellandese, M.H., Lerman, J.W., Gilbert, H.R.: An Acoustic Analysis of Excellent Female Esophageal, Tracheoesophageal, and Laryngeal Speakers. J. Speech Lang. Hear. Res. 44(6), 1315–1320 (2001)
2. Bocklet, T., Riedhammer, K., Nöth, E., Eysholdt, U., Haderlein, T.: Automatic intelligibility assessment of speakers after laryngeal cancer by means of acoustic modeling. J. Voice 26(3), 390–397 (2012)
3. Haderlein, T.: Automatic Evaluation of Tracheoesophageal Substitute Voices. Studien zur Mustererkennung, vol. 25. Logos Verlag, Berlin (2007)
4. Haderlein, T., Moers, C., Möbius, B., Rosanowski, F., Nöth, E.: Intelligibility Rating with Automatic Speech Recognition, Prosodic, and Cepstral Evaluation. In: Habernal, I., Matoušek, V. (eds.) TSD 2011. LNCS (LNAI), vol. 6836, pp. 195–202. Springer, Heidelberg (2011)
5. Haderlein, T., Zorn, D., Steidl, S., Nöth, E., Shozakai, M., Schuster, M.: Visualization of Voice Disorders Using the Sammon Transform. In: Sojka, P., Kopeček, I., Pala, K. (eds.) TSD 2006. LNCS (LNAI), vol. 4188, pp. 589–596. Springer, Heidelberg (2006)
6. International Phonetic Association (IPA): Handbook of the International Phonetic Association. Cambridge University Press, Cambridge (1999)

7. Maier, A., Exner, J., Steidl, S., Batliner, A., Haderlein, T., Nöth, E.: An Extension to the Sammon Mapping for the Robust Visualization of Speaker Dependencies. In: Sojka, P., Horák, A., Kopeček, I., Pala, K. (eds.) TSD 2008. LNCS (LNAI), vol. 5246, pp. 381–388. Springer, Heidelberg (2008)
8. Maier, A., Schuster, M., Eysholdt, U., Haderlein, T., Cincarek, T., Rosanowski, F., Steidl, S., Batliner, A., Wenhardt, S., Nöth, E.: QMOS – A Robust Visualization Method for Speaker Dependencies with Different Microphones. Journal of Pattern Recognition Research 4(1), 32–51 (2009)
9. Maryn, Y., Roy, N., De Bodt, M., Van Cauwenberge, P., Corthals, P.: Acoustic measurement of overall voice quality: A meta-analysis. J. Acoust. Soc. Am. 126, 2619–2634 (2009)
10. Middag, C.: Automatic Analysis of Pathological Speech. Ph.D. thesis. Ghent University, Ghent, Belgium (2012)
11. Middag, C., Bocklet, T., Martens, J.-P., Nöth, E.: Combining phonological and acoustic ASR-free features for pathological speech intelligibility assessment. In: Proc. Interspeech, pp. 3005–3008. Int. Speech Comm. Assoc. (2011)
12. Middag, C., Saeys, Y., Martens, J.-P.: Towards an ASR-free objective analysis of pathological speech. In: Proc. Interspeech, pp. 294–297. Int. Speech Comm. Assoc. (2010)
13. Moerman, M.B.J., Pieters, G., Martens, J.P., van der Borgt, M.J., Dejonckere, P.H.: Objective evaluation of the quality of substitution voices. Eur. Arch. Otorhinolaryngol. 261(10), 541–547 (2004)
14. Parsa, V., Jamieson, D.G.: Acoustic discrimination of pathological voice: sustained vowels versus continuous speech. J. Speech Lang. Hear. Res. 44, 327–339 (2001)
15. Sammon Jr., J.: A Nonlinear Mapping for Data Structure Analysis. IEEE Trans. on Computers C-18(5), 401–409 (1969)
16. Shozakai, M., Nagino, G.: Analysis of Speaking Styles by Two-Dimensional Visualization of Aggregate of Acoustic Models. In: Proc. ICSLP, Jeju Island, Korea, pp. 717–720 (2004)
17. van As, C.J., Koopmans-van Beinum, F.J., Pols, L.C., Hilgers, F.J.M.: Perceptual evaluation of tracheoesophageal speech by naive and experienced judges through the use of semantic differential scales. J. Speech Lang. Hear. Res. 46(4), 947–959 (2003)

Using Suprasegmental Information in Recognized Speech Punctuation Completion

Marek Boháč and Karel Blavka

Institute of Information Technology and Electronics,
Studentská 2/1402, 461 17 Liberec, Czech Republic
{marek.bohac,karel.blavka}@tul.cz
https://www.ite.tul.cz/itee/

Abstract. We propose a scheme to determine punctuation of the text produced by an automatic speech recognizer. We deal with the addition of commas based on the recognized text and we propose a full stop detection scheme using both – the textual and prosody information. We also propose an expanded scheme which utilizes enriched audio document information (e.g. speaker diarization, language detection etc.) to improve the sentence boundary detection. We compare the above mentioned schemes and its accuracy in terms of (in)correctly estimated punctuation markers and its ability to mark the positions of sentence boundaries. Hence we want to show it is better to incorporate all the relevant information sources in one reasonable scheme than to split the document processing into independent layers. Proposed schemes are evaluated over a set of recordings from the Czech (and Czechoslovak) radio broadcasts.

Keywords: punctuation completion, fundamental frequency detection, comma, full stop, automatic speech recognition, document segmentation.

1 Introduction

An increasing number of automatic speech recognition (ASR) applications aims to provide the access to different media sources and make it searchable. Some typical examples are on-line media monitoring systems [1], audio archive indexing engines [2] or lecture streaming [3]. Common feature of all mentioned applications is an interface presenting the ASR textual output to the user. When the presented plain text fragment is longer than one sentence, reading becomes very demanding and uncomfortable. As some languages (e.g. Slavic languages) have very loose form – especially their spoken form – it becomes essential to complement the ASR results with appropriate post-processing and punctuation.

We propose an approach to supplement the recognized text with punctuation markers and we show the importance of utilizing more information sources than to perform each post-processing level individually. From the whole post-processing task we focus on the modules estimating the commas and full stops. The comma estimation is based on N-gram language modeling. The full stop estimation employs the recognized text, prosodic information (speech fundamental frequency and non-speech events) and the document segmentation. We evaluate the consequences of employing or not employing concrete information sources.

P. Sojka et al. (Eds.): TSD 2014, LNAI 8655, pp. 555–562, 2014.

The next section introduces all the employed modules with emphasis on the full stop determination. Section 3 describes the experimental setup and results. In Section 4 we make conclusions and propose the future work.

2 Proposed Scheme

As can be seen in Fig. 1 the audio document processing can be generalized into four functional blocks: Automatic Speech Recognition system (ASR), Document Segmentation (Doc Seg), Comma completion and Full Stop completion. The complexity of concrete blocks may highly differ accordingly to the structure of the audio document (e.g. single vs. multiple-language documents) and to the demands of the user (e.g. one-pass vs. multi-pass ASR). In the following paragraphs we introduce the employed functional blocks with emphasis on the Full Stop completion as the other blocks were already published.

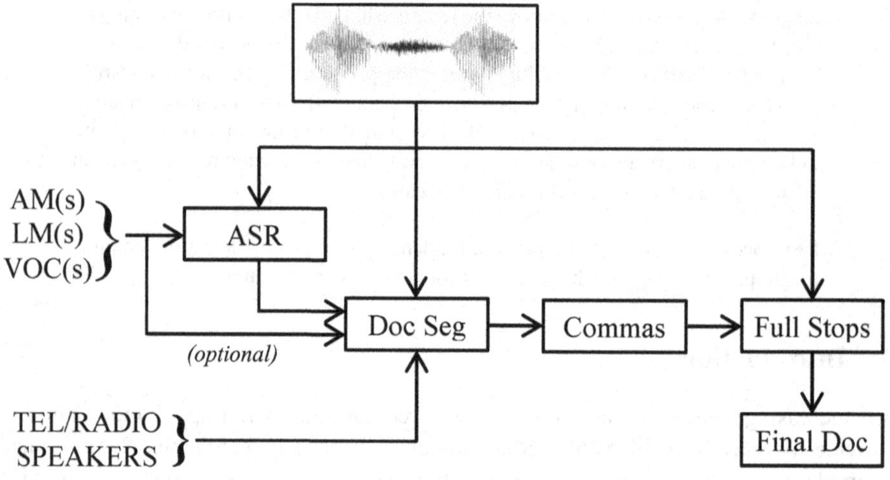

Fig. 1. Overall scheme of the audio document processing

2.1 ASR

The automatic speech recognizer (ASR) is our own one-pass time-synchronous Viterbi decoder. The HMM-based acoustic model contains speech phonemes as well as non-speech events (e.g. breathing, hesitation, click and cough). The features are 13-dimensional MFCCs with first and second derivate. The input audio presumes at least 16 kHz sampling frequency so the FFmpeg can convert it into standard PCM Wave format (mono, 16 bits per sample).

The acoustic models (AM), language models (LM) and vocabularies (VOC) are trained over different sets of data. The Czech vocabulary contains 550k items and the

AM is trained using 300 h of recordings. The Slovak vocabulary contains 320k items, its AM is trained on 100 h and the combined Czech-Slovak vocabulary contains 50k+50k items and the AM is trained on 100 h+100 h of recordings. The language models use different amounts of stored (precomputed) word-pairs that can be mixed and weighted according to the application topic domain. In the case of Czech-Slovak LM there are special features preferring to keep the current language over changing it.

2.2 Document Segmentation

The document segmentation module (Doc Seg) employs several modules – some of them optional. The base and mandatory module is the speaker diarization. The optional modules are the channel classification (e.g. telephone vs. radio speech), the speaker identification and the jingle/song detection. All these partial information sources are passed to a logical layer which outputs smoothed language, speaker and channel homogeneous document segments (paragraphs).

Speaker Diarization. The diarization module includes usual layers: i) voice activity detection (VAD), ii) speaker turn detection and iii) speaker clustering.

The VAD module is substituted by the usage of ASR output which already exists. This brings few benefits. In the case of noisy recordings the ASR-based speech detection grants higher robustness than a standalone solution. We can also limit the speaker turn points only to the borders of words so it decreases the computational demands and synchronizes the ASR output and the diarization.

Speaker turn point detection is performed by a variable-length window sliding along the frames of the parameterized recording. The test (derived from the Bayesian Information Criterion - BIC) compares the speech-labeled frames on the left and on the right from the examined turn point (for details see [4]). If the test exceeds the given threshold the speaker turn point is marked (we prefer over-splitting as it can be corrected in the clustering phase).

The speaker clustering is performed in two hierarchical steps. Firstly the segments are pre-clustered by the BIC-based classifier [5], secondly i-vector [6] representation of clusters with cosine similarity measure finalizes the clustering.

As the audio document diarization is a complex task and detailed description was already published, we kindly recommend to see [4,5,6] for details.

Speaker Identification and Channel and Gender Classification. The tools performing the document diarization were adjusted so the document segments can be classified with regards to different criteria. If the document contains utterances of already known speakers we can train i-vector based speaker models and identify the speaker. The score returned by the tool can be used to find a threshold so we can verify the speaker (we can detect if the speaker was not observed before). Second tool is a GMM-based classifier used to distinguish between standard recordings and narrow-band recordings (typically telephone transmissions). The GMM models are also used to obtain the speaker's gender.

Language Identification and Document Segmentation Smoothing. If there can be more than one language present in the recording (e.g. historical Czech and Slovak news) we apply special combined LM, AM and VOC in the first-pass recognition. For every speaker-and-channel homogenous segment the dominant language is determined (by the word count of the languages). If we can identify concrete speakers (not only diarization clusters), the language decision is smoothed over all speaker's utterances in the document.

The document segmentation is enhanced by the detection of jingles and strongly noised segments (e.g. music, terrain recordings). It is found by a sliding window traversing through the ASR transcription. Segments we want to detect have very high share of certain non-speech events (for details see [7]). Finally we smooth the document clustering with regards to the speakers' ID, gender, channel and language but also preventing over-segmentation. Segments are optionally recognized in the second pass using appropriate LM, AM and VOC [4].

2.3 Comma Completion

Our comma completion module is based on the textual input only. For the training we used a hand-made corpora (109 MB of Czech texts and 130 MB of Slovak texts) containing both – spontaneous (interviews, talk shows) and prepared speech (news broadcasts, public speeches). We searched the corpora for the most frequent words/phrases preceding and following the occurrence of commas. Then we carried out a statistical analysis of these rules so we defined the rules for Czech and Slovak comma completion (some of them produce the comma while others forbid it). These rules are word N-grams where $N \leq 3$ (e.g. prepositions, conjunctions and some common phrases). The application of the rules is performed via Weighted Finite State Transducers – WFST. One of the biggest advantages is the WFST ability to determine N-best solutions so it solves the cases of overlapping rules. The Czech rules are 1,243 phrases after comma (with 1,883 negatives) and 130 rules before comma (with 518 negatives). The Slovak rules are 2,518 rules after comma (with 5,071 negatives) and 333 rules before comma (with 5,752). The negatives prevent false commas by longer rule restricting the shorter one (which generates the comma).

2.4 Full Stop Determination

In the following paragraphs, we propose a prosody-based scheme for the full stop determination. It employs the ASR output to localize words and non-speech events. Speech melody (F0) is estimated using STFT and a dynamic programming-based decoder, choosing the most likely F0, that is searched for the potential full stop points. We also show how the ASR output and the diarization can be utilized to improve the full stop (and sentence boundary) determination.

Speech Localization and STFT. To localize the audio segments containing speech we use the ASR output. As it provides the time stamps for all the words and non-speech events in the recording, the detection of speech segments is a straightforward task. The recording is processed word by word – inside every time span we compute the Short Time Fourier Transform (STFT). The STFT is computed within 20ms frames, 10ms overlap, zero-padded to 4096 samples and windowed with the Hamming window. From every frame we choose 5 most significant components (magnitude local maximums) in range 60–600 Hz (this interval should cover all – children, male and female speakers as shown in [8]). The detected components are given weights according to their significance (in our case 10, 9, 7, 5, 5) so we obtain a spectrogram with a kind of histogram equalization which is passed to the prosody decoder.

Fundamental Frequency Decoding. The F0 decoder solves few tasks at once. It chooses the best fitting F0 according to the spectrogram and it performs a kind of F0 smoothing also. The algorithm is based on dynamic programming and optimizes a path between averaged three last frames and three first frames. Between frames inside the word borders it can pass directly between the significant components or can keep the component from the preceding frame. Passing between the significant components is favored by its weights (so the most significant components will most probably form the prosody). Big steps between the components are penalized (even a trained singer has very limited speed of prosody change) as well as the keeping of preceding component has its penalization. The best path through the spectrogram matrix is the detected prosody.

Prosody-based Sentence Boundary Determination. As the Czech and Slovak prosody is not very distinctive (when compared for example with English or French native speakers) we can extract weaker cues to detect the sentence borders. Generally, we observed that pitch declines as the sentence end approaches and the new sentence starts at a higher pitch. The last word of the sentence has usually very changing pitch while the words inside usually keep flatter pitch trend. We decided to detect this behavior by two features – the mean pitch of the word $\overline{P}$ and a normalized difference between maximum and minimum pitch P of the word as shown in (1).

The sentence boundaries are proposed as subsequent word pairs where the mean pitch declines and the normalized pitch difference of the second word exceeds a given threshold. Such a word pair must be followed by a word with a higher mean pitch or by a non-speech event (e.g. breath, laughing).

$$NormDiff = \frac{\max(P) - \min(P)}{\overline{P}} \tag{1}$$

ASR and Segmentation Utilization. To decide if the proposed sentence boundary induces a full stop we firstly check if there is a comma. Secondly we check the ASR output around the proposed sentence boundary – some words do not occur at the sentence end/begin (we found these words statistically in the previously mentioned corpora). If the previous fulfilled we place the full stop.

As the recordings, we process suffer from background music and noises, the prosody information is not as reliable as for clear recordings. Hence we prefer a setup with high precision and lower recall. Thus we need to detect more full stops using other information sources. Segmentation information is a natural choice. We place full stops at the ends of document segments – paragraphs.

The last source of full stops is the ASR output. We detect non-speech events within long sequences without a sentence delimiter. We place a full stop to every non-speech event which constitutes a sentence longer than 12 words (as it is the average lengths of Czech sentences – see [7]).

3 Experimental Evaluation

In this section we describe the evaluation data and define the metrics. As we want to show the impact of using different information sources, we evaluate the punctuation scheme in five setups. All the schemes have the same comma detection module but they differ in the full stop determination stage: i) using the prosody and ASR output (fs_pros), ii) using prosody, ASR and speaker-turn points (fs_turns), iii) using all – the prosody, ASR, speaker-turn point and heuristics splitting of too long sentences (fs_full), iv) using only the speaker-turn points (fs_trn) and v) using speaker-turn points and heuristic splitting (fs_trhe).

3.1 Evaluation Data and Metrics

The evaluation data consists of 21 radio broadcasts recorded within years 1971 – 2005. Total duration of the recordings is 9 hours 21 minutes. In 10 of them Czech and Slovak occurs (Slovak forms 10%–40% of the concrete recordings). Some parts are recorded outside (e.g. telephone entry, street interview, etc.) so there are strongly noised parts as well as jingles and background music.

The reference transcripts were made manually. As there can be some differences between the reference and ASR-recognized text, we aligned it using the word time stamps obtained via ASR and forced alignment of the references [9]. However there are still two minor drawbacks of this approach. First one is possible misalignment between the reference and the ASR. We manually checked the data and the error is negligible. The second is that different annotators usually mark the same positions as sentence boundaries but the inter-annotator agreement on concrete punctuation marker is low – see [7]. As every document was transcribed by one annotator we must consider it the ground truth.

We use these evaluation metrics: accuracy (2), precision (3), recall (4), detection rate (5) and false alarm rate (6), where *TP* stands for true positives (correctly marked positions), *FP* stands for false positives (false alarms), *FN* stands for false negatives (missing markers) and *CF* stands for confused markers (e.g. annotator makes a full stop and the system generates comma). Subscript *com* denotes commas and *fs* stands for full stops.

$$ACC = \frac{TP}{TP + FP + FN} \tag{2}$$

$$PRC = \frac{TP}{TP + FP} \tag{3}$$

$$REC = \frac{TP}{TP + FN} \tag{4}$$

$$DR = \frac{TP_{fs} + TP_{com} + CF}{TP_{fs} + FN_{fs} + TP_{com} + FN_{com}} \tag{5}$$

$$FA = \frac{FP_{fs} + FP_{com}}{TP_{fs} + FN_{fs} + TP_{com} + FN_{com}} \tag{6}$$

3.2 Experimental Results

As the commas and full stops occupy the same set of positions we evaluate the experiments together – see Table 1. This is clearly shown by the results of comma detection. Results of the same module differ as *CF* is not the same (there are full stops instead of missing commas).

The comma completion results show over 80% precision (low false alarm rate). The problem is with lower recall (we can mark approx. 50% of the positions). Similar is the situation with the prosody based full stop detection. It has very good precision (over 85%) but low recall (approx. 25%). If we presume the application of "full scheme", this is what we need – place the markers where we are sure and pass those which can be placed by other knowledge sources.

The results of employing the document segmentation are predictable – it improves the full stop determination. The decrease of precision is caused by over-segmentation of the document (some longer paragraphs are interrupted). The additional "heuristic" completion of commas naturally carries decrease of precision but the impact on the recall and sentence boundary detection far outweights it. This can be clearly seen in the increase of sentence boundaries detection rate. Our experience also proves that slightly over-segmented text is more reader-friendly than a text with very long sentences.

Table 1. Experimental results – full stops, commas and sentence boundaries

	Full Stops			Commas			Sentence Boundaries	
scheme	ACC [%]	PRC [%]	REC [%]	ACC [%]	PRC [%]	REC [%]	DR [%]	FA [%]
fs_pros	23.32	86.85	24.18	43.78	83.12	48.06	40.90	7.03
fs_turns	29.65	81.42	31.81	44.90	83.18	49.40	44.94	8.74
fs_full	48.53	61.02	70.35	50.86	83.18	56.70	74.47	27.95
fs_trn	16.88	78.41	17.70	43.71	83.28	47.92	36.50	7.35
fs_trhe	43.93	59.21	62.98	49.80	83.28	55.14	70.06	27.68

4 Conclusions and Future Work

We presented a scheme for comma and full stop completion. Our results show the advantage of combining all the available knowledge (text, prosody, segmentation)

against separated independent layers. Our punctuation scheme is sufficient to make the document easily readable although there are some reserves.

A closer view showed us that to mark the missing comma positions we would have to carry out the semantic analysis of the text. Another future work is to redefine the stop-lists (words implying that the sentence continues) using more linguistic knowledge (not only the statistics). Our main future interest lies in better definition of heuristics for additional full stop placing – especially in utilizing verb detection in the text (and so preventing the false alarms).

Acknowledgments. This work was supported by Czech Ministry of Culture (project DF11P01OVV013 in program NAKI).

References

1. Pawlaczyk, L., Bosky, P.: Skrybot – a system for automatic speech recognition of polish language. In: Cyran, K.A., Kozielski, S., Peters, J.F., Stańczyk, U., Wakulicz-Deja, A. (eds.) Man-Machine Interactions. AISC, vol. 59, pp. 381–387. Springer, Heidelberg (2009)
2. Ordelman, R., de Jong, F., Huijbregts, M., van Leeuwen, D.: Robust audio indexing for dutch spoken-word collections. In: XVIth International Conference of the Association for History and Computing, pp. 215–223. KNAW, Amsterdam (2005)
3. Cerva, P., Silovsky, J., Zdansky, J., Nouza, J., Malek, J.: Real-time lecture transcription using asr for czech hearing impaired or deaf students. In: 13th Annual Conference of the International-Speech-Communication-Association, pp. 762–765. ISCA, Portland (2012)
4. Cerva, P., Silovsky, J., Zdansky, J., Nouza, J., Seps, L.: Speaker-adaptive speech recognition using speaker diarization for improved transcription of large spoken archives. Speech Communication 55, 1033–1046 (2013)
5. Chen, S., Gopalakrishnan, P.: Speaker, environment and channel change detection and clustering via the bayesian information criterion. In: Proc. DARPA Broadcast News Transcription and Understanding Workshop, Virginia, USA, pp. 127–132 (1998)
6. Dehak, N., Kenny, P., Dehak, R., Dumouchel, P., Ouellet, P.: Front-end factor analysis for speaker verification. IEEE Transactions on Audio, Speech, and Language Processing 19, 788–798 (2011)
7. Bohac, M., Blavka, K., Kucharova, M., Skodova, S.: Post-processing of the recognized speech for web presentation of large audio archive. In: 2012 35th International Conference on Telecommunications and Signal Processing, pp. 441–445 (2012)
8. Atassi, H.: Metody detekce základního tónu řeči - methods for speech pitch detection. Elektrorevue (2008)
9. Boháč, M., Blavka, K.: Text-to-speech alignment for imperfect transcriptions. In: Habernal, I. (ed.) TSD 2013. LNCS (LNAI), vol. 8082, pp. 536–543. Springer, Heidelberg (2013)

Two-Layer Semantic Entity Detection and Utterance Validation for Spoken Dialogue Systems

Adam Chýlek, Jan Švec, and Luboš Šmídl*

University of West Bohemia, Faculty of Applied Sciences
New Technologies for the Information Society
Univerzitní 22, 306 14 Plzeň, Czech Republic
{chylek,honzas,smidl}@kky.zcu.cz

Abstract. In this paper we present a novel method for semantic entity detection in a limited domain for spoken language understanding. The target domain of this method is a dialogue system for an interactive training of air traffic controllers (ATC). The method comprises of two layers of detection. First layer uses formerly proposed method for semantic entity detection to extract domain-dependent set of semantic entities. This semantic entities are modelled using context-free grammars. To detect mispronounced words or words which do not comply with the ATC radio-telephony rules we use the second layer of semantic entity detection. Together with that, we assign a semantic meaning to the utterance. We also discuss the possibility of using this approach for semantic-based correction of an utterance. The experiments were performed on transcribed data as well as on an output from speech recognizer.

Keywords: semantic entity detection, spoken language understanding, finite state machines.

1 Introduction

The need for a novel approach for semantic entity detection and utterance validation comes from the development of a spoken dialogue system for interactive ATC training. Such training comprises of ATC-to-pilot communication lessons, where an ATC trainee controls virtual airspace in order to learn the rules and phrases that apply in air control environment. An ATC trainee communicates with several human pseudo-pilots that respond to commands or initiate the communication based on time plan created by an instructor. Pseudo-pilots process the commands into an input for virtual aircraft, which is visible on both pseudo-pilot's and trainee's radar screen.

The goal of this system is to provide a means of communication for ATC trainees mimicking real air traffic communication with pilots, without the actual need for human pseudo-pilots using dialogue system with automatic speech recognition (ASR), spoken language understanding (SLU) and text-to-speech (TTS) capability. At present, the dialogue system uses only semantic entity detection to highlight important semantic entities for human pseudo-pilot who then relays commands to several virtual aircraft

* This work was supported by the Technology Agency of the Czech Republic project No. TE01020197 and by an internal grant SGS-2013-032.

P. Sojka et al. (Eds.): TSD 2014, LNAI 8655, pp. 563–570, 2014.

under his control according to instructions from ATC. Such entities are *flight levels, communication frequencies, headings, clearances*, etc. We propose a method to not only extract those entities, but also assign another semantic meaning to the whole instruction, allowing us to manage virtual aircraft according to ATC's commands as a part of dialogue system.

Our method is also designed to validate user's utterance, because the domain of ATC radio communication is based on rules strictly defined in phraseology guides and as a part of their training, ATC trainees have to follow the rules without any deviation. The dialogue system can then use the information about utterance validity in order to give a feedback to its user on whether (and where) was the input incorrect. We also propose a method for utterance correction, giving the system an opportunity to not only validate against a set of rules, but also to offer the closest (in terms of Levenshtein distance) valid sequence of words and semantic entities the user might want to say instead.

Given that ATC communication is restricted by rules, we have chosen knowledge-based approach of spoken language understanding. The expert knowledge is represented as in many current commercial dialogue system by probabilistic context free grammars (PCFGs) which are used for both speech recognition and understanding. The developer of PCFG grammar does not need to be a dialogue system expert to write a grammar with good coverage of target semantic entities. On the other hand, the expansion probabilities are very hard to determine using knowledge only. Because of that, the probabilistic nature of PCFG is very rarely used and a large portion of recognition and understanding grammars are virtually deterministic context free grammars (CFG). The fact that CFG for a given semantic entity can be transferred and shared between many dialogue systems allows for rapid development of a SLU module for a new dialogue domain.

2 Semantic Entity Detection

A widely used approach in SLU systems is preprocessing of input sentences with respect to lexical classes. In the Semantic Tuple Classifier model [4], prior to training and decoding, lexical classes are replaced with class labels. The description of lexical classes consist of a list of lexical units pertaining to a given class. Having more complicated structures such as aircraft callsigns or time entities it is usual to replace only parts of the semantic entity with the corresponding lexical class. This paper uses an approach which allows to detect semantic entities described using a grammar (or finite state transducer) in a generic weighted finite state transducer (WFST). Our in-house ASR decoder [8] is able to compute directly the transition probabilities in such WFST. In other words, the lattices are normalized, the probabilities of all hypotheses sums to one. For manipulation with WFST we are using OpenFST framework [1].

We have considered using word confusion networks (WCN) for optimised representation of ASR output, appearing for example in [2]. Their results promise the same detection performance as with word lattices, but allow higher processing speeds. We have encountered problems with this approach. Using a simple lattice F (Fig. 1) we've created WCN F' (Fig. 2) using an algorithm from [2], where the best path is chosen as the pivot path. Consider a sequence of words ab. We can clearly see that in the original

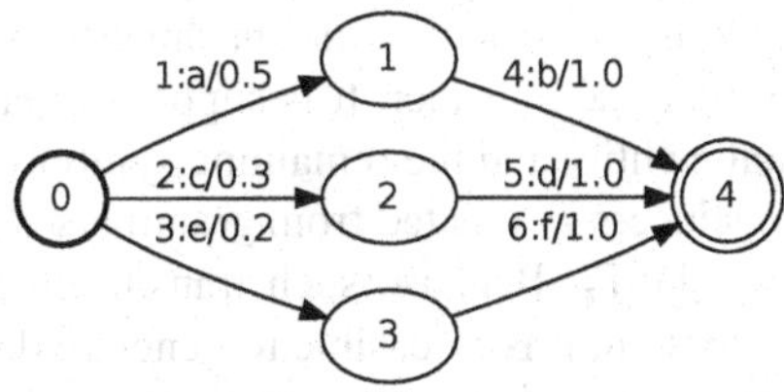

Fig. 1. Lattice example

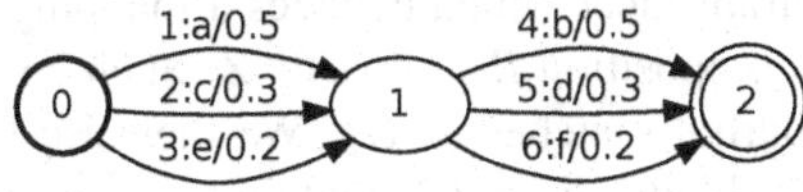

Fig. 2. Word confusion network example

lattice the probability of this sequence was 0.5, whereas the probability in WCN was reduced to 0.25. Experiments in [10] pointed out that the dominance of WCN stands only for single-word entities. Because our entities consist mostly of multiple words, we have chosen to use word lattices instead.

2.1 First Layer

Each type of semantic entity z has a corresponding CFG G_z. Generally, the CFG is parsed using the pushdown automaton with unlimited stack depth. In the case where CFG is not recursive, the stack depth is limited and such an automaton can be converted into a FST where the input symbols correspond to CFG terminal symbols and output symbols are the so called tags assigned by the CFG. If the CFG is recursive, it can be approximated by FST using for example algorithm presented in [7].

In this work, we use the standardized W3C speech recognition grammar specification (SRGS) [3] notation which allows us to use tags inside the rules definitions (Fig. 3). The use of a knowledge-based CFGs does not imply that the method is not probabilistic – it allows to assign posterior probabilities to every semantic entity detected by CFGs.

```
$number = ($d | ten{10} | twenty{20}[$d]);
$d = (one {1} | two {2} | three {3});
$runway = runway {RWY} $number;
```

Fig. 3. Example of grammar with tags for semantic entity detection

Considering our ASR hypotheses are represented as a WFST $\tilde{u}$, we first compile the knowledge base expressed as set of CFGs G_z into a transducer which accepts a string of symbols representing exactly one semantic entity and transduces this string onto a sequence of semantic tags e_i. The output symbols of T_z directly form the entity type and interpretation and allow the construction of "machine readable" objects (database entries, time objects) in a dialogue manager. It is supposed that the first symbol in e_i indicates the type of semantic entity and the remaining symbols are its interpretation.

We represent all the transducers T_z created from grammars G_z as a single transducer by making an union $Z = \bigoplus_z T_z$. Because such transducer does not model all the words an input lattice can contain, it is infeasible to generate desired output by simple composition $\tilde{u} \circ Z$. As an alternative to creation of filler model which would match any of the words unseen in grammars, we create a factor automaton $F(\tilde{u})$ from the input lattice $\tilde{u}$. The factor automaton accepts all subpaths of the original lattice $\tilde{u}$ [6].

We can then generate transducer which encodes a mapping from a lattice subpath to a semantic entity by a composition $R = F(\tilde{u}) \circ Z$. Such composition also contains paths representing only partial matches of G_z. We then use heuristics of maximum unambiguous coverage to obtain subset F^* containing only unambiguously assigned entities from the set of all ambiguous semantic entities F. From the set F we use the subset F^* where each transition in the lattice has assigned at most one semantic entity and the number of transitions with assigned semantic entities is maximal.

An algorithm which takes F^* and time alignment of semantic entities to produce a lexical-semantic lattice F_e where each path encodes one sequence e consisting of semantic entities and words that did not match any grammar G_z is described in [10]. The weights of F_e correspond to posterior probability distribution $P(E = e|W = \tilde{u})$.

2.2 Second Layer for Validation

After creating a lexical-semantic lattice F_e in the first layer, we can now use a set of rules G_v to validate that the sequence of words and semantic entities correspond to phraseology rules. Because most of the phraseology rules have also a semantic information associated with it (for example *change of flight level, report of heading*, we chose to design this validating layer to not only emit information about validity, but also assign semantic meaning to the utterance.

Because the input for the second layer is a lexical-semantic lattice, we are using an lexical-semantic context free grammar in SRGS format made by expert using phraseology guide as the source of rules. These rules have also assigned tags to them to allow the dialogue manager to decide on further actions. An example of such rules are shown in Fig. 4, where the first rule states that path "climbing to _FL_ _CS_" in lexical-semantic lattice means the system should react to a "change of flight level confirmed from pilot.". We use "_" at the beginning and end of the names of semantic entities in order to distinguish them from words.

```
{air_response} climbing to _FL_ _CS_ {level_change} |
{air_response} maintaining _FL_ _CS_ {level_report}
```

Fig. 4. Example of grammar for utterance validation and further semantic annotation with semantic entities surrounded by "_"

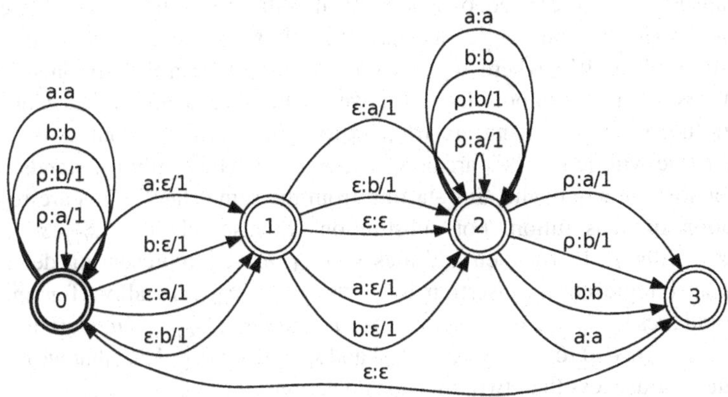

Fig. 5. Finite state transducer E for evaluation of edit distance between transducer F_1, F_2 using composition $F_1 \circ E \circ F_2$ with symbol table containing symbols a and b with only 2 consecutive deletions or insertions allowed

The set of rules G_v is then converted into FST F_v. Because of the way the grammar was created, there is guaranteed to be no recursion. That allows us to convert to FST without any need of approximation as in Sec. 2.1. F_v is created in such way that semantic entities from first layer are considered only on the level of entity types, independent of their interpretation (i.e. entity `_FL_:2:3:0` with type `_FL_` and interpretation `230` matches any `_FL_` in rules G_v).

We are able to validate user's utterance by composing $R_v = F_e \circ F_v$ with lexical-semantic lattice F_e from first layer. If the resulting transducer R_v is empty, then we consider the input invalid, because we didn't find any rules the utterance would match. Otherwise we consider the input valid and we can find the semantic information assigned to it by traversing transducer R_v and noting the output symbols on its arcs.

As this layer does not use any probabilistic approach (rules in G_v do not have any probability assigned to them), the cost of each path is determined only from the score given by previous layer. In case of multiple hypotheses is the best hypotheses based solely on those scores.

2.3 Second Layer for Utterance Validation and Correction

To further enhance our utterance-validating layer, we extend the final step of composing $F_e \circ F_v$ by using intermediary FST that allows us to correct user's input. The algorithm produces lattice where can all paths be considered as valid (based on the same grammar created by an expert as in Sec. 2.2), with cost of the paths being a measure of how many correction had to be made to make them valid.

We are using Levenshtein distance to find the closest valid input interpretation. Concretely, we are using a finite state transducer E to compute edit distance between two transducers with restrictions on number of consequent deletions or insertions c_{id} based on [5].

The transducer E is created over a tropical semiring with $c_{id} + 2$ states, with state s_0 as a start state and $s_{c_{id}+1}$ as final state. For each symbol w from a set of symbols $W = W_e \cup W_v$, where W_e is a set of symbols from the first layer's output F_e and W_v is a set of symbols from F_v, we create several arcs in E. Considering s_0 as a starting state, we create arc from s_0 to s_0 transcribing input w to ρ with a weight of 1 (we will denote this arc as $s_0, s_0, w, \rho, 1$). The symbol ρ stands for "any symbol not seen on any of the arcs starting from the same state". This arc represents the operation of substitution. For substitution we also add arcs $s_{c_{id}}, s_{c_{id}}, w, \rho, 1$ and $s_{c_{id}}, s_{c_{id}+1}, w, \rho, 1$. We then add arcs $s_i, s_{i+1}, w, \epsilon, 1$ representing deletion and $s_i, s_{i+1}, \epsilon, w, 1$ representing insertion for $i \in \{0, \ldots, c_{id}\}$. To allow for unchanged words we add arcs $s_0, s_0, w, w, 0$,$s_{c_{id}}, s_{c_{id}}, w, w, 0$ and $s_{c_{id}}, s_{c_{id}+1}, w, w, 0$. Finally, we add arcs $s_j, s_{j+1}, \epsilon, \epsilon, 0$ for $j \in \{1, \ldots, c_{id}\}$ and $s_{c_{id}+1}, s_0, \epsilon, \epsilon, 1$, so that we can repeat whole pattern or use less than two insertions/deletions.

An example of such transducer allowing for computation of edit distance over a small set of symbols $W = \{a, b\}$ is shown in Fig. 5. Restriction on two deletions or insertions in a row can be seen between states number 0 and 2.

Composing input transducer, edit distance transducer and grammar transducer results in FST $R = F_e \circ E \circ F_v$. An important step before the composition itself is unweighting of transducer F_e, because weights are being used to compute the edit distance. We are using tropical semiring in order for the resulting distance between strings α and β (with insertion, deletion and substitution being penalized by 1) to be obtained as a sum of costs of transitions transcribing string α into β.

It is possible that transducer R contains paths where an edit distance (represented as cost of the path) is too high to be of any use for utterance correction. We can prune the FST R with a reasonably low threshold r, leaving us with transducer R_r that contains only paths with cost lower than the cost of the shortest path increased by r and the dialogue system can then offer all the closest valid variations to the user. We can also choose to use only n-best hypotheses instead of pruning, leaving us with guaranteed number of suggestions to the user.

The algorithm marks the input utterance as valid if there is a path in R_r with cost $\bar{0}$ (i.e. with edit distance equal to 0). Finding this path is a task of finding best path in R_r and then summing up the weights.

Table 1. Percentage of utterances corresponding exactly to the phraseological rules (referred to as valid) in test set using two-layer validation with precision, recall and F1-measure against reference

	1-best	raw	reference
valid	25.17%	29.29%	23.22%
precision	0.76	0.72	–
recall	0.83	0.91	–
F1	0.79	0.80	–

3 Experiments

The experiments were performed on data [9] that were obtained from real communication between pilots and ATCs. The test set contains 1434 sentences. The ASR recognizer vocabulary contains 10.3k words with accuracy of ASR being 83.9%. We were using the following representation of ASR hypothesis: the best hypothesis (1-best) and the raw unoptimized lattice. We've also used reference transcriptions made by human annotators.

The grammars used in our experiments were defined by phraseology guide for semantic entities in first layer as well as for validity rules in second layer.

In our experiments, we did not evaluate the number of semantic entities and validation rules being covered by expert-defined grammars. Based on fixed set of grammars, we evaluated the influence of different ASR hypotheses representations on the validation performance. As a reference for validation, we used results of semantic entity detection and validation on the manually annotated transcription. We consider utterance valid, if it corresponds exactly to the rules.

Results can be seen in Table 1. The number of valid utterances was increased by 14% by using raw lattices instead of 1-best hypothesis. Confusion matrices are in Table 2 for 1-best hypotheses and Table 3 for raw lattices.

We assume the difference between valid sentences in reference transcription and ASR output is caused by the fact that language model in ASR tries to create sequences based on n-grams seen in training data, whereas human annotators that do not know the common word order in a pilot-to-ATC communication (hence do not posses any training data) prefer to mark words only as unintelligible instead of trying to guess.

Table 2. Confusion matrix for validation results using 1-best ASR hypotheses

	valid in reference	invalid in reference
valid from 1-best	275	86
invalid from 1-best	58	1015

Table 3. Confusion matrix for validation results using raw lattice ASR hypotheses

	valid in reference	invalid in reference
valid from lattice	303	117
invalid from lattice	30	984

4 Conclusion and Future Work

We presented a novel algorithm for semantic entity detection with utterance validation and its extension to utterance correction. Using this algorithm we were able to determine the percentage of valid sentences in our data set and assign semantic information to user's input which will be useful in further dialogue system development.

The results showed that by using a raw lattice from an ASR we are not only able to extract a valid hypothesis, but also increase the percentage of valid utterances our system recognized in comparison to the ones retrieved from 1-best hypothesis.

Further work can be done in utterance correction. Our approach using unweighted input lattices causes a loss of information about hypothesis probability obtained from language and acoustic model. This can be improved by using other than tropical semirings, for example lexicographic semiring would allow us to store both the edit distance and the probability from an ASR encoded into weights of transitions.

References

1. Allauzen, C., Riley, M., Schalkwyk, J., Skut, W., Mohri, M.: OpenFst: A general and efficient weighted finite-state transducer library. In: Holub, J., Žd'árek, J. (eds.) CIAA 2007. LNCS, vol. 4783, pp. 11–23. Springer, Heidelberg (2007), `http://www.openfst.org`
2. Hakkani-Tür, D., Béchet, F., Riccardi, G., Tur, G.: Beyond asr 1-best: Using word confusion networks in spoken language understanding. Computer Speech & Language 20(4), 495–514 (2006)
3. Hunt, A., McGlashan, S.: Speech recognition grammar specification version 1.0. In: World Wide Web Consortium, Recommendation REC-speech-grammar-20040316 (March 2004)
4. Mairesse, F., Gasic, M., Jurcicek, F., Keizer, S., Thomson, B., Yu, K., Young, S.: Spoken language understanding from unaligned data using discriminative classification models. In: IEEE International Conference on Acoustics, Speech and Signal Processing, ICASSP 2009, pp. 4749–4752 (April 2009)
5. Mohri, M.: Edit-distance of weighted automata. In: Champarnaud, J.-M., Maurel, D. (eds.) CIAA 2002. LNCS, vol. 2608, pp. 1–23. Springer, Heidelberg (2003)
6. Mohri, M., Moreno, P., Weinstein, E.: Factor automata of automata and applications. In: Holub, J., Žd'árek, J. (eds.) CIAA 2007. LNCS, vol. 4783, pp. 168–179. Springer, Heidelberg (2007)
7. Mohri, M., Nederhof, M.J.: Regular approximation of context-free grammars through transformation. In: Robustness in Language and Speech Technology, pp. 153–163. Springer (2001)
8. Pražák, A., Psutka, J.V., Hoidekr, J., Kanis, J., Müller, L., Psutka, J.: Automatic online subtitling of the Czech parliament meetings. In: Sojka, P., Kopeček, I., Pala, K. (eds.) TSD 2006. LNCS (LNAI), vol. 4188, pp. 501–508. Springer, Heidelberg (2006)
9. Šmídl, L.: Air traffic control communication corpus. Published in LINDAT/CLARING repository (2012), available under CC BY-NC-ND 3.0 from `http://hdl.handle.net/11858/00-097C-0000-0001-CCA1-0`
10. Švec, J., Ircing, P., Šmídl, L.: Semantic entity detection from multiple ASR hypotheses within the WFST framework. In: 2013 IEEE Workshop on Automatic Speech Recognition and Understanding (ASRU), pp. 84–89 (December 2013)

Ontology Based Strategies for Supporting Communication within Social Networks

Ivan Kopeček, Radek Ošlejšek, and Jaromír Plhák

Faculty of Informatics, Masaryk University
Botanická 68a, 602 00 Brno, Czech Republic
{kopecek,oslejsek,xplhak}@fi.muni.cz

Abstract. In this paper, ontology based dialogue strategies are presented in connection with the concept of communicative images. Communicative images are graphical objects integrated with a dialogue interface and linked to an associated knowledge database which stores the semantics of the objects depicted. The relevant pieces of information can be linked to the external knowledge distributed in a social network. Exploiting a formal ontology approach facilitates the process of deriving information from relevant texts that can be found in the social network and it simultaneously forms a suitable framework for supporting dialogue communication in natural language. This approach is discussed and illustrated with various examples in this paper.

1 Introduction

Current technologies enable us to associate many useful pieces of information with images that can be used to support users in their retrieval of relevant and interesting information from pictures. Some data are stored automatically when the image is made, typically the date and time of the snapshot and its geographic coordinates, while other data can be retrieved subsequently through the use of various image recognition strategies, social interaction (e.g. people identification in social networks) or manual annotation (e.g. assigning a list of keywords to images within various photo organizers). Regardless of the way the information is obtained it can subsequently be used to support dialogue-based interaction. We refer to images with the ability to communicate with the user in natural language as communicative images.

The basic principles of communicative images have been proposed in [1]. A single communicative image consists of three data structures: (a) graphical content, (b) localization marks that determine the approximate location of the depicted objects and (c) semantic data, as shown in Figure 1. Our approach exploits the Scalable Vector Graphics (SVG) format [2] to encode all these data structures in a single file. The semantics are encoded as Web Ontology Language (OWL) ontologies [3] which provide a suitable formalism for information retrieval and machine-generated dialogues.

For example, let us imagine a holiday photo in JPEG format, as shown in Figure 1: My wife Jane in front of a castle. To make this photo communicative, it has to be converted into SVG format and stored with the original JPEG data. The supporting OWL ontology that is linked from the SVG, should define terms like "castle" and "wife". The annotation data, which is also embedded in the SVG file, links specific

P. Sojka et al. (Eds.): TSD 2014, LNAI 8655, pp. 571–578, 2014.

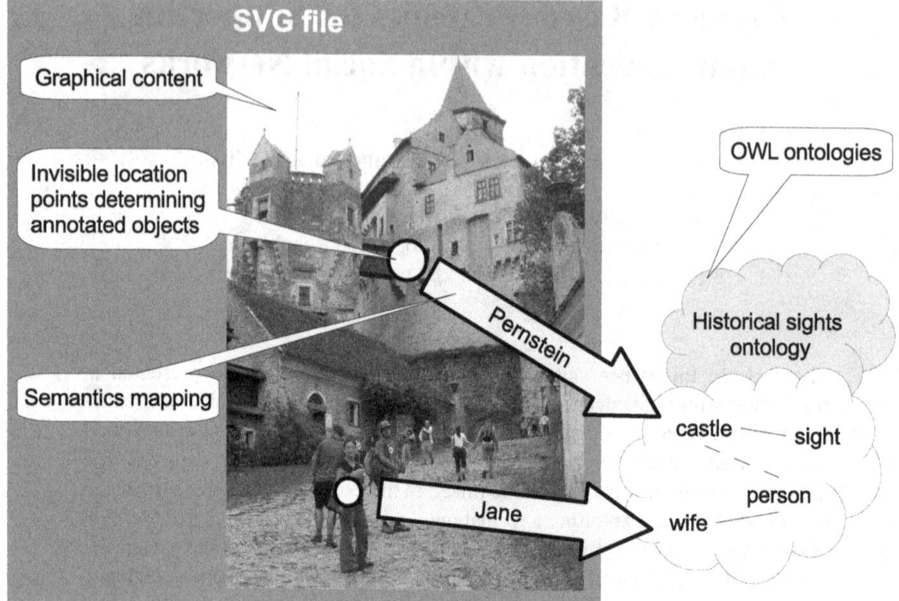

Fig. 1. Structure of a communicative image

graphical elements with terms in the ontology. Provided that the annotation data include the information that the wife has the name Jane and the castle is Pernštejn, the image is able to answer questions like "What is the castle behind Jane?". Moreover, connecting to a general ontology of historical sights, the communicative image would be able to provide the user with details about the castle.

Dialogue-based communication is essentially interactive. It enables us to control the complexity of the information and it prevents the user from being overburdened by superficial information. However, the ability of an image to communicate informatively depends on the ability to associate sufficient relevant information with the image. But making picture annotations and creating ontologies by hand is very labor-intensive and tedious. Furthermore, gathering such data by means of existing image recognition and auto detection techniques cannot always solve the problem because they are still far from being able to fully describe an analyzed picture. Our goal is therefore to enable images to actively learn from the communication. New pieces of information are provided either directly by the user or indirectly by the social community being active in a social network.

Consider the holiday photo discussed above and assume that this photo has no semantic data assigned so far. The following fragment of a dialogue demonstrates the ideas and principles of active learning.

Example 1 (*U = User, P = Picture, // comment*):

U: What is the castle behind Jane?
// At this moment, the communicative picture can assume that it depicts

some castle and some object called Jane. The picture takes over the initiative to learn more about these two things.
P: Who or what is Jane?
U: Jane is my wife.
// Picture semantics can be extended with the fact that there is a woman, the wife of the user, called Jane.
P: Based on geographic coordinates the castle would be Pernštejn. Is it possible?
// An explore external shared knowledge database covering geographic data of sights is now needed to identify the castle according to its position.
U: Yes, it is Pernštejn.
// The user recalled the trip and confirmed the information. The picture extends its semantics with this fact for later re-use.
U: When was Pernštejn built?
// At this moment, the picture tries to retrieve details about Pernštejn castle, either searching the available ontologies or asking the social community for help.
P: I'm sorry but I don't know.
// No information was found at this moment and then the picture asked the social community for assistance. It takes some time to get response but next time the picture should know the answer.

This dialogue demonstrates a way of overcoming the lack of information when neither the user nor the associated knowledge databases are able to provide valid semantic data. In this case, the user question can be stored in the image and shared with social community. This process has to be carefully managed so as to prevent violating the user's privacy and to protect the community from being overburdened with such questions. To optimize this task, question similarity detection, well-formed question verification and answer credibility need to be taken into account.

2 Related Work

Our approach to communicative images is based on formal modelling through ontologies. Ontologies are considered one of the pillars of the semantic web. Many papers have been published dealing with ontology-based intelligent data retrieval from databases [4], the annotation of graphical data [5,6] and from e-learning [7,8]. In [9], Chai et al. proposed an intelligent photo album enabling collections of family photos to be organized and searched by means of ontologies.

In the context of communicative images, dialogue management has to be able to achieve high efficiency in managing the image ontology and provide suitable dialogue strategies. Some approaches proposed in [10,11,12] that enhance the efficiency of the dialogue manager exploiting the knowledge bases have been presented. The dialogue management is separated from the domain knowledge management. In [13], it has been argued that this separation reduces the complexity of the systems and enhances further extensions.

3 Dialogue Strategies for Communicative Images within Social Networks

Communicative images represent dialogue systems that support providing information about non-textual data and other related modes of communication. The pieces of information that are exploited in this process are of the following types:

- *Type 1*: The information directly integrated with the object at the moment of creation, e.g. date and time the picture was shot, geographic coordinates, technical data of the camera, etc.
- *Type 2*: Basic semantic information about the picture which is incorporated directly into its SVG format. It is possible that there is no such information when the picture is created – it is provided by the users during the process of communicating with the picture.
- *Type 3*: The global information situated in the ontology forms in the cloud. Because this ontology saves all possible pieces of information based on the communication within the network, it can be too large to be integrated with the format of the picture. The structure of the ontology, however, makes finding relevant information or providing basic inference processes feasible.
- *Type 4*: The pieces of information spread across the network and internet in various forms, mainly as text. Such information can be converted into the format of the ontology, which enables more flexible use. But if this conversion is not feasible, it can be presented in its original form.

Based on this typology, the process of supporting communication with the user can be characterized as follows.

- Queries about Type 1 information are answered based on the integrated graphical ontology (see [14]).
- Queries about Type 2 information are answered based on the ontology which is integrated with the format of the picture.
- Queries about Type 3 information are answered based on the ontology in the cloud.
- Queries about Type 4 information are answered based on the ontology which is dynamically created in real time during the communication. If it appears that the conversion of the pieces of information into the ontology is neither feasible nor efficient, they can be presented in their original forms.
- Within the social network, the users can add comments, descriptions and answers to the questions of other users about the picture at any time.
- If the query cannot be understood by the system, the system either provides metacommunication with the user or passes the query to other users of the social network to be answered by them. In such cases, it is assumed that they will not be answered in real time, and it is possible that they will remain unanswered.

4 Testbed Implementation

The concept of ontology-based communication strategies is tested and verified by means of a testbed application, which has been developed to deal with communicative images. Its client-server architecture enables us to develop many specialized clients as well as to share the knowledge on the server side.

Clients, typically web applications or plugins to social networks, handle a user's interactions with the image and redirect these interactions to the remote server.

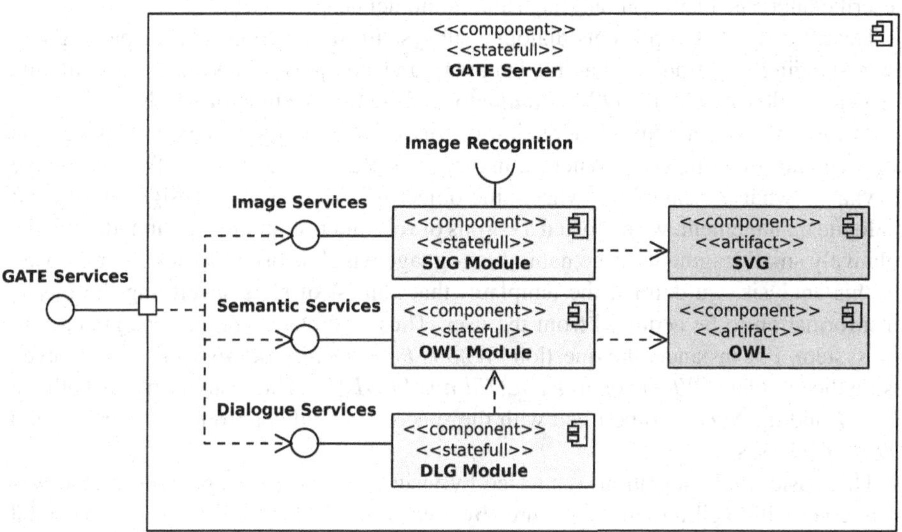

Fig. 2. Component architecture of GATE server

GATE server is designed as a component-based Java enterprise application, as shown in Fig. 2. It provides high-level session-oriented remote services via *SOAP* or *RESTfull* APIs. These services enable clients to upload an image and then to explore it by means of natural language. The server consists of several lower-level components providing services for the graphical content management, semantics management and dialogue subsystem.

The *SVG Module* handles graphical content of a single uploaded image. Provided services enables to explore the SVG Document Object Model (DOM). Either SVG or raster image can be uploaded at the beginning of the communication. Raster images are automatically wrapped with initial SVG content. When connected to external *Image Recognition* services, this module can automatically analyze the image and get initial semantic data by employing domain-specific algorithms, e.g. face recognition [15,16] or similarity search algorithms in large image collections [17,18,19].

The *OWL Module* is responsible for managing the semantic data encoded in OWL format. Annotation data and ontologies encoded in the uploaded communicative image are gathered and processed automatically. Additional ontologies can be added during the communication. Provided services cover ontology management, low-level traversal of the OWL DOM tree, ontology reasoning and information filtering.

Several OWL ontologies have been integrated into this module so far. A *Graphical ontology* [14] can be used to express significant or unusual visual features of the objects depicted as well as their location and mutual position, enabling the annotator to describe the scene in terms like "Abnormally big cat in the upper left corner." or "Oval pool in front of a house." (expressions defined by the ontology are underlined). A *Family ontology* can be used to classify people or to automatically infer their family relationships. A *Sights* ontology provides the vocabulary and background knowledge to describe interests, historical buildings and monuments.

The *DLG Module* represents dialogue subsystem. It is responsible for parsing and understanding questions in natural language and composing answers. This module cooperates closely with the *OWL Module* to analyze the meaning of words.

At present, only a simplified version of this module is implemented. This version supports questions in What-Where Language (WWL) [6], having the format "where is what", "what is where" or "what some object is". Moreover, to design an efficient dialogue management, we analyzed a corpus of relevant user utterances and identified a relatively small fragment of the natural lan- guage which is used by most users. Based on this analysis, we defined the templates that consist of slots specifying the pieces of information to be acquired from the user. These templates are also integrated into the system. For instance, the question "*Who is between me and my wife?*" is resolved using the template "*Who is between SLOT1 and SLOT2?*". The system expects both the *SLOT1* and the *SLOT2* to be filled with the specific individuals from the "*Person*" and "*Object*" classes.

This basic implementation was tested by visually impaired people whose goal was to interact with well-annotated pictures by means of WWL and give us their personal feedback. Although the current functionality is still very limited, there is no social network integration nor any generation of ontologies from texts at the moment. Despite this, the preliminary responses of users are promising and show that the concept of communicative images is feasible and seems to have potential in many application domains.

5 Application Domains

Precisely annotated communicative images are applicable to e-learning systems where the tutor prepares study materials thoroughly by hand. In this case, connecting the depicted objects with a broader knowledge database, e.g. historical, can provide students with a "window" into the whole topic of a field of study. For instance, a well-annotated picture of the battle of Austerlitz enables the students to discuss scene and naturally learn who Napoleon was, when the battle took place and who won.

Communicative images also support the accessibility of graphics for people with special needs. A dialogue with the image held in natural language makes the graphical data accessible especially to visually impaired people. The users are not limited to a simple summary of the image's content. Since the data is structured and related to different parts, objects and aspects of the image, a complex dialogue can be undertaken, ultimately leading to a more natural and fulfilling experience for the users.

Any user can benefit from the ease of access to photo albums. The user can browse family pictures while being reminded of the age and names of the people in the photos, their birthdays, the time and occasion the picture was taken. Moreover, ontologies used in communicative images could support efficient searching for specific images. The search query referring to specific objects or relations depicted in a source image can be specified either in natural language or using an intelligent graphical user interface. Pictures that have satisfy a query can be grouped into thematic albums creating and maintaining some order in large collections of images.

6 Conclusions and Future Work

In this paper, we have outlined basic principles of communicative images as well as the general architecture of the system. Currently, the concept of communication images suffers from the problems associated with the automatic gathering of semantics from available sources. The aim of this paper has been to propose a formal approach of obtaining semantic data from relevant texts. Nevertheless, the concept of communicative images has still many problems yet to be addressed. For instance, there is a gap between semantic models and dialogue strategies. We have to carefully prepare and fine-tune the dialogue subsystem for each concrete domain ontology by hand, instead of generating dialogue strategies automatically from the internal structure of the ontology provided. Also continual enhancement and enlargement of knowledge bases together with automatic learning from dialogues pose a great challenge. In spite of these obstacles, the preliminary results show that our approach is implementable and promises important application to many domains.

References

1. Kopecek, I., Oslejsek, R.: Communicative images. In: Dickmann, L., Volkmann, G., Malaka, R., Boll, S., Krüger, A., Olivier, P. (eds.) SG 2011. LNCS, vol. 6815, pp. 163–173. Springer, Heidelberg (2011)
2. Dahlstóm, E., et al.: Scalable vector graphics (svg) 1.1, 2nd edn. (2011)
3. Lacy, L.W.: OWL: Representing information using the Web Ontology Language. Trafford Publishing, Victoria BC (2005)
4. Muñoz, A., Aguilar, J.: Ontological scheme for intelligent database. In: Proc. of the 11th WSEAS Int. Conf. on Computers, ICCOMP 2007, pp. 1–6 (2007)
5. Schreiber, A.T.G., Dubbeldam, B., Wielemaker, J., Wielinga, B.: Ontology-based photo annotation. IEEE Intelligent Systems 16, 66–74 (2001)
6. Kopeček, I., Ošlejšek, R.: Gate to accessibility of computer graphics. In: Miesenberger, K., Klaus, J., Zagler, W., Karshmer, A.I. (eds.) ICCHP 2008. LNCS, vol. 5105, pp. 295–302. Springer, Heidelberg (2008)
7. Henze, N., Dolog, P., Nejdl, W.: Reasoning and ontologies for personalized e-learning in the semantic web. Educational Technology & Society 7, 82–97 (2004)
8. Lemnitzer, L., et al.: Using a domain-ontology and semantic search in an e-learning environment. In: Innovative Techniques in Instruction Technology, E-learning, E-assessment, and Education, pp. 279–284. Springer, Netherlands (2008)
9. Chai, Y., Xia, T., Zhu, J., Li, H.: Intelligent digital photo management system using ontology and swrl. In: Proc. of the 2010 Int. Conf. on Computational Intelligence and Security, pp. 18–22. IEEE Computer Society (2010)
10. Nyrkko, S., et al.: Ontology-based knowledge in interactive maintenance guide. In: Proc. of the 40th Annual Hawaii Int. Conf. on System Sciences. IEEE Computer Society (2007)

11. Pérez, G., et al.: Integrating owl ontologies with a dialogue manager. Technical report, Spoken Language Technology Workshop 2006, pp. 134–137. IEEE, Phanouriou (2006)
12. Araki, M., Funakura, Y.: Impact of semantic web on the development of spoken dialogue systems. In: Lee, G.G., Mariani, J., Minker, W., Nakamura, S. (eds.) IWSDS 2010. LNCS (LNAI), vol. 6392, pp. 144–149. Springer, Heidelberg (2010)
13. Flycht-Eriksson, A.: Design and Use of Ontologies in Information-providing Dialogue Systems. Dissertation, Linköping studies in science and technology, thesis no. 874, School of Engineering at Linköping University (2004)
14. Ošlejšek, R.: Annotation of pictures by means of graphical ontologies. In: Proc. Int. Conf. on Internet Computing ICOMP, pp. 296–300 (2009)
15. Bartlett, M.S., Movellan, J.R., Sejnowski, T.J.: Face recognition by independent component analysis. IEEE Transactions on Neural Networks 13, 1450–1464 (2002)
16. Haddadnia, J., Ahmadi, M.: N-feature neural network human face recognition. Image and Vision Computing 22, 1071–1082 (2004)
17. Batko, M., Dohnal, V., Novák, D., Sedmidubský, J.: Mufin: A multi-feature indexing network. In: Proceedings of the 2009 Second International Workshop on Similarity Search and Applications, pp. 158–159. IEEE Computer Society (2009)
18. Jaffe, A., Naaman, M., Tassa, T., Davis, M.: Generating summaries and visualization for large collections of geo-referenced photographs. In: Proceedings of the 8th ACM International Workshop on Multimedia Information Retrieval, pp. 89–98. ACM (2006)
19. Abbasi, R., Chernov, S., Nejdl, W., Paiu, R., Staab, S.: Exploiting flickr tags and groups for finding landmark photos. In: Boughanem, M., Berrut, C., Mothe, J., Soule-Dupuy, C. (eds.) ECIR 2009. LNCS, vol. 5478, pp. 654–661. Springer, Heidelberg (2009)

A Factored Discriminative Spoken Language Understanding for Spoken Dialogue Systems

Filip Jurčíček, Ondřej Dušek, and Ondřej Plátek

Charles University in Prague
Faculty of Mathematics and Physics
Institute of Formal and Applied Linguistics
{jurcicek, odusek, oplatek}@ufal.mff.cuni.cz
https://ufal.mff.cuni.cz

Abstract. This paper describes a factored discriminative spoken language understanding method suitable for real-time parsing of recognised speech. It is based on a set of logistic regression classifiers, which are used to map input utterances into dialogue acts. The proposed method is evaluated on a corpus of spoken utterances from the Public Transport Information (PTI) domain. In PTI, users can interact with a dialogue system on the phone to find intra- and inter-city public transport connections and ask for weather forecast in a desired city. The results show that in adverse speech recognition conditions, the statistical parser yields significantly better results compared to the baseline well-tuned handcrafted parser.

Keywords: spoken language understanding, dialogue systems, meaning representation.

1 Introduction

Semantic parsing is a key component of any spoken dialogue system. Its purpose is to map natural language to a formal meaning representation – semantics, which can be defined either by a grammar, e.g. LR grammar for the GeoQuery domain [1], or by frames and slots, e.g. the TownInfo domain [2], or dialogue acts [3]. In this work, dialogue acts are used to represent the meaning. A dialogue act (DA) is composed of one or more dialogue act items (DAI). A dialogue act item represents basic intents (such as `inform`, `request`, etc.) and optionally the semantic content, also referred to as slots, in the input utterance (e.g. `vehicle=bus`, `time=1:30`). In some cases, the value of a slot can be omitted, for example, where the intention is to query it, as in `request(arrival_time)`. Table 1 shows examples of the dialogue acts in the public transport information (PTI) domain. As dialogue managers commonly use semantics in the form of frames and slots [3,4], the presented approach learns to map directly from natural language into the frame and slot semantics.

This paper describes a probabilistic discriminative Spoken Language Understanding (SLU) based on Dialogue Act Item Classifiers using Logistic Regression (DAICLR), where logistic regression classifiers are used to map input utterances into dialogue acts. To obtain a compact probabilistic representation, the predicted dialogue acts are factored according to dialogue act items, each associated with a corresponding item

P. Sojka et al. (Eds.): TSD 2014, LNAI 8655, pp. 579–586, 2014.

Table 1. Examples of the PTI semantics

1.	ZE STANICE NÁDRAŽÍ HOLEŠOVICE *from the NÁDRAŽÍ HOLEŠOVICE station* `inform(from_stop="Nádraží Holešovice")`
2.	JÁ BYCH CHTĚLA JET V OSM HODIN RÁNO *I would like to leave at 8 am* `inform(ampm="morning")&inform(departure_time="8:00")`

3.	JEDE TO NA ZVONAŘKU
	`is it going to ZVONAŘKA`
	`confirm(to_stop="Zvonařka")`

4.	V KOLIK HODIN TO BUDE NA STANICI VELETRŽNÍ PALÁC *what time does it arrive to the VELETRŽNÍ PALÁC stop* `inform(to_stop="Veletržní palác")&request(arrival_time)`

Top to bottom for each statement: Czech user utterance, English literal translation, dialogue act.

marginal probability instead of representing the uncertainty of dialogue acts in N-best lists. The proposed DAICLR method is evaluated on a corpus of spoken utterances from the Public Transport Information (PTI) domain. In PTI, users can interact in Czech language with a telephone-based dialogue system to find intra- and inter-city public transport connections and ask for weather forecast in a desired city. The PTI system is publicly available at a toll-free phone number and covers virtually all cities in the Czech Republic [5].

A successful SLU component in a spoken dialogue system must be robust to recognition errors, easy to build, computationally efficient, and provide accurate predictive probabilities. As the DAICLR parser is directly trained on the output of the ASR component which includes recognition errors, the DAICLR is capable of learning to deal with systematic recognition errors. The parser learns from data which has no alignment between words and semantics. It can efficiently parse not only 1-best ASR output but also N-best lists with ASR hypotheses. In addition, it learns a small set of classifiers that allows real-time parsing. Finally, the output probability estimates accurately model the chance of the predicted items being correct due to the use of discriminatively trained logistic regression classifiers. This is particularly important in probabilistic dialogue state tracking components [6].

In the next section, related work on mapping natural language into formal meaning representations is described. Section 3 presents the proposed DAICLR parser and describes the training process. Section 4 compares the proposed DAICLR parser to a well-tuned handcrafted parser developed for the PTI domain. Finally, Section 5 concludes this work.

2 Related Work

There has already been a substantial amount of work on data-driven approaches to SLU. This section briefly describes some of the main contributions to this in literature.

The Hidden Vector State (HVS) technique has been used to model an approximation of a pushdown automaton with semantic concepts as non-terminal symbols [7,8]. From the output parse trees, a deterministic algorithm was used to recover slot names and their values.

A probabilistic parser using Combinatory Categorical Grammar (PCCG) has been used to map utterances to lambda calculus [10]. The combinatory categorical grammar is converted into a probabilistic model by learning a log-linear model. An online learning algorithm updates weights of features representing the parse tree of an input utterance. However, apart from using the lexical categories (city names, airport names, etc.) readily available from the ATIS corpus [11], this method also needs a considerable number of handcrafted entries in its initial lexicon.

Markov Logic Networks (MLN) have been used to extract slot values by combining probabilistic graphical models and first-order logic [12]. In this approach, weights are attached to first-order clauses which represent the relationship between slot names and their values. Such weighted clauses are used as templates for features of Markov networks.

Semantic Tuple Classifiers (STC) based on support vector machines have been used to build semantic trees by recursively calling classifiers that predict fragments of the semantic representation from N-gram features [2].

The domain-independent semantic role labelling was used as a form of preprocessing to reduce complexity of mapping to domain-dependent meaning representations [14].

Machine translation techniques [15] have been used with a translation model based on synchronous context-free grammars.

Inductive logic programming [16] has been used to incrementally develop a theory including a set of predicates. In each iteration, the predicates were generalised from predicates in the theory and predicates automatically constructed from examples.

The DAICLR approach is similar to the STC parser; however, its implementation is more straightforward and consequently more computationally efficient.

3 Methodology

This section describes the DAICLR parsers. First, a description of an utterance and dialogue act abstraction using the in-domain gazetteers is provided. Second, the method of training dialogue act item classifiers is described. Third, features used in the dialogue act classifiers are detailed.

3.1 Utterance and Dialogue Act Abstraction

The DAICLR model uses a set of independent classifiers for dialogue act items that can appear in the output dialogue act. Since the number of possible slot values for each slot is generally very high, this would lead to a very large set of classifiers specialised to classify individual combinations of dialogue act type, slot name, and slot value in the input utterance. Consequently, the training process would suffer from severe data sparsity since most of the slot values are never seen in training data. Therefore, to reduce the data sparsity, a form generalisation using gazetteers with surface forms for slot categories such as city and stop is implemented [7,10,12,13]. A simple deterministic procedure is used to abstract an utterance and associated dialogue act. This procedure has two variants: the first one is used in training, the second one in decoding. In training, surface forms of slot values found in the dialogue act are replaced by their category labels in both the utterance and the dialogue act. This is demonstrated in the next example:

chtěla bych jet z Anděla ⇒ `inform(from_stop="Anděl")`
i want to leave from Anděl (Eng. lit. tran.)

is abstracted to

chtěla bych jet z STOP ⇒ `inform(from_stop="STOP")`
i want to leave from STOP (Eng. lit. tran.)

Then an abstract classifier for `inform(from_stop="STOP")` is trained. However, gazetteers are not always accurate and do not include all possible surface forms. Therefore, when no surface forms for a given slot value can be found in the utterance, the slot value is left un-abstracted and a specialised classifier just for this specific slot value is trained. To prevent creation of too many specialised classifiers, only classifiers for values with occurence counts larger than a certain threshold are created.

In decoding, the utterance is first abstracted by replacing the surface forms of slot values by their category labels. Each substitution of a surface form is recorded together with the corresponding slot value. Then, dialogue act item classification is performed using both abstract and specialised classifiers. Finally, outputs predicted by abstract classifiers are converted back, replacing the category labels with the corresponding slot values from the substitution records. A similar approach to generalisation was used in [2].

3.2 DAICLR Model

The DAICLR dialogue act predictive model is factored according to dialogue act items as follows:

$$p(d|u) = \prod_i p(i|u), \tag{1}$$

where d is a dialogue act, u is the input utterance, i is the i-th item of the dialogue act d, and each probability $p(i|w)$ is modelled by a logistic regression classifier. Let the feature function $f(i,u) \in R^d$, defined as a d-dimensional vector, represent features extracted from the input utterance u for the dialogue act item i. Let be the $w \in R^d$ a parameter vector. Then the probability of a dialogue act item i is defined as

$$p(i|u) = \frac{1}{1 + \exp\{-w_i \cdot f(i,u))\}}. \tag{2}$$

In this work, the logistic regression classifier training was performed using the Scikit-Learn [18][1] software package. However, any other tool could be used. Other types of

[1] `http://scikit-learn.org/stable/`

classifiers such as support vector machines and kernelised logistic regression [9] were evaluated as well. As these approaches provided similar performance on the evaluation task, their evaluation is omitted in this work. The source code of the DAICLR parser is available on GitHub[2]. In the next Section, the features used in the classifiers are detailed.

3.3 Feature Extraction

To make the DAICLR parser computationally efficient, only lexical N-gram features extracted from the input utterance are used. In informal experiments, it was observed that N-grams for N up to 4 bring consistent improvement in accuracy. In addition, skipping bigrams, which can skip up to 3 words, were used. The skipping bigrams have a large span, yet they do not suffer from data sparsity as much as high-order N-grams. To prevent overfitting, simple feature pruning based on counts of occurrences in the training data was introduced. The threshold was set to the size of the N-gram plus a fixed small constant, which was tuned on development data. While abstraction was used to generate abstract features (see Section 3.1), the final feature set also includes examples of original surface forms. This enables classifiers to adjust its classification to common values, such as Prague (the capital city of the Czech Republic) or Anděl (a stop with high public transport traffic). For example, the features extracted for the sentence "chtěla bych jet z Anděla" include N-grams 'chtěla', 'z Anděla', 'chtěla * Anděla', 'z STOP', 'chtěla * STOP', and so on.

So far, extraction from text or 1-best ASR hypothesis was described. To process ASR N-best lists, the same feature extraction process is performed for each hypothesis in the list. The final feature set is then a weighted combination of features extracted from individual ASR hypotheses where the weights correspond to the probabilities of the hypotheses.

4 Experiments

In this section, the DAICLR parser is evaluated on the corpus of user interactions with a statistical spoken dialogue system in the PTI domain [5].

4.1 Data

The PTI corpus consist of approximately 1800 dialogue call logs, which amount to about 11870 user utterances. All audio recordings in the data were transcribed by professional transcribers. The transcriptions are orthographic and capture several kinds of non-speech events as well as incompletely pronounced words and foreign words used in Czech discourse. To obtain semantic annotation, a semi-automatic transcription process was employed. A handcrafted parser was built by an expert in an iterative manner: The parser was first used to obtain semantic annotations by processing human

[2] `https://github.com/UFAL-DSG/alex/blob/master/alex/components/slu/dailrclassifier.py`

transcriptions, these automatic annotations were then verified by an independent expert and identified errors were corrected in the handcrafted parser.

This approach seems to be appropriate since most of advantages of a trainable SLU come from the ability to adapt to ASR errors. In addition, obtaining semantic annotation for Czech data is relatively slow and complicated; using crowdsourcing is not a possibility due to lack of speakers of Czech on platforms such as Amazon Mechanical Turk[3].

The data were divided into training, development, and test sections, where the corresponding data sizes were 9,496, 1,188, 1,188 utterances respectively. Apart from manual transcriptions, the data includes the 1-best and 10-best lists of ASR hypotheses, which allows us to evaluate the robustness of our models to recognition errors. Any tuneable parameters such as pruning thresholds were set on the development data and the reported results were obtained using the test set.

To obtain ASR hypotheses, we used the Google cloud-based ASR[4]. The main advantage of Google ASR is that it is fast, can be used off-the-shelf without any additional modifications, and provides state-of-the-art quality for many tasks [17]. This setup yields performance of 45.2% WER (Word Error Rate) on our data. The high WER is presumably caused by adverse acoustic conditions presented in the PTI speech data, e.g. street noise, and mismatch between the Google's language model and the PTI domain.

Table 2. Dialogue act item precision, recall and F-measure for the PTI test set

Parser	Precision	Recall	F-measure
1-best ASR output			
handcrafted parser	50.83	47.39	49.12
DAICLR	68.39	67.52	67.95
N-best ASR output			
handcrafted parser	50.89	47.49	49.13
DAICLR	68.59	67.74	68.16

4.2 Results

The results for the PTI test data are shown in Table 2. The model accuracy is measured in terms of precision, recall, and F-measure (harmonic mean of precision and recall) on dialogue act items. A dialogue act item is correct only if the dialogue act type, slot, and value are correct.

The DAICLR parser significantly outperforms the baseline handcrafted parser. Regarding the 1-best ASR output, Table 2 shows that DAICLR produces 67.95% of F-measure, which represents a 18.83% improvement over the handcrafted parser. Concerning the N-best ASR output, results shows only modest improvement over the

[3] `https://www.mturk.com/mturk/welcome`

[4] The API is located at `https://www.google.com/speech-api/v1/recognize` for the Czech version of the service, and its use is described in a blog post at `http://mikepultz.com/2013/07/google-speech-api-full-duplex-php-version/`.

1-best results. Presumably, the probabilities in the N-best ASR output do not reflect the accuracy of the ASR hypotheses well. On manual inspection of the data, we observed that only a small portion of the total probability mass was distributed among alternative hypotheses on average.

The DAICLR parser is very computationally efficient on domains such as PTI because the final number of abstracted and specialised classifiers is small. There are 24 unique dialogue act types and 21 unique slots in the PTI domain and the total number of learnt classifiers was 135. On a standard workstation, the DAICLR parser can process one utterance in under 50 milliseconds.

5 Conclusion

This paper presents a novel spoken language understanding parser based on a factored probabilistic model with individual factors modelled using logistic regression classifiers. This approach learns a small set of abstract and specialised classifiers which generalise or specialise to any slot value in a domain. The concept of a factored model for dialogue act item classification and a small set of classifiers is the core of the parser's computational efficiency. It was verified that in adverse acoustic conditions, the trainable DIACLR parser significantly outperforms a well-tuned handcrafted parser and can significantly mitigate the problem of inaccurate speech recognition.

Acknowledgments. This research was partly funded by the Ministry of Education, Youth and Sports of the Czech Republic under the grant agreement LK11221, core research funding and grant GAUK 2058214 of Charles University in Prague. This work has been using language resources distributed by the LINDAT/CLARIN project of the Ministry of Education, Youth and Sports of the Czech Republic (project LM2010013).

References

1. Kate, R.J., Wong, Y.W., Mooney, R.J.: Learning to Transform Natural to Formal Languages. In: Proceedings of AAAI, pp. 1062–1068 (2005)
2. Mairesse, F., Gasic, M., Jurčíček, F., Keizer, S., Thomson, B., Yu, K., Young, S.: Spoken language understanding from unaligned data using discriminative classification models. In: Proceedings of IEEE International Conference on Acoustics, Speech and Signal Processing, pp. 4749–4752 (2009)
3. Thomson, B., Gašić, M., Keizer, S., Mairesse, F., Schatzmann, J., Yu, K., Young, S.: User study of the Bayesian update of dialogue state approach to dialogue management. In: Proceedings of Interspeech, pp. 483–486 (2008)
4. Williams, J., Young, S.: Partially observable Markov decision processes for spoken dialog systems. Computer Speech and Language 21(2), 393–422 (2007)
5. Public Transport Information System for Czech Republic (2014), `https://ufal.mff.cuni.cz/alex-dialogue-systems-framework/ptics`
6. Žilka, L., Marek, D., Korvas, M., Jurčíček, F.: Comparison of Bayesian Discriminative and Generative Models for Dialogue State Tracking. In: SIGDIAL 2013: Proc. of the 14th Annual Meeting of the Special Interest Group on Discourse and Dialogue, Metz, France, pp. 452–457 (2013)

7. He, Y., Young, S.: Semantic processing using the Hidden Vector State model. Computer Speech & Language 19(1), 85–106 (2005)
8. Jurčíček, F., Švec, J., Müller, L.: Extension of the HVS semantic parser by allowing left-right branching. In: Proceedings of ICASSP, pp. 4993–4996 (2008)
9. Zhu, J., Hastie, T.: Kernel logistic regression and the import vector machine. Journal of Computational and Graphical Statistics 14(1), 109–185 (2005)
10. Zettlemoyer, L.S., Collins, M.: Online learning of relaxed CCG grammars for parsing to logical form. In: Proceedings of the 2007 Joint Conference on Empirical Methods in Natural Language Processing and Computational Natural Language Learning, pp. 678–687 (2007)
11. Dahl, D.A., Bates, M., Brown, M., Fisher, W., Hunicke-Smith, K., Pallett, D., Pao, C., Rudnicky, A., Shriberg, E.: Expanding the scope of the ATIS task: The ATIS-3 corpus. In: Proceedings of the ARPA HLT Workshop, pp. 43–48 (1994)
12. Meza-Ruiz, I.V., Riedel, S., Lemon, O.: Spoken Language Understanding in dialogue systems, using a 2-layer Markov Logic Network: Improving semantic accuracy. In: Proceedings of Londial (2008)
13. Tür, G., Hakkani-Tür, D.Z., Hillard, D., Celikyilmaz, A.: Unsupervised Spoken Language Understanding: Exploiting Query Click Logs for Slot Filling. In: Proceedings of Interspeech, pp. 1293–1296 (2011)
14. Henderson, J.: Semantic Decoder which Exploits Syntactic-Semantic Parsing, for the TownInfo Task. In: CLASSiC Project Deliverable 2.2 (2009)
15. Wong, Y.W., Mooney, R.J.: Learning for Semantic Parsing with Statistical Machine Translation. In: Proceedings of HLT/NAACL, pp. 439–446 (2006)
16. Tang, L.R., Mooney, R.J.: Using multiple clause constructors in inductive logic programming for semantic parsing. In: Flach, P.A., De Raedt, L. (eds.) ECML 2001. LNCS (LNAI), vol. 2167, p. 466. Springer, Heidelberg (2001)
17. Morbini, F., Audhkhasi, K., Sagae, K., Arstein, R., Can, D., Georgiou, P.G., Narayanan, S.S., Leuski, A., Traum, D.: Which ASR should I choose for my dialogue system? In: Proc. of SIGDIAL, Metz, France, pp. 394–403 (2013)
18. Pedregosa, F., et al.: Scikit-learn: Machine Learning in Python. JMLR 12, 2825–2830 (2011)

Alex: A Statistical Dialogue Systems Framework

Filip Jurčíček, Ondřej Dušek, Ondřej Plátek, and Lukáš Žilka

Charles University in Prague
Faculty of Mathematics and Physics
Institute of Formal and Applied Linguistics
{jurcicek,odusek,oplatek,zilka}@ufal.mff.cuni.cz
https://ufal.mff.cuni.cz

Abstract. This paper describes the Alex Dialogue Systems Framework (ADSF). The ADSF currently includes mature components for public telephone network connectivity, voice activity detection, automatic speech recognition, statistical spoken language understanding, and probabilistic belief tracking. The ADSF is used in a real-world deployment within the Public Transport Information (PTI) domain. In PTI, users can interact with a dialogue system on the phone to find intra- and inter-city public transport connections and ask for weather forecast in a desired city. Based on user responses, vast majority of the system users are satisfied with the system performance.

Keywords: automatic speech recognition, spoken language understanding, dialogue state tracking, dialogue systems.

1 Introduction

This paper introduces the Alex Dialogue Systems Framework (ADSF). The ADSF is freely available for download on GitHub[1] and designed for easy adaptation to new domains and languages. The ambition of the ADSF is to provide a modular platform for experimenting with statistical methods in the area of spoken dialogue systems. As of now, the system already allows the use of different data-driven components for automatic speech recognition, spoken language understanding, and dialogue state tracking. Since the framework follows a modular design, the system allows for easy replacement of individual components.

The ADSF is used for the development of our experimental Czech spoken dialogue system in the Public Transit Information (PTI) domain [6]. In PTI, the dialogue system provides information about all kinds of public transport in the Czech Republic and is publicly available at a toll-free 800 telephone number. It was launched into public use as soon as a first minimal working version was developed. Nine months after the launch, we have collected nearly 900 calls from the general public. The domain supported by the system includes transport information among ca. 5,000 towns and cities in the whole Czech Republic, plus weather and time information.

In the next section, the overall structure of our ADSF is described. Section 3 describes the PTI domain in detail and analyses its public deployment. Finally, Section 4 concludes this work.

[1] https://github.com/UFAL-DSG/alex

P. Sojka et al. (Eds.): TSD 2014, LNAI 8655, pp. 587–594, 2014.

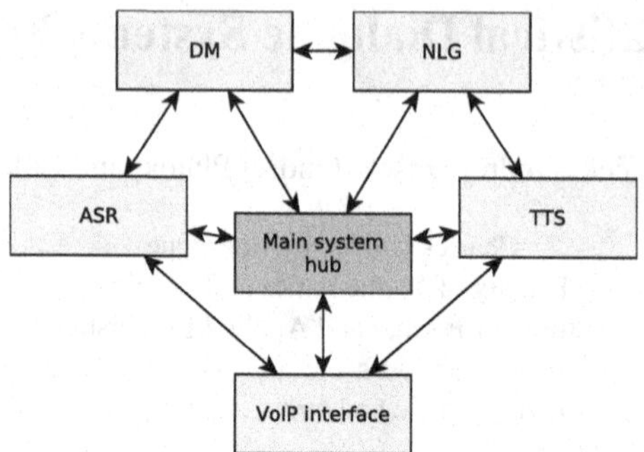

Fig. 1. The structure of our SDS framework

2 Framework Structure and the Main Components

The basic architecture of our system is modular and consists of the traditional Spoken Dialogue System (SDS) components: an automatic speech recogniser (ASR) including a voice activity detector, spoken language understanding (SLU), dialogue manager (DM), natural language generator (NLG), and a text-to-speech (TTS) module.

The framework is developed in Python, a high level programing language. If needed due to performance issues, C/C++ extensions can be easily integrated with Python. To ensure real-time responsiveness of the whole system, each of the components runs in a separate process on a multi-core machine and communicates with other components via operating system pipes. The central component is a hub which coordinates all components used within the framework. All components have access to a common logging process to allow for synchronised logging of events and data. The logged data includes input and output audio streams, speech segments identified by voice activity detector, ASR hypotheses, SLU hypotheses, dialogue state for each dialogue turn, the responses of the in the form of dialogue acts, and the output of the NLG module. A schema of the system is shown in Figure 1.

In the remainder of the section, the VoIP, ASR, SLU and DM components are described in detail as they already reached maturity. Concerning the NLG and TTS components, NLG uses a simple template-based approach and TTS uses web-based TTS services such as SpeechTech[2] or VoiceRSS[3].

2.1 Voice-Over-IP Interface

The Voice-Over-IP interface is based on the freely available PJSIP 2.1 library[4]. To bring real-time processing of voice audio into Python, a fork of the project[5] was created. This fork enables applications written in Python to perform in-memory audio signal recording and playing, which was not possible with the standard Python module

[2] http://www.speechtech.cz
[3] http://www.voicerss.org/
[4] http://www.pjsip.org/
[5] https://github.com/UFAL-DSG/pjsip

provided by the PJSIP 2.1 library. The interface allows for standard telephone network functionality such as answering incoming calls or dialling outgoing calls.

2.2 Automatic Speech Recognition

Same as in most traditional dialogue systems, the ADSF uses voice activity detection to split the incoming audio signal stream into speech and silence segments. To perform voice activity detection, the ADSF implements two algorithms: Gaussian Mixture Models and Feed-Forward Neural Networks. Both approaches are trained from data force-aligned by the HTK[6]. Once speech segments are identified, they are sent to an ASR component to be recognised. In ADSF, two ASR systems are currently supported: the Google cloud-based ASR and Kaldi OnlineLatgenRecogniser.

Google ASR. The main advantage of the Google cloud-based ASR[7] is that it is relatively fast and can be used off-the-shelf without any additional modifications. However, the Word Error Rate (WER) of the Google ASR in the PTI domain is about 45% and its average latency is well above 400 ms. The high WER is likely caused by a mismatch between Google's acoustic and language models and the test data, and the high latency is presumably caused by the batch processing of audio data and network latency.

Kaldi OnlineLatgenRecogniser. The OnlineLatgenRecogniser is our own extension of the state-of-the-art Kaldi automatic speech recognition toolkit [5]. The OnlineLatgenRecogniser recogniser may use acoustic models trained using techniques such as Linear Discriminant Analysis (LDA), Maximum Likelihood Linear Transform (MLLT), Boosted Maximum Mutual Information (BMMI), or Minimum Phone Error (MPE). In the PTI domain, OnlineLatgenRecogniser yields a performance of 18.54% WER when trained on VYSTADIAL 2013 – Czech telephone speech data[8] and the PTI speech data using BMMI discriminative training with LDA and MLLT feature transformations. A class based N-gram language model with a vocabulary consisting of approximately 52k words was estimated only from the PTI speech data transcriptions. Informal experiments show that the OnlineLatgenRecogniser has a latency of about 60 ms on average. The scripts used for training the acoustic models are available in the ADSF repository and a detailed description of the data and training procedure is given in [12]. The OnlineLatgenRecogniser is distributed under the Apache 2.0 license and the source code is available in the Kaldi repository [9].

---

[6] `http://htk.eng.cam.ac.uk/`

[7] The API is accessible at `https://www.google.com/speech-api/v1/recognize`, and its use is described in a blog post at `http://mikepultz.com/2013/07/google-speech-api-full-duplex-php-version/`.

[8] `http://hdl.handle.net/11858/00-097C-0000-0023-4670-6`

[9] `https://sourceforge.net/p/kaldi/code/HEAD/tree/sandbox/oplatek2/src/dec-wrap/`

Czech utterance: "v kolik hodin to bude na stanici veletržní palác"
English translation: *what time does it arrive to the veletržní palác stop*
Dialogue act: `inform(to_stop="Veletržní palác")&request(arrival_time)`

Fig. 2. Example dialogue act representation of an utterance

2.3 Spoken Language Understanding

The meaning extracted from the recognised utterances is represented using dialogue acts. A dialogue act (DA) is composed of one or more dialogue act items, where each dialogue act item represents basic intents (such as `inform`, `request`, etc.) and optionally the semantic content, also referred to as slots, in the input utterance (e.g. `vehicle=bus`, `time=1:30`). In some cases, the value of a slot can be omitted, for example when the intention is to query it, as in `request(arrival_time)`. An example of this is shown in Figure 2.

In ADSF, trainable spoken language understanding is provided by a probabilistic discriminative module based on dialogue act item classifiers using logistic regression (DAICLR). Since the number of possible slot values for each slot is generally very high, a form generalisation using gazetteers with surface forms for slots categories such as city and stop is implemented [8,11,16,17]. The DAICLR parser may be directly trained on the output of the ASR component to learn to deal with systematic speech recognition errors. In the DAICLR parser, only lexical N-gram features are extracted from the input utterance. To prevent overfitting, simple feature pruning based on counts of occurrences in the training data was introduced. To process ASR N-best lists, the same feature extraction process is performed for each hypothesis in the list. However, the final feature set is a weighted combination of features extracted from individual ASR hypotheses where the weights correspond to the probabilities of the hypotheses. The DAICLR parser is very computationally efficient on domains such as PTI because the typical number of classifiers is small. There are 24 unique dialogue act types and 21 unique slots in the PTI domain and the total number of classifiers trained is 135. The source code of the DAICLR parser is available in the ADSF repository[10].

2.4 Dialogue Management

Dialogue management in ADSF is split into two components: dialogue state tracking and dialogue policy.

The purpose of state tracking is to monitor dialogue progress and provide a compact representation of the past user input and system output in the form of a dialogue state. To represent the uncertainty in the estimate of the dialogue state, statistical dialogue systems maintain a probability distribution over all dialogue states called the *belief state* [4].

The dialogue policy is then responsible for decisions on how to continue in a conversation. This can be implemented, for example, using handcrafted rule-based policies or data driven policies optimised by reinforcement learning.

[10] `https://github.com/UFAL-DSG/alex/blob/master/alex/components/slu/dailrclassifier.py`

Belief tracking. In ADSF, we use a probabilistic discriminative tracker [7] which achieves near state-of-the-art performance while remaining completely parameter-free. This is possible because it uses a simple deterministic state transition probability distribution. Interestingly, this discriminative model gives performance comparable to more complex trackers; yet it is significantly more computationally efficient.

Dialogue policy. As of now, the ADSF only supports handcrafted dialogue policies. However, it can take take advantage of uncertainty estimated by the belief tracker. The main logic of the dialogue policy implemented for PTI is similar to that of [18]. First, it implements a set of domain-independent actions, such as: dialogue opening, closing, and restart, implicit confirmation of changed slots with high probability of the most probable value, explicit confirmation for slots with a lower probability of the most probable value, or a choice among two similarly probable values. Second, domain-specific actions are implemented for the PTI domain, such as: informing about a connection, informing about weather forecast, informing about current time.

3 Public Deployment: Transport Information Domain

The ADSF is used for the development of our experimental Czech spoken dialogue system in the PTI domain. In the PTI domain, a user can ask for information about public transport connections in Czech Republic, weather forecast, and current time. Users can control conversation by:

- asking for help,
- asking to “restart” the dialogue and start a new conversation,
- ending the call - for example, by saying “good-bye”,
- asking for a repetition of the last system utterance,
- confirming or rejecting system questions.

When asking for transport connections, the user provides his/her origin and destination and the application finds a connection. The user may specify departure or arrival time in absolute or relative terms (“in ten minutes”, “tomorrow morning”, “at 6 pm.”, “at 8:35” etc.) and request more details about the connection: number of transfers, journey duration, departure and arrival time. The user may travel not only among public transport stops within one city, but also among multiple cities or towns. The system is able to infer the city name from stop name and use a default stop for the given city. As of now, the system supports ca. 44,000 stops in 5,000 cities and towns using the Czech national public transport database provided by CHAPS[11].

Weather information service provided by OpenWeatherMap[12] covers all Czech cities. The user may ask for weather at the given time or on the whole day.

The system is accessible at a public toll-free 800 number. We advertised the service at our university, among friends, and via Facebook. We cooperate with the Czech

[11] `http://www.idos.cz`

[12] `http://openweathermap.org/`

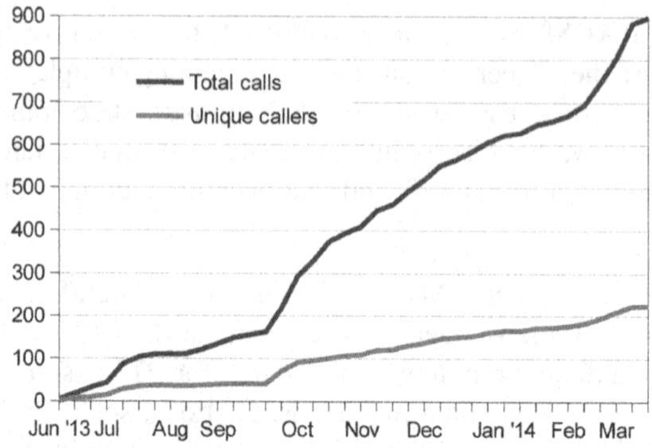

Fig. 3. Cumulative number of calls and unique callers from the public by weeks

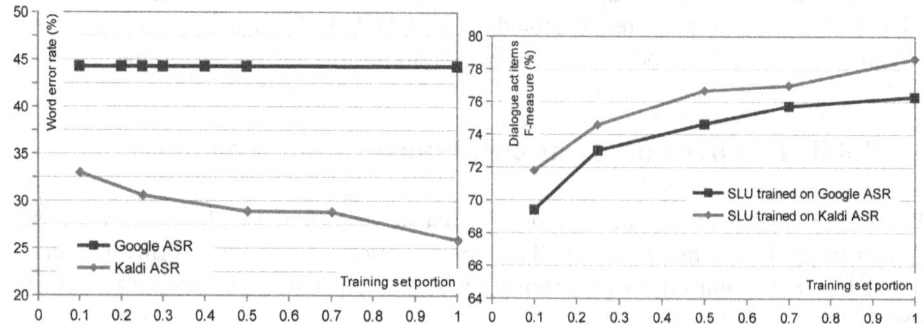

Fig. 4. ASR word error rate and SLU performance (F-measure on dialogue act items) depending on the size of in-domain training data

Blind United association[13], promoting our system among its members and receiving comments about its use.

We record and collect all calls to the system, including our own testing calls, to obtain training data for the satistical components of our system. The number of calls and unique users (caller phone numbers) grows steadily; so far, over 200 users from the public have made nearly 900 calls to the system (see Figure 3)[14]. Information was provided to the user in the vast majority of calls (740 out of 900). We included a measure of the subjective user success rate into our system. After the users says good-bye, the system asks them if they received the information they were looking for. By looking at the transcriptions of responses to this question, we recognise about 90% of them as rather positive ("Yes", "Nearly" etc.).

While the system was deployed, we have been gradually improving our ASR and SLU components. A performance comparison of Google ASR with Kaldi ASR trained on our data is shown in Figure 4 (left). One can see that the Kaldi ASR improves as

[13] `http://www.sons.cz`

[14] We only count calls with at least one valid user utterance, disregarding calls where users hang up immediately.

more training data is available over time and that it outperforms Google ASR very early. Figure 4 (right) shows that the performance of the statistical SLU module improves with more training data. Interestingly, the trainable SLU helps significantly to mitigate the high WER of Google ASR.

4 Conclusion

This paper presents the Alex Dialogue Systems Framework (ADSF). Although its development is in an early stage, it already includes Voice-over-IP interface, various automatic speech recognition options, statistical spoken language understanding, and probabilistic belief tracking. The ADSF is an extensible dialogue system framework and is available for download under the Apache 2.0 license.

The ADSF is currently used in a dialogue system within the public transport information domain. The system is publicly available on a toll-free phone number. We have already collected nearly 900 calls with real users and this is currently increasing by 10 calls per day on average. The analysis of our call logs shows that our system is able to provide information in the vast majority of cases. Success rating provided by the users themselves is very positive.

Future plans include development of data-driven models for belief tracking, dialogue policy, and natural language generation. The PTI domain is expected to serve as an evaluation platform with real users.

Acknowledgments. This research was partly funded by the Ministry of Education, Youth and Sports of the Czech Republic under the grant agreement LK11221, core research funding, grant GAUK 2058214, and 2076214 of Charles University in Prague. This work has been using language resources distributed by the LINDAT/CLARIN project of the Ministry of Education, Youth and Sports of the Czech Republic (project LM2010013).

References

1. Kate, R.J., Wong, Y.W., Mooney, R.J.: Learning to Transform Natural to Formal Languages. In: Proceedings of AAAI, pp. 1062–1068 (2005)
2. Mairesse, F., Gasic, M., Jurčíček, F., Keizer, S., Thomson, B., Yu, K., Young, S.: Spoken language understanding from unaligned data using discriminative classification models. In: Proceedings of IEEE International Conference on Acoustics, Speech and Signal Processing, pp. 4749–4752 (2009)
3. Thomson, B., Gašić, M., Keizer, S., Mairesse, F., Schatzmann, J., Yu, K., Young, S.: User study of the Bayesian update of dialogue state approach to dialogue management. In: Proceedings of Interspeech, pp. 483–486 (2008)
4. Williams, J., Young, S.: Partially observable Markov decision processes for spoken dialog systems. Computer Speech and Language 21(2), 393–422 (2007)
5. Povey, D., et al.: The Kaldi speech recognition toolkit. In: Proc. ASRU, Hawaii, US, pp. 1–4 (December 2011)

6. Public Transport Information System for Czech Republic (2014), https://ufal.mff.cuni.cz/alex-dialogue-systems-framework/ptics
7. Žilka, L., Marek, D., Korvas, M., Jurčíček, F.: Comparison of Bayesian Discriminative and Generative Models for Dialogue State Tracking. In: SIGDIAL 2013: Proc. of the 14th Annual Meeting of the Special Interest Group on Discourse and Dialogue, Metz, France, pp. 452–457 (2013)
8. He, Y., Young, S.: Semantic processing using the Hidden Vector State model. Computer Speech & Language 19(1), 85–106 (2005)
9. Jurčíček, F., Švec, J., Müller, L.: Extension of the HVS semantic parser by allowing left-right branching. In: Proceedings of ICASSP, pp. 4993–4996 (2008)
10. Zhu, J., Hastie, T.: Kernel logistic regression and the import vector machine. Journal of Computational and Graphical Statistics 14(1), 1081–1088 (2005)
11. Zettlemoyer, L.S., Collins, M.: Online learning of relaxed CCG grammars for parsing to logical form. In: Proceedings of the 2007 Joint Conference on Empirical Methods in Natural Language Processing and Computational Natural Language Learning, pp. 678–687 (2007)
12. Korvas, M., Plátek, O., Dušek, O., Žilka, L., Jurčíček, F.: Free English and Czech telephone speech corpus shared under the CC-BY-SA 3.0 license. In: Proceedings of International Conference on Language Resources and Evaluation, pp. 4423–4428 (2014)
13. The Kaldi ASR toolkit (2014), http://sourceforge.net/projects/kaldi
14. The Alex Dialogue Systems Framework (2014), https://github.com/UFAL-DSG/alex
15. Dahl, D.A., Bates, M., Brown, M., Fisher, W., Hunicke-Smith, K., Pallett, D., Pao, C., Rudnicky, A., Shriberg, E.: Expanding the scope of the ATIS task: The ATIS-3 corpus. In: Proceedings of the ARPA HLT Workshop, pp. 43–48 (1994)
16. Meza-Ruiz, I.V., Riedel, S., Lemon, O.: Spoken Language Understanding in dialogue systems, using a 2-layer Markov Logic Network: Improving semantic accuracy. In: Proceedings of Londial (2008)
17. Tür, G., Hakkani-Tür, D.Z., Hillard, D., Celikyilmaz, A.: Unsupervised Spoken Language Understanding: Exploiting Query Click Logs for Slot Filling. In: Proceedings of Interspeech, pp. 1293–1296 (2011)
18. Jurčíček, F., Thomson, B., Young, S.: Reinforcement learning for parameter estimation in statistical spoken dialogue systems. Computer Speech & Language 26(3), 168–192 (2012)

Speech Synthesis and Uncanny Valley*

Jan Romportl

Department of Cybernetics, Faculty of Applied Sciences
University of West Bohemia, Pilsen, Czech Republic
rompi@kky.zcu.cz

Abstract. The paper discusses a hypothesis relating high quality text-to-speech (TTS) synthesis in spoken dialogue systems with the concept of "uncanny valley". It introduces a "Wizard-of-Oz" experiment with 30 volunteers engaged in conversations with two synthetic voices of different naturalness. The results of the experiment are summarized and interpreted, leading to the conclusion that the TTS uncanny valley effect in dialogue systems can probably be superseded and inverted by a positive attitude of the systems' users toward new technologies.

Keywords: text-to-speech synthesis, spoken dialogue system, uncanny valley, experiment.

1 Introduction

The concept of *uncanny valley* [1] has been originally introduced by Masahiro Mori in 1970 for description of a typical emotional effect that near-human artifacts elicit in human – for example seeing a very accurate prosthetic hand can provoke feelings of eeriness, whereas seeing an artificial robotic "hand" is as neutral as seeing a real human hand (indeed in its proper functioning conditions as a part of a healthy human body, not e.g. amputated).

Analogically, an archetypical human-style robot, such as Number 5 from the movie Short Circuit or C-3PO from Star Wars, is often perceived almost as cute, whereas a highly accurate near-human robot like Geminoid F, especially when moving, is usually assessed as literally creepy. Mori also notes that the uncanny valley effect is stronger when the artifact is moving.[1] The dimension in which uncanny valley occurs and can be "measured" (at least informally) is called *shinwakan*, a Japanese neologism coined by Mori for what can be translated as "familiarity", or probably better as "affinity" [2], i.e. affinity of a human towards the artifact.

There have been many studies reporting various findings and theories on uncanny valley. Their brief overview is for example in Roger Moore's paper [3] where he also proposes a probabilistic framework for uncanny valley formalization, based on

* This work was supported by the European Regional Development Fund (ERDF), project "New Technologies for Information Society" (NTIS), European Centre of Excellence, ED1.1.00/02.0090.

[1] Due to quite strict space limitation we do not reprint the notoriously known Mori's uncanny valley curve which very well illustrates the effect. However, it is easily accessible e.g. here http://en.wikipedia.org/wiki/Uncanny_valley.

P. Sojka et al. (Eds.): TSD 2014, LNAI 8655, pp. 595–602, 2014.

the presumption that uncanny valley is a specific manifestation of a more general psychological phenomenon called "perceptual magnet effect".

The goal of our paper is to explore (at least to some preliminary extent) links between uncanny valley and text-to-speech (TTS) synthesis used in spoken man-machine dialogue systems. It has been inspired by our previous involvement in the FP6 project Companions where we worked on a highly natural and expressive Embodied Conversational Agent (ECA) for seniors. As a part of this work, we have recorded an audio-visual corpus of 60 hours of spoken dialogues between seniors and a simplified graphical ECA with a rather artificial TTS voice controlled by hidden human operators – so called "Wizard-of-Oz" (WoZ) method [4].

The initial expectations were that the users (seniors) would not truly enjoy the conversations and immerse into them unless ECA is equipped with a highly natural expressive and emotional TTS voice, which was not the case in the aforementioned corpus acquisition. However, to our surprise, the users enjoyed the conversations very much and often got highly emotionally involved in them (no matter their rather artificial conversation partner), as was recently shown by our M.Sc. student Pavlína Heiderová who analyzed in her master's thesis [5] emotional responses of the users of this WoZ-simulated ECA.

As a result of these findings, we posed a question whether it could actually be the *artificial* ("robotic") voice itself that helps the users engage in a *natural* and pleasant conversation with an obviously *artificial* agent. In other words: whether it is possible that by using a highly natural state-of-the-art emotional speech synthesis the system would actually "drag" the users into the uncanny valley, degrading their user experience and conversational comfort by exposing them to a mismatch between different sensory cues (natural voice possessed by otherwise fully artificial entity), distorting their categorization in the aforementioned perceptual magnet effect, just as it is with the contrast between C-3PO and Geminoid F, only this time with speech synthesis. Therefore, we have performed a set of initial experiments aimed at answering this question.

2 Experiments

The experiments were conducted by our M.Sc. student Daniela Tisarová as a part of her master's thesis [6]. The goal of the experiments was to acquire empirical data for supporting or rejecting the aforementioned hypothesis about TTS and uncanny valley.

The experimental protocol was based on "Wizard-of-Oz" simulation of a fictional AI-based small talk dialogue system (chatterbot) having spoken Skype conversation with research volunteers (probands henceforth) over several neutral casual topics, such as public transportation, weather, etc. The dialogue was immediately followed by a structured questionnaire.

The group of the probands consisted of 30 individuals with the average age of 23.5 years, selected mostly from university students of a technical and a philosophical faculty. The group consisted of 15 female and 15 male probands.

2.1 Experimental Protocol

Prior to the experiments, all the probands were briefly introduced to the field of AI-based dialogue systems, their state of the art, problems and challenges. However, they were not explicitly introduced to the field of TTS synthesis and its evaluation, so as to eliminate a potential bias resulting from their knowledge of which particular speech synthesis method is used in the experiment.

The probands were then instructed that they would go through a Skype call with an AI-based spoken dialogue system that will casually chat with them over two photos (a bus station and a railway station) about public transportation. The probands were told that the system has two female identities represented by two different voices and that the Skype call will be divided into two separate conversations, each of them with a different voice and over a different photo. The probands were intentionally given a false idea that the dialogues are for testing purposes of the actual AI-based system performance and that the system is equipped with a state of the art ASR, NLP, dialogue manager and TTS, whereas in fact TTS was the only automatic component really present; the rest was simulated by the experimenter Daniela ("wizard").

The two voices used in the experiment will be henceforth denoted as "Voice A" and "Voice B". Both voices were synthesized by our TTS system ARTIC. Voice A is an old single unit instance TD-PSOLA voice, judged ex post by the probands as being very "robotic". Voice B is a state of the art highly natural unit selection (with no acoustic signal modifications) voice based on a 10+k-sentence corpus. The order of the voices' engagement with the probands was randomized – in 15 cases Voice A was taking part in the first conversation, in 15 cases it was Voice B. The order of the photographs giving a background for the dialogues was randomized along the same lines, too.

The probands did not know that their communication partner is actually the experimenter who was using the TTS system with Voice A/B for synthesizing the turns of the fictional chatterbot and who was mimicking typical behavior of contemporary spoken dialogue systems, such as inappropriate timing of responses (either too long pauses or badly timed barge-ins), lack of common sense (or somehow caricatured common sense), problems with semantical and pragmatical interpretations, etc. – simply, the experimenter mimicked the stereotypes that most contemporary spoken dialogue systems meet and that are usually expected from them by general public. The probands were thus talking to the supposed AI system and were receiving synthesized replies generated by the hidden experimenter.

The experimenter indeed tried to keep the content of all the dialogues across the probands as similar as the individual situations allowed, so that the experiment is not influenced by uncontrolled differences among the dialogues.

2.2 Assessment of Dialogues

Immediately after the Skype call when both conversations ended, the probands were asked to complete a structured online questionnaire. The most important questions were:

1. Which *conversation* was less unpleasant? (with Voice A / with Voice B / no preference)
2. Which *voice* did you like less? (Voice A / Voice B / no preference)

3. Were the conversations interesting for you? (absolutely / mostly yes / mostly no / not at all)
4. How would you assess the Voice A and B, respectively? (robotic or artificial / usual widespread synthetic voice / close to human voice / same as human voice)

Then the questionnaire comprised questions about the probands (age, education, etc.) and also several unstructured questions, such as "How would you assess the dialogue by one word?", "Were you surprised by anything?", "What was the most/least pleasant aspect of the dialogue?", etc. These are, however, beyond the scope of the present paper.

The questions (1) and (2) were formulated in the negative voice because we expected the probands to feel some discomfort or unease in any case (based on our own subjective experience; and it was also an objective of the experiment to build up little tension, otherwise there would be no space for uncanny valley), and so asking them "Which conversation/voice did you like *more*?" could psychologically lead to more frequent frustrated reply "none", meaning simply "I don't like talking to machines at all."

The first two questions show an apparent effort to filter out the probands' opinion on which voice sounds "more/less natural" – we did not ask for naturalness, instead wanted to hear which voice and which conversation caused them more troubles, strangeness or discomfort, which voice in that particular situation "dragged" them more to the uncanny valley. Moreover, we wanted to see if there is a difference in the probands' assessment of their attitude towards the particular *voice* and towards the more complex concept of *conversation*.

3 Results and Their Discussion

The most obvious question that we would like to get answered is indeed whether the probands disliked significantly more Voice A, or Voice B. If Voice B is disliked unequivocally, then there is a clear indication of uncanny valley because the highly natural Voice B does not match properly the typical "machine-like" behavior of the system in all other aspects of the conversation. On the other hand, if Voice A is disliked unequivocally, then the uncanny valley hypothesis for this kind of TTS application can be rejected. However, as Table 1 shows (based on questions (1) and (2)), the results are not unequivocal at all. They are somewhat in favor of Voice B (and rejection of the uncanny valley hypothesis) but they are definitely not convincing. It means that this major question still remains open, as we will discuss further.

The table also shows there is a very accurate complementary relation between "(dis)liking the voice" and "(dis)liking the conversation", which indirectly supports our assumption that the content of all the dialogues is coherent and that the only aspect making the difference is the voice.

We also checked if there is any significant difference between the voice preferences in male and female probands, and by Pearson's chi-squared tests we confirmed the voice preferences are independent on the probands' gender (with $p = 0.74$).

3.1 Order of Conversations

Since the preferences (almost one-fourth) in favor of the robotic Voice A must not be neglected, we have investigated more factors that could explain them differently than as an inherent aspect of psychology and cognition of the respective probands.

Table 1. Overall preferences of the voices and conversations

	this voice is less pleasant	conversation with this voice is less unpleasant
Voice A	22	6
Voice B	7	20
no preferences	1	4

One of the hypothesis was that the preference could be influenced by the order of the voices that engaged in two conversations with each proband; e.g., if a proband hears the robotic Voice A in the second conversation after being engaged with the natural Voice B for some time, he/she might subjectively feel disappointment purely with the voice qualities, that cannot be cognitively separated from the much more complex attribute of affinity (or Mori's *shinwakan*) towards the voice. However, the chi-squared test shows (again with $p = 0.74$) that we cannot reject the null hypothesis of independence between the voice preference and the order of its respective conversation (the table is thus not necessary here because the values quite closely follow the distribution in Table 1). Therefore, it is quite reasonable to assume that the voice preference is independent on the order of its engagement with the proband.

3.2 Background of Probands

It is quite clear that the background of the probands can influence their affinity towards man-machine conversation – those who are familiar with new technologies, are in active daily contact with them or even are professionally involved in their development are quite likely to react differently when verbally exposed to an AI than those who have either a priori lukewarm attitude towards technology or simply have not been in touch with it.

We have presupposed (among others on the basis of our prior teaching experience at various faculties) that such background differences can very roughly be captured by the field of the proband's study/work – either "rather technical" (including economics), or "rather humanities". Our expectations were quite confirmed by the results of the experiment, as illustrated by Table 2. This contingency table shows the relation between the probands' field of study/work (i.e. field of expertise) and their voice preference. When we stated the null hypothesis as the independence of the voice preference on the field of expertise, we had in this case a reason to reject it by the chi-squared test with calculated $p = 0.05$.

Such a borderline value gives us somewhat medium presumption against the null hypothesis but we must keep in mind that the total number of the probands (here 29 because we have excluded the proband with no preference for/against any of the voices) is still significantly lower than what the rule of thumb says for the Pearson's chi-squared test (often said to be 50), and therefore the statistical results are not much robust. In any case, it at least points out a very important factor that can be addressed in future experiments.

What is, however, clear is that in this particular experiment the probands with technical expertise quite explicitly disliked conversations with the robotic Voice A. At

Table 2. The relation between the probands' field of expertise and their voice preference

	rather technical	rather humanities	*total*
dislike Voice A	11 (92 %)	10 (59 %)	21
dislike Voice B	1 (8 %)	7 (41 %)	8
total	12	17	29

this moment, we can only speculate about the cause of such an effect – one of the speculations can be that the probands with more technical background easily identified the technical shortcomings of Voice A and that their psychologically default modus operandi can be informally paraphrased as "the more technical shortcomings a thing has, the worse it is". On the other hand, the probands from the field of humanities are more likely to be spared from such technophillic assessments and their preference is driven more by their unconscious affinity towards the communication partner. Such a speculation can thus lead to two (maybe not disjoint) conclusions: 1) people without technical background are more likely to be "dragged into the uncanny valley" of TTS; or 2) people with technical background pay less attention to their unconscious affinity because it is superseded by their technophillic enthusiasm. However, in order to prove or falsify these statements, much more elaborate and extensive experiments are needed.

Since we have found that the voice preference most likely depends on the field of expertise, we wanted to see if the probands' interest in the conversation (question (3) – 15 answers "absolutely", 14 answers "mostly yes", 1 did not answer) could make a difference too. The conclusion is that the null hypothesis "the voice preference is independent on the probands' interest" cannot be rejected by the Pearson's chi-squared test (with $p = 0.40$).

On the other hand, there is perhaps some form of dependence between the probands' field of expertise and their interest in the conversations. We have low presumption against the null hypothesis "the probands' interest is independent on their field of expertise" (with $p = 0.09$), which at least means (given the aforementioned low robustness of the statistics) that this aspect should be further explored in detail. At this moment, we can at least speculate that this again fits well to the image pictured by the previous tests (and also the intuitive stereotypical thinking about the technology users). We do not present the respective tables here due to lack of space.

3.3 Duration of Conversations

Another hypothesis that emerged together with the experiment was a relation between the probands' preference and the duration of their engagement with the voices. What if the probands were getting more frustrated with the robotic Voice A as the conversation took longer? Or the other way around – what if they were getting used to the robotic Voice A in the course of the conversation, while noticing more and more "uncanny glitches" in Voice B?

Therefore, we have calculated the following quantitative parameters for each proband: a) duration of the whole Skype call, given as mm:ss (average 16:15, standard deviation 03:37); b) duration of the conversation with Voice A (avg. 07:13, stdev.

02:12) and Voice B respectively (avg. 08:02, stdev. 02:39); c) number of turns Voice A (avg. 28.6, stdev. 8.6) and Voice B respectively (avg. 30.7, stdev. 6.5) had in each conversation.

For each parameter and each proband, we have categorized the respective conversation/call into one of three groups: Short, Medium, Long. The Medium category was delimited by the interval of the respective average value minus/plus its standard deviation. The Short category was then indeed everything below this interval and the Long category above. We do not present the tables with the frequencies of each category here mainly due to space limitation and the fact that the most interesting information is given by the aforementioned moments.

We formulated and tested (again by the Pearson's chi-squared test) the following list of null hypotheses:

1. The voice preference is independent on the duration of the Skype call; $p = 0.54$; cannot be rejected.
2. The voice preference is independent on the duration of the Voice A conversation; $p = 0.25$; cannot be rejected.
3. The voice preference is independent on the duration of the Voice B conversation; $p = 0.58$; cannot be rejected.
4. The voice preference is independent on the number of the system's turns in the Voice A conversation; $p = 0.25$; cannot be rejected.
5. The voice preference is independent on the number of the system's turns in the Voice B conversation; $p = 0.98$; cannot be rejected.
6. The Skype call duration is independent on the proband's field of expertise; $p = 0.54$; cannot be rejected.

The conclusion for this point is thus quite clear: the probands' voice preference most likely did not depend in any way on the duration of the conversations.

4 Conclusion

As we have already discussed in the previous section, we have not received unequivocal results. The majority (about three quarters) of the probands preferred the more natural synthetic voice, which speaks against the initial hypothesis of uncanny valley related to "too natural" speech synthesis in spoken dialogue systems. However, still a significant number of probands had the opposite preference, which also cannot be ignored.

Our most important finding here is quite a remarkable influence of the field of expertise of the probands, which leads to our new speculative hypothesis that the "technophillic attitude" of a significant part of the probands covered and superseded their primary *affinity* towards their artificial partner in conversation. We will address this hypothesis in our future experiments that will be aimed at statistically more robust, balanced and extensive group of probands (especially laypeople outside academic environment).

Moreover, we have shown that the conversation preference is quite well reducible to the voice preference and that the voice preference does not depend on the order of their respective conversations, nor does it depend on the duration of the conversations. The only aspect that made the difference was the background of the probands. This will help future experiments as well.

References

1. Mori, M.: The uncanny valley (translated by MacDorman, K.F., Kageki, N.). IEEE Robotics & Automation Magazine 19(2), 98–100 (2012)
2. Bartneck, C., Kanda, T., Ishiguro, H., Hagita, N.: Is the uncanny valley an uncanny cliff? In: 16th IEEE International Symposium on Robot and Human Interactive Communication, Jeju, Korea, pp. 368–373 (2007)
3. Moore, R.K.: A Bayesian explanation of the 'Uncanny Valley' effect and related psychological phenomena. Nature Scientific Reports 2(864) (2012)
4. Romportl, J., Zovato, E., Santos, R., Ircing, P., Relaño Gil, J., Danieli, M.: Application of expressive TTS synthesis in an advanced ECA system. In: Proceedings of the ISCA Tutorial and Research Workshop on Speech Synthesis, Kyoto, Japan, pp. 120–125 (2010)
5. Heiderová, P.: Perspektivy řečové komunikace mezi člověkem a strojem (Perspectives of Speech Communication Between Human and Machine). Master's thesis, University of West Bohemia, Pilsen (2012)
6. Tisarová, D.: Hypotéza "uncanny valley" ve vztahu k syntetické řeči (The Uncanny Valley Hypothesis in Relation to Synthetic Speech). Master's thesis, University of West Bohemia, Pilsen (2014)

Integration of an On-line Kaldi Speech Recogniser to the Alex Dialogue Systems Framework

Ondřej Plátek and Filip Jurčíček

Charles University in Prague,
Faculty of Mathematics and Physics,
Institute of Formal and Applied Linguistics
{oplatek,jurcicek}@ufal.mff.cuni.cz
https://ufal.mff.cuni.cz

Abstract. This paper describes the integration of an on-line Kaldi speech recogniser into the Alex Dialogue Systems Framework (ADSF). As the Kaldi *OnlineLatgenRecogniser* is written in C++, we first developed a Python wrapper for the recogniser so that the ADSF, written in Python, could interface with it. Training scripts for acoustic and language modelling were developed and integrated into ADSF, and acoustic and language models were build. Finally, optimal recogniser parameters were determined and evaluated. The dialogue system Alex with the new speech recogniser is evaluated on Public Transport Information (PTI) domain.

Keywords: automatic speech recognition, Kaldi, Alex, dialogue systems.

1 Introduction

The Alex Dialogue Systems Framework (ADSF) is used for development of our experimental Czech spoken dialogue system in the Public Transport Information (PTI) domain [8]. In PTI, the dialogue system provides information about all kinds of public transport in the Czech Republic, publicly available at a toll-free 800 899 998 number.

This paper describes and evaluates the integration of the Kaldi *OnlineLatgenRecogniser* in the ADSF. Important goal of ADSF is to process the speech incrementally. Incremental speech processing recognise the incoming speech while a user speaks and the Kaldi *OnlineLatgenRecogniser* implements on-line interface which allows incremental audio processing. As a result, the ASR output can be obtained with minimal latency.

In the next section, the Python interface of *OnlineLatgenRecogniser* is described and alternative ASR implementations are discussed. Integration of *PyOnlineLatgenRecogniser* into Alex is described in Section 3. Section 4 details the PTI domain, acoustic and language model training. The evaluation itself is presented in Section 5. Finally, Section 6 concludes this work.

2 Automatic Speech Recognition Implementations

There are various options for automatic speech recognition. Among the most popular are: OpenJulius and RWTH decoder.

P. Sojka et al. (Eds.): TSD 2014, LNAI 8655, pp. 603–610, 2014.

The OpenJulius decoder can be used with custom-built acoustic and language models and for on-line speech recognition [2]. However, OpenJulius suffers from software instability when producing lattices and confusion networks. Therefore, it is not suitable for practical use. The RWTH decoder is not free software and a license must be purchased for commercial applications [6].

In the following text, we present ASR implementations which are integrated in ADSF. As of now, Google cloud based speech recognition service and *OnlineLatgen-Recogniser* are available in ADSF.

2.1 Google ASR

The main advantage of Google's cloud-based ASR is that it can be used off-the-shelf without any additional modifications and is relatively fast. However, the Word Error Rate (WER) is rather high on the PTI domain. The Google ASR API is available at `https://www.google.com/speech-api/v1/recognize`[1].

2.2 PyOnlineLatgenRecogniser

The *PyOnlineLatgenRecogniser* is a thin Python wrapper of the Kaldi *OnlineLatgen-Recogniser*, which is implemented in C++ [12]. The *OnlineLatgenRecogniser* implements on-line interface to speech parametrisation, feature transformations, and Kaldi *LatticeFasterDecoder*. It supports state of the art acoustic models for non-speaker adaptive recognition. Namely, we are able to use MFCC speech parametrisation, $\Delta - \Delta\Delta$ or LDA+MLLT feature transformation, generative training and MMI or MPE discriminative training [7].

The *OnlineLatgenRecogniser* implements the following interface:

- *AudioIn* – queueing new audio for pre-processing,
- *Decode* – decoding a fixed number of audio frames,
- *PruneFinal* – preparing internal data structures for lattice extraction,
- *GetLattice* – extracting a word posterior lattice and returning log likelihood of processed audio,
- *Reset* – preparing the recogniser for a new utterance,

The minimalistic Python example in Listing 1.1 shows usage of the *PyOnlineLatgen-Recogniser* and the decoding of a single utterance. The audio is passed to the recogniser in small chunks (line 4), so the decoding (line 5 and 8) can be performed as user speaks. When no more audio data is available a likelihood and a word posterior lattice is extracted from the recogniser (line 10).

The word posterior lattices are returned as instances of the OpenFST [3] class. The OpenFST implementation for Python is provided by *pyfst* library [13]. In the ADSF we implemented conversion of the word posterior lattices to an n-best list. The implementation is efficient since the OpenFST shortest path algorithm is used on small lattices.

[1] Its use is described in a blog at `http://mikepultz.com/2013/07/google-speech-api-full-duplex-php-version/`

Listing 1.1. The snippet implements incremental recognition with *PyOnlineLatgenRecogniser's* interface abstracting from anaudio source and parameters details.

```
d = PyGmmLatgenWrapper()
d.setup(argv)
while audio_to_process():
    d.audio_in(get_audio_chunks())
    dec_t = d.decode(max_frames=10)
    while dec_t > 0:
        decoded_frames += dec_t
        dec_t = d.decode(max_frames=10)
d.prune_final()
lik, lat = d.get_lattice()
```

3 Integration of PyOnlineLatgenRecogniser into ADSF

The integration of the Kaldi real-time recognizer into the Alex framework consisted of implementing the following features:

1. The *KaldiASR* as subclass of *ASRInterface*, so that the Alex's ASR component can use *PyOnlineLatgenRecogniser*.
2. The training scripts for acoustic and language models.
3. The scripts for evaluation and decoding graph creation so that various settings, language and acoustic models can be tested.

This section discuss the implementation of *KaldiASR* class which uses *PyOnlineLatgenRecogniser's* functionality. The acoustic model training and the evaluation are described in next sections.

The ASR component in the ADSF runs as a separate process, and speech recognition is triggered by Voice Activity Detection (VAD) decisions. The ADSF uses voice activity detection to split a stream of incoming audio signal into speech and silence segments. Once speech segments are identified, they are sent to an ASR component to be recognised.

The ADSF defines abstract interface for the ASR unit as illustrated in Listing 1.2. The two most important methods are *rec_in* and *hyp_out*.

Listing 1.2. ASRInterface

```
class ASRInterface(object):

    def rec_in(self, frame):

    def flush(self):

    def hyp_out(self):

    def rec_wav(self, pcm):
        self.rec_in(pcm)
        return self.hyp_out()
```

The *rec_wav* method in Listing 1.2 nicely illustrates how the two methods *rec_in* and *hyp_out* are used for decoding. The *rec_in* queues in audio input and decodes the audio using beam search [5]. The *hyp_out* method extract ASR hypothesis. Since the *rec_wav* method is used only for testing purposes, it sends all input audio to the speech recogniser at once. However, the audio is passed in small chunks to *PyOnlineLatgenRecogniser* in real-time applications, so the *rec_in* method decodes the audio as a user speaks.

When VAD recognises the end of speech, the *hyp_out* method is called in order to extract ASR hypothesis. *PyOnlineLatgenRecogniser* extracts word posterior lattices where its probabilities are computed using the forward-backward algorithm. The resulting lattice is converted to an n-best list and returned to hyp_out method which is processed by spoken language understanding component.

The latency of the ASR unit depends on the time spent in *hyp_out* method. The *hyp_out* method itself spend most of the time on preparing a word lattice. The word lattice is extracted by Kaldi determinisation algorithm [5]. The *lattice-beam* affects a level of approximations. The higher value of the *lattice-beam* parameter the better determinised lattice is extracted with longer latency.

The hypothesis extraction takes a nearly constant amount of time for variable utterance length and fixed *lattice-beam*. We fixed the *lattice-beam* value so it takes around 60 ms on average.

The Alex dialogue system frequently handles several spoken requests immediately one after another. Nevertheless, at the end of each utterance the *hyp_out* method is called and the ASR hypothesis is extracted. Since the user already speaks when the lattice is extracted, *rec_in* cannot be called and the audio is buffered. Consequently, the *rec_in* should decode the buffered audio faster than the user speaks.

In extreme cases, the *flush* method may be used in order to throw away the buffered audio input and reset the decoding. Skipping some user requests is arguably a better strategy than baffling the user with responses to a request which was asked a long time ago.

4 Training Acoustic and Language Models for the PTI Domain

The *OnlineLatgenRecogniser* is evaluated on a corpus of audio data from the Public Transport Information (PTI) domain. In PTI, users can interact in Czech language with a telephone-based dialogue system to find public transport connections [8]. The PTI corpus consists of approximately 12,000 user utterances with a length varying between 0.4 s and 18 s with a median around 3 s. The data were divided into training, development, and test data sections where the corresponding data sizes were 9,496, 1,188, 1,188 respectively. For evaluation, a domain specific class-based language model with a vocabulary size of approximately 52,000 and 559,000 n-grams was estimated from the training data. Named entities e.g., cities or bus stops, in class-based language model are expanded before building a decoding graph which is used for beam search in *OnlineLatgenRecogniser*. The perplexity of the resulting language model evaluated on the development data is about 48.

Since the PTI acoustic data amounts to less then 5 hours, the acoustic training data was extended by additional 15 hours of telephone out-of-domain data from VYSTADIAL 2013 – Czech corpus [9]. The best acoustic models were obtained by BMMI discriminative training with LDA and MLLT feature transformations. A detailed description of the training procedure is given in [9].

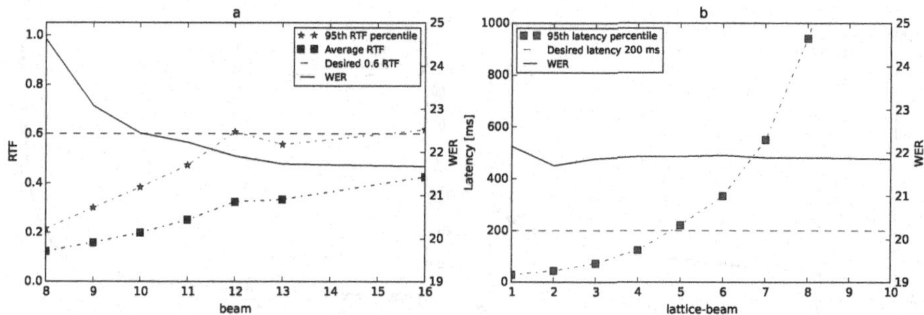

Fig. 1. The left graph (a) shows that WER decreases with increasing *beam* and the average RTF linearly grows with the beam. Setting the maximum number of active states to 2000 stops the growth of the 95th RTF percentile at 0.6, indicating that even in the worst case, we can guarantee an RTF around 0.6. The right graph (b) shows how latency grows in response to increasing *lattice-beam*.

5 Evaluation of *PyOnlineLatgenRecogniser* in ADSF

We focus on evaluating the speed of the *OnlineLatgenRecogniser* and its relationship to the accuracy of the decoder. We evaluate the following measures:

- Real Time Factor (RTF) of decoding – the ratio of the recognition time to the duration of the audio input,
- Latency – the delay between the end of utterance and the availability of the recognition results,
- Word Error Rate (WER).

The accuracy and speed of the *OnlineLatgenRecogniser* are controlled by the *max-active-states*, *beam*, and *lattice-beam* parameters [7]. *Max-active-states* limits the maximum number of active tokens during decoding. *Beam* is used during graph search to prune ASR hypotheses at the state level. *Lattice-beam* is used when producing word level lattices after the decoding is finished. It is crucial to tune these parameters optimally to obtain good results.

In general, one aims for a RTF smaller than 1.0. Moreover, it is useful in practice if the RTF is even smaller because other processes running on the machine can influence the amount of available computational resources. Therefore,we target in our setup the RTF of 0.6.

We used grid search on the test set to identify the optimal parameters. Figure 1 (a) shows the impact of *beam* on the WER and RTF measures. In this case, we set *max-active-states* to 2000 in order to limit the worst case RTF to 0.6. Based on the results shown in Figure 1 (a), we set *beam* to 13 for further experiments as this setting balances the WER. Figure 1 (b) shows the impact of *lattice-beam* on WER and latency when *beam* is fixed to 13. We set *lattice-beam* to 5 based on Figure 1 (b) to obtain the 95th latency percentile[2] of 200 ms, which is considered natural in a dialogue [1].

[2] For example, the 95th percentile is the value of a measure such that 95% of the data has the measure below that value.

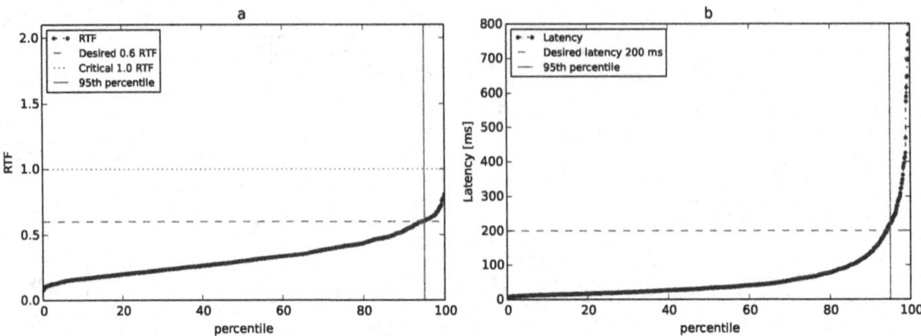

Fig. 2. The percentile graphs show RTF and Latency scores for test data for *max-active-sates*=2000, *beam*=13, *lattice-beam*=5. Note that 95 % of utterances were decoded with a latency lower than 200ms.

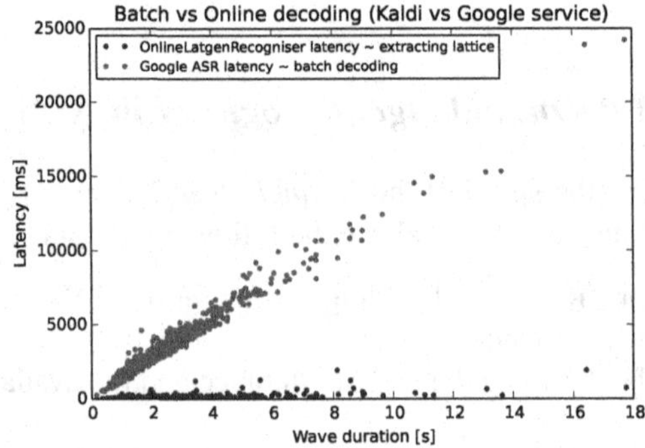

Fig. 3. Relation between latency and utterance length. Comparing on-line decoder (OnlineLatgenRecogniser) and batch decoding (Google cloud ASR service).

Lattice-beam does not affect WER, but larger *lattice-beam* improves the oracle WER of generated lattices [5]. Richer lattices may improve SLU performance.

Figure 2 shows the percentile graph of the RTF and latency measures over the test set. One can see from Figure 2 that 95% of test utterances is decoded with RTF under 0.6 and latency under 200 ms. The extreme values are typically caused by decoding long noisy utterances where uncertainty in decoding slows down the recogniser. Using this setting, the *OnlineLatgenRecogniser* decodes the test utterances with a WER of about 21%.

In addition, previously used ASR engine, Google ASR service, is evaluated. The Google ASR service decoded the test utterances from the PTI domain with a 95% latency percentile of 1900ms and reached WER about 48%. The high latency is presumably caused by the batch processing of audio data and network latency, and the high WER is likely caused by a mismatch between Google's acoustic and language models and the test data.

6 Conclusion

This work described the integration of the Kaldi *OnlineLatgenRecogniser* into the Alex Dialogue Systems Framework (ADSF). The *OnlineLatgenRecogniser* offers state-of-the-art accuracy as well as outstanding real-time and latency performance. The source code of the *OnlineLatgenRecogniser*, acoustic modelling scripts, and ADSF is available for download under the permissive Apache 2.0 license. The evaluation showed that the *OnlineLatgenRecogniser* significantly outperforms Google ASR service in terms of accuracy and latency in the PTI domain.

Overall, *OnlineLatgenRecogniser*, and the Kaldi toolkit offer an excellent alternative to Google ASR or OpenJulius when considered in the context of spoken dialogue systems.

Acknowledgments. We would also like to thank Daniel Povey and Ondřej Dušek for their useful comments and discussions. We also thank the anonymous reviewers for their helpful comments and suggestions.

This research was funded by the Ministry of Education, Youth and Sports of the Czech Republic under the grant agreement LK11221, by the core research funding of Charles University in Prague. The language resources presented in this work are stored and distributed by the LINDAT/CLARIN project of the Ministry of Education, Youth and Sports of the Czech Republic (project LM2010013).

References

1. Skantze, G., Schlangen, D.: Incremental dialogue processing in a micro-domain. In: Proc. ECACL, pp. 745–753 (2009)
2. Akinobu, L.: Open-Source Large Vocabulary CSR Engine Julius (2014), `http://julius.sourceforge.jp/en_index.php`
3. Allauzen, C., Riley, M., Schalkwyk, J., Skut, W., Mohri, M.: OpenFst: A general and efficient weighted finite-state transducer library. In: Holub, J., Žďárek, J. (eds.) CIAA 2007. LNCS, vol. 4783, pp. 11–23. Springer, Heidelberg (2007)
4. Huggins-Daines, D., Kumar, M., Chan, A., Black, A., Ravishankar, M., Rudnicky, A.: Pocketsphinx: A free, real-time continuous speech recognition system for hand-held devices. In: Proc. ICASSP, pp. I–I (December 2006)
5. D. Povey, M. Hannemann, G. Boulianne, L. Burget, A. Ghoshal, M. Janda, M. Karafiát, S. Kombrink, P. Motlicek, Y. Qian at al.: Generating exact lattices in the WFST framework. In Proc. ICASSP, pp. 4213–4216 (2012)
6. Rybach, D., Hahn, S., Lehnen, P., Nolden, D., Sundermeyer, M., Tüske, Z., Wiesler, S., Schlüter, R., Ney, H.: The RASR-The RWTH Aachen University open source speech recognition toolkit. In: Proc. IEEE Automatic Speech Recognition and Understanding Workshop (2011)
7. Povey, D., et al.: The Kaldi speech recognition toolkit. In: Proc. ASRU, Hawaii, US, pp. 1–4 (December 2011)
8. Public Transport Information System for Czech Republic, `https://ufal.mff.cuni.cz/alex-dialogue-systems-framework/ptics`
9. Korvas, M., Plátek, O., Dušek, O., Žilka, L., Jurčíček, F.: Free English and Czech telephone speech corpus shared under the CC-BY-SA 3.0 license. In: Proceedings of International Conference on Language Resources and Evaluation (to be published, 2014)

10. The Kaldi ASR toolkit (2014), http://sourceforge.net/projects/kaldi
11. The Alex Dialogue Systems Framework (2014), https://github.com/UFAL-DSG/alex
12. The OnlineLatgenRecogniser (2014), https://github.com/UFAL-DSG/pykaldi
13. The pyfst library: OpenFst in Python (2014), http://pyfst.github.com/

Author Index